NUMERICAL SOLUTION OF HYPERBOLIC PARTIAL DIFFERENTIAL EQUATIONS

This is a new type of graduate textbook, with both print and interactive electronic components (on CD). It is a comprehensive presentation of modern shock-capturing methods, including both finite volume and finite element methods, covering the theory of hyperbolic conservation laws and the theory of the numerical methods.

Classical techniques for judging the qualitative performance of the schemes, such as modified equation analysis and Fourier analysis, are used to motivate the development of classical higher-order methods (the Lax–Wendroff process) and to prove results such as the Lax Equivalence Theorem.

The range of applications (shallow water, compressible gas dynamics, magnetohydrodynamics, finite deformation in solids, plasticity, polymer flooding and water/gas injection in oil recovery) is broad enough to engage most engineering disciplines and many areas of applied mathematics.

The solution of the Riemann problems for these applications is developed, so that the reader can use the theory to develop test problems for the methods, especially to measure errors for comparisons of accuracy and efficiency. The numerical methods involve a variety of important approaches, such as MUSCL and PPM, TVD, wave propagation, Lax–Friedrichs (aka central schemes), ENO and discontinuous Galerkin; all of these are discussed in one and multiple spatial dimensions. Since many of these methods depend on Riemann solvers, there is extensive discussion of the basic design principles of approximate Riemann solvers, and several computationally useful techniques. The final chapter contains a discussion of adaptive mesh refinement via structured grids.

The accompanying CD contains a hyperlinked version of the text which provides access to computer codes for all of the text figures. Through this electronic text students can:

- See the codes and run them, choosing their own input parameters interactively
- View the online numerical results as movies
- Gain an appreciation for both the dynamics of the problem application, and the growth of numerical errors
- Download and modify the code for use with other applications
- Study the code to learn how to structure their programs for modularity and ease of debugging

John A. Trangenstein is Professor of Mathematics at Duke University, North Carolina.

NUMERICAL SOLUTION OF HYPERBOLIC PARTIAL DIFFERENTIAL EQUATIONS

JOHN A. TRANGENSTEIN

Department of Mathematics, Duke University
Durham, NC 27708-0320

CAMBRIDGE
UNIVERSITY PRESS

CAMBRIDGE UNIVERSITY PRESS
Cambridge, New York, Melbourne, Madrid, Cape Town, Singapore, São Paulo, Delhi

Cambridge University Press
The Edinburgh Building, Cambridge CB2 8RU, UK

Published in the United States of America by Cambridge University Press, New York

www.cambridge.org
Information on this title: www.cambridge.org/9780521877275

First published 2009

Printed in the United Kingdom at the University Press, Cambridge

A catalog record for this publication is available from the British Library

ISBN 978-0-521-87727-5 Hardback

To James A. Rowe

Contents

Preface *page* xvii

1 Introduction to Partial Differential Equations 1

2 Scalar Hyperbolic Conservation Laws 6
2.1 Linear Advection 6
2.1.1 Conservation Law on an Unbounded Domain 6
2.1.2 Integral Form of the Conservation Law 8
2.1.3 Advection–Diffusion Equation 9
2.1.4 Advection Equation on a Half-Line 10
2.1.5 Advection Equation on a Finite Interval 11
2.2 Linear Finite Difference Methods 12
2.2.1 Basics of Discretization 12
2.2.2 Explicit Upwind Differences 14
2.2.3 Programs for Explicit Upwind Differences 16
2.2.3.1 First Upwind Difference Program 16
2.2.3.2 Second Upwind Difference Program 17
2.2.3.3 Third Upwind Difference Program 18
2.2.3.4 Fourth Upwind Difference Program 20
2.2.3.5 Fifth Upwind Difference Program 21
2.2.4 Explicit Downwind Differences 23
2.2.5 Implicit Downwind Differences 24
2.2.6 Implicit Upwind Differences 25
2.2.7 Explicit Centered Differences 26
2.3 Modified Equation Analysis 30
2.3.1 Modified Equation Analysis for Explicit Upwind Differences 30

2.3.2 Modified Equation Analysis for Explicit Downwind Differences 31
2.3.3 Modified Equation Analysis for Explicit Centered Differences 32
2.3.4 Modified Equation Analysis Literature 33
2.4 Consistency, Stability and Convergence 35
2.5 Fourier Analysis of Finite Difference Schemes 38
2.5.1 Constant Coefficient Equations and Waves 39
2.5.2 Dimensionless Groups 40
2.5.3 Linear Finite Differences and Advection 41
2.5.4 Fourier Analysis of Individual Schemes 44
2.6 L^2 Stability for Linear Schemes 53
2.7 Lax Equivalence Theorem 55
2.8 Measuring Accuracy and Efficiency 69

3 Nonlinear Scalar Laws 81
3.1 Nonlinear Hyperbolic Conservation Laws 81
3.1.1 Nonlinear Equations on Unbounded Domains 81
3.1.2 Characteristics 82
3.1.3 Development of Singularities 84
3.1.4 Propagation of Discontinuities 85
3.1.5 Traveling Wave Profiles 89
3.1.6 Entropy Functions 92
3.1.7 Oleinik Chord Condition 95
3.1.8 Riemann Problems 97
3.1.9 Galilean Coordinate Transformations 99
3.2 Case Studies 102
3.2.1 Traffic Flow 102
3.2.2 Miscible Displacement Model 103
3.2.3 Buckley–Leverett Model 105
3.3 First-Order Finite Difference Methods 111
3.3.1 Explicit Upwind Differences 111
3.3.2 Lax–Friedrichs Scheme 112
3.3.3 Timestep Selection 117
3.3.4 Rusanov's Scheme 118
3.3.5 Godunov's Scheme 120
3.3.6 Comparison of Lax–Friedrichs, Godunov and Rusanov 124
3.4 Nonreflecting Boundary Conditions 125
3.5 Lax–Wendroff Process 129
3.6 Other Second Order Schemes 132

4 Nonlinear Hyperbolic Systems 135
4.1 Theory of Hyperbolic Systems 135
4.1.1 Hyperbolicity and Characteristics 135
4.1.2 Linear Systems 139
4.1.3 Frames of Reference 140
4.1.3.1 Useful Identities 141
4.1.3.2 Change of Frame of Reference for Conservation Laws 143
4.1.3.3 Change of Frame of Reference for Propagating Discontinuities 145
4.1.4 Rankine–Hugoniot Jump Condition 146
4.1.5 Lax Admissibility Conditions 150
4.1.6 Asymptotic Behavior of Hugoniot Loci 152
4.1.7 Centered Rarefactions 156
4.1.8 Riemann Problems 159
4.1.9 Riemann Problem for Linear Systems 159
4.1.10 Riemann Problem for Shallow Water 162
4.1.11 Entropy Functions 164
4.2 Upwind Schemes 176
4.2.1 Lax–Friedrichs Scheme 176
4.2.2 Rusanov Scheme 179
4.2.3 Godunov Scheme 179
4.3 Case Study: Maxwell's Equations 183
4.3.1 Conservation Laws 183
4.3.2 Characteristic Analysis 184
4.4 Case Study: Gas Dynamics 186
4.4.1 Conservation Laws 187
4.4.2 Thermodynamics 187
4.4.3 Characteristic Analysis 188
4.4.4 Entropy Function 190
4.4.5 Centered Rarefaction Curves 192
4.4.6 Jump Conditions 194
4.4.7 Riemann Problem 200
4.4.8 Reflecting Walls 205
4.5 Case Study: Magnetohydrodynamics (MHD) 208
4.5.1 Conservation Laws 208
4.5.2 Characteristic Analysis 209
4.5.3 Entropy Function 218
4.5.4 Centered Rarefaction Curves 218
4.5.5 Jump Conditions 220

4.6 Case Study: Finite Deformation in Elastic Solids 221
4.6.1 Eulerian Formulation of Equations of Motion for Solids 221
4.6.2 Lagrangian Formulation of Equations of Motion for Solids 222
4.6.3 Constitutive Laws 223
4.6.4 Conservation Form of the Equations of Motion for Solids 225
4.6.5 Jump Conditions for Isothermal Solids 226
4.6.6 Characteristic Analysis for Solids 227
4.7 Case Study: Linear Elasticity 233
4.8 Case Study: Vibrating String 235
4.8.1 Conservation Laws 235
4.8.2 Characteristic Analysis 237
4.8.3 Jump Conditions 238
4.8.4 Lax Admissibility Conditions 240
4.8.5 Entropy Function 240
4.8.6 Wave Families for Concave Tension 241
4.8.7 Wave Family Intersections 245
4.8.8 Riemann Problem Solution 249
4.9 Case Study: Plasticity 255
4.9.1 Lagrangian Equations of Motion 255
4.9.2 Constitutive Laws 256
4.9.3 Centered Rarefactions 258
4.9.4 Hugoniot Loci 259
4.9.5 Entropy Function 261
4.9.6 Riemann Problem 261
4.10 Case Study: Polymer Model 267
4.10.1 Constitutive Laws 268
4.10.2 Characteristic Analysis 269
4.10.3 Jump Conditions 270
4.10.4 Riemann Problem Solution 271
4.11 Case Study: Three-Phase Buckley–Leverett Flow 273
4.11.1 Constitutive Models 273
4.11.2 Characteristic Analysis 275
4.11.3 Umbilic Point 276
4.11.4 Elliptic Regions 276
4.12 Case Study: Schaeffer–Schechter–Shearer System 277
4.13 Approximate Riemann Solvers 283
4.13.1 Design of Approximate Riemann Solvers 283
4.13.2 Artificial Diffusion 290
4.13.3 Rusanov Solver 292
4.13.4 Weak Wave Riemann Solver 293

4.13.5 Colella–Glaz Riemann Solver 295
4.13.6 Osher–Solomon Riemann Solver 297
4.13.7 Bell–Colella–Trangenstein Approximate Riemann Problem Solver 298
4.13.8 Roe Riemann Solver 303
4.13.9 Harten–Hyman Modification of the Roe Solver 312
4.13.10 Harten–Lax–van Leer Scheme 314
4.13.11 HLL Solvers with Two Intermediate States 316
4.13.12 Approximate Riemann Solver Recommendations 319

5 Methods for Scalar Laws 326
5.1 Convergence 326
5.1.1 Consistency and Order 326
5.1.2 Linear Methods and Stability 328
5.1.3 Convergence of Linear Methods 330
5.2 Entropy Conditions and Difference Approximations 331
5.2.1 Bounded Convergence 331
5.2.2 Monotone Schemes 341
5.3 Nonlinear Stability 353
5.3.1 Total Variation 353
5.3.2 Total Variation Stability 354
5.3.3 Other Stability Notions 357
5.4 Propagation of Numerical Discontinuities 359
5.5 Monotonic Schemes 361
5.5.1 Smoothness Monitor 361
5.5.2 Monotonizations 362
5.5.3 MUSCL Scheme 364
5.6 Discrete Entropy Conditions 367
5.7 E-Schemes 368
5.8 Total Variation Diminishing Schemes 370
5.8.1 Sufficient Conditions for Diminishing Total Variation 370
5.8.2 Higher-Order TVD Schemes for Linear Advection 374
5.8.3 Extension to Nonlinear Scalar Conservation Laws 378
5.9 Slope-Limiter Schemes 382
5.9.1 Exact Integration for Constant Velocity 383
5.9.2 Piecewise Linear Reconstruction 385
5.9.3 Temporal Quadrature for Flux Integrals 387
5.9.4 Characteristic Tracing 388
5.9.5 Flux Evaluation 389
5.9.6 Non-Reflecting Boundaries with the MUSCL Scheme 390

5.10 Wave Propagation Slope Limiter Schemes 391
5.10.1 Cell-Centered Wave Propagation 391
5.10.2 Side-Centered Wave Propagation 394
5.11 Higher-Order Extensions of the Lax–Friedrichs Scheme 395
5.12 Piecewise Parabolic Method 402
5.13 Essentially Non-Oscillatory Schemes 408
5.14 Discontinuous Galerkin Methods 412
5.14.1 Weak Formulation 412
5.14.2 Basis Functions 413
5.14.3 Numerical Quadrature 414
5.14.4 Initial Data 415
5.14.5 Limiters 416
5.14.6 Timestep Selection 417
5.15 Case Studies 418
5.15.1 Case Study: Linear Advection 418
5.15.2 Case Study: Burgers' Equation 422
5.15.3 Case Study: Traffic Flow 426
5.15.4 Case Study: Buckley–Leverett Model 427

6 Methods for Hyperbolic Systems 432
6.1 First-Order Schemes for Nonlinear Systems 432
6.1.1 Lax–Friedrichs Method 432
6.1.2 Random Choice Method 433
6.1.3 Godunov's Method 433
6.1.3.1 Godunov's Method with the Rusanov Flux 434
6.1.3.2 Godunov's Method with the Harten–Lax–vanLeer (HLL) Solver 435
6.1.3.3 Godunov's Method with the Harten–Hyman Fix for Roe's Solver 436
6.2 Second-Order Schemes for Nonlinear Systems 438
6.2.1 Lax–Wendroff Method 438
6.2.2 MacCormack's Method 439
6.2.3 Higher-Order Lax–Friedrichs Schemes 439
6.2.4 TVD Methods 443
6.2.5 MUSCL 447
6.2.6 Wave Propagation Methods 448
6.2.7 PPM 450
6.2.8 ENO 452
6.2.9 Discontinuous Galerkin Method 453

6.3 Case Studies 456
6.3.1 Wave Equation 456
6.3.2 Shallow Water 456
6.3.3 Gas Dynamics 459
6.3.4 MHD 461
6.3.5 Nonlinear Elasticity 461
6.3.6 Cristescu's Vibrating String 461
6.3.7 Plasticity 464
6.3.8 Polymer Model 467
6.3.9 Schaeffer–Schechter–Shearer Model 470

7 Methods in Multiple Dimensions 474
7.1 Numerical Methods in Two Dimensions 474
7.1.1 Operator Splitting 474
7.1.2 Donor Cell Methods 476
7.1.2.1 Traditional Donor Cell Upwind Method 478
7.1.2.2 First-Order Corner Transport Upwind Method 479
7.1.2.3 Wave Propagation Form of First-Order Corner Transport Upwind 483
7.1.2.4 Second-Order Corner Transport Upwind Method 485
7.1.3 Wave Propagation 488
7.1.4 2D Lax–Friedrichs 489
7.1.4.1 First-Order Lax–Friedrichs 490
7.1.4.2 Second-Order Lax–Friedrichs 491
7.1.5 Multidimensional ENO 494
7.1.6 Discontinuous Galerkin Method on Rectangles 494
7.2 Riemann Problems in Two Dimensions 498
7.2.1 Burgers' Equation 498
7.2.2 Shallow Water 500
7.2.3 Gas Dynamics 503
7.3 Numerical Methods in Three Dimensions 506
7.3.1 Operator Splitting 506
7.3.2 Donor Cell Methods 508
7.3.3 Corner Transport Upwind Scheme 510
7.3.3.1 Linear Advection with Positive Velocity 513
7.3.3.2 Linear Advection with Arbitrary Velocity 517
7.3.3.3 General Nonlinear Problems 518
7.3.3.4 Second-Order Corner Transport Upwind 519
7.3.4 Wave Propagation 521

7.4 Curvilinear Coordinates 521
7.4.1 Coordinate Transformations 522
7.4.2 Spherical Coordinates 523
7.4.2.1 Case Study: Eulerian Gas Dynamics in Spherical Coordinates 527
7.4.2.2 Case Study: Lagrangian Solid Mechanics in Spherical Coordinates 529
7.4.3 Cylindrical Coordinates 533
7.4.3.1 Case Study: Eulerian Gas Dynamics in Cylindrical Coordinates 537
7.4.3.2 Case Study: Lagrangian Solid Mechanics in Cylindrical Coordinates 539
7.5 Source Terms 542
7.6 Geometric Flexibility 542

8 Adaptive Mesh Refinement 544
8.1 Localized Phenomena 544
8.2 Basic Assumptions 546
8.3 Outline of the Algorithm 547
8.3.1 Timestep Selection 548
8.3.2 Advancing the Patches 549
8.3.2.1 Boundary Data 549
8.3.2.2 Flux Computation 550
8.3.2.3 Time Integration 552
8.3.3 Regridding 553
8.3.3.1 Proper Nesting 553
8.3.3.2 Tagging Cells for Refinement 556
8.3.3.3 Tag Buffering 559
8.3.3.4 Logically Rectangular Organization 559
8.3.3.5 Initializing Data after Regridding 559
8.3.4 Refluxing 560
8.3.5 Upscaling 560
8.3.6 Initialization 561
8.4 Object Oriented Programming 561
8.4.1 Programming Languages 562
8.4.2 AMR Classes 563
8.4.2.1 Geometric Indices 563
8.4.2.2 Boxes 567
8.4.2.3 Data Pointers 569
8.4.2.4 Lists 569

8.4.2.5 FlowVariables 570
8.4.2.6 Timesteps 571
8.4.2.7 TagBoxes 571
8.4.2.8 DataBoxes 571
8.4.2.9 EOSModels 572
8.4.2.10 Patch 572
8.4.2.11 Level 573
8.5 ScalarLaw Example 573
8.5.1 ScalarLaw Constructor 576
8.5.2 initialize 576
8.5.3 stableDt 577
8.5.4 stuffModelGhost 577
8.5.5 stuffBoxGhost 578
8.5.6 computeFluxes 578
8.5.7 conservativeDifference 579
8.5.8 findErrorCells 579
8.5.9 Numerical Example 579
8.6 Linear Elasticity Example 580
8.7 Gas Dynamics Examples 581

Bibliography 584
Index 593

Preface

Hyperbolic conservation laws describe a number of interesting physical problems in diverse areas such as fluid dynamics, solid mechanics, and astrophysics. Our emphasis in this book is on nonlinearities in these problems, especially those that lead to the development of propagating discontinuities. These propagating discontinuities can appear as the familiar shock waves in gases (the "boom" from explosions or super-sonic airplanes), but share many mathematical properties with other waves that do not appear to be so "shocking" (such as steep changes in oil saturations in petroleum reservoirs). These nonlinearities require special treatment, usually by methods that are themselves nonlinear. Of course, the numerical methods in this book can be used to solve linear hyperbolic conservation laws, but our methods will not be as fast or accurate as possible for these problems. If you are only interested in *linear* hyperbolic conservation laws, you should read about spectral methods and multipole expansions.

This book grew out of a one-semester course I have taught at Duke University over the past decade. Quite frankly, it has taken me at least 10 years to develop the material into a form that I like. I may tinker with the material more in the future, because I expect that I will never be fully satisfied.

I have designed this book to describe both numerical methods and their applications. As a result, I have included substantial discussion about the analytical solution of hyperbolic conservation laws, as well as discussion about numerical methods. In this course, I have tried to cover the applications in such a way that the engineering students can see the mathematical structure that is common to all of these problem areas. With this information, I hope that they will be able to adapt new numerical methods developed for other problem areas to their own applications. I try to get the mathematics students to adopt one of the physical models for their computations during the semester, so that the numerical experiments can help them to develop physical intuition.

I also tried to discuss a variety of numerical methods in this text, so that students could see a number of competing ideas. This book does not try to favor any one particular numerical scheme, and it does not serve as a user manual to a software package. It does have software available, to allow the reader to experiment with the various ideas. But the software is not designed for easy application to new problems. Instead, I hope that the readers will learn enough from this book to make intelligent decisions on which scheme is best for their problems, as well as how to implement that scheme efficiently.

There are a number of very good books on related topics. LeVeque's *Finite Volume Methods for Hyperbolic Problems* [97] is one that covers the mathematics well, describes several important numerical methods, but emphasizes the wave propagation scheme over all. Other books are specialized for particular problem areas, such as Hirsch's *Numerical Computation of Internal and External Flows* [73], Peyret and Taylor's *Computational Methods for Fluid Flow* [131], Roache's *Computational Fluid Dynamics* [137] and Toro's *Riemann Solvers and Numerical Methods for Fluid Dynamics* [159]. These books contain very interesting techniques that are particular for fluid dynamics, and should not be ignored.

Because this text develops analytical solutions to several problems, it is possible to measure the errors in the numerical methods on interesting test problem. This relates to a point I try to emphasize in teaching the course, that it is essential in numerical computation to perform mesh refinement studies in order to make sure that the method is performing properly. Another topic in this text is that numerical methods can be compared for accuracy (error for a given mesh size) and efficiency (error for a given amount of computational time). Sometimes people have an innate bias toward higher-order methods, but this may not be the most cost-effective approach for many problems. Efficiency is tricky to measure, because subtle programming issues can drive up computational time. I do not claim to have produced the most efficient version of any of the schemes in this text, so the efficiency comparisons should be taken "with a grain of salt."

The numerical comparisons produced some surprises for me. For example, I was surprised that approximate Riemann problem solvers often produce better numerical results in Godunov methods than "exact" Riemann solvers. Another surprise is that there is no clear best scheme or worst scheme in this text (although I have omitted discussions of schemes that have fallen out of favor in the literature for good reasons). There are some schemes that generally work better than most and some that often are less efficient than most, but all schemes have their niche in which they perform well. The journal literature, of course, is full of examples of the latter behavior, since the authors get to choose computational examples that benefit their method.

During the past ten years, I have watched numerical methods evolve, computers gain amazing speed, and students struggle harder with programming. The evolution of the methods led me to develop the course material into a form that students could access online. In that way, I could insert additional text for ready access by the students. The speed of current desktop machines allows us to make some reasonably interesting computations during the semester, seeing in a few minutes what used to require overnight runs on supercomputers. During that time, however, the new operating systems have separated the students ever farther from programming details.

As I gained experience with online text generation, I started to ask if it would be possible to develop an interactive text. First, I wanted students to be able to view the example programs while they were reading the text online. Next, I wanted students to be able to examine links to information available on the web. Then, I decided that it would be really nice if students could perform "what if" experiments within the text, by running numerical methods with different parameters and seeing the results immediately. Because I continue to think that only "real" programming languages (*i.e.*, C, C^{++} and Fortran) should be used for the material such as this, I rejected suggestions that I rewrite the programs in Matlab or Java. Eventually, our department systems programmer, Andrew Schretter, found a way to make things work for me, provided that I arrange for all parameter entry through graphical user interfaces. Our senior systems programmer, Yunliang Yu, did a lot of the development of the early form of the graphical user interface. One of my former graduate students, Wenjun Ying, programmed carefully the many cases for the marching cubes algorithm for visualizing level surfaces in three dimensions. I am greatly indebted to Andrew, Wenjun and Yunliang for their help.

This text is being published in two forms: traditional paper copy and a PDF file on a companion CD. The electronic form of the text contains links between equation or theorem references and the original statements. Similar links lead to bibliography citations or to occurrences of key words in the index. There are electronic links in the online text to source code and executables on the CD. This allows students to view computer implementations of the algorithms developed in the book, and to perform "what if" experiments with program and model parameters. However, since the text is the same for both versions of the book, this means that the paper text contains instructions to click on electronic links.

The graphical user interface (GUI) makes it easy for students to change parameters (and, in fact, to see all of the input parameters). The GUI also complicates the online programs. There is a danger that students may think that they have to program GUIs in order to solve these problems. That is not my intent. I have provided several example programs in the online version of chapter 2 to show students how they can write simple programs (that produce data sets for post processing) or slightly more complex programs (that display numerical results during the computation to look like movies), or very sophisticated programs (that use GUIs for input parameters). I would be happy if all students could program successfully in the first style. After all, CLAWPACK is a very successful example of that simple and direct style of programming.

It is common that students in this class are taking it in order to learn programming in Fortran or C^{++}, as much as they want to learn about the numerical methods. Both of these languages have advantages and disadvantages. Fortran is very good with arrays (subscripts can start at arbitrary values, which is useful for "ghost cells" in many methods) and has a very large set of intrinsic functions (for example, max and min with more than two arguments for slope limiters). Fortran is not very good with memory allocation, or with pointers in general. I use C^{++} to perform all memory allocation, and for all interactive graphics, including GUIs. When users select numerical methods through a GUI, then I set values for function pointers and pass those as arguments to Fortran routines. I do not recommend such practices for novice programmers. On the other hand, students who want to expand their programming skills can find several interesting techniques in the codes.

I do try to emphasize **defensive programming** when I teach courses that involve scientific computing. By this term, I mean the use of programming practices that make it easier to prevent or identify programming errors. It is often difficult to catch the use of uninitialized variables, the access of memory out of bounds, or memory leaks. The mixed-language programs all use the following defensive steps. First, floating-point traps are enabled in unoptimized code. Second, floating-point array values are initialized to IEEE infinity. Third, a memory debugger handles all memory allocation by overloading `operator new` in C^{++}. When the program makes an allocation request, the memory debugger gets even more space from the heap, and puts special bit patterns into the space before and after the user memory. As a result, the programmer can ask the memory debugger to check individual pointers or all pointers for writes out of bounds. This memory debugger is very fast, and does not add significantly to the overall memory requirements. The memory debugger also informs the programmer about memory leaks, providing information about where the unfreed pointer was allocated.

Unfortunately, mixing Fortran and C^{++} allows the possibility of truly bizarre programming errors. For example, declaring a Fortran subroutine to have a return value in a C^{++} `extern "C"` block can lead to stack corruption. I don't have a good defensive programming technique for that error.

But this book is really about numerical methods, not programming. I became interested in hyperbolic conservation laws well after graduate school, and I am indebted to several people for helping me to develop that interest. John Bell and

Gregory Shubin were particularly helpful when we worked together at Exxon Production Research. At Lawrence Livermore National Laboratory, I learnt much about Godunov methods from both John Bell and Phil Colella, and about object oriented programming from Bill Crutchfield and Mike Welcome. I want to thank all of them for their kind assistance during our years together.

Finally, emotional support throughout a project of this sort is essential. I want to thank my wife, Becky, for all her love and understanding throughout our years together. I could not have written this book without her.

1
Introduction to Partial Differential Equations

Partial differential equations arise in a number of physical problems, such as fluid flow, heat transfer, solid mechanics and biological processes. These equations often fall into one of three types. **Hyperbolic equations** are most commonly associated with advection, and **parabolic equations** are most commonly associated with diffusion. **Elliptic equations** are most commonly associated with steady states of either parabolic or hyperbolic problems.

Not all problems fall easily into one of these three types. Advection–diffusion problems involve important aspects of both hyperbolic and parabolic problems. Almost all advection problems involve a small amount of diffusion.

It is reasonably straightforward to determine the type of a general second-order partial differential equation. Consider the equation

$$\sum_{j=1}^{d}\sum_{i=1}^{d}\mathbf{A}_{ij}\frac{\partial^2 u}{\partial \mathbf{x}_i \partial \mathbf{x}_j} + \sum_{i=1}^{d} b_i \frac{\partial u}{\partial \mathbf{x}_i} + cu = 0.$$

Without loss of generality, we can assume that $\mathbf{A}$ is symmetric, by averaging the coefficients of the i, j and j, i derivative terms. By performing a linear coordinate transformation

$$\xi = \mathbf{F}\mathbf{x}$$

we hope to transformation the equation into a simpler form. We will find a way to choose the transformation matrix $\mathbf{F}$ below.

Note that

$$\begin{aligned}
\frac{\partial \xi_i}{\partial \mathbf{x}_j} &= \mathbf{F}_{ij} \\
\frac{\partial u}{\partial \mathbf{x}_i} &= \sum_{j=1}^{d} \frac{\partial u}{\partial \xi_j}\frac{\partial \xi_j}{\partial \mathbf{x}_i} = \sum_{j=1}^{d} \frac{\partial u}{\partial \xi_j}\mathbf{F}_{ji} \\
\frac{\partial^2 u}{\partial \mathbf{x}_i \mathbf{x}_j} &= \sum_{\ell=1}^{d}\sum_{k=1}^{d} \frac{\partial \xi_k}{\partial \mathbf{x}_i}\frac{\partial^2 u}{\partial \xi_k \partial \xi_\ell}\frac{\partial \xi_\ell}{\partial \mathbf{x}_j} = \sum_{\ell=1}^{d}\sum_{k=1}^{d} \mathbf{F}_{ki}\frac{\partial^2 u}{\partial \xi_k \partial \xi_\ell}\mathbf{F}_{\ell j}
\end{aligned}$$

After the coordinate transformation, the differential equation takes the form

$$0 = \sum_{j=1}^{d} \sum_{i=1}^{d} \mathbf{A}_{ij} \left[\sum_{\ell=1}^{d} \sum_{k=1}^{d} \mathbf{F}_{ki} \frac{\partial^2 u}{\partial \xi_k \partial \xi_\ell} \mathbf{F}_{\ell j} \right] + \sum_{i=1}^{d} b_i \left[\sum_{j=1}^{d} \frac{\partial u}{\partial \xi_j} \mathbf{F}_{ji} \right] + cu$$

$$= \sum_{\ell=1}^{d} \sum_{k=1}^{d} \left[\sum_{j=1}^{d} \sum_{i=1}^{d} \mathbf{F}_{ki} \mathbf{A}_{ij} \mathbf{F}_{\ell j} \right] \frac{\partial^2 u}{\partial \xi_k \partial \xi_\ell} + \sum_{j=1}^{d} \left[\sum_{i=1}^{d} \mathbf{F}_{ji} b_i \right] \frac{\partial u}{\partial \xi_j} + cu.$$

We would like to choose the matrix $\mathbf{F}$ so that $\mathbf{D} = \mathbf{F}\mathbf{A}\mathbf{F}^\top$ is diagonal. Recall that we can diagonalize a symmetric matrix by means of an orthogonal change of variables. In other words, we can choose $\mathbf{F}$ to be an orthogonal matrix.

If $\mathbf{D}$ has nonzero diagonal entries all of the same sign, the differential equation is **elliptic**. The canonical example of an elliptic equation is the Laplace equation $\nabla_\mathbf{x} \cdot \nabla_\mathbf{x} u = 0$. If $\mathbf{D}$ has nonzero diagonal entries with one entry of different sign from the others, then the differential equation is **hyperbolic**. The canonical example of a hyperbolic equation is the wave equation $\frac{\partial^2 u}{\partial t^2} - \nabla_\mathbf{x} \cdot \nabla_\mathbf{x} u = 0$. We will discuss simple hyperbolic equations in Chapter 2, and general hyperbolic equations in Chapter 4. If $\mathbf{D}$ has one zero diagonal entry, the equation may be **parabolic**. The canonical example of a parabolic equation is the heat equation $\frac{\partial u}{\partial t} + \nabla_\mathbf{x} \cdot \nabla_\mathbf{x} u = 0$.

Example 1.0.1 Consider the differential equation

$$\frac{\partial^2 u}{\partial \mathbf{x}_1^2} + \frac{\partial^2 u}{\partial \mathbf{x}_2^2} - \frac{\partial^2 u}{\partial \mathbf{x}_3 \partial \mathbf{x}_4} = 0$$

which arises in the Khokhlov–Zabolotskaya–Kuznetsov (KZK) equation for biomedical imaging. In this case, the coefficient matrix is

$$\mathbf{A} = \begin{bmatrix} 1 & 0 & 0 & 0 \\ 0 & 1 & 0 & 0 \\ 0 & 0 & 0 & -1/2 \\ 0 & 0 & -1/2 & 0 \end{bmatrix}.$$

A coordinate transformation that diagonalizes $\mathbf{A}$ is given by

$$\mathbf{F} = \begin{bmatrix} 1 & 0 & 0 & 0 \\ 0 & 1 & 0 & 0 \\ 0 & 0 & 1/\sqrt{2} & 1/\sqrt{2} \\ 0 & 0 & -1/\sqrt{2} & 1/\sqrt{2} \end{bmatrix}$$

and the new coefficient matrix is

$$\mathbf{D} = \begin{bmatrix} 1 & 0 & 0 & 0 \\ 0 & 1 & 0 & 0 \\ 0 & 0 & 1/2 & 0 \\ 0 & 0 & 0 & -1/2 \end{bmatrix}.$$

In this case, we see that the KZK equation is hyperbolic.

This book will discuss analytical and numerical methods for solving hyperbolic equations. Our emphasis will be on numerical methods and nonlinear problems, but a knowledge of some analytical approaches will be very useful for computation. Generally, our hyperbolic equations will arise from a physical law describing the conservation of some quantity, such as mass, momentum or energy. These will take a special form, which we will build into our numerical methods, so that our computations conserve these physical quantities as well.

Here is an outline of the analytical approaches in this book, whether they are applied to problems or numerical methods. In Chapter 2 we will study linear hyperbolic conservation laws in a single unknown. We will learn how the solution of such problems depends on initial and boundary data, so that we can construct numerical methods that respect this dependence. We will also develop some simple methods for analyzing the behavior of the numerical methods. First, we will use calculus to see how the approximations in the numerical method cause us to be solving a differential equation that is slightly different from the problem that was posed. This approach, called a modified equation analysis, gives us a qualititative feel for how the method should perform in practice. Second, we will use Fourier analysis to see how the methods propagate waves, and use this analysis to develop the very important Lax equivalence theorem.

In Chapter 3 we will begin our study of nonlinear hyperbolic conservation laws. We will learn about the development and propagation of discontinuities, and see that an understanding of infinitesimal diffusive effects is essential to understanding how nature selects certain solutions to these problems. We will also begin to learn how to build numerical diffusion into our computational methods, so that we can expect to compute the physically correct solutions as well. This numerical diffusion will arise in subtle ways, depending on how how numerical schemes use upwinding and averaging techniques. Some approaches will concentrate on building important analytical information about the wave propagation into the method, while other schemes will assiduously avoid such analytical work. We will apply these methods to problems in traffic flow and oil recovery/contaminant cleanup.

Chapter 4 will discuss hyperbolic systems of conservation laws. This is where the discussion becomes most practical, because the physical applications are so interesting. Once we understand the basic principles underlying the analytical solution of hyperbolic systems, we will perform case studies of shallow water, compressible gas dynamics, magnetohydrodynamics, solid mechanics and flow in porous media. The analytical solution of the equations of motion for these problems for special initial data (Riemann problems) can be very useful in building some of our numerical methods. Unfortunately, this analytical information is often expensive to compute and difficult to program, when it is available. As a result, we will find methods to approximate the solution of Riemann problems.

Amazingly enough, several of these approximate Riemann solvers produce better numerical results than the analytical methods, and at far less cost with far simpler programs.

In Chapter 5 we will try to analyze the numerical methods, and use the analysis to design better methods. We will run into an obstacle due to Godunov: linear schemes that preserve monotonicity are at best first-order accurate. In order to achieve higher-order accuracy, we will design nonlinear schemes, even for use on linear problems. These schemes will be very useful for solving problems with propagating discontinuities. They will not be the most effective schemes for solving linear problems with smooth solutions. We will extend these higher-order numerical schemes to solve hyperbolic systems in Chapter 6, and to solve problems in multiple dimensions in Chapter 7.

But this book is not just about analysis of problems and methods. In each chapter, there are discussions of numerical results and comparisons of numerical methods. It is important that the student learn how to judge when a numerical method is working properly, sometimes by understanding its numerical stability, and often by performing mesh refinement studies to verify the correct order of convergence. Numerical methods can differ greatly in their achieved accuracy even when they have the same nominal order of accuracy. Methods can also differ greatly in their efficiency, meaning how much it costs us to achieve a given accuracy. Unfortunately, there is no best method in this book, that applies to all problems and is always most (or nearly most) efficient.

In order to assist the student in gaining knowledge about the design and performance of numerical methods, we have provided an interactive form of this book. Fortunately, you are currently reading that version. In this way, students can view computer programs to learn about code organization. Students can also run the programs from inside the book, and adjust parameters that control the numerical performance. Through the use of interactive graphics, the student can see the evolution of the numerical solution; this really helps in understanding instability and the spread of discontinuities due to numerical diffusion.

In order to execute programs from inside this book, it was necessary to use graphical user interfaces. These make the selection of program parameters easy once the code is written, but makes the example code somewhat larger than it needs to be just to solve the problem. In order to help the student here, we have provided a series of programs in Section 2.2.3 of Chapter 2. These programs start with short Fortran programs, proceed through more modular Fortran to mixed language programs, and end up with the more complicated program containing interactive graphics and graphical user interfaces. Students can write their own programs in any of these styles, as is appropriate for their experience or the expectations of their instructor.

If the student can learn about mixed language programming, then the discussion on adaptive mesh refinement in Chapter 8 should be interesting. This chapter describes the basic principles behind Marsha Berger's structured adaptive mesh refinement, and describes the basic ideas in the design of the author's adaptive mesh refinement program. The hope is that after study of the applications of adaptive mesh refinement to oil recovery, linear elasticity and gas dynamics, the student can apply adaptive mesh refinement to other research problems.

2

Scalar Hyperbolic Conservation Laws

In numerical analysis or scientific computing courses, it is common to examine ordinary differential equations and some basic numerical methods to solve these problems. In this chapter we will develop several basic numerical methods to solve initial value problems arising from a particular class of partial differential equations, namely scalar hyperbolic conservation laws. In some cases, we will be able to transform the solution of partial differential equations into ordinary differential equations. However, in many practical problems there are physical effects, such as diffusion, that prevent such analytical reductions. These ideas will be developed in Section 2.1.

The design of numerical methods for scalar conservation laws involves principles that are different from those commonly considered in the solution of ordinary differential equations. Some experimentation with obvious numerical discretizations in Section 2.2 will produce surprises, and illustrate the utility of interactive graphical displays in programming. Analysis of these basic numerical methods using Taylor series and Fourier transforms in Sections 2.3 and 2.5 will yield some basic numerical principles, and the limitations of the simple numerical methods.

2.1 Linear Advection

Linear advection describes the motion of some conserved quantity along a constant velocity field. This is the simplest conservation law, but it illustrates many of the important features we will see in more complicated conservation laws.

2.1.1 Conservation Law on an Unbounded Domain

The unbounded linear advection problem takes the form

$$\frac{\partial u}{\partial t} + \frac{\partial cu}{\partial x} = 0 \text{ for all } x \in \mathbf{R} \text{ for all } t > 0, \tag{2.1a}$$

$$u(x,0) = u_0(x) \text{ for all } x \in \mathbf{R}. \tag{2.1b}$$

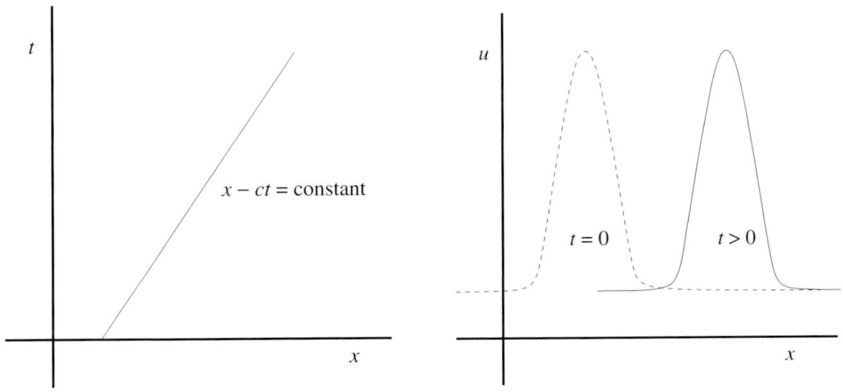

(a) Characteristic line (b) Evolution

Fig. 2.1 Characteristics in linear advection

In this initial-value problem, we assume that the velocity c is constant. Then the differential equation (2.1a) can be rewritten in the form

$$0 = \begin{bmatrix} 1, & c \end{bmatrix} \begin{bmatrix} \frac{\partial u}{\partial t} \\ \frac{\partial u}{\partial x} \end{bmatrix} \quad \text{for all } x \in \mathbf{R} \text{ for all } t > 0.$$

This equation says that the gradient of u is orthogonal to a constant vector. It follows that u is constant on lines parallel to that constant vector:

$$\text{for all } (x_0, t_0) \text{ for all } \tau \; u(x_0 + c\tau, t_0 + \tau) = \text{constant}.$$

Choosing $\tau = t' - t_0$ gives us

$$u(x_0 + c(t - t_0), t) = u(x_0 - ct_0, 0) \equiv u_0(x_0 - ct_0).$$

Given x, choose $x_0 = x - ct + ct_0$ to get

$$u(x, t) = u_0(x - ct).$$

This is a formula for the solution of problem (2.1). It is clear from this formula that the **characteristic** lines

$$x - ct = \text{constant}$$

are especially important. Along a characteristic line, the solution of the conservation law at time $t > 0$ is equal to the initial value at time $t = 0$. These ideas are illustrated in Figure 2.1.

There is an easy way to verify this solution. Suppose that we define the new variables

$$\xi = x - ct, \quad \tau = t \tag{2.2}$$

and the function

$$\tilde{u}(\xi, \tau) \equiv u(x, t). \tag{2.3}$$

Then the chain rule implies that

$$\frac{\partial u}{\partial t} = \frac{\partial \tilde{u}}{\partial \tau} - \frac{\partial \tilde{u}}{\partial \xi} c \quad \text{and} \quad \frac{\partial u}{\partial x} = \frac{\partial \tilde{u}}{\partial \xi}.$$

It follows that $\tilde{u}$ solves the initial value problem

$$0 = \frac{\partial u}{\partial t} + c\frac{\partial u}{\partial x} = \frac{\partial \tilde{u}}{\partial \tau},$$
$$\tilde{u}(\xi, 0) = u_0(\xi).$$

The differential equation for $\tilde{u}$ shows that $\tilde{u}$ is a function of ξ alone. In summary, after we change to characteristic coordinates the original partial differential equation becomes a system of ordinary differential equations, parameterized by ξ. Further, these ordinary differential equations have the trivial solution

$$\tilde{u}(\xi) = u_0(\xi).$$

2.1.2 Integral Form of the Conservation Law

In general, a **conservation law** in one dimension takes the form

$$\frac{\partial u}{\partial t} + \frac{\partial f}{\partial x} = 0. \tag{2.4}$$

Here u is the conserved quantity, and f is the flux. For example, the linear advection flux is $f = cu$.

There is a physical reason for calling Equation (2.4) a conservation law. By integrating over the space-time rectangle $(a, b) \times (0, t)$ and applying the divergence theorem, we obtain

$$\int_a^b u(x, t)\, dx = \int_a^b u(x, 0)\, dx + \int_0^t f(a, \tau)\, d\tau - \int_0^t f(b, \tau)\, d\tau. \tag{2.5}$$

We can interpret the conservation law (2.5) as follows. The quantity u represents a density, *i.e.*, the conserved quantity per length. Thus the spatial integrals represent the total conserved quantity in the interval (a, b) at some advanced time t and the initial time 0. The temporal integrals represent the total amount of the conserved quantity flowing through ends of the interval in space during the given interval in time. Thus equation (2.5) says that the total conserved quantity in the interval (a, b) at time t is equal to the total conserved quantity in the same interval initially, plus what flows into the interval on the left and minus what flows out on the right.

2.1.3 Advection–Diffusion Equation

Many physically realistic problems actually involve some amount of diffusion. For example, the miscible displacement problem described in Section 3.2.2 is a linear advection problem involving a physical diffusion. In general, one-dimensional linear advection with constant diffusion takes the form

$$\frac{\partial u}{\partial t} + \frac{\partial cu}{\partial x} = \frac{\partial}{\partial x}\left(\epsilon \frac{\partial u}{\partial x}\right) \text{ for all } x \in \mathbf{R} \text{ for all } t > 0, \tag{2.6a}$$

$$u(x,0) = u_0(x) \text{ for all } x \in \mathbf{R}. \tag{2.6b}$$

Here, we assume that the diffusion coefficient satisfies $\epsilon > 0$, so that the conservation law is well-posed. The need for this restriction on ϵ will become obvious in Equation (2.8) below.

Let us transform again to characteristic coordinate $\xi = x - ct$ and time $\tau = t$ as in Equation (2.2) and define the solution $\tilde{u}$ in terms of these coordinates as in (2.3). Then substitution into the advection–diffusion equation (2.6) leads to

$$\begin{aligned} \frac{\partial \tilde{u}}{\partial \tau} &= \epsilon \frac{\partial^2 \tilde{u}}{\partial \xi^2} \text{ for all } \xi \in \mathbf{R} \text{ for all } \tau > 0, \\ \tilde{u}(\xi,0) &= u_0(\xi) \text{ for all } \xi \in \mathbf{R}. \end{aligned} \tag{2.7}$$

This is the one-dimensional **heat equation** on an unbounded interval. If the initial data u_0 grow sufficiently slowly for large values of its argument, then it is well known that the analytical solution of this equation is

$$\tilde{u}(\xi,\tau) = \int_{-\infty}^{\infty} \frac{1}{\sqrt{4\pi\epsilon\tau}} e^{-(\xi-y)^2/(4\epsilon\tau)} u_0(y)dy \equiv \int_{-\infty}^{\infty} G(\xi - y, \tau)u_0(y)dy. \tag{2.8}$$

Here

$$G(\xi,\tau) = \frac{1}{\sqrt{4\pi\epsilon\tau}} e^{-\xi^2/(4\epsilon\tau)}$$

is called the **Green's function**. Because the diffusion constant ϵ is positive, the Green's function is real-valued. Here $\tilde{u}$ is smooth for $t > 0$ because derivatives of $\tilde{u}$ involve derivatives of the smooth Green's function G, and not derivatives of the initial data u_0.

It follows that the solution of the linear advection–diffusion problem (2.6) is

$$u(x,t) = \int_{-\infty}^{\infty} \frac{1}{\sqrt{4\pi\epsilon t}} e^{-(x-ct-y)^2/(4\epsilon t)} u_0(y)dy.$$

The lines $x - ct =$ constant are still important, in that they carry the bulk of the initial information for small diffusion, but they are no longer lines along which the solution u is constant.

Note that the Green's function G approaches a delta-function as the diffusion coefficient $\epsilon \to 0$. On the other hand, after sufficiently large time even a small diffusion will spread the effect of disturbances in the initial data over significant intervals in space. If the initial data is zero outside some bounded interval, then at very large times the solution $\tilde{u}$ will decay to zero. These observations are important, because in many practical situations we are interested in the solution of conservation laws obtained in the limit as the diffusion tends to zero. The study of the interplay between small diffusion and large times is an appropriate matter for asymptotics, and would take the current discussion too far astray.

It is sometimes useful to note that the linear advection–diffusion equation (2.6a) is a conservation law. In fact, we can rewrite it in the form

$$\frac{\partial u}{\partial t} + \frac{\partial}{\partial x}\left(cu - \epsilon \frac{\partial u}{\partial x}\right) = 0 \text{ for all } x \in \mathbf{R} \text{ for all } t > 0.$$

Here the flux $f(x,t) \equiv cu - \epsilon \frac{\partial u}{\partial x}$ is the difference of the advective flux cu and the diffusive flux $\epsilon \frac{\partial u}{\partial x}$. We can develop an integral form of this conservation law by using (2.5).

2.1.4 Advection Equation on a Half-Line

Here is another important modification to the problem (2.1). For both practical and computational purposes, we might be interested in solving a semi-infinite problem with boundary data:

$$\frac{\partial u}{\partial t} + \frac{\partial (cu)}{\partial x} = 0 \text{ for all } x > 0 \text{ for all } t > 0, \tag{2.9a}$$

$$u(0,t) = v(t) \text{ for all } t > 0, \tag{2.9b}$$

$$u(x,0) = u_0(x) \text{ for all } x > 0. \tag{2.9c}$$

If we transform to characteristic coordinates as in Equation (2.2), we see that the solution of (2.9) depends on the data $v(t)$ at the left-hand boundary only for $x - ct < 0$. If $c < 0$, this inequality cannot be satisfied for any (x,t) in the problem domain; in other words, no points in the problem domain will depend on the data at the left-hand boundary. Thus in this case we assume that the velocity c is positive: $c > 0$. Since the solution of (2.9) is constant along characteristics, we can easily solve to get

$$u(x,t) = \begin{cases} u_0(x - ct), & x - ct > 0 \\ v(t - x/c), & x - ct < 0. \end{cases}$$

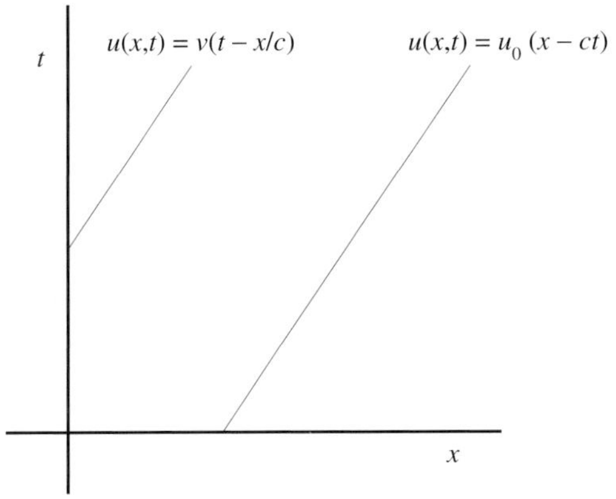

Fig. 2.2 Characteristics in linear advection on a quarter-plane

This solution is illustrated in Figure 2.2. In other words, the solution in part of the domain, namely points that can be reached by characteristics from the positive x-axis, is given by the initial data; the solution in the remainder of the domain is given by tracing back along characteristics to the boundary data on the positive t-axis.

2.1.5 Advection Equation on a Finite Interval

In practice, we will not want to compute on an unbounded domain. Instead, we will typically work with a problem on a bounded domain

$$\frac{\partial u}{\partial t} + \frac{\partial (cu)}{\partial x} = 0 \text{ for all } a < x < b \text{ for all } t > 0, \tag{2.10a}$$

$$u(a,t) = v(t) \text{ for all } t > 0, \tag{2.10b}$$

$$u(x,0) = u_0(x) \text{ for all } a < x < b. \tag{2.10c}$$

Here we have assumed for simplicity that the velocity c satisfies $c > 0$. Note that we do not specify a value for u at the right-hand boundary, since the characteristics show us that these values are determined by the initial data and the data on the left-hand boundary.

It is interesting to note that if we added diffusion to this problem, as in Equation (2.6a), then we would have to specify u at the right-hand side of the domain. In this case, the analytical solution of the advection–diffusion problem on a finite interval would typically involve a boundary layer at the right-hand side, unless the boundary data there is chosen very carefully.

Exercises for 2.1

2.1.1 Consider the variable coefficient conservation law

$$\frac{\partial u}{\partial t} + \frac{\partial (uc)}{\partial x} = 0, \quad \text{for all } x \in \mathbf{R} \text{ for all } t > 0$$
$$u(x,0) = u_0(x), \quad \text{for all } x \in \mathbf{R}$$

where c is a function of x. Assume that $c(x) \neq 0$ for all x.

(a) Let $f(x,t) = u(x,t)c(x)$ and show that f satisfies the partial differential equation

$$\frac{\partial f}{\partial t} + c\frac{\partial f}{\partial x} = 0.$$

(b) If $b(x)$ satisfies $db/dx = 1/c(x)$, let $\xi(x,t) = t - b(x)$ and show that $w(\xi(x,t),t) \equiv f(x,t)$ satisfies the partial differential equation

$$\frac{\partial w}{\partial t} = 0.$$

(c) If $b(x)$ has an inverse function, show that

$$u(x,t) = \frac{c(b^{-1}(b(x)-t))}{c(x)} u_0(b^{-1}(b(x)-t))$$

satisfies the original differential equation.

(d) Find the solution of the variable-coefficient linear advection problem if $c(x) = a + bx$ with constant a and constant $b \neq 0$.

2.2 Linear Finite Difference Methods

2.2.1 Basics of Discretization

In order to approximate the solution to the linear advection problem (2.10), we will discretize space by a finite increasing sequence of grid points

$$a = x_{-1/2} < x_{1/2} < \cdots < x_{I-\frac{3}{2}} < x_{I-1/2} = b$$

and time points

$$0 = t^0 < t^1 < \ldots < t^{N-1} < t^N = T.$$

We will define the computational grid **cells** to be the intervals $(x_{i-1/2}, x_{i+1/2})$, with cell widths

$$\triangle x_i \equiv x_{i+1/2} - x_{i-1/2} \quad \text{for all } 0 \leq i < I.$$

We will also define the **timesteps** to be

$$\triangle t^{n+1/2} \equiv t^{n+1} - t^n \quad \text{for all } 0 \leq n < N.$$

See Figure 2.3 for an illustration of spatial and temporal discretization.

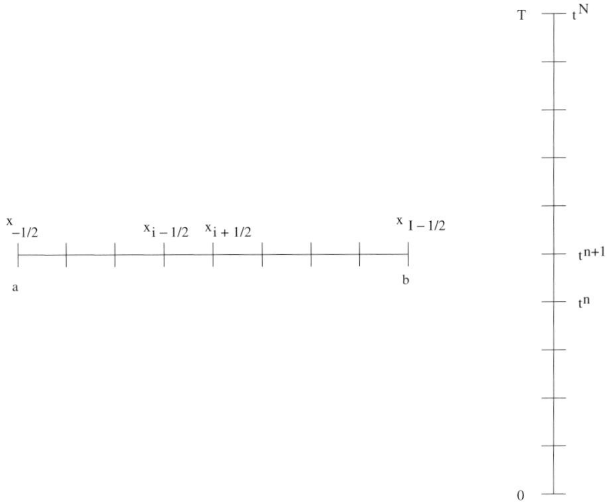

Fig. 2.3 Spatial and temporal discretization

As in section 2.1.2, we can integrate the differential equation (2.10a) over the space-time rectangle $(x_{i-1/2}, x_{i+1/2}) \times (t^n, t^{n+1})$ to find that for all $0 \leq i < I$ and all $0 \leq n < N$

$$\int_{x_{i-1/2}}^{x_{i+1/2}} u(x, t^{n+1})\, dx = \int_{x_{i-1/2}}^{x_{i+1/2}} u(x, t^n)\, dx + \int_{t^n}^{t^{n+1}} cu(x_{i-1/2}, t)\, dt - \int_{t^n}^{t^{n+1}} cu(x_{i+1/2}, t)\, dt. \quad (2.11)$$

This equation involves no approximations. For the more general nonlinear scalar conservation law (2.4) we obtain the similar equation

$$\int_{x_{i-1/2}}^{x_{i+1/2}} u(x, t^{n+1})\, dx = \int_{x_{i-1/2}}^{x_{i+1/2}} u(x, t^n)\, dx + \int_{t^n}^{t^{n+1}} f(u(x_{i-1/2}, t))\, dt - \int_{t^n}^{t^{n+1}} f(u(x_{i+1/2}, t))\, dt. \quad (2.12)$$

Equation (2.12) suggests that we construct numerical approximations to cell averages of the conserved quantity in the conservation law. Our numerical scheme will involve discrete quantities u_i^n that approximate the cell averages in the following sense:

$$u_i^n \approx \frac{1}{\triangle x_i} \int_{x_{i-1/2}}^{x_{i+1/2}} u(x, t^n)\, dx \quad \text{for all } 0 \leq i < I. \quad (2.13)$$

Similarly, we will work with discrete quantities $f_{i+1/2}^{n+1/2}$ that approximate time averages of the flux in the conservation law:

$$f_{i+1/2}^{n+1/2} \approx \frac{1}{\triangle t^{n+1/2}} \int_{t^n}^{t^{n+1}} f(u(x_{i+1/2}, t))\, dt \quad \text{for all } 0 \le i < I.$$

As suggested by Equation (2.12) we will require our discretization to satisfy

$$u_i^{n+1} = u_i^n - \frac{\triangle t^{n+1/2}}{\triangle x_i} \left[f_{i+1/2}^{n+1/2} - f_{i-1/2}^{n+1/2} \right] \quad \text{for all } 0 \le i < I. \tag{2.14}$$

Once we relate the fluxes $f_{i+1/2}^{n+1/2}$ to the solution values u_i^n and u_i^{n+1}, this will have the form of a recurrence relation.

If we sum equation (2.14) over all spatial intervals, we obtain a telescoping sum that simplifies to

$$\sum_{i=0}^{I-1} u_i^{n+1} \triangle x_i = \sum_{i=0}^{I-1} u_i^n \triangle x_i - \triangle t^{n+1/2} \left[f_{I-1/2}^{n+1/2} - f_{-1/2}^{n+1/2} \right] \quad \text{for all } 0 \le n < N.$$

If there is no flux at the boundaries (that is, if $f_{-1/2} = 0 = f_{I-1/2}^{n+1/2}$), then this equation shows that the total discrete amount of u is conserved. As a result, schemes of the form (2.14) are called **conservative schemes**. Conservative finite difference schemes for solving the conservation law are distinguished solely by their choices for the numerical fluxes $f_{i+1/2}^{n+1/2}$, for $0 \le i < I$ and $0 \le n < N$.

Initial values for our numerical method are chosen to be the cell averages of the initial data:

$$u_i^0 = \frac{1}{\triangle x_i} \int_{x_{i-1/2}}^{x_{i+1/2}} u_0(x)\, dx \quad \text{for all } 0 \le i < I.$$

At the left-hand boundary $x_{-1/2} = a$ we define the numerical fluxes by time averages of the boundary data:

$$f_{-1/2}^{n+1/2} = \frac{1}{\triangle t^{n+1/2}} \int_{t^n}^{t^{n+1}} f(v(t))\, dt \quad \text{for all } 0 \le n < N. \tag{2.15}$$

2.2.2 *Explicit Upwind Differences*

The simplest numerical approximation to the linear advection equation is the **explicit upwind difference** method

$$u_i^{n+1} = u_i^n - \left[u_i^n - u_{i-1}^n \right] \frac{c \triangle t^{n+1/2}}{\triangle x_i}, \quad 0 < i < I \tag{2.16a}$$

$$u_0^{n+1} = u_0^n - \left[c u_0^n - f_{-1/2}^{n+1/2} \right] \frac{\triangle t^{n+1/2}}{\triangle x_i}. \tag{2.16b}$$

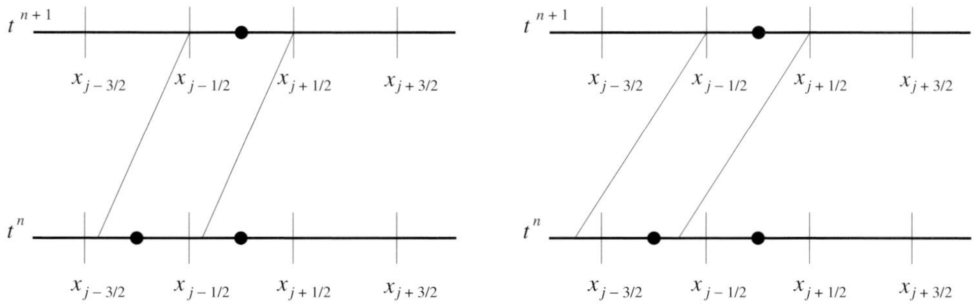

(a) CFL < 1

(b) CFL > 1

Fig. 2.4 Explicit upwind stencil

This is a conservative difference scheme in which the numerical fluxes are computed by $f_{i+1/2}^{n+1/2} = cu_i^n$ for all $0 \le i < I$, and by Equation (2.15) at the inflow boundary $i = 0$.

For explicit upwind differences away from the left boundary, the new solution u_i^{n+1} depends solely on the previous data u_i^n and u_{i-1}^n (the so-called **stencil** of the scheme). Note that these are averages over the intervals $(x_{i-\frac{3}{2}}, x_{i-1/2})$ and $(x_{i-\frac{1}{2}}, x_{i+1/2})$, respectively. Thus the domain of dependence of u_i^{n+1} is the interval $(x_{i-\frac{3}{2}}, x_{i+1/2})$. Recall that the physical domain of dependence is the interval $(x_{i-1/2} - c\triangle t^{n+1/2}, x_{i+1/2} - c\triangle t^{n+1/2})$, which corresponds to tracing the endpoints of the interval $(x_{i-1/2}, x_{i+1/2})$ backward along characteristics from t^{n+1} to t^n. It follows that the numerical domain of dependence contains the physical domain of dependence if and only if the timestep satisfies the **Courant–Friedrichs–Levy condition** (also known as the **CFL condition**)

$$c\triangle t^{n+1/2} \le \min_i \{\triangle x_i\}. \tag{2.17}$$

These ideas are illustrated in Figure 2.4.

The explicit upwind difference scheme (2.16) and the CFL condition (2.17) suggest that we define the dimensionless Courant–Friedrichs–Levy number, usually abbreviated to **CFL number**, namely

$$\gamma_i^{n+1/2} \equiv \frac{c\triangle t^{n+1/2}}{\triangle x_i}. \tag{2.18}$$

If the timestep is chosen so that the CFL number satisfies $\gamma_i^{n+1/2} \le 1$ for all i, then the explicit upwind scheme can be rewritten as the weighted average

$$u_i^{n+1} = u_i^n(1 - \gamma_i^{n+1/2}) + u_{i-1}^n \gamma_i^{n+1/2}, \quad 0 < i < I.$$

Because the new solution is an average of the previous values when the CFL number is at most one, the extreme value of the solution at the new time lies between the extreme values of the previous solution. This implies that the upwind difference scheme is stable when the CFL condition (2.17) is satisfied.

Finally, let us note that the explicit upwind scheme depends on the assumption that the velocity c is positive. In general, the explicit upwind scheme chooses the numerical flux to be

$$f_{i+1/2}^{n+1/2} = \begin{cases} cu_i^n, & c \geq 0 \text{ for all } 0 \leq i < I, \\ cu_{i+1}^n, & c < 0 \text{ for all } -1 \leq i < I-1. \end{cases}$$

Furthermore, if $c < 0$ then we have to change the boundary condition (2.10b) and define fluxes $f_{I-1/2}^{n+1/2}$ similar to Equation (2.15).

2.2.3 Programs for Explicit Upwind Differences

Explicit upwind differences are easy to program. Nevertheless, in order to assist the student with code organization, visualization and debugging, we have provided five example programs. These programs will increase the complexity in the main program and makefile, but typically share subroutines that compute the solution of the conservation law. When we begin to experiment with different integration schemes and differential equations, we will use the the last of these programs.

To prepare to obtain copies of these codes, perform the following steps:

(i) Type "`cd`" to return to your home directory.
(ii) Type "`mkdir scalar_law`" to make a directory to contain the program code in this chapter.
(iii) Type "`cd scalar_law`" to enter the new directory.
(iv) Download **Program 2.2-1: tarfile** from the web page.
(v) Type "`tar -xvf tarfile`" to unbundle the codes in your new `scalar_law` directory.
(vi) Type "`rm tarfile`" to remove the code bundle in your `scalar_law` directory. The unbundled code will remain.

2.2.3.1 First Upwind Difference Program

The first program is designed to be as simple as possible. It consists of the single main program **Program 2.2-2: main.f**, written in Fortran. This program is specifically designed to solve the linear advection problem with a positive advection velocity. It also uses a uniform mesh, specifies a fixed value for the solution at the left boundary, and uses piecewise-constant initial data. The time evolution of the solution is terminated by exceeding either a specified number of timesteps or a specified simulation time. The program prints the final results for use by a separate plotting program. Instructions for using the program can be found in the accompanying **Program 2.2-3: README**.

To run this code, perform the following steps:

(i) Type "`cd`" to return to your home directory.
(ii) Type "`cd scalar_law/PROGRAM0`" to enter the directory for the first program.
(iii) Type "`g77 main.f`" build the program executable.
(iv) Type "`a.out > output`" to run the program and direct the output to the file names "output."
(v) Type "`xmgrace output`" to plot the computational results.

This information is contained in the `README` file.

2.2.3.2 Second Upwind Difference Program

The second program is designed to be more modular than the first program. This program consists of several pieces:

- **Program 2.2-4: linaddmain.f** Fortran main program for solving a linear advection problem over some specified time or number of timesteps.
- **Program 2.2-5: riemprob.f** Fortran routines for initializing the solution and mesh, and handling boundary conditions;
- **Program 2.2-6: linearad.f** Fortran routines for computing characteristic speeds and solving Riemann problems;
- **Program 2.2-7: upwind.f** Fortran routine to compute a numerical approximation to the time integral of the flux at a cell side;
- **Program 2.2-8: consdiff.f** Fortran routine to apply conservative differences;
- **Program 2.2-9: linearad.i** Fortran common block for parameters used in the linear advection model and Riemann problem;
- **Program 2.2-10: const.i** Fortran common block for machine dependent parameters, and parameter statements for some common constants;
- **Program 2.2-11: GNUmakefile** Makefile to compile and load the Fortran files.

It is strongly suggested that the student maintain this basic style of organization for the code. The separation of initial and boundary conditions from the time integration will make the experimentation with alternative numerical schemes easier. It will also make it easier for us to apply the methods to a variety of differential equations, or to different initial values or boundary conditions. Furthermore, the modular organization will assist the transition to adaptive mesh refinement in Chapter 8.

File `riemprob.f` contains three routines, `initsl`, `bcmesh` and `bccells`. Subroutine `initsl` initializes the conserved quantity and mesh for a scalar law, such as linear advection. Subroutine `bcmesh` sets values for the mesh outside the physical domain, in case these are need for boundary conditions. Subroutine `bccells` sets values for the conserved quantity outside the physical domain, using the user-specified physical boundary condition. The only tricky aspect of this routine is the array addressing. In order to simplify the treatment of boundary conditions, we may begin array addresses with negative indices, determined by the value of the integer arguments (`fc,lc` *etc.*) to the routine. This simplifies the treatment of boundary conditions because index 0 in the Fortran array corresponds to the first cell inside the domain; this corresponds to the way in which we have written the difference scheme in this text.

File `linearad.f` contains two routines, `fluxderv` and `riemann`. Initially, we will not use either of these routines. The subroutine `fluxderv` computes the characteristic speeds, and `riemann` solves a Riemann problem. We will discuss these issues later.

File `upwind.f` uses upwind differencing to compute the time integrals of the numerical flux at the cell sides. File `consdiff.f` applies the conservative difference to compute the solution of the differential equation at the new time.

Note that these Fortran files have been designed to organize the various problem-dependent parts of the code. File `consdiff.f` should be the same for all scalar conservation laws in one dimension. File `upwind.f` could be replaced to change the scheme without changing the differential equation. File `riemprob.f` would have to be changed if we change the initial or boundary conditions, or change the differential equation.

To run a copy of this code, perform the following steps:

(i) Type "`cd`" to return to your home directory.
(ii) Type "`cd scalar_law/PROGRAM1`" to enter the directory for this program.
(iii) Type "`make`" to compile the program files and make the executable `flinearad`.
(iv) Type "`flinearad > output`"; `flinearad` runs the program and `> output` redirects the results to the file `output`.
(v) Type "`xmgrace output`" to plot the computational results.

The final step will show a graph of the numerical solution plotted as a function of space, at the final time in the simulation.

There are several difficulties with this simple Fortran code. One is that whenever we want to change the input parameters, such as the number of grid cells or timesteps, we have to recompile the main program. Another problem is that the arrays have to be given a fixed size, because Fortran 77 does not perform dynamic memory allocation. We will fix these problems with the next program.

2.2.3.3 Third Upwind Difference Program

Our third program is more sophisticated, employing a mixture of C^{++} and Fortran. This program consists of several pieces that we have already seen, namely riemprob.f, linearad.f, upwind.f, consdiff.f, linearad.i and const.i. However, we have a different main program, a different make file and a new input file:

- **Program 2.2-12: LinearAdvectionMain.C** C^{++} main program;
- **Program 2.2-13: GNUmakefile** Makefile to compile and load the mixed-language program;
- **Program 2.2-14: input** the input file for executing the program.

The C^{++} file, LinearAdvectionMain.C, has several important features. Since C^{++} is strongly typed, this file contains function prototypes for the Fortran routines; these can be found in the extern "C" block. Next, this C^{++} file also defines C^{++}structures for the Fortran common blocks, namely linearad_common and machine_common; these allow us to refer to the data in the Fortran common blocks from within the main program.

Inside the main program itself, we provide values for the machine-dependent constants in the Fortran common block machine, and default values for the problem-dependent constants in the linearad common block. Afterward, we read the parameters from the input file.

After this preliminary work, the main program is prepared for computation. It allocates memory for the computational arrays, and defines array bounds for the Fortran subroutine calls. Next, the main program initializes the array entries to IEEE infinity; if the program uses an entry before it is given a proper value, then the resulting values will be obviously wrong. This initialization is useful in debugging; think of it as **defensive programming**.

Now that the problem parameters are known and the data arrays have been allocated, the main program calls initsl to set the initial values, and bcmesh to set boundary values for the mesh. Since the characteristic speed is fixed in linear advection, the main program calls stabledt once to compute the stable step size. At the end of the computation, the main program writes out the final results.

To run a copy of this code, perform the following steps:

(i) Type "cd" to return to your home directory.
(ii) Type "cd scalar_law/PROGRAM2" to enter the directory for this program.
(iii) Type "make" to compile the program files and make the executable linearad.
(iv) Type "linearad input > output" to run the program and redirect the results to the file output.
(v) Type "xmgrace output" to plot the computational results.

There are still difficulties with this program. Since we cannot see the numerical results during execution, it is difficult to see the time evolution ofthe computation.

Further, we are not fully able to perform other important aspects of defensive programming that we will introduce in the next program.

2.2.3.4 Fourth Upwind Difference Program

Our fourth program is even more sophisticated, employing a mixture of C^{++} and Fortran, together with some references to external libraries. This program consists of several pieces that we have already seen, namely `riemprob.f`, `linearad.f`, `upwind.f`, `consdiff.f`, `linearad.i` and `const.i`. However, the main program, input file and make file are different:

- **Program 2.2-15: LinearAdvectionMain.C** C^{++} main program;
- **Program 2.2-16: GNUmakefile** Makefile to compile the mixed-language program and link with libraries;
- **Program 2.2-17: input** the input file for executing the program.

The first change in `LinearAdvectionMain.C` is that we construct a `MemoryDebugger` to watch for out-of-bounds writes and unfreed pointers. Later, we define `InputParameters` for everything we would like to read from our `input` file. Each `InputParameter` knows the location of the variable to be assigned, a character string identifier and lower/upper bounds on permissible values. After defining the `InputParameters`, we read the parameters from the `input` file.

The biggest change to the main program is our use of interactive graphics to plot the solution. To do this, we compute the upper and lower bounds on the mesh and the solution. Then we construct an `XYGraphTool` that will plot our results. The arguments to the `XYGraphTool` constructor are the title to appear on the graphics window, the user coordinates for the window, a pointer to the colormap, and the desired size of the window as a fraction of the screen size. Next, we set the colors for the background and foreground, and draw the axes. Afterward, we draw plus signs at the cell centers for the numerical solution. During the loop over timesteps, we also plot the new results.

The makefile is necessarily complicated, because we are linking with other libraries for memory debugging, graphics, and graphical user interfaces. At the beginning of makefile, we include `macros.gnu`, which contains machine-dependent macros to describe the compiler names and options. Next, we set some internal macros for compiling and linking. Afterward, we make a list of routines needed by our program.

The trickiest part of the makefile is how we provide different targets to construct code for debugging or optimized performance. We can choose whether we will make debug or optimized code by setting the `OPT_OPTIONS` flag in `GNUmakefile`. The choice `d` will generate code for debugging with no optimization, while the choice `o` will generate optimized code. During code development, you will want to work with debug code. Once your code has been tested, you can create optimized code for greater execution speed.

To run a copy of this code, perform the following steps:

(i) Type "`cd`" to return to your home directory.
(ii) Type "`cd scalar_law/PROGRAM3`" to enter the directory for this program.
(iii) Type "`make`" to compile the program files and make the executable `1d/linearad`.
(iv) Type "`1d/linearad input`" to run the program.

When the program is run, the user will see a movie of the simulation, showing the conserved quantity plotted as a function of space at each time in the movie.

The directory `1d` refers to code written in one dimension, for debugging. Optimized code is compiled and loaded in directory `1o`. Figure 2.5 contains some example results with this program, at the final time in each simulation.

In order to capture the graphics into a file for printing, you can create a shell script, such as **Program 2.2-18: eps4paper** This command first copies the contents of a window to a `.gif` file, then converts that file to `.pdf` form.

2.2.3.5 Fifth Upwind Difference Program

Our fifth and final version of our upwind finite difference program is designed to be run from within this book. For this purpose, it is necessary that the user be able to change all input parameters interactively, from within a graphical user interface. This program consists of several pieces that we have already seen, namely `riemprob.f`, `linearad.f`, `consdiff.f`, `linearad.i` and `const.i`. However, the main program, input file and make file are different:

- **Program 2.2-19: GUILinearAdvectionMain.C** C^{++} main program and C^{++} auxiliary procedures;
- **Program 2.2-20: GNUmakefile** Makefile to compile the mixed-language program and link with libraries;
- **Program 2.2-21: input** the input file for executing the program.

In order to work with the graphical user interface, the main program basically performs some preliminary work before entering an event loop. The event loop calls various routines in response to user interaction with the graphical user interface. One of these callback routines is `runMain` in `GUILinearAdvectionMain.C`; this routine contains most of the statements that appeared in the main program of the previous example. The event loop allows the user to perform one simulation, adjust the input parameters, and then perform another simulation, all in the same run of the program. However, because of the separate threads used for the events, such a program is more difficult to debug than the previous examples.

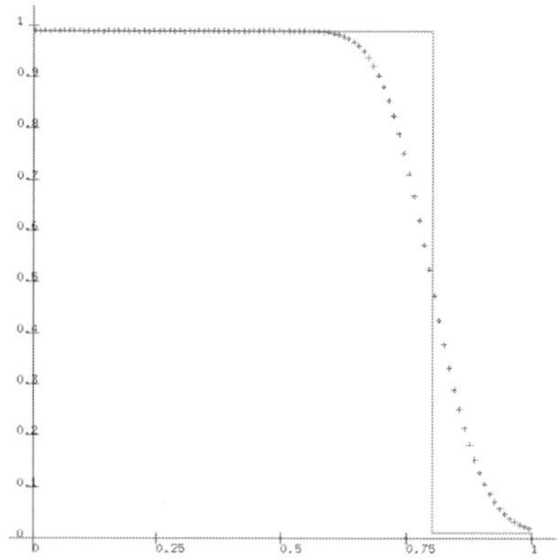

(a) CFL = 0.1

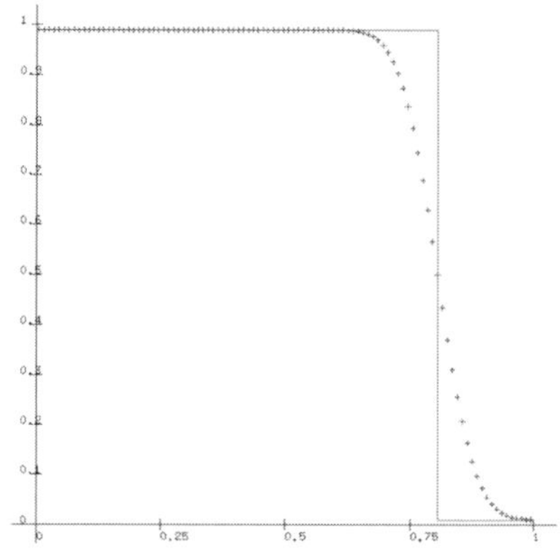

(b) CFL = 0.5

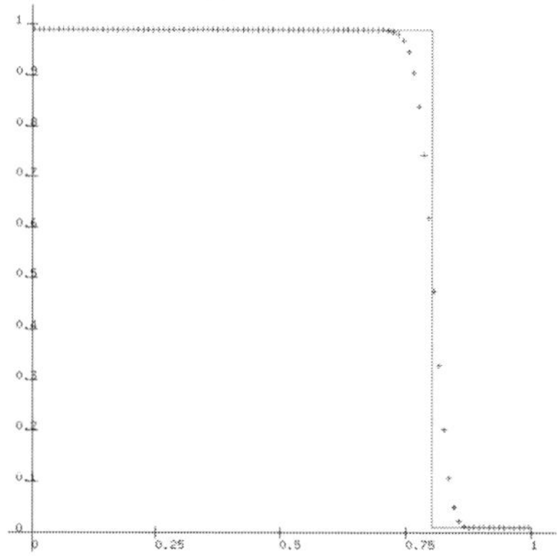

(c) CFL = 0.0

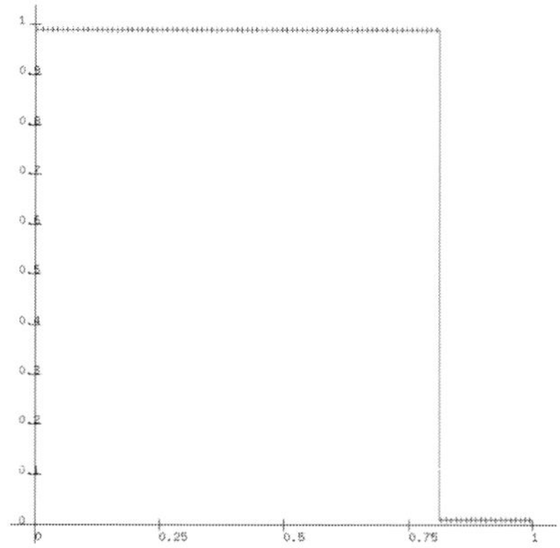

(d) CFL = 1.0

Fig. 2.5 Explicit upwind for linear advection (red = exact, blue = numerical solution)

To run a copy of this code, perform the following steps:

(i) Type "`cd`" to return to your home directory.
(ii) Type "`cd scalar_law`" to enter the directory for this program.
(iii) Type "`make`" to compile the program files and make the executable `1d/guilinearad`.
(iv) Type "`1d/guilinearad input`" to run the program.

The directory `1d` refers to code written in one dimension, for debugging. Optimized code is compiled and loaded in directory `1o`. You can also run the executable by clicking on the following: **Executable 2.2-1: guilinearad** The latter will use a graphical user interface for parameter input. Pull down on "`View`" and release the mouse on "`Main`". Click on any of the arrows to see current values of either the "`Riemann Problem Parameters`", "`Linear Advection`

`Parameters`", "`Numerical Method Parameters`" or "`Graphics`" parameters. After selecting your values, click on "`Start Run Now`" in the original graphical user interface. As with the executable "`1d/guilinearad`", you will get a window displaying a movie of the conserved quantity plotted as a function of space during simulation time.

2.2.4 Explicit Downwind Differences

Now that we have examined one useful scheme for integrating the linear advection equation, let us consider the development of alternative numerical schemes. The explicit upwind difference scheme (2.16a) can be viewed as a finite difference approximation to the linear advection equation (2.10a). In fact, we can rewrite explicit upwind differences as

$$\frac{u_i^{n+1} - u_i^n}{\triangle t^{n+1/2}} + \frac{cu_i^n - cu_{i-1}^n}{\triangle x_i} = 0, \quad 0 < i < I.$$

Note that the spatial difference is a first-order approximation to $\partial cu/\partial x$.

When we view the numerical method solely in terms of order of approximation of difference quotients to derivatives, we do not have any reason to prefer one first-order difference to another. We might be tempted to try the finite difference approximation

$$\frac{u_i^{n+1} - u_i^n}{\triangle t^{n+1/2}} + \frac{cu_{i+1}^n - cu_i^n}{\triangle x_i} = 0, \quad 0 < i < I,$$

which uses a different first-order approximation to the spatial derivative in the linear advection equation. We can rewrite this **explicit downwind difference** scheme in the form

$$u_i^{n+1} = u_i^n - [u_{i+1}^n - u_i^n]\frac{c\triangle t^{n+1/2}}{\triangle x_i}, \quad 0 \le i < I.$$

In other words, the explicit downwind difference scheme is a conservative difference scheme in which the numerical flux is chosen to be $f_{i+1/2}^{n+1/2} = cu_{i+1}^n$.

If we look carefully at this scheme, we can see that it ignores the boundary data at the left, and does not know how to compute the new solution in the last grid cell on the right. An optimist might hope that these flaws could be overcome by special treatment. Actually, these are indicators of a much more serious flaw.

It is easy to see that in the explicit downwind scheme, the new cell average u_u^{n+1} depends on the cell averages u_i^n and u_{i+1}^n. Thus, the domain of dependence of u_i^{n+1} is the interval $(x_{i-1/2}, x_{i+\frac{3}{2}})$. Recall that the physical domain of dependence for the solution at (x, t^{n+1}) is the point $(x - c\triangle t^{n+1/2}, t^n)$, so the physical domain of dependence of the cell average $\int_{x_{i-1/2}}^{x_{i+1/2}} u(x, t^{n+1})\, dx$ is the interval $(x_{i-1/2} - c\triangle t^{n+1/2}, x_{i+1/2} - c\triangle t^{n+1/2})$. Thus for downwind differences, the numerical domain of dependence *never* contains the physical domain of dependence, no matter what the size of the timestep $\triangle t^{n+1/2}$ may be. This indicates that, in general, the explicit downwind difference scheme cannot converge to the physical solution.

If the CFL number $\gamma_i^{n+1/2}$ is given by (2.18), then the downwind scheme can be rewritten in terms of the CFL number $\gamma_i^{n+1/2} = c\triangle t^{n+1/2}/\triangle x_i$ as follows:

$$u_i^{n+1} = u_i^n + \frac{c\triangle t^{n+1/2}}{\triangle x_i}[u_i^n - u_{i-1}^n] = u_i^n(1 + \gamma_i^{n+1/2}) - u_{i-1}^n \gamma_i^{n+1/2}, \ 0 \le i < I - 1.$$

Because the new solution involves amplification of u_i^n, the upwind scheme allows for instability to develop. We will present two different discussions of this instability in Sections 2.3 and 2.5 below.

Figure 2.6 contains some example results with this scheme. These results can be obtained by running Executable 2.2-1 with `scheme` set to `explicit downwind`. Since this scheme is unstable, the program should be run with a very small number of timesteps.

2.2.5 *Implicit Downwind Differences*

Let us continue to experiment with first-order discretizations of the space and time derivatives in the linear advection equation (2.10). The explicit downwind scheme uses first-order temporal differencing and first-order downwind spatial differencing at the old time. If instead we evaluate the spatial downwind difference at the new time, we obtain the **implicit downwind difference** scheme

$$\frac{u_i^{n+1} - u_i^n}{\triangle t^{n+1/2}} + \frac{cu_{i+1}^{n+1} - cu_i^{n+1}}{\triangle x_i} = 0, \quad 0 \le i < I - 1.$$

This is a conservative difference scheme in which the numerical fluxes are chosen to be $f_{i+1/2}^{n+1/2} = cu_{i+1}^{n+1}$.

Note that we can rewrite the scheme in the form

$$(1 - \gamma_i^{n+1/2})u_i^{n+1} + \gamma_i^{n+1/2} u_{i+1}^{n+1} = u_i^n, \quad 0 \le i < I - 1.$$

This gives us a right-triangular system of equations for the new solution. Back-solution of this linear system shows that u_i^{n+1} depends on cell averages u_j^n at the previous time for all $j \ge i$. Thus the domain of dependence of u_i^{n+1} is the union of the cells $(x_{j-1/2}, x_{j+1/2}$ for all $j \ge i$, or in other words the interval $(x_{i-1/2}, x_{I-1/2}) = (x_{i-1/2}, b)$, at the previous time. This interval does not contain the physical domain of dependence $(x_{i-1/2} - c\triangle t^{n+1/2}, x_{i+1/2} - c\triangle t^{n+1/2})$ for any $\triangle t^{n+1/2} > 0$. As a result, implicit downwind differences cannot be convergent.

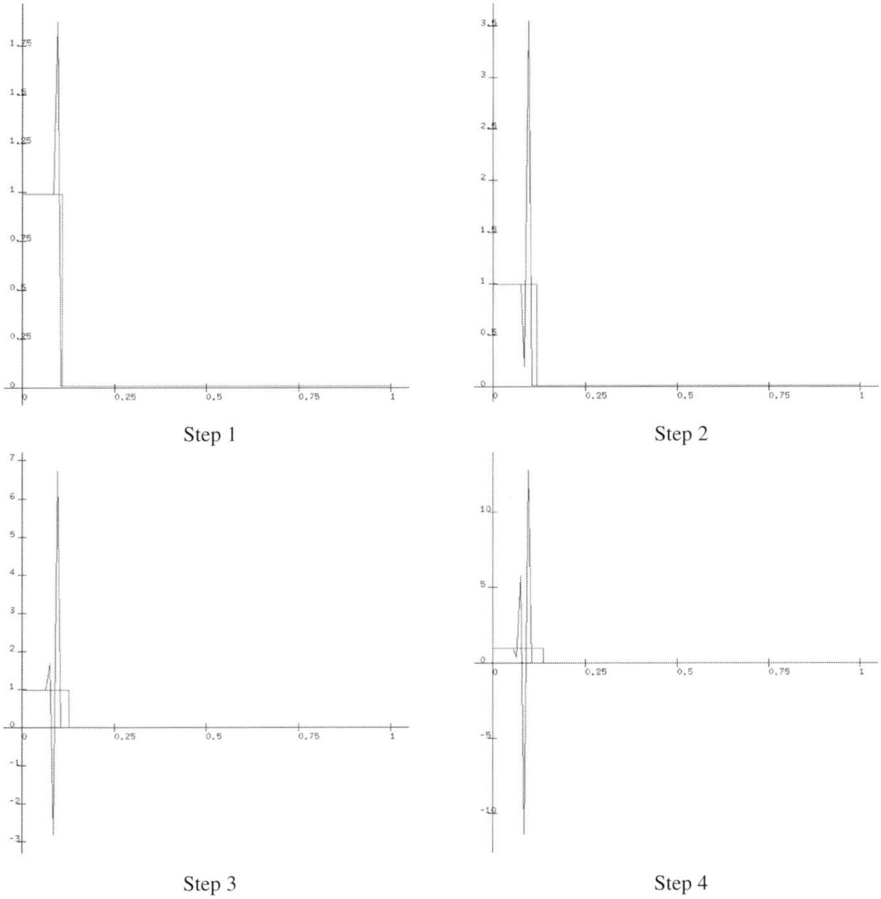

Fig. 2.6 Explicit downwind for linear advection (monotone curve = exact, oscillatory curve = numerical solution): first four steps at CFL = 0.9

The implicit downwind scheme does not use the boundary data at the left, and has trouble defining the solution at the right-hand (outflow) boundary. These observations are further indication of trouble with this scheme. However, it is possible to show (see Section 2.5.4 below) that this scheme is stable for sufficiently large timesteps, namely those for which the CFL numbers satisfy $\gamma_i^{n+1/2} \geq 1$. Thus stability is not the only consideration in choosing a numerical method; convergence to the physically correct solution is also important.

2.2.6 Implicit Upwind Differences

Our next example of a fully first-order discretization of the linear advection equation involves evaluating the upwind spatial difference at the new time. This **implicit**

upwind difference scheme can be written

$$\frac{u_i^{n+1} - u_i^n}{\triangle t^{n+1/2}} + \frac{cu_i^{n+1} - cu_{i-1}^{n+1}}{\triangle x_i} = 0, \quad 0 < i < I. \tag{2.19a}$$

$$u_i^{n+1} = u_i^n - [cu_i^{n+1} - f_{i-1/2}^{n+1/2}]\frac{\triangle t^{n+1/2}}{\triangle x_i}, \quad i = 0. \tag{2.19b}$$

This is a conservative difference scheme in which the numerical flux is chosen to be $f_{i+1/2}^{n+1/2} = cu_i^{n+1}$ for $0 \le i < I$.

Note that we can rewrite the scheme in the form

$$(1 + \gamma_i^{n+1/2})u_i^{n+1} - \gamma_i^{n+1/2}u_{i-1}^{n+1} = u_i^n, \quad 0 < i < I.$$

This gives us a left-triangular system of equations for the new solution. As a result, u_i^{n+1} depends on u_j^n for $j \le i$. It follows that the domain of dependence of u_i^{n+1} is $(x_{-1/2}, x_{i+1/2}) = (a, x_{i+1/2})$, plus the boundary data on the left. This interval contains the physical domain of dependence for all $\triangle t^{n+1/2} > 0$.

In Section 2.5.4 below we will see that this scheme is unconditionally stable. In comparison, remember that the explicit upwind scheme is stable for $\gamma_i^{n+1/2} \le 1$. If stability were the only consideration, we would clearly prefer the implicit upwind scheme. However, we will see in Sections 2.3 and 2.5.4 that this scheme introduces more numerical diffusion than explicit upwind differences, and involves greater numerical cost (namely the cost of solving a linear system of equations for u_i^{n+1}).

Figure 2.7 contains some example results with this scheme. These results can be obtained by running Executable 2.2-1 with the `scheme` set to `implicit upwind`. Note that the spreading of the numerical discontinuity increases as the Courant number increases. In other words, taking larger timesteps decreases the accuracy of the implicit upwind scheme; this result stands in contrast to our experiments with the explicit upwind scheme, in which the accuracy improved as the size of the timestep was increased. This is important to remember, because the temptation is to take larger timesteps with the implicit upwind method in order to decrease the cost of the scheme.

2.2.7 Explicit Centered Differences

Our final example of finite difference schemes for linear advection will use a second-order spatial difference. In order to determine this second-order spatial approximation, we will use the cell averages u_i^n to form a quadratic approximation to the integral of u, and differentiate that quadratic at a cell side. We construct a divided difference table by first forming columns, of spatial positions

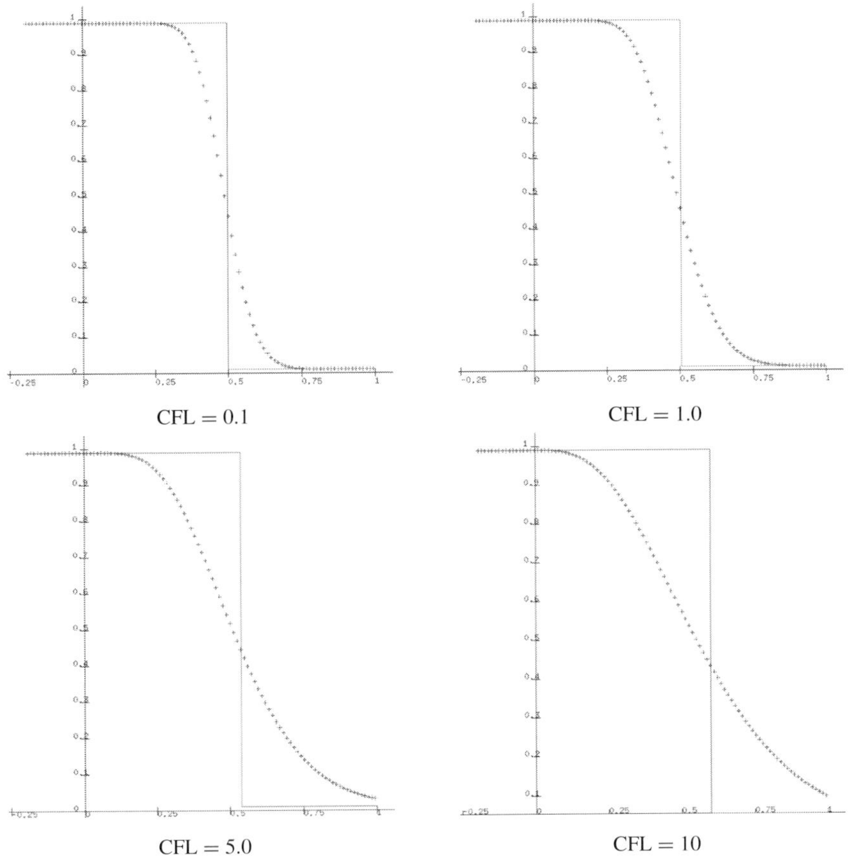

Fig. 2.7 Implicit upwind for linear advection (red = exact, blue = numerical solution)

$x_{j+1/2}$ and corresponding spatial integrals $\sum_{k=i}^{j} u_k^n \triangle x_k$, and then computing divided differences:

$$\begin{array}{ccccc}
 & x_{i-1/2} & \sum_{k=i}^{i-1} u_k^n \triangle x_k & & \\
\triangle x_i & & & u_i^n & \\
 & x_{i+1/2} & \sum_{k=i}^{i} u_k^n \triangle x_k & & u[x_i, x_{i+1}] \\
\triangle x_{i+1} & & & u_{i+1}^n & \\
 & x_{i+3/2} & \sum_{k=i}^{i+1} u_k^n \triangle x_k & &
\end{array}$$

Here we have used divided difference notation for one term in the table, namely

$$u[x_i, x_{i+1}] \equiv \frac{u_{i+1}^n - u_i^n}{\triangle x_{i+1} + \triangle x_i},$$

even though this difference quotient is not the same as that found in numerical analysis textbooks. This gives us the quadratic interpolation

$$\int_{x_{i-1/2}}^{x} u(x)\,dx \approx u_i^n(x - x_{i-1/2}) + u[x_i, x_{i+1}](x - x_{i-1/2})(x - x_{i+1/2}).$$

If we differentiate this quadratic and evaluate it at $x_{i+1/2}$, we obtain an approximation for u at the cell side:

$$\begin{aligned}\frac{d}{dx}\int_{x_{i-1/2}}^{x} u(x)\,dx|_{x_{i+1/2}} &\approx u_i^n + u[x_i, x_{i+1}]\triangle x_i\\ &= u_i^n \frac{\triangle x_{i+1}}{\triangle x_i + \triangle x_{i+1}} + u_{i+1}^n \frac{\triangle x_i}{\triangle x_i + \triangle x_{i+1}}.\end{aligned}$$

In the **explicit centered difference** scheme for linear advection, we evaluate the fluxes for a conservative difference by using our approximation to u at the cell side:

$$f_{i+1/2}^{n+1/2} = c\frac{u_i^n \triangle x_{i+1} + u_{i+1}^n \triangle x_i}{\triangle x_i + \triangle x_{i+1}}.$$

The new solution is then computed by the conservative difference

$$u_i^{n+1} = u_i^n - \frac{\triangle t^{n+1/2}}{\triangle x_i}\left[f_{i+1/2}^{n+1/2} - f_{i-1/2}^{n+1/2}\right].$$

This can be rewritten in the form of a weighted average

$$\begin{aligned}u_i^{n+1} = {}& u_{i-1}^n \frac{c\triangle t^{n+1/2}}{\triangle x_i + \triangle x_{i+1}} + u_i^n\left[1 + \frac{c\triangle t^{n+1/2}}{\triangle x_i + \triangle x_{i+1}}\frac{\triangle x_{i-1} - \triangle x_{i+1}}{\triangle x_i}\right]\\ &- u_{i+1}^n \frac{c\triangle t^{n+1/2}}{\triangle x_i + \triangle x_{i+1}}.\end{aligned} \tag{2.20}$$

On a uniform grid, the fluxes simplify to the more common formula $f_{i+1/2}^{n+1/2} = c(u_i^n + u_{i+1}^n)/2$, and the new solution can be evaluated by

$$u_i^{n+1} = u_i^n - \frac{c\triangle t^{n+1/2}}{\triangle x_i}[u_{i+1}^n - u_{i-1}^n].$$

Note that there is a difficulty in determining the solution u_{I-1}^{n+1} at the right-hand boundary; if this were the only problem with this scheme, it might be overcome by using explicit upwind differences there.

We do not gain useful information by examining the domain of dependence of the explicit centered difference scheme. Note that u_i^{n+1} depends on u_{i-1}^n, u_i^n and u_{i+1}^n. Thus the numerical domain of dependence is $(x_{i-\frac{3}{2}}, x_{i+\frac{3}{2}})$. This domain of dependence contains the physical domain of dependence if and only if the timestep is chosen so that $|c\triangle t^{n+1/2}/\triangle x_{i-1}| \leq 1$ for every grid cell. This seems to indicate that the explicit centered difference scheme could be conditionally convergent.

We can get some indication that this scheme is unstable by examining the weighted average form of the scheme (2.20). Note that the coefficient of u^n_{i+1} is always negative; this will lead to instability. In fact, we will demonstrate in sections 2.3.3 and 2.5.4 that this scheme is unconditionally unstable.

Explicit centered differences are easy to program, except for the treatment of the downstream boundary. This fact, together with the lure of second-order spatial accuracy, is so compelling that careless people are sometimes attracted to this scheme.

Exercises for 2.2

2.2.1 Consider the linear advection problem on the interval $(a, b) = (0, 1)$ with initial data $u(x, 0) = 2$ for $x < 0.1$ and $u(x, 0) = 1$ for $x > 0.1$ and boundary data $u(0, t) = 2$ for $t > 0$. Determine the analytical solution to this problem, and write a program to plot this analytical solution as a function of x for any $t > 0$.

2.2.2 Run the explicit upwind scheme for CFL numbers 0.1, 0.5, 1.0 and 1.1 with $c = 1$, $(a, b) = (0, 1)$ and $\triangle x = 0.01$. Stop each simulation at time $t = 0.55$. Plot the numerical results together with the analytical solution from the previous exercise, labeling each CFL number case carefully. Describe the qualitative differences in the numerical results. Which results look sharp? Which results look unstable?

2.2.3 Program the explicit downwind scheme for the problem in the first exercise. In order to treat boundary data, take $u = 2$ on the left, and $u = 1$ on the right. Plot the numerical results after 5, 10 and 20 timesteps with each of the CFL numbers in the previous exercise. Describe the qualitative differences in the numerical results. How does the CFL number affect the instability?

2.2.4 Program the implicit upwind scheme for the problem in the first exercise. Plot the numerical results together with the analytical solution, labeling each CFL number case carefully. Describe the qualitative differences in the numerical results. How does the CFL number affect the sharpness of the results?

2.2.5 There are several very interesting test problems for numerical schemes applied to linear advection, suggested by Zalesak [180]:

square pulse for $0.1 \le x \le 0.2$ $u(x, 0) = 2$, otherwise $u(x, 0) = 1$;
triangular pulse for $0.1 \le x \le 0.2$ $u(x, 0) = 2 - 20|x - 0.15|$, otherwise $u(x, 0) = 1$;
smooth Gaussian pulse for $0.1 \le x \le 0.2$ $u(x, 0) = 1 + \exp(-10^4(x - 0.15)^2) - \exp(-25)$, otherwise $u(x, 0) = 1$;
quadratic pulse for $0.1 \le x \le 0.2$ $u(x, 0) = 1 + \sqrt{1 - 400(x - 0.15)^2}$, otherwise $u(x, 0) = 1$.

Each problem should be solved with 100 cells on a uniform grid, so that the initial disturbance is described in a fixed number of grid cells. Solve each of these problems by the explicit upwind scheme, using CFL = 1.0, 0.9, 0.5 and 0.1. Plot the analytical solution with a continuous curve, and the numerical solution with discrete markers. Be sure to compute the initial data for the scheme as the cell average of the given data.

2.2.6 Repeat the previous exercise using the implicit upwind scheme. How do the results compare, both for accuracy and computational speed?

2.3 Modified Equation Analysis

The examples in section 2.2 indicate that we need to develop some methods for assessing the qualitative behavior of finite difference schemes. In this section, we will discover that the numerical solution of linear advection by finite differences actually solves a partial differential equation that is slightly different from the original problem. We will examine a heuristic method for determining which partial differential equation the numerical method is actually approximating, and what effect the modification might have.

Please note that we will continue to assume that the advection velocity satisfies $c > 0$.

2.3.1 Modified Equation Analysis for Explicit Upwind Differences

Lemma 2.3.1 *Suppose that the discrete values u_i^n satisfy the explicit upwind difference*

$$u_i^{n+1} = u_i^n - \frac{c\triangle t}{\triangle x}[u_i^n - u_{i-1}^n].$$

Further suppose that

$$u_i^n = \tilde{u}(i\triangle x, n\triangle t) + o(\triangle t^2) + o(\triangle x^2) + o(\triangle t\triangle x),$$

where $\tilde{u}$ is twice continuously differentiable in x and t, and $\tilde{u}$ satisfies a modified equation of the form

$$\frac{\partial \tilde{u}}{\partial t} + \frac{\partial c\tilde{u}}{\partial x} = e = O(\triangle t) + O(\triangle x). \tag{2.21}$$

Then the modification e satisfies

$$e = \frac{c\triangle x}{2}\left(1 - \frac{c\triangle t}{\triangle x}\right)\frac{\partial^2 \tilde{u}}{\partial x^2} + o(\triangle t) + o(\triangle x).$$

Proof Since $\tilde{u}$ is twice continuously differentiable,

$$\begin{aligned}
u_i^{n+1} &= \tilde{u}(x, t + \triangle t) + o(\triangle t^2) + o(\triangle x^2) + o(\triangle t\triangle x)\\
&= \tilde{u}(x,t) + \frac{\partial \tilde{u}}{\partial t}\triangle t + \frac{1}{2}\frac{\partial^2 \tilde{u}}{\partial t^2}\triangle t^2 + o(\triangle t^2) + o(\triangle x^2) + o(\triangle t\triangle x),\\
u_{i-1}^n &= \tilde{u}(x - \triangle x, t) + o(\triangle t^2) + o(\triangle x^2) + o(\triangle t\triangle x)\\
&= \tilde{u}(x,t) - \frac{\partial \tilde{u}}{\partial x}\triangle x + \frac{\partial^2 \tilde{u}}{\partial x^2}\frac{\triangle x^2}{2} + o(\triangle t^2) + o(\triangle x^2) + o(\triangle t\triangle x).
\end{aligned}$$

In these expressions, the partial derivatives are all evaluated at (x, t). Since $\tilde{u}$ satisfies the modified equation (2.21),

$$\frac{\partial^2 \tilde{u}}{\partial t^2} = \frac{\partial}{\partial t}\left(e - \frac{\partial c\tilde{u}}{\partial x}\right) = \frac{\partial e}{\partial t} - c\frac{\partial}{\partial x}\left(e - \frac{\partial c\tilde{u}}{\partial x}\right) = c^2\frac{\partial^2 \tilde{u}}{\partial x^2} + \frac{\partial e}{\partial t} - c\frac{\partial e}{\partial x}.$$

When we substitute the Taylor series approximations into the explicit upwind difference scheme, we get the equation

$$\begin{aligned} 0 &= \frac{u_i^{n+1} - u_i^n}{\triangle t} + c\frac{u_i^n - u_{i-1}^n}{\triangle x} = \frac{\partial \tilde{u}}{\partial t} + \frac{\partial c\tilde{u}}{\partial x} + \frac{\triangle t}{2}\frac{\partial^2 \tilde{u}}{\partial t^2} - \frac{c\triangle x}{2}\frac{\partial^2 \tilde{u}}{\partial x^2} + o(\triangle t) + o(\triangle x) \\ &= \frac{\partial \tilde{u}}{\partial t} + \frac{\partial c\tilde{u}}{\partial x} - \frac{c\triangle x}{2}\left(1 - \frac{c\triangle t}{\triangle x}\right)\frac{\partial^2 \tilde{u}}{\partial x^2} + \frac{\triangle t}{2}\left(\frac{\partial e}{\partial t} - c\frac{\partial e}{\partial x}\right) + o(\triangle t) + o(\triangle x) \\ &= \frac{\partial \tilde{u}}{\partial t} + \frac{\partial c\tilde{u}}{\partial x} - \frac{c\triangle x}{2}\left(1 - \frac{c\triangle t}{\triangle x}\right)\frac{\partial^2 \tilde{u}}{\partial x^2} + o(\triangle t) + o(\triangle x). \end{aligned}$$

This equation determines the error term to be

$$e = \frac{c\triangle x}{2}\left(1 - \frac{c\triangle t}{\triangle x}\right)\frac{\partial^2 \tilde{u}}{\partial x^2} + o(\triangle t) + o(\triangle x).$$

□

If $c > 0$ and $c\triangle t \le \triangle x$ the modified equation shows that the upwind scheme is actually solving an advection–diffusion equation with a small diffusion. This numerical diffusion is proportional to the product of the cell width and one minus the CFL number. It is interesting to note that the upwind scheme involves no diffusion if we choose $\gamma = c\triangle t/\triangle x = 1$. In this case, the explicit upwind scheme is exact (although we did not determine this fact from the modified equation analysis). In fact, the explicit upwind scheme itself shows that when $c\triangle t/\triangle x = 1$, then $u_i^{n+1} = u_{i-1}^n$. In other words, $u_i^n = u_{i-n}^0$; if the initial data for the scheme is chosen to be $u_i^0 = u_0(i\triangle x)$, then $u_i^n = u_0([i-n]\triangle x) = u_0(i\triangle x - cn\triangle t)$, and the scheme is exact.

2.3.2 Modified Equation Analysis for Explicit Downwind Differences

Lemma 2.3.2 *Suppose that the discrete values u_i^n satisfy the explicit downwind difference*

$$u_i^{n+1} = u_i^n - \frac{c\triangle t}{\triangle x}[u_{i+1}^n - u_i^n],$$

Further suppose that

$$u_i^n = \tilde{u}(i\triangle x, n\triangle t) + o(\triangle t^2) + o(\triangle x^2) + o(\triangle t\triangle x),$$

where $\tilde{u}$ is twice continuously differentiable in x and t, and $\tilde{u}$ satisfies a modified equation of the form (2.21). *Then the modification e satisfies*

$$e = -\frac{c\triangle x}{2}\left(1+\frac{c\triangle t}{\triangle x}\right)\frac{\partial^2 \tilde{u}}{\partial x^2} + o(\triangle t) + o(\triangle x).$$

Proof Using Taylor series and the modified equation assumption (2.21), we obtain

$$0 = \frac{u_i^{n+1} - u_i^n}{\triangle t} + c\frac{u_{i+1}^n - u_i^n}{\triangle x} = \frac{\partial \tilde{u}}{\partial t} + \frac{\partial c\tilde{u}}{\partial x} + \frac{\triangle t}{2}\frac{\partial^2 \tilde{u}}{\partial t^2} + \frac{c\triangle x}{2}\frac{\partial^2 \tilde{u}}{\partial x^2} + o(\triangle t) + o(\triangle x)$$
$$= \frac{\partial \tilde{u}}{\partial t} + \frac{\partial c\tilde{u}}{\partial x} + \frac{c\triangle x}{2}\left(1+\frac{c\triangle t}{\triangle x}\right)\frac{\partial^2 \tilde{u}}{\partial x^2} + o(\triangle t) + o(\triangle x).$$

In this case the error term in the modified equation is

$$e = -\frac{c\triangle x}{2}\left(1+\frac{c\triangle t}{\triangle x}\right)\frac{\partial^2 \tilde{u}}{\partial x^2} = -\frac{c\triangle x}{2}(1+\gamma)\frac{\partial^2 \tilde{u}}{\partial x^2} + o(\triangle t) + o(\triangle x).$$

□

Since $c > 0$, the downwind scheme is anti-diffusive for all CFL numbers $\gamma \equiv c\triangle t/\triangle x > 0$. This anti-diffusion leads to instability. In fact, the analytical solution for the advection-diffusion equation in Section 2.1.3 does not apply to this modified equation, because the diffusion coefficient is negative. Rather, an examination of the Fourier series solution for this modified equation would show that all wave numbers lead to growth, and the growth increases with the square of the wave number. We will perform a Fourier analysis of this scheme in Section 2.5.4.

2.3.3 Modified Equation Analysis for Explicit Centered Differences

The modified equation analyses of the explicit upwind and explicit downwind schemes were very similar. However, in order to study the explicit centered difference scheme we will need to make the additional assumption that the modified equation error has an asymptotic expansion.

Lemma 2.3.3 *Suppose that the discrete values u_i^n satisfy the explicit centered difference*

$$u_i^{n+1} = u_i^n - \frac{c\triangle t}{2\triangle x}[u_{i+1}^n - u_{i-1}^n].$$

Further suppose that

$$u_i^n = \tilde{u}(i\triangle x, n\triangle t) + o(\triangle t^2) + o(\triangle x^3) + o(\triangle t \triangle x^2),$$

where $\tilde{u}$ is twice continuously differentiable in x and t, and $\tilde{u}$ satisfies a modified equation of the form (2.21). *Then the modification e satisfies*

$$e = -\frac{c^2 \triangle t}{2} \frac{\partial^2 \tilde{u}}{\partial x^2} + o(\triangle t) + o(\triangle x^2).$$

Proof If we substitute Taylor series expansions into the linear advection equation, we obtain

$$\begin{aligned}
0 &= \frac{u_i^{n+1} - u_i^n}{\triangle t} + c \frac{u_{i+1}^n - u_{i-1}^n}{2 \triangle x} \\
&= \frac{\partial \tilde{u}}{\partial t} + \frac{\partial c \tilde{u}}{\partial x} + \frac{\triangle t}{2} \frac{\partial^2 \tilde{u}}{\partial t^2} + \frac{1}{3} c \triangle x^2 \frac{\partial^3 \tilde{u}}{\partial x^3} + o(\triangle t) + o(\triangle x^2) + o(\triangle t \triangle x) \\
&= \frac{\partial \tilde{u}}{\partial t} + \frac{\partial c \tilde{u}}{\partial x} + \frac{\triangle t}{2} \left(c^2 \frac{\partial^2 \tilde{u}}{\partial x^2} + \frac{\partial e}{\partial t} - c \frac{\partial e}{\partial x} \right) + \frac{1}{3} \triangle x^2 \frac{\partial^3 \tilde{u}}{\partial x^3} + o(\triangle t) + o(\triangle x^2) + o(\triangle t \triangle x).
\end{aligned}$$

It follows that

$$e = -\frac{\triangle t}{2} \left(c^2 \frac{\partial^2 \tilde{u}}{\partial x^2} + \frac{\partial e}{\partial t} - c \frac{\partial e}{\partial x} \right) + o(\triangle t) + o(\triangle x^2) + o(\triangle t \triangle x).$$

This defines e implicitly. The dominant term in the expansion for e is the first; it implies that

$$e \approx -\frac{c^2 \triangle t}{2} \frac{\partial^2 \tilde{u}}{\partial x^2} + o(\triangle t) + o(\triangle t \triangle x) + o(\triangle x^2).$$

□

Note that the dominant term in the modified equation error e is anti-diffusive for all $\triangle t$, no matter what the sign of the velocity c. This anti-diffusion indicates that the explicit centered difference scheme is unconditionally unstable.

2.3.4 Modified Equation Analysis Literature

There are several interesting papers on modified equation analysis. Hedstrom [70] showed that the modified equation analysis is valid for general diffusive difference approximations to scalar linear equations with discontinuous initial data. Majda and Ralston [115] used modified equation analysis for first-order schemes and weak shocks in nonlinear systems to provide necessary and sufficient conditions to guarantee physical (or non-physical) discrete shock profiles. Engquist and Osher [49] gave examples of steady wave profiles for Riemann problems for which the modified equation analysis fails. However, in this case the schemes involved are not diffusive when linearized around the sonic points. Goodman and Majda [60] proved that the modified equation analysis is valid for upwind differencing on scalar nonlinear conservation laws.

Exercises for 2.3

2.3.1 Perform a modified equation analysis of the implicit upwind scheme. Compare its numerical diffusion to that of the explicit upwind scheme.

2.3.2 Perform a modified equation analysis of the implicit downwind scheme. Under what circumstances is it diffusive?

2.3.3 Suppose that we want to perform an explicit upwind difference on a non-uniform grid. We assume that there are constants $\underline{c}$ and $\overline{c}$ and a scalar h so that

$$\underline{c}h \leq \triangle x_i \leq \overline{c}h \text{ for all } i$$

as the mesh widths decrease to zero.

(a) Perform a modified equation analysis of the scheme

$$u_i^{n+1} = u_i^n - \frac{c\triangle t}{\triangle x_i}[u_i^n - u_{i-1}^n].$$

to show that the error is $O(1)$. Note that

$$u_i^n - u_{i-1}^n \approx \frac{\partial u}{\partial x}\frac{\triangle x_i + \triangle x_{i-1}}{2}.$$

(b) Show that the scheme

$$u_i^{n+1} = u_i^n - \frac{2c\triangle t}{\triangle x_i + \triangle x_{i-1}}[u_i^n - u_{i-1}^n]$$

is first-order but not conservative.

(c) To construct a first-order scheme on a non-uniform grid, we can define the numerical fluxes to be given by a Newton interpolation to the flux at the cell centers:

$$f_{i+1/2}^n = f(u_i^n) + \frac{f(u_i^n) - f(u_{i-1}^n)}{\triangle x_i + \triangle x_{i-1}}\triangle x_i.$$

First show that

$$f_{i+1/2}^n \approx f_i^n + \frac{\partial f}{\partial x}\frac{\triangle x_i}{2} - \frac{\partial^2 f}{\partial x^2}\frac{(\triangle x_i + \triangle x_{i-1})\triangle x_i}{8}$$

and

$$f_{i-1/2}^n \approx f_i^n - \frac{\partial f}{\partial x}\frac{\triangle x_i}{2} - \frac{\partial^2 f}{\partial x^2}\alpha$$

where $\alpha = O(h^2)$. Perform a modified equation analysis of the conservative difference scheme

$$u_i^{n+1} = u_i^n - \frac{\triangle t}{\triangle x_i}[f_{i+1/2}^n - f_{i-1/2}^n]$$

to show that it is first-order in both space and time. Also show that on a uniform grid this scheme is second order in space.

2.3.4 Perform a modified equation analysis of **Leonard's scheme** [92] (without second-order time correction)

$$u_i^{n+1} = u_i^n - \frac{c\triangle t}{\triangle x}[u_{i+1/2}^n - u_{i-1/2}^n],$$

where

$$u_{i+1/2}^n = u_i^n + \frac{2u_{i+1}^n - u_i^n - u_{i-1}^n}{6}$$

to determine the order of the scheme. Under what circumstances is this scheme diffusive?

2.4 Consistency, Stability and Convergence

The modified equation analysis gives us a qualitative measurement of the behavior of numerical methods. With some minor modification of that analysis and some additional assumptions, we can prove the convergence to solutions of partial differential equations.

First, let us describe what we mean by linear explicit two-step schemes. We assume that the numerical method can be written

$$u^{n+1} = Q^n u^n$$

where Q^n is some operator on the solution vector. Note that the solution vector may be defined at an infinite number of points, for the purposes of this analysis. It will typically be convenient to use the shift operators

$$(S_+u)_i^n = u_{i+1}^n \quad \text{and} \quad (S_-u)_i^n = u_{i-1}^n,$$

to define Q^n in specific schemes.

Example 2.4.1 The explicit upwind scheme can be written

$$u_i^{n+1} = \left(1 - \gamma_i^{n+\frac{1}{2}}\right) u_i^n + \gamma_i^{n+\frac{1}{2}} u_{i-1}^n,$$

where $\gamma_i^{n+\frac{1}{2}} = c\triangle t^{n+\frac{1}{2}}/\triangle x_i$ is the CFL number. In this case, we have

$$Q^n = I\left(1 - \gamma_i^{n+\frac{1}{2}}\right) + S_-\gamma_i^{n+\frac{1}{2}}.$$

Example 2.4.2 The implicit upwind scheme can be written

$$\left(1 + \gamma_i^{n+\frac{1}{2}}\right) u_i^{n+1} - \gamma_i^{n+\frac{1}{2}} u_{i-1}^{n+1} = u_i^n.$$

In this case, we have

$$Q^n = \left[I(1 + \gamma_i^{n+\frac{1}{2}}) - S_- \gamma_i^{n+\frac{1}{2}} \right]^{-1}.$$

Next, let $\|\cdot\|$ represent some norm on the solution vector. For example, we could use the ℓ^∞ norm

$$\|u^n\| \equiv \sup_i |u_i^n|$$

or the ℓ^1 norm

$$\|u^n\| \equiv \sum_i |u_i^n| \triangle x_i.$$

The induced norm on Q^n is defined by

$$\|Q^n\| \equiv \sup_{u^n} \frac{\|Q^n u^n\|}{\|u^n\|}.$$

Example 2.4.3 If the CFL number satisfies $0 < \gamma_i^{n+\frac{1}{2}} = c\triangle t^{n+\frac{1}{2}}/\triangle x_i \leq 1$, then in the max norm the solution operator Q^n for the explicit upwind scheme satisfies

$$\begin{aligned}
\|Q^n u^n\| &= \max_i \left| u_i^n(1 - \gamma_i^{n+\frac{1}{2}}) + u_{i-1}^n \gamma_i^{n+\frac{1}{2}} \right| \\
&\leq \max \left\{ |u_i^n|(1 - \gamma_i^{n+\frac{1}{2}}) + |u_{i-1}^n| \gamma_i^{n+\frac{1}{2}} \right\} \\
&\leq \max_i\{|u^n|\} \max_i \left\{ \left(1 - \gamma_i^{n+\frac{1}{2}}\right) + \gamma_i^{n+\frac{1}{2}} \right\} = \max_i |u_i^n| \equiv \|u^n\|.
\end{aligned}$$

Thus in this case, $\|Q^n\| \leq 1$ whenever the CFL number is at most one.

Example 2.4.4 Recall that in the implicit upwind scheme we have

$$Q^n = \left[I\left(1 + \gamma_i^{n+\frac{1}{2}}\right) - S_- \gamma_i^{n+\frac{1}{2}} u_{i-1}^n \right]^{-1}.$$

From the definition of this scheme, we see that for any norm

$$\begin{aligned}
\|u^n\| &= \left\| \left(1 + \gamma_i^{n+\frac{1}{2}}\right) u_i^{n+1} - \gamma_i^{n+\frac{1}{2}} u_{i-1}^{n+1} \right\| \\
&\geq \|u^{n+1}\| \left(1 + \gamma_i^{n+\frac{1}{2}}\right) - \|S_- u^{n+1}\| \gamma_i^{n+\frac{1}{2}} \\
&= \|u^{n+1}\| \left(1 + \gamma_i^{n+\frac{1}{2}}\right) - \|u^{n+1}\| \gamma_i^{n+\frac{1}{2}} = \|u^{n+1}\|.
\end{aligned}$$

Thus $\|Q^n u^n\| \leq \|u\|$, so $\|Q^n\| \leq 1$ for all norms and all positive CFL numbers.

The following important theorem proves that linear explicit two-step schemes are convergent under reasonable assumptions, and provides an estimate for the error in the numerical solution.

Theorem 2.4.5 *(Lax) Assume that*

(i) $u^{n+1} = Q^n u^n$ *is a scheme to approximate the solution of some linear partial differential equation*

(ii) *given a final time T and a maximum number of timesteps $n > 0$, we perform n steps with this scheme, using timesteps $\triangle t^{m+\frac{1}{2}}$ satisfying*

$$\text{for all } T > 0 \text{ for all } n > 0 \sum_{m=0}^{n-1} \triangle t^{m+\frac{1}{2}} \leq T \text{ and}$$

$$\text{for all } T > 0 \text{ there exists } \alpha > 0 \text{ such that } n \max_{0 \leq m < n} \triangle t_{m+\frac{1}{2}} \leq T\alpha$$

(iii) *the scheme is* **stable**, *meaning that*

$$\text{there exists } C > 0 \text{ for all } n \ \|Q^n\| \leq 1 + C\triangle t^{n+\frac{1}{2}}$$

(iv) *the scheme has order p in time and order 1 in space, meaning that if u is the exact solution of the partial differential equation, then the* **local truncation error** ϵ^n *satisfies*

$$\text{there exists } C_t > 0 \text{ there exists } p > 0 \text{ there exists } C_x > 0 \text{ there exists } q > 0 \text{ for all } \triangle t^{n+\frac{1}{2}} \text{ for all } \triangle x_i$$
$$\epsilon^n \equiv \frac{1}{\triangle t^{n+1/2}} \|u(x_i, t^n) - Q^n u(\cdot, t^n)\| \leq C_t \left(\triangle t^{n+\frac{1}{2}}\right)^p + C_x (\triangle x_i)^q$$

Then the error in the approximate solution satisfies

$$\|u(\cdot, t^n) - u^n\| \leq e^{CT} \|u(\cdot, 0) - u^0\| + \alpha T e^{CT} \left[C_t \max_n (\triangle t^{n+\frac{1}{2}})^p + C_x \max_i (\triangle x_i)^q \right].$$

Proof For all timesteps satisfying the assumptions,

$$\begin{aligned} \|u(\cdot, t^n) - u^n\| &\leq \|u(\cdot, t^n) - Q^{n-1} u(\cdot, t^{n-1})\| + \|Q^{n-1} u(\cdot, t^n) - Q^{n-1} u^{n-1}\| \\ &\leq \epsilon^{n-1} + \|Q^{n-1}\| \|u(\cdot, t^{n-1}) - u^{n-1}\|. \end{aligned}$$

We can solve this recurrent inequality to get

$$
\begin{aligned}
\|u(\cdot,t^n)-u^n\| &\leq \left\{\prod_{k=0}^{n-1}\|Q^k\|\right\}\|u(\cdot,t^0)-u^0\| + \sum_{\ell=0}^{n-1}\left\{\prod_{k=\ell}^{n-1}\|Q^k\|\right\}\epsilon^\ell \\
&\leq \left\{\prod_{k=0}^{n-1}\|Q^k\|\right\}\|u(\cdot,t^0)-u^0\| + \max_\ell \epsilon^\ell \sum_{\ell=0}^{n-1}\prod_{k=\ell}^{n-1}\|Q^k\| \\
&\leq \|u(\cdot,t^0)-u^0\|\prod_{k=0}^{n-1}\left(1+C\triangle t^{k+\frac{1}{2}}\right) + n\max_\ell \epsilon^\ell \prod_{k=\ell}^{n-1}\left(1+C\triangle t^{\ell+\frac{1}{2}}\right) \\
&\leq \|u(\cdot,t^0)-u^0\| e^{C\sum_{k=0}^{n-1}\triangle t^{k+\frac{1}{2}}} \\
&\quad + e^{C\sum_{k=\ell}^{n-1}\triangle t^{\ell+\frac{1}{2}}} \max_{k,i}\left\{n\triangle t^{k+\frac{1}{2}}\left[C_t(\triangle t^{k+\frac{1}{2}})^p + C_x(\triangle x_i)^q\right]\right\} \\
&\leq e^{CT}\|u(\cdot,t^0)-u^0\| + \alpha T e^{CT}\left[C_t \max_n(\triangle t^{n+\frac{1}{2}})^p + C_x\max_i(\triangle x_i)^q\right].
\end{aligned}
$$

□

Example 2.4.6 The same Taylor series expansions that were used in the modified equation analysis of the explicit upwind scheme can be used to show that the local truncation error for the explicit upwind scheme is

$$
\begin{aligned}
\epsilon_i^n &\equiv u(x_i,t^{n+1}) - \left\{u(x_i,t^n) - \gamma_i^{n+\frac{1}{2}}[u(x_i,t^n)-u(x_{i-1},t^n)]\right\} \\
&= \triangle t^{n+\frac{1}{2}}\left\{-\frac{\partial^2 u}{\partial t^2}\frac{\triangle t^{n+\frac{1}{2}}}{2} + \frac{\partial^2 u}{\partial x^2}\frac{c\triangle x_i}{2}\right\}.
\end{aligned}
$$

Recall that in Example 2.4.1 we showed that the explicit upwind scheme is stable in the max norm provided that the timesteps are chosen so that the CFL number is always at most one. If the second partial derivatives of u are uniformly bounded for all states within the range of the problem of interest, then Theorem 2.4.5 proves that the explicit upwind scheme is first-order accurate in both space and time.

Exercise for 2.4

2.4.1 Prove that the implicit upwind scheme for linear advection is first-order accurate in both space and time.

2.5 Fourier Analysis of Finite Difference Schemes

In Section 2.3 we developed the modified equation analysis as a heuristic tool for understanding the qualitative behavior of finite difference schemes. We found that the modified equation analysis was useful in understanding the order of the scheme, and comparing the numerical diffusion employed by different schemes. The modified equation analysis is reasonably general, in that it can be applied to linear schemes for nonlinear differential equations on bounded domains.

In this section, we will develop a different tool for analyzing finite difference schemes. We will use Fourier transforms to study the dissipation and dispersion introduced by linear schemes in solving linear problems on unbounded domains. The Fourier analysis will give us useful information about the inter-relationship between dissipation and dispersion in controlling numerical oscillations. However, Fourier analysis can only be used to study linear schemes applied to linear problems.

For a second source of some of the information in this section, the reader can consult [151].

2.5.1 Constant Coefficient Equations and Waves

Let us consider the linear partial differential equation

$$\frac{\partial u}{\partial t} + c\frac{\partial u}{\partial x} = ru + d\frac{\partial^2 u}{\partial x^2} + f\frac{\partial^3 u}{\partial x^3} \quad \text{for all } x \in \mathbf{R} \text{ for all } t > 0 \tag{2.22a}$$

$$u(x,0) = u_0(x). \tag{2.22b}$$

In order to understand the behavior of this problem, we will define the **Fourier transform** of an integrable function to be

$$\hat{u}(\xi, t^n) = \int_{-\infty}^{\infty} u(x, t^n) e^{-\imath\xi x} dx.$$

It is well-known [139] that if both u and $\hat{u}$ are integrable in x and ξ, respectively, then the appropriate inversion formula for the Fourier transform is

$$u(x, t^n) = \frac{1}{2\pi}\int_{-\infty}^{\infty} \hat{u}(\xi, t^n) e^{\imath\xi x} d\xi \tag{2.23}$$

almost everywhere. If we take the Fourier transform in space of (2.22a), we obtain an ordinary differential equation that is parameterized by the Fourier variable ξ:

$$\frac{\partial \hat{u}}{\partial t} = \left(\left[r - d\xi^2\right] - i\left[c\xi + f\xi^3\right]\right)\hat{u}(\xi, t).$$

The solution of this ordinary differential equation is

$$\hat{u}(\xi, t) = e^{[r-d\xi^2]t - i[c\xi + f\xi^3]t}\hat{u}_0(\xi).$$

If the initial data has Fourier transform $\hat{u}(\xi, 0) = \alpha\delta(\xi - \beta)$, then the Fourier inversion formula (2.23) shows that $u(x, 0) = \frac{\alpha}{2\pi}e^{i\beta x}$. Thus the initial data consists of a single **wave number** β; a wave number is equal to 2π over the wavelength. It is also easy to use the inverse Fourier transform to find that the solution of (2.22) is

$$u(x,t) = e^{[r-d\beta^2]t - i[c\beta + f\beta^3]t} u(x,0).$$

It is common to define the **frequency**

$$\omega = -(\beta c + \beta^3 f) + i(-r + \beta^2 d).$$

so that $u(x,t) = e^{i\omega t} u(x,0)$. Note that the frequency ω has units of one over time.

We can provide several important examples of frequencies, by considering Equation (2.22a) with only one nonzero coefficient and initial data consisting of a single wave number.

advection: If $r = 0$, $d = 0$ and $f = 0$ then the frequency is $\omega = -c\beta$ and the solution of (2.22) is $u(x,t) = \alpha e^{i\beta(x-ct)}$. In this case, all wave numbers β travel with the same speed c.

reaction: If $c = 0$, $d = 0$ and $f = 0$ then the frequency of the wave is $\omega = -ir$, and the solution of (2.22) is $u(x,t) = \alpha e^{rt} e^{i\beta x}$. All wave numbers β remain stationary, and the amplitude of the wave either grows ($r > 0$) or decays ($r < 0$) in time.

diffusion: If $c = 0$, $r = 0$ and $f = 0$ then the frequency is $\omega = i\beta^2 d$ and the solution of (2.22) is $u(x,t) = \alpha e^{-\beta^2 dt} e^{i\beta x}$. In this case, all wave numbers remain stationary. If $d > 0$ all nonzero wave numbers decay, and large wave numbers decay faster than small wave numbers. If $d < 0$ then all nonzero wave numbers grow, and large wave numbers grow faster than small wave numbers.

dispersion: If $c = 0, r = 0$ and $d = 0$ then the frequency of the wave is $\omega = -\beta^3 f$ and the solution of (2.22) is $u(x,t) = \alpha e^{i\beta(x-\beta^2 ft)}$. This says that different wave numbers travel with different speeds, and high wave numbers travel faster than slow wave numbers.

2.5.2 Dimensionless Groups

Another collection of interesting partial differential equations involves the time derivative and two other terms in (2.22a). There are five interesting cases among the six possibilities:

convection–diffusion: Suppose that $r = 0$ and $f = 0$. Given some useful length L (such as the problem length or the grid cell width), we can define a dimensionless time coordinate $\tau = ct/L$ and a dimensionless spatial coordinate $\eta = (x - ct)/L$. We then change variables by defining $\tilde{u}(\eta, \tau) = u(x,t)$.

These lead to the transformed diffusion equation

$$\frac{\partial \tilde{u}}{\partial \tau} = \frac{d}{cL} \frac{\partial^2 \tilde{u}}{\partial \eta^2}.$$

Here the ratio cL/d of convection to diffusion is called the **Peclet number**.

convection–dispersion: If $r = 0$ and $d = 0$ we can define $\tau = ct/L$, $\eta = (x - ct)/L$ and $\tilde{u}(\eta, \tau) = u(x, t)$ to obtain the dispersion equation

$$\frac{\partial \tilde{u}}{\partial \tau} = \frac{f}{cL^2} \frac{\partial^3 \tilde{u}}{\partial \eta^3}.$$

The dimensionless ratio cL^2/f of convection to dispersion does not have a commonly used label.

convection–reaction: If $d = 0$ and $f = 0$ we can define $\tau = ct/L$, $\eta = (x - ct)/L$ and $\tilde{u}(\eta, \tau) = u(x, t)$ to obtain the system of ordinary differential equations (parameterized by η)

$$\frac{\partial \tilde{u}}{\partial \tau} = \frac{rL}{c} \tilde{u}.$$

reaction–diffusion: If $c = 0$ and $f = 0$ we can define $\tau = rt$, $\xi = x/L$ and $e^\tau \tilde{u}(\xi, \tau) = u(x, t)$ to obtain the diffusion equation

$$\frac{\partial \tilde{u}}{\partial \tau} = \frac{d}{rL^2} \frac{\partial^2 \tilde{u}}{\partial \xi^2}.$$

reaction–dispersion: If $c = 0$ and $d = 0$ we can define $\tau = rt$, $\xi = x/L$ and $u(x, t) = e^\tau \tilde{u}(\xi, \tau)$ to obtain the dispersion equation

$$\frac{\partial \tilde{u}}{\partial \tau} = \frac{f}{rL^3} \frac{\partial^3 \tilde{u}}{\partial \xi^3}.$$

2.5.3 Linear Finite Differences and Advection

Although Fourier analysis is applicable to general linear partial differential equations, in this section we are interested only in linear advection. Recall that the Fourier transform of the solution of the linear advection equation satisfies $\hat{u}(\xi, t) = e^{-ic\xi t} \hat{u}_0(\xi)$. Thus

$$\hat{u}(\xi, t + \triangle t) = e^{-i\xi c \triangle t} \hat{u}(\xi, t),$$

so the exact solution merely involves multiplying the Fourier transform by a fixed ratio. Let us define γ to be the (dimensionless) **CFL number**

$$\gamma = \frac{c \triangle t}{\triangle x},$$

and θ the (dimensionless) **mesh wave number**

$$\theta = \xi \triangle x.$$

Then the solution ratio is $e^{-i\xi c \Delta t} = e^{-i\theta\gamma}$.

Next, let us consider a linear finite difference scheme

$$\sum_k a_k u_{j+k}^{n+1} = \sum_k b_k u_{j+k}^n \tag{2.24}$$

on an equally spaced space-time mesh $t^n = n\triangle t$, $x_j = j\triangle x$. We will assume that the numerical scheme is nonzero only for $x \in (-a, a)$ over all timesteps of interest; as a result,

$$\text{for all } 0 \leq n \leq N = T/\triangle t, \quad \text{for all } |j| > J = a/\triangle x \quad u_j^n = 0.$$

We will define the **finite Fourier transform** of this discrete data to be

$$\hat{u}^n(\xi) = \sum_{j=-J}^{J} u_j^n e^{-\imath j \triangle x \xi} \triangle x.$$

This finite Fourier transform of the discrete data is a midpoint rule approximation to the Fourier transform of $u(x, t^n)$. Note that the corresponding inversion formula for the finite Fourier transform is

$$u_j^n = \frac{1}{2\pi} \int_{-\pi/\Delta x}^{\pi/\Delta x} \hat{u}^n(\xi) e^{\imath j \Delta x \xi} \, d\xi. \tag{2.25}$$

Also note that if we define the shift operators S_+ and S_- by

$$(S_+u)_j^n = u_{j+1}^n \text{ and } (S_-u)_j^n = u_{j-1}^n,$$

then it is easy to see that

$$\widehat{(S_+u)}^n = e^{\imath\xi\Delta x}\hat{u}^n \text{ and } \widehat{(S_-u)}^n = e^{-\imath\xi\Delta x}\hat{u}^n.$$

Thus, if we take the finite Fourier transform of the linear finite difference scheme (2.24), we obtain

$$\left[\sum_k a_k e^{ik\xi\Delta x}\right] \hat{u}^{n+1}(\xi) = \left[\sum_k b_k e^{ik\xi\Delta x}\right] \hat{u}^n(\xi).$$

Consequently, the finite Fourier transform at the new time satisfies

$$\hat{u}^{n+1}(\xi) = \frac{\sum_k b_k e^{ik\xi\Delta x}}{\sum_k a_k e^{ik\xi\Delta x}} \hat{u}^n(\xi).$$

We define the numerical solution ratio by

$$z(\theta) = \frac{\hat{u}^{n+1}(\xi)}{\hat{u}^n(\xi)} = \frac{\sum_k b_k e^{ik\theta}}{\sum_k a_k e^{ik\theta}}. \tag{2.26}$$

We will say that the scheme (2.24) is **dissipative** if and only if $|z| < 1$ for all $\theta \neq 0$ and **dispersive** if and only if $\arg(z)/\theta$ depends on θ. Recall that if the complex number z has the polar form $z = |z|e^{i\psi}$, then $\arg(z) \equiv \psi$.

In order to assess the cumulative effect of numerical dissipation and dispersion over several timesteps, we will compare the numerical solution to the analytical solution at the time required for the wave to cross a grid cell. The number of timesteps required for this crossing is $1/\gamma$. The analytical solution at this time is

$$\hat{u}(\xi, (n + 1/\gamma)\triangle t) = \hat{u}(\xi, n\triangle t)e^{-i\xi c\triangle t/\gamma} = \hat{u}(\xi, n\triangle t)e^{-i\theta},$$

and the numerical solution is

$$\hat{u}^{n+1/\gamma}(\xi) = \hat{u}^n(\xi)z(\theta)^{1/\gamma}.$$

These results give us quantitative measures of the errors introduced by numerical methods. The total **numerical dissipation error** in the time required for the wave to cross a grid cell is

$$|e^{-i\theta}| - |z(\theta)|^{1/\gamma} = 1 - |z(\theta)|^{1/\gamma}.$$

The total numerical dispersion is measured by the **phase error**

$$1 - \arg\left(z(\theta)^{1/\gamma}\right) / \arg\left(e^{-i\theta}\right) = 1 + \arg\left(z(\theta)^{1/\gamma}\right) /\theta.$$

The phase error measures the relative error in how fast information associated with a specific mesh wave number moves in a single timestep, while the dissipation error measures how much the amplitude of that information changes in one timestep. Positive phase errors indicate that information associated with the particular mesh wave number is moving slower than it should. Negative dissipation errors mean that the amplitude is larger than it should be, and indicates instability.

Numerical schemes can be related to rational trigonometric polynomial approximations of the form

$$z(\theta) \equiv \frac{\sum_k b_k e^{ik\theta}}{\sum_k a_k e^{ik\theta}} \approx e^{-i\gamma\theta}.$$

The **order** of the scheme is the power p such that

$$z(\theta) - e^{-i\gamma\theta} = O(\theta^{p+1}). \tag{2.27}$$

The order of the scheme contains information about the extent to which the scheme is consistent with the differential equation; when we say that a scheme has positive order we mean that its local truncation tends to zero as the mesh size approaches zero. However, a scheme can have positive order without being convergent.

An ideal scheme would have nearly no dispersion or dissipation over all wave numbers. Since this is impossible, it is common to look for schemes that have dissipation that matches dispersion in some useful sense, so that as waves are dispersed away from some wave-front and appear as oscillations, there is sufficient numerical dissipation to reduce the amplitude of these oscillations to acceptably small levels.

2.5.4 Fourier Analysis of Individual Schemes

Let us consider explicit upwind differences for linear advection:

$$u_j^{n+1} = u_j^n - \frac{c\triangle t}{\triangle x}[u_j^n - u_{j-1}^n] = (1-\gamma)u_j^n + \gamma u_{j-1}^n.$$

The Fourier transform of this scheme is

$$\hat{u}^{n+1}(\xi) = (1-\gamma)\hat{u}^n(\xi) + \gamma e^{-i\theta}\hat{u}^n(\xi).$$

Thus the solution ratio is

$$z(\theta) = 1 - \gamma + \gamma e^{-i\theta} = [1 - \gamma(1-\cos\theta)] - i\gamma\sin\theta. \tag{2.28}$$

Note that the graph of $z(\theta)$ in Figure 2.8 is a circle in the complex plane with center $1-\gamma$ and radius γ. This circle is contained within the unit circle if and only if $\gamma \leq 1$.

Alternatively, we can compute the modulus of the solution ratio

$$|z(\theta)|^2 = [1-\gamma(1-\cos\theta)]^2 + [\gamma\sin\theta]^2 = 1 - 4\gamma(1-\gamma)\sin^2(\theta/2).$$

Thus the scheme is dissipative (*i.e.*, for all θ we have $|z(\theta)| \leq 1$) if and only if $\gamma \leq 1$. The smallest solution ratio is associated with $\theta = \pi$. At this value of θ the wave number is $\xi = \pi/\triangle x$; in this case, the wavelength is on the order of the mesh width. For a fixed mesh wave number θ, the solution ratio goes to one as $\gamma \to 1$. However the dissipation error

$$1 - |z(\theta)|^{1/\gamma} = 1 - [1 - 4\gamma(1-\gamma)\sin^2(\theta/2)]^{1/(2\gamma)}$$

does not go to zero uniformly as $\gamma \to 0$.

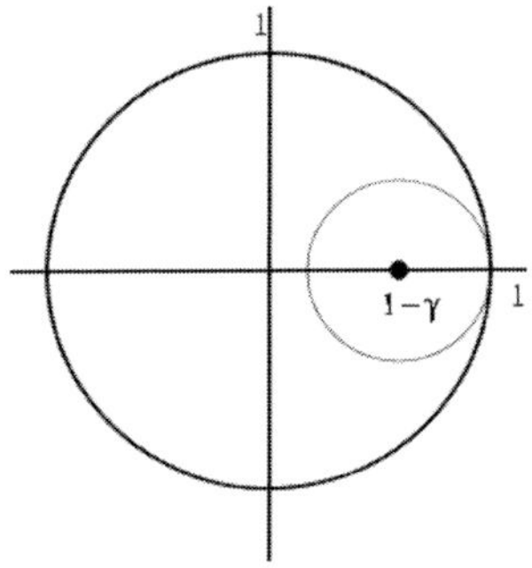

Fig. 2.8 Explicit upwind solution ratio

In order to determine the order of the explicit upwind scheme, we note that

$$\begin{aligned} z(\theta) - e^{-i\gamma\theta} &= 1 - \gamma + \gamma\cos\theta - i\gamma\sin\theta - \cos\gamma\theta + i\sin\gamma\theta \\ &= 1 - \gamma + \gamma\left[1 - \frac{1}{2}\theta^2 + O(\theta^4)\right] - i\gamma[\theta - O(\theta^3)] \\ &\quad - \left[1 - \frac{1}{2}\gamma^2\theta^2 + O(\gamma^4\theta^4)\right] + i[\gamma\theta - O(\gamma^3\theta^3)] \\ &= -\frac{1}{2}\gamma(1-\gamma)\theta^2 + O(\theta^3) \end{aligned}$$

This shows that the explicit upwind scheme has order 1 for $\gamma < 1$.

Next, let us investigate the dispersion in the explicit upwind scheme. Note that (2.28) shows that

$$\tan(\arg z(\theta)) = \frac{-\gamma\sin\theta}{1 - 2\gamma\sin^2(\theta/2)}.$$

For large wave numbers $\theta \to \pm\pi$ we have that $\tan z(\theta) \to 0$. This says that high wave numbers are nearly stationary. For small θ,

$$\arg z(\theta) = \tan^{-1}\left[\frac{-\gamma\sin\theta}{1 - 2\gamma\sin^2(\theta/2)}\right] \approx -\gamma\theta\left[1 - \frac{1}{6}(1-2\gamma)(1-\gamma)\theta^2\right] + o(\theta^3).$$

Thus for $\frac{1}{2} < \gamma < 1$ and small θ we have $\arg z(\theta) < -\gamma\theta$; this says that low wave numbers are dispersed behind for CFL greater than one half. For $\frac{1}{2} > \gamma > 0$ we have $\arg z(\theta) > -\gamma\theta$; this says that low wave numbers are dispersed ahead for CFL less than one half.

Two values of the CFL number are special. For $\gamma = 1$ we have $z(\theta) = e^{-i\theta}$. In this case, the explicit upwind scheme has zero dissipation and zero phase error; in fact, in this case the scheme is exact for these input data. For $\gamma = \frac{1}{2}$, we have

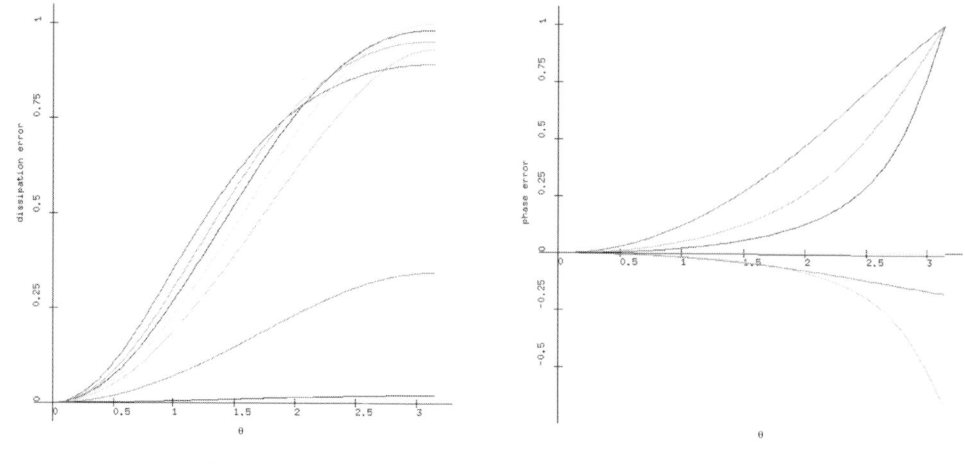

(a) Dissipation error (b) Phase error

Fig. 2.9 Fourier analysis for explicit upwind differences. CFL values for the curves are 0.9 (red), 0.7 (blue), 0.6 (green), 0.5 (yellow), 0.4 (cyan), 0.15 (magenta), 0.01 (black)

$z(\theta) = \frac{1}{2}(1 + e^{-i\theta})$. In this case, we have

$$\tan(\arg z(\theta)) = \frac{-\sin\theta}{1 + \cos\theta} = \tan(-\theta/2) = \tan(-\gamma\theta),$$

so again the scheme has zero phase error. These two choices of γ are very useful in computation. We can keep both numerical dissipation and dispersion small by taking γ to be slightly less than one; on the other hand, we can introduce numerical dissipation (and smear numerical fronts) without dispersion (numerical oscillations) by taking $\gamma = \frac{1}{2}$. For most problems, CFL numbers close to one give the best results.

Figure 2.9 shows the dissipation error $1 - |z(\theta)|^{1/\gamma}$ and phase error $1 + \arg(z(\theta)^{1/\gamma})/\theta$ for the explicit upwind scheme. This figure was produced by using the C^{++} main program **Program 2.5-22: fourierMain.C** and Fortran subroutine **Program 2.5-23: fourier.f**. Students can produce error curves for individual values of the CFL number by choosing `scheme` equal to 0 while running **Executable 2.5-2: guifourier.**

Next, recall explicit downwind differences for linear advection:

$$u_j^{n+1} = u_j^n - \frac{c\triangle t}{\triangle x}[u_{j+1}^n - u_j^n] = (1+\gamma)u_j^n - \gamma u_{j+1}^n.$$

In this case the solution ratio is

$$z(\theta) = 1 + \gamma - \gamma e^{i\theta}. \tag{2.29}$$

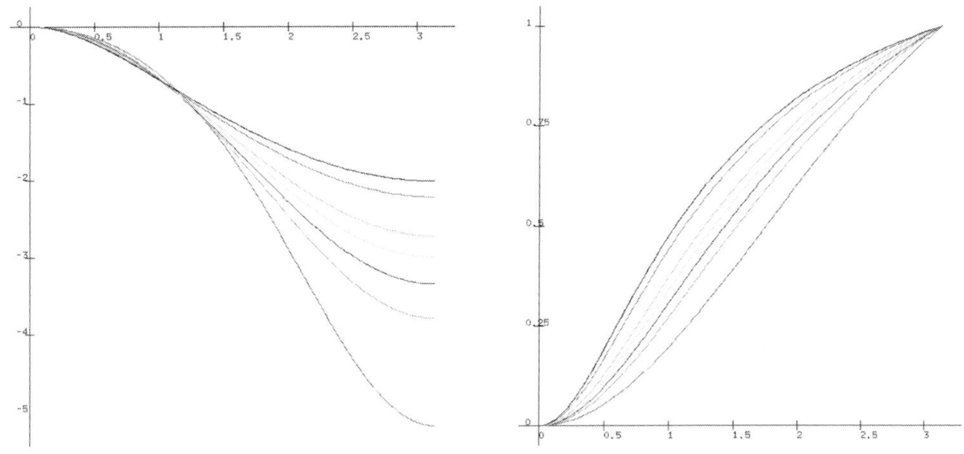

(a) Dissipation error

(b) Phase error

Fig. 2.10 Fourier analysis for explicit downwind differences. CFL values for the curves are 0.99, 0.85, 0.6, 0.5, 0.4, 0.3 and 0.1; these appear in order from top to bottom on the right in the dissipation errors, and from top to bottom in the middle of the phase errors. Negative dissipation errors indicate instability.

Note that $z(\theta)$ is the equation for a circle with center $1 + \gamma$ and radius γ. This circle is never contained inside the unit circle. As a result, the explicit downwind scheme is anti-dissipative.

This anti-dissipation appears as numerical instability. Numerical oscillations will grow until they become too large to represent on the machine. Such a method is not convergent, even though in this case the order of the scheme is 1. Here, it is useful to recall that the order of a scheme, as defined in Equation (2.27) measures only how the numerical solution ratio differs from the true solution ratio as the dimensionless mesh wave number tends to zero. Figure 2.10 shows the results of a Fourier analysis of the explicit downwind scheme. Note that the dissipation error is negative at all mesh wave numbers for all values of CFL; this indicates *unconditional* instability. Students can produce error curves for individual values of the CFL number by choosing `scheme` equal to 1 while running Executable 2.5-2 with the `scheme` set to `explicit downwind`.

Recall explicit centered differences for linear advection:

$$u_j^{n+1} = u_j^n - \frac{c\triangle t}{2\triangle x}[u_{j+1}^n - u_{j-1}^n] = \frac{1}{2}\gamma u_{j-1}^n + u_j^n - \frac{1}{2}\gamma u_{j+1}^n.$$

In this case the solution ratio is

$$z(\theta) = \frac{1}{2}\gamma e^{-i\theta} + 1 - \frac{1}{2}\gamma e^{i\theta} = 1 - i\gamma \sin\theta.$$

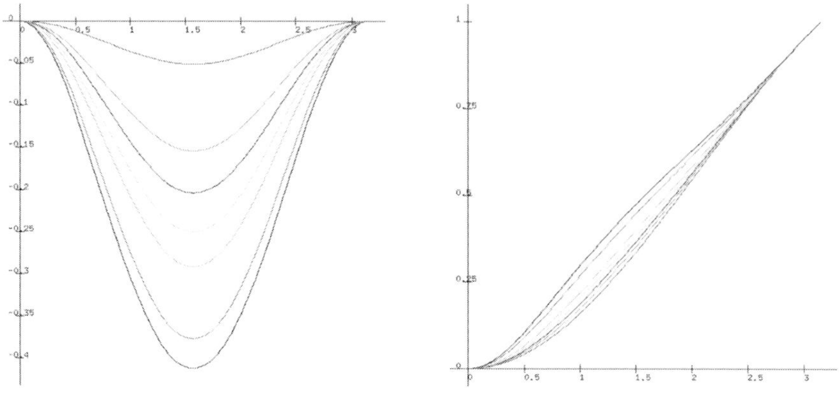

(a) Dissipation error (b) Phase error

Fig. 2.11 Fourier analysis for explicit centered differences. CFL values for the curves are 0.99, 0.85, 0.6, 0.5, 0.4, 0.3 and 0.1; these appear in order from bottom to top in the middle of the dissipation errors, and from top to bottom in the middle of the phase errors. Negative dissipation errors indicate instability.

Note that

$$|z(\theta)|^2 = 1 + \gamma^2 \sin^2 \theta.$$

Thus this scheme is anti-dissipative for all CFL numbers γ. Its order is 1. The Fourier analysis of the explicit centered difference scheme is shown in Figure 2.11. Students can produce error curves for individual values of the CFL number by choosing `scheme` equal to `explicit centered` while running Executable 2.5-2.

Let us consider implicit upwind differences for linear advection:

$$u_j^{n+1} = u_j^n - \frac{\lambda \triangle t}{\triangle x}[u_j^{n+1} - u_{j-1}^{n+1}].$$

This can be rewritten

$$(1+\gamma)u_j^{n+1} - \gamma u_{j-1}^{n+1} = u_j^n.$$

Thus the solution ratio is

$$z(\theta) = \frac{1}{1+\gamma - \gamma e^{-i\theta}}.$$

Note that $1/z(\theta)$ is the equation for a circle with center $1+\gamma$ and radius γ. Thus this circle lies outside the unit circle for all γ, so $z(\theta)$ lies inside the unit circle. It follows that the scheme is dissipative. The dissipation increases ($z(\theta)$ moves farther into the interior of the unit circle) as γ increases. It is possible to see that this scheme is more dissipative than explicit upwind differencing for small

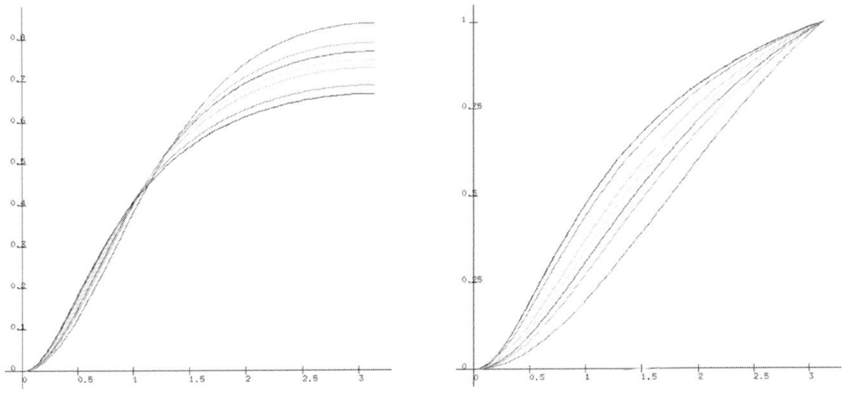

(a) Dissipation error (b) Phase error

Fig. 2.12 Fourier analysis for implicit upwind differences. CFL values for the curves are 0.99, 0.85, 0.6, 0.5, 0.4, 0.3 and 0.1; these appear in order from bottom to top on the right in the dissipation errors, and from top to bottom in the middle of the phase errors.

θ at the same CFL number γ. The Fourier analysis of this scheme is shown in Figure 2.12 Students can produce error curves for individual values of the CFL number by choosing `scheme` equal to `implicit upwind` while running Executable 2.5-2.

Let us consider implicit downwind differences for linear advection:

$$u_j^{n+1} = u_j^n - \frac{\lambda \triangle t}{\triangle x}[u_{j+1}^{n+1} - u_j^{n+1}].$$

This can be rewritten

$$\gamma u_j^{n+1} + (1-\gamma) u_j^{n+1} = u_j^n.$$

Thus the solution ratio is

$$z(\theta) = \frac{1}{1-\gamma+\gamma e^{i\theta}}.$$

Note that $1/z(\theta)$ is the equation for a circle with center $1-\gamma$ and radius γ. Thus this circle lies outside the unit circle for all $\gamma > 1$, so $z(\theta)$ lies inside the unit circle under these circumstances. It follows that the scheme is dissipative whenever $\gamma > 1$.

Let us consider the **Lax–Wendroff scheme** for linear advection:

$$\begin{aligned} u_j^{n+1} &= u_j^n - \frac{c\triangle t}{2\triangle x}[u_{j+1}^n - u_{j-1}^n] + \frac{1}{2}\left(\frac{c\triangle t}{\triangle x}\right)^2 [u_{j+1}^n - 2u_j^n + u_{j-1}^n] \\ &= \frac{1}{2}\gamma(1+\gamma)u_{j-1}^n + (1-\gamma^2)u_j^n - \frac{1}{2}\gamma(1-\gamma)u_{j+1}^n. \end{aligned}$$

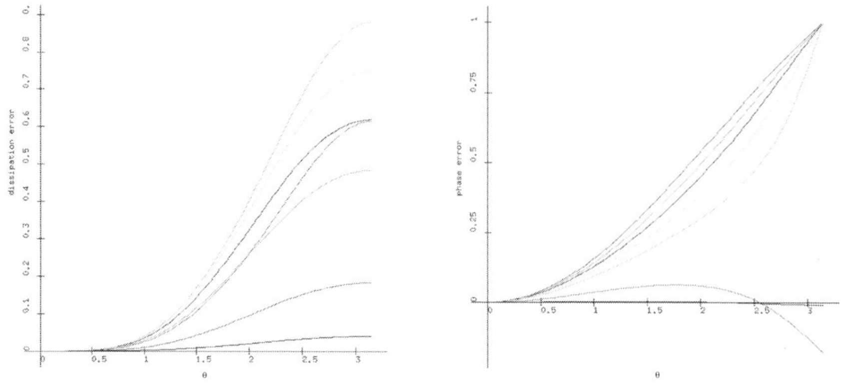

(a) Dissipation error (b) Phase error

Fig. 2.13 Fourier analysis for the Lax–Wendroff scheme. Dissipation error is small for all wave numbers with CFL $= 0.99$, and increases at large wave number as CFL decreases (0.85, 0.6, 0.5, 0.4), then decreases slightly at high wave number for smaller CFL (0.3, 0.1); phase error is essentially zero for CFL $= 0.99$, then becomes negative at large wave number (CFL $= 0.85$ and 0.6), then becomes zero for all wave numbers at CFL $= 0.5$, and becomes positive for smaller CFL

Thus the solution ratio is

$$z(\theta) = \frac{1}{2}\gamma(1+\gamma)e^{-i\theta} + 1 - \gamma^2 - \frac{1}{2}\gamma((1-\gamma)e^{i\theta}.$$

This implies that

$$|z(\theta)|^2 = 1 - 4\gamma^2(1-\gamma^2)(\sin\theta/2)^4$$

and the scheme is dissipative for $\gamma < 1$. The dissipation and phase errors for this scheme are shown in Figure 2.13. Note that for CFL near one, both the dissipation and phase errors are uniformly small for all mesh wave numbers. Students can produce error curves for individual values of the CFL number by choosing `scheme` equal to `lax wendroff` while running executable 2.5-2.

Finally, let us consider the **leap-frog scheme** for linear advection:

$$u_j^{n+1} = u_j^{n-1} - \frac{c\triangle t}{\triangle x}[u_{j+1}^n - u_{j-1}^n].$$

Thus

$$\mathbf{v^{n+1}} \equiv \begin{bmatrix} \hat{u}^{n+1} \\ \hat{u}^n \end{bmatrix} = \begin{bmatrix} -2\imath\gamma\sin\theta & 1 \\ 1 & 0 \end{bmatrix} \begin{bmatrix} \hat{u}^n \\ \hat{u}^{n-1} \end{bmatrix} \equiv A\mathbf{v^n}.$$

The matrix A in this expression has eigenvalues

$$\lambda = \pm\sqrt{1-\gamma^2\sin^2\theta} - \iota\gamma\sin\theta;$$

in fact,

$$\begin{aligned} AX &\equiv \begin{bmatrix} -2\iota\gamma\sin\theta & 1 \\ 1 & 0 \end{bmatrix} \begin{bmatrix} \sqrt{1-\gamma^2\sin^2\theta} - \iota\gamma\sin\theta & -\sqrt{1-\gamma^2\sin^2\theta} - \iota\gamma\sin\theta \\ 1 & 1 \end{bmatrix} \\ &= \begin{bmatrix} \sqrt{1-\gamma^2\sin^2\theta} - \iota\gamma\sin\theta & -\sqrt{1-\gamma^2\sin^2\theta} - \iota\gamma\sin\theta \\ 1 & 1 \end{bmatrix} \\ &\quad \begin{bmatrix} \sqrt{1-\gamma^2\sin^2\theta} - \iota\gamma\sin\theta & -\sqrt{1-\gamma^2\sin^2\theta} - \iota\gamma\sin\theta \\ 1 & 1 \end{bmatrix} \equiv X\Lambda. \end{aligned}$$

It follows that

$$\mathbf{v^n} = A^n\mathbf{v^0} = X\Lambda^n X^{-1}\mathbf{v^0}.$$

In other words,

$$\hat{u}^n = \left(\sqrt{1-\gamma^2\sin^2\theta} - \iota\gamma\sin\theta\right)^n \alpha + \left(-\sqrt{1-\gamma^2\sin^2\theta} - \iota\gamma\sin\theta\right)^n \beta$$

for some coefficients α and β that depend on the initial data for the scheme.

Since there is no well-defined solution ratio for the leap-frog scheme, we need to find other ways in which to discuss its performance as a result of its Fourier analysis. Note that for either eigenvalue of A,

$$|\lambda| = 1.$$

This result implies that if the initial data were chosen so that exactly one of α or β were nonzero, then $|\hat{u}^n|$ would be constant for all n. In this sense, the leap-frog scheme involves zero dissipation. On the other hand, this scheme does involve phase errors, which we could examine in the special cases where exactly one of α or β is nonzero. Students can produce error curves for individual values of the CFL number by choosing `scheme` equal to `leap frog` while running executable 2.5-2.

It is possible to use energy estimates [151] to show that the leap-frog scheme is stable for $\gamma < 1$.

Table 2.1 summarizes the results of our Fourier analyses.

Table 2.1 *Fourier Analysis Results*

	Dissipation error	Phase error	Order
explicit upwind	dissipative for $\gamma \in (0,1)$	$\begin{cases} \text{negative for } \gamma \in (\frac{1}{2},1) \\ \text{positive for } \gamma \in (0,\frac{1}{2}) \\ \text{zero for } \gamma = \frac{1}{2}, 1 \end{cases}$	1
explicit downwind	anti-dissipative for all γ	positive for $\gamma \in (0,1)$	1
explicit centered	anti-dissipative for all γ	positive for $\gamma \in (0,1)$	1
implicit upwind	dissipative for all γ	positive for $\gamma \in (0,1)$	1
implicit downwind	dissipative for $\gamma > 1$		1
Lax–Wendroff	dissipative for $\gamma \in (0,1)$	$\begin{cases} \text{negative for } \gamma \in (\frac{1}{2},1) \\ \text{positive for } \gamma \in (0,\frac{1}{2}) \\ \text{zero for } \gamma = \frac{1}{2}, 1 \end{cases}$	2
leap-frog	dissipative for $\gamma \in (0,1)$	positive for $\gamma \in (0,1)$	2

Exercises for 2.5

2.5.1 Plot the total dissipation $1 - |z(\theta)|^{1/\gamma}$ and the phase error $|1 + \arg(z(\theta)^{1/\gamma})/\theta|$ as functions of the mesh wave number θ for $-\pi \leq \theta \leq \pi$ and $\gamma = 0.9, 0.6, 0.5, 0.4$ and 0.1 for each of the following schemes
(a) explicit upwind
(b) explicit downwind
(c) explicit centered differences
Discuss which features of the plots indicate stability, and which features indicate order of convergence.

2.5.2 Repeat the previous exercise for **Eulerian-Lagrangian localized adjoint method** [25, 172]

$$\frac{1}{6}u_{i+1}^{n+1} + \frac{2}{3}u_i^{n+1} + \frac{1}{6}u_{i-1}^{n+1} = \frac{1}{6}(1-\alpha)^3 u_{i-\lfloor\gamma\rfloor-2} + \frac{1}{6}(4 - 6\alpha^2 + 3\alpha^3)u_{i-\lfloor\gamma\rfloor-1} + \frac{1}{6}(1 + 3\alpha + 3\alpha^2 - 3\alpha^3)u_{i-\lfloor\gamma\rfloor} + \frac{1}{6}\alpha^3 u_{i-\lfloor\gamma\rfloor+1}$$

where

$$\alpha = 1 - (\gamma - \lfloor\gamma\rfloor)$$

and

$$\gamma = \frac{c\triangle t}{\triangle x}.$$

Since this scheme is implicit, it is potentially competitive with explicit schemes only if it can take larger timesteps than the explicit schemes. Therefore, pay particular attention to the behavior of this scheme for CFL numbers $\gamma > 1$.

2.5.3 The **discontinuous Galerkin method** [134] for linear advection

$$\frac{\partial u}{\partial t} + \frac{\partial (cu)}{\partial x} = 0$$

with $c > 0$ is determined by the weak form

$$0 = \int_{x_{i-1/2}}^{x_{i+1/2}} w(x) \left[\frac{\partial u}{\partial t} + \frac{\partial (cu)}{\partial x} \right] dx$$
$$= \frac{d}{dt} \int_{x_{i-1/2}}^{x_{i+1/2}} w(x) u_i(x,t)\, dx + w(x_{i+1/2}) c u_i(x_{i+1/2}, t)$$
$$- w(x_{i-1/2}) c u_{i-1}(x_{i-1/2}, t) - \int_{x_{i-1/2}}^{x_{i+1/2}} \frac{dw}{dx} c u_i(x,t)\, dx.$$

Here $u_i(x,t)$ is a polynomial of degree at most p in x for each t, and $w(x)$ is a basis function for the set of all polynomials of degree at most p. In this formula, we used upwind values for the fluxes at cell sides.

(a) If $p = 0$, then we can take $w(x) = 1$ and $u_i(x,t) = u_i(t)$ and integrate in time by forward Euler. Show that the resulting scheme is explicit upwind.

(b) If $p = 1$, then $w_0(x) = 1$ and $w_1(x) = (x - x_i)/\triangle x_i$ are orthogonal basis functions for first-degree polynomials on $(x_{i-1/2}, x_{i+1/2})$. Use modified Euler time integration ($y' = f(y)$ approximated by $y^{n+1/2} = y^n + (\triangle t/2) f(y^n)$, and $y^{n+1} = y^n + \triangle t f(y^{n+1/2})$). Program the resulting scheme, and test it for piecewise constant initial data.

(c) Perform a Fourier analysis on the scheme, and plot its dissipation and dispersion. What do the results tell you about the scheme?

2.5.4 Choose one of either the **Lax–Wendroff scheme**, the **leap-frog scheme**, or the **Eulerian–Lagrangian localized adjoint method** described above, and perform the following:

(a) Modify the GNUmakefile so that **Program 2.5-24: fourierMain.C**. calls the Fortran subroutine `fourier` from whichever of **Program 2.5-25: lax_wendroff.f**, **Program 2.5-26: leap_frog.f** or **Program 2.5-27: ellam.f** is appropriate for your choice of scheme. Plot the fourier analysis of your scheme.

(b) Select interesting values of CFL for your scheme, and make runs with your scheme for Riemann problem initial data and at most 100 grid cells. Explain why you chose these values of CFL, and describe how your numerical results correspond to the results of the fourier analysis.

2.6 L^2 Stability for Linear Schemes

Let us recall **Parseval's identity** for the finite Fourier transform

$$\frac{1}{2\pi} \int_{-\pi/\Delta x}^{\pi/\Delta x} |\hat{u}^n(\xi)|^2\, d\xi = \sum_{j=-\pi/\Delta x}^{\pi/\Delta x} |u_j^n|^2 \triangle x \equiv \|u^n\|_{\triangle x}^2. \tag{2.30}$$

The corresponding Parseval identity for the Fourier transform is

$$\frac{1}{2\pi}\int_{-\infty}^{\infty} |\hat{u}^n(\xi)|^2 \, d\xi = \int_{-\infty}^{\infty} |u(x)|^2 \, dx. \tag{2.31}$$

Then the definition (2.26) of the numerical solution ratio and Parseval's identity (2.30) imply that

$$\begin{aligned}\sum_{j=-J}^{J} (u_j^{n+1})^2 \triangle x &= \frac{1}{2\pi}\int_{-\pi/\Delta x}^{\pi/\Delta x} |z(\xi \triangle x)\hat{u}^n(\xi)|^2 \, d\xi \\ &\leq \max_{\theta|\leq\pi}\{|z(\theta)|\}^2 \frac{1}{2\pi}\int_{-\pi/\Delta x}^{\pi/\Delta x} |\hat{u}^n(\xi)|^2 \, d\xi.\end{aligned}$$

If $\max_{|\theta|\leq\pi}\{|z(\theta)|\} \leq 1$, then we see that

$$\begin{aligned}\|u^{n+1}\|_2 \equiv \sqrt{\sum_{j=-J}^{J} (u_j^{n+1})^2 \triangle x} &\leq \max_{|\theta|\leq\pi}\{|z(\theta)|\} \frac{1}{2\pi}\int_{-\pi/\Delta x}^{\pi/\Delta x} |\hat{u}^n(\xi)|^2 \\ &\leq \frac{1}{2\pi}\int_{-\pi/\Delta x}^{\pi/\Delta x} |\hat{u}^n(\xi)|^2 = \sqrt{\sum_{j=-J}^{J} (u_j^n)^2 \triangle x} = \|u^n\|_2.\end{aligned}$$

This shows that the L^2 norm of the solution cannot grow whenever the modulus of z is bounded above by 1.

Fourier analysis is seldom used for studying the convergence of linear schemes on linear problems. However, it will be useful to examine how the dissipation and phase errors affect the error in the numerical solution. From our Fourier inversion formulas,

$$\begin{aligned}u(j\triangle x, t^{n+1}) - u_j^{n+1} &= \frac{1}{2\pi}\left[\int_{-\infty}^{\infty} \hat{u}(\xi, t^{n+1})e^{\imath x\xi} \, d\xi - \int_{-\pi/\Delta x}^{\pi/\Delta x} \hat{u}^{n+1}(\xi)e^{\imath j\Delta x\xi} \, d\xi\right] \\ &= \frac{1}{2\pi}\left[\int_{|\xi|>\pi/\Delta x} \hat{u}(\xi, t^{n+1})e^{\imath x\xi} \, d\xi\right. \\ &\quad \left. + \int_{-\pi/\Delta x}^{\pi/\Delta x} \{\hat{u}(\xi, t^n)e^{-\imath c\Delta t\xi}e^{\imath x\xi} - \hat{u}^n(\xi)z(\xi\triangle x)e^{\imath j\Delta x\xi}\} \, d\xi\right] \\ &= \frac{1}{2\pi}\left[\int_{|\xi|>\pi/\Delta x} \hat{u}(\xi, t^{n+1})e^{\imath x\xi} \, d\xi\right. \\ &\quad + \int_{-\pi/\Delta x}^{\pi/\Delta x} z(\xi\triangle x)\{\hat{u}(\xi, t^n)e^{-\imath c\Delta t\xi} - \hat{u}^n(\xi)e^{\imath j\Delta x\xi}\} \, d\xi \\ &\quad \left. + \int_{-\pi/\Delta x}^{\pi/\Delta x} \{e^{-\imath c\Delta t\xi} - z(\xi\triangle x)\}\hat{u}(\xi, t^n)e^{\imath x\xi} \, d\xi\right]\end{aligned}$$

The first term on the right is small if the solution involves very little high-frequency information. The second term is small for all timesteps if z has modulus less than one and the errors in the initial data for the numerical method are small. The third term is small if $z(\xi \triangle x)$ is close to $e^{-\imath c \triangle t \xi}$; for smooth initial data, only low wave number information is important. This motivates our definition (2.27) of the order of the scheme.

2.7 Lax Equivalence Theorem

For more information about the material in this section see, for example, Strikwerda [154].

In Section 2.5 we considered the Fourier analysis of linear schemes for linear advection. In this section, we will use Fourier analysis to examine more general linear partial differential equations of the form

$$Pu \equiv \frac{\partial u}{\partial t} - \mathcal{Q}\left(\frac{\partial}{\partial x}\right) u = 0. \tag{2.32}$$

Here $\mathcal{Q}$ is some linear operator. If we take the Fourier transform of equation (2.32) in space, we obtain

$$p\left(\xi, \frac{\partial}{\partial t}\right) \hat{u}(\xi, t) \equiv \frac{\partial \hat{u}}{\partial t} - q(\xi)\hat{u} = 0 \tag{2.33}$$

where $q(\xi)$ is whatever comes out of the Fourier transform of the spatial derivatives in the partial differential equation. The function $p(\xi, s) = s - q(\xi)$ is called the **symbol** in (2.32). Note that

$$p(\xi, q(\xi)) = 0.$$

It is also useful to note that the function $w_{\xi,s}(x, t) \equiv e^{st} e^{\imath \xi x}$ is an eigenfunction of the differential operator in (2.32); in other words,

$$P w_{\xi,s} = [s - q(\xi)] w_{\xi,s}.$$

Example 2.7.1 The symbol for the linear advection operator $\frac{\partial u}{\partial t} + c\frac{\partial u}{\partial x}$ is $p(\xi, s) = s + \imath c \xi$. In this case, $q(\xi) = -\imath c \xi$.

Example 2.7.2 The symbol for the linear diffusion operator $\frac{\partial u}{\partial t} - \frac{\partial}{\partial x}\left(d\frac{\partial u}{\partial x}\right) = 0$ is $p(\xi, s) = s + d\xi^2$. In this case, $q(\xi) = -d\xi^2$.

Next, suppose that we have a linear explicit two-step numerical scheme of the form

$$P_{\triangle x, \triangle t} u^n \equiv \sum_k a_k u_{j+k}^{n+1} - \sum_k b_k u_{j+k}^n = 0 \tag{2.34}$$

which is assumed to approximate the partial differential equation (2.32). If we take the finite Fourier transform of this scheme, we obtain

$$\left[\sum_k a_k e^{\imath k\xi\triangle x}\right]\hat{u}^{n+1}(\xi) - \left[\sum_k b_k e^{\imath k\xi\triangle x}\right]\hat{u}^n(\xi) = 0.$$

It is useful to define

$$w_j^n = w_{\xi,s}(j\triangle x, n\triangle t) = e^{sn\triangle t}e^{\imath\xi j\triangle x},$$

which are the mesh values of the eigenfunction of the differential operator in (2.32). Since

$$P_{\triangle x,\triangle t}w_j^n = \left\{\left[\sum_k a_k e^{\imath k\xi\triangle x}\right]e^{s\triangle t} - \left[\sum_k b_k e^{\imath k\xi\triangle x}\right]\right\}w_j^n \equiv p_{\triangle x,\triangle t}(\xi, s)w_j^n, \quad (2.35)$$

we see that the mesh function w_j^n is an eigenfunction of the difference operator $P_{\triangle x,\triangle t}$ in the numerical scheme (2.34). Here $p_{\triangle x,\triangle t}(\xi, s)$ is called the **symbol** of the numerical scheme. The **solution ratio** is the ratio

$$z(\xi\triangle x) \equiv \frac{\sum_k b_k e^{\imath k\xi\triangle x}}{\sum_k a_k e^{\imath k\xi\triangle x}} = \frac{\hat{u}^{n+1}(\xi)}{\hat{u}^n(\xi)}.$$

Note that $s = \ln z(\xi\triangle x)/\triangle t$ is a zero of the symbol $p_{\triangle x,\triangle t}$:

$$p_{\triangle x,\triangle t}\left(\xi, \frac{1}{\triangle t}\ln z(\xi\triangle x)\right) = 0.$$

Example 2.7.3 The symbol for the explicit upwind scheme applied to linear advection is $p_{\triangle x,\triangle t}(\xi, s) = e^{s\triangle t} - [(1-\gamma) + \gamma e^{\imath\xi\triangle x}]$ and the solution ratio is $z(\xi\triangle x) = (1-\gamma) + \gamma e^{\imath\xi\triangle x}$.

For computations that occur below, it will be useful to compute

$$\frac{\partial p_{\triangle x,\triangle t}}{\partial s}(\xi, s) = \triangle t e^{s\triangle t}\sum_k a_k e^{\imath k\xi\triangle x}.$$

In particular, it will be useful to note that

$$\frac{\partial p_{\triangle x,\triangle t}}{\partial s}\left(\xi, \frac{1}{\triangle t}\ln z(\xi\triangle x)\right) = \triangle t z(\xi\triangle x)\sum_k a_k e^{\imath k\xi\triangle x} = \triangle t\sum_k b_k e^{\imath k\xi\triangle x}.$$

Thus the zeros of $\frac{\partial p_{\triangle x,\triangle t}}{\partial s}(\xi, \frac{1}{\triangle t}\ln z(\xi\triangle x))$ are the zeros of the trigonometric polynomial $\sum_k b_k e^{\imath k\xi\triangle x}$, or equivalently, of the solution ratio $z(\xi\triangle x)$ considered as a function of ξ.

Example 2.7.4 The solution ratio for the explicit upwind scheme applied to linear advection with velocity c is $z(\xi\triangle x) = 1 - \gamma + \gamma e^{\imath\xi\triangle x}$, where $\gamma = c\triangle t/\triangle x$. In order for z to be zero, we must have $(1-\gamma)/\gamma = \pm 1$, since $|e^{-\imath\xi\Delta x}| = 1$. The only solution is $\gamma = \frac{1}{2}$ and $\xi = \pm\pi/\triangle x$.

Definition 2.7.5 We will say that the scheme $P_{\triangle x,\triangle t}u^n = 0$ is **consistent** with the partial differential equation $Pu = 0$ if and only if

$$\text{for all } \phi \in C^\infty \text{ for all } j \in \mathbf{Z} \text{ for all } n \in \mathbf{Z}_+ \text{ for all } \epsilon > 0 \text{ there exists } \triangle x_0 > 0$$
$$\text{there exists } \triangle t_0 > 0 \text{ for all } \triangle x \in (0, \triangle x_0] \text{ for all } \triangle t \in (0, \triangle t_0]$$
$$|P_{\triangle x,\triangle t}\phi(j\triangle x, n\triangle t) - (P\phi)(j\triangle x, n\triangle t)| < \epsilon.$$

Definition 2.7.6 We will say that the scheme $P_{\triangle x,\triangle t}u^n = 0$ has order α in time and order β in space if and only if the **local truncation error** satisfies

$$\text{there exists } \alpha > 0 \text{ there exists } \beta > 0 \text{ for all } \phi \in C^\infty \text{ for all } j \in \mathbf{Z} \text{ for all } n \in \mathbf{Z}_+$$
$$\text{there exists } C_\alpha > 0 \text{ there exists } C_\beta > 0 \text{ there exists } \triangle x_0 > 0 \text{ there exists } \triangle t_0 > 0$$
$$\text{for all } \triangle x \in (0, \triangle x_0] \text{ for all } \triangle t \in (0, \triangle t_0]$$
$$|P_{\triangle x,\triangle t}\phi(j\triangle x, n\triangle t) - (P\phi)(j\triangle x, n\triangle t)| \le C_\alpha \triangle t^\alpha + C_\beta \triangle t^\beta.$$

We expect that if the scheme is consistent with the partial differential equation, then the zero $s = \frac{1}{\Delta t}\ln z(\xi\triangle x)$ of the symbol $p_{\triangle x,\triangle t}(\xi, s)$ of the numerical scheme should be close to the zero $s = q(\xi)$ of the symbol $p(\xi, s)$ of the partial differential equation. The following lemma discusses one sense in which this is true.

Lemma 2.7.7 *Suppose that the scheme $P_{\triangle x,\triangle t}u^n = 0$ is consistent with the partial differential equation $Pu = 0$. Further, suppose that the symbol $p_{\triangle x,\triangle t}(\xi, s)$ of the scheme is continuously differentiable in s, and*

$$\frac{\partial p_{\triangle x,\triangle t}}{\partial s}\left(\xi, \frac{1}{\Delta t}\ln z(\xi\triangle x)\right) \neq 0.$$

Then $e^{q(\xi)\Delta t} - z(\xi\triangle x) = o(\triangle t)$; in other words,

$$\text{for all } \xi \text{ such that } \frac{\partial p_{\triangle x,\triangle t}}{\partial s}\left(\xi, \frac{1}{\triangle t}\ln z(\xi\triangle x)\right) \neq 0 \text{ for all } \delta > 0 \text{ there exists } \triangle x_0 > 0 \text{ there exists } \triangle t_0 > 0$$
$$\text{for all } \triangle x \in (0, \triangle x_0] \text{ for all } \triangle t \in (0, \triangle t_0]\ |e^{q(\xi)\Delta t} - z(\xi\triangle x)| \le \delta\triangle t.$$

Proof Since $s = q(\xi)$ is a zero of the symbol p of the differential equation, the definition of consistency 2.7.5 with $\phi(x, t) = e^{\imath x\xi}e^{q(\xi)t}$ implies that

$$\text{for all } j \in \mathbf{Z} \text{ for all } n \in \mathbf{Z}_+ \text{ for all } \delta > 0 \text{ there exists } \triangle x_0 > 0$$
$$\text{there exists } \triangle t_0 > 0 \text{ for all } \triangle x \in (0, \triangle x_0] \text{ for all } \triangle t \in (0, \triangle t_0]$$
$$\delta > |p_{\triangle x,\triangle t}(\xi, q(\xi))e^{\imath j\xi\Delta x}e^{q(\xi)n\Delta t} - p(\xi, q(\xi))e^{\imath j\xi\Delta x}e^{q(\xi)n\Delta t}|$$
$$= |p_{\triangle x,\triangle t}(\xi, q(\xi))|e^{q(\xi)n\Delta t}.$$

Since $s = \frac{1}{\triangle t} \ln z(\xi \triangle x)$ is a zero of the symbol $p_{\triangle x, \triangle t}$ of the scheme,

$$\begin{aligned} e^{-q(\xi)n\triangle t}\delta &> \left| p_{\triangle x,\triangle t}(\xi, q(\xi)) - p_{\triangle x,\triangle t}\left(\xi, \frac{1}{\triangle t}\ln z(\xi \triangle x)\right)\right| \\ &= \left| \int_{\ln z(\xi\triangle x)/\triangle t}^{q(\xi)} \frac{\partial p_{\triangle x,\triangle t}}{\partial s}(\xi, s)\, ds\right| \\ &\geq \left| q(\xi) - \frac{1}{\triangle t}\ln z(\xi\triangle x)\right| \min_{s\in \operatorname{int}(\ln z(\xi\triangle x)/\triangle t, q(\xi))} \left|\frac{\partial p_{\triangle x,\triangle t}}{\partial s}(\xi, s)\right| . \end{aligned}$$

It follows that

$$\begin{aligned} \left|\frac{e^{q(\xi)\triangle t} - z(\xi\triangle x)}{\triangle t}\right| &= \frac{1}{\triangle t}\left|\int_{\ln z(\xi\triangle x)}^{q(\xi)\triangle t} e^s\, ds\right| \\ &\leq \left| q(\xi) - \frac{1}{\triangle t}\ln z(\xi\triangle x)\right| e^{\max\{q(\xi)\triangle t, \ln z(\xi\triangle x)\}} \\ &< \delta \frac{e^{-q(\xi)n\triangle t} e^{\max\{q(\xi)\triangle t, \ln z(\xi\triangle x)\}}}{\min_{s\in \operatorname{int}(\ln z(\xi\triangle x)/\triangle t, q(\xi))} \left|\frac{\partial p_{\triangle x,\triangle t}}{\partial s}(\xi, s)\right|} . \end{aligned}$$

Now choose ξ so that $\frac{\partial p_{\triangle x,\triangle t}}{\partial s}$ is nonzero, and choose ϵ. The continuity of $\frac{\partial p_{\triangle x,\triangle t}}{\partial s}$ implies that

$$\text{there exists } \gamma > 0 \text{ there exists } \triangle x_0 > 0 \text{ there exists } \triangle t_0 \text{ for all } \triangle x \in (0, \triangle x_0] \text{ for all } \triangle t \in (0, \triangle t_0] \quad \min_{s\in \operatorname{int}(\ln z(\xi\triangle x)/\triangle t, q(\xi))} \left|\frac{\partial p_{\triangle x,\triangle t}}{\partial s}(\xi, s)\right| < \gamma .$$

Further, the continuity of q and z implies that

$$\begin{aligned} &\text{there exists } \beta > 0 \ \text{ there exists } n > 0 \ \text{ there exists } \triangle x_0 > 0 \ \text{ there exists } \triangle t_0 \\ &\text{for all } \triangle x \in (0, \triangle x_0] \ \text{ for all } \triangle t \in (0, \triangle t_0] \ e^{-q(\xi)n\triangle t} e^{\max\{q(\xi)\triangle t, \ln z(\xi\triangle x)\}} < \beta . \end{aligned}$$

Since δ is arbitrary, we can choose $\delta < \epsilon\gamma/\beta$ so that the conclusion of the lemma is satisfied. □

Corollary 2.7.8 *Suppose that the scheme $P_{\triangle x,\triangle t}u^n = 0$ is consistent with the partial differential equation $Pu = 0$ of order α in time and β in space. Further, suppose that the symbol $p_{\triangle x,\triangle t}(\xi, s)$ of the scheme is continuously differentiable in s, and*

$$\frac{\partial p_{\triangle x,\triangle t}}{\partial s}\left(\xi, \frac{1}{\triangle t}\ln z(\xi\triangle x)\right) \neq 0.$$

Then $[e^{q(\xi)\Delta t} - z(\xi \Delta x)]/\Delta t = O(\Delta t^\alpha) + O(\Delta x^\beta)$*; in other words,*

$$\text{for all } \xi \text{ such that } \frac{\partial p_{\Delta x,\Delta t}}{\partial s}\left(\xi, \frac{1}{\Delta t}\ln z(\xi\Delta x)\right) \neq 0 \text{ there exists } C_\alpha > 0 \text{ there exists } C_\beta > 0$$
$$\text{there exists } \Delta x_0 > 0 \text{ there exists } \Delta t_0 > 0 \text{ for all } \Delta x \in (0, \Delta x_0] \text{ for all } \Delta t \in (0, \Delta t_0]$$
$$\left|\frac{e^{q(\xi)\Delta t} - z(\xi\Delta x)}{\Delta t}\right| \le C_\alpha \Delta t^\alpha + C_\beta \Delta x^\beta.$$

Proof Replace ϵ in the previous proof with $C_\alpha \Delta t^\alpha + C_\beta \Delta x^\beta$. □

Now that we have discussed consistency, let us turn to stability.

Definition 2.7.9 We will say that the scheme $P_{\Delta x,\Delta t} u^n = 0$ is a **stable** finite difference approximation to the partial differential equation $Pu = 0$ if and only if

$$\text{there exists } \Delta x_0 > 0 \text{ there exists } \Delta t_0 > 0 \text{ for all } T > 0 \text{ there exists } C_T > 0$$
$$\text{for all } \Delta t \in (0, \Delta t_0] \text{ for all } n\Delta t \in [0, T] \text{ for all } \Delta x \in (0, \Delta x_0]$$
$$\|u^n\|^2_{\Delta x} \equiv \Delta x \sum_{j=-\infty}^{\infty} |u^n_j|^2 \le C_T \Delta x \sum_{j=-\infty}^{\infty} |u^0_j|^2 \equiv C_T \|u^0\|^2_{\Delta x}. \tag{2.36}$$

We expect that if the scheme is stable, then the solution ratio is bounded close to one.

Lemma 2.7.10 *Suppose that the scheme* $P_{\Delta x,\Delta t} u^n = 0$ *is a finite difference approximation to the partial differential equation* $Pu = 0$*, and that the solution ratio* $z(\xi \Delta x)$ *for the scheme is continuous. Then the scheme is a stable finite difference approximation to the partial differential equation if and only if* $z(\xi \Delta x)$ *bounded close to one in the following sense:*

$$\text{there exists } K > 0 \text{ there exists } \Delta x_0 > 0 \text{ there exists } \Delta t_0 > 0 \text{ for all } \Delta t \in (0, \Delta t_0]$$
$$\text{for all } \Delta x \in (0, \Delta x_0] \text{ for all } \theta\ |z(\theta)| \le 1 + K\Delta t. \tag{2.37}$$

Proof First, we will prove that the bounded solution ratio condition implies stability. By Parseval's identity (2.30) and the fact that $\hat{u}^{n+1}(\xi) = z(\xi\Delta x)\hat{u}^n(\xi)$, inequality (2.37) implies

$$\begin{aligned}\|u^n\|^2_{\Delta x} &= \int_{-\pi/\Delta x}^{\pi/\Delta x} |z(\xi\Delta x)|^{2n} |\hat{u}^0(\xi)|^2 \, d\xi \le (1 + K\Delta t)^{2n} \|\hat{u}^0\|^2_{\Delta x} \\ &\le \left[(1 + K\Delta t)^{T/\Delta t}\right]^2 \|\hat{u}^0\|^2_{\Delta x} \le e^{2KT} \|\hat{u}^0\|^2_{\Delta x}\end{aligned}$$

This shows that the scheme is stable.

Next, we will show that if inequality (2.37) cannot be satisfied, then the scheme is not stable. The negation of (2.37) is

$$\text{for all } K > 0 \text{ for all } \triangle x_0 > 0 \text{ for all } \triangle t_0 > 0 \text{ there exists } 0 < \triangle t \leq \triangle t_0$$
$$\text{there exists } 0 < \triangle x \leq \triangle x_0 \text{ there exists } \theta \text{ such that } \quad |z(\theta)| > 1 + K\triangle t.$$

Since z is continuous,

$$\text{for all } K > 0 \text{ for all } \triangle x_0 > 0 \text{ for all } \triangle t_0 > 0 \text{ there exists } 0 < \triangle t \leq \triangle t_0$$
$$\text{there exists } 0 < \triangle x \leq \triangle x_0 \text{ there exists } \theta_1 < \theta_2 \text{ for all } \theta_1 < \theta < \theta_2 \; |z(\theta)| > 1 + K\triangle t.$$

Given K and the corresponding $\triangle x$, θ_1 and θ_2, define the initial data in terms of its finite Fourier transform by

$$\hat{u}^0(\xi) = \begin{cases} \sqrt{\triangle x/(\theta_2 - \theta_1)}, & \theta_1 < \xi\triangle x < \theta_2 \\ 0, & \text{otherwise.} \end{cases}$$

Note that Parseval's identity (2.30) implies that

$$\|u^0\|_{\triangle x}^2 = \|\hat{u}^0\|^2 = \int_{\theta_1/\triangle x}^{\theta_2/\triangle x} \frac{\triangle x}{\theta_2 - \theta_1}\, d\xi = 1.$$

For any $T > 0$ and for $n\triangle t$ near T we have

$$\begin{aligned} \|u^n\|_{\triangle x}^2 &= \int_{-\pi/\triangle x}^{\pi/\triangle x} |z(\xi\triangle x)|^{2n} |\hat{u}^0(\xi)|^2 \, d\xi = \int_{\theta_1/\triangle x}^{\theta_2/\triangle x} |z(\xi\triangle x)|^{2n} \frac{\triangle x}{\theta_2 - \theta_1}\, d\xi \\ &\geq (1 + K\triangle t)^{2n} \geq \frac{1}{2} e^{2KT} = \frac{1}{2} e^{2KT} \|u^0\|_{\triangle x}^2. \end{aligned}$$

Since K is arbitrary, this shows that the negation of the stability definition (2.36) holds, namely

$$\text{for all } \triangle x_0 > 0 \text{ for all } \triangle t_0 > 0 \text{ there exists } T > 0 \text{ for all } C_T > 0 \text{ there exists } 0 < \triangle t \leq \triangle t_0$$
$$\text{there exists } 0 \leq n\triangle t \leq T \text{ there exists } 0 < \triangle x \leq \triangle x_0 \text{ such that } \|u^n\|_{\triangle x}^2 > C_T \|u^0\|_{\triangle x}^2$$

is satisfied with $C_T < \frac{1}{2} e^{2KT}$. □

Our next goal will be to study the connections between consistency, stability and convergence. In order to do so, we will make use of two new devices. The **interpolation operator** $I_{\triangle x} : L^2(\mathbf{Z}\triangle x) \to L^2(\mathbf{R})$ is defined for any grid function v_j in terms of its finite Fourier transform $\hat{v}$ by

$$(I_{\triangle x} v)(x) = \frac{1}{2\pi} \int_{-\pi/\triangle x}^{\pi/\triangle x} e^{\imath x\xi}\, \hat{v}(\xi)\, d\xi. \tag{2.38}$$

Note that the interpolation operator takes a grid function and returns a function of space that agrees in value with the grid function at the cell centers; in this way, it interpolates the values at the cell centers at all points in space. This definition and the Fourier inversion formula (2.25) implies that

$$\widehat{I_{\triangle x} v}(\xi) = \begin{cases} \hat{v}(\xi), & |\xi| \leq \pi/\triangle x \\ 0, & |\xi| > \pi/\triangle x. \end{cases}$$

Also, the **truncation operator** $T_{\triangle x} : L^2(\mathbf{R}) \to L^2(\mathbf{Z}\triangle x)$ is defined for any L^2 function $u(x)$ in terms of its Fourier transform $\hat{u}$ by

$$(T_{\triangle x} u)_j = \frac{1}{2\pi} \int_{-\pi/\Delta x}^{\pi/\Delta x} e^{\imath j \triangle x \xi} \hat{u}(\xi)\, d\xi.$$

The truncation operator takes a function of space and returns a grid function that agrees in value with its argument at the cell centers. Note that the finite Fourier transform of $T_{\triangle x} u$ satisfies

$$\text{for all } |\xi| \leq \frac{\pi}{\triangle x} \quad \widehat{T_{\triangle x} u}(\xi) = \hat{u}(\xi).$$

Both the interpolation operator and the truncation operator are linear.

These definitions lead to the following simple lemmas.

Lemma 2.7.11 *If u_j^n is a grid function, then the interpolation operator* (2.38) *satisfies*

$$\|I_{\triangle x} u^n\| = \|u^n\|_{\triangle x}.$$

Proof Parseval's identities (2.30) and (2.31) imply that

$$\begin{aligned} \|I_{\triangle x} u^n\|^2 &= \int_{-\infty}^{\infty} |(I_{\triangle x} u^n)(x)|^2 \, dx = \frac{1}{2\pi} \int_{-\infty}^{\infty} |\widehat{I_{\triangle x} u^n}(\xi)|^2(\xi)\, d\xi \\ &= \frac{1}{2\pi} \int_{-\pi/\Delta x}^{\pi/\Delta x} |\hat{u}(\xi)|^2(\xi)\, d\xi = \sum_{j=-\pi/\Delta x}^{\pi/\Delta x} |u_j|^2 \triangle x = \|u\|_{\triangle x}^2. \end{aligned}$$

□

Lemma 2.7.12 *Suppose that $u \in L^2(\mathbf{R})$ and the grid function v_j are given. Then the truncation operator* (2.7) *satisfies*

$$\text{for all } \triangle x > 0, \quad \|T_{\triangle x} u - v\|_{\triangle x} \leq \|u - I_{\triangle x} v\|.$$

Proof Using Parseval's identities (2.30) and (2.31), we compute

$$
\begin{aligned}
\|T_{\triangle x}u - v\|^2_{\triangle x} &= \triangle x \sum_{j=-\pi/\triangle x}^{\pi/\triangle x} |(T_{\triangle x}u)_j - v_j|^2 = \int_{-\pi/\triangle x}^{\pi/\triangle x} |\widehat{T_{\triangle x}u}(\xi) - \hat{v}(\xi)|^2 \, d\xi \\
&= \int_{-\pi/\triangle x}^{\pi/\triangle x} |\hat{u}(\xi) - \hat{v}(\xi)|^2 \, d\xi \\
&\leq \int_{-\pi/\triangle x}^{\pi/\triangle x} |\hat{u}(\xi) - \hat{v}(\xi)|^2 \, d\xi + \int_{|\xi|>\pi/\triangle x} |\hat{u}(\xi)|^2 \, d\xi \\
&= \int_{-\infty}^{\infty} |\hat{u}(\xi) - \widehat{I_{\triangle x}v}(\xi)|^2 \, d\xi = \|u - I_{\triangle x}v\|^2
\end{aligned}
$$

□

Lemma 2.7.13 *Suppose that $u \in L^2(\mathbf{R})$. Then Fourier interpolation applied to truncation approaches the original function as the mesh is refined:*

$$\textit{for all } \epsilon > 0 \ \textit{ there exists } \triangle x_0 > 0 \ \textit{ for all } 0 < \triangle x \leq \triangle x_0, \quad \|u - I_{\triangle x}(T_{\triangle x}u)\| < \epsilon.$$

Proof We compute

$$
\begin{aligned}
\|u - I_{\triangle x}(T_{\triangle x}u)\|^2 &= \int_{-\infty}^{\infty} |\widehat{u}(\xi) - I_{\triangle x}(\hat{T}_{\triangle x}u)(\xi)|^2 \, d\xi \\
&= \int_{-\pi/\triangle x}^{\pi/\triangle x} |\hat{u}(\xi) - \widehat{T_{\triangle x}u}(\xi)|^2 \, d\xi + \int_{|\xi|>\pi/\triangle x} |\hat{u}(\xi)|^2 \, d\xi \\
&= \int_{|\xi|>\pi/\triangle x} |\hat{u}(\xi)|^2 \, d\xi.
\end{aligned}
$$

Since $u \in L^2(\mathbf{R})$, the right-hand side of this inequality tends to zero as $\triangle x \to 0$.

□

These results lead us to the following important theorem.

Theorem 2.7.14 *(Lax Equivalence Part I: Stability Implies Convergence) Suppose that the scheme $P_{\triangle x,\triangle t}u^n = 0$ is defined by (2.34), and is consistent with the partial differential equation $Pu = 0$, which is defined by (2.32). Further, suppose that the symbol $p_{\triangle x,\triangle t}(\xi, s)$ of the scheme is defined by (2.35), is continuously differentiable in s, and*

$$\frac{\partial p_{\triangle x,\triangle t}}{\partial s}\left(\xi, \frac{1}{\triangle t} \ln z(\xi \triangle x)\right) \neq 0.$$

We also assume that the solution ratio $z(\xi \triangle x)$ for the scheme is continuous. We assume that the symbol of the partial differential equation P is defined by (2.33), and has the form $p(\xi, s) = s - q(\xi)$ where $q(\xi)$ is continuous. In addition, we *assume that the partial differential equation $Pu = 0$ is stable, in the sense that*

$$\text{for all } T > 0 \text{ there exists } C_T > 0 \text{ for all } 0 \le t \le T \text{ for all } \xi, |e^{q(\xi)t}| \le C_T. \quad (2.39)$$

Finally, we assume that the initial data for the scheme is convergent to the true initial data, in the sense that the interpolation operator (given by (2.38)) applied to the initial data for the scheme approaches the true data in the following sense:

$$\text{for all } u_0 \in L^2(\mathbf{R}) \text{ for all } \delta > 0 \text{ there exists } \triangle x_0 > 0 \text{ for all } 0 < \triangle x \le \triangle x_0$$
$$\|u_0 - I_{\triangle x} u^0\| < \delta \quad (2.40)$$

Under these conditions, if the scheme is stable (see condition (2.36)) then it is convergent, meaning that

$$\text{for all } u_0 \in L^2(\mathbf{R}) \text{ for all } \epsilon > 0 \text{ for all } T > 0 \text{ there exists } \triangle x_0 > 0 \text{ there exists } \triangle t_0 > 0$$
$$\text{for all } 0 < \triangle t \le \triangle t_0 \text{ for all } 0 \le n\triangle t \le T \text{ for all } 0 < \triangle x < \triangle x_0 \;\; \|u(\cdot, n\triangle t) - I_{\triangle x} u^n\| < \epsilon. \quad (2.41)$$

Proof Given any $u_0 \in L^2(\mathbf{R})$, suppose that the grid function w_j^n satisfies the scheme $P_{\Delta x,\Delta t} w^n = 0$ with initial data $w^0 = T_{\Delta x} u_0$. Using the finite Parseval identity (2.30) we compute

$$\begin{aligned}
&\|u(\cdot, n\triangle t) - I_{\triangle x} w^n\|^2 \\
&= \frac{1}{2\pi} \int_{-\infty}^{\infty} |\hat{u}(\cdot, n\triangle t)(\xi) - \widehat{I_{\Delta x} w^n}(\xi)|^2 \, d\xi \\
&= \frac{1}{2\pi} \int_{-\pi/\Delta x}^{\pi/\Delta x} |e^{q(\xi)n\Delta t} - z(\xi\triangle x)^n|^2 |\hat{u}_0(\xi)|^2 \, d\xi + \int_{|\xi| > \pi/\Delta x} |e^{q(\xi)n\Delta t} \hat{u}_0(\xi)|^2 \, d\xi \\
&\equiv \frac{1}{2\pi} \int_{-\infty}^{\infty} \phi_{\triangle x}(\xi) \, d\xi.
\end{aligned}$$

Here we have defined

$$\phi_{\triangle x}(\xi) \equiv \begin{cases} |e^{q(\xi)n\Delta t} - z(\xi\triangle x)^n|^2 |\hat{u}_0(\xi)|^2, & |\xi| < \pi/\triangle x \\ |e^{q(\xi)n\Delta t}|^2 |\hat{u}_0(\xi)|^2, & |\xi| \ge \pi/\triangle x. \end{cases}$$

Since the scheme is stable, Lemma 2.7.10 implies that the solution ratio is bounded in the following sense:

$$\text{there exists } K > 0 \text{ there exists } \triangle x_0 > 0 \text{ there exists } \triangle t_0 > 0 \text{ for all } 0 < \triangle t \le \triangle t_0$$
$$\text{for all } 0 < \triangle x \le \triangle x_0 \text{ for all } \theta \quad |z(\theta)| \le 1 + K\triangle t.$$

This may place our first restrictions on $\triangle x$ and $\triangle t$. Since the scheme is consistent, Lemma 2.7.7 implies that

$$\text{for all } \xi \text{ such that } \frac{\partial p_{\triangle x,\triangle t}}{\partial s}\left(\xi, \frac{1}{\triangle t}\ln z(\xi\triangle x)\right) \neq 0$$
$$\text{for all } \delta > 0 \text{ there exists } \triangle x_0 > 0 \text{ there exists } \triangle t_0 > 0$$
$$\text{for all } 0 < \triangle x \leq \triangle x_0 \text{ for all } 0 < \triangle t \leq \triangle t_0 \; |e^{q(\xi)\triangle t} - z(\xi\triangle x)| \leq \delta\triangle t.$$

This places further restrictions on $\triangle x_0$ and $\triangle t$, and possibly a restriction on ξ. These last two inequalities imply that

$$\text{there exists } K > 0 \text{ for all } \delta > 0 \text{ for all } \xi \text{ such that } \frac{\partial p_{\triangle x,\triangle t}}{\partial s}\left(\xi, \frac{1}{\triangle t}\ln z(\xi\triangle x)\right) \neq 0$$
$$\text{there exists } \triangle x_0 > 0 \text{ there exists } \triangle t_0 > 0 \text{ for all } 0 < \triangle x \leq \triangle x_0 \text{ for all } 0 < \triangle t \leq \triangle t_0$$
$$|e^{q(\xi)\triangle t}| \leq |e^{q(\xi)\triangle t} - z(\xi\triangle x)| + |z(\xi\triangle x)| \leq 1 + (K+\delta)\triangle t.$$

The bounds on $|z(\theta)|$ and $|e^{q(\xi)\triangle t}|$ imply

$$\text{there exists } K > 0$$
$$\text{for all } \xi \text{ such that } \frac{\partial p_{\triangle x,\triangle t}}{\partial s}\left(\xi, \frac{1}{\triangle t}\ln z(\xi\triangle x)\right) \neq 0 \text{ for all } n \geq 0$$
$$\text{there exists } \triangle x_0 > 0$$
$$\text{there exists } \triangle t_0 > 0$$
$$\text{for all } \triangle x \in (0, \triangle x_0] \text{ for all } \triangle t \in (0, \triangle t_0]$$
$$\begin{aligned} |e^{q(\xi)n\triangle t} - z(\xi\triangle x)^n| &= \left|(e^{q(\xi)\triangle t} - z(\xi\triangle x))\sum_{k=0}^{n-1} e^{q(\xi)(n-k)\triangle t} z(\xi\triangle x)^k\right| \\ &\leq |e^{q(\xi)\triangle t} - z(\xi\triangle x)| n[1 + (K+\delta)\triangle t]^n [1 + K\triangle t]^n \\ &\leq \delta n\triangle t e^{(2K+\delta)n\triangle t}. \end{aligned}$$

Thus $\phi_{\triangle x}(\xi) \to 0$ almost everywhere as $\triangle x, \triangle t \to 0$:

$$\text{there exists } K > 0 \text{ for all } \xi \text{ such that } \frac{\partial p_{\triangle x,\triangle t}}{\partial s}\left(\xi, \frac{1}{\triangle t}\ln z(\xi\triangle x)\right) \neq 0 \text{ for all } n \geq 0$$
$$\text{there exists } \triangle x_0 > 0 \text{ there exists } \triangle t_0 > 0$$
$$\text{for all } \triangle x \in (0, \triangle x_0] \text{ for all } \triangle t \in (0, \triangle t_0]$$
$$\phi_{\triangle x}(\xi) \leq \begin{cases} (\delta n\triangle t)^2 e^{2(2K+\delta)n\triangle t}|\hat{u}_0(\xi)|^2, & |\xi| < \pi/\triangle x \\ e^{2(K+\delta)n\triangle t}|\hat{u}_0(\xi)|^2, & |\xi| \geq \pi/\triangle x. \end{cases}$$

Lebesgue's dominated convergence theorem implies that

$$\text{for all } \delta > 0 \text{ for all } n > 0 \text{ there exists } \triangle x_0 > 0 \text{ there exists } \triangle t_0 > 0$$
$$\text{for all } 0 < \triangle x \leq \triangle x_0 \text{ for all } 0 < \triangle t \leq \triangle t_0$$
$$\|u(\cdot, n\triangle t) - I_{\triangle x} w^n\|^2 = \int_{-\infty}^{\infty} \phi_{\triangle x}(\xi)\, d\xi < \delta. \tag{2.42}$$

For the general scheme with initial data u_j^0 we use the triangle inequality

$$\|u(\cdot, n\triangle t) - I_{\triangle x}u^n\| \leq \|u(\cdot, n\triangle t) - I_{\triangle x}w^n\| + \|I_{\triangle x}w^n - I_{\triangle x}u^n\|. \tag{2.43}$$

Lemma 2.7.11 implies that the second term on the right is $\|I_{\triangle x}w^n - I_{\triangle x}u^n\| = \|w^n - u^n\|_{\triangle x}$. Since both w_j^n and u_j^n are grid functions generated by a stable linear scheme

there exists $K > 0$ for all $n > 0$ there exists $\triangle x_0 > 0$ there exists $\triangle t_0 > 0$

for all $0 < \triangle t \leq \triangle t_0$ for all $0 < \triangle x \leq \triangle x_0$

$$\begin{aligned}
\|w^n - u^n\|_{\triangle x}^2 &= \frac{1}{2\pi}\int_{-\pi/\Delta x}^{\pi/\Delta x} |\hat{w}^n(\xi) - \hat{u}^n(\xi)|^2\, d\xi \\
&= \frac{1}{2\pi}\int_{-\pi/\Delta x}^{\pi/\Delta x} |z(\xi\triangle x)|^{2n}|\hat{w}^0(\xi) - \hat{u}^0(\xi)|^2\, d\xi \\
&\leq \frac{(1 + K\triangle t)^{2n}}{2\pi}\int_{-\pi/\Delta x}^{\pi/\Delta x} |\hat{w}^0(\xi) - \hat{u}^0(\xi)|^2\, d\xi \\
&\leq \frac{e^{2Kn\Delta t}}{2\pi}\int_{-\pi/\Delta x}^{\pi/\Delta x} |\hat{w}^0(\xi) - \hat{u}^0(\xi)|^2\, d\xi \\
&= e^{2Kn\Delta t}\|w^0 - u^0\|_{\triangle x}^2.
\end{aligned}$$

Since the grid function w_j^n uses initial data $w^0 = T_{\triangle x}u_0$, Lemma 2.7.12 together with inequalities (2.43), (2.42) and (2.40) implies that

there exists $K > 0$ for all $n > 0$ there exists $\triangle x_0 > 0$ there exists $\triangle t_0 > 0$

for all $0 < \triangle t \leq \triangle t_0$ for all $0 < \triangle x \leq \triangle x_0$

$$\begin{aligned}
\|u(\cdot, n\triangle t) - I_{\triangle x}u^n\| &\leq \|u(\cdot, n\triangle t) - I_{\triangle x}w^n\| + e^{Kn\Delta t}\|T_{\triangle x}u_0 - u^0\|_{\triangle x} \\
&\leq \|u(\cdot, n\triangle t) - I_{\triangle x}w^n\| + e^{Kn\Delta t}\|u_0 - I_{\triangle x}u^0\|_{\triangle x}.
\end{aligned}$$

We showed in inequality (2.42) that for any initial data and any $\epsilon > 0$ and any $n > 0$ we can choose $\triangle x$ and $\triangle t$ so that the first of the two terms on the right hand side is less than $\epsilon/2$. Since we assumed in inequality (2.40) that the error in the initial data can be chosen to be small, for any initial data and any $\epsilon > 0$ we can further restrict $\triangle x$ so that the second of these two terms is less than $\epsilon/2$. This proves the conclusion (2.41) of our theorem, that stability implies convergence. □

Here is the second part of the Lax equivalence theorem.

Theorem 2.7.15 *(Lax Equivalence Part II: Convergence implies stability) Suppose that the scheme $P_{\Delta x,\Delta t}u^n = 0$ is defined by* (2.34)*, and is consistent with the partial differential equation $Pu = 0$, which is defined by* (2.32)*. Further, suppose that the symbol $p_{\Delta x,\Delta t}(\xi, s)$ of the scheme is defined by* (2.35)*, is continuously differentiable* *in s, and*

$$\frac{\partial p_{\triangle x,\triangle t}}{\partial s}\left(\xi,\frac{1}{\triangle t}\ln z(\xi\triangle x)\right)\neq 0.$$

We also assume that the solution ratio $z(\xi \triangle x)$ for the scheme is continuous. We assume that the symbol of the partial differential equation P is defined by (2.33)*, and has the form $p(\xi, s) = s - q(\xi)$ where $q(\xi)$ is continuous. In addition, we assume that the partial differential equation $Pu = 0$ is stable, in the sense that condition* (2.39) *holds. Finally, we assume that the initial data for the scheme is convergent to the true initial data, in the sense that the interpolation operator (given by* (2.38)*) applied to the initial data for the scheme approaches the true data in the following sense that condition* (2.40) *holds. Under these conditions, if the scheme is not stable then it is not convergent.*

Proof We will prove this by constructing initial data $u_0(x)$ so that the numerical solution w_j^n, satisfying $P_{\triangle x,\triangle t} w^n = 0$ and $w_j^0 = T_{\triangle x} u_0$, does not converge to $u(x, t)$.

Note that the negation of the stability condition in Lemma 2.7.10 says that the solution ratio satisfies

$$\begin{aligned}&\text{for all } K>0 \text{ for all } \triangle x_0>0 \text{ for all } \triangle t_0>0 \text{ there exists } 0<\triangle t\le\triangle t_0\\&\text{there exists } 0<\triangle x\le\triangle x_0 \text{ there exists } \theta \text{ such that } |z(\theta)|>1+K\triangle t.\end{aligned}$$

Since the solution ratio z is assumed to be continuous, this negation of stability implies that

$$\begin{aligned}&\text{for all } K>0 \text{ for all } \triangle x_0>0 \text{ for all } \triangle t_0>0 \text{ there exists } 0<\triangle t\le\triangle t_0\\&\text{there exists } 0<\triangle x\le\triangle x_0 \text{ there exists } \xi_K \text{ there exists } \eta_K>0 \text{ for all } |\xi-\xi_K|\le\eta_K\\&\qquad |z(\xi\triangle x)|>1+\frac{1}{2}K\triangle t.\end{aligned}$$

In particular, we may further restrict the choices as follows:

$$\begin{aligned}&\text{for all } K\in\mathbf{Z}_+ \text{ there exists } 0<\triangle t_K<\triangle t_{K-1} \text{ there exists } 0<\triangle x_K<\triangle x_{K-1} \text{ there exists } \xi_K\\&\text{there exists } 0<\eta_K\le 1/K^2 \text{ for all } |\xi-\xi_K|\le\eta_K\\&\qquad |z(\xi\triangle x)|>1+\frac{1}{2}K\triangle t.\end{aligned}\tag{2.44}$$

We now claim that for $K > 1$, the interval $\Omega_K = [\xi_K - \eta_K, \xi_K + \eta_K]$ can be chosen to be disjoint from the previous intervals $\Omega_1, \ldots, \Omega_{K-1}$. Note that this claim is obviously satisfied for $K = 1$. We will prove the claim is true by induction and contradiction. Suppose that $K > 1$ is the first so that I_K cannot be disjoint from the previous intervals. In other words,

$$\begin{aligned}&\text{there exists } K \in \mathbf{Z}_+ \text{ there exists } 0 < \triangle t_K < \triangle t_{K-1} \text{ there exists } 0 < \triangle x_K < \triangle x_{K-1}\\ &\text{if there exists } \xi_K \text{ there exists } \eta_K > 0 \text{ for all } \xi \in [\xi_K - \eta_K, \xi_K + \eta_K]\ |z(\xi \triangle x)| > 1 + \frac{1}{2} K \triangle t\\ &\text{then there exists } J < K \text{ such that } [\xi_K - \eta_K, \xi_K + \eta_K] \subset [\xi_J - \eta_J, \xi_J + \eta_J]\end{aligned}$$

Then we must have the following bound on z outside the union of the previous intervals:

$$\begin{aligned}&\text{there exists } K \in \mathbf{Z}_+ \text{ there exists } 0 < \triangle t_K < \triangle t_{K-1}\\ &\text{there exists } 0 < \triangle x_K < \triangle x_{K-1} \text{ such that for all } \xi \notin \cup_{N<K} I_N,\\ &\qquad |z(\xi \triangle x)| \leq 1 + K \triangle t_K.\end{aligned}$$

Since the scheme is consistent, Lemma 2.7.7 implies that

$$\begin{aligned}&\text{for all } \xi \text{ such that } \frac{\partial p_{\triangle x, \triangle t}}{\partial s}\left(\xi, \frac{1}{\triangle t} \ln z(\xi \triangle x)\right) \neq 0\\ &\text{for all } \epsilon > 0 \text{ there exists } \triangle x_* > 0 \text{ there exists } \triangle t_* > 0 \text{ such that}\\ &\text{for all } 0 < \triangle x \leq \triangle x_* \text{ for all } 0 < \triangle t \leq \triangle t_*\\ &\left| \frac{e^{q(\xi)\triangle t} - z(\xi \triangle x)}{\triangle t} \right| \leq \epsilon.\end{aligned}$$

As we noted above, the exclusions on ξ at the zeros of $\frac{\partial p_{\triangle x, \triangle t}}{\partial s}$ are identical with excluding ξ at zeros of z; these can be ignored for $\xi \in \bigcup_{N<K} I_N$, because z is large there. Since $\bigcup_{N<K} I_N$ is a union of closed bounded intervals and since q and z are continuous,

$$\begin{aligned}&\text{there exists } \triangle x_* < \triangle x_K \text{ there exists } \triangle t_* < \triangle t_K\\ &\text{there exists } C_* > 0 \text{ for all } \xi \in \cup_{N<K} I_N \text{ for all } 0 < \triangle x \leq \triangle x_*\\ &\qquad \text{for all } 0 < \triangle t \leq \triangle t_* \quad \left| \frac{e^{q(\xi)\triangle t} - z(\xi \triangle x)}{\triangle t} \right| \leq C_*.\end{aligned}$$

Since the partial differential equation is assumed to be stable, inequality (2.39) implies that

$$\begin{aligned}&\text{for all } T > 0 \text{ there exists } C_T \geq 1 \text{ for all } \triangle t > 0\\ &\text{for all } 0 \leq n \triangle t \leq T \text{ for all } M \geq (C_T^{1/n} - 1)/\triangle t \text{ for all } \xi\\ &\qquad |e^{q(\xi)\triangle t}| = |e^{q(\xi) n \triangle t}|^{1/n} \leq C_T^{1/n} \leq 1 + M \triangle t.\end{aligned}$$

Thus for sufficiently fine mesh and $\xi \notin \cup_{N<K} I_N$, $|z(\xi \triangle x)|$ is bounded by $1 + K \triangle t$, while for $\xi \in \cup_{N<K} I_N$ we can bound

$$|z(\xi \triangle x)| \leq |e^{q(\xi)\Delta t}| + \triangle t \left| \frac{e^{q(\xi)\Delta t} - z(\xi \triangle x)}{\triangle t} \right| \leq (1 + M \triangle t) + C_* \triangle t.$$

Thus

$$\begin{aligned} &\text{there exists } \triangle x_* < \triangle x_K \;\; \text{there exists } \triangle t_* < \triangle t_K \;\; \text{there exists } C_* > 0 \\ &\text{for all } \xi \;\; \text{for all } 0 < \triangle x \leq \triangle x_* \;\; \text{for all } 0 < \triangle t \leq \triangle t_* \\ &\qquad\qquad |z(\xi \triangle x)| \leq 1 + \max\{K, C_* + M\} \triangle t. \end{aligned}$$

This contradicts our assumption that the scheme is unstable. Thus the intervals can be chosen to be disjoint.

Next, let us use these disjoint intervals to define initial data for the partial differential equation. We choose

$$u_0(x) = \sum_{K=1}^{\infty} w_K(x)$$

where the Fourier transform of w_K is given by

$$\hat{w}_K(\xi) = \begin{cases} \frac{1}{K\sqrt{\eta_K}}, & |\xi - \xi_K| \leq \eta_K \\ 0, & \text{otherwise.} \end{cases}$$

Note that

$$\int_{-\infty}^{\infty} |u_0(x)|^2 \, dx = \sum_{K=1}^{\infty} \frac{1}{2\pi} \int_{-\infty}^{\infty} |\hat{w}_K(\xi)|^2 \, d\xi = \frac{1}{\pi} \sum_{K=1}^{\infty} \frac{1}{K^2 \eta_K} \eta_K = \frac{1}{\pi} \frac{\pi^2}{3} = \frac{\pi}{3},$$

so $u_0 \in L^2(\mathbf{R})$.

We now claim that the scheme does not converge for these initial data. First, we note that

$$\text{for all } T > 0 \;\; \text{there exists } n > 0 \;\; \text{there exists } K \geq 1 \;\; \frac{T}{2} \leq n \; \triangle t_K \leq T \text{ and } \frac{C_T - 1}{K} \leq \frac{T}{8}.$$

Next, note that for all $\xi \in [\xi_K - \eta_K, \xi_K + \eta_K]$, inequality (2.39) and then inequality (2.44) imply that

$$|e^{q(\xi)n\Delta t} - z(\xi \triangle x_K)^n| \geq |z(\xi \triangle x_K)|^n - C_T \geq \left(1 + \frac{1}{2} K \triangle t_K\right)^n - C_T.$$

Lemma 2.7.12 and the inequality $(1+x)^n \geq 1+nx$ (which holds for all $x > 0$ and $n \geq 1$) imply that

$$\begin{aligned}
\|I_{\triangle x}u^n - u(\cdot,t^n)\|^2 &\geq \|u^n - T_{\triangle x}u(\cdot,t^n)\|^2_{\triangle x} \\
&= \frac{1}{2\pi}\int_{-\infty}^{\infty} |z(\xi\triangle x)^n - e^{q(\xi)n\Delta t}|^2 |\hat{u}_0(\xi)|^2 \, d\xi \\
&= \frac{1}{2\pi}\sum_{N=1}^{\infty}\int_{-\infty}^{\infty} |z(\xi\triangle x)^n - e^{q(\xi)n\Delta t}|^2 |\hat{w}_N(\xi)|^2 \, d\xi \\
&> \frac{1}{2\pi}\int_{\xi_K-\eta_K}^{\xi_K+\eta_K} |z(\xi\triangle x)^n - e^{q(\xi)n\Delta t}|^2 |\hat{w}_K(\xi)|^2 \, d\xi \\
&= \frac{1}{2\pi}\left[\left(1+\frac{1}{2}K\triangle t_K\right)^n - C_T\right]^2 \frac{1}{K^2\eta_K}2\eta_K \\
&= \frac{1}{\pi}\left[\frac{(1+\frac{1}{2}K\triangle t_K)^n - C_T]^2}{K}\right]^2 \\
&\geq \frac{1}{\pi}\left[\frac{1+\frac{1}{2}Kn\triangle t_K - C_T]^2}{K}\right]^2 \geq \frac{1}{\pi}(T/8)^2.
\end{aligned}$$

We have shown that there exists initial data u_0 so that for any time $T > 0$ there is an error tolerance $\epsilon = T^2/(64\pi)$ so that for all sufficiently fine mesh the error in the numerical solution at time at most T is greater than ϵ. This proves that instability implies non-convergence. □

The Lax equivalence theorem explains the importance of stability in the design of convergent numerical methods. However, it is also important for the student to remember the limitations of the assumptions in the theorem. In particular, the theorem only applies to linear schemes for linear partial differential equations in one spatial dimension. On the other hand, the theorem is very general, because it makes no assumption about the type of the differential equation; it applies equally well to linear advection and diffusion equations.

2.8 Measuring Accuracy and Efficiency

Different numerical schemes have different convergence properties, even when they have the same order of convergence. It is important to compare the performance of numerical schemes, in order to construct efficient numerical methods. For our purposes, we will measure efficiency by comparing the computational time required to achieve a specified numerical accuracy. This means that we will have to determine how to measure the accuracy of numerical methods.

The first difficulty we face in measuring the accuracy of finite difference methods is that our numerical results have point values on a grid, while the solution of the differential equation is defined on an interval in space. We could overcome this problem by restricting the solution of the differential equation to points on the grid, or by extending the numerical solution to all of the problem domain, and then applying standard norms. The truncation operator in (2.7) and interpolation operator in (2.38) served these purposes in section 2.7, in combination with L^2 norms in space or on a grid. However, the use of L^2 norms for studying hyperbolic equations is uncommon (especially for nonlinear problems). Also, L^∞ norms (*i.e.*, max norms) are uncommon. Instead, we will typically use L^1 norms in our comparisons. These are still not ideal; in fact, theoreticians have not yet determined the norms that are appropriate for proving existence or uniqueness of multidimensional hyperbolic systems of conservation laws.

In Section 2.2 we developed conservative finite difference methods by constructing various approximations to the time integrals of the flux in the integral form of the conservation law. The numerical solution values were taken to be approximations to the cell averages of the solution in equation (2.13). Thus it seems reasonable to define the average of a function $w(x)$ over a cell $(x_{i-1/2}, x_{i+1/2})$ by

$$A(w)_i \equiv \frac{1}{x_{i+1/2} - x_{i-1/2}} \int_{x_{i-1/2}}^{x_{i+1/2}} w(x)\, dx$$

and use the L^1 norm

$$\begin{aligned} \|u^n - A(u(\cdot, t^n))\|_1 &\equiv \sum_{i=0}^{I-1} \left| u_i^n - \frac{1}{\Delta x_i} \int_{x_{i-1/2}}^{x_{i+1/2}} u(x, t^n)\, dx \right| \Delta x_i \\ &= \sum_{i=0}^{I-1} \left| \int_{x_{i-1/2}}^{x_{i+1/2}} u_i^n - u(x, t^n)\, dx \right|, \end{aligned} \tag{2.45}$$

or the L^∞ norm

$$\begin{aligned} \|u^n - A(u(\cdot, t^n))\|_\infty &\equiv \max_{0 \le i < I} \left| u_i^n - \frac{1}{\Delta x_i} \int_{x_{i-1/2}}^{x_{i+1/2}} u(x, t^n)\, dx \right| \Delta x_i \\ &= \max_{0 \le i < I} \left| \int_{x_{i-1/2}}^{x_{i+1/2}} u_i^n - u(x, t^n)\, dx \right| \frac{1}{\Delta x_i}, \end{aligned} \tag{2.46}$$

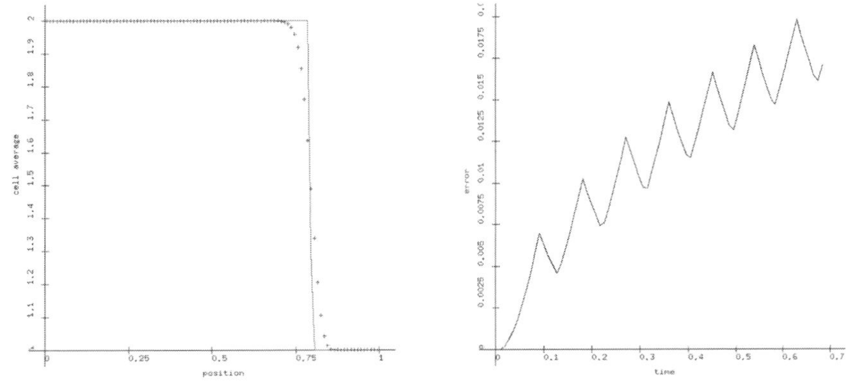

(a) Cell average vs. position (b) Discrete L^1 error in cell average vs. time

Fig. 2.14 Explicit upwind scheme for linear advection at CFL = 0.9, velocity = 1.0, 100 grid cells, Riemann problem initial data with jump at $x = 0.1$

or the L^2 norm

$$\|u^n - A(u(\cdot, t^n))\|_2^2 \equiv \sum_{i=0}^{I-1} \left| u_i^n - \frac{1}{\triangle x_i} \int_{x_{i-1/2}}^{x_{i+1/2}} u(x, t^n)\, dx \right|^2 \triangle x_i$$

$$= \sum_{i=0}^{I-1} \left| \int_{x_{i-1/2}}^{x_{i+1/2}} u_i^n - u(x, t^n)\, dx \right|^2 \frac{1}{\triangle x_i}. \qquad (2.47)$$

Alternatively, we could define dimensionless relative errors by dividing the norms above by the corresponding norms of the solution. These norms will determine the accuracy of our methods.

Let us examine the use of these norms for the explicit upwind difference scheme. In Figure 2.14 we show the numerical solution computed with a CFL number of 0.9 and 100 grid cells at time 0.5, superimposed with the true cell averages, in the left-hand image. In the right-hand image of the same figure, we also show the L^1 norm of the error, defined by Equation (2.45), versus time. Generally speaking, the error increases with time, but does not increase monotonically. This is because the L^1 norm of the error is principally determined by the errors in just a couple of the grid cells, and the positioning of the front within these grid cells. The results in Figure 2.15 were computed with a CFL number of 0.1. Note that the resolution of the propagating discontinuity is worse in this simulation, and the L^1 norm of the error in the computed results is larger. We expect the explicit upwind scheme to be more accurate as the CFL number approaches one. These results were obtained by running **Executable 2.8-3: guilinearerror** with `initial_data` equal to `riemann`, `scheme` equal to `explicit upwind` and `cfl` equal to 0.9 or 0.1. Students will find

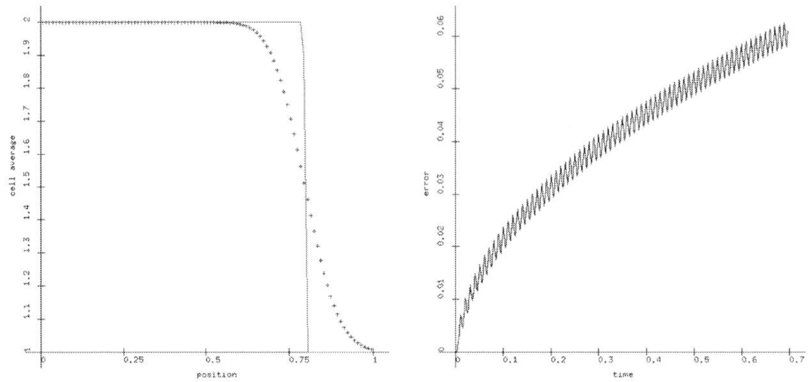

(a) Cell average vs. position (b) Discrete L^1 error in cell average vs. time

Fig. 2.15 Explicit upwind scheme for linear advection at CFL = 0.1, velocity = 1.0, 100 grid cells, Riemann problem initial data with jump at $x = 0.1$

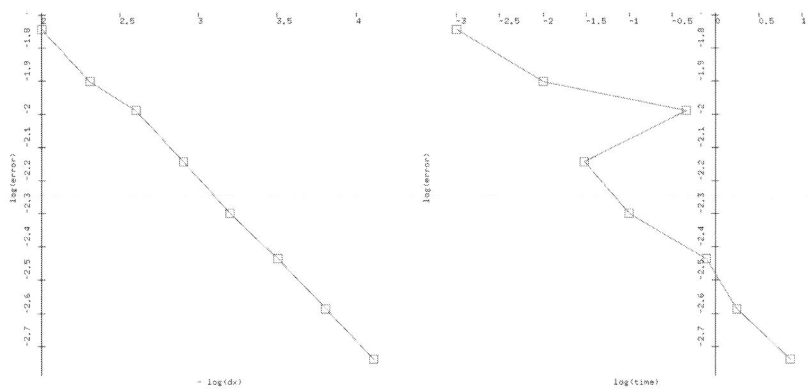

(a) Log_{10} Error vs. -Log_{10} Cell Width (b) Log_{10} Error vs. Log_{10} Time

Fig. 2.16 Refinement Study with Explicit Upwind Scheme for Linear Advection at CFL = 0.9

similar numerical results for `square pulse`, `triangular pulse`, `smooth Gaussian` or `quadratic pulse` initial data. The plots of error versus time are somewhat smoother when more grid cells are used (*e.g.* `ncells` = 1000).

It is also useful to examine how the computational errors behave as the mesh is refined. In Figure 2.16 we show the results of a mesh refinement study for the explicit upwind scheme. These results used Riemann problem initial data, and a CFL number of 0.9. One interesting observation is that the error is roughly proportional to $\sqrt{\triangle x}$, even though the explicit upwind scheme is supposedly first-order accurate. The plot of error versus computational time shows somewhat erratic behavior for

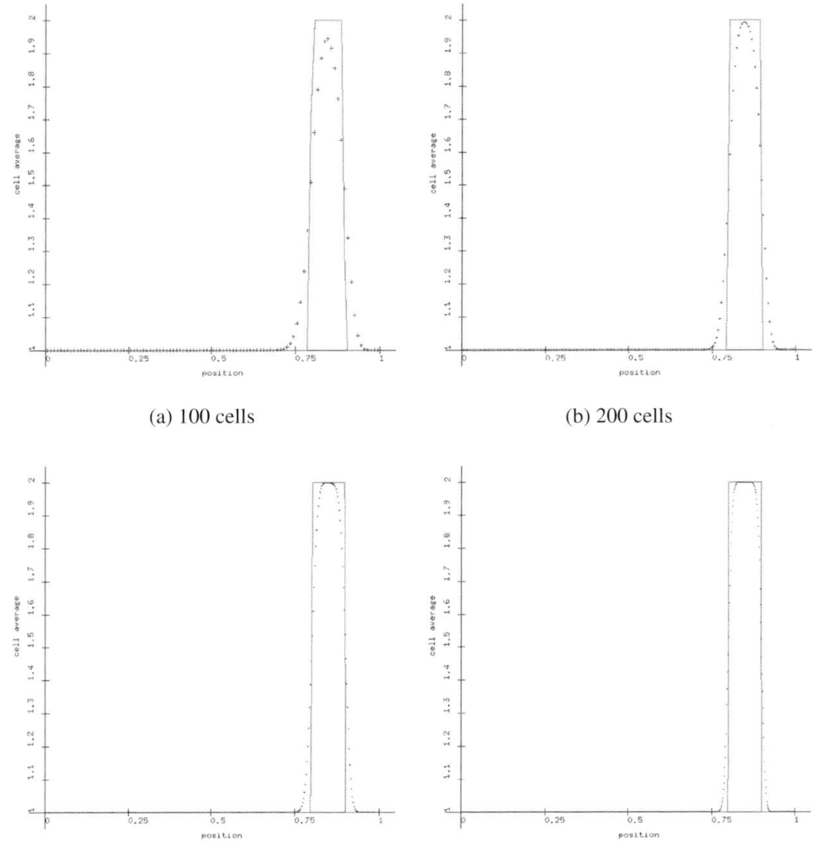

(a) 100 cells

(b) 200 cells

(c) 400 cells

(d) 800 cells

Fig. 2.17 Refinement Study with Explicit Upwind Scheme for Linear Advection Square Pulse at CFL $= 0.9$

coarse mesh, due to the inherent inaccuracy in the available system timing routines. For more refined computations, however, this figure seems to indicate that the error is roughly proportional to the computational time raised to the power 0.25. These results were obtained by running executable 2.8-3 with `initial_data` equal to `riemann`, `scheme` equal to `explicit upwind`, `cfl` equal to 0.9 and `ncells` equal to 0. This seemingly nonsensical value for the number of grid cells is a signal for the executable to perform a mesh refinement study.

Figure 2.17 shows the results with the explicit upwind scheme for linear advection of a square pulse. This figure will give the student an idea of how this scheme converges during mesh refinement. Figure 2.18 shows how the implicit upwind scheme converges for linear advection of a square pulse at CFL $= 2$. Figure 2.19 shows computational results for the implicit upwind scheme for linear advection of a square pulse with various choices for CFL. Note that the results are significantly

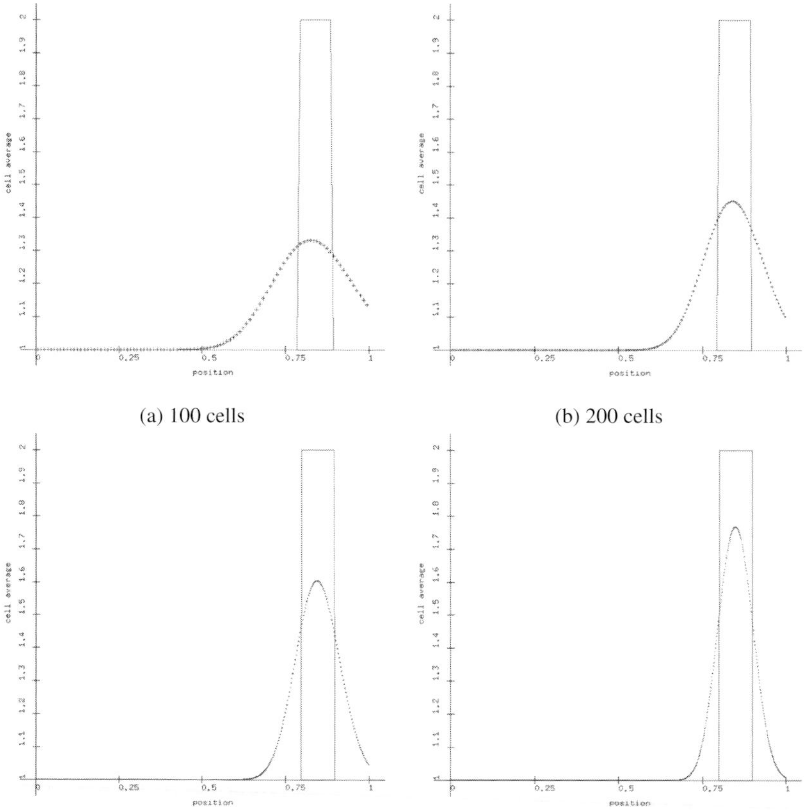

(a) 100 cells (b) 200 cells

(c) 400 cells (d) 800 cells

Fig. 2.18 Refinement study with implicit upwind scheme for linear advection square pulse at CFL = 2.0

smeared for all values of CFL; the differences show up in the computational time, because large values of CFL correspond to larger timesteps, which means fewer timesteps and less computational time.

Figure 2.20 shows the results of mesh refinement for linear advection of a square pulse with the Lax–Wendroff scheme. Note that the resolution of the pulse is better than with either explicit or implicit upwind, but there are significant oscillations to the left of each discontinuity in the pulse.

Figure 2.21 shows the error in the explicit upwind scheme, plotted against computational time for several values of the CFL number. From this picture, it is clear that the explicit upwind scheme becomes more efficient as the CFL number approaches 1. In other words, the schemes requires less computational time to reach a given level of accuracy as the CFL number is increased. The efficiency of the implicit upwind scheme for this linear advection problem seems to be greatest for CFL

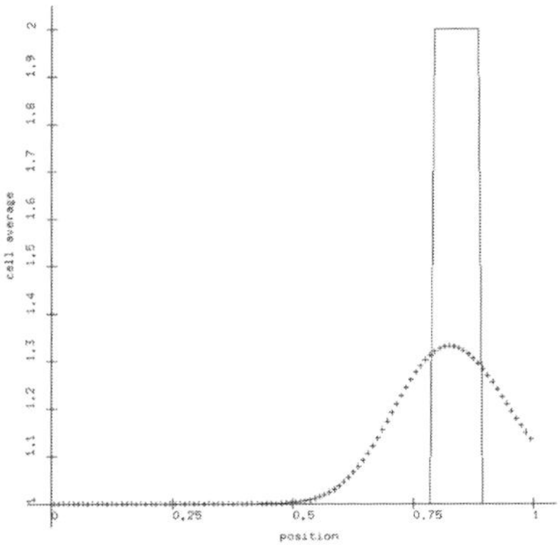

(a) CFL = 0.9

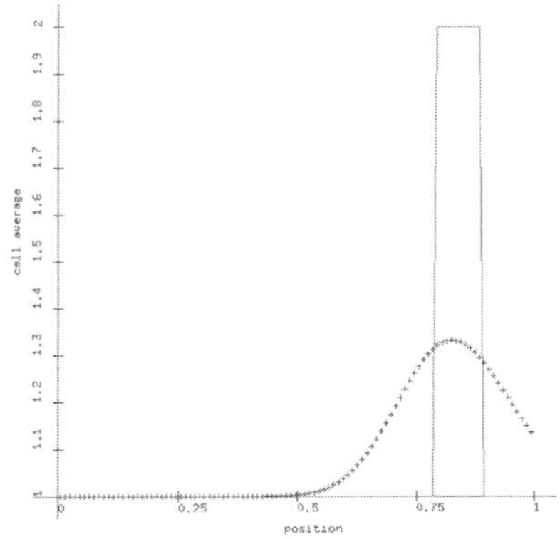

(b) CFL = 2.0

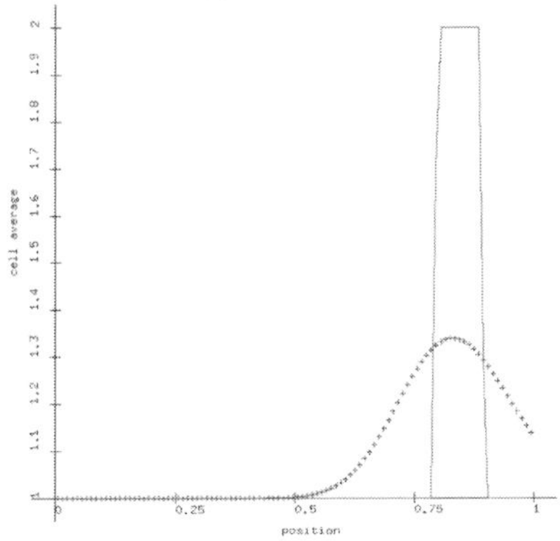

(c) CFL = 5.0

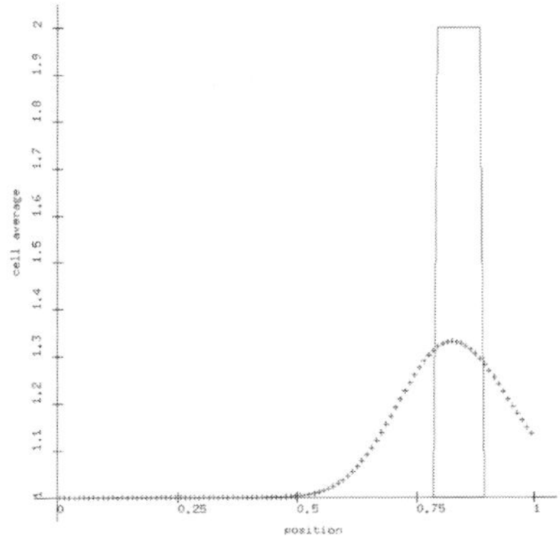

(d) CFL = 10.0

Fig. 2.19 Study with implicit upwind scheme for linear advection square pulse (100 cells)

numbers between 0.5 and 2. Low CFL numbers increase the cost of the implicit upwind scheme while reducing the numerical spreading of the discontinuities. On the other hand, high CFL numbers reduce the cost of the scheme while increasing the numerical spreading of the discontinuities. Figure 2.22 shows the error refinement study for the explicit upwind scheme at an efficient CFL number of 0.9, together with the error refinement study for implicit upwind at its efficient CFL number of 1. This figure indicates that the explicit upwind scheme is more efficient than the implicit upwind scheme for the linear advection problem with square pulse initial data.

Our previous examples have involved linear advection with discontinuous initial data. Since the analytical solution is not smooth, the numerical methods do not reach their expected order of accuracy. It is reasonable to ask if the methods would perform differently on a smooth problem. Figure 2.23 shows numerical results

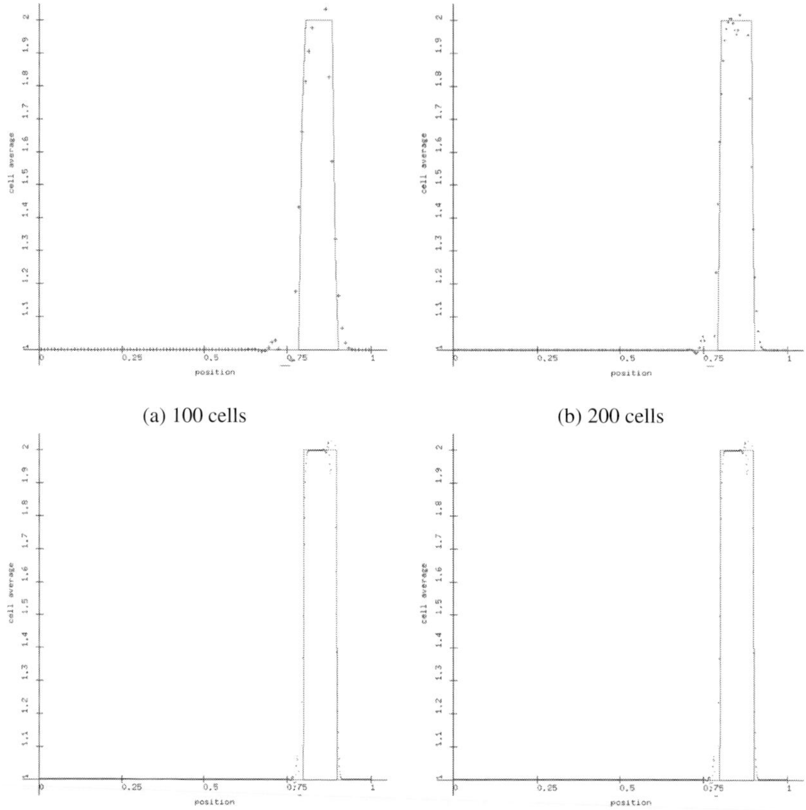

(a) 100 cells

(b) 200 cells

(c) 400 cells

(d) 800 cells

Fig. 2.20 Refinement study with Lax–Wendroff scheme for linear advection square pulse at CFL = 0.9

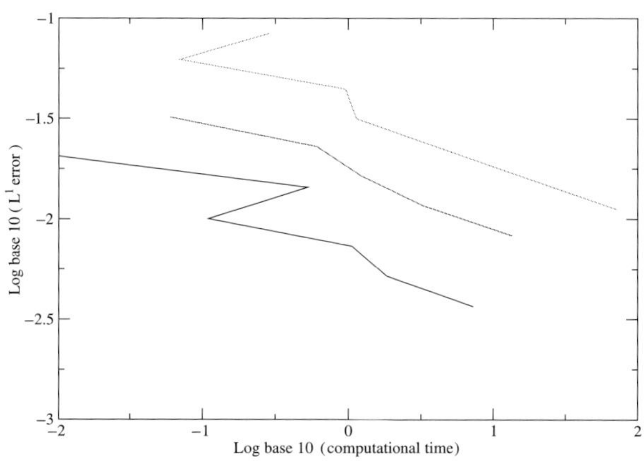

Fig. 2.21 Refinement study with explicit upwind scheme for linear advection, square pulse initial data, bottom: CFL = 0.9; middle: CFL = 0.5; top: CFL = 0.1

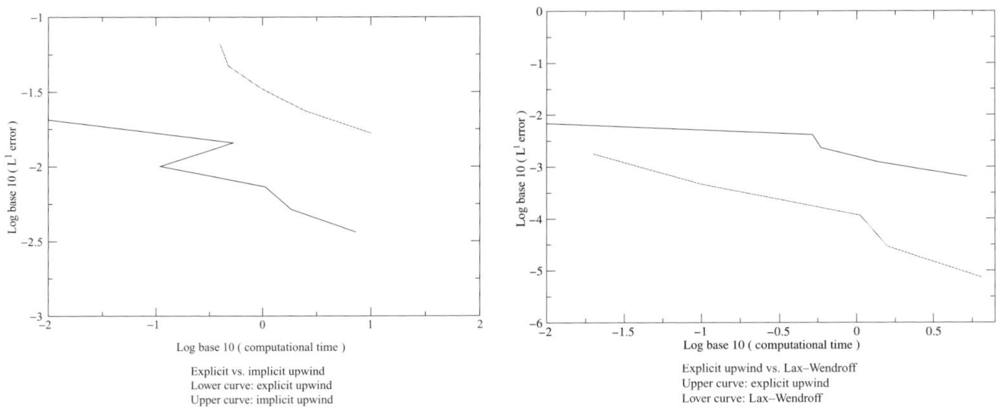

Fig. 2.22 Refinement studies comparing explicit upwind scheme for linear advection, square pulse initial data

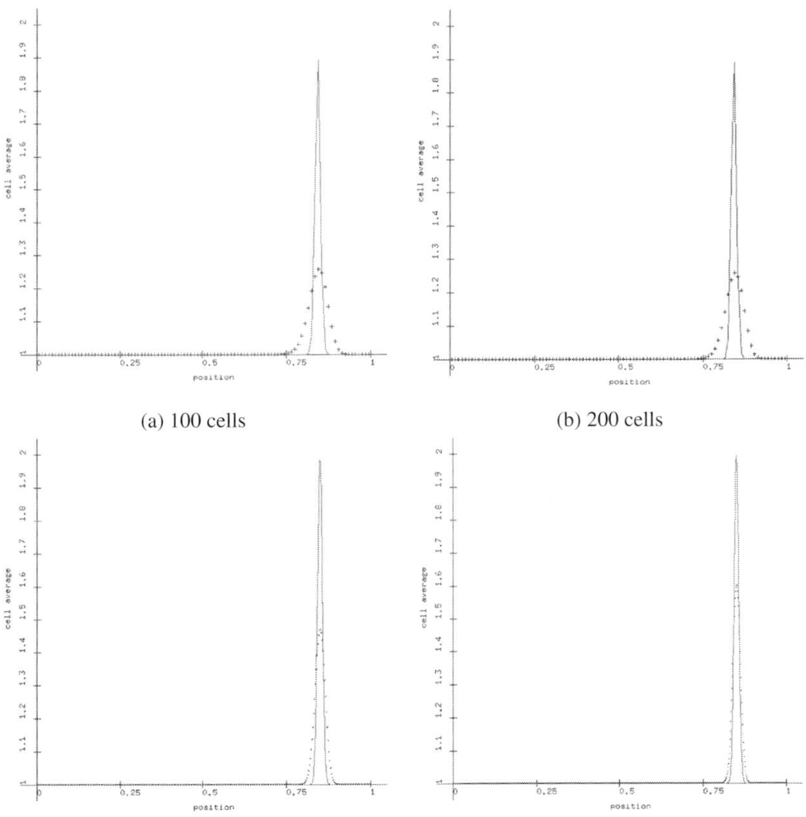

(a) 100 cells

(b) 200 cells

(c) 400 cells

(d) 800 cells

Fig. 2.23 Refinement study with explicit upwind scheme for linear advection smooth Gaussian at CFL = 0.9

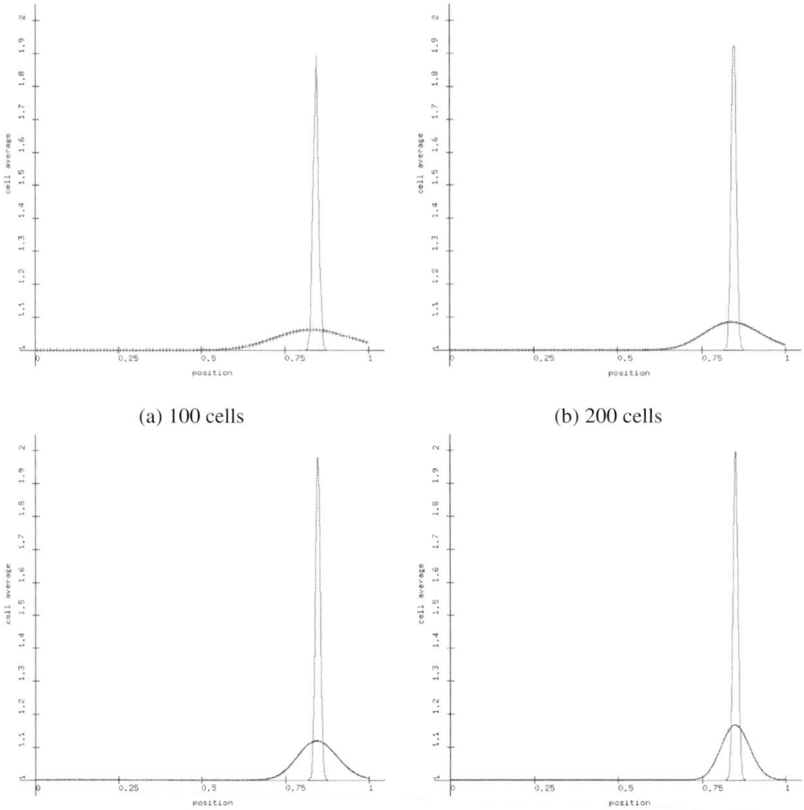

(a) 100 cells (b) 200 cells

(c) 400 cells (d) 800 cells

Fig. 2.24 Refinement study with implicit upwind scheme for linear advection smooth Gaussian at CFL = 2.0

for the explicit upwind scheme applied to linear advection with initial data given by a narrow smooth Gaussian. Note that the peak value of the solution is not very accurate in these simulations. However, the explicit upwind scheme does significantly better than the implicit upwind scheme, for which the results are shown in Figure 2.24. As expected, the Lax–Wendroff scheme does much better than either of these schemes; see Figure 2.25. So, it is not surprising that when we plot the errors for these schemes in Figure 2.26, we see that the Lax–Wendroff scheme is the most accurate and efficient of the three. Furthermore, the accuracy figure shows that the Lax–Wendroff scheme is indeed second-order, and the explicit upwind scheme is first-order. If the mesh were refined even more, we would see that the implicit upwind scheme is first-order accurate for the smooth Gaussian, as well.

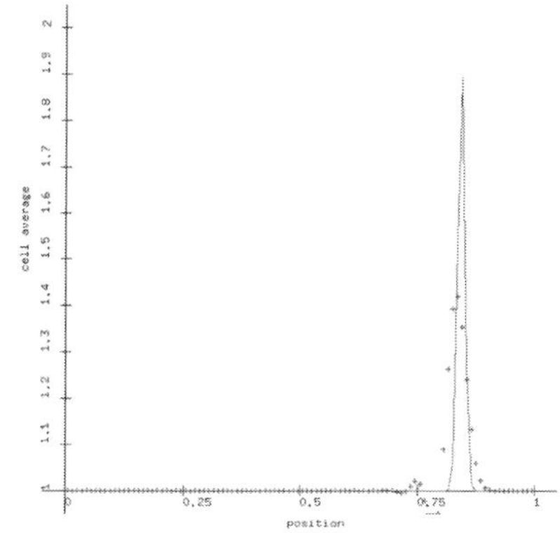

(a) 100 cells

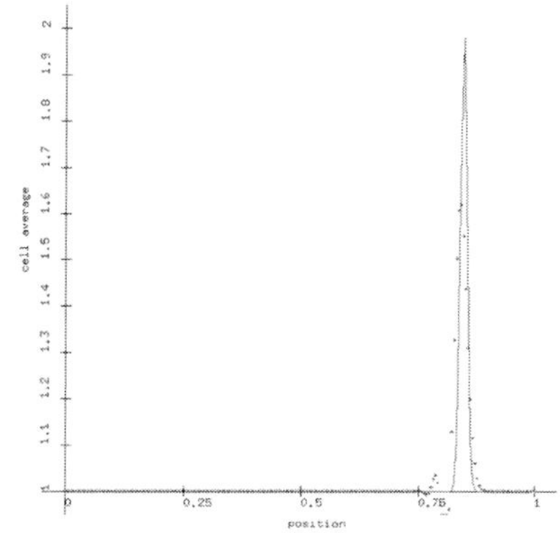

(b) 200 cells

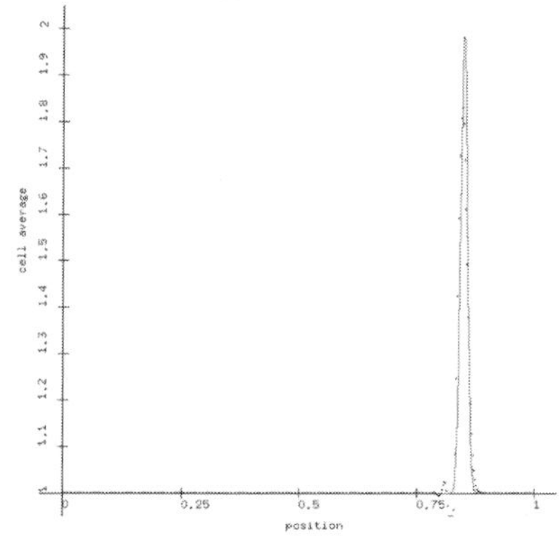

(c) 400 cells

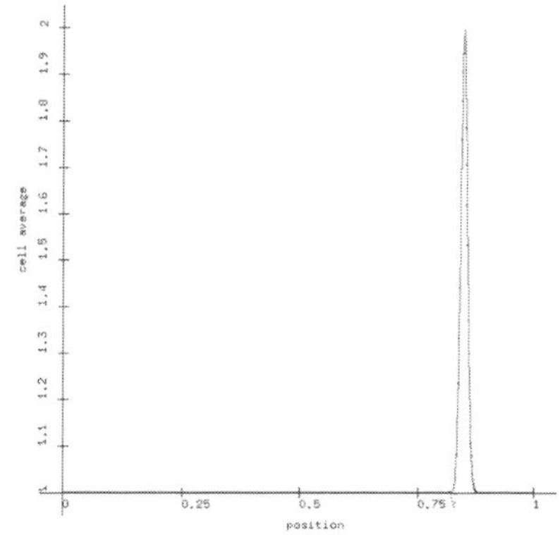

(d) 800 cells

Fig. 2.25 Refinement study with Lax–Wendroff scheme for linear advection of a smooth Gaussian at CFL = 0.9

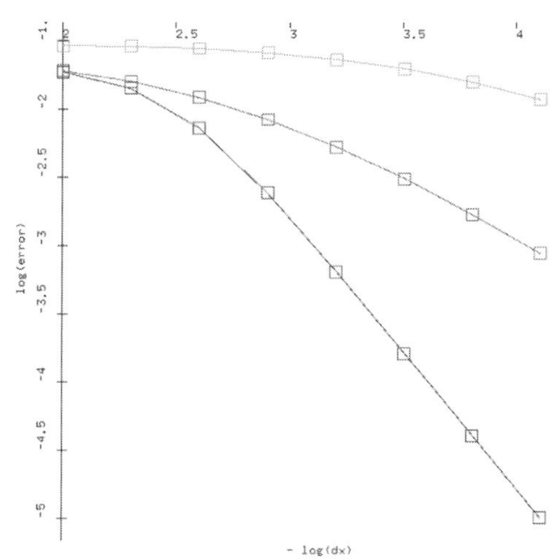

Log(error) vs. Log(number cells)

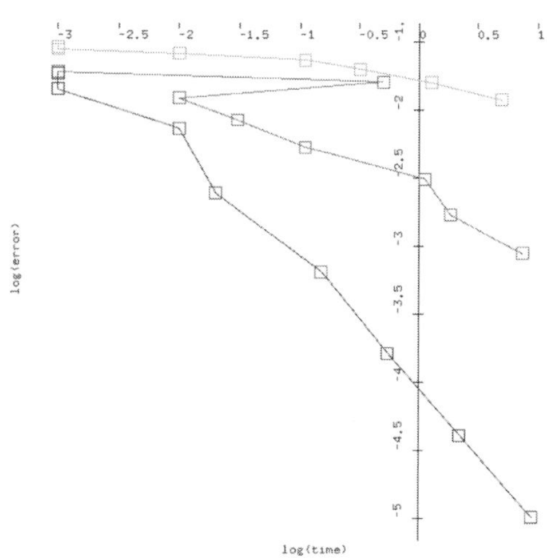

Log(error) vs. Log(computer time)

Fig. 2.26 Refinement studies comparing schemes for linear advection, smooth Gaussian initial data (lower curves: Lax–Wendroff, middle curves: explicit upwind, upper curves: implicit upwind

It is tricky to compare numerical schemes for efficiency. The parameters that make an individual scheme operate efficiently cannot be assumed to be the best parameters for another scheme. Computational times can be affected by programming care and the choice of computing machinery.

Some general observations may apply. It is reasonable to expect that implicit numerical schemes are more efficient than explicit numerical schemes only if the former can take timesteps much larger than the latter for a given level of accuracy. This is because the implicit numerical schemes involve greater numerical cost in solving the linear systems for the implicit treatment.

It is also reasonable to expect that high-order numerical schemes should be more efficient than low-order numerical schemes when high accuracy is required. This is because the low-order scheme produces small errors only by using small mesh widths. Of course, this observation is problem-dependent. For example, the Lax–Wendroff scheme is $O(\sqrt{\triangle x})$ for linear advection problems involving propagating discontinuities, first-order for problems with continuous but not continuously differentiable initial data, and second-order accurate for smooth initial data. These results were obtained by running executable 2.8-3 with `scheme` equal to `lax wendroff`. Figure 2.22 also shows the results of two mesh refinement studies for the explicit upwind scheme and the Lax–Wendroff scheme for smooth Gaussian initial data. Both schemes were run at CFL $= 0.9$. The results show that the second-order Lax–Wendroff scheme is more efficient than the first-order explicit upwind scheme for this problem at any of the mesh sizes in the study.

Exercise for 2.8

2.8.1 The modified equation analysis in Section 2.3.1 showed that the explicit upwind difference scheme is approximately solving a diffusion equation with diffusion coefficient proportional to the mesh width. The discussion of convection–diffusion equations in Section 2.5.2 showed how to transform a convection diffusion equation into a diffusion equation. Note that the analytical solution of the diffusion equation involves a Green's function that is proportional to one over the square root of the diffusion coefficient. Use the analytical solution of the diffusion equation to explain why the numerical solution of the linear advection problem with Riemann problem initial data should have an error that decreases proportional to the square root of the mesh width.

3
Nonlinear Scalar Laws

The linear advection equation is very easy to solve analytically, but somewhat tricky to solve well numerically. We could design high-quality numerical methods that would work particularly well for linear advection; for example, we could transform to characteristic coordinates and solve ordinary differential equations. A number of high-order pseudo-spectral [61] and collocation methods also work well for linear advection. However, we will want to solve other more complicated problems. The most challenging physical problems are nonlinear, because these develop propagating discontinuities known as shocks. The mathematical foundations of shock formation and propagation are discussed in Section 3.1. Several practical examples of nonlinear scalar conservation laws are presented in Section 3.2. The chapter concludes with several important numerical methods that are useful for solving nonlinear scalar conservation laws.

3.1 Nonlinear Hyperbolic Conservation Laws

An excellent reference for the material in this section is the monograph by Lax [89]. An alternative reference is LeVeque's book [96], which also covers the theory of numerical methods.

3.1.1 Nonlinear Equations on Unbounded Domains

The general nonlinear scalar conservation law on an unbounded domain takes the form:

$$\frac{\partial u}{\partial t} + \frac{\partial f(u)}{\partial x} = 0 \quad \text{for all } t > 0 \quad \text{for all } x \in \mathbf{R}, \tag{3.1a}$$

$$u(x, 0) = u_0(x) \quad \text{for all } x \in \mathbf{R}. \tag{3.1b}$$

In these equations, u represents the density of some conserved quantity, and $f(u)$ represents the flux of that conserved quantity. Note that $f(u)$ must have units of velocity times units of u.

Example 3.1.1 The most common example of a nonlinear scalar conservation law is **Burgers' equation** [24], for which the flux function is

$$f(u) = \frac{1}{2}u^2. \tag{3.2}$$

This conservation law is not terribly important in practice, but it is useful in illustrating important concepts.

If we integrate Equation (3.1a) over a region $a < x < b$, we obtain

$$\frac{d}{dt}\int_a^b u(x,t)\,dx = -f(u(b,t)) + f(u(a,t)). \tag{3.3}$$

This equation says that the rate of change of the conserved quantity in the fixed region (a,b) is equal to the net flux into the interval. This equation should be viewed as a fundamental physical principle, which is assumed to hold even in the presence of discontinuities inside the interval (a,b). Actually, Equation (3.3) is the most careful statement of a physical conservation law, and the partial differential Equation (3.1a) is actually derived from the integral form (3.3) where appropriate. We will discuss the meaning of "appropriate" in Section 3.1.4 below.

In order to develop numerical schemes, we often integrate this conservation law over an interval (t_1, t_2) in time to obtain

$$\int_a^b u(x,t_2)\,dx = \int_a^b u(x,t_1)\,dx - \int_{t_1}^{t_2} f(u(b,t))\,dt + \int_{t_1}^{t_2} f(u(a,t))\,dt.$$

This equation says that the total conserved quantity at the new time t_2 is equal to the total at the old time, plus the total flux into the interval minus the total flux out.

3.1.2 Characteristics

The partial differential equation (3.1a) has **quasilinear form**

$$0 = \frac{\partial u}{\partial t} + \frac{df}{du}\frac{\partial u}{\partial x} = \begin{bmatrix} 1 & \frac{df}{du} \end{bmatrix} \begin{bmatrix} \frac{\partial u}{\partial t} \\ \frac{\partial u}{\partial x} \end{bmatrix}. \tag{3.4}$$

Let us use the notation

$$\lambda(u) = \frac{df}{du}.$$

Note that λ has units of velocity. We will call λ the **characteristic speed**. If $u(x,t)$ has differentiable initial data $u_0(x)$, then the quasilinear form of the conservation law (3.4) shows that the gradient of u is orthogonal to the vector $[1 \quad \lambda(u)]$. This fact implies that u is constant along trajectories $x = x(t)$ that propagate with speed $\frac{dx}{dt} = \lambda(u(x,t))$, since

$$\frac{d}{dt}u(x(t),t) = \frac{\partial u}{\partial t} + \frac{\partial u}{\partial x}\frac{dx}{dt} = \frac{\partial u}{\partial t} + \frac{\partial u}{\partial x}\frac{\partial f}{\partial u} = 0.$$

As a result, a solution of the differential equation (3.4) is

$$u(x,t) = u_0(x - t\lambda(u(x,t))). \tag{3.5}$$

Note that this equation defines $u(x,t)$ implicitly.

Lemma 3.1.2 *If u_0 is continuously differentiable in x, and the solution $u(x,t)$ of the initial value problem*

$$\frac{\partial u}{\partial t} + \frac{\partial f(u)}{\partial x} = 0 \ \text{for all } x \in \mathbf{R} \ \text{for all } t > 0$$
$$u(x,0) = u(x) \ \text{for all } x \in \mathbf{R}$$

is continuously differentiable in x and t, then u is defined implicitly by (3.5), where $\lambda(u) \equiv \frac{df}{du}$.

Proof To check the solution (3.5) to problem (3.4), we will use implicit differentiation:

$$\frac{\partial u}{\partial t} = \left\{-\lambda - t\frac{\partial \lambda}{\partial u}\frac{\partial u}{\partial t}\right\}u_0' \Longrightarrow \frac{\partial u}{\partial t} = \frac{-\lambda u_0'}{1 + tu_0'\frac{\partial \lambda}{\partial u}}, \tag{3.6a}$$

$$\frac{\partial u}{\partial x} = \left\{1 - t\frac{\partial \lambda}{\partial u}\frac{\partial u}{\partial x}\right\}u_0' \Longrightarrow \frac{\partial u}{\partial x} = \frac{u_0'}{1 + tu_0'\frac{\partial \lambda}{\partial u}}. \tag{3.6b}$$

These equations imply that the solution (3.5) satisfies the quasilinear form (3.4), provided that the initial data u_0 is differentiable, and the denominator $1 + tu_0'\frac{\partial \lambda}{\partial u}$ is nonzero for all time t up to the time of interest. □

Example 3.1.3 The linear advection equation (2.1a) has flux $f(u) = cu$ where c is a constant. In this case, the characteristic speed is $\lambda = \frac{df}{du} = c$. Since the characteristic speed is constant, it is easy to see that the solution of the conservation law is

$$u(x,t) = u_0(x - ct).$$

Example 3.1.4 Recall that **Burgers' equation** has flux $f(u) = \frac{1}{2}u^2$. In this case, the characteristic speed is $\lambda(u) = u$, so the solution

$$u(x,t) = u_0(x - tu(x,t))$$

is implicitly defined through the initial data u_0.

3.1.3 Development of Singularities

Since the solution (3.5) generally defines u implicitly, we need to find circumstances under which we can solve this equation for u. We have used (3.6) to solve for the partial derivatives of u when the initial data is differentiable. These equations allow us to make several observations.

convex flux: Suppose that $\frac{d\lambda}{du} > 0$ for all u. In this case, Equations (3.6) show that $\frac{\partial u}{\partial t}$ and $\frac{\partial u}{\partial x}$ are bounded for all t if and only if $u_0'(x) \geq 0$ for all x. In other words, for convex flux functions, the temporal and spatial derivatives of the solution are bounded for all time if and only if the initial data is a nondecreasing function of x.

concave flux: Next, suppose that $\frac{\partial \lambda}{\partial u} < 0$ for all u. In this case, $\frac{\partial u}{\partial t}$ and $\frac{\partial u}{\partial x}$ are bounded for all t if and only if $u_0'(x) \leq 0$ for all x. In other words, for concave flux functions, the temporal and spatial derivatives of the solution are bounded for all time if and only if the initial data is a nonincreasing function of x.

linear flux: Next, suppose that $\frac{\partial \lambda}{\partial u} = 0$ for all u. In this case, $\frac{\partial u}{\partial t} = -\lambda u_0'$ and $\frac{\partial u}{\partial x} = u_0'$ are bounded for all t.

Example 3.1.5 Let us consider Burgers' equation again. Here the flux function is $f(u) = \frac{1}{2}u^2$, so the characteristic speed is $\lambda(u) = u$ and $\frac{\partial \lambda}{\partial u} = 1 > 0$ for all u. Thus the Burgers' flux is convex. We expect the first-order partial derivatives of u to be bounded for all t if and only if the initial data $u_0(x)$ is non-decreasing.

Suppose we are given the nonincreasing initial data

$$u_0(x) = \begin{cases} 1, & x \leq 0 \\ 1 - x, & 0 \leq x \leq 1 \\ 0, & 1 \leq x, \end{cases}$$

as shown in Figure 3.1a. For $x < 0$, the characteristic lines intersecting the x axis are given by $x - t \cdot 1 = const < 0$; along these curves the solution is $u(x, t) = 1$. For $x > 1$, the characteristic lines intersecting the x axis are given by $x - t \cdot 0 = const > 1$; along these curves the solution is $u(x, t) = 0$. For $0 < x < 1$ the characteristic lines intersecting the x axis are given by $x - t \cdot (1 - x_0) = x_0 \in (0, 1)$, which can be rewritten $x(t) = x_0(1 - t) + t$. At $t = 1$, we have $x(1) = 1$ for any initial value $x_0 \in (0, 1)$. Thus all of these characteristic lines intersect at $x = 1$, $t = 1$. Along these curves the solution of the conservation law is $u(x(t), t) = 1 - x(t) = (1 - x_0)(1 - t)$ for $0 \leq x_0 \leq 1$ and $0 \leq t \leq 1$. The characteristics are shown in Figure 3.1b, and the solution at a time $0 < t < 1$ is depicted in Figure 3.1c.

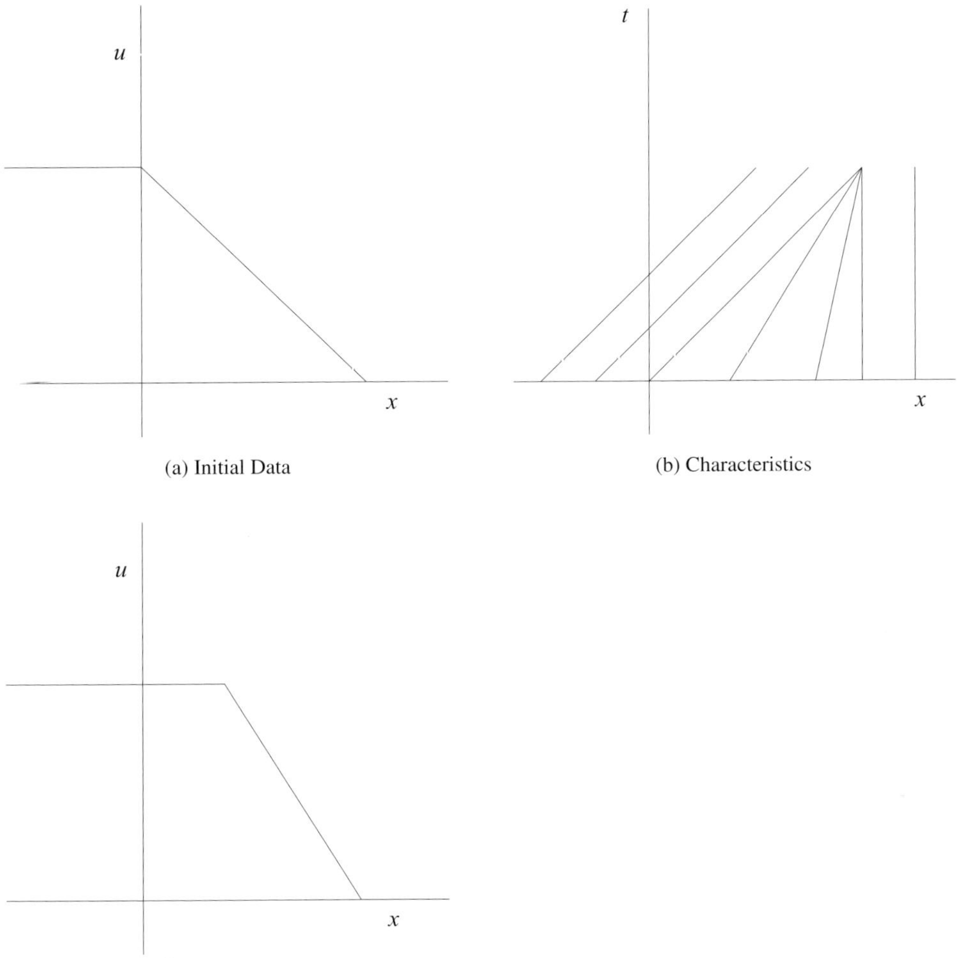

(a) Initial Data

(b) Characteristics

(c) Solution at time $t < 1$

Fig. 3.1 Decreasing data for Burgers' equation

For $0 < x < 1$ we have that $u_0' = -1$ and $\frac{\partial \lambda}{\partial u} = 1$; it follows that $1 + u_0' t \frac{\partial \lambda}{\partial u} = 1 - t$. Equation (3.6) shows that the partial derivatives of u become infinite at $t = 1$. It follows that the solution (3.5) cannot be continuous for $t \geq 1$.

3.1.4 Propagation of Discontinuities

As we have seen, hyperbolic conservation laws can develop discontinuities, even when provided with continuous initial data. We would like to determine how a discontinuity will propagate once it is formed. In developing the formula for the propagation of a discontinuity, we will use the notation $\lim_{x \downarrow z} f(x)$ for the one-sided limit from the right of f at z, and $\lim_{x \uparrow z} f(x)$ for the one-sided limit from the left of f at z.

Lemma 3.1.6 *Suppose that $u(x,t)$ satisfies the integral form of the conservation law*

$$\frac{d}{dt}\int_a^b u(x,t)\,dx = f(u(a,t)) - f(u(b,t)) \text{ for all } 0 < t < T.$$

If u is discontinuous along the space-time curve $(z(t),t)$ that moves with speed $\frac{dz}{dt}$, then the jumps across the discontinuity satisfy the **Rankine–Hugoniot jump condition**

$$\begin{aligned}&\text{for all } 0 < t < T \quad \lim_{x \downarrow z(t)} f(u(x,t)) - \lim_{x \uparrow z(t)} f(u(x,t)) \\ &= \left\{\lim_{x \downarrow z(t)} u(x,t) - \lim_{x \uparrow z(t)} u(x,t)\right\}\frac{dz}{dt}.\end{aligned} \tag{3.7}$$

Proof We compute

$$\begin{aligned}
&f(u(a,t)) - f(u(b,t)) \\
&= \frac{d}{dt}\int_a^b u(x,t)\,dx \\
&= \frac{d}{dt}\int_a^{z(t)} u(x,t)\,dx + \frac{d}{dt}\int_{z(t)}^b u(x,t)\,dx \\
&= \int_a^{z(t)} \frac{\partial u}{\partial t}\,dx + \lim_{x \uparrow z(t)} u(x,t)\frac{dz}{dt} + \int_{z(t)}^b \frac{\partial u}{\partial t}\,dx - \lim_{x \downarrow z(t)} u(x,t)\frac{dz}{dt} \\
&= \int_a^{z(t)} -\frac{\partial f}{\partial x}\,dx + \int_{z(t)}^b -\frac{\partial f}{\partial x}\,dx + \left\{\lim_{x \uparrow z(t)} u(x,t) - \lim_{x \downarrow z(t)} u(x,t)\right\}\frac{dz}{dt} \\
&= \left\{f(u(a,t)) - \lim_{x \uparrow z(t)} f(u(x,t))\right\} + \left\{\lim_{x \downarrow z(t)} f(u(x,t)) - f(u(b,t))\right\} \\
&\quad + \left\{\lim_{x \uparrow z(t)} u(x,t) - \lim_{x \downarrow z(t)} u(x,t)\right\}\frac{dz}{dt} \\
&= \{f(u(a,t)) - f(u(b,t))\} + \left\{\lim_{x \downarrow z(t)} f(u(x,t)) - \lim_{x \uparrow z(t)} f(u(x,t))\right\} \\
&\quad + \left\{\lim_{x \uparrow z(t)} u(x,t) - \lim_{x \downarrow z(t)} u(x,t)\right\}\frac{dz}{dt}.
\end{aligned}$$

After canceling the expression on the original left-hand side across the equation to the final right-hand side, we obtain the claimed result. □

The **Rankine–Hugoniot jump condition** says that the jump in the flux across the discontinuity is equal to the jump in the density of the conserved quantity times the speed of the discontinuity. It is customary to represent the jump in some quantity by square brackets, and the discontinuity speed by σ. Thus for some function $w(x,t)$ with a jump at $(z(t),t)$,

$$[w] \equiv \lim_{x \downarrow z(t)} w(x,t) - \lim_{x \uparrow z(t)} w(x,t),$$

and a discontinuity speed $\sigma = \frac{dz}{dt}$, the Rankine–Hugoniot jump condition (3.7) can be written in the terse form

$$[f] = [u]\sigma.$$

Example 3.1.7 Consider the Burgers' equation in example 3.1.5. We know that at time $t = 1$ the solution of this equation must develop a discontinuity at $x = 1$. On the right side of this discontinuity, the solution of the conservation law is $u = 0$, as determined by tracing characteristics back to the initial condition at $x > 1$. On the left side of the discontinuity, the solution of the conservation law is $u = 1$. According to the Rankine–Hugoniot condition, the speed of the discontinuity once it forms is

$$\sigma = \frac{[f]}{[u]} = \frac{f(0) - f(1)}{0 - 1} = \frac{1}{2}.$$

It follows that for $0 \le t \le 1$ the solution of the conservation law is

$$u(x,t) = \begin{cases} 1, & t \ge x \\ \frac{1-x}{1-t}, & t \le x \le 1 \\ 0, & 1 \le x, \end{cases}$$

and for $1 \le t$ the solution is

$$u(x,t) = \begin{cases} 1, & 1 + t > 2x \\ 0, & 1 + t < 2x. \end{cases}$$

Example 3.1.8 Next, consider Burgers' equation with nondecreasing discontinuous initial data

$$u_0(x) = \begin{cases} 0, & x < 0 \\ 1, & x > 0. \end{cases}$$

Figure 3.2 shows that the characteristics for these initial data do not intersect. If we trace back to the axis at $x < 0$, the characteristics have the form $x(t) = x_0 < 0$. If we trace back to the other half of the axis at $x > 0$, the characteristics have the form $x(t) - t = x_0 > 0$. Thus the characteristics do not provide any information about the solution of the conservation law in the region $0 < x < t$. It appears that the solution of this problem might not be unique. For example, the initial discontinuity

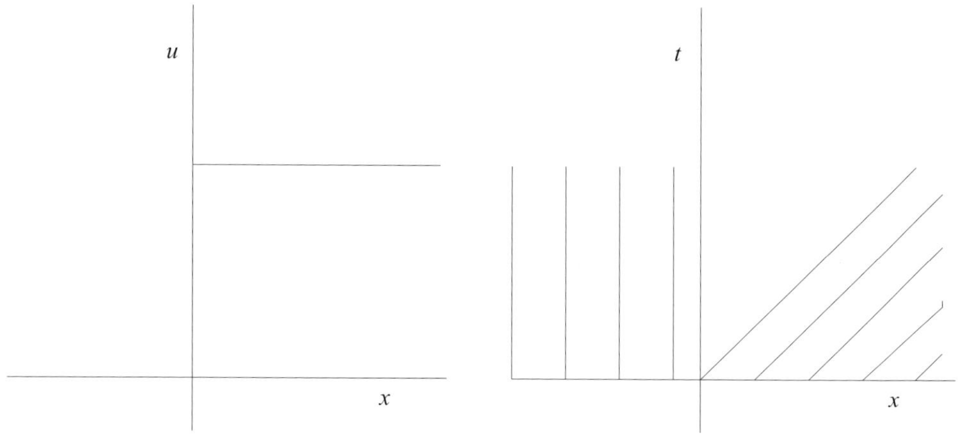

(a) Initial data (b) Characteristics

Fig. 3.2 Discontinuous decreasing data for Burgers' equation

could continue to propagate with speed $\sigma = [f]/[u] = \frac{1}{2}$:

$$u(x,t) = \begin{cases} 0, & t > 2x \\ 1, & t < 2x. \end{cases}$$

On the other hand, the initial discontinuity could evolve into a continuous solution

$$u(x,t) = \begin{cases} 0, & x \le 0 \\ x/t, & 0 \le x \le t \\ 1, & x \ge t. \end{cases}$$

The form of this solution for $0 \le x \le t$ can be determined as follows. The solution must be constant along characteristics through the origin, and the speed of the trajectory is $\lambda = \frac{df}{du} = u$. Thus the characteristic has the form $x - u(x,t)t = 0$; this implies that $u(x,t) = x/t$.

Example 3.1.9 Consider the following two forms of Burgers' equation:

$$\frac{\partial u}{\partial t} + \frac{\partial \frac{1}{2}u^2}{\partial x} = 0,$$

and

$$\frac{\partial (u^k/k)}{\partial t} + \frac{\partial (u^{k+1}/(k+1))}{\partial x} = 0,$$

both with discontinuous initial data

$$u_0(x) = \begin{cases} u_-, & x < 0 \\ u_+, & x > 0. \end{cases}$$

The second form of the conservation law can be obtained by multiplying the first form of Burgers' equation by u^{k-1}. Note that we get the same characteristic speeds for both problems:

$$\frac{\partial \frac{1}{2}u^2}{\partial u} = u = \frac{\partial\left(\frac{1}{k+1}u^{k+1}\right)}{\partial u}\left(\frac{\partial(\frac{1}{k}u^k)}{\partial u}\right)^{-1}.$$

On the other hand, we get different discontinuity speeds from the Rankine–Hugoniot jump conditions for these problems: the propagating discontinuity speeds are either

$$\frac{\left[\frac{1}{2}u^2\right]}{[u]} = \frac{1}{2}\frac{u_+^2 - u_-^2}{u_+ - u_-} = \frac{1}{2}(u_+ + u_-),$$

or

$$\frac{[u^{k+1}/(k+1)]}{[u^k/k]} = \frac{k}{k+1}\frac{u_+^{k+1} - u_-^{k+1}}{u_+^k - u_-^k} = \frac{k}{k+1}\sum_{j=0}^{k} u_+^j u_-^{k-j}.$$

Again, we do not yet know which information correctly specifies the solution to the problem.

3.1.5 Traveling Wave Profiles

The discussion in this section follows that in the book by Smoller [150].

As we mentioned in section 2.1.3, many hyperbolic conservation laws actually represent the limit of a diffusion equation, in the limit as the diffusion approaches zero. In order to determine when the solution of a conservation law involves a propagating discontinuity, and to find the correct speed of the discontinuity, we will return to the viscous form of the equation. The following lemma will determine which are the correct propagating discontinuities for use in solving our nonlinear conservation laws.

Lemma 3.1.10 *Consider the viscous conservation law*

$$\frac{\partial u_\epsilon}{\partial t} + \frac{\partial f(u_\epsilon)}{\partial x} = \epsilon\frac{\partial^2 u_\epsilon}{\partial x^2} \text{ for all } x \in \mathbf{R} \text{ for all } t > 0, \tag{3.8a}$$

$$u_\epsilon(x,t) = \begin{cases} u_-, & x < 0 \\ u_+, & x > 0. \end{cases} \tag{3.8b}$$

Here $\epsilon > 0$ is assumed to be small. Suppose that this problem has a traveling wave solution $u_\epsilon(x,t) = w(\frac{x-\sigma t}{\epsilon})$, and that this traveling wave tends to constants at large values of its argument:

$$\lim_{\xi\to-\infty} w(\xi) = u_-, \quad \lim_{\xi\to\infty} w(\xi) = u_+.$$

Then u_- and u_+ satisfy the Rankine–Hugoniot condition

$$f(u_+) - f(u_-) = (u_+ - u_-)\sigma \tag{3.9}$$

and the traveling wave satisfies the ordinary differential equation

$$w' = f(w) - f(u_-) - \sigma(w - u_-). \tag{3.10}$$

If, in addition, $u_\epsilon(x,t)$ tends to a solution $u(x,t)$ of the conservation law

$$\frac{\partial u}{\partial t} + \frac{\partial f(u)}{\partial x} = 0 \ \textit{for all}\ x \in \mathbf{R}\ \textit{for all}\ t > 0,$$

then u_- is an unstable stationary point of (3.10), *u_+ is a stable stationary point, and the Lax admissibility conditions*

$$f'(u_-) > \sigma > f'(u_+) \tag{3.11}$$

are satisfied.

Proof Because of the diffusion, u_ϵ is smooth and the partial differential equation (3.8) should be valid for all space and all positive time. We assume that we have a traveling wave solution of the form

$$u_\epsilon(x,t) = w\left(\frac{x - \sigma t}{\epsilon}\right).$$

In other words, w is a function of the variable $\xi = \frac{x-\sigma t}{\epsilon}$. If we substitute w into the differential equation (3.8) we obtain

$$-\frac{\sigma}{\epsilon}w' + \frac{1}{\epsilon}f'(w)w' = \epsilon\frac{1}{\epsilon^2}w''.$$

Since the equation involves the same power of ϵ in all terms, this suggests that the form of the traveling wave variable ξ is correct. If we cancel out ϵ and integrate once with respect to ξ, we obtain

$$w' = f(w) - \sigma w + C, \tag{3.12}$$

where C is independent of ξ.

In order for the traveling wave w to converge, as $\epsilon \to 0$, to a propagating discontinuity with value u_- on the left of $(x, 0)$ and u_+ on the right, we have required

$$\lim_{\xi\to-\infty} w(\xi) = u_-, \qquad \lim_{\xi\to\infty} w(\xi) = u_+.$$

Since w must tend to a constant for large ξ, we obtain

$$0 = \lim_{\xi\to-\infty} w'(\xi) = f(u_-) - \sigma u_- + C,$$

$$0 = \lim_{\xi\to\infty} w'(\xi) = f(u_+) - \sigma u_+ + C.$$

The former equation implies that

$$C = \sigma u_- - f(u_-), \tag{3.13}$$

and the latter equation then gives us an alternative derivation of the Rankine-Hugoniot condition

$$f(u_+) - f(u_-) = (u_+ - u_-)\sigma.$$

If we substitute the value of C from Equation (3.13) into the Equation (3.12) for the traveling wave derivative, we obtain (3.10).

Note that both $w = u_-$ and $w = u_+$ are stationary points of the ordinary differential equation (3.10). In order for w to tend to a solution of the inviscid conservation law with the prescribed values on either side of the discontinuity, we must find an orbit of the ordinary differential equation (3.10) such that

$$\lim_{\xi \to -\infty} w(\xi) = u_-, \qquad \lim_{\xi \to \infty} w(\xi) = u_+.$$

We also want u_- to be an unstable stationary point of this orbit, and we want u_+ to be a stable stationary point.

To determine the stability of stationary points for w, we will consider a perturbation of w:

$$\frac{d(w + y\delta)}{d\xi} = f(w + y\delta) - f(u_-) - \sigma(w + y\delta - u_-).$$

We can subtract the traveling wave equation for w (3.10) from this equation and take δ to be small, in order to obtain

$$\frac{dy}{d\xi} = (f'(w) - \sigma)y.$$

So that u_- is an unstable stationary point, and u_+ is a stable stationary point, of Equation (3.10), we must have

$$f'(u_-) - \sigma > 0, \quad f'(u_+) - \sigma < 0.$$

These imply the Lax admissibility condition (3.11). □

Definition 3.1.11 A propagating discontinuity that satisfies the Lax admissibility conditions (3.11) is called a **shock**.

The Lax admissibility conditions can be used to determine the correct solutions to conservation laws where characteristic information may not suffice.

Example 3.1.12 Let us return to Burgers' equation with nondecreasing discontinuous initial data

$$u_0(x) = \begin{cases} 0, & x < 0 \\ 1, & x > 0. \end{cases}$$

If this problem involves a propagating discontinuity, the speed of the discontinuity must be determined by the Rankine–Hugoniot condition:

$$\sigma = [f]/[u] = \frac{1}{2}.$$

We would like to determine if this propagating discontinuity is the limit as diffusion vanishes of a solution to the viscous Burgers' equation. Thus we must check the Lax admissibility conditions (3.11). Note that $u_- = 0$ and $f'(u_-) = u_- = 0$. Thus the left-hand admissibility condition in (3.11) is violated. This problem cannot involve a shock. Thus the initial discontinuity evolves into a continuous solution. We will determine the form of this continuous solution in Section 3.1.8 below.

Example 3.1.13 Recall the following two forms of Burgers' equation:

$$\frac{\partial u}{\partial t} + \frac{\partial \frac{1}{2}u^2}{\partial x} = 0,$$

and

$$\frac{\partial u^k/k}{\partial t} + \frac{\partial u^{k+1}/(k+1)}{\partial x} = 0,$$

with discontinuous initial data

$$u_0(x) = \begin{cases} u_-, & x < 0 \\ u_+, & x > 0. \end{cases}$$

Note that we cannot obtain the viscous form of the second equation by multiplying the viscous form of Burgers' equation by u^{k-1}. Since the two problems have different shock speeds, namely

$$\frac{\frac{1}{2}u_+^2 - \frac{1}{2}u_-^2}{u_+ - u_-} = \frac{u_+ + u_-}{2}$$

and

$$\frac{1}{k+1}\frac{u_+^{k+1} - u_-^{k+1}}{u_+ - u_-} = \frac{\sum_{j=0}^{k} u_+^{k-j} u_-^j}{k+1},$$

they also have different Lax admissibility conditions. Thus it is possible for the two problems to have different solutions.

3.1.6 Entropy Functions

Definition 3.1.14 Given a hyperbolic conservation law with flux $f(u)$, suppose that we can find a continuous function $s(u)$ and a continuous function $\psi(u)$ so that

$$\frac{d\psi}{du} = \frac{ds}{du}\frac{df}{du}. \tag{3.14}$$

Then $s(u)$ is called an **entropy function** and $\psi(u)$ is called the corresponding **entropy flux**.

Lemma 3.1.15 *If $s(u)$ is an entropy function for the conservation law*

$$\frac{\partial u}{\partial t} + \frac{\partial f(u)}{\partial x} = 0$$

and if the solution $u(x,t)$ of the conservation law is continuously differentiable, then s satisfies the conservation law

$$\frac{\partial s(u)}{\partial t} + \frac{\partial \psi(u)}{\partial x} = 0 \tag{3.15}$$

Proof We compute

$$\frac{\partial s}{\partial t} + \frac{\partial \psi}{\partial x} = \frac{ds}{du}\frac{\partial u}{\partial t} + \frac{d\psi}{du}\frac{\partial u}{\partial x} = \frac{ds}{du}\left\{\frac{\partial u}{\partial t} + \frac{\partial f}{\partial x}\right\} = 0.$$

□

Lemma 3.1.16 *Suppose that $s(u)$ is an entropy function for the conservation law*

$$\text{for all } x \in \mathbf{R} \text{ for all } t > 0 \; \frac{\partial u}{\partial t} + \frac{\partial f(u)}{\partial x} = 0$$

and $u_\epsilon(x,t)$ satisfies the viscous conservation law

$$\text{for all } x \in \mathbf{R} \text{ for all } t > 0 \; \frac{\partial u_\epsilon}{\partial t} + \frac{\partial f(u_\epsilon)}{\partial x} = \epsilon \frac{\partial^2 u_\epsilon}{\partial x^2},$$

with $\epsilon > 0$. Also suppose that almost everywhere in x and t, $u_\epsilon(x,t) \to u(x,t)$. If $s(u)$ is convex and $s(u(x,t))$ is bounded for all x and t, then

$$\text{for all } \phi(x,t) \geq 0, \; \phi \in C_0^\infty(\mathbf{R} \times \mathbf{R})$$
$$-\int_0^\infty \int_{-\infty}^\infty \frac{\partial \phi}{\partial t} s(u) + \frac{\partial \phi}{\partial x} \psi(u) \, dx \, dt - \int_{-\infty}^\infty \phi(x,0) u(x,0) \, dx \leq 0.$$

Proof The maximum principle shows that the viscous conservation law has at most one solution, but the conservation law may have multiple solutions; this is the reason for the assumption that u_ϵ converges to u. Also note that the solution u_ϵ of the viscous conservation law is smooth for any $\epsilon > 0$.

Since $s(u)$ is an entropy function with entropy flux $\psi(u)$ for our conservation law, we have that

$$\begin{aligned} 0 &= \frac{\partial s}{\partial u}\left[\frac{\partial u_\epsilon}{\partial t} + \frac{\partial f(u_\epsilon)}{\partial x} - \epsilon \frac{\partial^2 u_\epsilon}{\partial x^2}\right] \\ &= \frac{\partial s(u_\epsilon)}{\partial t} + \frac{\partial \psi(u_\epsilon)}{\partial x} - \epsilon \frac{\partial}{\partial x}\left(\frac{\partial s}{\partial u}\frac{\partial u_\epsilon}{\partial x}\right) + \epsilon \frac{\partial^2 s}{\partial u^2}\left(\frac{\partial u_\epsilon}{\partial x}\right)^2. \end{aligned}$$

Since the solution u_ϵ of the viscous conservation law is smooth, for any nonnegative smooth $\phi(x,t)$ we can compute

$$\int_0^\infty \int_{-\infty}^\infty \left[\frac{\partial s(u_\epsilon)}{\partial t} + \frac{\partial \psi(u_\epsilon)}{\partial x}\right] dx\,dt = -\int_0^\infty \int_{-\infty}^\infty \frac{\partial \phi}{\partial t} s(u_\epsilon) + \frac{\partial \phi}{\partial x}\psi(u_\epsilon)\,dx\,dt - \int_{-\infty}^\infty \phi(x,0)u_\epsilon(x,0)\,dx.$$

We can also compute

$$\int_0^\infty \int_{-\infty}^\infty \phi \frac{\partial}{\partial x}\left(\frac{\partial s}{\partial u}\frac{\partial u_\epsilon}{\partial x}\right) dx\,dt = -\int_0^\infty \int_{-\infty}^\infty \frac{\partial \phi}{\partial x}\frac{\partial s(u_\epsilon)}{\partial x}\,dx\,dt = \int_0^\infty \int_{-\infty}^\infty \frac{\partial^2 \phi}{\partial x^2} s(u_\epsilon)\,dx\,dt$$

and note that the convexity of s implies that

$$\int_0^\infty \int_{-\infty}^\infty \phi \frac{\partial^2 s}{\partial u^2}\left(\frac{\partial u_\epsilon}{\partial x}\right)^2 dx\,dt \geq 0.$$

Putting these results together, we obtain

$$0 \geq -\int_0^\infty \int_{-\infty}^\infty \frac{\partial \phi}{\partial t} s(u_\epsilon) + \frac{\partial \phi}{\partial x}\psi(u_\epsilon)\,dx\,dt - \int_{-\infty}^\infty \phi(x,0)u_\epsilon(x,0)\,dx - \epsilon \int_0^\infty \int_{-\infty}^\infty \frac{\partial^2 \phi}{\partial x^2} s(u_\epsilon)\,dx\,dt$$

Since $s(u)$ is bounded and $u_\epsilon \to u$ almost everywhere, the term involving a factor of ϵ tends to zero as $\epsilon \to 0$. Taking limits as $\epsilon \to 0$ now produces the claimed result. □

Often, the result of this lemma is written in the form of an inequality, which is said to hold "weakly":

$$\frac{\partial s(u)}{\partial t} + \frac{\partial \psi(u)}{\partial x} \leq 0. \tag{3.16}$$

Also note that similar results can be proved if the entropy function is concave; the obvious inequalities are reversed.

Next, let us discuss the behavior of the entropy at a discontinuity.

Lemma 3.1.17 *Suppose that the solution $u(x,t)$ of the conservation law*

$$\text{for all } x \in \mathbf{R} \text{ for all } t > 0 \quad \frac{\partial u}{\partial t} + \frac{\partial f(u)}{\partial x} = 0$$

is the limit, as the diffusion tends to zero, of the corresponding viscous conservation law. Further suppose that $u(x,t)$ involves an isolated discontinuity located at $x(t)$

and moving with speed $\sigma = \frac{dx}{dt}$. *If the entropy* $s(u)$ *is convex, then the jumps in the entropy flux and entropy function at* $x(t)$ *satisfy*

$$[\psi] \leq [s]\sigma.$$

Proof Suppose that $\phi(x,t) \in C_0^\infty((-\infty,\infty)\times(0,\infty))$ and $\phi(x(t),t) > 0$. From Lemma 3.1.16 we have

$$\begin{aligned}
0 \geq & -\int_0^\infty \int_{-\infty}^\infty \frac{\partial\phi}{\partial t}s + \frac{\partial\phi}{\partial x}\psi \, dx\, dt \\
= & -\int_0^\infty \int_{-\infty}^{x(t)} \left[\frac{\partial}{\partial x} \quad \frac{\partial}{\partial t}\right]\begin{bmatrix}\phi\psi \\ \phi s\end{bmatrix} dx\, dt + \int_0^\infty \int_{-\infty}^{x(t)} \phi\left[\frac{\partial s}{\partial t} + \frac{\partial\psi}{\partial x}\right] dx\, dt \\
& -\int_0^\infty \int_{x(t)}^\infty \left[\frac{\partial}{\partial x} \quad \frac{\partial}{\partial t}\right]\begin{bmatrix}\phi\psi \\ \phi s\end{bmatrix} dx\, dt + \int_0^\infty \int_{x(t)}^\infty \phi\left[\frac{\partial s}{\partial t} + \frac{\partial\psi}{\partial x}\right] dx\, dt
\end{aligned}$$

If the support of ϕ is chosen so that no other discontinuity of u lies in its support, then Lemma 3.1.15 shows that

$$0 \geq -\int_0^\infty \int_{-\infty}^{x(t)} \left[\frac{\partial}{\partial x} \quad \frac{\partial}{\partial t}\right]\begin{bmatrix}\phi\psi \\ \phi s\end{bmatrix} dx\, dt - \int_0^\infty \int_{x(t)}^\infty \left[\frac{\partial}{\partial x} \quad \frac{\partial}{\partial t}\right]\begin{bmatrix}\phi\psi \\ \phi s\end{bmatrix} dx\, dt$$

Using the divergence theorem, we can rewrite this result in the form

$$\begin{aligned}
0 \geq & -\int_0^\infty \left[1 \quad \frac{dx(t)}{dt}\right]\begin{bmatrix}\phi\psi \\ \phi s\end{bmatrix}\Big|_{x=x(t)-0} dt + \int_0^\infty \left[1 \quad \frac{dx(t)}{dt}\right]\begin{bmatrix}\phi\psi \\ \phi s\end{bmatrix}\Big|_{x=x(t)+0} dt \\
= & \int_0^\infty \phi\left\{[\psi] - [s]\frac{dx(t)}{dt}\right\} dt.
\end{aligned}$$

The result follows by noting that the support of ϕ is arbitrary. □

3.1.7 Oleinik Chord Condition

The following result is due to Kruzkov [84].

Lemma 3.1.18 *For any scalar conservation law*

$$\frac{\partial u}{\partial t} + \frac{\partial f(u)}{\partial x} = 0$$

and any constant c,

$$s_c(u) = |u - c|$$

is a convex entropy function with entropy flux

$$\psi_c(u) = [f(u) - f(c)]sign(u - c).$$

Proof Note that s_c and ψ_c are continuous, s_c is convex, and

$$\frac{d\psi_c}{du} = -\frac{df}{du}\mathrm{sign}(u-c) = \frac{df}{du}\frac{ds_c}{du}.$$

□

The following lemma is very useful in describing the solutions of hyperbolic conservation laws. The reader can find some pictures of this result in Figures 3.3 and 3.4 below.

Lemma 3.1.19 *Suppose that $u(x,t)$ solves the conservation law*

$$\frac{\partial u}{\partial t} + \frac{\partial f(u)}{\partial x} = 0$$

and that u is the limit of the solution of the viscous conservation law as the diffusion tends to zero. If $u(x,t)$ has a propagating discontinuity at $x(t)$, with $u_L(t) = \lim_{x \uparrow x(t)} u(x,t)$ and $u_R(t) = \lim_{x \downarrow x(t)} u(x,t)$, then the **Oleinik chord condition** *is satisfied:*

$$\frac{f(u)-f(u_L)}{u-u_L} \geq \sigma \equiv \frac{f(u_R)-f(u_L)}{u_R-u_L} \geq \frac{f(u_R)-f(u)}{u_R-u} \text{ for all } u \text{ between } u_- \text{ and } u_+. \tag{3.17}$$

Proof Lemma 3.1.17 shows that

$$\begin{aligned}
0 &\geq \{\psi_c(u_R) - \psi_c(u_L)\} - \{s_c(u_R) - s_c(u_L)\}\sigma \\
&= \{f(u_R) - f(c)\}\mathrm{sign}(u_R - c) - \{f(u_L) - f(c)\}\mathrm{sign}(u_L - c) \\
&\quad - \{|u_R - c| - |u_L - c|\}\sigma.
\end{aligned}$$

In the case $u_L < c < u_R$, we obtain

$$\begin{aligned}
0 &\leq -\{f(u_R) - f(c)\}\mathrm{sign}(u_R - c) + \{f(u_L) - f(c)\}\mathrm{sign}(u_L - c) \\
&\quad + \{|u_R - c| - |u_L - c|\}\sigma \\
&= 2f(c) - f(u_R) - f(u_L) - \{2c - u_R - u_L\}\sigma.
\end{aligned}$$

By adding the Rankine–Hugoniot jump condition $0 = f(u_R) - f(u_L) - \{u_R - r_L\}\sigma$ to this result, we obtain

$$0 \leq 2\{f(c) - f(u_L) - (c - u_L)\sigma\};$$

by subtracting the Rankine–Hugoniot condition we obtain

$$0 \leq 2\{f(c) - f(u_R) - (c - u_R)\sigma\}.$$

By choosing c to be an intermediate state u, we obtain (3.17). A similar argument can be used in the case $u_R < u_L$. □

3.1.8 Riemann Problems

As we have seen, it is interesting to consider conservation laws with piecewise-constant initial data:

$$\frac{d}{dt}\int_a^b u(x,t)\,dx + f(u(b,t)) - f(u(a,t)) = 0 \text{ for all } a < b \text{ for all } t > 0, \quad (3.18a)$$

$$u(x,0) = \begin{cases} u_-, & x < 0 \\ u_+, & x > 0. \end{cases} \quad (3.18b)$$

Initial value problems for conservation laws on the entire real line with initial data given by two constant states are called **Riemann problems**. We would like to understand the analytical solutions to these problems.

The solutions of Riemann problems are **self-similar**. By this, we mean that $u(x,t)$ is a function of x/t; as a result, we will be able to write the solution of the Riemann problem in the form

$$u(x,t) = \mathcal{R}\left(u_-,\ u_+;\ \frac{x}{t}\right).$$

This fact will require some explanation.

If the self-similar solution w is differentiable, then

$$\frac{\partial u}{\partial t} = -w'\frac{x}{t^2},\quad \frac{\partial u}{\partial x} = w'\frac{1}{t}.$$

In order for $u(x,t) = w(x/t)$ to satisfy the conservation law (3.18a), we must have

$$0 = -w'\frac{x}{t^2} + \lambda w'\frac{1}{t} = \left(\lambda - \frac{x}{t}\right)w'\frac{1}{t}.$$

Here $\lambda = f'(w)$ is the characteristic speed. It follows that the **centered rarefaction** $w(x/t)$ will satisfy the conservation law (3.18a) whenever x/t is equal to a characteristic speed. Note that in order to have a continuous solution w the characteristics must not collide; in other words, a centered rarefaction requires that $f'(u_-) < f'(u) < f'(u_+)$ for all states u between u_- and u_+. If $u_- < u_+$, the existence of a single centered rarefaction connecting the left and right states requires the flux function to be convex on (u_-, u_+); if $u_- > u_+$, a centered rarefaction between u_- and u_+ requires the flux function to be concave in the interval (u_+, u_-).

Example 3.1.20 Let us return to Burgers' equation with nondecreasing discontinuous initial data

$$u_0(x) = \begin{cases} 0, & x < 0 \\ 1, & x > 0. \end{cases}$$

We saw in Example 3.1.12 that this problem cannot involve a shock. Thus the initial discontinuity evolves into a continuous solution. This continuous solution must be self-similar, with $x/t = \frac{df}{du} = u$ inside any centered rarefaction. Thus the solution of this Riemann problem is

$$u(x,t) = \begin{cases} 0, & x \le 0 \\ x/t, & 0 \le x/t \le 1 \\ 1, & 1 \le x/t. \end{cases}$$

We can always construct a discontinuous self-similar solution to a Riemann problem:

$$w(x/t) = \begin{cases} u_-, & x < \sigma t \\ u_+, & x > \sigma t. \end{cases}$$

Here $\sigma = [f]/[u]$ is the Rankine–Hugoniot jump speed. In order for this discontinuity to be a shock, the shock speed must satisfy the Lax admissibility conditions (3.11), *i.e.*,

$$f'(u_-) > \sigma > f'(u_+).$$

In order for this single discontinuity to remain coherent, the discontinuity must satisfy the **Oleinik chord condition** (3.17).

This gives a simple rule for solving Riemann problems for general scalar flux functions.

Algorithm 3.1.21 *Scalar conservation law Riemann problem solution*

(i) *If $u_- < u_+$, then the solution of the Riemann problem is determined by the convex hull of the flux function on (u_-, u_+). (The convex hull is the largest convex function less than or equal to the given flux function.) Wherever the convex hull is equal to a continuous portion of f, we have a centered rarefaction, and wherever the convex hull jumps across points on f, we have a(n admissible) shock.*

(ii) *On the other hand, if $u_- > u_+$, then the solution of the Riemann problem is determined by the concave hull of the flux function on (u_+, u_-). (The concave hull is the smallest concave function greater than or equal to the given flux function.) Wherever the concave hull is equal to a continuous portion of f, we have a centered rarefaction, and wherever the concave hull jumps across points on f, we have a(n admissible) shock.*

Example 3.1.22 Consider the general flux function in Figure 3.3. This flux function is neither convex nor concave between the left and right states for the Riemann problem. In this case, the left state is less than the right state, so the solution of the Riemann problem is determined by the convex hull of the flux function between the two states in the Riemann problem. This leads to a solution involving a rarefaction moving rapidly to the left, followed by a shock, a transonic rarefaction (meaning that the characteristic speeds change sign), and a shock moving to the right behind a fast rarefaction. It is most appropriate to plot the solution of the Riemann problem versus the self-similar coordinate x/t. When we plot the characteristic speeds for the solution of this problem, we find that $\lambda = x/t$ in rarefactions, so these curves follow lines with unit slope through the origin. Characteristic speeds must not

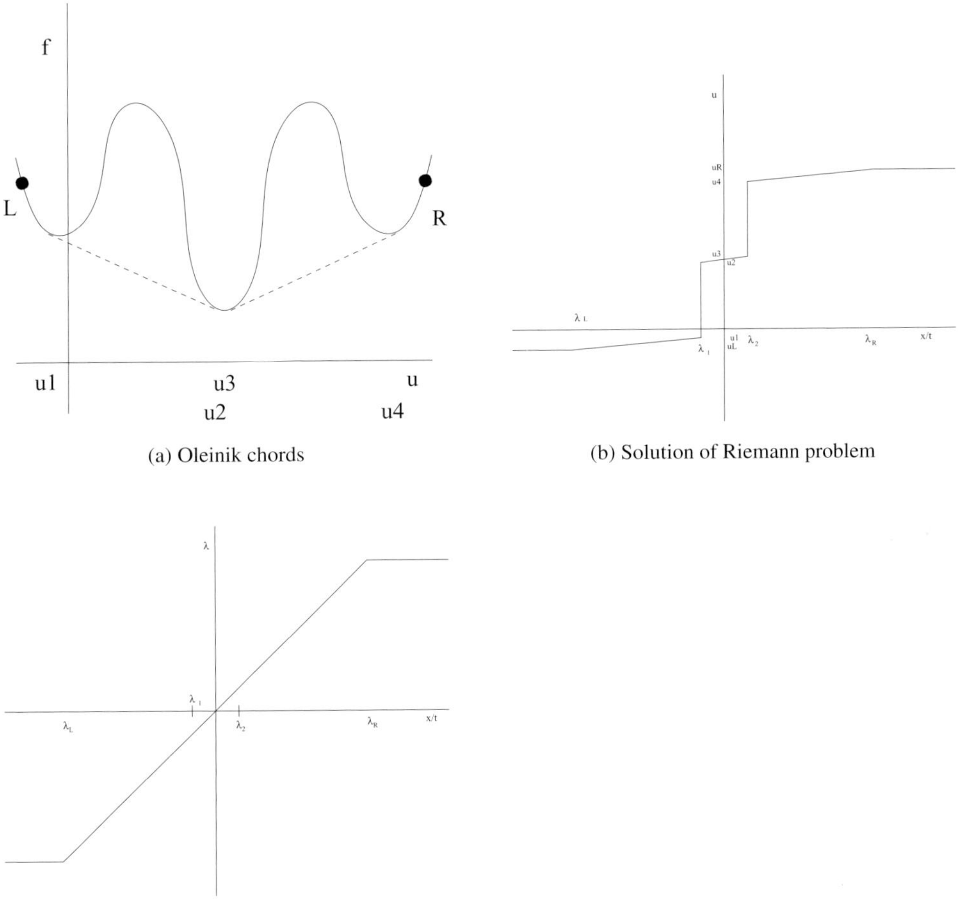

(a) Oleinik chords

(b) Solution of Riemann problem

(c) Characteristic speeds

Fig. 3.3 Riemann problem for general flux function

decrease from left to right in a rarefaction, otherwise we could not draw a picture of the conserved quantity as a *function* of x/t.

Example 3.1.23 We can also consider the same flux function as in the previous example, but reverse the left and right states. This leads to Figure 3.4. In this case, the left state is greater than the right state, so the solution follows the concave hull of the flux function.

3.1.9 Galilean Coordinate Transformations

Occasionally it is useful to study a conservation law in a moving frame of reference. Suppose that we have a coordinate system (ξ, τ) moving at a fixed velocity c with

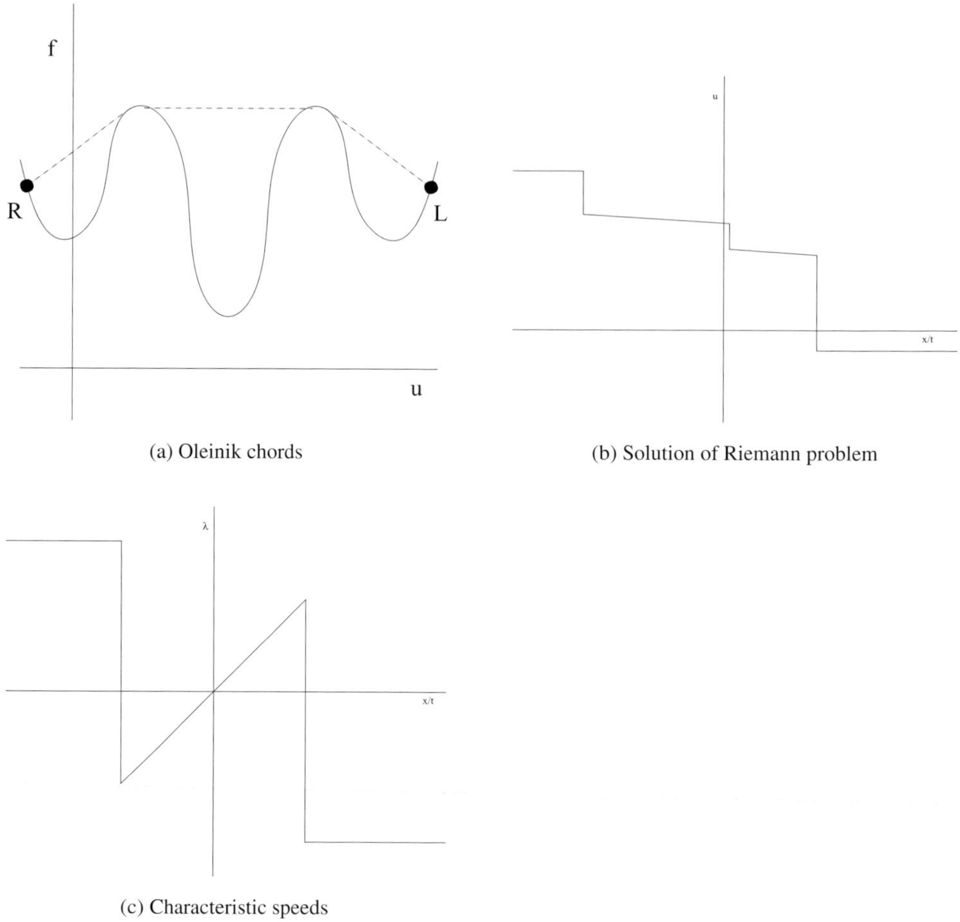

(a) Oleinik chords

(b) Solution of Riemann problem

(c) Characteristic speeds

Fig. 3.4 Riemann problem for general flux function

respect to the coordinate system (x, t):

$$\xi = x - ct, \;\; \tau = t.$$

Let $\tilde{u}(\xi, \tau) \equiv u(x, t)$ where u satisfies the conservation law

$$\frac{\partial u}{\partial t} + \frac{\partial f(u)}{\partial x} = 0.$$

Then it is easy to see that

$$\frac{\partial f(u)}{\partial x} = \frac{\partial f(\tilde{u})}{\partial \xi} \quad \text{and} \quad \frac{\partial u}{\partial t} = \frac{\partial \tilde{u}}{\partial \tau} - c\frac{\partial \tilde{u}}{\partial \xi}.$$

Thus in the moving frame of reference, the conservation law takes the form

$$\frac{\partial \tilde{u}}{\partial \tau} + \frac{\partial (f(\tilde{u}) - c\tilde{u})}{\partial \xi} = 0.$$

This suggests that we define the flux in the moving frame of reference to be $\tilde{f}(\tilde{u}) = f(\tilde{u}) - c\tilde{u}$. In the moving frame of reference, the characteristic speeds are

$$\tilde{\lambda} = \frac{d\tilde{f}}{d\tilde{u}} = f'(\tilde{u}) - c = \lambda - c$$

and the speeds of propagating discontinuities are

$$\tilde{\sigma} = \frac{[\tilde{f}]}{\tilde{u}]} = \frac{[f] - c[\tilde{u}]}{[\tilde{u}]} = \frac{[f]}{[u]} - c = \sigma - c.$$

Thus, we can arbitrarily adjust the speed c so that specific points in the solution of a problem move at some desired speed. For example, for a Riemann problem with a shock we can choose a Galilean transformation that transforms to a problem with a stationary discontinuity.

Exercises for 3.1

3.1.1 Suppose that we want to solve Burgers' equation with initial data $u_0(x) = x^2$ for $x \geq 0$. Show that the solution is

$$u(x, t) = \frac{1}{2t^2}[1 + 2xt - \sqrt{1 + 4xt}].$$

3.1.2 Show that in a Riemann problem for Burgers' equation, the flux at the state that moves with zero speed is

$$f = \begin{cases} \frac{1}{2}\max\{u_L, \min\{u_R, 0\}\}^2, & u_L < u_R \\ \frac{1}{2}\max\{|u_L|, |u_r|\}^2, & u_L \geq u_R. \end{cases}$$

3.1.3 Show that the solution to the scalar Riemann problem is $u(x, t) = w(x/t)$ where

$$f(w(\xi)) - \xi w(\xi) = \begin{cases} \min_{u_- \leq v \leq u_+}[f(v) - \xi v], & u_- \leq u_+ \\ \max_{u_+ \leq v \leq u_-}[f(v) - \xi v], & u_- \geq u_+. \end{cases}$$

Here we use the notation $\xi = x/t$. (Hint: use a Galilean transformation to reduce the problem to finding the state in the solution to the Riemann problem that moves with zero speed.)

3.1.4 Show that in the sense of distributions, the self-similar solution to the scalar Riemann problem satisfies

$$w(\xi) = \begin{cases} -\frac{d}{d\xi}(\min_{u_- \leq v \leq u_+}[f(v) - \xi v]), & u_- \leq u_+ \\ -\frac{d}{d\xi}(\max_{u_+ \leq v \leq u_-}[f(v) - \xi v]), & u_- \geq u_+. \end{cases}$$

3.1.5 Determine the analytical solution to Burgers' equation with initial data [59]

$$u_0(x) = \begin{cases} -0.5, & x < 0.5 \\ 0.2 + 0.7cos(2\pi x), & x > 0.5. \end{cases}$$

3.2 Case Studies

Many interesting applications of hyperbolic conservation laws are nonlinear. We have already seen one example of a nonlinear conservation law, namely Burgers' equation, in the series of Examples 3.1.1, 3.1.5, 3.1.7, 3.1.8, 3.1.9, 3.1.12 and 3.1.13. This nonlinear conservation law is useful in certain simplified models of gas dynamics. In this section, we would like to present some other models of nonlinear conservation laws.

3.2.1 Traffic Flow

We will consider a simple model of traffic flow. Of course, highway traffic is composed of discrete particles (the vehicles) that do not respond identically (some are impatient, sluggish or inattentive). Nevertheless, we will use a continuum model to describe the large-scale behavior of traffic. If averaged over time and applied to reasonably heavy traffic, the models can be pretty good.

Let $\rho(x, t)$ be the density of vehicles (say vehicles per mile), and let $\mathrm{v}(\rho)$ be the velocity of the vehicles. The flux of vehicles is $\mathrm{v}\rho$. Assuming that we are watching a section of the highway with no entrances or exits, conservation of mass of the vehicles can be written

$$\frac{\partial \rho}{\partial t} + \frac{\partial \rho \mathrm{v}}{\partial x} = 0.$$

In order to specify the problem completely, we need to describe the vehicle velocity function.

One simple model for the vehicle velocity is

$$\mathrm{v}(\rho) = \mathrm{v}_{\max}(1 - \rho/\rho_{\max}).$$

Here $\rho_{\max}$ is the maximum density of vehicles, and $\mathrm{v}_{\max}$ is the maximum velocity. The latter may be determined at times of low density, either by the topology of the highway or the attentiveness of law enforcement. The former may be determined at rush hour, especially when a collision impedes the flow. Thus at $\rho = 0$ the vehicle speed is $\mathrm{v}_{\max}$, but the speed decreases linearly with density after that.

Another model for the velocity takes

$$\mathrm{v}(\rho) = a \ln(\rho_{\max}/\rho). \tag{3.19}$$

Note that negative values of the density are unphysical, as are values of density greater than $\rho_{\max}$. In particular, in this model for the velocity, negative values for the density can make it impossible to compute the velocity. We will not want our numerical methods to produce negative densities (or velocities) in such cases.

3.2.2 Miscible Displacement Model

The **miscible displacement model** describes the flow in a porous medium of a fluid consisting of a single incompressible phase but multiple chemical components. This problem occurs in modeling the flow of water-soluble contaminants in aquifers, and of solvent-enhanced recovery of oil. For simplicity, we will assume that the fluid is composed of two components, water and a tracer. It is assumed that the tracer is inert; in other words, there are no chemical reactions that would transform the water and tracer into other chemicals. Further, the tracer is transported entirely with the water, and does not adsorb onto the surface of the porous medium.

We will denote the concentration of the tracer by c; by definition, c is the volume of tracer divided by the total volume of the fluid in some region in space. It follows that the concentration of water is $1 - c$. Because the tracer concentration can vary, the fluid density ρ can vary; we will assume that density is a function of tracer concentration. Similarly, the fluid viscosity is $\mu(c)$.

The fluid moves through tiny holes in the rock. The ratio of the volume of these holes to the total rock volume is called the porosity $\phi(x)$. Thus porosity is dimensionless. Since the rock is incompressible, ϕ is independent of time, but may vary in space.

The holes must be connected for the fluid to move through the rock. This is measured by the permeability $K(x)$ of the rock. It turns out that permeability has units of area. Typically the permeability is independent of time, but varies in space. Neither the permeability nor the porosity need be continuous functions. The velocity of the fluid is typically modeled by **Darcy's law**. This takes the form

$$\mathbf{v} = \mathbf{K}\left[-\nabla_{\mathbf{x}} p + \mathbf{g}\rho\right]\frac{1}{\mu},$$

where p is the pressure in the fluid and g is the gravity vector.

The flux of the components is represented in two parts. One is due to the macro-scale flow from Darcy's law; this part of the flux takes the form $c\mathbf{v}$ for the tracer, and $(1 - c)\mathbf{v}$ for water. The second part of the flux represents smaller scale convective mixing of the components as they flow through irregular pore channels, and molecular diffusion. Typically, this part of the flux is represented by **Fick's law**. The resulting flux for the tracer is

$$\mathbf{f}_t(c) = c\mathbf{v} - \left[\mathbf{v}\frac{\alpha_\ell}{\|\mathbf{v}\|}\mathbf{v}^\top + (I\mathbf{v}^\top\mathbf{v} - \mathbf{v}\mathbf{v}^\top)\frac{\alpha_t}{\|\mathbf{v}\|} + I\frac{\phi\delta_c}{\tau}\right]\nabla_{\mathbf{x}} c.$$

Here, α_ℓ and α_t are the longitudinal and transverse mixing lengths, δ_c is the diffusivity of the tracer and τ is the tortuosity of the rock. The equation for the flux of water is similar: we replace c by $1 - c$ to get

$$\mathbf{f}_w(c) = (1 - c)\mathbf{v} + \left[\mathbf{v}\frac{\alpha_\ell}{\|\mathbf{v}\|}\mathbf{v}^\top + (I\mathbf{v}^\top\mathbf{v} - \mathbf{v}\mathbf{v}^\top)\frac{\alpha_t}{\|\mathbf{v}\|} + I\frac{\phi\delta_c}{\tau}\right]\nabla_{\mathbf{x}} c.$$

We must have equations representing the conservation of mass for water and the tracer. It is easy to see that the volume of tracer per bulk (rock) volume is $c\phi$, and the volume of water per bulk volume is $(1-c)\phi$. Thus conservation of the tracer and water can be written

$$\frac{\partial c\phi}{\partial t} + \nabla_{\mathbf{x}} \cdot \mathbf{f}_t = 0,$$
$$\frac{\partial (1-c)\phi}{\partial t} + \nabla_{\mathbf{x}} \cdot \mathbf{f}_w = 0.$$

If desired, an equation for the pressure can be determined by adding together the two mass conservation laws. This leads to an elliptic partial differential equation for pressure:

$$0 = \nabla_{\mathbf{x}} \cdot \mathbf{v} = \nabla_{\mathbf{x}} \cdot \left\{ \mathbf{K}\left[-\nabla_{\mathbf{x}} p + \mathbf{g}\rho(c)\right] \frac{1}{\mu(c)} \right\}. \tag{3.20}$$

It is interesting to note that the coefficients $\rho(c)$ and $\mu(c)$ in this problem are functions of the variable tracer concentration. As a result, since the tracer concentration c can change in time, so can the pressure p. This completes the description of the miscible displacement model.

Next, let us restrict our discussion to one dimension. We need to manipulate these equations into a conservation law. We can add the two mass conservation equations to get

$$\frac{\partial \mathbf{v}}{\partial \mathbf{x}} = 0.$$

This is says that the total fluid velocity is independent of position in one dimension. Further, the transverse mixing length has no effect in one dimension, so the tracer flux simplifies to

$$\mathbf{f}_t = c\mathbf{v} - \left[|\mathbf{v}|\alpha_\ell + \frac{\phi\delta_c}{\tau} \right] \frac{\partial c}{\partial \mathbf{x}}.$$

Given the total fluid velocity, the two mass conservation equations become redundant. It will suffice to work with conservation of the tracer:

$$\frac{\partial c\phi}{\partial t} + \frac{\partial \mathbf{v}c}{\partial \mathbf{x}} = \frac{\partial}{\partial \mathbf{x}} \left\{ \left[|\mathbf{v}|\alpha_\ell + \frac{\phi\delta_c}{\tau} \right] \frac{\partial c}{\partial \mathbf{x}} \right\}. \tag{3.21}$$

In the typical case, we specify the total fluid velocity $\mathbf{v}(t)$ and the tracer concentration $c(x_L, t)$ at inflow. In this case, the equations for conservation of mass of the tracer (3.21) and for the fluid pressure (3.20) decouple; the solution of the pressure equation can be determined by specifying the fluid pressure at outflow. Suppose that we are given $p(x_R, t) = p_R(t)$ at the right-hand boundary. Since the one-dimensional pressure equation says that

$$\mathbf{K}\left[-\frac{\partial p}{\partial \mathbf{x}} + \mathbf{g}\rho\right]\frac{1}{\mu} = \mathbf{v}$$

is equal to the inflow fluid velocity, the solution of the pressure equation is

$$p(\mathbf{x}, t) = p_R(t) - g\int_{\mathbf{x}}^{\mathbf{x}_R} \rho(c(\mathbf{x}, t))\, d\mathbf{x} - \mathbf{v}(t)\int_{\mathbf{x}}^{\mathbf{x}_r} \frac{\mu(c(\mathbf{x}, t))}{\mathbf{K}(\mathbf{x})}\, d\mathbf{x}.$$

Some common test problems choose the fluid velocity to be around 30 centimeters per day, the porosity to be 0.25, and the longitudinal mixing length α_L to be 0.01 times the problem length. Normally, molecular diffusion is negligible unless we are working on very fine scales (close to the scale of the rock pores), meaning that $\delta_c \ll |\mathbf{v}|\tau\alpha_\ell/\phi$.

For our purposes in this chapter, we will not need to compute the pressure, so we will not need to describe the density or viscosity. Rather, we will fix the fluid velocity $\mathbf{v}$ and solve equation (3.21).

3.2.3 Buckley–Leverett Model

The **Buckley–Leverett model** for flow of two immiscible incompressible phases in a porous medium is important to models of oil reservoirs and contaminated aquifers. In this model, we assume that the fluid consists of two distinct phases, oil and water. It is assumed that the chemicals forming these two phases do not interact or move from one phase to the other. Since the fluid is incompressible and the chemical composition of the phases is fixed, the phase densities ρ_o and ρ_w are constants. Since the chemical composition of each phase remains constant, the viscosities μ_w and μ_o are constant.

The fluid moves through tiny holes in the rock. The ratio of the volume of these holes to the total rock volume is called the porosity $\phi(x)$. Thus porosity is dimensionless. Since the rock is incompressible, ϕ is a constant in time, but may vary in space.

The holes must be connected for the fluid to move through the rock. This is measured by the permeability $\mathbf{K}(\mathbf{x})$ of the rock. It turns out that permeability has units of area. Typically the permeability is a constant in time, but varies in space. Neither the permeability nor the porosity need be continuous functions.

The saturations of the phases s_o and s_w are the ratios of the phase volumes to the fluid volume. By definition,

$$s_w + s_o = 1. \tag{3.22}$$

Typically, water is the *wetting* phase, meaning that it prefers to move along the surface of the rock pores. Thus oil is the *non-wetting* phase, and prefers to sit as disconnected droplets in the center of cell pores, or move as ganglia when the droplets can connect. Thus the presence of both oil and water reduces the flow of the other. This effect is often modeled by a relative permeability, which is a dimensionless modification to the total permeability **K**. Typically, relative permeability of a phase is chosen to be an empirical function of that phase saturation. Thus the relative permeability of oil is $\kappa_{ro}(s_o)$, and the relative permeability of water is $\kappa_{rw}(s_w)$. We must have

$$\kappa_{ro}(0) = 0 = \kappa_{rw}(0),$$

because neither phase can flow if it occupies no volume in the fluid. We must also have

$$\kappa_{ro}(1) \le 1 \quad \text{and} \quad \kappa_{rw}(1) \le 1,$$

because neither phase can flow more easily than the total permeability permits. Finally, we must have

$$\kappa_{ro}'(s_o) \ge 0 \;\; \text{for all } s_o \in [0,1] \quad \text{and} \quad \kappa_{rw}'(s_w) \ge 0 \;\; \text{for all } s_w \in [0,1],$$

because an increase in relative volume of a phase makes it easier for that phase to flow.

The velocities of the phases are typically modeled by Darcy's law. In Section 3.2.2 for a single phase, Darcy's law was written

$$\mathbf{v} = \mathbf{K}\left[-\nabla_{\mathbf{x}} p + \mathbf{g}\rho\right]\frac{1}{\mu},$$

where p is the pressure in the fluid and $\mathbf{g}$ is the gravity vector. For two-phase flow, Darcy's law is usually modified to take the forms

$$\mathbf{v}_o = \mathbf{K}\left[-\nabla_{\mathbf{x}} p_o + \mathbf{g}\rho_o\right]\frac{\kappa_{ro}}{\mu_o}, \qquad \mathbf{v}_w = \mathbf{K}\left[-\nabla_{\mathbf{x}} p_w + \mathbf{g}\rho_w\right]\frac{\kappa_{rw}}{\mu_w}.$$

For two-phase flow the pressures in the phases are not necessarily the same. Instead,

$$p_o = p_w + P_c(s_w),$$

where $P_c(s_w)$ is the capillary pressure between the phases. Capillary pressure arises from the interfacial tension between the phases and the narrow flow paths available to the fluids. Typically, capillary pressure is a strictly decreasing function of water saturation, and a very large pressure is required to drive the water saturation to zero. See Figure 3.5 for a typical capillary pressure curve.

It will simplify notation if we define the phase mobilities

$$\lambda_o(s_o) \equiv \frac{\kappa_{ro}(s_o)}{\mu_o}, \qquad \lambda_w(s_w) \equiv \frac{\kappa_{rw}(s_w)}{\mu_w}.$$

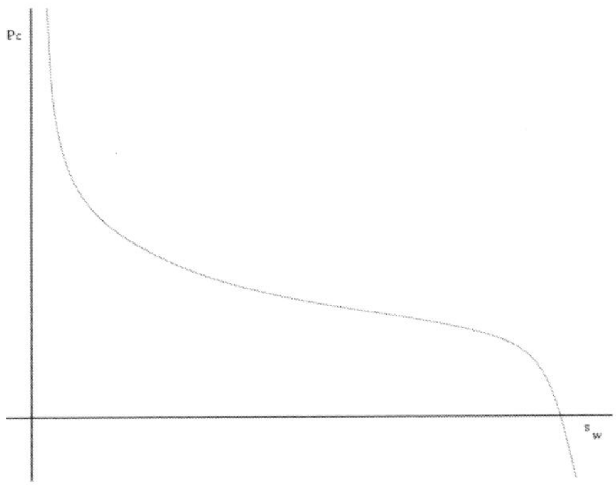

Fig. 3.5 Capillary pressure curve

Then the two-phase modification of Darcy's law can be written

$$\mathbf{v}_o = \mathbf{K}[-\nabla_{\mathbf{x}} p_o + \mathbf{g}\rho_o]\lambda_o, \quad \mathbf{v}_w = \mathbf{K}[-\nabla_{\mathbf{x}} p_w + \mathbf{g}\rho_w]\lambda_w.$$

Finally, we must have equations representing the conservation of mass for oil and water. It is easy to see that the mass of oil per bulk (rock) volume is $\rho_o s_o \phi$, and the mass of water per bulk volume is $\rho_w s_w \phi$. The volumetric flux of oil is $\rho_o \mathbf{v}_o$, and the flux of water is $\rho_w \mathbf{v}_w$. Thus conservation of oil and water can be written

$$\frac{\partial \rho_o s_o \phi}{\partial t} + \nabla_{\mathbf{x}} \cdot (\rho_o \mathbf{v}_o) = 0, \quad \frac{\partial \rho_w s_w \phi}{\partial t} + \nabla_{\mathbf{x}} \cdot (\rho_w \mathbf{v}_w) = 0.$$

This completes the description of the Buckley–Leverett model.

Next, we need to manipulate these equations into a conservation law in one dimension. We can divide the mass conservation equations by the constant phase densities to get

$$\frac{\partial s_o \phi}{\partial t} + \frac{\partial \mathbf{v}_o}{\partial \mathbf{x}} = 0, \quad \frac{\partial s_w \phi}{\partial t} + \frac{\partial \mathbf{v}_w}{\partial \mathbf{x}} = 0.$$

Next, we can add these two equations and use equation (3.22) to get

$$\frac{\partial \mathbf{v}_o + \mathbf{v}_w}{\partial \mathbf{x}} = 0.$$

This is an expression of the incompressibility of the flow: the total fluid velocity has zero divergence. Given the total fluid velocity, the two mass conservation equations become redundant. It will suffice to work with conservation of oil.

We can use the two-phase flow modification of Darcy's law to represent the total fluid velocity as a function of the water phase pressure and the oil saturation:

$$\mathbf{v}_T \equiv \mathbf{v}_o + \mathbf{v}_w = \mathbf{K}\left[-\frac{\partial (p_w + P_c)}{\partial \mathbf{x}}\lambda_o - \frac{\partial p_w}{\partial \mathbf{x}}\lambda_w + \mathbf{g}(\rho_o\lambda_o + \rho_w\lambda_w)\right].$$

Table 3.1 *Density and Viscosity of Fluids*

Fluid	Density (g/cc)	Viscosity (gm/sec/cm)
water	0.998	0.0114
diesel fuel	0.729	0.0062
kerosene	0.839	0.0230
prudhoe bay crude	0.905	0.6840

Since the total fluid velocity is divergence-free, in one dimension it is a constant in space. Thus this equation can be viewed as providing a relationship between the gradient of the water-phase pressure and the oil saturation:

$$-\frac{\partial p_w}{\partial \mathbf{x}} = \frac{\mathbf{v}_T/\mathbf{K} - \mathbf{g}(\rho_o\lambda_o + \rho_w\lambda_w) + \lambda_o \partial P_c/\partial \mathbf{x}}{\lambda_o + \lambda_w}.$$

This equation allows us to eliminate the pressure gradient from the expression for the Darcy velocity for oil

$$\mathbf{v}_o = \left[\mathbf{v}_T + \mathbf{Kg}\lambda_w(\rho_o - \rho_w) - \mathbf{K}\frac{\partial P_c}{\partial \mathbf{x}}\lambda_w\right]\frac{\lambda_o}{\lambda_o + \lambda_w}.$$

This means that conservation of oil can be written

$$\frac{\partial s_o\phi}{\partial t} + \frac{\partial}{\partial \mathbf{x}}\left\{[\mathbf{v}_T + \mathbf{K}g\lambda_w(\rho_o - \rho_w)]\frac{\lambda_o}{\lambda_o + \lambda_w}\right\} = \frac{\partial}{\partial \mathbf{x}}\left(\frac{\mathbf{K}\lambda_o\lambda_w}{\lambda_o + \lambda_w}\frac{\partial P_c}{\partial \mathbf{x}}\right). \quad (3.23)$$

This has the form of a conservation law. The capillary pressure term introduces a physical diffusion; this diffusion term is nonlinear. In most oil recovery problems, the capillary pressure gradient is small compared to the fluid pressure gradient.

Some common test problems choose the relative permeabilities to be

$$\kappa_{ro}(s_o) = s_o^2, \ \kappa_{rw}(s_w) = s_w^2.$$

It is common to take the capillary pressure to be zero in oil recovery problems. Oil is less dense than water and more viscous. A porosity $\phi = 0.25$ is typical. Problem lengths might be 100 meters, with a total fluid velocity $\mathbf{v}_T = 0.3$ meters/day. The permeability $\mathbf{K}$ can be defined in terms of the **gravity number** $\mathbf{Kg}(\rho_o - \rho_w)/(\mu_w\mathbf{v}_T)$. Some typical values for the densities and viscosities are contained in Table 3.1

The Buckley–Leverett flux function is neither convex nor concave. With zero total fluid velocity, the flux function formed solely by the action of gravity is shaped like a script "V". This model is especially interesting when both the total fluid velocity and gravity are nonzero. Figure 3.6 shows some examples of the Buckley–Leverett flux function.

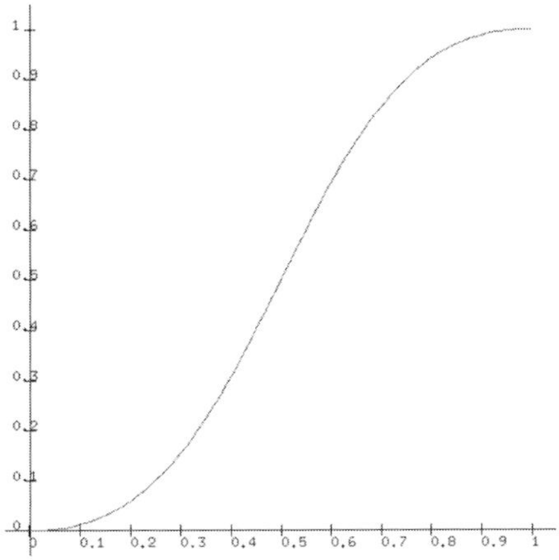

(a) $v_T = 1, g = 0$

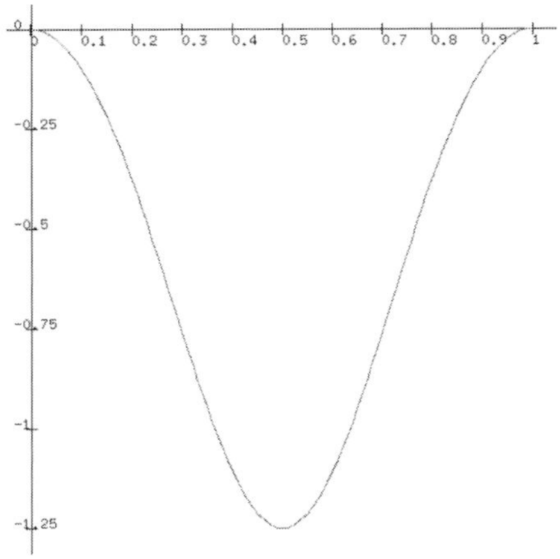

(b) $v_t = 0, g = -10$

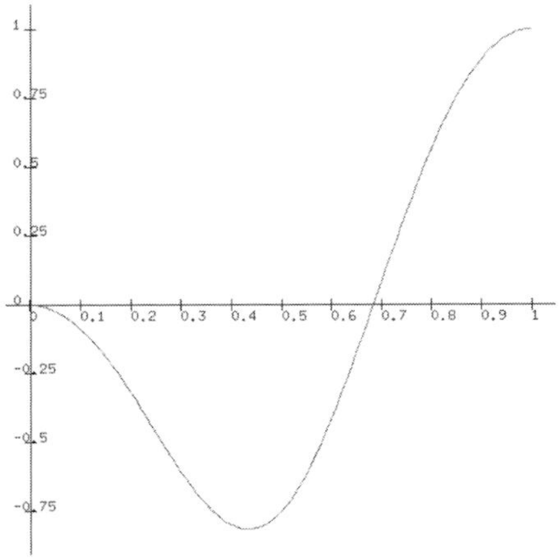

(c) $v_T = 1, g = -10$

Fig. 3.6 Buckley–Leverett flux function: $f(s) = (v_T + g(1-s)^2)s^2/(s^2 + (1-s)^2)\mu_o/\mu_w$ with $\mu_o/\mu_w = 1$

Exercises for 3.2

3.2.1 Compute the characteristic speed for traffic flow with velocity function given by Equation (3.19). Plot the characteristic speed as a function of traffic density.

3.2.2 Solve the Riemann problem for traffic flow with velocity function given by equation (3.19). In particular, find a general formula in terms of the left and right densities ρ_L and ρ_R for the state ρ that moves with zero speed in the solution of the Riemann problem.

3.2.3 Consider the traffic flow with velocity function given by Equation (3.19). Take $\rho_{\max} = 1$ and $a = 100$.

(a) Suppose that we want to approximate the solution of the Riemann problem with $\rho = \rho_{max}/\sqrt{e}$ on the left and $\rho = 0$ on the right. Show that the solution of this Riemann problem involves a rarefaction with all characteristic speeds nonnegative.

(b) Program the following scheme for the Riemann problem just posed:

$$\rho_i^{n+1} = \rho_i^n - \frac{\triangle t^{n+1/2}}{\triangle x_i}[(\rho \mathrm{v})_i^n - (\rho \mathrm{v})_{i-1}^n].$$

Choose the timestep to be less than the cell width $\triangle x$ divided by the maximum characteristic speed in the problem. Use 100 grid cells and a problem length of 1 for your calculations. Plot the traffic density and characteristic speed versus position divided by time.

(c) Suppose that we want to approximate the solution of the Riemann problem with $\rho = \rho_{max}/\sqrt{e}$ on the right and $\rho = \rho_{max}$ on the left. Show that the solution of this Riemann problem involves a rarefaction with all characteristic speeds nonpositive.

(d) Program the following scheme for the Riemann problem just posed:

$$\rho_i^{n+1} = \rho_i^n - \frac{\triangle t^{n+1/2}}{\triangle x_i}[(\rho \mathrm{v})_{i+1}^n - (\rho \mathrm{v})_i^n].$$

Choose the timestep to be less than the cell width $\triangle x$ divided by the maximum characteristic speed in the problem. Use 100 grid cells and a problem length of 1 for your calculations. Plot the traffic density and characteristic speed versus position divided by time.

(e) Suppose that we want to approximate the solution of the Riemann problem with $\rho = 0$ on the right and $\rho = \rho_{max}$ on the left. Show that the solution of this Riemann problem involves a rarefaction with negative and positive characteristic speeds.

(f) What happens if you try to solve the problem just posed with the upwind method given for the problem with all nonnegative characteristic speeds? Run your program and describe your results.

(g) What happens if you try to solve the problem just posed with the upwind method given for the problem with all nonpositive characteristic speeds? Run your program and describe your results.

3.2.4 Compute the characteristic speed for the miscible displacement model (3.21). Ignore the diffusive terms (convective mixing and molecular diffusion) when you compute the characteristic speed. Also find a formula for the Peclet number, defined in Section 2.5.2.

3.2.5 Typically, people measure time in miscible displacement problems in terms of "pore volumes injected." The total pore volume is

$$\int_{x_L}^{x_R} \phi(x)\, dx$$

and the total fluid injected is

$$\int_0^T \mathrm{v}(t)\, dt.$$

If the porosity ϕ is constant and the injection rate v is constant, find a formula for the time at which the volume of the total fluid injected is half the total pore volume.

3.2.6 Program explicit upwind differences for the miscible displacement problem. Ignore convective mixing and molecular diffusion. Choose the problem length to be 10 meters, and use 100 grid cells. Assume that the porous medium initially has no tracer concentration, and that you inject a fixed concentration of 0.01 on the left for all time. Plot your numerical results when one-half pore volume has been injected. Describe how you chose your timestep.

3.2.7 Determine the characteristic speed for the Buckley–Leverett equation (3.23). Use Prudhoe Bay crude, and choose the permeability so that the gravity number is -100. Plot the oil flux f_o and characteristic speed versus the oil saturation s_o. (Hint: if we ignore capillary pressure, the quasilinear form of the equation is

$$\phi \frac{\partial s_o}{\partial t} + \frac{\partial f_o}{\partial s_o} \frac{\partial s_o}{\partial x} = 0.$$

The characteristic speed can be determined from this equation.)

3.2.8 Discuss the solution of the Riemann problem for the Buckley–Leverett problem, using the permeability in the previous exercise. Which kinds of flow problems produce multiple shocks? Which kinds of flow problems correspond to injecting water to produce oil from an oil reservoir with (nearly) no water?

3.2.9 Choose left and right states for a Buckley–Leverett Riemann problem so that all of the characteristic speeds are positive. Program the explicit upwind scheme for this problem. Use 100 grid cells, and plot the numerical results at 0.5 pore volumes injected. Also plot the characteristic speed. Describe how you chose your timestep.

3.3 First-Order Finite Difference Methods

3.3.1 Explicit Upwind Differences

Suppose that we want to approximate the solution of

$$\frac{\partial u}{\partial t} + \frac{\partial f(u)}{\partial x} = 0 \tag{3.24}$$

by explicit upwind differences. In the special case when $\frac{df}{du} > 0$ for all u, we can generalize explicit upwind differences as follows:

$$u_i^{n+1} = u_i^n - \frac{\triangle t^{n+1/2}}{\triangle x_i} \left[f(u_i^n) - f(u_{i-1}^n) \right].$$

This is a conservative difference scheme with numerical fluxes defined by $f_{i+1/2}^{n+1/2} = f(u_i^n)$, and for which all timesteps are chosen so that

$$\frac{\partial f}{\partial u}(u_i^n) \triangle t^{n+1/2} \leq \gamma \triangle x_i$$

for some $0 < \gamma < 1$. Similarly, if $\partial f / \partial u < 0$ for all u, the explicit upwind difference method takes the form

$$u_i^{n+1} = u_i^n - \frac{\triangle t^{n+1/2}}{\triangle x_i} \left[f\left(u_{i+1}^n\right) - f\left(u_i^n\right) \right].$$

This is a conservative difference scheme with numerical fluxes defined by $f_{i+1/2}^{n+1/2} = f(u_{i+1}^n)$, and with all timesteps chosen so that

$$-\frac{\partial f}{\partial u}\left(u_i^n\right) \triangle t^{n+1/2} \le \gamma \triangle x_i$$

for some $0 < \gamma < 1$. With negative characteristic speed, the upwind state is on the right.

However, we will have to be more clever to develop schemes for problems in which the sign of the characteristic speed is not known. If we can find a bound c on the absolute value of the characteristic speeds, so that for all states of interest in our problem

$$-c \le \frac{\partial f}{\partial u} \le c,$$

then we can employ upwind methods to solve

$$\frac{\partial \tilde{u}}{\partial t} + \frac{\partial f(\tilde{u}) + c\tilde{u}}{\partial x} = 0.$$

This, of course, is a Galilean transformation of the original problem. (See Section 3.1.9.) This leads to the conservative difference

$$\tilde{u}_i^{n+1} = \tilde{u}_i^n - \frac{\triangle t^{n+1/2}}{\triangle x_i} [(f(\tilde{u}_i^n) + c\tilde{u}_i^n) - (f(\tilde{u}_{i-1}^n) + c\tilde{u}_{i-1}^n)],$$

and to the timestep restriction

$$[f'(\tilde{u}_i^n) + c] \triangle t^{n+1/2} \le \gamma \triangle x_i.$$

Choosing c to be too large can lead to unnecessarily small timesteps.

3.3.2 Lax–Friedrichs Scheme

A very popular scheme for general nonlinear flux functions f is the **Lax–Friedrichs scheme**. This scheme depends on three basic assumptions.

(i) the computational results are replaced with the piecewise constant cell averages at times t^n and $t^n + \frac{1}{2} \triangle t^{n+1/2}$;
(ii) there is an upper bound λ on the characteristic speed so that for all u $|df/du| \le \lambda$ (we will have to clarify just what we mean by the quantifier on u),
(iii) the timestep is chosen so that

$$\text{for all } i, \; \lambda \triangle t^{n+1/2} < \triangle x_i. \tag{3.25}$$

Let us introduce the notation

$$x_i = \frac{1}{2}\left(x_{i-1/2} + x_{i+1/2}\right)$$

for the grid cell centers. We will apply the divergence theorem to the conservation law (3.24) over the space-time rectangle $(x_i, x_{i+1}) \times (t^n, t^n + \frac{1}{2}t^{n+1/2})$. Since the solution is piecewise constant at time t^n and the timestep is chosen so that waves from the constant states do not reach the cell centers by time $t^n + \triangle t^{n+1/2}/2$, the fluxes are constant in time at the cell centers in this application of the divergence theorem:

$$\begin{aligned}
0 &= \int_{t^n}^{t^n+\triangle t^{n+1/2}/2} \int_{x_i}^{x_{i+1}} \frac{\partial u}{\partial t} + \frac{\partial f(u)}{\partial x} dx\, dt \\
&= \int_{x_i}^{x_{i+1}} u\left(x, t^n + \frac{1}{2}\triangle t^{n+1/2}\right) dx - \int_{x_i}^{x_{i+1}} u(x, t^n)\, dx \\
&\quad + \int_{t^n}^{t^n+\triangle t^{n+1/2}/2} f\left(u(x_{i+1}, t)\right) dt - \int_{t^n}^{t^n+\triangle t^{n+1/2}/2} f(u(x_i, t))\, dt \\
&= \int_{x_i}^{x_{i+1}} u\left(x, t^n + \frac{1}{2}\triangle t^{n+1/2}\right) dx - \left[u_i^n \frac{\triangle x_i}{2} + u_{i+1}^n \frac{\triangle x_{i+1}}{2}\right] \\
&\quad + f(u_{i+1}^n)\frac{\triangle t^{n+1/2}}{2} - f(u_i^n)\frac{\triangle t^{n+1/2}}{2}
\end{aligned}$$

We replace the numerical results at the half-time with the cell averages. A second half-step is similar, applying the divergence theorem over the rectangle $(x_{i-1/2}, x_{i+1/2}) \times (t^n + \frac{1}{2}t^{n+1/2}, t^n + \triangle t^{n+1/2})$. We obtain the following formulas for the two half-steps:

$$u_{i+1/2}^{n+1/2} = \left\{u_i^n \triangle x_i + u_{i+1}^n \triangle x_{i+1} - \left[f\left(u_{i+1}^n\right) - f\left(u_i^n\right)\right] \triangle t^{n+1/2}\right\} \frac{1}{\triangle x_i + \triangle x_{i+1}}, \tag{3.26a}$$

$$u_i^{n+1} = \left\{u_{i-1/2}^{n+1/2} + u_{i+1/2}^{n+1/2} - \left[f\left(u_{i+1/2}^{n+1/2}\right) - f\left(u_{i-1/2}^{n+1/2}\right)\right] \frac{\triangle t^{n+1/2}}{\triangle x_i}\right\} \frac{1}{2}. \tag{3.26b}$$

These are conservative differences. For example, summing the conserved quantity at the half-time leads to the mass at the old time plus a telescoping sum of fluxes:

$$\begin{aligned}
\sum_i u_{i+1/2}^{n+1/2} \frac{\triangle x_i + \triangle x_{i+1}}{2} &= \frac{1}{2}\sum_i u_i^n \triangle x_i + \frac{1}{2}\sum_i u_{i+1}^n \triangle x_{i+1} \\
&\quad - \triangle t^{n+1/2} \sum_i \left[f\left(u_{i+1}^n\right) - f\left(u_i^n\right)\right] \\
&= \sum_i u_i^n \triangle x_i + \triangle t^{n+1/2}\left[f\left(u_0^n\right) - f\left(u_I^n\right)\right].
\end{aligned}$$

The second step is similarly conservative.

The only approximations in these equations are the replacement of the solution at each new time by the cell averages. Of course, it is necessary to modify the calculation at the boundary of the domain, in order to prevent the application of the divergence theorem over a rectangle that does not lie entirely inside the problem domain. Except for certain boundary conditions (periodic boundaries, first-order non-reflecting boundaries and reflecting boundaries), this is a delicate topic. For example, specifying the the flux on a boundary can make it difficult to determine the solution at the half-time boundaries.

Example 3.3.1 For linear advection on a uniform grid, the Lax–Friedrichs scheme can be written

$$\begin{aligned} u_{i+1/2}^{n+1/2} &= \frac{1}{2}\left(u_i^n + u_{i+1}^n\right) - \frac{\lambda \triangle t}{2\triangle x_i}\left[u_{i+1}^n - u_i^n\right], \\ u_i^{n+1} &= \frac{1}{2}\left(u_{i-1/2}^{n+1/2} + u_{i+1/2}^{n+1/2}\right) - \frac{\lambda \triangle t}{2\triangle x}\left[u_{i+1/2}^{n+1/2} - u_{i-1/2}^{n+1/2}\right]. \end{aligned}$$

In other words, if $\gamma = \lambda \triangle t / \triangle x$ then

$$\begin{aligned} u_i^{n+1} &= \frac{1}{2}\left[(1-\gamma) u_{i+1/2}^{n+1/2} + (1+\gamma) u_{i-1/2}^{n+1/2}\right] \\ &= \frac{1}{4}(1-\gamma)^2 u_{i+1}^n + \frac{1}{2}(1-\gamma^2) u_i^n + \frac{1}{4}(1+\gamma)^2 u_{i-1}^n. \end{aligned}$$

A Fourier analysis shows that the solution ratio is

$$\begin{aligned} z &= \frac{1}{4}(1-\gamma)^2 e^{i\theta} + \frac{1}{2}(1-\gamma^2) + \frac{1}{4}(1+\gamma)^2 e^{-i\theta} \\ &= \frac{1}{2}(1+\gamma^2)\cos\theta - i\gamma\sin\theta + \frac{1}{2}(1-\gamma^2) \\ &= \cos^2\frac{\theta}{2} - \gamma^2\sin^2\frac{\theta}{2} - 2i\gamma\sin\frac{\theta}{2}\cos\frac{\theta}{2}. \end{aligned}$$

Thus

$$\begin{aligned} |z|^2 &= \left[\cos^4\frac{\theta}{2} - 2\gamma^2\cos^2\frac{\theta}{2}\sin^2\frac{\theta}{2} + \gamma^4\sin^4\frac{\theta}{2}\right] + 4\gamma^2\sin^2\frac{\theta}{2}\cos^2\frac{\theta}{2} \\ &= \left[\cos^2\frac{\theta}{2} + \gamma^2\sin\frac{\theta}{2}\right]^2. \end{aligned}$$

It follows that the Lax–Friedrichs scheme is dissipative for $\gamma < 1$.

A modified equation analysis of the Lax–Friedrichs scheme for linear advection on a uniform grid shows that

$$\begin{aligned} u_i^n + \frac{\partial \tilde{u}}{\partial t}\triangle t + \frac{1}{2}\frac{\partial^2 \tilde{u}}{\partial t^2}\triangle t^2 \approx u_i^{n+1} &= \frac{1}{4}(1-\gamma)^2 u_{i+1}^n + \frac{1}{2}(1-\gamma^2)u_i^n + \frac{1}{4}(1+\gamma)^2 u_{i-1}^n. \\ &\approx \frac{1}{4}(1-\gamma)^2\left[u_i^n + \frac{\partial \tilde{u}}{\partial x}\triangle x + \frac{1}{2}\frac{\partial^2 \tilde{u}}{\partial x^2}\triangle x^2\right] + \frac{1}{2}(1-\gamma^2)u_i^n \\ &\quad + \frac{1}{4}(1+\gamma)^2\left[u_i^n - \frac{\partial \tilde{u}}{\partial x}\triangle x + \frac{1}{2}\frac{\partial^2 \tilde{u}}{\partial x^2}\triangle x^2\right] \\ &= u_i^n - \frac{\partial \tilde{u}}{\partial x}\gamma \triangle x + \frac{\partial^2 \tilde{u}}{\partial x^2}(1+\gamma^2)\frac{\triangle x}{4}. \end{aligned}$$

This can be rewritten in the form

$$\frac{\partial \tilde{u}}{\partial t} + \lambda \frac{\partial \tilde{u}}{\partial x} = -\frac{\partial^2 \tilde{u}}{\partial t^2}\frac{\triangle t}{2} + \frac{\partial^2 \tilde{u}}{\partial x^2}(1+\gamma^2)\frac{\triangle x^2}{4\triangle t} \equiv e.$$

It follows from this equation that

$$\begin{aligned} \frac{\partial^2 \tilde{u}}{\partial t^2} &= \frac{\partial}{\partial t}\left(-\lambda \frac{\partial \tilde{u}}{\partial x} + e\right) = -\lambda \frac{\partial}{\partial x}\left(-\lambda \frac{\partial \tilde{u}}{\partial x} + e\right) + \frac{\partial e}{\partial t} \\ &= \lambda^2 \frac{\partial^2 \tilde{u}}{\partial x^2} + \frac{\partial e}{\partial t} - \lambda \frac{\partial e}{\partial x}. \end{aligned}$$

Thus the modified equation for the Lax–Friedrichs scheme is

$$\frac{\partial \tilde{u}}{\partial t} + \lambda \frac{\partial \tilde{u}}{\partial x} \approx \left[-\frac{\lambda^2 \triangle t}{2} + \frac{\triangle x^2 + \lambda^2 \triangle t^2}{4\triangle t}\right]\frac{\partial^2 \tilde{u}}{\partial x^2} = (1-\gamma^2)\frac{\triangle x^2}{4\triangle t}\frac{\partial^2 \tilde{u}}{\partial x^2}.$$

This result shows that the scheme is diffusive for $\gamma < 1$, and that the diffusion becomes infinite as $\triangle t \to 0$.

A program to implement the Lax-Friedrichs scheme can be found in either **Program 3.3-28: laxFriedrichs.f** or **Program 3.3-29: Schemes.C** By deleting the file name from the browser window that displays the current web address, the user can see a list of all of the files in the directory. In particular, the `GNUmakefile` will describe which files are used to make a particular executable. Figure 3.7 shows numerical results for the Lax-Friedrichs scheme applied to Burgers' equation. The numerical solution at a fixed time is plotted versus x/t. Thus it is possible to read the shock speed from the horizontal axis. Note that the discontinuity is spread out more as we decrease the CFL number. We obtain similar results for a rarefaction in Figure 3.8. In this case, since the conserved quantity u is equal to the characteristic speed for Burgers' equation, the exact solution in the middle of the rarefaction should be a line going through the origin with slope 1. Students can also execute

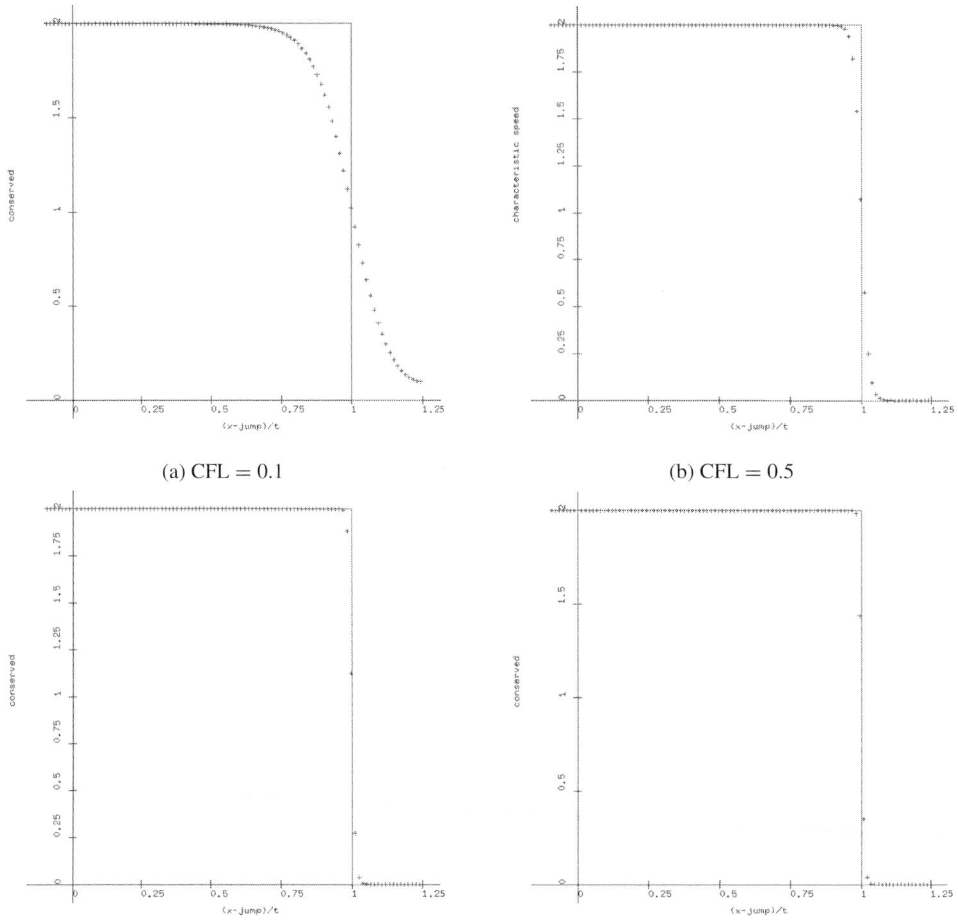

(a) CFL = 0.1

(b) CFL = 0.5

(c) CFL = 0.9

(d) CFL = 1.0

Fig. 3.7 Lax–Friedrichs scheme for Burgers' shock: u vs. x/t (solid = exact, plus = numerical solution)

the Lax Friedrichs scheme by clicking on the web link **Executable 3.3-4: guiconvex1** and selecting `scheme` to be `lax friedrichs` in the `View` pulldown menu, inside the `Numerical Method Parameters` group.

The advantage of the Lax–Friedrichs scheme is that it avoids the need to represent the wave interactions (Riemann problems) arising from the piecewise-constant initial data. The disadvantages of the Lax–Friedrichs scheme are that it must work on a staggered grid using half-interval timesteps, and must use some special calculations at boundaries. Let us note that the staggered mesh in the second half-step can be avoided by averaging the staggered cell averages back onto the original mesh [156]. The timestep must still be chosen so that waves from cell sides cannot

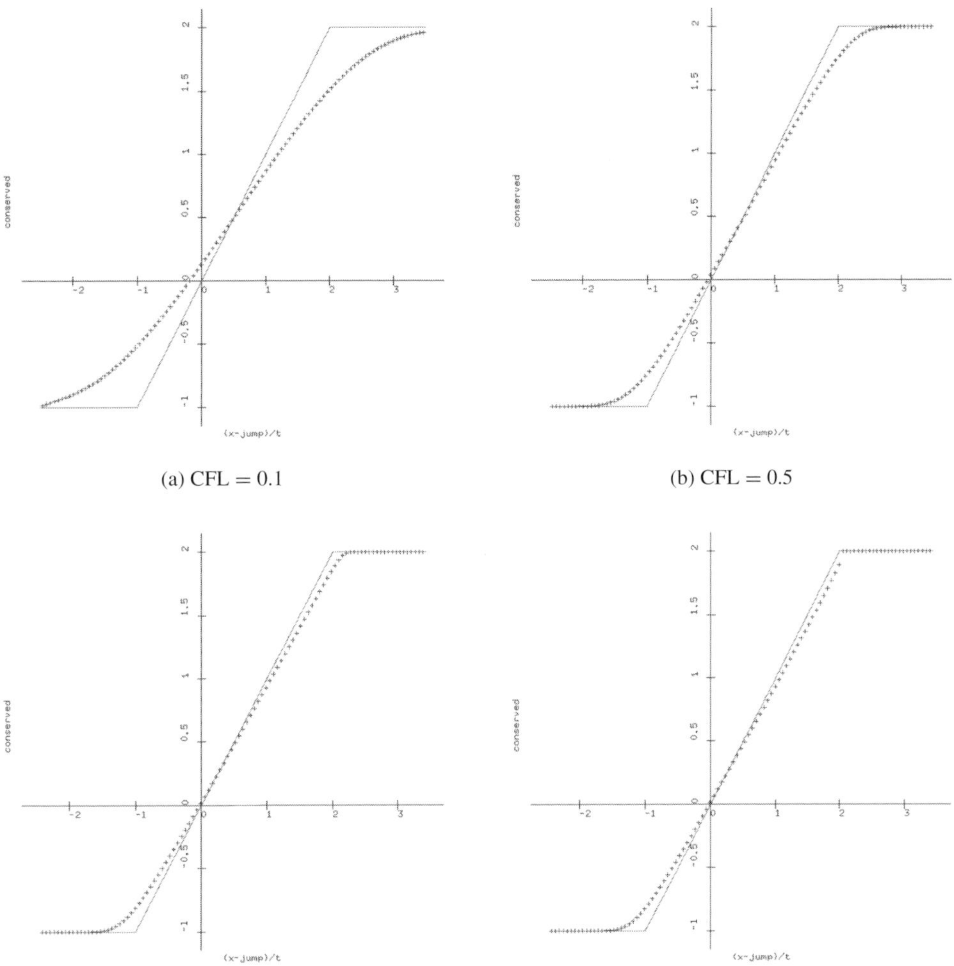

(a) CFL = 0.1

(b) CFL = 0.5

(c) CFL = 0.9

(d) CFL = 1.0

Fig. 3.8 Lax–Friedrichs scheme for Burgers' rarefaction: u vs. x/t (solid = exact, plus = numerical solution)

reach cell centers, and the additional averaging step introduces additional numerical diffusion.

3.3.3 Timestep Selection

In all of the schemes for integrating conservation laws, we will identify a stability condition of the form

$$\lambda \triangle t^{n+1/2} < \alpha \triangle x_i \quad \text{for all } i$$

where α is some stability constant, and λ is some upper bound on the largest characteristic speed found in the problem. For the Lax–Friedrichs scheme we found that $\alpha = 1$. The value $\alpha = 1$ will work for most of the schemes we will consider.

In many cases, it is difficult for us to compute a strict upper bound λ on the largest characteristic speed. Typically, we will sample the discrete values and approximate

$$\lambda = \max_i \left| \frac{df}{du}(u_i^n) \right|.$$

Because of the discrete sampling, the computed λ may be less than the analytical value. In order to protect the integration, we usually reduce the timestep size by some fixed factor ω, called the **CFL number**. The revised timestep selection would take the form

$$\lambda \triangle t^{n+1/2} < \omega \triangle x_i \quad \text{for all } i$$

where $\omega \leq \alpha$.

Typically we will choose $\omega = 0.9$ for schemes that are stable with $\alpha = 1$. For some schemes applied to difficult problems we may choose $\omega = 0.5$. This will be desirable when the scheme has zero phase error at at CFL number of 0.5, and the problem has strong discontinuities. The choice of the CFL number can be guided by the Fourier analysis of the scheme applied to linear advection.

3.3.4 Rusanov's Scheme

In some cases, we would like to avoid the half-step complications of the Lax-Friedrichs scheme. Suppose that for each cell side we can find an upper bound $\lambda_{i+1/2}$ so that

$$\text{for all } u \text{ between } u_i^n \text{ and } u_{i+1}^n, \left| \frac{\partial f}{5u}(u) \right| \leq \lambda_{i+1/2}.$$

Further, let

$$\lambda = \max_i \{\lambda_{i+1/2}\}.$$

Note that $f_{i+1/2}^+(u) \equiv f(u) + \lambda_{i+1/2} u$ is such that $\partial f_{i+1/2}^+(u)/\partial u \geq 0$ for all u between u_i^n and u_{i+1}^n, and that $f_{i+1/2}^-(u) \equiv f(u) - \lambda_{i+1/2} u$ is such that $\partial f_{i+1/2}^-(u)/\partial u \leq 0$ for all u between u_i^n and u_{i+1}^n. Rusanov's scheme uses the upwind flux evaluation for each of these two parts of the flux:

$$f_{i+1/2}^{n+1/2} = \frac{1}{2}\left[f_{i+1/2}^+(u_i^n) + f_{i+1/2}^-(u_{i+1}^n)\right] = \frac{1}{2}\left[f(u_i^n) + f(u_{i+1}^n) - \lambda_{i+\frac{1}{2}}(u_{i+1}^n - u_i^n)\right].$$

Thus the Rusanov flux is the explicit centered differences flux plus an additional artificial diffusion.

Example 3.3.2 For linear advection

$$\frac{\partial u}{\partial t} + c\frac{\partial u}{\partial x} = 0$$

and with $\lambda \geq |c|$, the Rusanov flux is

$$f_{i+1/2}^{n+1/2} = \frac{1}{2}\left[cu_i^n + cu_{i+1}^n - \lambda\left(u_{i+1}^n - u_i^n\right)\right] = \frac{\lambda + c}{2}u_i^n - \frac{\lambda - c}{2}u_{i+1}^n.$$

It follows that Rusanov's scheme for linear advection can be written

$$\begin{aligned} u_i^{n+1} &= u_i^n - \frac{\triangle t^{n+1/2}}{\triangle x_i}\left[\frac{\lambda + c}{2}u_i^n - \frac{\lambda - c}{2}u_{i+1}^n - \frac{\lambda + c}{2}u_{i-1}^n + \frac{\lambda - c}{2}u_i^n\right] \\ &= \frac{\triangle t^{n+1/2}}{\triangle x_i}\frac{\lambda + c}{2}u_{i-1}^n + \left(1 - \lambda\frac{\triangle t^{n+1/2}}{\triangle x_i}\right)u_i^n + \frac{\triangle t^{n+1/2}}{\triangle x_i}\frac{\lambda - c}{2}u_{i+1}^n. \end{aligned}$$

Note that if $\lambda \pm c \geq 0$, then u_i^{n+1} is a weighted average of the solution at the previous time; this lends stability to the scheme.

A modified equation analysis of Rusanov's scheme for linear advection shows that

$$\begin{aligned} u_i^n + \frac{\partial \tilde{u}}{\partial t}\triangle t + \frac{1}{2}\frac{\partial^2 \tilde{u}}{\partial t^2}\triangle t^2 &\approx \frac{\triangle t}{\triangle x}\frac{\lambda + c}{2}\left[u_i^n - \frac{\partial \tilde{u}}{\partial x}\triangle x + \frac{1}{2}\frac{\partial^2 \tilde{u}}{\partial x^2}\triangle x^2\right] \\ &\quad + \left(1 - \lambda\frac{\triangle t}{\triangle x}\right)u_i^n + \frac{\triangle t}{\triangle x}\frac{\lambda - c}{2}\left[u_i^n + \frac{\partial \tilde{u}}{\partial x}\triangle x + \frac{1}{2}\frac{\partial^2 \tilde{u}}{\partial x^2}\triangle x^2\right] \\ &= u_i^n - c\triangle t\frac{\partial \tilde{u}}{\partial x} + \frac{\lambda \triangle t \triangle x}{2}\frac{\partial^2 \tilde{u}}{\partial x^2}. \end{aligned}$$

It follows that the modified equation is

$$\frac{\partial \tilde{u}}{\partial t} + \frac{\partial \tilde{u}}{\partial x} \approx \frac{|c|\triangle x}{2}\left(\frac{\lambda}{|c|} - \frac{|c|\triangle t}{\triangle x}\right)\frac{\partial^2 \tilde{u}}{\partial x^2}.$$

This indicates that Rusanov's scheme is diffusive if $\lambda > \frac{c^2 \triangle t}{\triangle x}$. However, a Fourier stability analysis leads to solution ratio

$$z = \frac{1}{2}(\gamma + \gamma_c)e^{-i\theta} + 1 - \gamma + \frac{1}{2}(\gamma - \gamma_c)e^{i\theta},$$

where $\gamma = \lambda \triangle t / \triangle x$ is the CFL number and $\gamma_c = c\triangle t/\triangle x$ is the natural CFL number. It follows that

$$|z|^2 = 1 - 2\gamma + \gamma^2(1 + \cos^2\theta) + \gamma_c^2 \sin^2\theta.$$

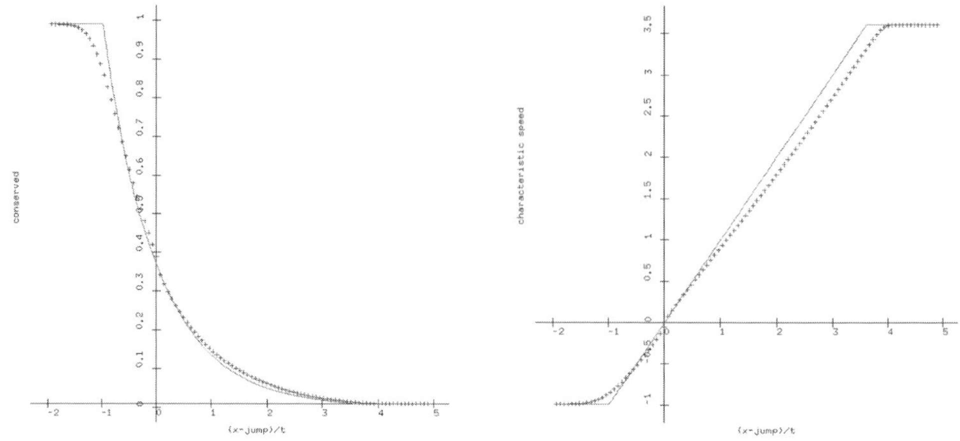

(a) Density vs. x/t (b) Characteristic speed vs. x/t

Fig. 3.9 Rusanov scheme for traffic flow: $\mathrm{v}(\rho) = -\log(\rho)$, CFL $= 0.9$ (solid = exact, plus = numerical solution)

Since $\lambda \geq |c|$, we see that $\gamma \geq |\gamma_c|$. It follows that

$$1 - |z|^2 = 2\gamma - \gamma^2(1 + \cos^2\theta) - \gamma_c^2 \sin^2\theta \leq 2\gamma(1 - \gamma).$$

This result indicates that in order for the Rusanov scheme to be dissipative, it is sufficient that the CFL number satisfy $0 < \gamma < 1$.

A program to implement the Rusanov scheme can be found in **Program 3.3-30: rusanov.f** or in **Program 3.3-31: Schemes.C** Figure 3.9 shows some numerical results with the Rusanov scheme for the traffic flow problem with $\mathrm{v}(\rho) = -\rho\log(\rho)$. This is a Riemann problem for which the analytical solution is a rarefaction. Note that we plot both the solution ρ and the characteristic speed $d(\rho\mathrm{v}(\rho))/d\rho$ versus x/t. Inside a rarefaction, the characteristic speed should be identical to x/t. Thus, the graph of the characteristic speed gives us a way to check that the numerical solution is (approximately) correct. It is useful to plot the characteristic speed versus x/t for all Riemann problems, as an *a posteriori* check on the numerical results. Students can also execute the Lax–Friedrichs scheme by clicking on the web link **Executable 3.3-5: guiconvex1** and selecting `scheme` to be `rusanov` in the `View` pulldown menu, inside the `Numerical Method Parameters` group.

3.3.5 Godunov's Scheme

Like the Lax–Friedrichs scheme, **Godunov's scheme** considers the solution at each timestep to be piecewise-constant. Unlike the Lax–Friedrichs scheme, Godunov's scheme purposely uses information from Riemann problems to determine the numerical fluxes.

Godunov's scheme applies the divergence theorem to the conservation law (3.24) over the space-time rectangle $(x_{i-1/2}, x_{i+1/2}) \times (t^n, t^{n+1})$ to get

$$\begin{aligned} u_i^{n+1} &\equiv \frac{1}{\triangle x_i} \int_{x_{i-1/2}}^{x_{i+1/2}} u(x, t^{n+1})\, dx \\ &= \frac{1}{\triangle x_i} \int_{x_{i-1/2}}^{x_{i+1/2}} u(x, t^n)\, dx \\ &\quad - \frac{1}{\triangle x_i} \left[\int_{t^n}^{t^{n+1}} f\big(\mathcal{R}(u_i^n, u_{i+1}^n; 0)\big)\, dt - \int_{t^n}^{t^{n+1}} f\big(\mathcal{R}(u_{i-1}^n, u_i^n; 0)\big)\, dt \right] \\ &\equiv u_i^n - \frac{\triangle t^{n+1/2}}{\triangle x_i} \left[f_{i+1/2}^{n+1/2} - f_{i-1/2}^{n+1/2} \right]. \end{aligned}$$

Recall that $\mathcal{R}(u_L, u_R; \lambda)$ represents the state that moves with speed λ in the solution of the Riemann problem with left state u_L and right state u_R. Godunov's scheme does not require that we find the complete solution to the Riemann problem; it only requires the flux at the stationary state in the solution to the Riemann problem.

Godunov's scheme is particularly useful for boundary conditions. For example, suppose that we want the boundary condition to represent an unbounded domain; then no waves should go from the boundary to the interior of the domain. This means that we expect that there are no characteristics going from inside the domain across the left boundary. This can be checked during the calculation of the Godunov flux at the left boundary; in particular, the flux at the boundary should be the flux at the state inside the domain at the boundary.

Suppose that $u_i^n < u_{i+1}^n$. Then the fastest characteristics from the individual Riemann problems at the cell sides do not intersect other sides if the convex hull $\underline{f}$ of the flux function f between the two states satisfies

$$\max_{u_i^n \le u \le u_{i+1}^n} \left\{ \left| \frac{d\underline{f}}{du}(u) \right| \right\} \triangle t^{n+1/2} \le \min\{\triangle x_i, \triangle x_{i+1}\} \quad \text{for all } i.$$

Similarly, if $u_i^n > u_{i+1}^n$ then the fastest characteristics from the individual Riemann problems at the cell sides do not intersect other sides if the concave hull $\overline{f}$ of the flux function f satisfies

$$\max_{u_{i+1}^n \le u \le u_i^n} \left\{ \left| \frac{\partial \overline{f}}{\partial u}(u) \right| \right\} \triangle t^{n+1/2} \le \min\{\triangle x_i, \triangle x_{i+1}\} \quad \text{for all } i.$$

However, it is possible that the interaction of the fastest waves from two Riemann problems at neighboring sides could produce an even faster characteristic speed. In order to avoid this difficult situation, and in order to avoid the construction of the convex or concave hull of the flux function at each step, we typically select the timestep so that

$$\max_{u \text{ between } u_i^n \text{ and } u_{i+1}^n} \left\{ \left| \frac{df}{du}(u) \right| \right\} \triangle t^{n+1/2} \leq \min\{\triangle x_i, \triangle x_{i+1}\} \text{ for all } i. \tag{3.27}$$

With this choice, it may be possible to pre-compute the inflection points of the flux to find the largest possible characteristic speeds.

Example 3.3.3 Consider Godunov's method for linear advection,

$$\frac{\partial u}{\partial t} + \frac{\partial \lambda u}{\partial x} = 0.$$

If the velocity satisfies $\lambda > 0$, then $\mathcal{R}(u_L, u_R; 0) = u_L$. In this case, Godunov's scheme is

$$u_i^{n+1} = u_i^n - \frac{\triangle t^{n+1/2}}{\triangle x_i} \left[\lambda u_i^n - \lambda u_{i-1}^n \right],$$

which is identical to explicit upwind differencing. Similarly, if $\lambda < 0$, Godunov's scheme is

$$u_i^{n+1} = u_i^n - \frac{\triangle t^{n+1/2}}{\triangle x_i} \left[\lambda u_{i+1}^n - \lambda u_i^n \right],$$

which again is upwind differencing. Thus Godunov's scheme should be viewed as a generalization of upwind differencing to nonlinear problems. Also note that for linear advection, Godunov's scheme is dissipative if $|\lambda| \triangle t^{n+1/2} < \triangle x_i$ for all cells i.

Example 3.3.4 Consider Godunov's scheme for Burgers' equation. For Godunov's method, we need to find the state that moves with zero speed in the solution of the Riemann problem. It is easy to see that the flux at that state is

$$f(\mathcal{R}(u_L, u_R; 0)) = \begin{cases} \frac{1}{2} \max\{u_L, \min\{u_R, 0\}\}^2, & u_L < u_R \\ \frac{1}{2} \max\{|u_L|, |u_r|\}^2, & u_L \geq u_R. \end{cases}$$

Thus Godunov's scheme is easy to implement for Burgers' equation.

Note that the Godunov flux $f(\mathcal{R}(u_L, u_R; 0))$ is not a continuously differentiable function of u_L and u_R. This may cause difficulty in the computation of steady-state solutions, stationary waves or transonic rarefactions. It is also interesting to note the following formulation of Godunov's scheme. Suppose that the timestep is chosen so that (3.27) is satisfied. If we are given piecewise constant initial data, we can approximate the the solution to the conservation law by the cell averages of the analytical solution to the conservation law. Like the Lax-Friedrichs scheme, the only approximation error in the method is due to replacing the analytical solution by piecewise constant averages.

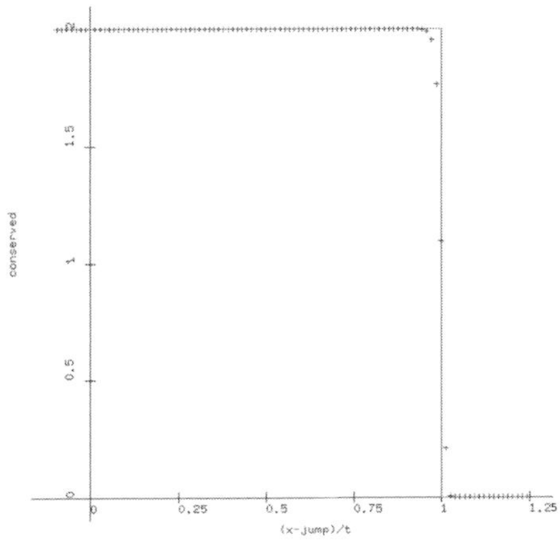

(a) CFL = 0.1

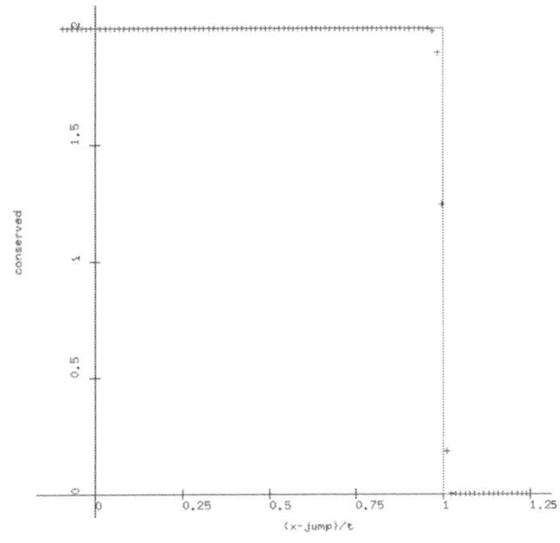

(b) CFL = 0.5

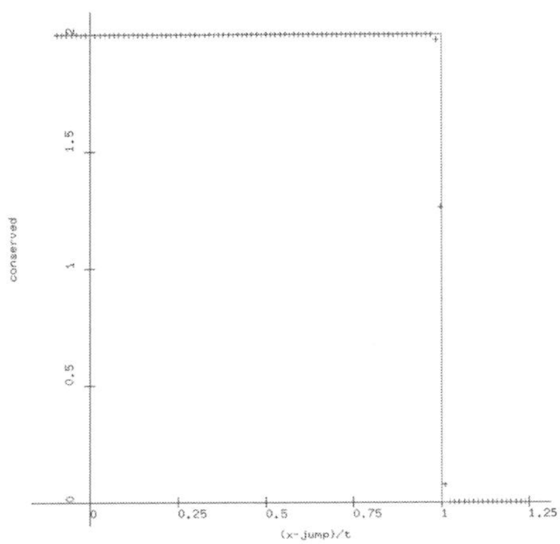

(c) CFL = 0.9

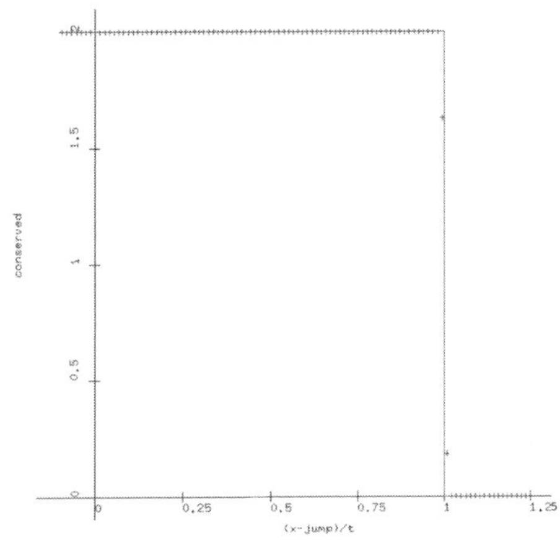

(d) CFL = 1.0

Fig. 3.10 Godunov scheme for Burgers' shock: u vs. x/t (solid = exact, plus = numerical solution)

A program to implement the Godunov scheme can be found in **Program 3.3-32: godunov.f** or in **Program 3.3-33: Schemes.C** The numerical flux evaluation in Godunov's scheme requires the solution of Riemann problems, which can be found in **Program 3.3-34: burgers.f** or in routine `solveRiemann` in **Program 3.3-35: GUIConvexRiemannProblem1.C**. These routines for solving scalar law Riemann problems are valid only for convex or concave flux functions. Figures 3.10 and 3.11 show numerical results for Godunov's scheme applied to a shock and a rarefaction in Burgers' equation. Note that the numerical solution is not smeared as much as it was with the Lax-Friedrichs scheme. Students can perform these computations interactively

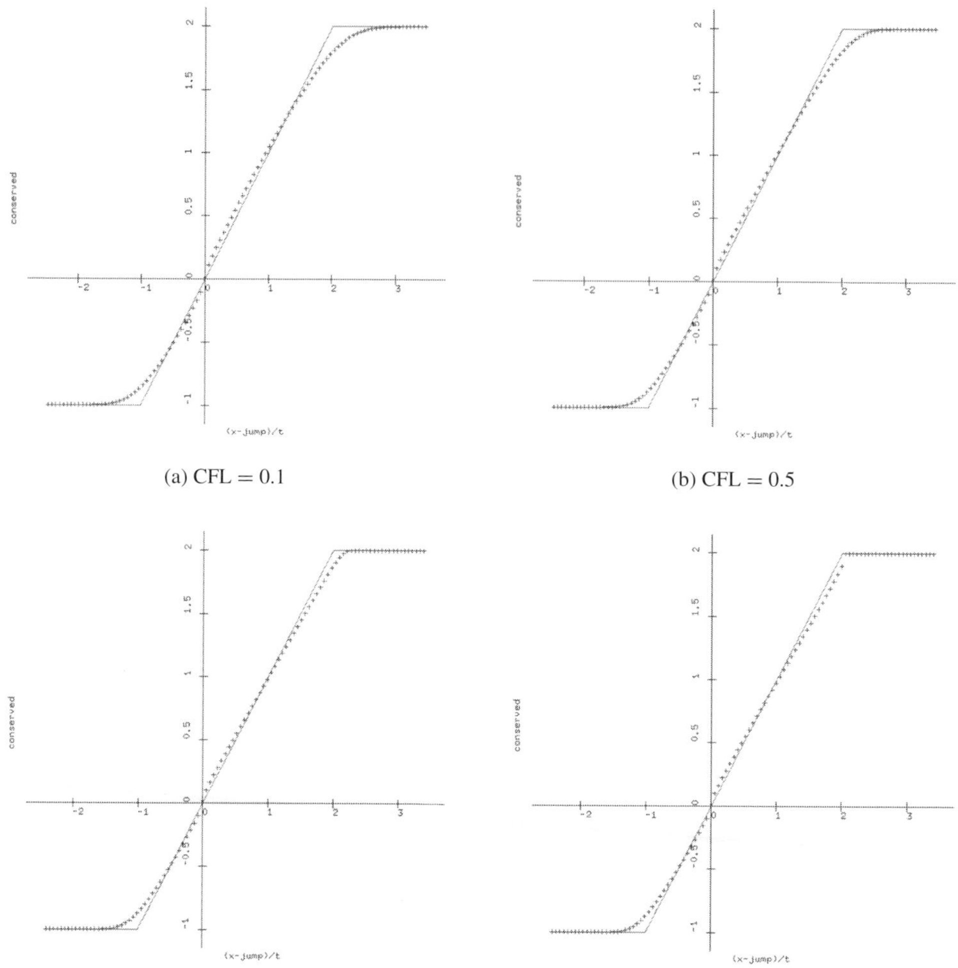

(a) CFL $= 0.1$ (b) CFL $= 0.5$

(c) CFL $= 0.9$ (d) CFL $= 1.0$

Fig. 3.11 Godunov Scheme for Burgers' Rarefaction: u vs. x/t (solid = exact, plus = numerical solution)

by clicking on **Executable 3.3-6: guiconvex1** and selecting `scheme` to be `rusanov` in the `View` pulldown menu, inside the `Numerical Method Parameters` group.

3.3.6 Comparison of Lax–Friedrichs, Godunov and Rusanov

Of the three schemes we have just examined, the Rusanov scheme is probably the easiest to implement, and the Godunov scheme is the least diffusive. For linear advection, the modified equation analyses gave us

$$\frac{\partial u}{\partial t} + \frac{\partial cu}{\partial x} \approx \frac{c\triangle x}{2}(1-\gamma)\frac{\partial^2 u}{\partial x^2} \equiv e_G$$

for the Godunov scheme, which is equivalent to explicit upwind in this case. The modified equation analysis for Lax–Friedrichs gave us

$$\frac{\partial u}{\partial t} + \frac{\partial cu}{\partial x} \approx \frac{\triangle x^2}{4\triangle t}(1-\gamma^2)\frac{\partial^2 u}{\partial x^2} \equiv e_G \frac{1+\gamma}{\gamma}$$

and the modified equation analysis for Rusanov gave us

$$\frac{\partial u}{\partial t} + \frac{\partial cu}{\partial x} \approx \frac{c\triangle x}{2}\left(\frac{\lambda}{c}-\gamma\right)\frac{\partial^2 u}{\partial x^2} \equiv e_G \frac{\lambda/c-\gamma}{\gamma}.$$

From these analyses, we expect that both Lax–Friedrichs and Rusanov will be more diffusive than Godunov.

Figures 3.12 and 3.13 show a comparison of the Lax–Friedrichs, Godunov, Rusanov and Marquina (exercise 3.4.3) schemes for a shock and a rarefaction with the logarithmic traffic model. All schemes were run with CFL = 0.9. Of these four schemes, the Godunov scheme generally produces results closest to the analytical solution, except for states close to the sonic point (where the characteristic speed is zero in the rarefaction).

Students can perform their own comparisons of these schemes for convex conservation laws by clicking on the web link **Executable 3.3-7: guiconvexerror1.** The main program for this executable can be viewed by clicking on **Program 3.3-36: GUIConvexErrorAnalysis1.C** This program allows the student to compare Lax-Friedrichs, Rusanov, Godunov and the Marquina scheme for problems involving Burgers' equation and traffic flow. For comparisons involving Burgers's equation and the Buckley-Leverett model, view **Program 3.3-37: GUIErrorAnalysis1.C** or run the executable **Executable 3.3-8: guierror1.** Individual schemes can be run for Burgers' equation and the Buckley-Leverett model by clicking on **Executable 3.3-9: guiriemann1.** The main program for this executable can be viewed at **Program 3.3-38: GUIRiemannProblem1.C**

3.4 Nonreflecting Boundary Conditions

Although boundary conditions can take many forms, there is one condition in particular that we would like to be able to treat. This condition represents a **nonreflecting boundary**. Such a boundary represents an infinite medium; waves that cross this boundary are supposed to continue on and never be seen again.

Suppose that our numerical method assumes that we have piecewise constant data at the end of each timestep. We also assume that our timestep is chosen so that

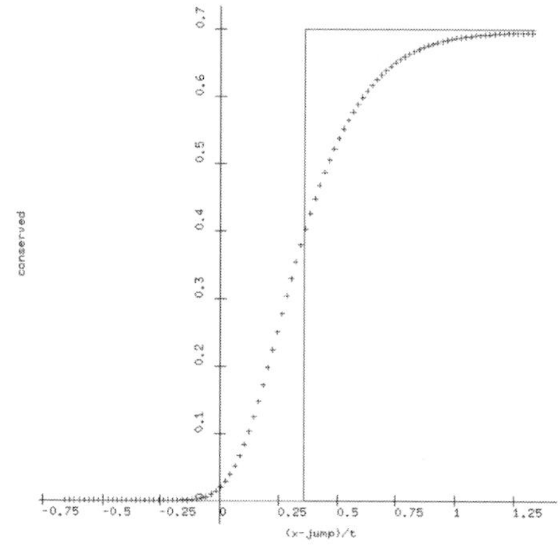

(a) Lax–Friedrichs

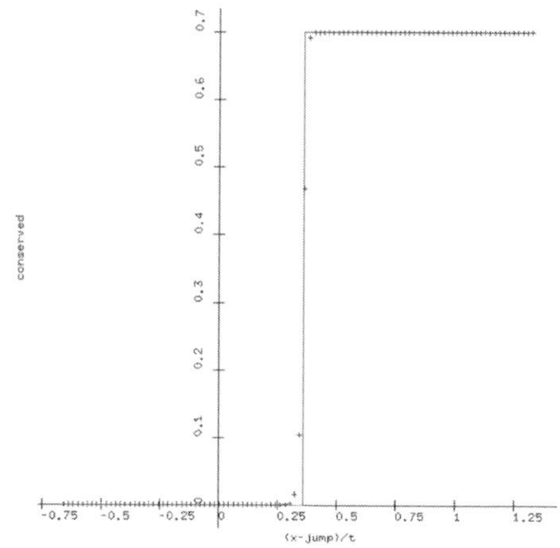

(b) Rusanov

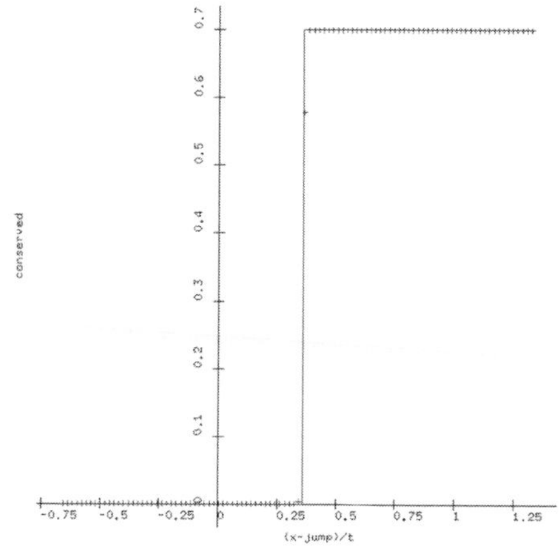

(c) Godunov

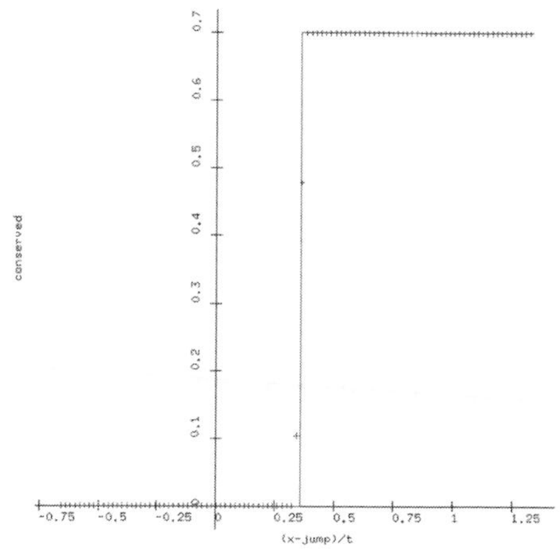

(d) Marquina

Fig. 3.12 Comparison of schemes for log traffic shock: u vs. x/t, CFL $= 0.9$, $u_L = 0$, $u_R = 0.7$

the interaction of waves at individual cell sides cannot cross a cell in less than one timestep. For simplicity, let us assume that the non-reflecting boundary is at the right-hand side of the domain. Given the solution at the previous time, the analytical solution of the conservation law at the right-hand boundary is

$$w^n(x) = u_j^n$$

where j is the index of the last cell. In order that information only move to the right out of the domain, we must have

$$f'\left(u_j^n\right) \geq 0.$$

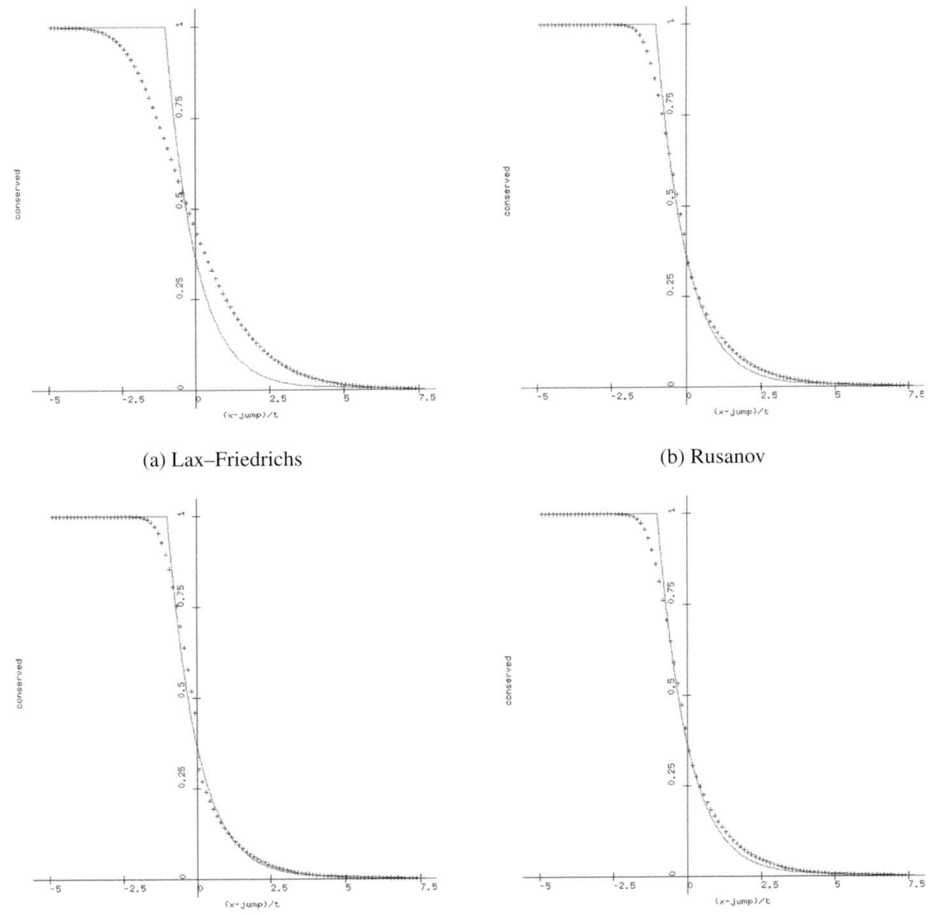

(a) Lax–Friedrichs

(b) Rusanov

(c) Godunov

(d) Marquina

Fig. 3.13 Comparison of schemes for log traffic rarefaction: u vs. $x/t, u_L = 1$, $r_R = 0$

In this case, we can use **ghost cells** to cause the numerical method to compute the correct flux at the right-hand boundary. We carry extra cells $j+1, \ldots, j+k$ where k is the width of the finite difference stencil, and set

$$u^n_{j+i} = u^n_j, \; 1 \leq i \leq k$$

Similarly, at a nonreflecting boundary on the left, we can set the ghost cell values to the value of the solution in the first cell inside the domain.

Higher-order methods typically involve a higher-order representation of the solution than the piecewise-constant function described in the previous paragraph. These methods will usually require a more elaborate treatment of a nonreflecting boundary to preserve their order at the boundary.

Exercises for 3.4

3.4.1 Many texts (for example [96, page 125] and [72, page 315]) incorrectly write the Lax–Friedrichs scheme on a uniform grid in the form

$$u_i^{n+1} = \frac{1}{2}(u_{i-1}^n + u_{i+1}^n) - \frac{\triangle t^{n+1/2}}{2\triangle x}[f(u_{i+1}^n) - f(u_{i-1}^n)].$$

(a) Show that this is a conservative difference scheme with flux

$$f_{i+1/2}^n = \frac{f(u_{i+1}^n) + f(u_i^n)}{2} - \frac{\triangle x}{2\triangle t}[u_{i+1}^n - u_i^n].$$

(b) Show that this incorrect form of the Lax–Friedrichs scheme is such that odd-indexed values of the solution at the new time depend only on even-indexed values of the solution at the old time.

(c) Program this scheme and the correct Lax–Friedrichs scheme for linear advection, and compare the results.

3.4.2 Program Lax-Friedrichs, Godunov and Rusanov for Burgers' equation. Compare the numerical results for the following Riemann problems:

(a) a shock with $u_L = 2$, $u_R = 0$ on $x \in (-0.1, 1.0)$ and $0 < t \leq 0.9$

(b) a shock with $u_L = 0$, $u_R = -2$ on $x \in (-1.0, 0.1)$ and $0 < t \leq 0.9$

(c) a rarefaction with $u_L = 1$, $u_R = 2$ on $x \in (-0.2, 2.0)$ and $0 < t \leq 0.9$

(d) a transonic rarefaction, with $u_L = -1$, $u_R = 1$ on $x \in (-1.0, 1.0)$ and $0 < t \leq 0.9$.

When comparing the schemes, do the following steps:

(a) use 100 grid cells in each calculation;

(b) place the initial discontinuity in the interior of your grid at $x = 0$;

(c) run each scheme for CFL numbers 0.9 and 0.5;

(d) plot the solution at the final time, but scale the horizontal axis by dividing the spatial coordinates by the final time. (This is what is meant by "plotting the solution versus x/t".)

3.4.3 **Marquina's flux formula** [45] for scalar laws involves using the Rusanov scheme to approximate the flux in transonic Riemann problems, and Godunov's scheme otherwise. In other words, Marquina's flux is given by

$$f_{i+1/2}^{n+1/2} = \begin{cases} f(u_L), & f'(u) > 0 \text{ for all } u \in \text{int}[u_L, u_R] \\ f(u_R), & f'(u) < 0 \text{ for all } u \in \text{int}[u_L, u_R] \\ \frac{1}{2}[f(u_L) + f(u_R) - (u_R - u_L)\max_{u \in int[u_L, u_r]} |f'(u)|], & \text{otherwise.} \end{cases}$$

Test this scheme on the computations in the previous problem.

3.4.4 Program Lax–Friedrich, Godunov and Rusanov for the traffic flow problem with density given by (3.19). Compare the numerical results (see the previous problem for the instructions in the comparison) for the following Riemann problems:

(a) a shock with $\rho_R = \rho_{\max}/e$, $\rho_L = 0$, $a = 1$ on $x \in (-0.1, 1.0)$ and $0 < t \leq 0.9$

(b) a shock with $\rho_R = \rho_{\max}$, $\rho_L = \rho_{\max}/e$, $a = e - 1$ on $x \in (-1.0, 0.1)$ and $0 < t \leq 0.9$

(c) a rarefaction with $\rho_L = \rho_{\max}/e^2$, $\rho_L = \rho_{\max}/e^3$, $a = 1$ on $x \in (-0.1, 2.0)$ and $0 < t \leq 0.9$

(d) a transonic rarefaction, with $\rho_L = \rho_{\max}$, $\rho_L = \rho_{\max}/e^2$, $a = 1$ on $x \in (-1.0, 1.0)$ and $0 < t \leq 0.9$

In addition, plot the characteristic speeds versus x/t for each of these Riemann problems.

3.4.5 Program Lax–Friedrichs, Godunov and Rusanov for the miscible displacement problem with tracer concentration given by (3.21). In order to discretize the diffusive term, use explicit centered differences:

$$\frac{\partial}{\partial x}\left(D\frac{\partial c}{\partial x}\right) \approx \frac{1}{\Delta x^2}\left[D_{i+1/2}(c_{i+1}^n - c_i^n) - D_{i-1/2}(c_i^n - c_{i-1}^n)\right].$$

(a) Assuming that the velocity v, longitudinal mixing length α_ℓ, porosity ϕ, diffusivity δ_c and tortuosity τ are all constant, rewrite the miscible displacement problem as a convection-diffusion problem and determine a formula for the cell **Peclet number**.

(b) Perform a Fourier analysis of explicit upwind differences for the miscible displacement problem to determine conditions on the timestep so that the scheme is diffusive.

(c) Write the upwind difference scheme for the miscible displacement problem with explicit centered differences for the diffusive terms as a conservative difference scheme.

(d) Program the upwind difference scheme for miscible displacement. Experiment with values of the Peclet number to determine when the numerical diffusion dominates the physical diffusion.

3.4.6 Program Lax–Friedrichs, Godunov and Rusanov for the Buckley–Leverett problem.

3.4.7 We can plot approximate trajectories for numerical solutions of conservation laws using a contour plotter. Suppose that we compute the numerical solution of the conservation law

$$\frac{\partial u}{\partial t} + \frac{\partial f(u)}{\partial x} = 0$$

and store the numerical solution at all grid cell centers x_i and all timesteps t^n. This gives us a 2D lattice of numerical values u_i^n. Then we use a contour plotter to plot lines of constant values of u.

(a) Explain why this contour plot would approximately produce the trajectories of $u(x, t)$.

(b) Use this approach to plot the trajectories of the solution of Burgers' equation with initial data

$$u_0(x) = \begin{cases} -0.5, & x < 0.5 \\ 0.2 + 0.7\cos(2\pi x), & x > 0.5 \end{cases}$$

The difficulty with this approach for plotting trajectories is that it does not plot trajectories in regions where the solution is constant.

3.5 Lax–Wendroff Process

Modified equation analyses of any of the previous schemes we have studied indicate that the error term is first-order in either Δt or Δx. In this section, we will see how we can use the modified equation analysis to construct a second-order scheme.

Consider the explicit centered difference scheme on a uniform grid

$$u_i^{n+1} = u_i^n - \frac{\triangle t^{n+1/2}}{2\triangle x}\left[f\left(u_{i+1}^n\right) - f\left(u_{i-1}^n\right)\right].$$

This is a conservative difference scheme in which the numerical fluxes are chosen to be

$$f_{i+1/2}^{n+1/2} = \frac{1}{2}\left[f\left(u_{i+1}^n\right) + f\left(u_i^n\right)\right].$$

As we have seen, the modified equation analysis of the explicit centered difference method for linear advection shows that the error term is $O(\triangle x^2) + O(\triangle t)$. We have also seen that the scheme is unconditionally unstable for linear advection.

Nevertheless, let us carry out a modified equation analysis for the general nonlinear scalar conservation law. We can see that

$$u_i^{n+1} \approx u_i^n + \frac{\partial \tilde{u}}{\partial t}\triangle t + \frac{1}{2}\frac{\partial^2 \tilde{u}}{\partial t^2}\triangle t^2,$$

and

$$f\left(u_{i\pm 1}^n\right) \approx f\left(u_i^n\right) \pm \frac{\partial f}{\partial x}\triangle x + \frac{1}{2}\frac{\partial^2 f}{\partial x^2}\triangle x^2 \pm \frac{1}{6}\frac{\partial^3 f}{\partial x^3}\triangle x^3.$$

We assume that the modified equation is

$$\frac{\partial \tilde{u}}{\partial t} + \frac{\partial f(\tilde{u})}{\partial x} = e.$$

This implies that

$$\begin{aligned}
\frac{\partial^2 \tilde{u}}{\partial t^2} &= \frac{\partial e}{\partial t} - \frac{\partial}{\partial t}\left(\frac{\partial f(\tilde{u})}{\partial x}\right) = \frac{\partial e}{\partial t} - \frac{\partial}{\partial x}\left(\frac{df(\tilde{u})}{d\tilde{u}}\frac{\partial \tilde{u}}{\partial t}\right) \\
&= \frac{\partial e}{\partial t} - \frac{\partial}{\partial x}\left(\frac{\partial f(\tilde{u})}{\partial \tilde{u}}e\right) + \frac{\partial}{\partial x}\left(\frac{df(\tilde{u})}{d\tilde{u}}\frac{\partial f(\tilde{u})}{\partial x}\right).
\end{aligned}$$

Thus the difference equation leads to

$$\begin{aligned}
&\frac{u_i^{n+1} - u_i^n}{\triangle t} + \frac{f\left(u_{i+1}^n\right) - f\left(u_{i-1}^n\right)}{2\triangle x} \\
&\approx \frac{\partial \tilde{u}}{\partial t} + \frac{1}{2}\frac{\partial^2 \tilde{u}}{\partial t^2}\triangle t + \frac{\partial f(\tilde{u})}{\partial x} + \frac{1}{6}\frac{\partial^3 f}{\partial x^3}\triangle x^2 \\
&= e + \frac{1}{6}\frac{\partial^3 f}{\partial x^3}\triangle x^2 + \left\{\frac{\partial e}{\partial t} - \frac{\partial}{\partial x}\left(\frac{df(\tilde{u})}{d\tilde{u}}e\right) + \frac{\partial}{\partial x}\left(\frac{df(\tilde{u})}{d\tilde{u}}\frac{\partial f(\tilde{u})}{\partial x}\right)\right\}\frac{\triangle t}{2}.
\end{aligned}$$

Since the error term e in the modified equation analysis is $O(\triangle t)$, the dominant term of order $\triangle t$ is $\frac{\partial}{\partial x}(\frac{df(\tilde{u})}{d\tilde{u}}\frac{\partial f(\tilde{u})}{\partial x})\frac{\triangle t}{2}$. We will approximate this term by finite differences in order to obtain a higher-order method, called the **Lax–Wendroff method**

$$0 = \frac{u_i^{n+1} - u_i^n}{\triangle t^{n+1/2}} + \frac{f\left(u_{i+1}^n\right) - f\left(u_{i-1}^n\right)}{2\triangle x}$$
$$- \frac{\triangle t^{n+1/2}}{\triangle x}\left[\left(\frac{df}{du}\right)_{i+1/2}^n \frac{f\left(u_{i+1}^n\right) - f\left(u_i^n\right)}{2\triangle x} - \left(\frac{df}{du}\right)_{i-1/2}^n \frac{f\left(u_i^n\right) - f\left(u_{i-1}^n\right)}{2\triangle x}\right].$$

On a non-uniform mesh, we should take the Lax-Wendroff flux to be

$$f_{i+1/2}^{n+1/2} = \frac{f\left(u_i^n\right)\triangle x_{i+1} + f\left(u_{i+1}^n\right)\triangle x_i - \triangle t^{n+1/2}\frac{df}{du}\left[f\left(u_{i+1}^n\right) - f\left(u_i^n\right)\right]}{\triangle x_i + \triangle x_{i+1}}. \tag{3.28}$$

In this formula, we can evaluate the partial derivative of f as

$$\left(\frac{df}{du}\right)_{i+1/2}^n \equiv \frac{df}{du}\bigg|_{\frac{1}{2}(u_{i+1}^n + u_i^n)}.$$

In other words, the Lax–Wendroff scheme is a conservative difference scheme with numerical flux

$$f_{i+1/2}^{n+1/2} = \frac{1}{2}\left[f\left(u_{i+1}^n\right) + f\left(u_i^n\right)\right] - \left(\frac{df}{du}\right)_{i+1/2}^n \frac{f\left(u_{i+1}^n\right) - f\left(u_i^n\right)}{2\triangle x}\triangle t^{n+1/2}. \tag{3.29}$$

It turns out that the resulting scheme is second-order in both space and time, and linearly stable for

$$\left|\frac{df}{du}\right|\triangle t \le \triangle x.$$

However, this scheme does not work well for general nonlinear problems, as the next example shows.

Example 3.5.1 The Lax–Wendroff scheme computes the wrong solution for transonic rarefactions. For example, if we consider the Riemann problem initial data for Burgers' equation with initial data

$$u(x,0) = \begin{cases} -1, & x < 0 \\ 1, & x < 0 \end{cases}$$

then the Lax–Wendroff fluxes in (3.29) will be $f_{i+1/2}^{n+1/2} = \frac{1}{2}$ at all cell sides. The conservative difference (2.14) will produce no change in the solution for all timesteps.

It is more common to implement the Lax–Wendroff scheme in two steps. A classical approach is the **Richtmyer two-step Lax–Wendroff scheme**

$$u_{i+1/2}^{n+1/2} = \frac{u_{i+1}^n\triangle x_i + u_i^n\triangle x_{i+1} - \triangle t^{n+1/2}\left[f\left(u_{i+1}^n\right) - f\left(u_i^n\right)\right]}{\triangle x_i + \triangle x_{i+1}} \tag{3.30a}$$

$$u_i^{n+1} = u_i^n - \frac{\triangle t^{n+1/2}}{\triangle x_i}\left[f\left(u_{i+1/2}^{n+1/2}\right) - f\left(u_{i-1/2}^{n+1/2}\right)\right]. \tag{3.30b}$$

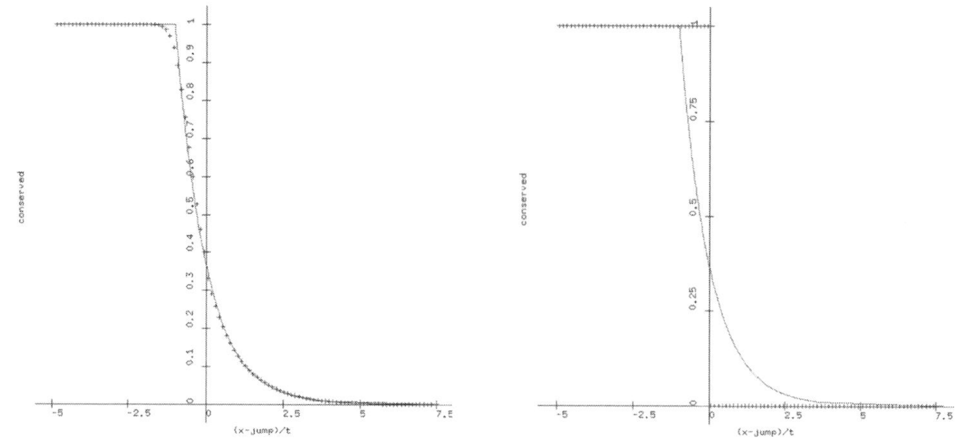

(a) Lax–Wendroff (b) MacCormack

Fig. 3.14 Second-order schemes for traffic flow rarefaction: CFL = 0.9

The first step is similar to the Lax–Friedrich scheme, and the second step is a conservative difference.

Figure 3.14 shows some numerical results with the Richtmyer two-step Lax–Wendroff scheme for a rarefaction with the traffic flow problem. Note that the Lax–Wendroff results are significantly more accurate than the first-order schemes in Figure 3.9. A program to implement the Lax–Wendroff scheme can be found in **Program 3.5-39: Schemes.C** Students can also execute the Lax–Wendroff scheme by clicking on the web link **Executable 3.5-10: guiconvex2** and selecting `scheme` to be `lax wendroff` in the `View` pulldown menu, inside the `Numerical Method Parameters` group.

3.6 Other Second Order Schemes

There are several other classical second-order methods for hyperbolic equations. A popular scheme is the **MacCormack scheme** [114]:

$$\tilde{u}_i^{n+1} = u_i^n - \frac{\triangle t^{n+1/2}}{\triangle x_i}\left[f\left(u_{i+1}^n\right) - f\left(u_i^n\right)\right]; \tag{3.31a}$$

$$\tilde{\tilde{u}}_i^{n+2} = \tilde{u}_i^{n+1} - \frac{\triangle t^{n+1/2}}{\triangle x_i}\left[f\left(\tilde{u}_i^{n+1}\right) - f\left(\tilde{u}_{i-1}^{n+1}\right)\right]); \tag{3.31b}$$

$$u_i^{n+1} = \frac{1}{2}\left[u_i^n + \tilde{\tilde{u}}_i^{n+2}\right]. \tag{3.31c}$$

An alternative form of this scheme uses backward differencing in the predictor, and forward differencing the the corrector. It is also common to alternate the forward and backward differencing between successive steps of this scheme. Figure 3.14 shows some numerical results with the MacCormack scheme for a rarefaction with the traffic flow problem. Note that the MacCormack computes the wrong solution to this problem. (In classical applications of these methods, artificial diffusion was added to reduce the likelihood of such behavior.) A program to implement the MacCormack scheme can be found in **Program 3.6-40: Schemes.C** Students can also execute the MacCormack scheme by clicking on the web link **Executable 3.6-11: guiconvex2** and selecting `scheme` to be `mac cormack` in the `View` pulldown menu, inside the `Numerical Method Parameters` group.

There are some other second-order schemes that differ in their choice of higher-order approximations to $f_{i+1/2}^{n+1/2}$ and the related second-order time correction. The **Beam–Warming scheme** [173]: uses the solution values u_j^n for $j = i-2, i-1, i$ to construct a second-order approximation to the flux at $x_{i+1/2}$. This scheme needs to be properly upwinded and combined with artificial diffusion to work properly. We will discuss this scheme for linear advection in Section 5.5.2.

The **Fromm scheme** [51] is basically the average of the Lax–Wendroff and Beam–Warming methods. This scheme also needs to be properly upwinded and combined with artificial diffusion to work properly. We will discuss this scheme for linear advection in Section 5.5.2.

The **leap-frog scheme** uses data from two previous timesteps to compute

$$u_i^{n+1} = u_i^{n-1} - \frac{\triangle t^{n+1/2} + \triangle t^{n-1/2}}{2\triangle x_i}\left[f\left(u_{i+1}^n\right) - f\left(u_{i-1}^n\right)\right].$$

This scheme is neutrally stable for linear advection problems [72, p. 313]. It works well for linear advection problems with smooth initial data, but does not work well for nonlinear problems.

Students can perform their own comparisons of these schemes for convex conservation laws by clicking on the web links **Executable 3.6-12: guiconvex2 Executable 3.6-13: guiconvexerror2, Executable 3.6-14: guiriemann2,** and **Executable 3.6-15: guierror2.** The main programs for these executables are **Program 3.6-41: GUIConvexRiemannProblem2.C, Program 3.6-42: GUIConvexErrorAnalysis2.C Program 3.6-43: GUIRiemannProblem2.C** and **Program 3.6-44: GUIErrorAnalysis2.C.**

Exercises for 3.6

3.6.1 Perform Fourier analyses for the Lax–Wendroff, MacCormack, and leap-frog applied to linear advection. Plot the total dissipation and dispersion for each. Show that the leap-frog scheme has no dissipation.

3.6.2 Perform modified equation analyses for the Lax–Wendroff, MacCormack, and leap-frog applied to linear advection. Determine the dominant error terms in each.

3.6.3 Program Lax–Wendroff, MacCormack, and leap-frog for linear advection. Test the schemes for different values of the CFL number, and compare to the results of the Fourier analysis. For example, compare the location of the oscillations in the Lax–Wendroff scheme to the dispersion results from the Fourier analysis.

3.6.4 Program the Lax–Wendroff, MacCormack, and leap-frog schemes for Burgers' equation. Compare the results to Godunov's scheme for a Riemann problem with shock moving at speed 1, and for a stationary shock.

3.6.5 Program the Lax–Wendroff, MacCormack, and leap-frog schemes for the traffic flow problem with logarithmic density. Compare the results to Godunov's scheme for a Riemann problem with shock moving at speed 1, and for a stationary shock. (Note that numerical oscillations from these schemes could cause trouble for the logarithmic density in the traffic flow problem.)

3.6.6 Develop a second-order treatment of non-reflecting boundary conditions for the Lax–Wendroff method.

4

Nonlinear Hyperbolic Systems

Many important problems in mathematical physics involve systems of conservation laws. For example, gas dynamics involves equations for conservation of mass, momentum and energy. In order to design effective numerical methods for the solutions of conservation laws, we will need to examine the mathematical features of the equations of motion.

4.1 Theory of Hyperbolic Systems

There are several important theoretical issues that we need to understand in order to develop effective numerical methods for solving nonlinear hyperbolic systems of conservation laws. In order to assist the student in understanding these issues, we will present applications of these issues to the shallow water equations throughout the discussion of the general theory. Later, we will present case studies of other important physical systems, and discuss all of the theory with regard to each single application.

4.1.1 Hyperbolicity and Characteristics

Systems of conservation laws take the form

$$\frac{d}{dt}\int_{\Omega}\mathbf{u}(\mathbf{x},t)\,d\mathbf{x}+\int_{\partial\Omega}\mathbf{F}(\mathbf{u}(\mathbf{x},t))\mathbf{n}\,ds=\int_{\Omega}\mathbf{b}(\mathbf{x},t)\,d\mathbf{x}, \tag{4.1}$$

where $\Omega \subset \mathbf{R}^k$, $\mathbf{u} \in \mathbf{R}^m$ is the vector of conserved quantities, $\mathbf{F} \in \mathbf{R}^{m\times k}$ is the array of fluxes of the conserved quantities, $\mathbf{b} \in \mathbf{R}^m$ is the vector of body forces and $\mathbf{n}$ is the outward normal on $\partial\Omega$.

Example 4.1.1 In a body of water with zero slope angle and zero friction coefficient, the equations for the motion of shallow water are [175, page 84]

$$\frac{d}{dt}\int_{\Omega}\begin{bmatrix} h \\ \mathbf{v}h \end{bmatrix} d\mathbf{x} + \int_{\partial\Omega}\begin{bmatrix} h\mathbf{v}\cdot\mathbf{n} \\ \mathbf{v}h\mathbf{v}\cdot\mathbf{n}+\mathbf{n}\frac{1}{2}gh^2 \end{bmatrix} ds = 0. \tag{4.2}$$

Here h is the height of the water, $\mathbf{v}$ is the velocity, g is the vertical component of the acceleration due to gravity.

The following definition is important to the discussion of systems of conservation laws.

Definition 4.1.2 The system of conservation laws (4.1) is **hyperbolic** if and only if for any $\mathbf{u}$ and for any fixed unit vector $\mathbf{n}$ the matrix $\partial\mathbf{F}(\mathbf{u})\mathbf{n}/\partial\mathbf{u}$ has real eigenvalues. In other words, (4.1) is hyperbolic if and only if for each $\mathbf{u}$ and $\mathbf{n}$ there is a nonsingular matrix $\mathbf{X}_{(\mathbf{n})}$ and Jordan canonical form $\Lambda_{(\mathbf{n})}$ with real diagonal entries so that

$$\frac{\partial\mathbf{F}\mathbf{n}}{\partial\mathbf{u}}\mathbf{X}_{(\mathbf{n})} = \mathbf{X}_{(\mathbf{n})}\Lambda_{(\mathbf{n})}.$$

The diagonal entries of $\Lambda_{(\mathbf{n})}$ are called the **characteristic speeds**. If the characteristic speeds are distinct, then the system is called **strictly hyperbolic**.

In order to simplify the notation in the definition of characteristic speeds, we will typically suppress the dependence of $\mathbf{X}$ and Λ on $\mathbf{n}$, and write

$$\frac{\partial\mathbf{F}\mathbf{n}}{\partial\mathbf{u}}\mathbf{X} = \mathbf{X}\Lambda.$$

In regions of smooth flow, the system of conservation laws can be written as a system of partial differential equations:

$$\frac{\partial\mathbf{u}}{\partial t} + \sum_{i=1}^{k}\frac{\partial\mathbf{F}\mathbf{e}_i}{\partial\mathbf{x}_i} = \mathbf{b}(\mathbf{x},t). \tag{4.3}$$

In many cases, it is useful to consider both $\mathbf{u}$ and $\mathbf{F}$ as functions of another set of variables $\mathbf{w}$, which we will call the "flux variables." The **quasilinear form** of the conservation law in smooth flow will then be written

$$\frac{\partial\mathbf{u}}{\partial\mathbf{w}}\frac{\partial\mathbf{w}}{\partial t} + \sum_{i=1}^{k}\frac{\partial\mathbf{F}\mathbf{e}_i}{\partial\mathbf{w}}\frac{\partial\mathbf{w}}{\partial\mathbf{x}_i} = \mathbf{b}(\mathbf{x},t). \tag{4.4}$$

Lemma 4.1.3 *Suppose that the conservation law* (4.1) *has a continuously differentiable solution* $\mathbf{u}$ *at some point* $(\mathbf{x}, t)$. *Suppose that* $\mathbf{w}$ *is a vector of flux variables for this equation; in other words,* $\frac{\partial\mathbf{u}}{\partial\mathbf{w}}$ *is nonsingular. Finally, suppose that we can* *find a nonsingular matrix* $\mathbf{Y}$ *and a Jordan canonical form* Λ *so that*

$$\left(\frac{\partial \mathbf{u}}{\partial \mathbf{w}}\right)^{-1} \frac{\partial \mathbf{Fn}}{\partial \mathbf{w}} \mathbf{Y} = \mathbf{Y}\Lambda. \tag{4.5}$$

Then the diagonal entries of Λ *are the characteristic speeds for* (4.1), *and the columns of* $\mathbf{X} = \frac{\partial \mathbf{u}}{\partial \mathbf{w}} \mathbf{Y}$ *are the characteristic directions.*

Proof Since

$$\frac{\partial \mathbf{Fn}}{\partial \mathbf{u}} = \frac{\partial \mathbf{Fn}}{\partial \mathbf{w}} \left(\frac{\partial \mathbf{u}}{\partial \mathbf{w}}\right)^{-1}$$

it follows that

$$\frac{\partial \mathbf{Fn}}{\partial \mathbf{u}} \mathbf{X} \equiv \frac{\partial \mathbf{Fn}}{\partial \mathbf{u}} \left(\frac{\partial \mathbf{u}}{\partial \mathbf{w}} \mathbf{Y}\right) = \left(\frac{\partial \mathbf{u}}{\partial \mathbf{w}} \mathbf{Y}\right) \Lambda \equiv \mathbf{X}\Lambda.$$

□

Example 4.1.4 For the shallow water equations (4.2), it is natural to take the array of flux variables to be

$$\mathbf{w} = \begin{bmatrix} h \\ \mathbf{v} \end{bmatrix}.$$

Then the array of derivatives of the conserved quantities is

$$\frac{\partial \mathbf{u}}{\partial \mathbf{w}} = \begin{bmatrix} 1 & 0 \\ \mathbf{v} & \mathbf{I}h \end{bmatrix}$$

and the array of flux derivatives is

$$\frac{\partial \mathbf{Fn}}{\partial \mathbf{w}} = \begin{bmatrix} \mathbf{v} \cdot \mathbf{n} & h\mathbf{n}^\top \\ \mathbf{v}\mathbf{v} \cdot \mathbf{n} + \mathbf{n}gh & \mathbf{I}h\mathbf{v} \cdot \mathbf{n} + \mathbf{v}h\mathbf{n}^\top \end{bmatrix}.$$

In order to determine if the shallow water conservation laws are hyperbolic, Definition 4.1.2 requires that we find the eigenvalues of

$$\left(\frac{\partial \mathbf{u}}{\partial \mathbf{w}}\right)^{-1} \frac{\partial \mathbf{Fn}}{\partial \mathbf{w}} = \begin{bmatrix} \mathbf{v} \cdot \mathbf{n} & h\mathbf{n}^\top \\ \mathbf{n}g & \mathbf{I}\mathbf{v} \cdot \mathbf{n} \end{bmatrix}.$$

Note that $\frac{\partial \mathbf{u}}{\partial \mathbf{w}}$ is singular at $h = 0$, so we restrict our discussion to problems in which $h > 0$. Let us define the speed of sound by $c = \sqrt{gh}$. Then

$$\left\{\left(\frac{\partial \mathbf{u}}{\partial \mathbf{w}}\right)^{-1} \frac{\partial \mathbf{Fn}}{\partial \mathbf{w}}\right\} \mathbf{Y} \equiv \begin{bmatrix} \mathbf{v} \cdot \mathbf{n} & h\mathbf{n}^\top \\ \mathbf{n}g & \mathbf{I}\mathbf{v} \cdot \mathbf{n} \end{bmatrix} \begin{bmatrix} -h/c & 0 & h/c \\ \mathbf{n} & \mathbf{N} & \mathbf{n} \end{bmatrix}$$

$$= \begin{bmatrix} -h/c & 0 & h/c \\ \mathbf{n} & \mathbf{N} & \mathbf{n} \end{bmatrix} \begin{bmatrix} \mathbf{v} \cdot \mathbf{n} - c & 0 & 0 \\ 0 & \mathbf{I}\mathbf{v} \cdot \mathbf{n} & 0 \\ 0 & 0 & \mathbf{v} \cdot \mathbf{n} + c \end{bmatrix} \equiv \mathbf{Y}\Lambda.$$

Here $\begin{bmatrix}\mathbf{n} & \mathbf{N}\end{bmatrix}$ is an orthogonal matrix with first column equal to $\mathbf{n}$. Thus the system is hyperbolic. The characteristic speeds are $\mathbf{v}\cdot\mathbf{n} \pm c$, and $\mathbf{v}\cdot\mathbf{n}$ in multiple dimensions $k > 1$. The characteristic directions are the columns of

$$\mathbf{X} = \frac{\partial \mathbf{u}}{\partial \mathbf{w}}\mathbf{Y} = \begin{bmatrix} 1 & 0 \\ \mathbf{v} & \mathbf{I}h \end{bmatrix} \begin{bmatrix} -h/c & 0 & h/c \\ \mathbf{n} & \mathbf{N} & \mathbf{n} \end{bmatrix} = \begin{bmatrix} -h/c & 0 & h/c \\ (\mathbf{n} - \mathbf{v}/c)h & \mathbf{N}h & (\mathbf{n} + \mathbf{v}/c)h \end{bmatrix}.$$

In many applied problems, the characteristic speeds are distinct, so the problem has a full set of characteristic directions. Shallow water in one dimension is an example of such a system. Many physical problems have a full set of characteristic directions even if the characteristic speeds are not distinct. Shallow water in multiple dimensions is an example. We will also see interesting problems in which some characteristic speeds are equal along special degenerate curves (as in the vibrating string of Section 4.8. below). In other cases (for example, polymer flooding in Section 4.10), Λ is not always diagonal.

The following definitions will be useful.

Definition 4.1.5 Given a unit vector $\mathbf{n}$, an eigenvalue $\lambda_i = \mathbf{e}_i^\top \Lambda \mathbf{e}_i$ of $\frac{\partial \mathbf{F}\mathbf{n}}{\partial \mathbf{u}}$ is **genuinely nonlinear** if and only if

$$\text{for all } \mathbf{u} \ \frac{\partial \lambda_i}{\partial \mathbf{u}} \mathbf{X}\mathbf{e}_i \neq 0.$$

On the other hand, λ_i is **linearly degenerate** if and only if

$$\text{for all } \mathbf{u} \ \frac{\partial \lambda_i}{\partial \mathbf{u}} \mathbf{X}\mathbf{e}_i = 0.$$

Linearly degenerate waves are sometimes called contact discontinuities.

When flux variables are used, a somewhat easier test for genuine nonlinearity is $\frac{\partial \lambda_i}{\partial \mathbf{w}} \mathbf{Y}\mathbf{e}_i \neq 0$, where $\mathbf{Y}$ is given by (4.5).

Example 4.1.6 To see that the shallow water characteristic speeds $\mathbf{v}\cdot\mathbf{n} \pm \sqrt{gh}$ are genuinely nonlinear, we compute

$$\frac{\partial \lambda_i}{\partial \mathbf{w}} \mathbf{Y}\mathbf{e}_i = \frac{\partial(\mathbf{v}\cdot\mathbf{n} \pm \sqrt{gh})}{\partial \mathbf{w}} \begin{bmatrix} \pm\sqrt{h/g} \\ \mathbf{n} \end{bmatrix} = \begin{bmatrix} \pm\frac{1}{2}\sqrt{g/h} & \mathbf{n}^\top \end{bmatrix} \begin{bmatrix} \pm\sqrt{h/g} \\ \mathbf{n} \end{bmatrix} = \frac{3}{2}.$$

To see that $\mathbf{v}\cdot\mathbf{n}$ is linearly degenerate in multiple dimensions, we compute

$$\frac{\partial \mathbf{v}\cdot\mathbf{n}}{\partial \mathbf{w}} \begin{bmatrix} 0 \\ \mathbf{N} \end{bmatrix} = \begin{bmatrix} 0 & \mathbf{n}^\top \end{bmatrix} \begin{bmatrix} 0 \\ \mathbf{N} \end{bmatrix} = 0.$$

4.1.2 Linear Systems

Some interesting physical problems, such as Maxwell's equations, are linear with constant coefficients. It is easy to solve such problems with continuously differentiable initial data in one dimension, as the next lemma shows.

Lemma 4.1.7 *Consider the linear hyperbolic system for* $\mathbf{u} \in \mathbf{R}^m$ *in a single spatial variable* x*, namely*

$$\frac{\partial \mathbf{u}}{\partial t} + \mathbf{A}\frac{\partial \mathbf{u}}{\partial x} = 0,$$
$$\mathbf{u}(x, 0) = \mathbf{u}_0(x),$$

where $\mathbf{A}$ *is some fixed matrix. We assume that* $\mathbf{A}$ *is diagonalizable; in other words,*

$$\mathbf{A}\mathbf{X} = \mathbf{X}\Lambda$$

where $\mathbf{X} \in \mathbf{R}^{m\times m}$ *is nonsingular and* $\Lambda \in \mathbf{R}^{m\times m}$ *is diagonal. Then the solution to this initial value problem is*

$$\mathbf{u}(x, t) = \sum_j \mathbf{X}\mathbf{e}_j\mathbf{e}_j^\top \mathbf{X}^{-1}\mathbf{u}_0(x - \lambda_j t).$$

Proof Define the characteristic expansion coefficients to be $\mathbf{c}(x, t) = \mathbf{X}^{-1}\mathbf{u}(x, t)$. Then we can rewrite the conservation law in the form

$$\frac{\partial \mathbf{c}}{\partial t} + \Lambda\frac{\partial \mathbf{c}}{\partial x} = 0,$$
$$\mathbf{c}(x, 0) = \mathbf{X}^{-1}\mathbf{u}_0(x).$$

In this equation, we have decoupled the system of m conservation laws into m separate conservation laws for the components of $\mathbf{c}$:

$$\frac{\partial \mathbf{c}_j}{\partial t} + \lambda_j\frac{\partial \mathbf{c}_j}{\partial x} = 0,$$
$$\mathbf{c}_j(x, 0) = \mathbf{e}_j^\top \mathbf{X}^{-1}\mathbf{u}_0(x).$$

We can solve for the individual components to get $\mathbf{c}_j(x, t) = \mathbf{e}_j^\top \mathbf{X}^{-1}\mathbf{u}_0(x - \lambda_j t)$. Then we can recombine these values to write the solution of the original equation in the form

$$\mathbf{u}(x, t) = \mathbf{X}\mathbf{c}(x, t) = \sum_j \mathbf{X}\mathbf{e}_j\mathbf{c}_j(x, t) = \sum_j \mathbf{X}\mathbf{e}_j\mathbf{e}_j^\top \mathbf{X}^{-1}\mathbf{u}_0(x - \lambda_j t).$$

□

Just for fun, we will examine the solution of a linear hyperbolic system of two conservation laws involving a nontrivial Jordan canonical form.

Lemma 4.1.8 *Suppose that*

$$\mathbf{J} = \begin{bmatrix} \lambda & 1 \\ 0 & \lambda \end{bmatrix}.$$

is a nontrivial Jordan block, and that $\mathbf{u}_0(x)$ *is continuously differentiable. Then the solution of the linear hyperbolic system in one spatial variable*

$$\frac{\partial \mathbf{u}}{\partial t} + \mathbf{J}\frac{\partial \mathbf{u}}{\partial x} = 0, \quad \mathbf{u}(x,0) = \mathbf{u}_0(x),$$

is

$$\mathbf{u}(x,t) = \begin{bmatrix} \mathbf{e}_1^\top \mathbf{u}_0(x-\lambda t) - t\, \mathbf{e}_2^\top \mathbf{u}_0'(x-\lambda t) \\ \mathbf{e}_2^\top \mathbf{u}_0(x-\lambda t) \end{bmatrix}.$$

Proof If we transform to characteristic coordinates $\xi = x - \lambda t$ and $\tau = t$, then $\tilde{\mathbf{u}}(\xi,\tau) \equiv \mathbf{u}(x,t)$ satisfies

$$\frac{\partial \tilde{\mathbf{u}}}{\partial \tau} + \begin{bmatrix} 0 & 1 \\ 0 & 0 \end{bmatrix} \frac{\partial \tilde{\mathbf{u}}}{\partial \xi} = 0.$$

It follows that $\mathbf{e}_2 \cdot \tilde{\mathbf{u}}(\xi,\tau) = \mathbf{e}_2 \cdot \mathbf{u}_0(\xi)$, and that $\mathbf{e}_1 \cdot \tilde{\mathbf{u}}(\xi,\tau) = \mathbf{e}_1 \cdot \mathbf{u}_0(\xi) - \tau \mathbf{e}_2 \cdot \mathbf{u}_0'(\xi)$. After transforming back to the original coordinates x and t, we obtain the claimed result. □

It is straightforward to generalize this solution to larger Jordan blocks. Note that the solution of this conservation law can involve polynomial growth in time, if the initial data is continuously differentiable; if the initial data is not continuously differentiable, then the solution can blow up immediately.

On the other hand, it is more difficult to extend the results of either of these two lemmas to multiple dimensions. A principal difficulty in this extension is the complication due to wave propagation on characteristic cones; see [54] for more details. The importance of characteristic cones for the multidimensional wave equation is well-known; see [153] for more details.

4.1.3 Frames of Reference

In developing the conservation laws for physical various models, it will be useful to view the fluid in various frames of reference. The **Lagrangian frame** views the problem in terms of the original position of the material particles. On the other hand, the **Eulerian frame** views the fluid in terms of the current position of the material particles. The material particles move in the Eulerian frame of reference, but are fixed in the Lagrangian frame.

Following the notation in Fung's solid mechanics book [52], we will represent the Lagrangian coordinates by $\mathbf{a} \in \Omega_0$, and the Eulerian coordinates by $\mathbf{x} \in \Omega_t$. Thus the current position of a particle originally at position $\mathbf{a}$ is $\mathbf{x}(\mathbf{a}, t)$. More detailed discussion of frames of references can be found in Truesdell's book [167], but the notation involves the use of many fonts for the letter "x."

4.1.3.1 Useful Identities

Because conservation laws in multiple dimensions involve partial derivatives of arrays and possible changes in frame of reference, it will be useful to provide some identities from calculus and linear algebra.

If $\mathbf{b}, \mathbf{c} \in \mathbf{R}^m$ then

$$\mathbf{b}^\top \mathbf{c} = \text{tr}\,(\mathbf{b}\mathbf{c}^\top), \tag{4.6}$$

where "tr" is the trace of a square matrix (the sum of the diagonal entries). Similarly, if $\mathbf{y} \in \mathbf{R}^k$ and $\mathbf{w}(\mathbf{y}) \in \mathbf{R}^k$, then it is easy to see that

$$\nabla_\mathbf{y} \cdot \mathbf{w} = \text{tr}\,(\nabla_\mathbf{y} \mathbf{w}^\top) = \text{tr}\left(\frac{\partial \mathbf{w}}{\partial \mathbf{y}}\right). \tag{4.7}$$

There are several identities based on the product rule for differentiation. If $\mathbf{y} \in \mathbf{R}^k$, $\mathbf{b}(\mathbf{y}) \in \mathbf{R}^k$ and $\mathbf{c}(\mathbf{y}) \in \mathbf{R}^m$ then

$$\nabla_\mathbf{y} \cdot (\mathbf{b}\mathbf{c}^\top) = (\nabla_\mathbf{y} \cdot \mathbf{b})\mathbf{c}^\top + \mathbf{b}^\top(\nabla_\mathbf{y}\mathbf{c}^\top) = (\nabla_\mathbf{y} \cdot \mathbf{b})\mathbf{c}^\top + \mathbf{b}^\top \left(\frac{\partial \mathbf{c}}{\partial \mathbf{y}}\right)^\top. \tag{4.8}$$

Similarly, if $\mathbf{y} \in \mathbf{R}^k$, $\mathbf{B}(\mathbf{y}) \in \mathbf{R}^{k\times m}$ and $\mathbf{c}(\mathbf{y}) \in \mathbf{R}^m$ then

$$\nabla_\mathbf{y} \cdot (\mathbf{B}\mathbf{c}) = (\nabla_\mathbf{y}^\top \mathbf{B})\mathbf{c} + (\mathbf{B}^\top \nabla_\mathbf{y})^\top \mathbf{c} = (\nabla_\mathbf{y}^\top \mathbf{B})\mathbf{c} + \text{tr}\left(\frac{\partial \mathbf{c}}{\partial \mathbf{y}}\mathbf{B}\right). \tag{4.9}$$

If $t \in \mathbf{R}$ and $\mathbf{F}(t) \in \mathbf{R}^{m\times m}$ is invertible, then $\mathbf{F}\mathbf{F}^{-1} = \mathbf{I}$ implies that

$$\frac{d\mathbf{F}^{-1}}{dt} = -\mathbf{F}^{-1}\frac{d\mathbf{F}}{dt}\mathbf{F}^{-1}. \tag{4.10}$$

Next, we will state some useful identities based on the chain rule for partial differentiation. In order to clarify the independent variables in the formulas, we will use subscript $_L$ for dependence on $(\mathbf{a}, t)$, and subscript $_E$ for dependence on $(\mathbf{x}, t)$. We define the **deformation gradient** by

$$\mathbf{J}_L(\mathbf{a}, t) \equiv \frac{\partial \mathbf{x}}{\partial \mathbf{a}}. \tag{4.11}$$

If $\mathbf{u}_E(\mathbf{x}, t) \in \mathbf{R}^m$, let

$$\mathbf{u}_L(\mathbf{a}, t) \equiv \mathbf{u}_E(\mathbf{x}(\mathbf{a}, t), t).$$

Then the chain rule implies that

$$\frac{\partial \mathbf{u}_L}{\partial \mathbf{a}} = \frac{\partial \mathbf{u}_E}{\partial \mathbf{x}} \mathbf{J}_L. \tag{4.12}$$

We can transpose this equation to obtain $\nabla_{\mathbf{a}} \mathbf{u}_L^\top \mathbf{J}_L^\top (\nabla_{\mathbf{x}} \mathbf{u}_E^\top)$, or take the trace to obtain $\nabla_{\mathbf{a}} \cdot \mathbf{u}_L = \operatorname{tr}(\frac{\partial \mathbf{u}_E}{\partial \mathbf{x}} \mathbf{J}_L)$.

We define the **velocity** by

$$\mathbf{v}_L(\mathbf{a}, t) = \frac{\partial \mathbf{x}}{\partial t}. \tag{4.13}$$

By interchanging the order of differentiation it is easy to see that

$$\frac{\partial \mathbf{J}_L}{\partial t} = \frac{\partial \mathbf{v}_L}{\partial a}. \tag{4.14}$$

The velocity and deformation gradient can also be defined in the Eulerian frame of reference by $\mathbf{v}_E(\mathbf{x}(\mathbf{a}, t), t) \equiv \mathbf{v}_L(\mathbf{a}, t)$ and $\mathbf{J}_E(\mathbf{x}(\mathbf{a}.t), t) \equiv \mathbf{J}_L(\mathbf{a}, t)$. Equations (4.14) and (4.12) applied to the velocity vector show that

$$\frac{\partial \mathbf{J}_L}{\partial t} = \frac{\partial \mathbf{v}_E}{\partial \mathbf{x}} \mathbf{J}_L. \tag{4.15}$$

Using the multilinearity of the determinant, then equation (4.15) and finally the fact that a matrix with one row a scalar multiple of another has zero determinant, we compute

$$\frac{\partial \det \mathbf{J}_L}{\partial t} = \sum_{i=1}^{k} \det \begin{bmatrix} \mathbf{e}_1^\top \mathbf{J}_L \\ \vdots \\ \mathbf{e}_i^\top \frac{\partial \mathbf{J}_L}{\partial t} \\ \cdots \\ \mathbf{e}_k^\top \mathbf{J}_L \end{bmatrix} = \sum_{i=1}^{k} \det \begin{bmatrix} \mathbf{e}_1^\top \mathbf{J}_L \\ \vdots \\ \sum_{\ell=1}^{k} \mathbf{e}_i^\top \frac{\partial \mathbf{v}_E}{\partial \mathbf{x}} \mathbf{e}_\ell \mathbf{e}_\ell^\top \mathbf{J}_L \\ \vdots \\ \mathbf{e}_k^\top \mathbf{J}_L \end{bmatrix}$$

$$= \sum_{i=1}^{k} \det \begin{bmatrix} \mathbf{e}_1^\top \mathbf{J}_L \\ \vdots \\ \mathbf{e}_i^\top \frac{\partial \cdot \mathbf{v}_E}{\partial \mathbf{x}} \mathbf{e}_i \mathbf{e}_i^\top \mathbf{J}_L \\ \vdots \\ \mathbf{e}_k^\top \mathbf{J}_L \end{bmatrix} = \left(\sum_{i=1}^{k} \mathbf{e}_i^\top \frac{\partial \mathbf{v}_E}{\partial \mathbf{x}} \mathbf{e}_i \right) \det \mathbf{J}_L.$$

This result can be written

$$\frac{\partial |\mathbf{J}_L|}{\partial t} = |\mathbf{J}_E| \nabla_{\mathbf{x}} \cdot \mathbf{v}_E. \tag{4.16}$$

The chain rule can also be used to derive the formula for the **material derivative** (which is sometimes called the **substantial derivative**): if $\mathbf{u}_L(\mathbf{a}, t) =$

$\mathbf{u}_E(\mathbf{x}(\mathbf{a}, t), t) \in \mathbf{R}^m$ then

$$\frac{\partial \mathbf{u}_L}{\partial t} = \frac{\partial \mathbf{u}_E}{\partial t} + \frac{\partial \mathbf{u}_E}{\partial \mathbf{x}} \mathbf{v}_E. \tag{4.17}$$

Equality of mixed partial derivatives can be used to prove that

$$\nabla_{\mathbf{a}} \cdot \left(|\mathbf{J}_L| \mathbf{J}_L^{-1} \right) = 0. \tag{4.18}$$

If $\mathbf{B}_L(\mathbf{a}, t) = \mathbf{B}_E(\mathbf{x}(a, t), t) \in \mathbf{R}^{k \times m}$ then we can use the product rule (4.9), equality of mixed partial derivatives (4.18), and the chain rule (4.12) to prove that

$$\nabla_{\mathbf{a}} \cdot \left(|\mathbf{J}_L| \mathbf{J}_L^{-1} \mathbf{B}_L \right) = |\mathbf{J}_E| \nabla_{\mathbf{x}} \cdot \mathbf{B}_E. \tag{4.19}$$

4.1.3.2 Change of Frame of Reference for Conservation Laws

We can use a number of identities from the previous subsection to prove the following lemma.

Lemma 4.1.9 *Suppose that* $\mathbf{u}_E(\mathbf{x}, t) \in \mathbf{R}^m$ *satisfies the (Eulerian) conservation law*

$$\frac{d}{dt} \int_{\Omega_{\mathbf{x}}} \mathbf{u}_E(\mathbf{x}, t)\, d\mathbf{x} + \int_{\partial \Omega_{\mathbf{x}}} \mathbf{F}_E(\mathbf{u}_E(\mathbf{x}, t)) \mathbf{n}_{\mathbf{x}}\, ds_{\mathbf{x}} = \int_{\Omega_{\mathbf{x}}} \mathbf{b}_E(\mathbf{x}, t)\, d\mathbf{x},$$

Also assume that $\mathbf{x}(\mathbf{a}, t) \in \mathbf{R}^k$ *and* $\mathbf{a} \in \mathbf{R}^k$, $\mathbf{v}_L = \frac{\partial \mathbf{x}}{\partial t}$, $\mathbf{J}_L = \frac{\partial \mathbf{x}}{\partial \mathbf{a}}$, $\Omega_{\mathbf{x}} = \{\mathbf{x}(\mathbf{a}, t) : \mathbf{a} \in \Omega_{\mathbf{a}}\}$, $\mathbf{F}_L(\mathbf{a}, t) = \mathbf{F}_E(\mathbf{x}(\mathbf{a}, t), t)$, *and* $\mathbf{b}_L(\mathbf{a}, t) = \mathbf{b}_E(\mathbf{x}(\mathbf{a}, t), t)$. *Then* $\mathbf{u}_L(\mathbf{a}, t) \equiv \mathbf{u}_E(\mathbf{x}(\mathbf{a}, t), t)$ *satisfies the (Lagrangian) conservation law*

$$\frac{d}{dt} \int_{\Omega_{\mathbf{a}}} \mathbf{u}_L(\mathbf{a}, t) |\mathbf{J}_L|\, d\mathbf{a} + \int_{\partial \Omega_{\mathbf{a}}} [(\mathbf{F}_L - \mathbf{u}_L \mathbf{v}_L^\top) \mathbf{J}_L^{-\top} |\mathbf{J}_L| \mathbf{n}_{\mathbf{a}}\, ds_{\mathbf{a}} = \int_{\Omega_{\mathbf{a}}} \mathbf{b}_L(\mathbf{a}, t) |\mathbf{J}_L|\, d\mathbf{a}.$$

Proof The formula for change of variables in integration implies that

$$\int_{\Omega_{\mathbf{x}}} \mathbf{b}_E(\mathbf{x}, t)\, d\mathbf{x} = \int_{\Omega_{\mathbf{a}}} \mathbf{b}_L(\mathbf{a}, t) |\mathbf{J}_L|\, d\mathbf{a}.$$

Further, the divergence theorem implies that for any fixed vector z,

$$\int_{\partial \Omega_{\mathbf{x}}} z^\top \mathbf{F}_E(\mathbf{u}(\mathbf{x}, t)) \mathbf{n}_{\mathbf{x}}\, ds_{\mathbf{x}} = \int_{\Omega_{\mathbf{x}}} (\nabla_{\mathbf{x}} \cdot \mathbf{F}_E^\top z)\, d\mathbf{x} = \int_{\Omega_{\mathbf{x}}} \operatorname{tr} \left(\frac{\partial \mathbf{F}_E^\top z}{\partial \mathbf{x}} \right) d\mathbf{x}$$

then change of variables in integration implies that

$$= \int_{\Omega_{\mathbf{a}}} \operatorname{tr} \left(\mathbf{J}_L^{-1} \frac{\partial \mathbf{F}_L^\top z}{\partial \mathbf{a}} \right) |\mathbf{J}_L|\, d\mathbf{a}$$

then the product rule (4.9) with $\mathbf{B} = \mathbf{J}_L^{-1}|\mathbf{J}_L|$ and $\mathbf{c} = \mathbf{F}_L^\top z$, together with equality of mixed partial derivatives (4.18) imply that

$$= \int_{\Omega_\mathbf{a}} \operatorname{tr}\left(\frac{\partial \mathbf{J}_L^{-1}\mathbf{F}_L^\top z|\mathbf{J}_L|}{\partial \mathbf{a}}\right) d\mathbf{a} = \int_{\Omega_\mathbf{a}} \nabla_\mathbf{a} \cdot \left(\mathbf{J}_L^{-1}\mathbf{F}_L^\top z|\mathbf{J}_L|\right) d\mathbf{a}$$

and finally the divergence theorem implies that

$$= \int_{\partial\Omega_\mathbf{a}} z^\top \mathbf{F}_L \mathbf{J}_L^{-\top} \mathbf{n}_\mathbf{a} |\mathbf{J}_L| \, ds_\mathbf{a}.$$

Finally, the formula (4.17) for the material derivative implies

$$\int_{\Omega_\mathbf{x}} \frac{\partial \mathbf{u}_E}{\partial t} \, d\mathbf{x} = \int_{\Omega_\mathbf{x}} \frac{\partial \mathbf{u}_L}{\partial t} - \frac{\partial \mathbf{u}_E}{\partial \mathbf{x}} \mathbf{v}_E \, d\mathbf{x}$$

then change of variables in integration and the chain rule (4.12) imply

$$= \int_{\Omega_\mathbf{a}} \left[\frac{\partial \mathbf{u}_L}{\partial t} - \frac{\partial \mathbf{u}_L}{\partial \mathbf{a}} \mathbf{J}_L^{-1} \mathbf{v}_L\right] |\mathbf{J}| \, d\mathbf{a}$$

then the formula (4.16) for the rate of change of the Jacobian, together with the chain rule (4.12), imply that

$$= \int_{\Omega_\mathbf{a}} \frac{\partial \mathbf{u}_L}{\partial t} |\mathbf{J}_L| + \mathbf{u}_L \left[\frac{\partial |\mathbf{J}_L|}{\partial t} - \operatorname{tr}\left(\frac{\partial \mathbf{v}_L}{\partial \mathbf{a}} \mathbf{J}_L^{-1} |\mathbf{J}_L|\right)\right] - \frac{\partial \mathbf{u}_L}{\partial \mathbf{a}} \mathbf{J}_L^{-1} \mathbf{v}_L |\mathbf{J}_L| \, d\mathbf{a}$$

then the product rule for time differentiation, Equation (4.18), and the product rule (4.9) with $\mathbf{B} = \mathbf{J}_L^{-1}|\mathbf{J}_L|$ and $\mathbf{c} = \mathbf{v}_L$ imply

$$= \int_{\Omega_\mathbf{a}} \frac{\partial \mathbf{u}_L |\mathbf{J}_L|}{\partial t} - \mathbf{u}_L \, \nabla_\mathbf{a} \cdot \left(\mathbf{J}_L^{-1} \mathbf{v}_L |\mathbf{J}_L|\right) - \frac{\partial \mathbf{u}_L}{\partial \mathbf{a}} \mathbf{J}_L^{-1} \mathbf{v}_L |\mathbf{J}_L| \, d\mathbf{a}$$

and finally the transpose of (4.8) with $\mathbf{b} = \mathbf{J}_L^{-1}\mathbf{v}_L|\mathbf{J}_L|$ and $\mathbf{c} = \mathbf{u}_L$ implies that

$$= \int_{\Omega_\mathbf{a}} \frac{\partial \mathbf{u}_L |\mathbf{J}_L|}{\partial t} - \left[\nabla_\mathbf{a}^\top \left(\mathbf{J}_L^{-1} \mathbf{v}_L |\mathbf{J}_L| \mathbf{u}_L^\top\right)\right]^\top \, d\mathbf{a}.$$

We can put these three equations together to obtain the claimed result. □

Note that in smooth flow, this lemma says that the conservation law

$$\frac{\partial \mathbf{u}_E}{\partial t} + \sum_{i=1}^{k} \frac{\partial \mathbf{F}_E \mathbf{e}_i}{\partial \mathbf{x}_i} = \mathbf{b}_E$$

is equivalent to the conservation law

$$\frac{\partial \mathbf{u}_L |\mathbf{J}_L|}{\partial t} + \sum_{i=1}^{k} \frac{\partial \left(\mathbf{F}_L - \mathbf{u}_L \mathbf{v}_L^\top\right) \mathbf{J}_L^{-\top} \mathbf{e}_i |\mathbf{J}_L|}{\partial \mathbf{a}_i} = \mathbf{b}_L |\mathbf{J}_L|. \tag{4.20}$$

4.1.3.3 Change of Frame of Reference for Propagating Discontinuities

Next, we would like to determine how discontinuity speeds and normals change when we change the frame of reference.

Lemma 4.1.10 *[63] Suppose that at some point in the domain, a discontinuity propagates with speed σ_E in direction $\mathbf{n}_E$ with respect the Eulerian coordinate system $\mathbf{x}(\mathbf{a}, t)$, and with speed σ_L and normal $\mathbf{n}_L$ in the Lagrangian coordinate system $\mathbf{a}$. Then*

$$\mathbf{n}_E = \mathbf{J}_L^{-\top} \mathbf{n}_L \frac{1}{\|\mathbf{J}_L^{-\top} \mathbf{n}_L\|}, \qquad \sigma_E - \mathbf{n}_E \cdot \mathbf{v} = \sigma_L \|\mathbf{J}_L^\top \mathbf{n}_L\|$$

where $\mathbf{v} = \frac{\partial \mathbf{x}}{\partial t}$, $\mathbf{J}_L = \frac{\partial \mathbf{x}}{\partial \mathbf{a}}$.

Proof Suppose that we have a propagating discontinuity surface described as the level set of a function ϕ_E in the Eulerian frame of reference: $\phi_E(\mathbf{x}, t) = 0$. Then the normal to the surface is $\mathbf{n}_E = \nabla_\mathbf{x} \phi_E / \|\nabla_\mathbf{x} \phi_E\|$. The velocity of the surface in this normal direction must be continuous across the surface. Since $\phi_E(\mathbf{x}, t) = 0$ gives us a 1-parameter representation of a point $\mathbf{x}$ on the surface as a function of time t, let us define that point $\mathbf{x}_t(t)$ by the equation $\phi_E(\mathbf{x}_t(t), t) = 0$ and the requirement that the point move along the normal to the surface $\frac{d\mathbf{x}_t}{dt} = \mathbf{n}_E \sigma_E$. Then we can differentiate the level set equation in time to get

$$0 = \frac{\partial \phi_E}{\partial \mathbf{x}} \frac{\partial \mathbf{x}_t}{\partial t} + \frac{\partial \phi_E}{\partial t} = \nabla_\mathbf{x} \phi_E \cdot \mathbf{n}_E \sigma_E + \frac{\partial \phi_E}{\partial t}.$$

This implies that the normal speed of the surface is $\sigma_E = -\frac{\partial \phi_E}{\partial t} \frac{1}{\|\nabla_\mathbf{x} \phi_E\|}$.

We can also write the discontinuity surface in the Lagrangian frame as $\phi_L(\mathbf{a}, t) = \phi_E(\mathbf{x}(\mathbf{a}, t), t) = 0$. Thus the Lagrangian normal to the discontinuity surface is $\mathbf{n}_L = \nabla_\mathbf{a} \phi_L / \|\nabla_\mathbf{a} \phi_L\|$. Note that the chain rule for partial differentiation implies that $\nabla_\mathbf{a} \phi_L = \mathbf{J}_L^\top \nabla_\mathbf{x} \phi_E$. where $\mathbf{J}_L$ is the deformation gradient $\mathbf{J}_L \equiv \partial \mathbf{x} / \partial \mathbf{a}$. Thus the normal directions in the two frames of reference are related as follows:

$$\mathbf{n}_L = \nabla_\mathbf{a} \phi_L \frac{1}{\|\nabla_\mathbf{a} \phi_L\|} = \mathbf{J}_L^\top \nabla_\mathbf{x} \phi_E \frac{1}{\|\nabla_\mathbf{a} \phi_L\|} = \mathbf{J}_L^\top \mathbf{n}_E \frac{\|\nabla_\mathbf{x} \phi_E\|}{\|\nabla_\mathbf{a} \phi_L\|}.$$

Taking norms of both sides of this equation leads to $1 = \|\mathbf{J}_L^\top \mathbf{n}_E\| \|\nabla_\mathbf{x} \phi_E\| / \|\nabla_\mathbf{a} \phi_L\|$. This result allows us to write

$$\mathbf{n}_L = \mathbf{J}_L^\top \mathbf{n}_E \frac{1}{\|\mathbf{J}_L^\top \mathbf{n}_E\|} \quad \text{or} \quad \mathbf{n}_E = \mathbf{J}_L^{-\top} \mathbf{n}_L \frac{1}{\|\mathbf{J}_L^{-\top} \mathbf{n}_L\|}.$$

Of course, the equation $\phi_L(\mathbf{a}, t) = 0$ gives us a one-parameter representation of the motion of points $\mathbf{a}$ in the surface. Let $\mathbf{a}_t(t)$ be the trajectory of a point on the discontinuity surface moving along the normal to the surface. Since

$$0 = \frac{\partial \phi_L}{\partial t} + \frac{\partial \phi_L}{\partial \mathbf{a}} \frac{d\mathbf{a}_t}{dt},$$

we see as before that the normal velocity of points on the surface is $\frac{d\mathbf{a}_t}{dt} = \mathbf{n}_L \sigma_L$, where the normal speed of the discontinuity surface in the Lagrangian frame is

$$\sigma_L = -\frac{\partial \phi_L}{\partial t} \frac{1}{\|\nabla_{\mathbf{a}} \phi_L\|}.$$

The Eulerian and Lagrangian trajectories are related by the equation $\mathbf{x}_t(t) = \mathbf{x}(\mathbf{a}_t(t), t)$, from which it follows that

$$\frac{d\mathbf{x}_t}{dt} = \frac{\partial \mathbf{x}}{\partial \mathbf{a}} \frac{d\mathbf{a}_t}{dt} + \frac{\partial \mathbf{x}}{\partial t} = \mathbf{J}_L \mathbf{n}_L \sigma_L + \mathbf{v}_L.$$

Now we can see that the Eulerian and Lagrangian normal discontinuity speeds are related by

$$\sigma_E = -\frac{\partial \phi_E}{\partial t} \frac{1}{\|\nabla_{\mathbf{x}} \phi_E\|} = \frac{1}{\|\nabla_{\mathbf{x}} \phi_E\|} (\nabla_{\mathbf{x}} \phi_E) \cdot \frac{d\mathbf{x}_t}{dt} = \mathbf{n}_E \cdot [\mathbf{J}_L \mathbf{n}_L \sigma_L + \mathbf{v}]$$

$$= \frac{\mathbf{n}_E^\top \mathbf{J}_L \mathbf{J}_L^\top \mathbf{n}_E}{\|\mathbf{J}_L^\top \mathbf{n}_E\|} \sigma_L + \mathbf{n}_E^\top \mathbf{v} = \|\mathbf{J}_L^\top \mathbf{n}_E\| \sigma_L + \mathbf{n}_E \cdot \mathbf{v} = \frac{\|\nabla_{\mathbf{a}} \phi_L\|}{\|\nabla_{\mathbf{x}} \phi_E\|} \sigma_L + \mathbf{n}_E \cdot \mathbf{v}.$$

□

4.1.4 Rankine–Hugoniot Jump Condition

It is valid to write the conservation laws (4.1) as partial differential equations (4.3) in regions of smooth flow. However, these equations could become invalid because of a propagating discontinuity. In order to understand how the conservation laws operate at a propagating discontinuity, we will prove the following important result, which generalizes our previous Lemma 4.1.9 in one dimension.

Lemma 4.1.11 *Suppose that we have a fixed domain $\Omega \subset \mathbf{R}^k$ which is divided into two sub-domains Ω_L and Ω_R by a single propagating discontinuity on a surface D, associated with the conservation law*

$$\frac{d}{dt} \int_\Omega \mathbf{u} \, d\mathbf{x} + \int_{\partial\Omega} \mathbf{F}\mathbf{n} \, ds = \int_\Omega \mathbf{b} \, d\mathbf{x}. \tag{4.21}$$

We will assume that the normal $\mathbf{n}$ to the discontinuity is oriented to point from Ω_L to Ω_R. At any point on the discontinuity surface, let $\mathbf{u}_L$ and $\mathbf{u}_R$ be the values of $\mathbf{u}$ as we approach the point from inside Ω_L and Ω_R, respectively. Similarly, let $\mathbf{F}_L$ and $\mathbf{F}_R$ be *the values of* **F** *associated with the two domains on either side of the discontinuity. Finally, let σ be the speed of the discontinuity, oriented corresponding to the normal* **n**. *Then the* **Rankine–Hugoniot jump condition** *holds:*

$$[\mathbf{F}_R - \mathbf{F}_L]\mathbf{n} = [\mathbf{u}_R - \mathbf{u}_L]\sigma. \tag{4.22}$$

Proof If $\mathbf{y}(\mathbf{x})$ is the velocity at a point $\mathbf{x}$ on the discontinuity surface, then the well-known formula for the derivative of an integral leads to the equations

$$\frac{d}{dt}\int_{\Omega_L} \mathbf{u}\, d\mathbf{x} = \int_{\Omega_L} \frac{\partial \mathbf{u}}{\partial t}\, d\mathbf{x} + \int_D \mathbf{u}_L\, \mathbf{n}\cdot\mathbf{y}\, ds$$

and

$$\frac{d}{dt}\int_{\Omega_R} \mathbf{u}\, d\mathbf{x} = \int_{\Omega_R} \frac{\partial \mathbf{u}}{\partial t}\, d\mathbf{x} - \int_D \mathbf{u}_R\, \mathbf{n}\cdot\mathbf{y}\, ds.$$

Away from the discontinuity surface, the conservation law can be written as the partial differential equation

$$\frac{\partial \mathbf{u}}{\partial t} + \sum_{i=1}^{k} \frac{\partial \mathbf{F}\mathbf{e}_i}{\partial \mathbf{x}_i} = \mathbf{b}.$$

We use the fact that $\mathbf{n}\cdot\mathbf{y} = \sigma$ is the normal speed of the discontinuity to get

$$\begin{aligned}\frac{d}{dt}\int_{\Omega} \mathbf{u}\, d\mathbf{x} &= \int_{\Omega_L} \mathbf{b} - \sum_{i=1}^{k} \frac{\partial \mathbf{F}\mathbf{e}_i}{\partial \mathbf{x}_i}\, d\mathbf{x} + \int_{\Omega_R} \mathbf{b} - \sum_{i=1}^{k} \frac{\partial \mathbf{F}\mathbf{e}_i}{\partial \mathbf{x}_i}\, d\mathbf{x}\\ &\quad + \int_D \mathbf{u}_L \sigma\, ds - \int_D \mathbf{u}_R \sigma\, ds\end{aligned}$$

then we apply the divergence theorem to get

$$= \int_{\Omega} \mathbf{b}\, d\mathbf{x} - \int_{\partial\Omega_L} \mathbf{F}\mathbf{n}\, ds - \int_{\partial\Omega_R} \mathbf{F}\mathbf{n}\, ds - \int_D [\mathbf{u}_R - \mathbf{u}_L]\sigma\, ds.$$

then we use the fact that $\partial\Omega_L \cup \partial\Omega_R = \partial\Omega \cup D$ to get

$$= \int_{\Omega} \mathbf{b}\, d\mathbf{x} - \int_{\partial\Omega} \mathbf{F}\mathbf{n}\, ds + \int_D [\mathbf{F}_R - \mathbf{F}_L]\mathbf{n}\, ds - \int_D [\mathbf{u}_R - \mathbf{u}_L]\sigma\, ds.$$

Subtracting the original form (4.21) for the conservation law, we obtain

$$\int_D [\mathbf{F}_R - \mathbf{F}_L]\mathbf{n}\, ds = \int_D [\mathbf{u}_R - \mathbf{u}_L]\sigma\, ds.$$

By shrinking D and Ω around a point, we obtain the Rankine–Hugoniot jump condition (4.22). □

This lemma says that the jump in the normal component of the flux is equal to the jump in the conserved quantities times the normal velocity of the discontinuity. Note that the normal $\mathbf{n}$ appears in the definition of the discontinuity speed σ; thus reversing the sign of the normal leads to the same result.

Example 4.1.12 The Rankine–Hugoniot jump conditions for the shallow water Equations (4.2) are

$$[h\mathbf{v}\cdot\mathbf{n}] = [h]\sigma$$
$$\left[\mathbf{v}h\mathbf{v}\cdot\mathbf{n} + \frac{1}{2}\mathbf{n}gh^2\right] = [\mathbf{v}h]\sigma.$$

Let $\xi = \sigma - \mathbf{v}\cdot\mathbf{n}$ be the velocity of the discontinuity relative to the fluid velocity. Then the jump conditions can be rewritten

$$[h\xi] = 0$$
$$\left[\frac{1}{2}\mathbf{n}gh^2 - \mathbf{v}h\xi\right] = 0.$$

The second of these two jump equations can be rewritten in the separate forms

$$\left[\frac{1}{2}gh^2 - \mathbf{v}\cdot\mathbf{n}h\xi\right] = 0,$$
$$[\mathbf{v}\cdot\mathbf{n}^{\perp}h\xi] = 0$$

where $\mathbf{n}^{\perp}$ is any vector orthogonal to $\mathbf{n}$. Note that $[\mathbf{v}\cdot\mathbf{n}] = -[\xi]$ and $[h\xi] = 0$, so

$$[\mathbf{v}\cdot\mathbf{n}h\xi] = (h_R\xi_R)[\mathbf{v}\cdot\mathbf{n}] = -(h_R\xi_R)[\xi] = -[h\xi^2];$$

thus we can further modify the jump conditions to get

$$[h\xi] = 0$$
$$\left[\frac{1}{2}gh^2 + h\xi^2\right] = 0$$
$$[\mathbf{v}\cdot\mathbf{n}^{\perp}h\xi] = 0.$$

We will consider two cases, $[h] = 0$ and $[h] \neq 0$. If $[h] = 0$ and $h_L = h_R > 0$, then $[\xi] = 0$. Thus $[\mathbf{v}\cdot\mathbf{n}] = 0$. We must have $[\mathbf{v}\cdot\mathbf{n}^{\perp}] \neq 0$, otherwise there is no jump at all. Then $\xi_R = \xi_L = 0$, so the speed of the discontinuity is $\sigma = \mathbf{v}\cdot\mathbf{n}$. This is a contact discontinuity.

On the other hand, suppose that $[h] \neq 0$. There are four variables in the two equations

$$[h\xi] = 0$$
$$\left[\frac{1}{2}gh^2 + h\xi^2\right] = 0,$$

namely ξ_L, ξ_R, h_L and h_R. We need to specify two variables to determine a solution. Suppose that we are given $h_L > 0$ and $h_R > 0$. Define the relative jump in the water height to be $z = (h_R - h_L)/h_L$. Then we can rewrite the right height in terms of the left height and z to get $h_R = h_L(1+z)$ and $\xi_R = \xi_L/(1+z)$. We can use this expression for ξ_R to rewrite the second jump condition in the form

$$\frac{1}{2}gh_L(1+z)^2 + \frac{\xi_L^2}{(1+z)} = \frac{1}{2}gh_L + \xi_L^2.$$

If $z \neq 0$, we can solve this equation to obtain

$$\xi_L = \pm\sqrt{\frac{1}{2}gh_R(2+z)(1+z)} = \pm\sqrt{\frac{1}{2}gh_R(h_R+h_L)/h_L}.$$

At this point, given h_L and h_R, we know how to determine ξ_L and ξ_R.

If in addition we specify $\mathbf{v}_L$, then it follows that the discontinuity speed has two possible values:

$$\sigma = \mathbf{v}_L \cdot \mathbf{n} + \xi_L = \mathbf{v}_L \cdot \mathbf{n} \pm \sqrt{g\frac{h_L + h_R}{2}\frac{h_R}{h_L}}.$$

It is interesting to note that as $h_R \to h_L$, we find that $\sigma \to \mathbf{v}_L \cdot \mathbf{n} \pm \sqrt{gh_L} = \lambda$; in other words, the speed of infinitesimal discontinuities is the same as a characteristic speed. Alternatively, we could specify $\mathbf{v}_R$ and compute

$$\sigma \equiv \mathbf{v}_R \cdot \mathbf{n} + \xi_R = \mathbf{v}_R \cdot \mathbf{n} + \frac{\xi_L}{(1+z)} = \mathbf{v}_R \cdot \mathbf{n} \pm \sqrt{\frac{1}{2}gh_L(h_R+h_L)/h_R}.$$

The locus of points satisfying the Rankine–Hugoniot jump conditions consists of two curves in $\mathbf{w}$-space. For example, we could fix the left state $(h_L, \mathbf{v}_L)$ and determine $\mathbf{v}_R(h_R)$ as well as $\sigma(h_R)$:

$$\mathbf{v}_R = \mathbf{v}_L \pm \mathbf{n}(h_R - h_L)\sqrt{\frac{g}{2}\left(\frac{1}{h_L} + \frac{1}{h_R}\right)}.$$

Alternatively, we could fix the right state $(h_R, \mathbf{v}_R)$ and determine $\mathbf{v}_L(h_L)$ and $\sigma(h_L)$.

Summary 4.1.13 *Given a left state $\mathbf{w}_L = (h_L, \mathbf{v}_L)$, the points $\mathbf{w} = (h, \mathbf{v})$ on the slow Rankine–Hugoniot locus for the shallow water equations* (4.2) *satisfy*

$$\mathbf{v}(h) = \mathbf{v}_L - \mathbf{n}(h - h_L)\sqrt{\frac{g}{2}\left(\frac{1}{h_L} + \frac{1}{h}\right)} \text{ where } h > h_L > 0$$

with discontinuity speed

$$\sigma(h) = \mathbf{v}_L \cdot \mathbf{n} - \sqrt{g\frac{h_L + h}{2}\frac{h}{h_L}}.$$

Similarly, given a right state $\mathbf{w}_R = (h_R, \mathbf{v}_R)$, the points $\mathbf{w} = (h, \mathbf{v})$ on the fast Rankine-Hugoniot locus satisfy

$$\mathbf{v}(h) = \mathbf{v}_R - \mathbf{n}(h - h_R)\sqrt{\frac{g}{2}\left(\frac{1}{h_R} + \frac{1}{h}\right)} \text{ where } h > h_R > 0$$

with discontinuity speed

$$\sigma(h) = \mathbf{v}_R \cdot \mathbf{n} + \sqrt{g\frac{h_R + h}{2}\frac{h}{h_R}}.$$

Across a contact discontinuity, both the water height h and the normal velocity $\mathbf{v} \cdot \mathbf{n}$ are continuous, and the speed of a contact discontinuity is $\sigma = \mathbf{v} \cdot \mathbf{n}$.

4.1.5 Lax Admissibility Conditions

Our goal here is to find conditions that will enable us to determine uniquely the states $\mathbf{u}$ associated with propagating discontinuities.

Definition 4.1.14 Suppose that for all states $\mathbf{u}$ and all directions $\mathbf{n}$, the characteristic speeds obtained from $\partial \mathbf{Fn}/\partial \mathbf{u}$ are real and either genuinely nonlinear or linear degenerate. Further assume that genuinely nonlinear characteristic speeds are distinct, both from other genuinely nonlinear characteristic speeds and from any linearly degenerate characteristic speeds. Suppose that the characteristic speeds have been ordered so that

$$\lambda_1 \leq \lambda_2 \leq \ldots \leq \lambda_m. \tag{4.23a}$$

Then a discontinuity moving with speed σ is a **shock** if and only if there is an index j associated with a genuinely nonlinear characteristic speed λ_j so that

$$(\lambda_j)_L > \sigma > (\lambda_j)_R \quad \text{and} \quad (\lambda_{j-1})_L < \sigma < (\lambda_{j+1})_R. \tag{4.23b}$$

These inequalities require that $m - j + 1$ characteristics enter the shock on the left (namely those with indices $j, \ldots, m$) and j characteristics enter the shock on the right (namely those with indices $1, \ldots, j$). Note that the Rankine–Hugoniot conditions

$$[\mathbf{F}(\mathbf{u}_R) - \mathbf{F}(\mathbf{u}_L)]\mathbf{n} = [\mathbf{u}_R - \mathbf{u}_L]\sigma$$

involve $2m + 1$ unknowns, namely $\mathbf{u}_L$, $\mathbf{u}_R$ and σ. The Rankine-Hugoniot jump conditions give us m equations to determine these unknowns. The Lax admissibility conditions (4.23) provide the remaining conditions needed to completely specify the left and right states and the shock speed.

To see why the Lax admissibility conditions (4.23) are important, we will generalize the analysis in section 3.1.5 to a system of conservation laws in one spatial dimension. Suppose that

$$\frac{\partial \mathbf{u}}{\partial t} + \frac{\partial \mathbf{f}(\mathbf{u})}{\partial x} = 0, \quad \mathbf{u}(x, 0) = \begin{cases} \mathbf{u}_L, & x < 0 \\ \mathbf{u}_R, & x > 0 \end{cases}$$

where the initial data satisfies the Rankine–Hugoniot jump conditions

$$[\mathbf{f}(\mathbf{u}_R) - \mathbf{f}(\mathbf{u}_L)] = [\mathbf{u}_R - \mathbf{u}_L]\sigma.$$

Also suppose that the system is hyperbolic with a full set of characteristic directions; then by Definition 4.1.2,

$$\frac{\partial \mathbf{f}}{\partial \mathbf{u}}\mathbf{X} = \mathbf{X}\Lambda,$$

where Λ is diagonal, and its diagonal entries satisfy the Lax admissibility conditions (4.23). Consider the viscous modification of the conservation law

$$\frac{\partial \mathbf{u}_\epsilon}{\partial t} + \frac{\partial \mathbf{f}(\mathbf{u}_\epsilon)}{\partial x} = \epsilon \frac{\partial^2 \mathbf{u}_\epsilon}{\partial x^2},$$

where $\epsilon > 0$. We will look for traveling wave solutions of the form

$$\mathbf{u}_\epsilon(x, t) = \mathbf{w}\left(\frac{x - \sigma t}{\epsilon}\right).$$

As in the case of a single conservation law, if $\mathbf{u}_\epsilon$ satisfies the viscous conservation law, then $\mathbf{w}$ satisfies the system of ordinary differential equations

$$\left(\frac{\partial \mathbf{f}}{\partial \mathbf{u}} - \mathbf{I}\sigma\right)\mathbf{w}' = \mathbf{w}''.$$

We can integrate this equation once to obtain $\mathbf{w}' - \mathbf{f}(\mathbf{w}) + \mathbf{w}\sigma = \mathbf{c}$ where the vector $\mathbf{c}$ is constant. Taking the limit as $\xi \to -\infty$ implies that $\mathbf{u}_\epsilon \to \mathbf{u}_L$, and consequently that $\mathbf{w}' \to 0$. It follows that $\mathbf{c} = -\mathbf{f}(\mathbf{u}_L) + \mathbf{u}_L\sigma$, and that the ordinary differential equations can be written

$$\mathbf{w}' = \mathbf{f}(\mathbf{w}) - \mathbf{f}(\mathbf{u}_L) - (\mathbf{w} - \mathbf{u}_L)\sigma.$$

Because the initial states $\mathbf{u}_L$ and $\mathbf{u}_r$ satisfy the Rankine–Hugoniot condition, both of these states are stationary states of this system of ordinary differential equations for $\mathbf{w}$. Let us consider the stability of these stationary states. Note that if $\mathbf{w} = \mathbf{u}_L$ or $\mathbf{w} = \mathbf{u}_R$ and $\mathbf{w} + \mathbf{y}\delta$ is a perturbed solution of the ordinary differential equations, then

$$\frac{d(\mathbf{w} + \mathbf{y}\delta)}{dt} = \mathbf{f}(\mathbf{w} + \mathbf{y}\delta) - \mathbf{f}(\mathbf{u}_L) - (\mathbf{w} + \mathbf{y}\delta - \mathbf{u}_L)\sigma.$$

As $\delta \to 0$, we obtain

$$\frac{d\mathbf{y}}{dt} = \frac{\partial \mathbf{f}}{\partial \mathbf{u}}(\mathbf{w})\mathbf{y} - \mathbf{y}\sigma.$$

Thus

$$\mathbf{X}^{-1}\frac{d\mathbf{y}}{dt} = (\Lambda - \mathbf{I}\sigma)\mathbf{X}^{-1}\mathbf{y}.$$

The Lax admissibility conditions (4.23) imply that the jth entry of $\mathbf{X}^{-1}\mathbf{w}$ is linearly unstable at $\mathbf{u}_L$ and linearly stable at $\mathbf{u}_R$. Smoller [150] shows that there must be an orbit in the system of ordinary differential equations for $\mathbf{w}$ that connects $\mathbf{u}_L$ to $\mathbf{u}_R$.

Example 4.1.15 The slow discontinuity in the shallow water equations is admissible as long as $\mathbf{v}_L \cdot \mathbf{n} - \sqrt{gh_L} > \sigma > \mathbf{v}_R \cdot \mathbf{n} - \sqrt{gh_R}$. We can rewrite these inequalities in the form

$$\begin{aligned}\mathbf{v}_L \cdot \mathbf{n} - \sqrt{gh_L} &> \mathbf{v}_L \cdot \mathbf{n} - \sqrt{g\frac{h_R + h_L}{2}\frac{h_R}{h_L}} \\ &> \mathbf{v}_L \cdot \mathbf{n} - \sqrt{g\frac{h_R + h_L}{2}}\left[\sqrt{\frac{h_R}{h_L}} - \sqrt{\frac{h_L}{h_R}}\right] - \sqrt{gh_R}.\end{aligned}$$

These inequalities imply that $h_L < \sqrt{\frac{h_L+h_R}{2}h_R}$ and $\sqrt{\frac{h_L+h_R}{2}h_L} < h_R$. These in turn can be written $0 < (h_R - h_L)(h_R + 2h_L)$ and $0 < (h_R - h_L)(h_L + 2h_R)$. Thus the slow shock is admissible for $h_R > h_L$. A similar analysis for the fast shock shows that it is admissible for $h_R < h_L$.

4.1.6 Asymptotic Behavior of Hugoniot Loci

The following discussion has been taken from Lax [89]. We will restrict the discussion in this section to one dimension. The multidimensional case is more complicated. See, for example, [37] for a discussion of shock reflection, or [104, 146] for a discussion of Riemann problems for two-dimensional gas dynamics.

Lemma 4.1.16 *(Lax[89]) Suppose that the hyperbolic system of m conservation laws*

$$\frac{\partial \mathbf{u}(\mathbf{w})}{\partial t} + \frac{\partial \mathbf{f}(\mathbf{w})}{\partial x} = 0$$

has distinct characteristic speeds Λ satisfying

$$\frac{\partial \mathbf{f}(\mathbf{w})}{\partial \mathbf{w}} \mathbf{Y} = \frac{\partial \mathbf{u}}{\partial \mathbf{w}} \mathbf{Y} \Lambda,$$

and that the characteristic speeds satisfy the Lax admissibility conditions (4.23). *Suppose that the jth characteristic speed is genuinely nonlinear. Given a left state $\mathbf{w}_L$, the locus of right states $\mathbf{w}_R(\epsilon)$ that can be connected to $\mathbf{w}_L$ by an admissible shock in the jth wave family, and the associated characteristic speed and shock speed along this locus satisfy*

$$\lambda_j(\epsilon) = \lambda_j(0) + \epsilon + O(\epsilon^2) \tag{4.24a}$$

$$\sigma(\epsilon) = \lambda_j(0) + \epsilon/2 + O(\epsilon^2) \tag{4.24b}$$

$$\frac{d\mathbf{w}_R}{d\epsilon} = \mathbf{Y}\mathbf{e}_j + \left.\frac{d\mathbf{Y}\mathbf{e}_j}{d\epsilon}\right|_{\epsilon=0} \epsilon + O(\epsilon^2). \tag{4.24c}$$

Proof Given a state $\mathbf{w}_L$, we would like to determine the flux variables $\mathbf{w}_R$ that can be connected to $\mathbf{w}_L$ by a shock in the jth wave family. We claim that these states form a one-parameter curve $\mathbf{w}_R(\epsilon)$ with $\mathbf{w}_R(0) = \mathbf{w}_L$. To see this fact, note that given $\mathbf{w}_L$, the Rankine–Hugoniot conditions

$$[\mathbf{f}(\mathbf{w}_R) - \mathbf{f}(\mathbf{w}_L)] = [(\mathbf{u}(\mathbf{w}_R) - \mathbf{u}(\mathbf{w}_L)]\sigma$$

involve m equations for $m + 1$ unknowns, namely $\mathbf{w}_R$ and σ. We will let ϵ represent the remaining degree of freedom, and write $\mathbf{w}_R(\epsilon)$ and $\sigma(\epsilon)$ to represent the one-parameter curve of solutions to the Rankine-Hugoniot conditions.

If we differentiate the Rankine–Hugoniot conditions with respect to ϵ, we get

$$\left.\frac{\partial \mathbf{f}(\mathbf{w})}{\partial \mathbf{w}}\right|_{\mathbf{w}_R} \frac{d\mathbf{w}_R}{d\epsilon} = \left.\frac{\partial \mathbf{u}}{\partial \mathbf{w}}\right|_{\mathbf{w}_R} \frac{d\mathbf{w}_R}{d\epsilon}\sigma(\epsilon) + [\mathbf{u}(\mathbf{w}_R(\epsilon)) - \mathbf{u}(\mathbf{w}_L)]\frac{d\sigma}{d\epsilon}. \tag{4.25}$$

At $\epsilon = 0$, these equations say that

$$\left.\frac{\partial \mathbf{f}(\mathbf{w})}{\partial \mathbf{w}}\right|_{\mathbf{w}_L} \left.\frac{d\mathbf{w}_R}{d\epsilon}\right|_{\epsilon=0} = \left.\frac{\partial \mathbf{u}}{\partial \mathbf{w}}\right|_{\mathbf{w}_L} \left.\frac{d\mathbf{w}_R}{d\epsilon}\right|_{\epsilon=0} \sigma(0).$$

The eigenvector equation for the characteristic speeds shows that, after adjusting by a scalar multiple if necessary, there is some index j so that $\frac{d\mathbf{w}_R}{d\epsilon}|_{\epsilon=0}$ is a scalar multiple of $\mathbf{Y}e_j$, and $\sigma(0) = \lambda_j$.

We will assume that the genuinely nonlinear eigenvectors $\mathbf{Y}e_i$ have been normalized so that for all genuinely nonlinear waves

$$\frac{\partial \lambda_j}{\partial \mathbf{w}}\bigg|_{\mathbf{w}_L} \mathbf{Y}\mathbf{e}_j = 1$$

and that for all linearly degenerate waves $\|\mathbf{Y}\mathbf{e}_i\| = 1$. This will imply that for genuinely nonlinear waves j,

$$\frac{d\lambda_j(\mathbf{w}_R(\epsilon))}{d\epsilon}\bigg|_{\epsilon=0} = \frac{\partial \lambda_j}{\partial \mathbf{w}}\bigg|_{\mathbf{w}_L} \frac{\partial \mathbf{w}_R}{\partial \epsilon}\bigg|_{\epsilon=0} = \frac{\partial \lambda_j}{\partial \mathbf{w}}\bigg|_{\mathbf{w}_L} \mathbf{Y}\mathbf{e}_j = 1.$$

Let us differentiate Equation (4.25) once more with respect to ϵ, and then set $\epsilon = 0$:

$$\begin{aligned}\frac{d}{d\epsilon}\left(\frac{\partial \mathbf{f}}{\partial \mathbf{w}}\bigg|_{\mathbf{w}_R}\right)\bigg|_{\epsilon=0} \mathbf{Y}\mathbf{e}_j + \frac{\partial \mathbf{f}}{\partial \mathbf{w}}\bigg|_{\mathbf{w}_L} \frac{d^2 \mathbf{w}_R}{d\epsilon^2}\bigg|_{\epsilon=0} \\ = \frac{d}{d\epsilon}\left(\frac{\partial \mathbf{u}}{\partial \mathbf{w}}\bigg|_{\mathbf{w}_R}\right)\bigg|_{\epsilon=0} \mathbf{Y}\mathbf{e}_j\lambda_j + \frac{\partial \mathbf{u}}{\partial \mathbf{w}}\bigg|_{\mathbf{w}_L} \frac{d^2 \mathbf{w}_R}{d\epsilon^2}\bigg|_{\epsilon=0} \lambda_j + \frac{\partial \mathbf{u}}{\partial \mathbf{w}}\bigg|_{\mathbf{w}_L} \mathbf{Y}\mathbf{e}_j \frac{d\sigma}{d\epsilon}\bigg|_{\epsilon=0} 2.\end{aligned} \tag{4.26}$$

We can also differentiate the definition of the jth characteristic direction and speed,

$$\frac{\partial \mathbf{f}}{\partial \mathbf{w}}\bigg|_{\mathbf{w}_R(\epsilon)} \mathbf{Y}(\mathbf{w}_R(\epsilon))\mathbf{e}_j = \frac{\partial \mathbf{u}}{\partial \mathbf{w}} \mathbf{Y}(\mathbf{w}_R(\epsilon))\mathbf{e}_j \lambda_j(\mathbf{w}_R(\epsilon)),$$

to get

$$\begin{aligned}\frac{d}{d\epsilon}\left(\frac{\partial \mathbf{f}}{\partial \mathbf{w}}\bigg|_{\mathbf{w}_R}\right)\bigg|_{\epsilon=0} \mathbf{Y}\mathbf{e}_j + \frac{\partial \mathbf{F}\mathbf{n}}{\partial \mathbf{w}}\bigg|_{\mathbf{w}_L} \frac{d\mathbf{Y}\mathbf{e}_j}{d\epsilon}\bigg|_{\epsilon=0} \\ = \frac{d}{d\epsilon}\left(\frac{\partial \mathbf{u}}{\partial \mathbf{w}}(\mathbf{w}_R)\right)\bigg|_{\epsilon=0} \mathbf{Y}\mathbf{e}_j\lambda_j + \frac{\partial \mathbf{u}}{\partial \mathbf{w}}(\mathbf{w}_L) \frac{d\mathbf{Y}\mathbf{e}_j}{d\epsilon}\bigg|_{\epsilon=0} \lambda_j + \frac{\partial \mathbf{u}}{\partial \mathbf{w}}(\mathbf{w}_L)\mathbf{Y}\mathbf{e}_j \frac{d\lambda_j}{\partial \epsilon}\bigg|_{\epsilon=0}.\end{aligned}$$

If we subtract this equation from (4.26), we obtain

$$\begin{aligned}\frac{\partial \mathbf{f}}{\partial \mathbf{w}}\bigg|_{\mathbf{w}_L} \left\{ \frac{d^2 \mathbf{w}_R}{d\epsilon^2}\bigg|_{\epsilon=0} - \frac{d\mathbf{Y}\mathbf{e}_j}{d\epsilon}\bigg|_{\epsilon=0} \right\} \\ = \frac{\partial \mathbf{u}}{\partial \mathbf{w}}\bigg|_{\mathbf{w}_L} \left\{ \frac{d^2 \mathbf{w}_R}{d\epsilon^2}\bigg|_{\epsilon=0} - \frac{d\mathbf{Y}\mathbf{e}_j}{d\epsilon}\bigg|_{\epsilon=0} \right\} \lambda_j + \frac{\partial \mathbf{u}}{\partial \mathbf{w}}\bigg|_{\mathbf{w}_L} \mathbf{Y}\mathbf{e}_j \left\{ 2\frac{d\sigma}{d\epsilon}\bigg|_{\epsilon=0} - \frac{d\lambda_j}{\partial \epsilon}\bigg|_{\epsilon=0} \right\}.\end{aligned} \tag{4.27}$$

We can multiply this equation by $\mathbf{e}_j^\top \mathbf{Y}^{-1}(\frac{\partial \mathbf{u}}{\partial \mathbf{w}})^{-1}$ to get

$$\begin{aligned}\mathbf{e}_j^\top \Lambda \mathbf{Y}^{-1} \left\{ \frac{d^2 \mathbf{w}_R}{d\epsilon^2}\bigg|_{\epsilon=0} - \frac{d\mathbf{Y}\mathbf{e}_j}{d\epsilon}\bigg|_{\epsilon=0} \right\} \\ = \mathbf{e}_j^\top \mathbf{Y}^{-1} \left\{ \frac{d^2 \mathbf{w}_R}{d\epsilon^2}\bigg|_{\epsilon=0} - \frac{d\mathbf{Y}\mathbf{e}_j}{d\epsilon}\bigg|_{\epsilon=0} \right\} \lambda_j + 2\frac{d\sigma}{d\epsilon}\bigg|_{\epsilon=0} - \frac{d\lambda_j}{\partial \epsilon}\bigg|_{\epsilon=0}.\end{aligned}$$

After we cancel equal terms on the two sides of this equation, we obtain

$$\left.\frac{d\lambda_j}{d\epsilon}\right|_{\epsilon=0} = 2\left.\frac{d\sigma}{d\epsilon}\right|_{\epsilon=0}.$$

Thus at the left state $\mathbf{w}_L$, the characteristic speed is changing twice as fast as the shock speed along the Hugoniot locus.

Note that Equation (4.27) can be rewritten

$$\left\{\left.\frac{\partial \mathbf{f}}{\partial \mathbf{w}}\right|_{\mathbf{w}_L} - \left.\frac{\partial \mathbf{u}}{\partial \mathbf{w}}\right|_{\mathbf{w}_L}\lambda_j\right\}\left\{\left.\frac{d^2\mathbf{w}_R}{d\epsilon^2}\right|_{\epsilon=0} - \left.\frac{d\mathbf{Y}\mathbf{e}_j}{d\epsilon}\right|_{\epsilon=0}\right\}$$
$$= \frac{\partial \mathbf{u}}{\partial \mathbf{w}}(\mathbf{w}_L)\mathbf{Y}\mathbf{e}_j\left\{2\left.\frac{d\sigma}{d\epsilon}\right|_{\epsilon=0} - \left.\frac{d\lambda_j}{\partial\epsilon}\right|_{\epsilon=0}\right\} = 0.$$

Thus

$$\left.\frac{d^2\mathbf{w}_R}{d\epsilon^2}\right|_{\epsilon=0} - \left.\frac{d\mathbf{Y}\mathbf{e}_j}{d\epsilon}\right|_{\epsilon=0} = \mathbf{Y}\mathbf{e}_j\beta$$

for some scalar β. Because the left-hand side in this equation involves derivatives of different order, we can parameterize ϵ so that $\beta = 0$. □

The last equation in (4.24) says that the shock curve is continuous to second order with the integral curve of the characteristic direction. Recall that the Lax admissibility condition requires for $\epsilon \neq 0$,

$$\lambda_j(\mathbf{w}_L) = \lambda_j(\mathbf{w}_R(0)) > \sigma(\epsilon) > \lambda_j(\mathbf{w}_R(\epsilon)) \approx \lambda_j(\mathbf{w}_L) + \epsilon.$$

Thus the Hugoniot locus $\mathbf{w}_R(\epsilon)$ only involves parameters $\epsilon < 0$. In other words, the Hugoniot locus $\mathbf{w}_R(\epsilon)$ must go out from $\mathbf{w}_L$ in the direction of decreasing characteristic speed.

Example 4.1.17 Let us perform an asymptotic analysis of the jump conditions for the shallow water equations. Suppose that

$$\sigma = v_L \mp \sqrt{gh_L} - \frac{1}{2}\epsilon,$$

so that the shock speed is a small perturbation of one of the characteristic speeds. Then the speed of the shock relative to the fluid velocity on the left is

$$\xi_L = \sigma - v_L = \mp\sqrt{gh_L} - \frac{1}{2}\epsilon.$$

From the equation for h_R in Example 4.1.12 we can also see that the height of the fluid to the right of the shock is

$$h_R = \sqrt{\frac{1}{4}h_L^2 + \frac{2h_L}{g}\left(\frac{1}{2}\epsilon \pm \sqrt{gh_L}\right)^2} - \frac{1}{2}h_L \approx \sqrt{\frac{9}{4}h_L^2 \pm 2h_L\sqrt{h_L/g}\epsilon} - \frac{1}{2}h_L$$
$$\approx \frac{3}{2}h_L\left[1 \pm \frac{4}{9}\frac{\epsilon}{\sqrt{gh_L}}\right] - \frac{1}{2}h_L = h_L \pm \frac{2}{3}\sqrt{\frac{h_L}{g}}\epsilon.$$

From this, we find that the relative shock speed on the right is

$$\xi_R = \frac{h_L\xi_L}{h_R} \approx \frac{h_L\left[\mp\sqrt{gh_L} - \frac{1}{2}\epsilon\right]}{h_L \pm \frac{2}{3}\sqrt{\frac{h_L}{g}}\epsilon} \approx \mp\left[\sqrt{gh_L} \pm \frac{1}{2}\epsilon\right]\left[1 \pm \frac{2}{3}\frac{\epsilon}{\sqrt{h_L g}}\epsilon\right]$$
$$\approx \mp\left[\sqrt{gh_L} \pm \frac{1}{2}\epsilon\right]\left[1 \mp \frac{2}{3}\frac{\epsilon}{\sqrt{h_L g}}\epsilon\right] \approx \mp\sqrt{gh_L} + \frac{1}{6}\epsilon.$$

Also, the fluid velocity to the right of the shock is

$$v_R = \sigma - \xi_R \approx \left[v_L \mp \sqrt{gh_L} - \frac{1}{2}\epsilon\right] - \left[\mp\sqrt{gh_L} + \frac{1}{6}\epsilon\right] = v_L - \frac{2}{3}\epsilon.$$

It follows that the characteristic speed to the right of the shock is

$$v_R - \sqrt{gh_R} \approx v_L - \frac{2}{3}\epsilon \mp \sqrt{gh_L \pm \frac{2}{3}\sqrt{gh_L}\epsilon} \approx v_L - \frac{2}{3}\epsilon \mp \sqrt{gh_L}\sqrt{1 \mp \frac{2}{3}\frac{\epsilon}{\sqrt{gh_L}}}$$
$$\approx v_L - \frac{2}{3}\epsilon \mp \sqrt{gh_L}\left(1 \mp \frac{1}{3}\frac{\epsilon}{\sqrt{gh_L}}\right) = v_L \mp \sqrt{gh_L} - \epsilon.$$

This result is consistent with the general result we obtained above. We also note that

$$\mathbf{w}_R = \begin{bmatrix} h_R \\ v_R \end{bmatrix} \approx \begin{bmatrix} h_L \\ v_L \end{bmatrix} - \begin{bmatrix} \mp\sqrt{h_L/g} \\ 1 \end{bmatrix}\frac{2}{3}\epsilon$$

is perturbed by the appropriate eigenvector of the system.

4.1.7 Centered Rarefactions

Suppose that the conservation law $\frac{\partial \mathbf{u}}{\partial t} + \sum_{i=1}^{k} \frac{\partial \mathbf{Fe}_i}{\partial \mathbf{x}_i} = 0$ has a continuously differentiable **self-similar** solution involving **flux variables** $\mathbf{w}(\mathbf{x}, t) = \tilde{\mathbf{w}}(\mathbf{x} \cdot \mathbf{n}/t)$. Then

$$\frac{\partial \mathbf{w}}{\partial t} = -\tilde{\mathbf{w}}'\frac{\mathbf{x} \cdot \mathbf{n}}{t^2} \text{ and } \frac{\partial \mathbf{w}}{\partial \mathbf{x}} = \tilde{\mathbf{w}}'\frac{1}{t}\mathbf{n}^\top.$$

We can substitute these values into the conservation law to get

$$0 = \frac{\partial \mathbf{u}}{\partial \mathbf{w}} \frac{\partial \mathbf{w}}{\partial t} + \sum_{i=1}^{k} \frac{\partial \mathbf{F} \mathbf{e}_i}{\partial \mathbf{w}} \frac{\partial \mathbf{w}}{\partial \mathbf{x}_i} = \left\{ \frac{\partial \mathbf{F} \mathbf{n}}{\partial \mathbf{w}} - \frac{\partial \mathbf{u}}{\partial \mathbf{w}} \frac{\mathbf{x} \cdot \mathbf{n}}{t} \right\} \tilde{\mathbf{w}}' \frac{1}{t}.$$

It follows that $\tilde{\mathbf{w}}'$ is an eigenvector of $(\frac{\partial \mathbf{u}}{\partial \mathbf{w}})^{-1} \frac{\partial \mathbf{F} \mathbf{n}}{\partial \mathbf{w}}$, and $\mathbf{x} \cdot \mathbf{n}/t$ is the corresponding eigenvalue. In other words, for some index j and some scalar α $\tilde{\mathbf{w}}' = \mathbf{Y} \mathbf{e}_j \alpha$ and $\mathbf{x} \cdot \mathbf{n}/t = \lambda_j$. Since we have normalized the genuinely nonlinear eigenvectors so that $\frac{\partial \lambda_j}{\partial \mathbf{w}} \mathbf{Y} \mathbf{e}_j = 1$, we have

$$\frac{\partial \lambda_j}{\partial \mathbf{w}} \tilde{\mathbf{w}}' = \frac{\partial \lambda_j}{\partial \mathbf{w}} \mathbf{Y} \mathbf{e}_j \alpha = \alpha.$$

These results motivate the following definition.

Definition 4.1.18 Consider the hyperbolic system of conservation laws

$$\frac{\partial \mathbf{u}(\mathbf{w})}{\partial t} + \sum_{i=1}^{k} \frac{\partial \mathbf{F}(\mathbf{w}) \mathbf{e}_i}{\partial \mathbf{x}_i} = 0$$

where $\mathbf{u}, \mathbf{w} \in \mathbf{R}^m$ and for any fixed unit vector $\mathbf{n}$

$$\frac{\partial \mathbf{F} \mathbf{n}}{\partial \mathbf{w}} \mathbf{Y} = \mathbf{Y} \Lambda.$$

Then the function $\mathbf{w}(\mathbf{x}, t) = \tilde{\mathbf{w}}(\mathbf{x} \cdot \mathbf{n}/t)$ is a **centered rarefaction** if and only if there is some index $1 \le j \le m$ such that

$$\tilde{\mathbf{w}}' = \mathbf{Y}(\tilde{\mathbf{w}}) \mathbf{e}_j \alpha \tag{4.28}$$

for some function α of the similarity variable $\mathbf{x} \cdot \mathbf{n}/t$.

According to the Lax admissibility condition, we solve this ordinary differential equation in the direction of increasing characteristic speed.

Note that we cannot have a centered rarefaction in a linearly degenerate wave family. Since the linearly degenerate wavespeeds are constant along integral curves of the corresponding characteristic direction, we cannot solve (4.28) for $\tilde{\mathbf{w}}'$, so we cannot find an ordinary differential equation for $\tilde{\mathbf{w}}$.

For some problems, it will be possible to find quantities that are constant along the rarefaction curves.

Definition 4.1.19 Consider the hyperbolic system of conservation laws

$$\frac{\partial \mathbf{u}(\mathbf{w})}{\partial t} + \sum_{i=1}^{k} \frac{\partial \mathbf{F}(\mathbf{w})\mathbf{e}_i}{\partial \mathbf{x}_i} = 0$$

where $\mathbf{u}, \mathbf{w} \in \mathbf{R}^m$ and for any unit vector $\mathbf{n}$

$$\frac{\partial \mathbf{F}\mathbf{n}}{\partial \mathbf{w}} \mathbf{Y} = \mathbf{Y}\Lambda.$$

Then $r_\ell(\mathbf{w})$ is a **Riemann invariant** of this system if and only if

$$\text{for all } \mathbf{n} \text{ for all } j \neq \ell \;\; \frac{\partial r_\ell}{\partial \mathbf{w}} \mathbf{Y}\mathbf{e}_j = 0.$$

If we can find a full set of Riemann invariants, then we can alternatively describe the jth centered rarefaction curve as the locus of points $\mathbf{w}$ where the Riemann invariants $r_\ell(\mathbf{w})$ are constant for $\ell \neq j$.

Example 4.1.20 For the shallow water equations (4.2), centered rarefactions satisfy

$$\frac{d}{d\mathbf{y}} \begin{bmatrix} h \\ \mathbf{v} \end{bmatrix} = \begin{bmatrix} \mp\sqrt{h/g} \\ \mathbf{n} \end{bmatrix} \frac{2}{3}.$$

Note that we obtained these equations for the flux variables in lemma 4.1.4 under the assumption that $h > 0$. This implies that $\frac{d\mathbf{v}\cdot\mathbf{n}}{dh} = \mp\sqrt{\frac{g}{h}}$. We can integrate this ordinary differential equation to get

$$2\sqrt{h_R} - 2\sqrt{h_L} = \mp \frac{\mathbf{v}_R \cdot \mathbf{n} - \mathbf{v}_L \cdot \mathbf{n}}{\sqrt{g}},$$

which is equivalent to $\mathbf{v}_R \cdot \mathbf{n} \pm 2\sqrt{gh_R} = \mathbf{v}_L \cdot \mathbf{n} \pm 2\sqrt{gh_L}$. In other words, the quantities $\mathbf{v} \cdot \mathbf{n} \pm 2\sqrt{gh}$ are constant along the centered rarefaction curves.

Summary 4.1.21 *The Riemann invariants for the shallow water equations* (4.2) *are* $r_\pm = \mathbf{v} \cdot \mathbf{n} \pm 2\sqrt{gh}$. *On fast centered rarefactions* r_- *is constant, and on slow centered rarefactions* r_+ *is constant. Any transverse components of velocity* (i.e. *in multiple spatial dimensions) are also constant in slow and fast centered rarefactions.*

Note that centered rarefaction curves cannot be constructed as functions in space unless the characteristic speeds are increasing from left to right. In order for $\mathbf{u}_L$ to lie in space to the left of $\mathbf{u}_R$, characteristic speed λ_L with which it moves must be less than λ_R.

4.1.8 Riemann Problems

Many interesting test problems for numerical methods arise from one-dimensional **Riemann problems**. These take the form

$$\frac{\partial \mathbf{u}}{\partial t} + \frac{\partial \mathbf{f}}{\partial x} = 0 \text{ for } x \in \mathbf{R},\ t > 0$$
$$\mathbf{u}(x,0) = \begin{cases} \mathbf{u}_L, & x < 0 \\ \mathbf{u}_R, & x > 0. \end{cases}$$

These problems are **self-similar**, meaning that the solution is a function of x/t only. For problems satisfying the hypotheses of the Lax admissibility condition, the solutions of the Riemann problems involve a combination of centered rarefaction waves, Hugoniot loci and contact discontinuities, which are discontinuities in linearly degenerate wave families.

To illustrate the solution of a Riemann problem, let us assume that we have two genuinely nonlinear characteristic speeds, and that these two characteristic speeds are the largest and smallest speeds. We construct curves in the state space given by the flux variables $\mathbf{w}$ as follows. From the left state $\mathbf{w}_L$ we construct the centered rarefaction wave in the direction of increasing smallest characteristic speed, and we construct the Hugoniot locus in the direction of decreasing smallest characteristic speed. Note that the admissibility conditions on the shock require that the shock curve be used only in the direction of decreasing characteristic speed. Further, so that the rarefaction can represent a continuous solution the characteristics associated with the slowest wave must not intersect; this implies that the rarefaction curve must be drawn in the direction of increasing characteristic speed. From the right state $\mathbf{w}_R$ we construct the centered rarefaction wave in the direction of decreasing largest characteristic speed, and we construct the Hugoniot locus in the direction of increasing largest characteristic speed. We assume that all of state space $\mathbf{w}$ can be completely parameterized by these curves and the curves $\lambda_i =$ constant for the linearly degenerate characteristic speeds λ_i. This assumption leads to a unique path from $\mathbf{w}_L$ to $\mathbf{w}_R$; this path is the solution of the Riemann problem.

The solution of two-dimensional Riemann problems is more difficult. See [104, 105, 145, 146].

4.1.9 Riemann Problem for Linear Systems

For linear hyperbolic systems, the solution of Riemann problems takes a simple form.

Lemma 4.1.22 *Suppose that* $\mathbf{u}(x,t) \in \mathbf{R}^m$ *solves the linear constant coefficient system*

$$\frac{\partial \mathbf{u}}{\partial t} + \mathbf{A}\frac{\partial \mathbf{u}}{\partial x} = 0, \quad \mathbf{u}(x,0) = \begin{cases} \mathbf{u}_L, & x < 0 \\ \mathbf{u}_R, & x > 0, \end{cases}$$

where the eigenvectors and eigenvalues of $\mathbf{A}$ *are given by* $\mathbf{AX} = \mathbf{X}\Lambda$ *and* Λ *is diagonal with real entries. Then if* $x/t \neq \lambda_i$ *for any* $1 \leq i \leq m$*, the solution of this Riemann problem is* $u(x,t) = \mathcal{R}(\mathbf{u}_L, \mathbf{u}_R, x/t)$ *where*

$$\begin{aligned} \mathcal{R}(\mathbf{u}_L, \mathbf{u}_R; \xi) &= \mathbf{u}_L + \frac{1}{2}[\mathbf{I} - \text{sign}(\mathbf{A} - \mathbf{I}\xi)](\mathbf{u}_R - \mathbf{u}_L) \\ &= \mathbf{u}_R - \frac{1}{2}[\mathbf{I} + \text{sign}(\mathbf{A} - \mathbf{I}\xi)](\mathbf{u}_R - \mathbf{u}_L) \\ &= (\mathbf{u}_R + \mathbf{u}_L)\frac{1}{2} - \text{sign}(\mathbf{A} - \mathbf{I}\xi)(\mathbf{u}_R - \mathbf{u}_L)\frac{1}{2}. \end{aligned}$$

Further, the flux evaluated at the solution of this Riemann problem is

$$\begin{aligned} \mathbf{A}\mathcal{R}(\mathbf{u}_L, \mathbf{u}_R; \xi) &= \mathbf{Au}_L + \frac{1}{2}[\mathbf{A} - \mathbf{A}\ \text{sign}(\mathbf{A} - \mathbf{I}\xi)](\mathbf{u}_R - \mathbf{u}_L) \\ &= \mathbf{Au}_R - \frac{1}{2}[\mathbf{A} + \mathbf{A}\ \text{sign}(\mathbf{A} - \mathbf{I}\xi)](\mathbf{u}_R - \mathbf{u}_L) \\ &= \mathbf{A}(\mathbf{u}_R + \mathbf{u}_L)\frac{1}{2} - \mathbf{A}\ \text{sign}(\mathbf{A} - \mathbf{I}\xi)(\mathbf{u}_R - \mathbf{u}_L)\frac{1}{2}. \end{aligned}$$

In particular, even if zero is an eigenvalue of $\mathbf{A}$*, the flux at the state moving with zero speed in the solution of the Riemann problem is*

$$\mathbf{A}\mathcal{R}(\mathbf{u}_L, \mathbf{u}_R; 0) = \mathbf{A}^+\mathbf{u}_L + \mathbf{A}^-\mathbf{u}_R,$$

where

$$\mathbf{A}^+ \equiv \frac{1}{2}(\mathbf{A} + |\mathbf{A}|) \text{ and } \mathbf{A}^- \equiv \frac{1}{2}(\mathbf{A} - |\mathbf{A}|).$$

Proof Let $\phi(\xi)$ be continuously differentiable and such that $\phi(\xi) = 0$ for $\xi < 0$ and $\phi(\xi) = 1$ for $\xi > 1$. Given left and right states $\mathbf{u}_L$ and $\mathbf{u}_R$ for the linear conservation law, consider the continuously differentiable initial data

$$\mathbf{u}_\epsilon(x,0) = \mathbf{u}_L + \sum_{j=1}^{m} \mathbf{X}\mathbf{e}_j\phi\left(\frac{x - 2(j-1)\epsilon}{\epsilon}\right)\mathbf{e}_j^\top \mathbf{X}^{-1}(\mathbf{u}_R - \mathbf{u}_L).$$

Note that $\mathbf{u}_\epsilon(x,0) = \mathbf{u}_L$ for $x < 0$ and $\mathbf{u}_\epsilon(x,0) = \mathbf{u}_R$ for $x > (2m-1)\epsilon$. Then Lemma 4.1.7 shows that the solution of the linear conservation law with initial data $\mathbf{u}_\epsilon$ is

$$\mathbf{u}_\epsilon(x,t) = \mathbf{u}_L + \sum_{j=1}^{m} \mathbf{X}\mathbf{e}_j \phi\left(\frac{x - \lambda_j t - 2(j-1)\epsilon}{\epsilon}\right) \mathbf{e}_j^\top \mathbf{X}^{-1}(\mathbf{u}_R - \mathbf{u}_L).$$

The solution of the original Riemann problem follows by taking the limit as $\epsilon \to 0$.

We compute the characteristic expansion coefficients $\mathbf{y}$ for the jump in the solution by solving $\mathbf{X}\mathbf{y} = \mathbf{u}_R - \mathbf{u}_L$. Each of these characteristic expansion coefficients are associated with discontinuities moving at the different characteristic speeds. Provided that $\xi \notin \{\lambda_j\}$, the part of the solution of the Riemann problem that moves with speed ξ can be written

$$\begin{aligned}
\mathcal{R}(\mathbf{u}_L, \mathbf{u}_R; \xi) &= \mathbf{u}_L + \sum_{j:\lambda_j < \xi} \mathbf{X}\mathbf{e}_j \mathbf{e}_j^\top \mathbf{X}^{-1}(\mathbf{u}_R - \mathbf{u}_L) \\
&= \mathbf{u}_L + \frac{1}{2}\sum_j \mathbf{X}\mathbf{e}_j[1 - \operatorname{sign}(\lambda_j - \xi)]\mathbf{e}_j^\top \mathbf{X}^{-1}(\mathbf{u}_R - \mathbf{u}_L) \\
&= \mathbf{u}_L + \frac{1}{2}[\mathbf{I} - \operatorname{sign}(\mathbf{A} - \mathbf{I}\xi)](\mathbf{u}_R - \mathbf{u}_L)
\end{aligned}$$

or

$$\begin{aligned}
\mathcal{R}(\mathbf{u}_L, \mathbf{u}_R; \xi) &= \mathbf{u}_R - \sum_{j:\lambda_j > \xi} (\mathbf{X}\mathbf{e}_j)(\mathbf{e}_j^\top \mathbf{y}) \\
&= \mathbf{u}_R - \sum_{j:\lambda_j > \xi} \mathbf{X}\mathbf{e}_j \mathbf{e}_j^\top \mathbf{X}^{-1}(\mathbf{u}_R - \mathbf{u}_L) \\
&= \mathbf{u}_R - \frac{1}{2}\sum_j \mathbf{X}\mathbf{e}_j[\operatorname{sign}(\lambda_j - \xi) + 1]\mathbf{e}_j^\top \mathbf{X}^{-1}(\mathbf{u}_R - \mathbf{u}_L) \\
&= \mathbf{u}_R - \frac{1}{2}[\mathbf{I} + \operatorname{sign}(\mathbf{A} - \mathbf{I}\xi)](\mathbf{u}_R - \mathbf{u}_L).
\end{aligned}$$

We can average these two results to get

$$\mathcal{R}(\mathbf{u}_L, \mathbf{u}_R; \xi) = (\mathbf{u}_R + \mathbf{u}_L)\frac{1}{2} - \operatorname{sign}(\mathbf{A} - \mathbf{I}\xi)(\mathbf{u}_R - \mathbf{u}_L)\frac{1}{2}.$$

Provided that $\xi \notin \{\lambda_j\}$, the flux $\mathbf{f}(\mathbf{u}) = \mathbf{A}\mathbf{u}$ at the solution to the Riemann problem is

$$\begin{aligned}
\mathbf{f}(\mathcal{R}(\mathbf{u}_L, \mathbf{u}_R; \xi)) &= \mathbf{A}\mathbf{u}_L + \frac{1}{2}[\mathbf{A} - \mathbf{A}\operatorname{sign}(\mathbf{A} - \mathbf{I}\xi)](\mathbf{u}_R - \mathbf{u}_L) \\
&= \mathbf{A}\mathbf{u}_R - \frac{1}{2}[\mathbf{A} + \mathbf{A}\operatorname{sign}(\mathbf{A} - \mathbf{I}\xi)](\mathbf{u}_R - \mathbf{u}_L) \\
&= \mathbf{A}(\mathbf{u}_L + \mathbf{u}_R)\frac{1}{2} - [\mathbf{A}\operatorname{sign}(\mathbf{A} - \mathbf{I}\xi)](\mathbf{u}_R - \mathbf{u}_L)\frac{1}{2}.
\end{aligned}$$

In the special case where $\xi = 0$, we have

$$\begin{aligned}\mathbf{f}(\mathcal{R}(\mathbf{u}_L, \mathbf{u}_R; 0)) &= \mathbf{A}(\mathbf{u}_L + \mathbf{u}_R)\frac{1}{2} - |\mathbf{A}|(\mathbf{u}_R - \mathbf{u}_L)\frac{1}{2} \\ &= \frac{1}{2}[\mathbf{A} + |\mathbf{A}|]\mathbf{u}_L + \frac{1}{2}[\mathbf{A} - |\mathbf{A}|]\mathbf{u}_R \\ &= \mathbf{A}^+\mathbf{u}_L + \mathbf{A}^-\mathbf{u}_R.\end{aligned}$$

Here we have defined

$$\begin{aligned}\mathbf{A}^+ &= \frac{1}{2}[\mathbf{A} + |\mathbf{A}|] = \sum_j \mathbf{X}\mathbf{e}_j \frac{\lambda_j + |\lambda_j|}{2}\mathbf{e}_j^\top \mathbf{X}^{-1} = \sum_j \mathbf{X}\mathbf{e}_j \max\{\lambda_j, 0\}\mathbf{e}_j^\top \mathbf{X}^{-1} \\ &= \sum_{j:\lambda_j > 0} \mathbf{X}\mathbf{e}_j \lambda_j \mathbf{e}_j^\top \mathbf{X}^{-1}\end{aligned}$$

and

$$\begin{aligned}\mathbf{A}^- &= \frac{1}{2}[\mathbf{A} - |\mathbf{A}|] = \sum_j \mathbf{X}\mathbf{e}_j \frac{\lambda_j - |\lambda_j|}{2}\mathbf{e}_j^\top \mathbf{X}^{-1} = \sum_j \mathbf{X}\mathbf{e}_j \min\{\lambda_j, 0\}\mathbf{e}_j^\top \mathbf{X}^{-1} \\ &= \sum_{j:\lambda_j < 0} \mathbf{X}\mathbf{e}_j \lambda_j \mathbf{e}_j^\top \mathbf{X}^{-1}.\end{aligned}$$

Since the formula for the flux does not depend on the sign function when $\xi = 0$, it is valid even if one of the eigenvalues is zero. □

4.1.10 Riemann Problem for Shallow Water

We are now able to describe the solution of the Riemann problem for the shallow water equations. Given a left state $(h_L, \mathbf{v}_L)$ and a right state $(h_R, \mathbf{v}_R)$, define the slow wave curve

$$\mathbf{v}_-(h) \equiv \begin{cases} \mathbf{v}_L - \mathbf{n}(h - h_L)\sqrt{\frac{g}{2}\left(\frac{1}{h_L} + \frac{1}{h}\right)}, & h > h_L \\ \mathbf{v}_L + \mathbf{n}(h_L - h)\frac{2g}{\sqrt{gh_L} + \sqrt{gh}}, & h \le h_L \end{cases}$$

and the fast wave curve

$$\mathbf{v}_+(h) \equiv \begin{cases} \mathbf{v}_R + \mathbf{n}(h - h_R)\sqrt{\frac{g}{2}\left(\frac{1}{h_R} + \frac{1}{h}\right)}, & h > h_R \\ \mathbf{v}_R - \mathbf{n}(h_R - h)\frac{2g}{\sqrt{gh_R} + \sqrt{gh}}, & h \le h_R. \end{cases}$$

If $\mathbf{n} \cdot v_-(h_*) = \mathbf{n} \cdot v_+(h_*)$, let

$$\xi_L = \begin{cases} \mathbf{n} \cdot \mathbf{v}_L - \sqrt{g\frac{h_L + h_*}{2}\frac{h_*}{h_L}}, & h_* > h_L \\ \mathbf{n} \cdot \mathbf{v}_L - \sqrt{gh_L}, & h_* \le h_L \end{cases}$$

and

$$\xi_- = \begin{cases} \mathbf{n} \cdot \mathbf{v}_L - \sqrt{g \frac{h_L + h_*}{2} \frac{h_*}{h_L}}, & h_* > h_L \\ \mathbf{n} \cdot \mathbf{v}_-(h_*) - \sqrt{g h_*}, & h_* \le h_L \end{cases}$$

be the wave speeds at the beginning and the end of the slow wave, and

$$\xi_+ = \begin{cases} \mathbf{n} \cdot \mathbf{v}_R + \sqrt{g \frac{h_R + h_*}{2} \frac{h_*}{h_R}}, & h_* > h_R \\ \mathbf{n} \cdot \mathbf{v}_+(h_*) + \sqrt{g h_*}, & h_* \le h_R \end{cases}$$

and

$$\xi_R = \begin{cases} \mathbf{n} \cdot \mathbf{v}_R + \sqrt{g \frac{h_R + h_*}{2} \frac{h_*}{h_R}}, & h_* > h_R \\ \mathbf{n} \cdot \mathbf{v}_R + \sqrt{g h_R}, & h_* \le h_R \end{cases}$$

be the wave speeds at the beginning and the end of the fast wave.

Given a left state $(h_L, \mathbf{v}_L)$, the Rankine-Hugoniot jump conditions satisfy the Lax admissibility condition on the shock in the direction of increasing h: $h_R > h_L$; this implies decreasing $\mathbf{n} \cdot \mathbf{v}$ along the Hugoniot locus. The slow rarefaction curve proceeds out of the left state $(h_L, \mathbf{v}_L)$ in the direction of decreasing h and increasing $\mathbf{n} \cdot \mathbf{v}$. Similarly, the fast wave family has characteristic speed $\mathbf{n} \cdot \mathbf{v} + \sqrt{gh}$ and Riemann invariant $\mathbf{n} \cdot \mathbf{v} - 2\sqrt{gh}$. Given a right state $(h_R, \mathbf{v}_R)$, the Rankine–Hugoniot jump conditions satisfy the Lax admissibility condition on the shock in the direction of increasing h: $h_L > h_R$; this implies increasing $\mathbf{n} \cdot \mathbf{v}$ along the Hugoniot locus. The fast rarefaction curve proceeds out of the right state $(h_R, \mathbf{v}_R)$ in the direction of decreasing h and decreasing $\mathbf{n} \cdot \mathbf{v}$. Thus the slow wave family has negative slope $d\mathbf{n} \cdot \mathbf{v}/dh$ and the fast wave family always has positive slope $d\mathbf{n} \cdot \mathbf{v}/dh$. If

$$\mathbf{n} \cdot \mathbf{v}_L + 2\sqrt{g h_L} > \mathbf{n} \cdot \mathbf{v}_R - 2\sqrt{g h_R},$$

then the slow wave curve out of $(h_L, \mathbf{v}_L)$ must intersect the water height and normal velocity of the fast wave curve out of $(h_R, \mathbf{v}_R)$, at some point where $(h_*, \mathbf{n} \cdot \mathbf{v}_-(h_*)) = (h_*, \mathbf{n} \cdot \mathbf{v}_+(h_*))$. The rest of the Riemann problem solution follows from the results in the discussions in examples 4.1.12 and 4.1.20.

Summary 4.1.23 *Given a left state $(h_L, \mathbf{v}_L)$ and a right state $(h_R, \mathbf{v}_R)$, the solution of the one-dimensional Riemann problem for the shallow water equations* (4.2) *involves (in multiple dimensions) the contact discontinuity jumping from the slow wave curve $v_-(h)$ and the fast wave curve $v_+(h)$. There are four different structural forms for the solution of the Riemann problem: either a slow shock or a slow rarefaction, a contact discontinuity, and then either a fast shock or a fast rarefaction.*

Further, the state that moves with speed ξ in the Riemann problem is

$$(h_\xi, \mathbf{v}_\xi) = \begin{cases} (h_L, \mathbf{v}_L), & \xi < \xi_L \\ ([\xi - \xi_L - 3\sqrt{gh_L}]^2/(9g), \mathbf{v}_-(h_\xi), & \xi_L < \xi < \xi_- \\ (h_*, \mathbf{v}_-(h_*)), & \xi_- < \xi < \mathbf{n} \cdot \mathbf{v}_-(h_*) \\ (h_*, \mathbf{v}_+(h_*)), & \mathbf{n} \cdot \mathbf{v}_+(h_*) < \xi < \xi_+ \\ ([\xi - \xi_R + 3\sqrt{gh_R}]^2/(9g), \mathbf{v}_+(h_\xi), & \xi_+ < \xi < \xi_R \\ (h_R, \mathbf{v}_R), & \xi_R < \xi \end{cases}$$

A program to solve the Riemann problem for the shallow water equations can be found in **Program 4.1-45: Riemann Solver for Shallow Water Equations** This program contains a function `slowwavesw` to find points on the slow wave curve given the left state, and a function `fastwavesw` to find points on the fast wave curve given the right state. The procedure `solve_riemann_sw` solves the Riemann problem by using Newton's method to find the water height h at which the two wave curves intersect (*i.e.*, have the same velocity).

You can also execute this program by clicking on **Executable 4.1-16: guiShallow Water** Once you have selected the left state for the Riemann problem with the mouse, you can click and drag in the image to select a right state for the Riemann problem, and see how the solution of the Riemann problems varies as a function of the right state.

In Figure 4.1 we show the analytical solution of the shallow water Riemann problem for a problem involving two rarefactions. Note that in a rarefaction, the characteristic speed associated with that rarefaction increases from left to right; further, in the plot of characteristic speed versus x/t, the relevant characteristic speed for a rarefaction plots as a straight line with slope one with zero intercept. Figure 4.2 shows the solution of a shallow water Riemann problem involving a rarefaction and a shock. Note that the characteristic speed decreases from left to right in the wave family associated with the shock. Figures 4.3 and 4.4 show solutions to shallow water Riemann problems involving a shock and a rarefaction, or two shocks.

4.1.11 Entropy Functions

In many cases, there are physically meaningful functions associated with the selection of the admissible discontinuities. We will call these functions **entropy functions**, although they may not necessarily have the physical units of entropy.

Given a conservation law

$$\frac{\partial \mathbf{u}(\mathbf{w})}{\partial t} + \sum_{i=1}^{k} \frac{\partial \mathbf{F}(\mathbf{w})\mathbf{e}_i}{\partial \mathbf{x}_i} = 0$$

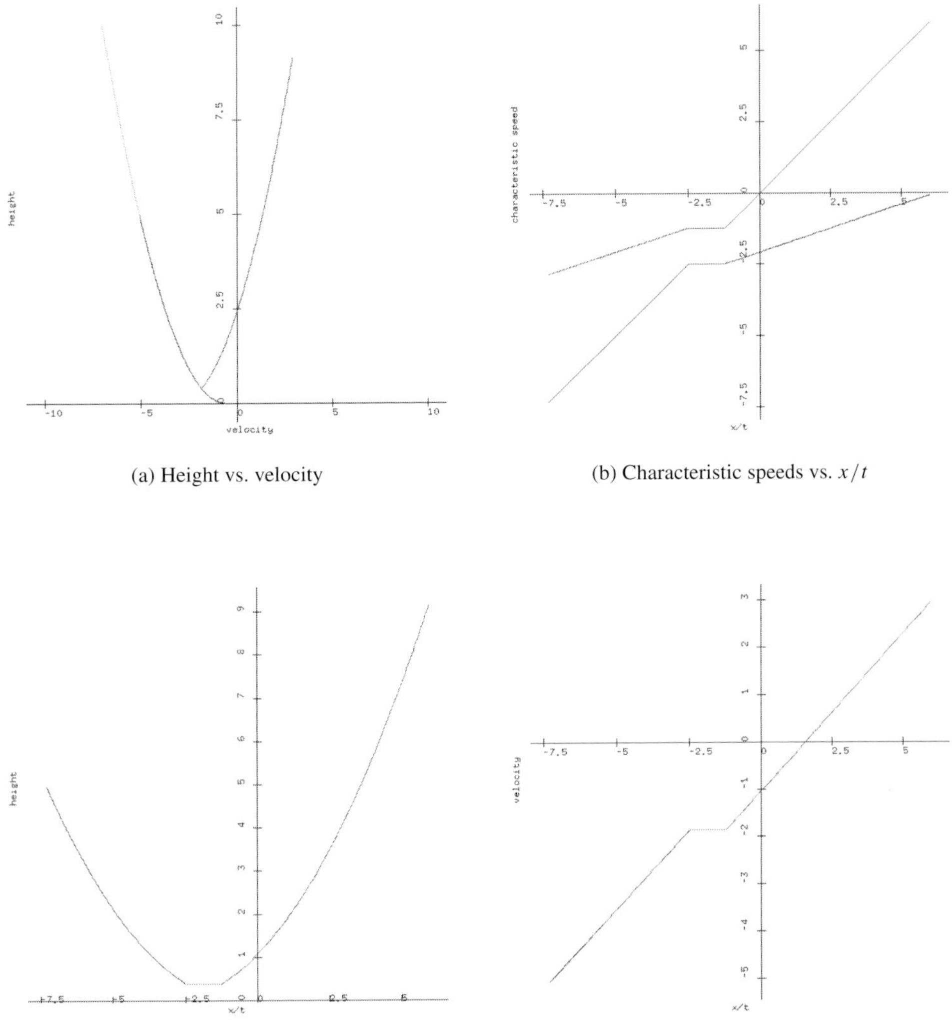

(a) Height vs. velocity

(b) Characteristic speeds vs. x/t

(c) Height vs. x/t

(d) Velocity vs. x/t

Fig. 4.1 Shallow water Riemann problem (rarefaction–rarefaction)

where $\mathbf{w}$ is a vector of flux variables, we want to find an entropy function $S(\mathbf{w})$ and entropy flux $\Psi(\mathbf{w})$ so that

$$\text{for all } \mathbf{x} \text{ for all } t > 0 \text{ for all } \mathbf{n} \text{ fixed, } \frac{\partial \mathbf{n}^\top \Psi}{\partial \mathbf{w}} = \frac{\partial S}{\partial \mathbf{w}} \left(\frac{\partial \mathbf{u}}{\partial \mathbf{w}} \right)^{-1} \frac{\partial \mathbf{F}\mathbf{n}}{\partial \mathbf{w}}. \tag{4.29}$$

If this equality is satisfied and $\mathbf{u}$ is continuously differentiable, then entropy is

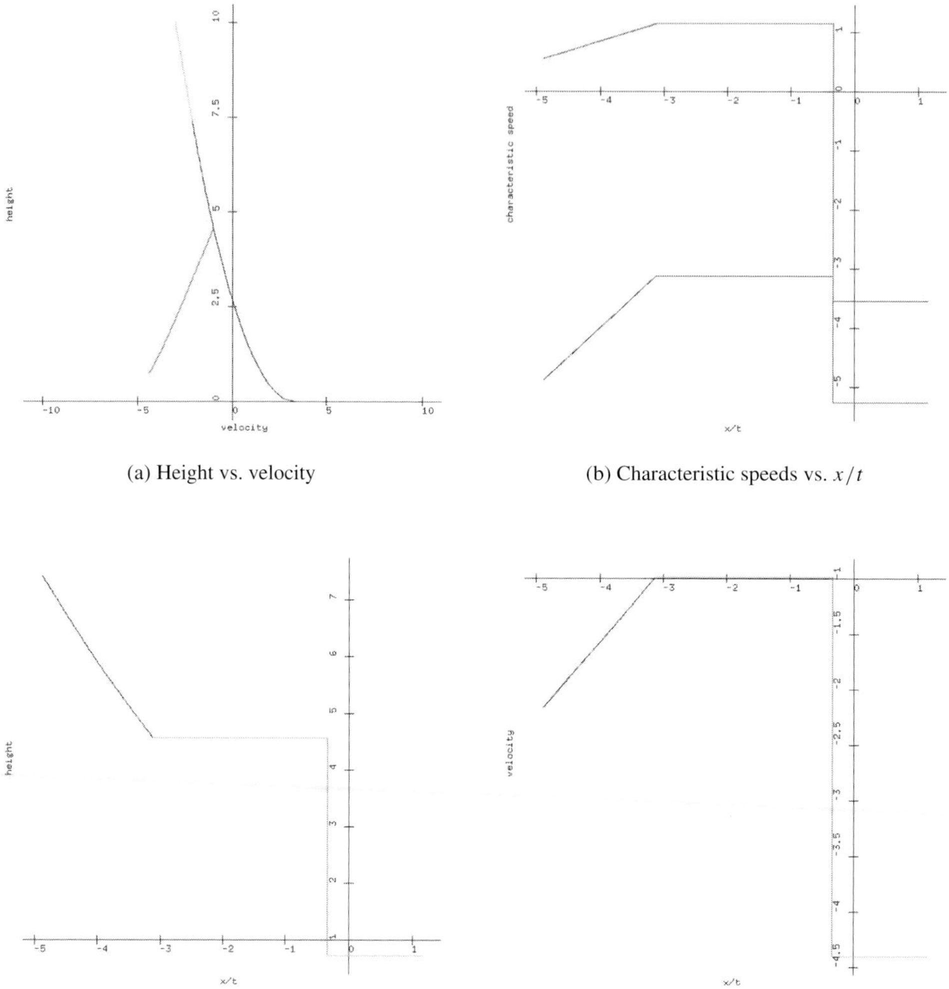

(a) Height vs. velocity

(b) Characteristic speeds vs. x/t

(c) Height vs. x/t

(d) Velocity vs. x/t

Fig. 4.2 Shallow water Riemann problem (rarefaction–shock)

conserved:

$$0 = \frac{\partial S}{\partial \mathbf{w}} \left(\frac{\partial \mathbf{u}}{\partial \mathbf{w}} \right)^{-1} \left[\frac{\partial \mathbf{u}}{\partial \mathbf{w}} \frac{\partial \mathbf{w}}{\partial t} + \sum_{i=1}^{k} \frac{\partial \mathbf{F}(\mathbf{w})\mathbf{e}_i}{\partial \mathbf{w}} \frac{\partial \mathbf{w}}{\partial \mathbf{x}_i} \right] = \frac{\partial S}{\partial \mathbf{w}} \frac{\partial \mathbf{w}}{\partial t} + \sum_{i=1}^{k} \frac{\mathbf{e}_i^\top \partial \Psi}{\partial \mathbf{w}} \frac{\partial \mathbf{w}}{\partial \mathbf{x}_i}$$

$$= \frac{\partial S}{\partial t} + \sum_{i=1}^{k} \frac{\partial \mathbf{e}_i^\top \Psi}{\partial \mathbf{x}_i}. \tag{4.30}$$

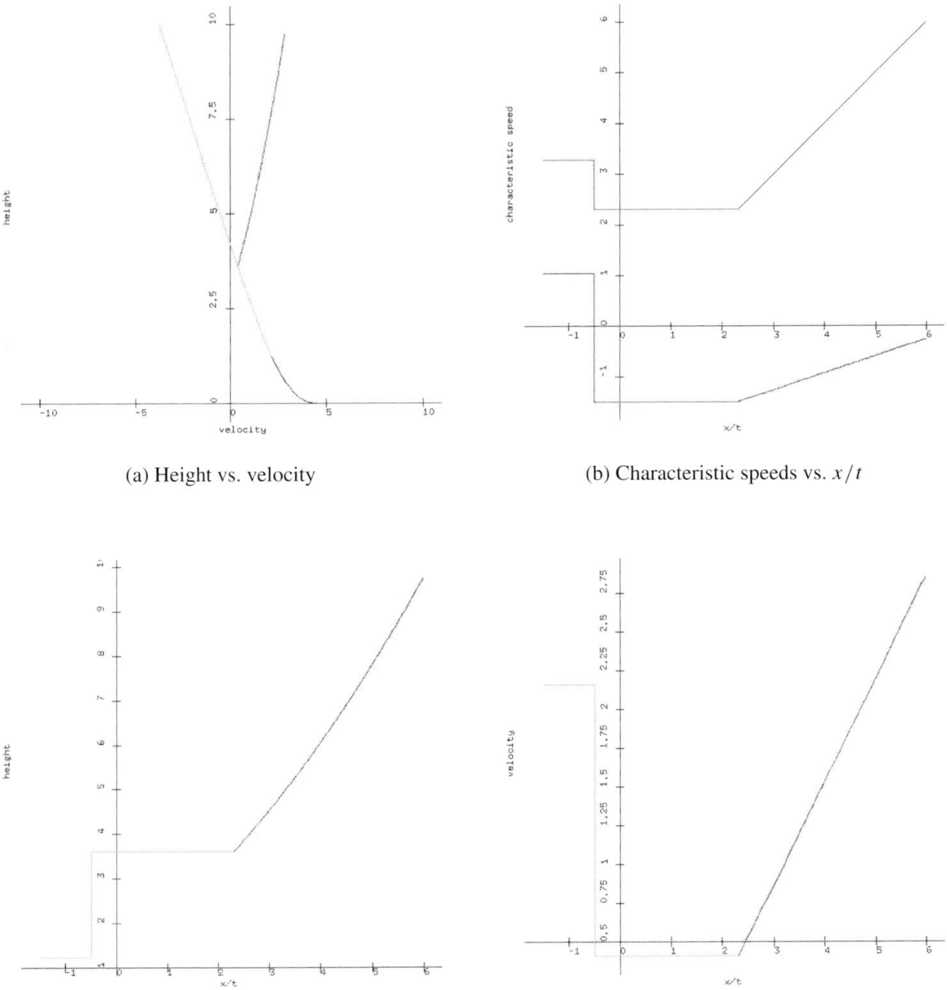

(a) Height vs. velocity

(b) Characteristic speeds vs. x/t

(c) Height vs. x/t

(d) Velocity vs. x/t

Fig. 4.3 Shallow water Riemann problem (shock–rarefaction)

Example 4.1.24 For the shallow water equations (4.2) the total energy is $E = \frac{1}{2}\rho(h\mathbf{v}\cdot\mathbf{v} + gh^2)$ and the energy flux is $\Psi = \rho(\frac{1}{2}h\mathbf{v}\cdot\mathbf{v} + gh^2)\mathbf{v}^\top$. We can compute

$$\frac{\partial E}{\partial \mathbf{w}} = \rho\left[gh + \tfrac{1}{2}\mathbf{v}\cdot\mathbf{v}, \quad h\mathbf{v}^\top\right]$$

and

$$\frac{\partial \Psi}{\partial \mathbf{w}} = \rho\left[(\tfrac{1}{2}\mathbf{v}\cdot\mathbf{v} + 2gh)\mathbf{v}\cdot\mathbf{n}, \quad gh^2\mathbf{n}^\top + h\mathbf{v}\cdot\mathbf{n}\mathbf{v}^\top + \tfrac{1}{2}h\mathbf{v}\cdot\mathbf{v}\mathbf{n}^\top\right].$$

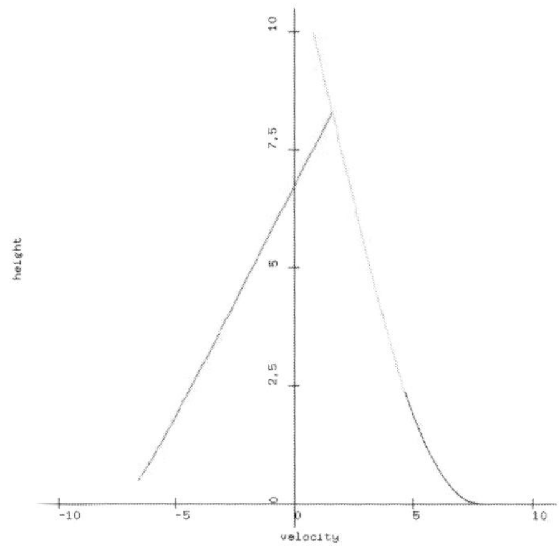

(a) Height vs. velocity

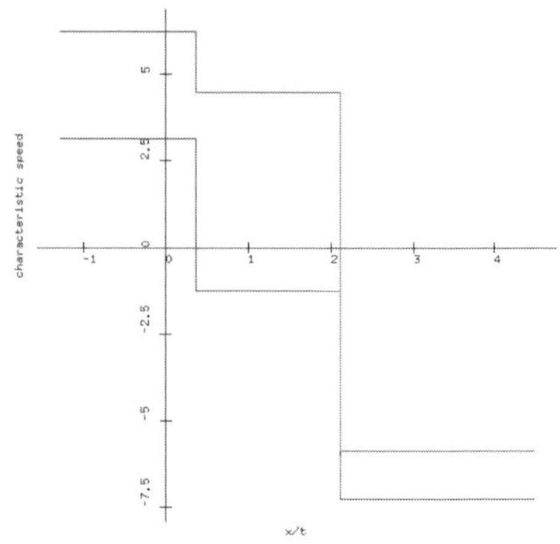

(b) Characteristic speeds vs. x/t

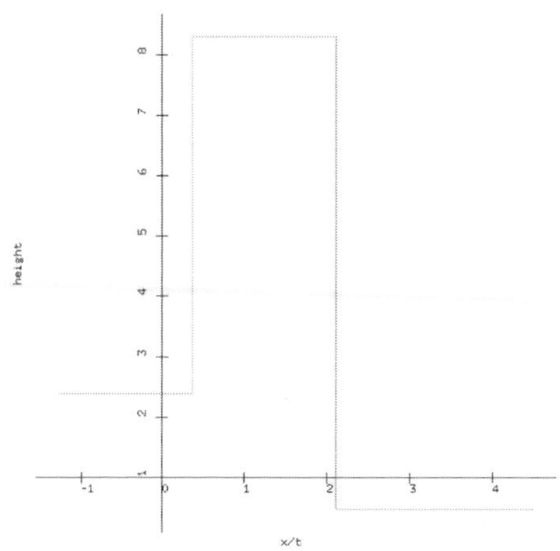

(c) Height vs. x/t

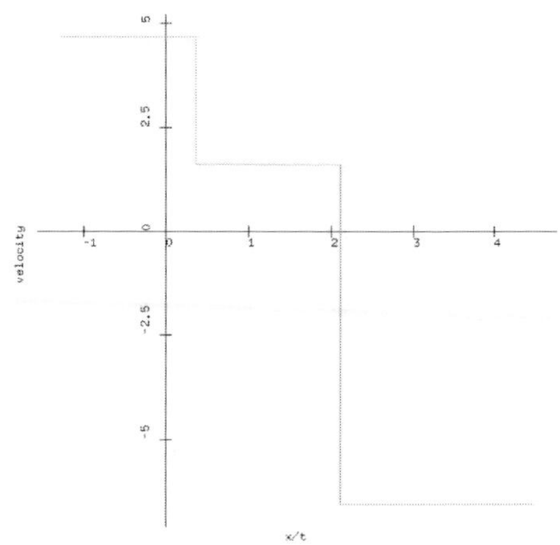

(d) Velocity vs. x/t

Fig. 4.4 Shallow water Riemann problem (shock–shock)

Thus

$$\frac{\partial E}{\partial \mathbf{w}} \left(\frac{\partial \mathbf{u}}{\partial \mathbf{w}} \right)^{-1} \frac{\partial \mathbf{F}}{\partial \mathbf{w}} = \rho \left[gh + \tfrac{1}{2} \mathbf{v} \cdot \mathbf{v}, \quad h\mathbf{v}^\top \right] \begin{bmatrix} \mathbf{v} \cdot \mathbf{n} & h\mathbf{n}^\top \\ \mathbf{n}g & \mathbf{I}\mathbf{v} \cdot \mathbf{n} \end{bmatrix}$$

$$= \rho \left[(\tfrac{1}{2} \mathbf{v} \cdot \mathbf{v} + 2gh) \mathbf{v} \cdot \mathbf{n}, \quad gh^2 \mathbf{n}^\top + h\mathbf{v} \cdot \mathbf{n} \mathbf{v}^\top + \tfrac{1}{2} h \mathbf{v} \cdot \mathbf{v} \mathbf{n}^\top \right] = \frac{\partial \Psi}{\partial \mathbf{w}}.$$

This says that the total energy function is an entropy function for the shallow water equations.

To see whether the total energy function is convex, we compute the matrix of second derivatives

$$\frac{\partial}{\partial \mathbf{w}}\left(\frac{\partial E}{\partial \mathbf{w}}\right)^\top = \begin{bmatrix} g & \mathbf{v}^\top \\ \mathbf{v} & \mathbf{I}h \end{bmatrix} \rho .$$

Attempting a Cholesky factorization of this matrix shows that it is positive definite if and only if $\mathbf{v} \cdot \mathbf{v} < gh$.

Recall that at a propagating discontinuity the Rankine-Hugoniot conditions require

$$[h\mathbf{v} \cdot \mathbf{n}] = [h]\sigma$$
$$\left[\mathbf{v}h\mathbf{v} \cdot \mathbf{n} + \frac{1}{2}\mathbf{n}gh^2\right] = [\mathbf{v}h]\sigma .$$

It follows that

$$\begin{aligned}[E] &= \frac{1}{2}\rho\left[gh^2 + h\mathbf{v}\cdot\mathbf{v}\right] = \frac{1}{2}\rho\left[h(\mathbf{v}\cdot\mathbf{n})^2 + \frac{1}{2}gh^2\right] + \frac{1}{4}\rho\left[gh^2\right] + \frac{1}{2}\rho[h(\|v\|^2 - |\mathbf{v}\cdot\mathbf{n}|^2)] \\ &= \frac{1}{2}\rho[h]\sigma^2 + \frac{1}{2}\rho g[h]\frac{h_L + h_R}{2} + [h]\|v^\perp\|^2 \\ &= \frac{1}{2}\rho[h]\left(\sigma^2 + g\frac{h_L + h_R}{2} + \|v^\perp\|^2\right).\end{aligned}$$

For a slow shock, the Lax admissibility condition requires that $[h] > 0$, which implies that $[E] > 0$. In this case, the velocities of the water relative to the shock are

$$\mathbf{n}\cdot\mathbf{v}_L - \sigma = \sqrt{g\frac{h_L + h_R}{2}\frac{h_R}{h_L}} \quad \text{and} \quad \mathbf{n}\cdot\mathbf{v}_R - \sigma = \sqrt{g\frac{h_L + h_R}{2}\frac{h_L}{h_R}}.$$

In both cases, the velocity of water relative to the shock is positive. Thus in this case the total energy E decreases from the pre-shock state (the right state) to the post-shock state (the left state). Similarly, for a fast wave we have $[h] < 0$ and then $[E] < 0$. On both sides of the shock the velocity of the water relative to the shock is negative, so this condition says that the total energy E decreases from the pre-shock state (the left state) to the post-shock state (the right state).

Next, let us discuss how the entropy function and flux are related, even if $\mathbf{u}$ is not continuously differentiable.

Lemma 4.1.25 *(Lax [89]) Suppose that* $\mathbf{u}(\mathbf{x}, t) \in \mathbf{R}^m$ *solves the hyperbolic conservation law*

$$\frac{\partial \mathbf{u}}{\partial t} + \sum_{i=1}^{k} \frac{\partial \mathbf{F}(\mathbf{u})\mathbf{e}_i}{\partial \mathbf{x}_i} = 0$$

in the limit of vanishing diffusion, meaning that $\mathbf{u}_\epsilon \to \mathbf{u}$ *weakly where* $\mathbf{u}$ *solves the viscous conservation law*

$$\frac{\partial \mathbf{u}_\epsilon}{\partial t} + \sum_{i=1}^{k} \frac{\partial \mathbf{F}(\mathbf{u}_\epsilon)\mathbf{e}_i}{\partial \mathbf{x}_i} = \epsilon \sum_{i=1}^{k} \frac{\partial^2 \mathbf{u}_\epsilon}{\partial \mathbf{x}_i^2}.$$

Further, suppose that there is a concave entropy function $S(\mathbf{u})$ *with entropy flux* $\Psi(\mathbf{u})$ *so that for all fixed directions* $\mathbf{n}$

$$\frac{\partial \mathbf{n}\cdot\Psi(\mathbf{u})}{\partial \mathbf{u}} = \frac{\partial S}{\partial \mathbf{u}} \frac{\partial \mathbf{F}(\mathbf{u})\mathbf{n}}{\partial \mathbf{u}}.$$

If $S(\mathbf{u})$ *is locally bounded for all* $\mathbf{x}$ *and* t*, then*

for all $\phi(\mathbf{x},t) \in C_0^\infty(\mathbf{R}^k \times \mathbf{R})$, $\phi(\mathbf{x},t) \geq 0$,

$$-\int_0^\infty \int_{\mathbf{R}^k} \frac{\partial \phi}{\partial t} S(\mathbf{u}) + \sum_{i=1}^{k} \frac{\partial \phi}{\partial \mathbf{x}_i} \Psi(\mathbf{u}) \cdot \mathbf{e}_i \, d\mathbf{x}\, dt - \int_{\mathbf{R}^k} \phi(\mathbf{x},0) S(\mathbf{u})(\mathbf{x},0)\, d\mathbf{x} \geq 0.$$

If S *is convex, then we replace* $\geq$ *with* $\leq$ *in this inequality.*

Proof As in the one-dimensional case, the maximum principle shows that the viscous conservation law has at most one solution, but the conservation law may have multiple solutions; this is the reason for the assumption that $\mathbf{u}_\epsilon$ converges to $\mathbf{u}$. Also note that the solution $\mathbf{u}_\epsilon$ of the viscous conservation law is smooth for any $\epsilon > 0$.

Since $S(\mathbf{u})$ is an entropy function with entropy flux $\Psi(\mathbf{u})$, we have that

$$\begin{aligned} 0 &= \frac{\partial S}{\partial \mathbf{u}} \left[\frac{\partial \mathbf{u}_\epsilon}{\partial t} + \sum_{i=1}^{k} \frac{\partial \mathbf{F}(\mathbf{u}_\epsilon)\mathbf{e}_i}{\partial \mathbf{x}_i} - \epsilon \sum_{i=1}^{k} \frac{\partial^2 \mathbf{u}_\epsilon}{\partial \mathbf{x}_i^2} \right] \\ &= \frac{\partial S(\mathbf{u}_\epsilon)}{\partial t} + \sum_{i=1}^{k} \frac{\partial \mathbf{e}_i \cdot \Psi(\mathbf{u}_\epsilon)}{\partial \mathbf{x}_i} - \epsilon \sum_{i=1}^{k} \frac{\partial}{\partial \mathbf{x}_i} \left(\frac{\partial S}{\partial \mathbf{u}} \frac{\partial \mathbf{u}_\epsilon}{\partial \mathbf{x}_i} \right) \\ &\quad + \epsilon \sum_{i=1}^{k} \left(\frac{\partial \mathbf{u}_\epsilon}{\partial \mathbf{x}_i} \right)^\top \frac{\partial^2 S}{\partial \mathbf{u}^2} \left(\frac{\partial \mathbf{u}_\epsilon}{\partial \mathbf{x}_i} \right). \end{aligned}$$

Since the solution $\mathbf{u}_\epsilon$ of the viscous conservation law is smooth, for any nonnegative smooth $\phi(\mathbf{x},t)$ we can compute

$$\begin{aligned} &\int_0^\infty \int_{\mathbf{R}^k} \phi \left[\frac{\partial S(\mathbf{u}_\epsilon)}{\partial t} + \sum_{i=1}^{k} \frac{\partial \mathbf{e}_i \cdot \Psi(\mathbf{u}_\epsilon)}{\partial \mathbf{x}_i} \right] d\mathbf{x}\, dt \\ &= -\int_0^\infty \int_{\mathbf{R}^k} \frac{\partial \phi}{\partial t} S(\mathbf{u}_\epsilon) + \sum_{i=1}^{k} \frac{\partial \phi}{\partial \mathbf{x}_i} \mathbf{e}_i \cdot \Psi(\mathbf{u}_\epsilon)\, d\mathbf{x}\, dt \\ &\quad - \int_{\mathbf{R}^k} \phi(\mathbf{x},0) S(\mathbf{u}_\epsilon)(\mathbf{x},0)\, d\mathbf{x} \end{aligned}$$

We can also compute

$$\int_0^\infty \int_{\mathbf{R}^k} \phi \sum_{i=1}^k \frac{\partial}{\partial \mathbf{x}_i}\left(\frac{\partial S}{\partial \mathbf{u}}\frac{\partial \mathbf{u}_\epsilon}{\partial \mathbf{x}_i}\right) d\mathbf{x}\, dt = -\int_0^\infty \int_{\mathbf{R}^k} \sum_{i=1}^k \frac{\partial \phi}{\partial \mathbf{x}_i}\frac{\partial S(\mathbf{u}_\epsilon)}{\partial \mathbf{x}_i} d\mathbf{x}\, dt$$
$$= \int_0^\infty \int_{\mathbf{R}^k} \sum_{i=1}^k \frac{\partial^2 \phi}{\partial \mathbf{x}_i^2} S(\mathbf{u}_\epsilon)\, d\mathbf{x}\, dt$$

and note that the concavity of S and nonnegativity of ϕ imply that

$$\int_0^\infty \int_{\mathbf{R}^k} \phi \left(\frac{\partial \mathbf{u}_\epsilon}{\partial \mathbf{x}}\right)^\top \frac{\partial^2 S}{\partial \mathbf{u}^2}\left(\frac{\partial \mathbf{u}_\epsilon}{\partial \mathbf{x}}\right) d\mathbf{x}\, dt \le 0.$$

Putting these results together, we obtain

$$0 \le -\int_0^\infty \int_{\mathbf{R}^k} \frac{\partial \phi}{\partial t} S(\mathbf{u}_\epsilon) + \sum_{i=1}^k \frac{\partial \phi}{\partial \mathbf{x}_i}\Psi(\mathbf{u}_\epsilon)\mathbf{e}_i\, d\mathbf{x}\, dt - \int_{\mathbf{R}^k} \phi(\mathbf{x},0)S(\mathbf{u}_\epsilon)(\mathbf{x},0)\, d\mathbf{x}$$
$$-\,\epsilon \int_0^\infty \int_{\mathbf{R}^k}^\infty \sum_{i=1}^k \frac{\partial^2 \phi}{\partial \mathbf{x}_i^2} S(\mathbf{u}_\epsilon)\, d\mathbf{x}\, dt.$$

Since $S(\mathbf{u})$ is bounded and $\mathbf{u}_\epsilon \to \mathbf{u}$ almost everywhere, the term involving a factor of ϵ tends to zero as $\epsilon \to 0$. Taking limits as $\epsilon \to 0$ now produces the claimed result. □

This lemma has the following useful corollary.

Corollary 4.1.26 *Suppose that $\mathbf{u}(\mathbf{x}, t) \in \mathbf{R}^m$ solves the hyperbolic conservation law*

$$\frac{\partial \mathbf{u}}{\partial t} + \sum_{i=1}^k \frac{\partial \mathbf{F}(\mathbf{u})\mathbf{e}_i}{\partial \mathbf{x}_i} = 0$$

in the limit of vanishing diffusion, that there is a corresponding concave entropy function $S(\mathbf{u})$ with entropy flux $\Psi(\mathbf{u})$, and that $S(\mathbf{u})$ is locally bounded for all $\mathbf{x}$ and t. Then for all rectangles $R = (a_1, b_1) \times \ldots \times (a_k, b_k) \subset \mathbf{R}^k$ and all intervals $(t^{[1]}, t^{[2]})$ we have

$$\int_R S(\mathbf{u}(\mathbf{x}, t^{[2]}))\, d\mathbf{x} - \int_R S(\mathbf{u}(\mathbf{x}, t^{[1]}))\, d\mathbf{x} + \int_{t^{[1]}}^{t^{[2]}} \int_{\partial R} \mathbf{n}^\top \Psi(\mathbf{u}(\mathbf{x}, t))\, ds\, dt \ge 0, \qquad (4.31)$$

where $\mathbf{n}$ is the outer normal to ∂R and ds is surface measure on ∂R.

Proof Consider the continuous piecewise linear functions

$$\phi_{0,h}(t) = \max\left\{0, \min\left\{1, 1 - \frac{t^{[1]} - t}{h}, 1 - \frac{t - t^{[2]}}{h}\right\}\right\}$$

$$\phi_{i,h}(\mathbf{x}_i) = \max\left\{0, \min\left\{1, 1 - \frac{a_i - \mathbf{x}_i}{h}, 1 - \frac{\mathbf{x}_i - b_i}{h}\right\}\right\}.$$

Let $\phi_h(\mathbf{x}, t) = \phi_{0,h}(t)\phi_{1,h}(\mathbf{x}_1)\ldots\phi_{k,h}(\mathbf{x}_h)$. This function is nonnegative and continuous, but not C^∞. However, it can be approximated arbitrarily well by a nonnegative smooth function with compact support. First, suppose that $t^{[1]} > 0$. Then lemma 4.1.25 implies that

$$\begin{aligned}
0 \leq & -h\int_0^\infty \int_{\mathbf{R}^k} \frac{\partial \phi}{\partial t} S(\mathbf{u}) + \sum_{i=1}^k \frac{\partial \phi}{\partial \mathbf{x}_i} \mathbf{e}_i^\top \Psi(\mathbf{u})\, d\mathbf{x}\, dt \\
= & -\int_{t^{[1]}-h}^{t^{[1]}} \int_{a_k-h}^{b_k+h} \cdots \int_{a_1-h}^{b_1+h} \phi_{1,h}(\mathbf{x}_1)\ldots\phi_{k,h}(\mathbf{x}_k) S(\mathbf{u}(\mathbf{x}, t))\, d\mathbf{x}_1 \ldots\, d\mathbf{x}_k\, dt \\
& + \int_{t^{[2]}}^{t^{[2]}+h} \int_{a_k-h}^{b_k+h} \cdots \int_{a_1-h}^{b_1+h} \phi_{1,h}(\mathbf{x}_1)\ldots\phi_{k,h}(\mathbf{x}_k) S(\mathbf{u}(\mathbf{x}, t))\, d\mathbf{x}_1 \ldots\, d\mathbf{x}_k\, dt \\
& - \int_{t^{[1]}-h}^{t^{[2]}+h} \int_{a_k-h}^{b_k+h} \cdots \int_{a_2-h}^{b_2+h} \int_{a_1-h}^{a_1} \phi_{0,h}(t)\phi_{2,h}(\mathbf{x}_2)\ldots\phi_{k,h}(\mathbf{x}_k)\mathbf{e}_1^\top \Psi(\mathbf{u}(\mathbf{x}, t))\, d\mathbf{x}_1 \ldots d\mathbf{x}_k\, dt \\
& + \int_{t^{[1]}-h}^{t^{[2]}+h} \int_{a_k-h}^{b_k+h} \cdots \int_{a_2-h}^{b_2+h} \int_{b_1}^{b_1+h} \phi_{0,h}(t)\phi_{2,h}(\mathbf{x}_2)\ldots\phi_{k,h}(\mathbf{x}_k)\mathbf{e}_1^\top \Psi(\mathbf{u}(\mathbf{x}, t))\, d\mathbf{x}_1 \cdots d\mathbf{x}_k\, dt \\
& + \cdots
\end{aligned}$$

As $h \to 0$, we get the claimed result. If $t^{[1]} = 0$, then we use the same test functions. In this case, the integral $\int_0^\infty \ldots dt$ does not include the positive slope in $\phi_{0,h}(t)$; the contribution we would expect from this term comes from the integral of the initial data in Lemma 4.1.25. □

Lemma 4.1.27 *(Lax,[89]) Suppose that* $\mathbf{u}(\mathbf{x}, t) \in \mathbf{R}^m$ *solves the hyperbolic conservation law*

$$\frac{\partial \mathbf{u}}{\partial t} + \sum_{i=1}^k \frac{\partial \mathbf{F}(\mathbf{u})\mathbf{e}_i}{\partial \mathbf{x}_i} = 0.$$

and is the limit, as the diffusion tends to zero, of the corresponding viscous conservation law. Further, suppose that there is a concave entropy function $S(\mathbf{u})$ *with entropy flux* $\Psi(\mathbf{u})$ *so that for all fixed directions* $\mathbf{n}$

$$\frac{\partial \mathbf{n} \cdot \Psi(\mathbf{u})}{\partial \mathbf{u}} = \frac{\partial S}{\partial \mathbf{u}} \frac{\partial \mathbf{F}(\mathbf{u})\mathbf{n}}{\partial \mathbf{u}}.$$

Suppose that $\mathbf{u}(\mathbf{x}, t)$ involves an isolated discontinuity surface $D(t)$ dividing $\mathbf{R}^k$ into two domains $\Omega_-(t)$ and $\Omega_+(t)$. Let $\mathbf{y}$ denote the velocity of a point on the discontinuity surface $D(t)$ and $\mathbf{n}$ denote the outer normal on $D(t)$ with respect to $\Omega_-(t)$. Further, assume that $\mathbf{y}$ points into $\Omega_+(t)$, and its normal velocity is $\sigma \equiv \mathbf{n} \cdot \mathbf{y} \geq 0$. Then at almost all points on the discontinuity surface we have

$$\mathbf{n} \cdot [\Psi_+ - \Psi_-] \geq [S_+ - S_-]\sigma. \tag{4.32}$$

The direction of this inequality is reversed if S is convex.

Proof Suppose that $\phi(\mathbf{x}, t) \in C_0^\infty(\mathbf{R}^k \times (0, \infty))$ and $\phi(\mathbf{x}(t), t) > 0$; also suppose that the support of ϕ is such that no other discontinuity surface of $\mathbf{u}$ lies in its support, and the support does not intersect any boundary of $\Omega_-(t)$ or $\Omega_+(t)$ other than $D(t)$. From Lemma 4.1.25 we have that

$$\begin{aligned}
0 \leq & - \int_0^\infty \int_{\mathbf{R}^k} \frac{\partial \phi}{\partial t} S + \sum_{i=1}^k \frac{\partial \phi}{\partial \mathbf{x}_i} \mathbf{e}_i \cdot \Psi \, d\mathbf{x} \, dt \\
= & - \int_0^\infty \int_{\Omega_-(t)} \begin{bmatrix} \nabla_{\mathbf{x}}^\top & \frac{\partial}{\partial t} \end{bmatrix} \begin{bmatrix} \phi \Psi \\ \phi S \end{bmatrix} d\mathbf{x} \, dt \\
& + \int_0^\infty \int_{\Omega_-(t)} \phi \left[\frac{\partial S}{\partial t} + \sum_{i=1}^k \frac{\partial \mathbf{e}_i \cdot \Psi}{\partial \mathbf{x}_i} \right] d\mathbf{x} \, dt \\
& - \int_0^\infty \int_{\Omega_+(t)} \begin{bmatrix} \nabla_{\mathbf{x}}^\top & \frac{\partial}{\partial t} \end{bmatrix} \begin{bmatrix} \phi \Psi \\ \phi S \end{bmatrix} d\mathbf{x} \, dt \\
& + \int_0^\infty \int_{\Omega_+(t)} \phi \left[\frac{\partial S}{\partial t} + \sum_{i=1}^k \frac{\partial \mathbf{e}_i \cdot \Psi}{\partial \mathbf{x}_i} \right] d\mathbf{x} \, dt
\end{aligned}$$

The entropy conservation law (4.30) shows that

$$0 \leq - \int_0^\infty \int_{\Omega_-(t)} \begin{bmatrix} \nabla_{\mathbf{x}}^\top & \frac{\partial}{\partial t} \end{bmatrix} \begin{bmatrix} \phi \Psi \\ \phi S \end{bmatrix} d\mathbf{x} \, dt - \int_0^\infty \int_{\Omega_+(t)} \begin{bmatrix} \nabla_{\mathbf{x}}^\top & \frac{\partial}{\partial t} \end{bmatrix} \begin{bmatrix} \phi \Psi \\ \phi S \end{bmatrix} d\mathbf{x} \, dt.$$

Using the divergence theorem, we can rewrite this result in the form

$$\begin{aligned}
0 \leq & - \int_0^\infty \int_{\partial\Omega_-(t)} \begin{bmatrix} \mathbf{n}_-^\top & \nu_- \end{bmatrix} \begin{bmatrix} \phi \Psi \\ \phi S \end{bmatrix} ds_L \, dt - \int_0^\infty \int_{\partial\Omega_+(t)} \begin{bmatrix} \mathbf{n}_+^\top & \nu_+ \end{bmatrix} \begin{bmatrix} \phi \Psi \\ \phi S \end{bmatrix} ds_R \, dt \\
= & - \int_{\{(\mathbf{x},t):\mathbf{x} \in D(t)\}} \begin{bmatrix} \mathbf{n}^\top & -\sigma \end{bmatrix} \begin{bmatrix} \phi \Psi_- \\ \phi S_- \end{bmatrix} ds - \int_{\{(\mathbf{x},t):\mathbf{x} \in D(t)\}} \begin{bmatrix} -\mathbf{n}^\top & \sigma \end{bmatrix} \begin{bmatrix} \phi \Psi_+ \\ \phi S_+ \end{bmatrix} ds \\
= & \int_{\{(\mathbf{x},t):\mathbf{x} \in D(t)\}} \mathbf{n} \cdot [\Psi_+ - \Psi_-] - [S_+ - S_-]\sigma \, ds
\end{aligned}$$

The result follows by shrinking the support of ϕ to a point on D. □

Entropy functions will be useful later in section 4.13.8 where we develop approximate Riemann solvers, and in section 5.2 where we discuss convergence of schemes to physically correct solutions of nonlinear conservation laws.

Exercises for 4.1

4.1.1 Consider the **wave equation**

$$\frac{\partial^2 \mathbf{u}}{\partial t^2} - c^2 \frac{\partial^2 \mathbf{u}}{\partial x^2} = 0 \text{ for all } x \in \mathbf{R} \text{ for all } t > 0$$

$$\mathbf{u}(x, 0) = \mathbf{u}_0(x), \ \frac{\partial \mathbf{u}}{\partial t}(x, 0) = v_0(x) \text{ for all } x \in \mathbf{R}$$

(a) Define $v(x,t) = \frac{\partial \mathbf{u}}{\partial t}(x,t)$ and $w(x,t) = \frac{\partial \mathbf{u}}{\partial x}(x,t)$. Explain why the wave equation can be written in the form of the linear system of conservation laws

$$\frac{\partial}{\partial t}\begin{bmatrix} v \\ w \end{bmatrix} + \begin{bmatrix} 0 & -c^2 \\ -1 & 0 \end{bmatrix} \frac{\partial}{\partial x}\begin{bmatrix} v \\ w \end{bmatrix} = 0 \text{ for all } x \in \mathbf{R} \text{ for all } t > 0$$

$$\begin{bmatrix} v \\ w \end{bmatrix}(x, 0) = \begin{bmatrix} v_0 \\ \frac{d\mathbf{u}_0}{dx} \end{bmatrix}(x) \text{ for all } x \in \mathbf{R}.$$

(b) Show that $\mathbf{AX} = \mathbf{X}\Lambda$ where

$$\mathbf{A} = \begin{bmatrix} 0 & -c^2 \\ -1 & 0 \end{bmatrix}, \ \mathbf{X} = \begin{bmatrix} c & c \\ 1 & -1 \end{bmatrix}, \ \Lambda = \begin{bmatrix} -c & 0 \\ 0 & c \end{bmatrix}.$$

(c) Use Lemma 4.1.7 to show that

$$\begin{bmatrix} v \\ w \end{bmatrix}(x, t) = \begin{bmatrix} c \\ 1 \end{bmatrix} \frac{1}{2c}\left\{v_0(x+ct) + c\frac{d\mathbf{u}_0}{dx}(x+ct)\right\}$$
$$+ \begin{bmatrix} c \\ -1 \end{bmatrix} \frac{1}{2c}\left\{v_0(x-ct) - c\frac{d\mathbf{u}_0}{dx}(x-ct)\right\}.$$

(d) Use this solution to derive **d'Alembert's formula**

$$\mathbf{u}(x,t) = \frac{1}{2}\{\mathbf{u}_0(x+ct) + \mathbf{u}_0(x-ct)\} + \frac{1}{2c}\int_{x-ct}^{x+ct} v_0(\xi)d\xi.$$

We remark that the solution of the wave equation in 3D is given by Kirchhoff's formula

$$v(\mathbf{x}, t) = \frac{1}{4\pi c^2 t}\int\int_{\|\xi - \mathbf{x}\| = ct} \frac{\partial v}{\partial t}(\xi, 0)\, ds$$
$$+ \frac{\partial}{\partial t}\left[\frac{1}{4\pi c^2 t}\int\int_{\|\xi - \mathbf{x}\| = ct} v(\xi, 0) ds\right]$$

and in 2D is given by

$$v(\mathbf{x},t) = \frac{1}{2\pi c}\int\int_{\|\xi-\mathbf{x}\|\le ct}\frac{\frac{\partial v}{\partial t}(\xi,0)}{\sqrt{c^2t^2-\|\xi-\mathbf{x}\|^2}}\,d\xi$$
$$+\frac{\partial}{\partial t}\left[\frac{1}{2\pi c}\int\int_{\|\xi-\mathbf{x}\|\le ct}\frac{v(\xi,0)}{\sqrt{c^2t^2-\|\xi-\mathbf{x}\|^2}}d\xi\right]$$

These correspond to the linear conservation law

$$\frac{\partial \mathbf{u}}{\partial t}+\sum_{i=1}^{k}\frac{\partial \mathbf{F}\mathbf{e}_i}{\partial \mathbf{x}_i}=0$$

where

$$\mathbf{u}=\begin{bmatrix} v \\ \mathbf{w}\end{bmatrix} \text{ and } \mathbf{F}(\mathbf{u})=\begin{bmatrix} c\mathbf{w}^\top \\ \mathbf{I}cv\end{bmatrix}.$$

In this case, the normal flux

$$\mathbf{F}(\mathbf{u})\mathbf{n}=\begin{bmatrix} 0 & c\mathbf{n}^\top \\ \mathbf{n}c & 0\end{bmatrix}\begin{bmatrix} v \\ \mathbf{w}\end{bmatrix}$$

is linear in its argument with a matrix that depends on $\mathbf{n}$. Although the eigenvalues of this matrix are independent of $\mathbf{n}$, its eigenvectors are not.

4.1.2 One easy application of Lemma 4.1.9 is to study the effect of a rotation of the coordinate system. For example, we might rotate the coordinate system so that the first coordinate direction is in the direction of a propagating discontinuity, or the jump in a Riemann problem. If $\mathbf{x}=\mathbf{Q}\mathbf{a}$ where $\mathbf{Q}$ is fixed, orthogonal and $\det(\mathbf{Q})=1$, show that the conservation law

$$\frac{\partial \mathbf{u}_E}{\partial t}+\sum_{i=1}^{k}\frac{\partial \mathbf{F}_E\mathbf{e}_i}{\partial \mathbf{x}_i}=\mathbf{b}_E$$

is equivalent to the conservation law

$$\frac{\partial \mathbf{u}_L}{\partial t}+\sum_{i=1}^{k}\frac{\partial \mathbf{F}_L\mathbf{Q}\mathbf{e}_i}{\partial \mathbf{a}_i}=\mathbf{b}_L.$$

In the latter equation, $\mathbf{u}_L(\mathbf{a},t)=\mathbf{u}_E(\mathbf{Q}\mathbf{a},t)$, and so on.

4.1.3 The shallow water equations were presented in the Eulerian frame of reference in Example 4.1.1.

(a) Show how to write the shallow water equations in the Lagrangian frame of reference. You may want to use equation (4.14) to have enough evolution equations to determine all of the variables in these equations.

(b) Perform a characteristic analysis of the Lagrangian form of the shallow water equations. Explain how the Lagrangian characteristic speeds relate to the Eulerian characteristic speeds.

4.1.4 Consider the one-dimensional system of constant coefficient equations

$$\frac{\partial \mathbf{u}}{\partial t} + \mathbf{A}\frac{\partial \mathbf{u}}{\partial \mathbf{x}} = 0 \quad \text{for all} \quad a < \mathbf{x} < b \quad \text{for all} \quad t > 0,$$

where $\mathbf{A}$ is diagonalizable with real eigenvalues, and $\mathbf{u}(\mathbf{x}, 0)$ is given. Describe what boundary data at $\mathbf{x} = a$ and $\mathbf{x} = b$ must be given for $t > 0$ in order to specify a unique solution to this problem.

4.2 Upwind Schemes

In Section 3.3 we presented several numerical methods for nonlinear scalar conservation laws. In this section, we will describe the application of three of these schemes to nonlinear systems.

4.2.1 Lax–Friedrichs Scheme

The Lax–Friedrichs scheme is applied to general one-dimensional nonlinear systems in the same way it is applied to scalar equations:

$$\mathbf{u}_{i+1/2}^{n+1/2} = \left\{\mathbf{u}_i^n \triangle x_i + \mathbf{u}_{i+1}^n \triangle x_{i+1} - [\mathbf{f}(\mathbf{u}_{i+1}^n) - \mathbf{f}(\mathbf{u}_i^n)]\triangle t^{n+1/2}\right\} \frac{1}{\triangle x_i + \triangle x_{i+1}} \tag{4.33a}$$

$$\mathbf{u}_i^{n+1} = \left\{\mathbf{u}_{i-1/2}^{n+1/2} + \mathbf{u}_{i+1/2}^{n+1/2} - [\mathbf{f}(\mathbf{u}_{i+1/2}^{n+1/2}) - \mathbf{f}(\mathbf{u}_{i-1/2}^{n+1/2})]\frac{\triangle t^{n+1/2}}{\triangle x_i}\right\} \frac{1}{2}. \tag{4.33b}$$

The principal complication is that the flux $\mathbf{f}$ is often really a function of the flux variables $\mathbf{w}$, so we have to decode the conserved quantities $\mathbf{u}$ to obtain $\mathbf{w}$ after each conservative difference.

Note that the Lax–Friedrichs scheme must choose the timestep $\triangle t^{\mathbf{n}+1/2}$ so that for all i we have $\triangle t^{\mathbf{n}+1/2}\lambda_i \leq \triangle x_i$. Here λ_i is an upper bound on all of the absolute values of the characteristic speeds.

Example 4.2.1 The Lax–Friedrichs scheme for shallow water begins with cell-centered values for the flux variables

$$\mathbf{w}_i^n = \begin{bmatrix} h \\ \mathbf{v} \end{bmatrix}_i^n$$

and the conserved quantities

$$\mathbf{u}_i^n = \begin{bmatrix} h \\ \mathbf{v}h \end{bmatrix}_i^n.$$

Note that the Lax–Friedrichs scheme must choose the timestep $\triangle t^{n+1/2}$ so that for all i $\triangle t^{n+1/2}\lambda_i \leq \triangle x_i$, where λ_i is an upper bound on the absolute values of the characteristic speeds. For shallow water, the least upper bound is $\lambda_i = |\mathbf{v}_i^n| + \sqrt{gh_i^n}$.

The first step of the Lax–Friedrichs scheme computes the cell-centered fluxes

$$\mathbf{f}(\mathbf{w}_i^n) = \begin{bmatrix} h\mathbf{v} \\ h\mathbf{v}^2 + \frac{1}{2}gh^2 \end{bmatrix}_i^n$$

and updates the conserved quantities by

$$\begin{aligned} h_{i+1/2}^{n+1/2} &= \left[h_i^n \triangle x_i + h_{i+1}^n \triangle x_{i+1} - \left\{h_{i+1}^n - h_i^n\right\} \triangle t^{n+1/2}\right] \frac{1}{\triangle x_i + \triangle x_{i+1}} \\ (\mathbf{v}h)_{i+1/2}^{n+1/2} &= \left[(\mathbf{v}h)_i^n \triangle x_i + (\mathbf{v}h)_{i+1}^n \triangle x_{i+1}\right. \\ &\quad \left. - \left\{\left(h\mathbf{v}^2 + \frac{1}{2}gh^2\right)_{i+1}^n - \left(h\mathbf{v}^2 + \frac{1}{2}gh^2\right)_i^n\right\} \triangle t^{n+1/2}\right] \frac{1}{\triangle x_i + \triangle x_{i+1}}. \end{aligned}$$

Before performing the second step of the Lax–Friedrichs scheme, it is necessary to decode the flux variables from the conserved quantities by computing $\mathbf{v}_{i+1/2}^{n+1/2} = (\mathbf{v}h)_{i+1/2}^{n+1/2} / h_{i+1/2}^{n+1/2}$. Then we compute the fluxes at the half-time

$$\mathbf{f}\left(\mathbf{w}_{i+1/2}^{n+1/2}\right) = \begin{bmatrix} h\mathbf{v} \\ h\mathbf{v}^2 + \frac{1}{2}gh^2 \end{bmatrix}_{i+1/2}^{n+1/2},$$

and determine the conserved quantities at the new time

$$\begin{aligned} h_i^{n+1} &= \frac{1}{2}\left[h_{i-1/2}^{n+1/2} + h_{i+1/2}^{n+1/2} - \left\{(h\mathbf{v})_{i+1/2}^{n+1/2} - (h\mathbf{v})_{i-1/2}^{n+1/2}\right\} \frac{\triangle t^{n+1/2}}{\triangle x_i}\right] \\ (\mathbf{v}h)_i^{n+1} &= \frac{1}{2}\left[(\mathbf{v}h)_{i-1/2}^{n+1/2} + (\mathbf{v}h)_{i+1/2}^{n+1/2}\right. \\ &\quad \left. - \left\{\left(h\mathbf{v}^2 + \frac{1}{2}gh^2\right)_{i+1/2}^{n+1/2} - \left(h\mathbf{v}^2 + \frac{1}{2}gh^2\right)_{i-1/2}^{n+1/2}\right\} \frac{\triangle t^{n+1/2}}{\triangle x_i}\right]. \end{aligned}$$

At this point, we need to determine the flux variables again: $\mathbf{v}_i^{n+1} = (\mathbf{v}h)_i^{n+1} / h_i^{n+1}$.

Figure 4.5 shows some numerical results with the Lax–Friedrichs scheme for the dam break problem. This is a Riemann problem with gravity $g = 1$ in which the left state is given by $h_L = 2$, $\mathbf{v}_L = 0$ and the right state is $h_R = 1$, $\mathbf{v}_R = 0$. The

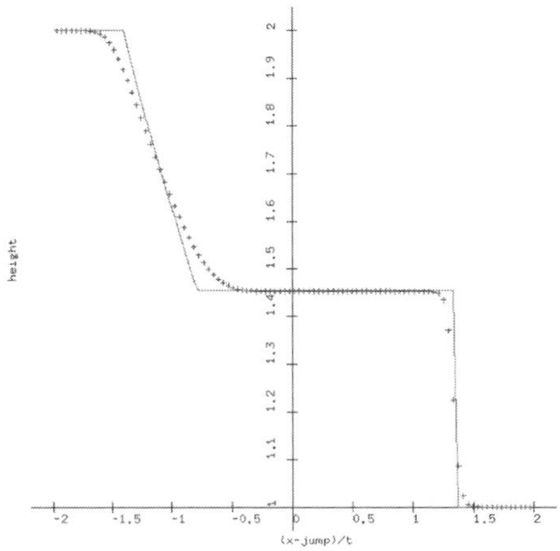

(a) Height vs. $\mathbf{x}/t$

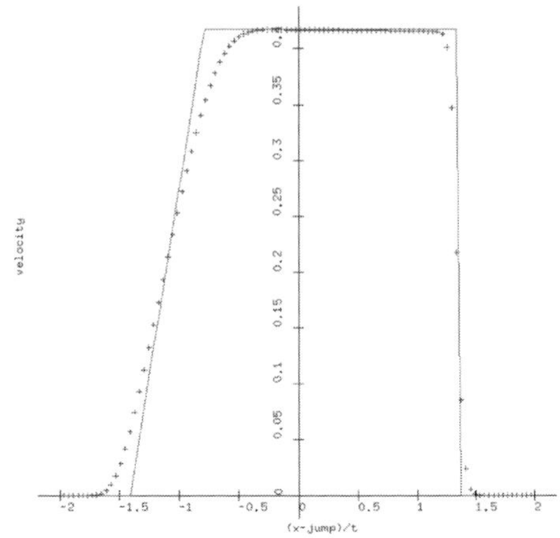

(b) Velocity vs. $\mathbf{x}/t$

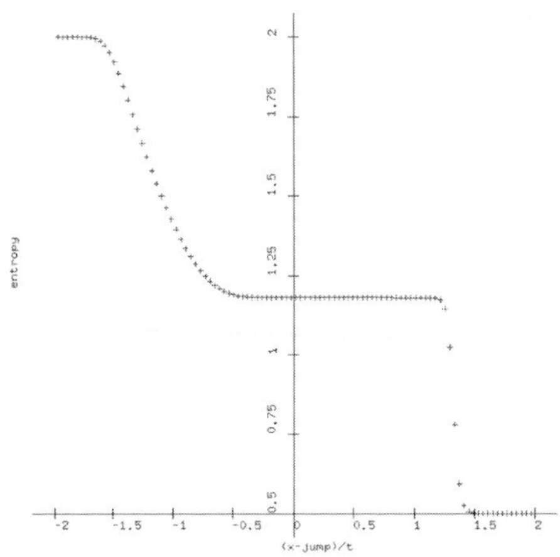

(c) Total energy vs. $\mathbf{x}/t$

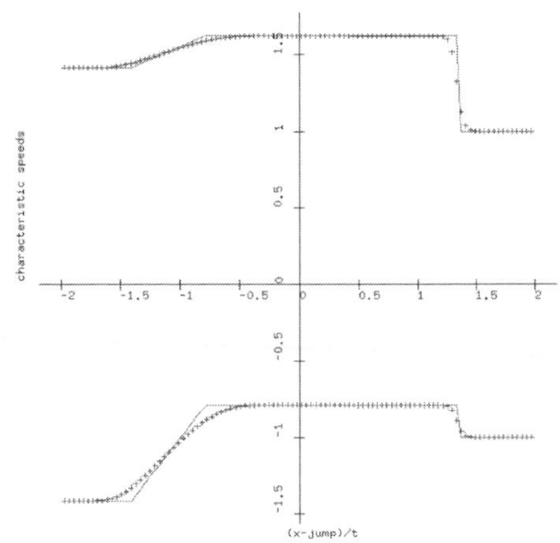

(d) Characteristic speed vs. $\mathbf{x}/t$

Fig. 4.5 Lax–Friedrichs Scheme for dam break problem: 100 grid cells, CFL $=$ 0.9, gravity $= 1$

numerical results are plotted versus $\mathbf{x}/t$. The solution involves a rarefaction moving to the left and a shock moving to the right. Nevertheless, the graph of the characteristic speeds indicates that the Lax–Friedrichs scheme is getting results consistent with the correct solution. In particular, note that the slow characteristic speed in the rarefaction nearly aligns with a line of slope 1 through the origin. Further, the shock involves a decrease in the fast characteristic speed as we move from left to right. Finally, note that the total energy increases as the shock passes.

4.2.2 Rusanov Scheme

For general nonlinear systems in one dimension, the Rusanov scheme takes the form

$$\mathbf{u}_i^{n+1} = \mathbf{u}_i^n - \frac{\triangle t^{n+1/2}}{\triangle x_i} \left[\mathbf{f}_{i+1/2}^{n+1/2} - \mathbf{f}_{i-1/2}^{n+1/2} \right]$$

where the fluxes are computed by

$$\mathbf{f}_{i+1/2}^{n+1/2} = \frac{1}{2} \left[\mathbf{f}(\mathbf{u}_i^n) + \mathbf{f}(\mathbf{u}_{i+1}^n) - \lambda_{i+1/2}(\mathbf{u}_{i+1}^n - \mathbf{u}_i^n) \right].$$

Here $\lambda_{i+1/2}$ is an upper bound for the absolute values of the characteristic speeds in either cell. The timestep is chosen so that $\lambda_{i+1/2} \triangle t^{n+1/2} \leq \min\{\triangle x_i, \triangle x_{i+1}\}$ for all cells i. Figure 4.6 shows some numerical results for the dam break problem using Rusanov's scheme. The quality of these results is similar to those with the Lax–Friedrichs scheme, but the computation is faster because the Rusanov scheme does not involve half-steps.

4.2.3 Godunov Scheme

Finally, we mention the Godunov scheme for nonlinear systems in one dimension. This scheme takes the form

$$\mathbf{u}_i^{n+1} = \mathbf{u}_i^n - \frac{\triangle t^{n+1/2}}{\triangle x_i} \left[\mathbf{f}_{i+1/2}^{n+1/2} - \mathbf{f}_{i-1/2}^{n+1/2} \right]$$

where the fluxes are computed by solving a Riemann problem:

$$\mathbf{f}_{i+1/2}^{n+1/2} = \mathbf{f}\left(\mathcal{R}(\mathbf{u}_i^n, \mathbf{u}_{i+1}^n; 0) \right).$$

The timestep is chosen as in the same fashion as the Lax–Friedrichs scheme or the Rusanov scheme. Figure 4.7 shows some numerical results with Godunov's method for the dam break problem. Note that the rarefaction is resolved better with Godunov's scheme than with either the Lax-Friedrichs scheme or the Rusanov scheme. Unfortunately, the use of the exact Riemann solver in the Godunov scheme increases both its computational time and programming difficulty.

Programs to perform the Lax-Friedrichs, Rusanov and Godunov schemes can be found in **Program 4.2-46: GUIRiemannProblem.C.** This program calls Fortran routines for many of the models to be discussed below. You can execute this Riemann problem solver by clicking on the link **Executable 4.2-17: guiRiemannProblem** You can select input parameters for the Riemann problem by pulling down on "View,"

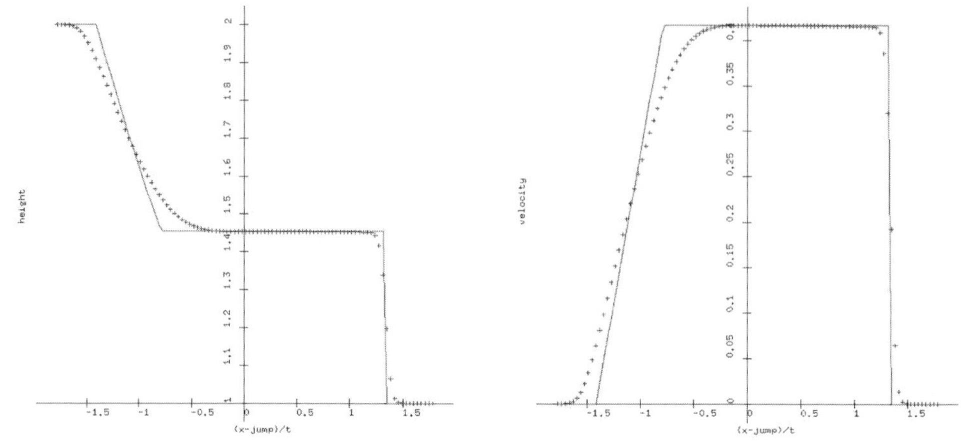

(a) Height vs. $\mathbf{x}/t$ (b) Velocity vs. $\mathbf{x}/t$

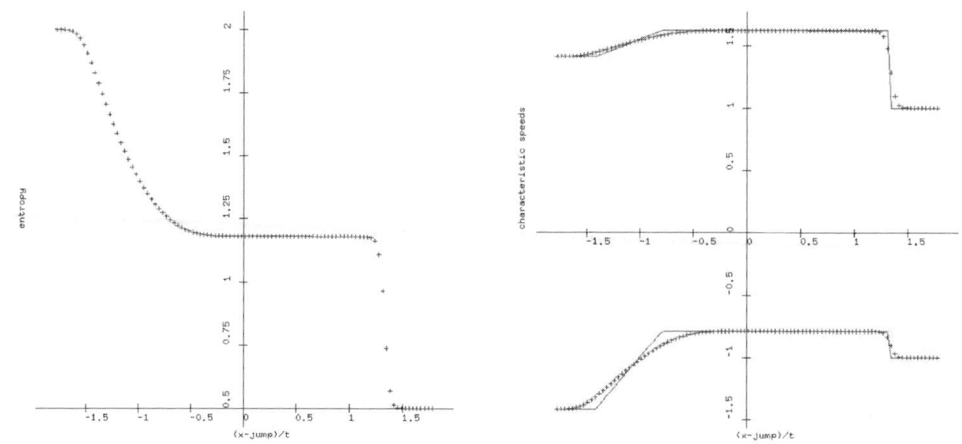

(c) Total Energy vs. $\mathbf{x}/t$ (d) Characteristic speed vs. $\mathbf{x}/t$

Fig. 4.6 Rusanov scheme for dam break problem: 100 grid cells, CFL $= 0.9$, gravity $= 1$

releasing on "Main," and then clicking on the arrow next to "Riemann Problem Parameters." The particular system of conservation laws can be selected by clicking on one of the radio buttons after "problem." Model parameters can be selected by clicking on the arrow next to the name of the model of interest. The particular numerical scheme can be selected under "Numerical Method Parameters." When you have selected all of your input parameters, click on "Start Run Now" in the window labeled "1d/guiRiemannProblem."

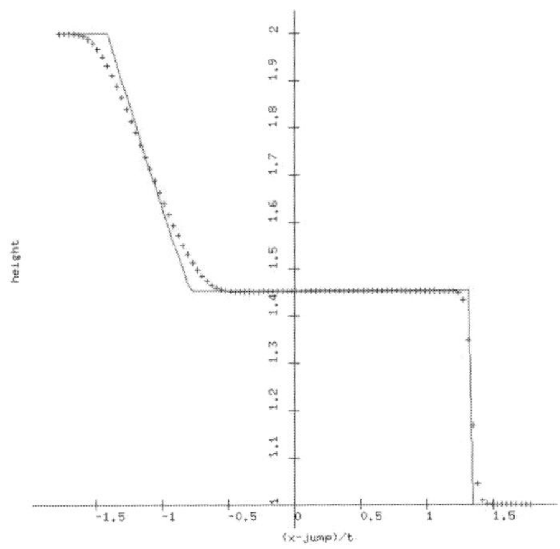

(a) Height vs. $\mathbf{x}/t$

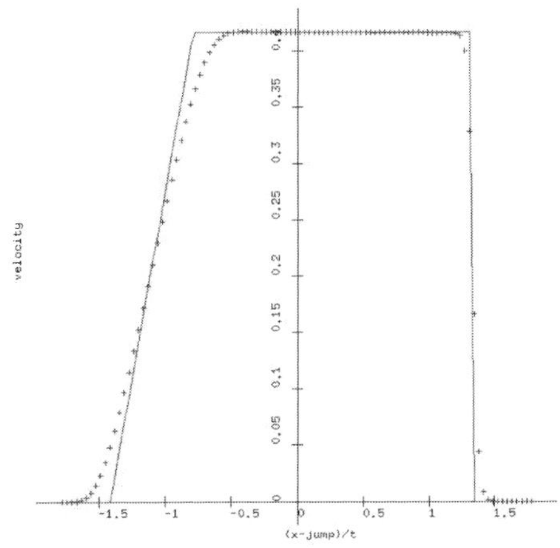

(b) Velocity vs. $\mathbf{x}/t$

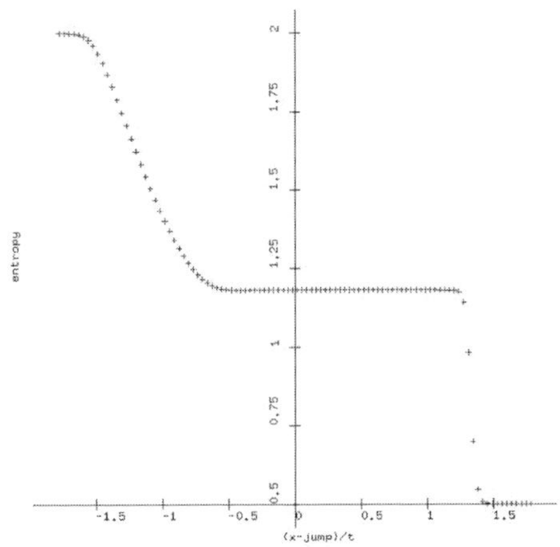

(c) Total Energy vs. $\mathbf{x}/t$

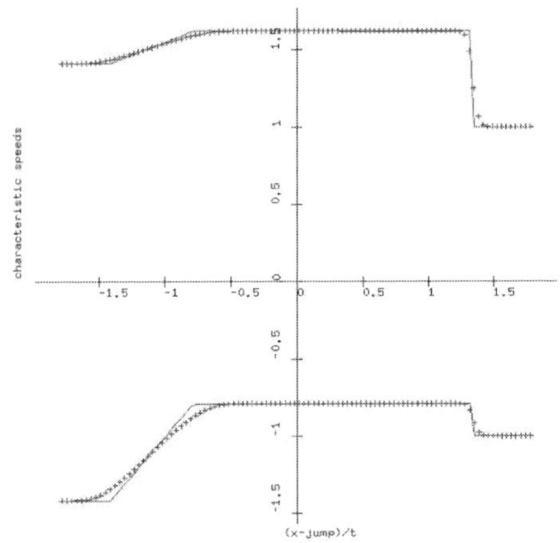

(d) Characteristic speed vs. $\mathbf{x}/t$

Fig. 4.7 Godunov scheme for dam break problem: 100 grid cells, CFL = 0.9, gravity = 1

The next lemma explains one nice feature of Godunov's scheme: whenever it converges, it converges to the correct solution.

Lemma 4.2.2 *If S is a convex (or concave) entropy function for some hyperbolic conservation law in one dimension, and Godunov's method for this conservation laws converges, then it converges to an entropy-satisfying solution of that conservation law.*

Proof Consider Godunov's method for a conservation law with a convex entropy function $S(\mathbf{u})$. Let $\tilde{\mathbf{u}}^n$ be the exact solution to the conservation law

$$\frac{\partial \mathbf{u}}{\partial t} + \frac{\partial \mathbf{f}(\mathbf{u})}{\partial \mathbf{x}} = 0$$

with piecewise constant initial data $\tilde{\mathbf{u}}^n(\mathbf{x}, t^n) = \mathbf{u}_i^n$ for $\mathbf{x} \in [\mathbf{x}_{i-1/2}, \mathbf{x}_{i+1/2}]$. Since $\tilde{\mathbf{u}}^n$ is the exact solution, the total entropy can only decrease in time:

$$\frac{1}{\triangle x_i}\int_{\mathbf{x}_{i-1/2}}^{\mathbf{x}_{i+1/2}} S(\tilde{\mathbf{u}}^n(\mathbf{x}, t^{n+1}))\, d\mathbf{x}$$
$$\leq \frac{1}{\triangle x_i}\int_{\mathbf{x}_{i-1/2}}^{\mathbf{x}_{i+1/2}} S(\mathbf{u}(\mathbf{x}, t^n))\, d\mathbf{x} - \frac{1}{\triangle x_i}\left[\int_{t^n}^{t^{n+1}} \Psi(\tilde{\mathbf{u}}^n(\mathbf{x}_{i+1/2}, t))\, dt - \int_{t^n}^{t^{n+1}} \Psi(\tilde{\mathbf{u}}^n(\mathbf{x}_{i-1/2}, t))\, dt\right].$$

In Godunov's method, it is natural to define the numerical entropy flux by $\tilde{\Psi}(\mathbf{w}_-, \mathbf{w}_+) = \Psi(\mathcal{R}(\mathbf{w}_-, \mathbf{w}_+; 0))$, where $\mathcal{R}(\mathbf{w}_-, \mathbf{w}_+; 0)$ is the state that moves with zero speed in the solution of the Riemann problem with left state $\mathbf{w}_-$ and right state $\mathbf{w}_+$. Since **Jensen's inequality** states that for any convex function S,

$$S\left(\frac{1}{\triangle x_i}\int_{\mathbf{x}_{i-1/2}}^{\mathbf{x}_{i+1/2}} \mathbf{w}(\mathbf{x})\, d\mathbf{x}\right) \leq \frac{1}{\triangle x_i}\int_{\mathbf{x}_{i-1/2}}^{\mathbf{x}_{i+1/2}} S(\mathbf{w}(\mathbf{x}))\, d\mathbf{x},$$

we have

$$\begin{aligned}
S(\mathbf{u}_i^{n+1}) &= S\left(\frac{1}{\triangle x_i}\int_{\mathbf{x}_{i-1/2}}^{\mathbf{x}_{i+1/2}} \tilde{\mathbf{u}}(\mathbf{x}, t^{n+1})\, d\mathbf{x}\right)\\
&\leq \frac{1}{\triangle x_i}\int_{\mathbf{x}_{i-1/2}}^{\mathbf{x}_{i+1/2}} S(\tilde{\mathbf{u}}(\mathbf{x}, t^{n+1}))\, d\mathbf{x} \leq \frac{1}{\triangle x_i}\int_{\mathbf{x}_{i-1/2}}^{\mathbf{x}_{i+1/2}} S(\mathbf{u}(\mathbf{x}, t^n))\, d\mathbf{x}\\
&\quad - \frac{1}{\triangle x_i}\left[\int_{t^n}^{t^{n+1}} \Psi(\tilde{\mathbf{u}}^n(\mathbf{x}_{i+1/2}, t))dt - \int_{t^n}^{t^{n+1}} \Psi(\tilde{\mathbf{u}}^n(\mathbf{x}_{i-1/2}, t))\, dt\right]\\
&= S(\mathbf{u}_i^n) - \frac{\triangle t^{n+1/2}}{\triangle x_i}[\Psi(\mathbf{u}_i^n, \mathbf{u}_{i+1}^n) - \Psi(\mathbf{u}_{i-1}^n, \mathbf{u}_i^n)].
\end{aligned}$$

It follows that in this case (Godunov's method for a conservation law with convex entropy function), whenever the numerical solution converges, it converges to a solution of the conservation law that satisfies the entropy inequality, similar to the "weak entropy inequality" (4.31). □

The difficulty with this scheme lies in computing the solution of the Riemann problem. We will discuss some examples of Riemann problem solutions in the case study sections that follow, and a number of ways to approximate the solution of Riemann problems in Section 4.13.

4.3 Case Study: Maxwell's Equations

Maxwell's equations for electromagnetic wave propagation are very important in electrical engineering and physics. Since these equations are linear, their solutions do not involve shocks. As a result, the shock-capturing techniques in this book are not particularly well-designed for this problem. We include this discussion because many students are interested in these equations, and because of the connections to the magnetohydrodynamics model in section 4.5 below.

4.3.1 Conservation Laws

In electromagnetic wave propagation, the electric displacement vector **D** is related to the electric field strength vector **E** by

$$\mathbf{D} = \mathbf{E}\epsilon,$$

and the magnetic induction vector **B** is related to the magnetic field strength vector **H** by

$$\mathbf{B} = \mathbf{H}\mu,$$

where ϵ is the permittivity and μ is the magnetic permeability. Here D has units of coulombs per square meter, E has units of volts per meter and ϵ has units of seconds per meter-ohm. Also **B** has units of volt-seconds per square meter, H has units of amps per meter and μ has units of ohm-seconds per meter. The induction current vector **J** is related to the electric field strength vector **E** by

$$\mathbf{J} = \mathbf{E}\sigma,$$

where σ is the conductivity. Here **J** has units of amps per square meter and $1/\sigma$ has units of ohm-meters. The electric charge density ρ satisfies

$$\rho = \nabla_{\mathbf{x}} \cdot \mathbf{D}.$$

Here ρ has units of coulombs per cubic meter. Since there are no free magnetic poles in nature,

$$0 = \nabla_{\mathbf{x}} \cdot \mathbf{B}.$$

The electromagnetic force on some closed circuit ∂S is

$$\int_{\partial S} \mathbf{E} \cdot \mathbf{t}\, ds = -\frac{d}{dt} \int_S \mathbf{B} \cdot \mathbf{n}\, dS$$

where $\mathbf{n}$ is the unit outer normal to the surface, ∂S is the closed curve that represents the boundary of the surface S, and $\mathbf{t}$ is the unit tangent vector to ∂S. Stokes theorem implies that

$$\int_S \left[\frac{\partial \mathbf{B}}{\partial t} + \nabla_{\mathbf{x}} \times \mathbf{E} \right] \cdot \mathbf{n} \, dS = 0.$$

The magnetomotive force around the circuit ∂S is

$$\int_{\partial S} \mathbf{H} \cdot \mathbf{t} \, ds = \int_S \left[\mathbf{J} + \frac{\partial \mathbf{D}}{\partial t} \right] \cdot \mathbf{n} \, dS.$$

Stokes theorem implies that

$$\int_S \left[\mathbf{J} + \frac{\partial \mathbf{D}}{\partial t} - \nabla_{\mathbf{x}} \times \mathbf{H} \right] \cdot \mathbf{n} \, dS = 0.$$

For simplicity, we will assume that the density of the electric charge is $\rho = 0$. Then we can eliminate $\mathbf{D}$ and $\mathbf{H}$ to obtain

$$\frac{\partial \mathbf{E}\epsilon}{\partial t} - \nabla_{\mathbf{x}} \times (\mathbf{B}/\mu) = -\mathbf{E}\sigma \tag{4.34a}$$

$$\frac{\partial \mathbf{B}}{\partial t} + \nabla_{\mathbf{x}} \times (\mathbf{E}) = 0. \tag{4.34b}$$

Note that if $\mathbf{E}$ is divergence-free at $t = 0$, then the former of these two equations implies that it is divergence-free for all $t > 0$.

4.3.2 Characteristic Analysis

We would like to determine if Maxwell's equations (4.34) are hyperbolic. We can rewrite the system in the form

$$\frac{\partial \mathbf{u}(\mathbf{w})}{\partial t} + \sum_{i=1}^{3} \frac{\partial \mathbf{F}(\mathbf{w})\mathbf{e}_i}{\partial \mathbf{x}_i} = \mathbf{r}$$

where

$$\mathbf{w} \equiv \begin{bmatrix} \mathbf{E}_1 \\ \mathbf{E}_2 \\ \mathbf{E}_3 \\ \mathbf{B}_1 \\ \mathbf{B}_2 \\ \mathbf{B}_3 \end{bmatrix}, \quad \mathbf{u}(\mathbf{w}) \equiv \begin{bmatrix} \mathbf{E}_1\epsilon \\ \mathbf{E}_2\epsilon \\ \mathbf{E}_3\epsilon \\ \mathbf{B}_1 \\ \mathbf{B}_2 \\ \mathbf{B}_3 \end{bmatrix},$$

$$\mathbf{F}(\mathbf{w}) \equiv \begin{bmatrix} 0 & -\mathbf{B}_3/\mu & \mathbf{B}_2/\mu \\ \mathbf{B}_3/\mu & 0 & -\mathbf{B}_1/\mu \\ -\mathbf{B}_2/\mu & \mathbf{B}_1/\mu & 0 \\ 0 & -\mathbf{E}_3 & \mathbf{E}_2 \\ \mathbf{E}_3 & 0 & -\mathbf{E}_1 \\ -\mathbf{E}_2 & \mathbf{E}_1 & 0 \end{bmatrix}, \quad \mathbf{r}(\mathbf{w}) \equiv \begin{bmatrix} -\mathbf{E}_1\sigma \\ -\mathbf{E}_2\sigma \\ -\mathbf{E}_3\sigma \\ 0 \\ 0 \\ 0 \end{bmatrix}.$$

We can compute

$$\frac{\partial \mathbf{u}}{\partial \mathbf{w}} = \begin{bmatrix} \mathbf{I}\epsilon & 0 \\ 0 & \mathbf{I} \end{bmatrix} \quad \text{and} \quad \frac{\partial \mathbf{Fn}}{\partial \mathbf{w}} = \begin{bmatrix} 0 & -\mathbf{N}\frac{1}{\mu} \\ \mathbf{N} & 0 \end{bmatrix},$$

where

$$\mathbf{N} \equiv \begin{bmatrix} 0 & -\mathbf{n}_3 & \mathbf{n}_2 \\ \mathbf{n}_3 & 0 & -\mathbf{n}_1 \\ -\mathbf{n}_2 & \mathbf{n}_1 & 0 \end{bmatrix}.$$

Note that $\mathbf{Nx} = \mathbf{n} \times \mathbf{x}$ for any vector $\mathbf{x}$. We can also compute

$$\frac{\partial \mathbf{Fn}}{\partial \mathbf{w}} \left(\frac{\partial \mathbf{u}}{\partial \mathbf{w}} \right)^{-1} = \begin{bmatrix} 0 & -\mathbf{N}\frac{1}{\mu} \\ \mathbf{N}\frac{1}{\epsilon} & 0 \end{bmatrix}.$$

The eigenvectors and eigenvalues of this matrix satisfy

$$\begin{bmatrix} 0 & -\mathbf{N}\frac{1}{\mu} \\ \mathbf{N}\frac{1}{\epsilon} & 0 \end{bmatrix} \begin{bmatrix} \mathbf{y} \\ \mathbf{z} \end{bmatrix} = \begin{bmatrix} \mathbf{y} \\ \mathbf{z} \end{bmatrix} \lambda.$$

These equations imply that

$$\mathbf{y}\epsilon\mu\lambda^2 = -\mathbf{Nz}\epsilon\lambda = \mathbf{z} \times \mathbf{n}\epsilon\lambda = (\mathbf{Ny}) \times \mathbf{n} = \mathbf{n} \times \mathbf{y} \times \mathbf{n} = (\mathbf{I} - \mathbf{nn}^\top)\mathbf{y}.$$

Since $(\mathbf{I} - \mathbf{nn}^\top)$ is the orthogonal projection onto the space of vectors orthogonal to $\mathbf{n}$, its eigenvectors are $\mathbf{n}$ (with eigenvalue 0), and any vector orthogonal to $\mathbf{n}$ (with eigenvalue 1). It follows that the eigenvalues of $\frac{\partial \mathbf{Fn}}{\partial \mathbf{w}} \left(\frac{\partial \mathbf{u}}{\partial \mathbf{w}} \right)^{-1}$ are 0 and $\pm 1/\sqrt{\epsilon\mu}$.

We can put this all together in the form of an array of eigenvectors and an array of eigenvalues of the flux derivatives. Let $\mathbf{n}^\perp$ be any (unit) vector that is orthogonal to $\mathbf{n}$. Define

$$\mathbf{Q} = \begin{bmatrix} -\mathbf{n}^\perp \times \mathbf{n} & -\mathbf{n}^\perp & -\mathbf{n} & \mathbf{n} & \mathbf{n}^\perp & \mathbf{n}^\perp \times \mathbf{n} \\ \mathbf{n}^\perp & \mathbf{n} \times \mathbf{n}^\perp & \mathbf{n} & \mathbf{n} & \mathbf{n} \times \mathbf{n}^\perp & \mathbf{n}^\perp \end{bmatrix} \frac{1}{\sqrt{2}},$$

$$\mathbf{X} = \begin{bmatrix} \mathbf{I}\sqrt{\epsilon} & 0 \\ 0 & \mathbf{I}\sqrt{\mu} \end{bmatrix} \mathbf{Q}$$

and

$$\Lambda = \begin{bmatrix} -\frac{1}{\sqrt{\epsilon\mu}} & 0 & 0 & 0 & 0 & 0 \\ 0 & -\frac{1}{\sqrt{\epsilon\mu}} & 0 & 0 & 0 & 0 \\ 0 & 0 & 0 & 0 & 0 & 0 \\ 0 & 0 & 0 & 0 & 0 & 0 \\ 0 & 0 & 0 & 0 & \frac{1}{\sqrt{\epsilon\mu}} & 0 \\ 0 & 0 & 0 & 0 & 0 & \frac{1}{\sqrt{\epsilon\mu}} \end{bmatrix}.$$

Then $\mathbf{Q}$ is an orthogonal matrix, and $\frac{\partial \mathbf{Fn}}{\partial \mathbf{w}} \left(\frac{\partial \mathbf{u}}{\partial \mathbf{w}}\right)^{-1} \mathbf{X} = \mathbf{X}\Lambda$.

Summary 4.3.1 *Suppose that the permittivity ϵ and magnetic permeability μ are constant. Then Maxwell's equations* (4.34) *are hyperbolic, with characteristic speeds 0 and $\pm 1/\sqrt{\epsilon\mu}$.*

The quantity $1/\sqrt{\epsilon\mu}$ is called the speed of light in the medium. Also note that the matrix $\mathbf{X}$ of eigenvectors is easy to invert because its 3×3 block entries are scalar multiples of orthogonal matrices. At any rate, we have shown that Maxwell's equations produce a hyperbolic system of conservation laws.

The Riemann problem for Maxwell's equation in 3D will be associated with a particular direction $\mathbf{n}$ only. The solution of the Riemann problem follows directly from the characteristic analysis in Lemma 4.3.1 and the solution of the linear Riemann problem in Lemma 4.1.22.

4.4 Case Study: Gas Dynamics

Some useful references for the material in this section are [27, 37, 96, 175]. The reasons for studying fluid dynamics are that its equations are derived by important physical considerations, the wave structure of gas dynamics illustrates many of the basic issues of hyperbolic conservation laws, and the most sophisticated numerical methods for conservation laws were originally developed for gas dynamics applications.

4.4.1 Conservation Laws

We will denote the velocity by $\mathbf{v}(\mathbf{x}, t)$, the pressure by $p(\mathbf{x}, t)$, the density by $\rho(\mathbf{x}, t)$, the specific internal energy (*i.e.* and the internal energy per mass) by $e(\mathbf{x}, t)$. The acceleration due to gravity is the constant vector $\mathbf{g}$.

In the Eulerian frame the arrays of conserved quantities, fluxes and body forces are

$$\mathbf{u} \equiv \begin{bmatrix} \rho \\ \mathbf{v}\rho \\ (e + \frac{1}{2}\mathbf{v} \cdot \mathbf{v})\rho \end{bmatrix}, \mathbf{F} \equiv \begin{bmatrix} \rho\mathbf{v}^\top \\ \mathbf{v}\rho\mathbf{v}^\top + \mathbf{I}p \\ \{\rho(e + \frac{1}{2}\mathbf{v} \cdot \mathbf{v}) + p\}\mathbf{v}^\top \end{bmatrix}, \mathbf{b} \equiv \begin{bmatrix} 0 \\ \mathbf{g}\rho \\ \mathbf{g} \cdot \mathbf{v}\rho \end{bmatrix}. \tag{4.35}$$

We obtain the system of conservation laws either in integral form

$$\frac{d}{dt} \int_{\Omega_t} \mathbf{u} \, d\mathbf{x} + \int_{\partial\Omega_t} \mathbf{F}\mathbf{n} \, ds = \int_{\Omega_t} \mathbf{b} \, d\mathbf{x}, \tag{4.36}$$

or (away from discontinuities) as a system of partial differential equations

$$\frac{\partial \mathbf{u}}{\partial t} + \sum_{i=1}^{k} \frac{\partial \mathbf{F} e_i}{\partial \mathbf{x}_i} = b. \tag{4.37}$$

The first conservation law in this system represents conservation of mass, the last represents conservation of energy, and the other equations represent conservation of momentum.

4.4.2 Thermodynamics

The material in this section has been taken from Courant and Friedrichs [37]. In order to save space, we will basically present their major results.

Let e denote the specific internal energy (internal energy per mass), S denote the specific entropy (entropy per mass) and T denote the temperature. Then the second law of thermodynamics for a closed reversible system is

$$de = T \, dS - p \, d(1/\rho).$$

This says that the change in internal energy is equal to the heat contributed plus the work done by pressure. If the internal energy is given as a function of specific volume and specific entropy, then the second law of thermodynamics says that

$$p = -\left.\frac{\partial e}{\partial 1/\rho}\right|_S, \; T = \left.\frac{\partial e}{\partial S}\right|_\rho.$$

For an **ideal gas**, $p = \rho R T$. Courant and Friedrichs show that this implies that the internal energy is a function of temperature alone. If in addition there is a constant c_v, called the **heat capacity** at constant volume, so that $e = c_v T$, then we have a **polytropic gas**. Courant and Friedrichs show that for a polytropic gas, the heat capacity at constant volume is given by $c_v = \frac{R}{\gamma-1}$. Here R is the **gas constant**, with value $R = 8.314 \times 10^7$ ergs / (gram-mole degree). The constant $1 \leq \gamma \leq 5/3$ is a property of the gas; for example, $\gamma = 7/5$ for air, $5/3$ for argon and $4/3$ for sulfur hexafluoride.

For a polytropic gas the specific internal energy is

$$e = \frac{p}{(\gamma - 1)\rho},$$

so the specific enthalpy is

$$h = e + \frac{p}{\rho} = \frac{\gamma}{\gamma - 1} \frac{p}{\rho} = \gamma e.$$

The second law of thermodynamics now implies that

$$dS = \frac{1}{T} de + \frac{p}{T\rho^2} d\rho = \frac{1}{T\rho(\gamma - 1)} dp - \frac{p\gamma}{T\rho^2(\gamma - 1)} d\rho.$$

This can be reinterpreted as saying that

$$\frac{\partial S}{\partial p} = \frac{1}{T\rho(\gamma - 1)} = \frac{R}{p(\gamma - 1)} = \frac{c_v}{p}$$

and

$$\frac{\partial S}{\partial \rho} = -\frac{p\gamma}{T\rho^2(\gamma - 1)} = -\frac{\gamma}{\gamma - 1} \frac{R}{\rho} = -\frac{\gamma c_v}{\rho}.$$

Thus the specific entropy is given by

$$S - S_0 = c_v \ln \left[\frac{p}{p_0} \left(\frac{\rho_0}{\rho} \right)^{\gamma} \right]. \tag{4.38}$$

The constant S_0 is arbitrary, and has no effect on the discussions to follow.

Thermodynamic stability requires that $dS \geq 0$. Away from discontinuities and in the absence of body forces and at a fixed material particle $\mathbf{a}$ the entropy is independent of time:

$$\left. \frac{dS}{dt} \right|_{\mathbf{a}} = 0.$$

4.4.3 Characteristic Analysis

In Lemma 4.1.3 we saw how to use the quasilinear form (4.4) of a conservation law to compute characteristic speeds and directions. In the case of Eulerian gas dynamics, note that **u**, **F** and **b**, defined in Equation (4.35), are functions of the flux variables

$$\mathbf{w} \equiv \begin{bmatrix} \rho \\ \mathbf{v} \\ p \end{bmatrix}. \tag{4.39}$$

These allow us to compute

$$\frac{\partial \mathbf{u}}{\partial \mathbf{w}} = \begin{bmatrix} 1 & 0 & 0 \\ \mathbf{v} & \mathbf{I}\rho & 0 \\ \frac{1}{2}\mathbf{v}\cdot\mathbf{v} & \rho\mathbf{v}^\top & \frac{1}{\gamma-1} \end{bmatrix},$$

and for any fixed unit vector $\mathbf{n}$

$$\frac{\partial \mathbf{Fn}}{\partial \mathbf{w}} = \begin{bmatrix} \mathbf{n}\cdot\mathbf{v} & \rho\mathbf{n}^\top & 0 \\ \mathbf{v}(\mathbf{n}\cdot\mathbf{v}) & \mathbf{I}\rho(\mathbf{n}\cdot\mathbf{v}) + \mathbf{v}\rho\mathbf{n}^\top & \mathbf{n} \\ \frac{1}{2}(\mathbf{v}\cdot\mathbf{v})(\mathbf{n}\cdot\mathbf{v}) & \rho(e+\frac{1}{2}\mathbf{v}\cdot\mathbf{v})\mathbf{n}^\top + p\mathbf{n}^\top + \rho(\mathbf{n}\cdot\mathbf{v})\mathbf{v}^\top & \frac{\gamma}{\gamma-1}(\mathbf{n}\cdot\mathbf{v}) \end{bmatrix}.$$

For any fixed unit vector $\mathbf{n}$,

$$\left(\frac{\partial \mathbf{u}}{\partial \mathbf{w}}\right)^{-1} \frac{\partial \mathbf{Fn}}{\partial \mathbf{w}} = \begin{bmatrix} \mathbf{n}\cdot\mathbf{v} & \rho\mathbf{n}^\top & 0 \\ 0 & \mathbf{I}(\mathbf{n}\cdot\mathbf{v}) & \mathbf{n}/\rho \\ 0 & \gamma p\mathbf{n}^\top & \mathbf{n}\cdot\mathbf{v} \end{bmatrix} \equiv \mathbf{A} + \mathbf{I}(\mathbf{n}\cdot\mathbf{v})$$

where $\mathbf{A}$ is the **acoustic tensor**

$$\mathbf{A} = \begin{bmatrix} 0 & \rho\mathbf{n}^\top & 0 \\ 0 & 0 & \mathbf{n}1/\rho \\ 0 & \gamma p\mathbf{n}^\top & 0 \end{bmatrix}.$$

It is easy to find the eigenvectors $\mathbf{Y}$ and eigenvalues Λ of $\mathbf{A}$. Let the matrix $[\mathbf{n},\ \mathbf{N}]$ be a rotation, and let c be the sound speed (*i.e.*, $c^2 = \gamma p/\rho$). Then

$$\begin{aligned} \mathbf{AY} &= \begin{bmatrix} 0 & \rho\mathbf{n}^\top & 0 \\ 0 & 0 & \mathbf{n}1/\rho \\ 0 & \gamma p\mathbf{n}^\top & 0 \end{bmatrix} \begin{bmatrix} \rho & 0 & 1 & \rho \\ \mathbf{n}c & \mathbf{N} & 0 & -\mathbf{n}c \\ \rho c^2 & 0 & 0 & \rho c^2 \end{bmatrix} \\ &= \begin{bmatrix} \rho & 0 & 1 & \rho \\ \mathbf{n}c & \mathbf{N} & 0 & -\mathbf{n}c \\ \rho c^2 & 0 & 0 & \rho c^2 \end{bmatrix} \begin{bmatrix} c & & & \\ & 0 & & \\ & & 0 & \\ & & & -c \end{bmatrix} = \mathbf{Y}\Lambda. \end{aligned}$$

As we saw in Lemma 4.1.3, the characteristic speeds are the eigenvalues of

$$\frac{\partial \mathbf{Fn}}{\partial \mathbf{u}} = \frac{\partial \mathbf{Fn}}{\partial \mathbf{w}} \left(\frac{\partial \mathbf{u}}{\partial \mathbf{w}}\right)^{-1} = \left(\frac{\partial \mathbf{u}}{\partial \mathbf{w}}\right) [\mathbf{A} + \mathbf{I}(\mathbf{n}\cdot\mathbf{v})] \left(\frac{\partial \mathbf{u}}{\partial \mathbf{w}}\right)^{-1}.$$

Thus the characteristic speeds for Eulerian gas dynamics are $\mathbf{n} \cdot \mathbf{v}$ and $\mathbf{n} \cdot \mathbf{v} \pm c$. Similarly, the characteristic directions are the eigenvectors of $\frac{\partial \mathbf{Fn}}{\partial \mathbf{u}}$, which we can compute as $\mathbf{X} = \frac{\partial \mathbf{u}}{\partial \mathbf{w}} \mathbf{Y}$.

The characteristic speeds are also functions of the flux variables ρ, $\mathbf{v}$ and p. Thus the test for genuine nonlinearity in Definition 4.1.5 can be written

$$\frac{\partial \lambda_i}{\partial \mathbf{u}} \mathbf{X} \mathbf{e}_i = \left\{ \frac{\partial \lambda_i}{\partial \mathbf{w}} \left(\frac{\partial \mathbf{u}}{\partial \mathbf{w}} \mathbf{Y} \right)^{-1} \right\} \left\{ \frac{\partial \mathbf{u}}{\partial \mathbf{w}} \mathbf{Y} \right\} \mathbf{e}_i = \frac{\partial \lambda_i}{\partial \mathbf{w}} \mathbf{Y} \mathbf{e}_i .$$

For example, the characteristic speed $\mathbf{n} \cdot \mathbf{v} + c$ is genuinely nonlinear, because

$$\begin{aligned} \frac{\partial (\mathbf{n} \cdot \mathbf{v} + c)}{\partial \mathbf{w}} \mathbf{Y} \mathbf{e}_1 &= \begin{bmatrix} -\frac{1}{2}\frac{c}{\rho} & \mathbf{n}^\top & \frac{1}{2}\frac{c}{p} \end{bmatrix} \begin{bmatrix} \rho \\ \mathbf{n}c \\ \rho c^2 \end{bmatrix} \\ &= -\frac{1}{2} c + c + \frac{1}{2} c^3 \frac{\rho}{p} = \frac{1}{2} c (1 + \gamma) \neq 0. \end{aligned}$$

Similarly, $\mathbf{n} \cdot \mathbf{v} - c$ is genuinely nonlinear:

$$\begin{aligned} \frac{\partial (\mathbf{n} \cdot \mathbf{v} - c)}{\partial \mathbf{w}} \mathbf{Y} \mathbf{e}_{k+2} &= \begin{bmatrix} \frac{1}{2}\frac{c}{\rho} & \mathbf{n}^\top & -\frac{1}{2}\frac{c}{p} \end{bmatrix} \begin{bmatrix} \rho \\ -\mathbf{n}c \\ \rho c^2 \end{bmatrix} \\ &= \frac{1}{2} c - c - \frac{1}{2} c^3 \frac{\rho}{p} = -\frac{1}{2} c (1 + \gamma) \neq 0. \end{aligned}$$

The other characteristic speeds, namely $\mathbf{n} \cdot \mathbf{v}$, are linearly degenerate because

$$\frac{\partial \mathbf{n} \cdot \mathbf{v}}{\partial \mathbf{w}} \begin{bmatrix} 0 & 1 \\ \mathbf{N} & 0 \\ 0 & 0 \end{bmatrix} = \begin{bmatrix} 0 & \mathbf{n}^\top & 0 \end{bmatrix} \begin{bmatrix} 0 & 1 \\ \mathbf{N} & 0 \\ 0 & 0 \end{bmatrix} = 0.$$

Summary 4.4.1 *The Eulerian equations* (4.37) *for a polytropic gas are hyperbolic with respect to any given direction* $\mathbf{n}$ *with characteristic speeds* $\mathbf{n} \cdot \mathbf{v}$ *and* $\mathbf{n} \cdot \mathbf{v} \pm c$*, where* $c = \sqrt{\gamma p / \rho}$ *is the speed of sound. The characteristic speed* $\mathbf{n} \cdot \mathbf{v}$ *is linearly degenerate, and the characteristic speeds* $\mathbf{n} \cdot \mathbf{v} \pm c$ *are genuinely nonlinear.*

4.4.4 Entropy Function

We derived the specific entropy for gas dynamics in equation (4.38). Thus the derivative of the specific entropy with respect to the vector of flux variables $\mathbf{w}$, defined by (4.39), is

$$\frac{\partial S}{\partial \mathbf{w}} = c_v \left[-\gamma/\rho, \quad 0, \quad 1/p \right].$$

This implies that the partial derivatives of the entropy per volume are

$$\frac{\partial S\rho}{\partial \mathbf{w}} = \left[S - c_v\gamma, \quad 0, \quad c_v\rho/p \right].$$

Next, we compute

$$\begin{aligned}
\frac{\partial S\rho}{\partial \mathbf{w}} \left(\frac{\partial \mathbf{u}}{\partial \mathbf{w}} \right)^{-1} \frac{\partial \mathbf{Fn}}{\partial \mathbf{w}} &= \left[S - c_v\gamma \quad 0 \quad c_v\rho/p \right] \begin{bmatrix} (\mathbf{n}^\top\mathbf{v}) & \rho\mathbf{n}^\top & 0 \\ 0 & \mathbf{I}(\mathbf{n}^\top\mathbf{v}) & \mathbf{n}1/\rho \\ 0 & \gamma p\mathbf{n}^\top & (\mathbf{n}^\top\mathbf{v}) \end{bmatrix} \\
&= \left[(S - c_v\gamma)(\mathbf{n}^\top\mathbf{v}) \quad S\rho\mathbf{n}^\top \quad c_v\tfrac{\rho}{p}(\mathbf{n}^\top\mathbf{v}) \right] \\
&= \frac{\partial S}{\partial \mathbf{w}} \rho\mathbf{n}^\top\mathbf{v} + S\mathbf{n}^\top \left[\mathbf{v} \quad \mathbf{I}\rho \quad 0 \right] = \frac{\partial S}{\partial \mathbf{w}} \rho\mathbf{n}^\top\mathbf{v} + S\mathbf{n}^\top \frac{\partial \mathbf{v}\rho}{\partial \mathbf{w}} \\
&= \frac{\partial S\rho(\mathbf{n}^\top\mathbf{v})}{\partial \mathbf{w}}.
\end{aligned}$$

The work in section 4.1.11 therefore shows that $S\rho\mathbf{n} \cdot \mathbf{v}$ is the entropy flux.

In order to see that the entropy function for gas dynamics is concave, we compute

$$\frac{\partial}{\partial \mathbf{w}} \left(\frac{\partial S\rho}{\partial \mathbf{w}} \right)^\top = \begin{bmatrix} -c_v\gamma/\rho & 0 & c_v/p \\ 0 & 0 & 0 \\ c_v/p & 0 & -c_v\rho/p^2 \end{bmatrix}.$$

It is easy to see that the eigenvalues of this matrix are either zero or

$$\lambda = -\frac{\rho}{2p^2} \left[1 + \frac{c^2}{\gamma} \pm \sqrt{\left(1 + \frac{c^2}{\gamma} \right)^2 - 4\frac{\gamma - 1}{\gamma} c^2} \right],$$

which are both nonpositive. Thus the entropy function is a concave function of ρ and p. Away from discontinuities, the Eulerian conservation law for entropy is

$$\begin{aligned}
\frac{\partial S\rho}{\partial t} + \nabla_{\mathbf{x}} \cdot (\mathbf{v}S\rho) &= \frac{\partial S\rho}{\partial \mathbf{u}} \frac{\partial \mathbf{u}}{\partial t} + \sum_{i=1}^{k} \frac{\partial S\rho\mathbf{e}_i \cdot \mathbf{v}}{\partial \mathbf{u}} \frac{\partial \mathbf{u}}{\partial \mathbf{x}_i} = \frac{\partial S\rho}{\partial \mathbf{u}} \left\{ \frac{\partial \mathbf{u}}{\partial t} + \sum_{i=1}^{k} \frac{\partial \mathbf{Fe}_i}{\partial \mathbf{u}} \frac{\partial \mathbf{u}}{\partial \mathbf{x}_i} \right\} \\
&= \frac{\partial S\rho}{\partial \mathbf{u}} \mathbf{b} = \left[S - c_v\gamma \quad 0 \quad c_v\rho/p \right] \begin{bmatrix} 0 \\ \mathbf{g}\rho \\ \mathbf{g} \cdot \mathbf{v}\rho \end{bmatrix} = c_v \frac{\mathbf{g} \cdot \mathbf{v}\rho^2}{p}.
\end{aligned}$$

Summary 4.4.2 *The specific entropy*

$$S = S_0 + c_v \left[\ln \frac{p}{p_0} - \gamma \ln \frac{\rho}{\rho_0} \right]$$

is an entropy function for gas dynamics, with corresponding flux $S\rho\mathbf{v}^\top$. *The specific entropy is a concave function of* p *and* ρ. *Finally, away from discontinuities, the Eulerian conservation law for entropy is*

$$\frac{\partial S\rho}{\partial t} + \nabla_{\mathbf{x}} \cdot (\mathbf{v}S\rho) = c_v \frac{\mathbf{g} \cdot \mathbf{v}\rho^2}{p}.$$

4.4.5 Centered Rarefaction Curves

Recall equation (4.28) for a centered rarefaction wave: $\tilde{\mathbf{w}}' = \mathbf{Y}(\tilde{\mathbf{w}})\mathbf{e}_j\alpha$, $\tilde{\mathbf{w}}(0) = \mathbf{w}_L$. Here $\mathbf{w}$ is the vector of flux variables, and $\mathbf{Y}$ is the matrix of eigenvectors of $(\frac{\partial \mathbf{u}}{\partial \mathbf{w}})^{-1}\frac{\partial \mathbf{Fn}}{\partial \mathbf{w}}$, which was computed in Section 4.4.3 We will use this equation to determine the centered rarefaction waves for the gas dynamics Equations (4.35) and (4.37).

The centered rarefaction corresponding to the slow characteristic speed $\mathbf{v} \cdot \mathbf{n} - c$ is the solution of the ordinary differential equation

$$\frac{d}{dy} \begin{bmatrix} \rho \\ \mathbf{v} \\ p \end{bmatrix} = \begin{bmatrix} \rho \\ -\mathbf{n}c \\ \rho c^2 \end{bmatrix} \alpha,$$

where y is some measure of distance along the rarefaction curve. Note that this system of ordinary differential equations says $\frac{dp}{d\rho} = c^2 = \frac{\gamma p}{\rho}$ and $\frac{d\mathbf{v}\cdot\mathbf{n}}{d\rho} = -\frac{c}{\rho}$. We can solve the former of these two equations to get $p = p_L(\frac{\rho}{\rho_L})^\gamma$. Note that this implies that the specific entropy $S = S_0 + c_v \ln\{\frac{p}{\rho^\gamma}\}$ is constant. We can solve the other ordinary differential equation to get

$$\begin{aligned} \mathbf{v}_L \cdot \mathbf{n} &= \mathbf{v} \cdot \mathbf{n} + \int_{\rho_L}^{\rho} \frac{c}{\rho} d\rho = \mathbf{v} \cdot \mathbf{n} + \int_{\rho_L}^{\rho} \sqrt{\frac{\gamma p}{\rho^3}} d\rho = \mathbf{v} \cdot \mathbf{n} + \int_{\rho_L}^{\rho} \sqrt{\frac{\gamma p_L}{\rho_L^\gamma}} \rho^{\gamma-3} d\rho \\ &= \mathbf{v} \cdot \mathbf{n} + \sqrt{\frac{\gamma p_L}{\rho_L^\gamma}} \frac{2}{\gamma-1} \rho^{\frac{1}{2}(\gamma-1)} \Big|_{\rho_L}^{\rho} = \mathbf{v} \cdot \mathbf{n} + \frac{2c}{\gamma-1} - \frac{2c_L}{\gamma-1}. \end{aligned}$$

It follows that $\mathbf{v} \cdot \mathbf{n} + 2c/(\gamma-1)$ is also constant on this centered rarefaction curve.

Since there are three flux variables for one-dimensional gas dynamics, and since the slow centered rarefaction curve involves a single degree of freedom, the centered rarefaction is completely described by the conditions that both the specific entropy S and $\mathbf{v} \cdot \mathbf{n} + 2c/(\gamma - 1)$ are constant. These quantities are called **Riemann invariants**. In multiple dimensions, the transverse components of velocity are also Riemann invariants.

It is useful to treat p as the independent variable in the description of the centered rarefaction curves. If $(\rho_L, \mathbf{v}_L, p_L)$ is on the centered rarefaction curve, then $\rho = \rho_L(\frac{p}{p_L})^{1/\gamma}$, $c = c_L(\frac{p}{p_L})^{\frac{1}{2}(1-1/\gamma)}$, $\mathbf{v} = \mathbf{v}_L - \mathbf{n}\frac{2}{\gamma-1}(c - c_L)$. Note that as $p \downarrow 0$, we have that $\rho \downarrow 0$, $c \downarrow 0$ and $\mathbf{v} \uparrow \mathbf{v}_L + \mathbf{n}\frac{2c_L}{\gamma-1}$. Also note that $2c\frac{dc}{dp} = \gamma\frac{dp/\rho}{dp} = \gamma(\frac{1}{\rho} - \frac{p}{\rho^2}\frac{d\rho}{dp}) = \frac{\gamma}{\rho^2}(\rho - p\frac{\rho}{\gamma p}) = \frac{\gamma-1}{\rho}$. Thus along the centered rarefaction curve the rate of change of the characteristic speed is $\frac{d(\mathbf{v}\cdot\mathbf{n}-c)}{dp} = -\frac{1}{\rho c} + \frac{\gamma-1}{2\rho c} = -\frac{3-\gamma}{2\rho c} < 0$, since section 4.4.2 gave us $1 < \gamma < 5/3$. Thus $\mathbf{v} \cdot \mathbf{n} - c$ increases as p decreases. Similar discussions apply to the centered rarefaction curves for $\mathbf{v} \cdot \mathbf{n} + c$.

Summary 4.4.3 *In a polytropic gas, with sound speed $c = \sqrt{\gamma p/\rho}$,*

(i) *a slow centered rarefaction, which is associated with characteristic speed $\mathbf{v} \cdot \mathbf{n} - c$, has for its Riemann invariants the specific entropy $S = c_v \ln(p/\rho^\gamma)$, the quantity $\mathbf{v} \cdot \mathbf{n} + 2c/(\gamma - 1)$, and (in multiple dimensions) the transverse components of velocity,*
(ii) *and a fast centered rarefaction, which is associated with characteristic speed $\mathbf{v} \cdot \mathbf{n} + c$, has for its Riemann invariants the specific entropy S, the quantity $\mathbf{v} \cdot \mathbf{n} - 2c/(\gamma - 1)$, and (in multiple dimensions) the transverse components of velocity.*

All states on

(i) *a slow rarefaction curve containing the state $(\rho_L, \mathbf{v}_L, p_L)$ satisfy the equations*

$$\rho = \rho_L(p/p_L)^{1/\gamma}, \quad c = c_L(p/p_L)^{(\gamma-1)/(2\gamma)}, \quad \mathbf{v} = \mathbf{v}_L - \mathbf{n}\frac{2}{\gamma - 1}(c - c_L), \tag{4.40}$$

(i) *and on a fast rarefaction containing the state $(\rho_R, \mathbf{v}_R, p_R)$ satisfy the equations*

$$\rho = \rho_R(p/p_R)^{1/\gamma}, \quad c = c_R(p/p_R)^{(\gamma-1)/2\gamma)}, \quad \mathbf{v} = \mathbf{v}_R + \mathbf{n}\frac{2}{\gamma - 1}(c - c_R).$$

Further,

(i) *the slow characteristic speed $\mathbf{v} \cdot \mathbf{n} - c$ increases as p decreases,*
(ii) *and the fast characteristic speed $\mathbf{v} \cdot \mathbf{n} + c$ increases as p increases.*

As $p \downarrow 0$,

(i) *along the slow rarefaction curve we have $\rho \downarrow 0$, $c \downarrow 0$ and $\mathbf{v} \uparrow v_L + \mathbf{n}\frac{2c_L}{\gamma-1}$,*
(ii) *and along the fast rarefaction curve we have $\rho \downarrow 0$, $c \downarrow 0$ and $\mathbf{v} \downarrow v_R - \mathbf{n}\frac{2c_L}{\gamma-1}$.*

4.4.6 Jump Conditions

If we apply the Rankine–Hugoniot jump conditions (4.22) to the Eulerian gas dynamics function (4.35), we obtain

$$[\rho \mathbf{v} \cdot \mathbf{n}] = [\rho]\sigma,$$
$$[\mathbf{v}\rho(\mathbf{v} \cdot \mathbf{n}) + \mathbf{n}p] = [\mathbf{v}\rho]\sigma$$
$$\left[\rho\left(e + \frac{1}{2}\mathbf{v} \cdot \mathbf{v}\right)(\mathbf{v} \cdot \mathbf{n}) + p\mathbf{v} \cdot \mathbf{n}\right] = \left[\left(e + \frac{1}{2}\mathbf{v} \cdot \mathbf{v}\right)\rho\right]\sigma.$$

Let $[\mathbf{n}, \mathbf{N}]$ be a rotation matrix with first column $\mathbf{n}$, and let $\nu = \mathbf{n} \cdot \mathbf{v} - \sigma$ be the normal velocity relative to the discontinuity speed. Then we can write the velocity in the form $\mathbf{v} = \mathbf{n}(\nu + \sigma) + \mathbf{N}\mathbf{v}^{\perp}$. If we multiply the normal component of the momentum jump condition by σ and subtract from the energy jump, we get

$$[\rho\nu] = 0, \tag{4.41a}$$
$$[p] = -[\rho\nu^2], \tag{4.41b}$$
$$[\mathbf{v}^{\perp}\rho\nu] = 0 \tag{4.41c}$$
$$\left[p\nu\frac{\gamma}{\gamma - 1} + \frac{1}{2}\rho\nu^3 + \frac{1}{2}\mathbf{v}^{\perp} \cdot \mathbf{v}^{\perp}\rho\nu\right] = 0. \tag{4.41d}$$

The mass jump condition (4.41a) implies that $\rho_R\nu_R = \rho_L\nu_L$. Let us consider the case in which $\rho_R\nu_R = \rho_L\nu_L$ is nonzero. Then $\rho_L > 0$, $\nu_R \neq 0$ and

$$\begin{aligned}
\mathbf{v}_R^{\perp} &= \mathbf{v}_L^{\perp}, \\
\nu_R &= \nu_L - \frac{p_R - p_L}{\rho_L\nu_L} = \frac{\rho_L\nu_L^2 - p_R + p_L}{\rho_L\nu_L} \\
\rho_R &= \frac{\rho_L\nu_L}{\nu_R} = \frac{(\rho_L\nu_L)^2}{\rho_L\nu_L^2 - p_R + p_L}, \\
0 &= p_R\nu_R\frac{\gamma}{\gamma - 1} + \frac{1}{2}\rho_L\nu_L\nu_R^2 - p_L\nu_L\frac{\gamma}{\gamma - 1} - \frac{1}{2}\rho_L\nu_L^3 \\
&= p_R\left\{\nu_L - \frac{p_R - p_L}{\rho_L\nu_L}\right\}\frac{\gamma}{\gamma - 1} + \frac{1}{2}\rho_L\nu_L\left\{\nu_L - \frac{p_R - p_L}{\rho_L\nu_L}\right\}^2 \\
&\quad - p_L\nu_L\frac{\gamma}{\gamma - 1} - \frac{1}{2}\rho_L\nu_L^3 \\
&= \frac{p_R - p_L}{2\rho_L\nu_L}\left\{\rho_L\nu_L^2\frac{2\gamma}{\gamma - 1} - p_R\frac{2\gamma}{\gamma - 1} - 2\rho_L\nu_L^2 + p_R - p_L\right\}.
\end{aligned}$$

Note that $p_R = p_L$ implies that $\nu_R = \nu_L$, which in turn implies that $\rho_R = \rho_L$ and therefore there is no jump. Thus we can divide the fourth of these jump conditions by $p_R - p_L$ to obtain

$$0 = \rho_L v_L^2 \left\{ \frac{2\gamma}{\gamma - 1} - 2 \right\} - p_R \frac{2\gamma}{\gamma - 1} + p_R - p_L = \rho_L v_L^2 \frac{2}{\gamma - 1} - p_R \frac{\gamma + 1}{\gamma - 1} - p_L .$$

This can be rewritten

$$v_L^2 = \frac{1}{2\rho_L} \{ p_L(\gamma - 1) + p_R(\gamma + 1) \}$$

This implies that the discontinuity speed is given by

$$\sigma = \mathbf{n} \cdot \mathbf{v}_L - v_L = \mathbf{n} \cdot \mathbf{v}_L \pm \sqrt{\frac{1}{2\rho_L} \{ p_L(\gamma - 1) + p_R(\gamma + 1) \}}.$$

Let us summarize these results.

Summary 4.4.4 *Suppose that* $\mathbf{n}$ *is the normal to a gas dynamics discontinuity propagating with speed* σ. *Also suppose that* $[\mathbf{n}, \ \mathbf{N}]$ *is a rotation. Given a left state* $\rho_L > 0$, $\mathbf{v}_L = \mathbf{n}\mathbf{n} \cdot \mathbf{v}_L + \mathbf{N}\mathbf{v}_L^\perp$, p_L *and the right pressure* p_R, *the gas dynamics jump conditions imply that* $\mathbf{v}_R = \mathbf{n}\mathbf{n} \cdot \mathbf{v}_R + \mathbf{N}\mathbf{v}_R^\perp$ *where*

$$\begin{aligned}
\mathbf{v}_R^\perp &= \mathbf{v}_L^\perp \\
\sigma &= \mathbf{n} \cdot \mathbf{v}_L - \sqrt{\frac{1}{2\rho_L} \{ p_L(\gamma - 1) + p_R(\gamma + 1) \}} \\
v_L &= \mathbf{n} \cdot \mathbf{v}_L - \sigma \\
v_R &= \mathbf{n} \cdot \mathbf{v}_R - \sigma = v_L - \frac{p_R - p_L}{\rho_L v_L} \\
\rho_R &= \frac{\rho_L v_L}{v_R}.
\end{aligned}$$

By symmetry, given $\rho_R > 0$, $\mathbf{v}_R$, p_R *and* p_L *the jump conditions imply*

$$\begin{aligned}
\mathbf{v}_L^\perp &= \mathbf{v}_R^\perp \\
\sigma &= \mathbf{n} \cdot \mathbf{v}_R + \sqrt{\frac{1}{2\rho_R} \{ p_R(\gamma - 1) + p_L(\gamma + 1) \}} \\
v_R &= \mathbf{n} \cdot \mathbf{v}_R - \sigma \\
v_L &= \mathbf{n} \cdot \mathbf{v}_L - \sigma = v_R - \frac{p_L - p_R}{\rho_R v_R} \\
\rho_L &= \frac{\rho_R v_R}{v_L}.
\end{aligned}$$

On the other hand, suppose that $\rho_L v_L = \rho_R v_R = 0$. Then the jump conditions (4.41) imply that $p_R = p_L$ but allow the transverse velocities $\mathbf{v}_R^\perp$ and $\mathbf{v}_L^\perp$ to be arbitrary. If $v_L = 0$, then (4.41d) implies that $p_R v_R = 0$; in this case, either $v_R = 0$ or both $\rho_R = 0$ and $p_R = 0$. The former of these two choices corresponds to a contact discontinuity.

Summary 4.4.5 *Suppose that* $\mathbf{n}$ *is the normal to a gas dynamics discontinuity, and that* $[\mathbf{n}, \mathbf{N}]$ *is a rotation. Suppose that we are given a left state* ρ_L, $\mathbf{v}_L = \mathbf{n}\mathbf{n} \cdot \mathbf{v}_L + \mathbf{N}\mathbf{v}_L^\perp$, *and* p_L. *If the discontinuity speed is* $\sigma = \mathbf{n} \cdot \mathbf{v}_L$ *then the gas dynamics jump conditions imply that*

$$\mathbf{v}_R = \mathbf{n}\mathbf{n} \cdot \mathbf{v}_L + \mathbf{N}\mathbf{v}_R^\perp$$
$$p_R = p_L$$

for any transverse velocity $\mathbf{v}_R^\perp$ *and any density* ρ_R. *In other words, across such a discontinuity the only jumps are in the transverse velocity and the density.*

In order to complete the solution of the gas dynamics jump conditions, we need to consider two more cases. In the case when $v_L = 0$ and $v_R \neq 0$ we have $\rho_R = 0$ and $p_L = p_R = 0$. Since both states are at zero pressure, this case is not very interesting. The other case has $\rho_L v_L = \rho_R v_R = 0$ with $v_L \neq 0$. This implies that $\rho_L = 0$. In order to have any gas particles in the problem, we would have to have $\rho_R > 0$, which would in turn imply that $v_R = 0$. This would lead to $0 = p_L = p_R$, so both states would be at zero pressure. Again, this case is not very interesting.

Sometimes, it is useful to discuss the jump conditions in terms of **shock strength**. Given a discontinuity propagating with speed $\sigma < \mathbf{n} \cdot \mathbf{v}_L$, we will define the shock strength by

$$z_L \equiv \frac{p_R - p_L}{p_L}.$$

Given $\rho_L > 0$, p_L and the shock strength z_L, Lemma 4.4.4 implies that the states on either side of the discontinuity satisfy

$$p_R = p_L(1 + z_L)$$
$$v_L = \sqrt{\frac{p_L}{\rho_L}\left\{\frac{\gamma - 1}{2} + \frac{1 + z_L}{2}(\gamma + 1)\right\}} = c_L\sqrt{1 + z_L\frac{\gamma + 1}{2\gamma}}$$
$$v_R = \frac{1}{\rho_L v_L}\left\{\rho_L v_L^2 - p_R + p_L\right\} = \frac{1}{\rho_L v_L}\left\{\rho_L c_L^2\left(1 + z_L\frac{\gamma + 1}{2\gamma}\right) - p_L(1 + z_L) + p_L\right\}$$
$$= \frac{c_L^2}{v_L}\left\{1 + z_L\frac{\gamma + 1}{2\gamma} - \frac{z_L}{\gamma}\right\} = \frac{c_L^2}{v_L}\left\{1 + z_L\frac{\gamma - 1}{2\gamma}\right\}$$
$$\rho_R = \frac{\rho_L v_L}{v_R} = \frac{\rho_L v_L^2}{c_L^2}\frac{1}{1 + z_L\frac{\gamma-1}{2\gamma}} = \rho_L\frac{1 + z_L\frac{\gamma+1}{2\gamma}}{1 + z_L\frac{\gamma-1}{2\gamma}}.$$

Given the same condition on the discontinuity speed, it is sometimes useful to define the **Mach number** by $M_L = \frac{v_L}{c_L}$. Given $\rho_L > 0$, p_L and the Mach number M_L, the states on either side of the discontinuity satisfy

$$
\begin{aligned}
v_L &= M_L c_L \\
p_R &= \frac{\gamma - 1}{\gamma + 1}\left\{\rho_L v_L^2 \frac{2}{\gamma - 1} - p_L\right\} = \frac{p_L}{\gamma + 1}\left\{\frac{2\rho_L v_L^2}{p_L} - \gamma + 1\right\} \\
&= \frac{p_L}{\gamma + 1}\left\{2\gamma M_L^2 - \gamma + 1\right\} = p_L\left\{\frac{2\gamma}{\gamma + 1}(M_L^2 - 1) + 1\right\} \\
v_R &= v_L - \frac{p_R - p_L}{\rho_L v_L} = v_L - \frac{p_L \frac{2\gamma}{\gamma+1}(M_L^2 - 1)}{\rho_L M_L c_L} \\
&= v_L - \frac{2c_L(M_L^2 - 1)}{M_L(\gamma + 1)} = c_L \frac{2 + M_L^2(\gamma - 1)}{M_L(\gamma + 1)} \\
\rho_R &= \frac{\rho_L v_L}{v_R} = \frac{\rho_L M_L c_L}{c_L} \frac{M_L(\gamma + 1)}{2 + M_L^2(\gamma - 1)} = \rho_L \frac{M_L^2(\gamma + 1)}{2 + M_L^2(\gamma - 1)}.
\end{aligned}
$$

Similar results can be obtained for a discontinuity propagating with speed $\sigma > \mathbf{n} \cdot \mathbf{v}_R$.

Finally, let us discuss thermodynamic stability for propagating discontinuities. If $0 < v_L \equiv \mathbf{n} \cdot \mathbf{v}_L - \sigma$, then we also have $v_R > 0$; in this case, gas particles move from the left (the pre-shock state) to the right (the post-shock state). Thermodynamic stability therefore requires that

$$
0 < \frac{S_R - S_L}{c_v} = \ln\left\{\frac{p_R}{p_L}\left(\frac{\rho_L}{\rho_R}\right)^\gamma\right\} = \ln\left\{(1 + z_L)\left(\frac{1 + z_L \frac{\gamma-1}{2\gamma}}{1 + z_L \frac{\gamma+1}{2\gamma}}\right)^\gamma\right\} \equiv \phi(z_L).
$$

Note that $\phi(0) = 1$ and

$$
\frac{d\phi}{dz} = \left(\frac{1 + z\frac{\gamma-1}{2\gamma}}{1 + z\frac{\gamma+1}{2\gamma}}\right)^\gamma \frac{z^2}{(1 + z\frac{\gamma+1}{2\gamma})(1 + z\frac{\gamma-1}{2\gamma})} \frac{\gamma^2 - 1}{4\gamma^2}.
$$

Thus $z > 0$ implies that $\phi(z) > 1$ and $\ln\phi(z) > 0$. Thus thermodynamic stability for a discontinuity propagating with speed $\sigma < \mathbf{n} \cdot \mathbf{v}_R$

requires

$$
\begin{aligned}
&p_R > p_L \\
&\frac{\rho_R}{\rho_L} > 1 \Longrightarrow \rho_R > \rho_L \\
&\frac{v_R}{v_L} > 1 \Longrightarrow \mathbf{n} \cdot \mathbf{v}_L > \mathbf{n} \cdot \mathbf{v}_R \\
&M_L = \frac{v_L}{c_L} = \sqrt{1 + z_L \frac{\gamma + 1}{2\gamma}} > 1 \Longrightarrow \mathbf{n} \cdot \mathbf{v}_L - c_L > \sigma \\
&c_R^2 = \frac{\gamma p_R}{\rho_R} = \frac{\gamma p_L (1 + z_L)}{\rho_L} \frac{1 + z_L \frac{\gamma-1}{2\gamma}}{1 + z_L \frac{\gamma+1}{2\gamma}} = c_L^2 (1 + z_L) \frac{1 + z_L \frac{\gamma-1}{2\gamma}}{1 + z_L \frac{\gamma+1}{2\gamma}} > c_L^2
\end{aligned}
$$

and

$$
\begin{aligned}
v_R - c_R &= \frac{c_L^2}{v_L} \left(1 + z_L \frac{\gamma - 1}{2\gamma} \right) - \sqrt{\frac{\gamma p_L}{\rho_L} (1 + z_L) \frac{1 + z_L(\gamma - 1)/(2\gamma)}{1 + z_L(\gamma + 1)/(2\gamma)}} \\
&= c_L \frac{1 + z_L(\gamma - 1)/(2\gamma)}{\sqrt{1 + z_L(\gamma + 1)/(2\gamma)}} - c_L \sqrt{(1 + z_L) \frac{1 + z_L(\gamma - 1)/(2\gamma)}{1 + z_L(\gamma + 1)/(2\gamma)}} \\
&= c_L \sqrt{\frac{1 + z_L(\gamma - 1)/(2\gamma)}{1 + z_L(\gamma + 1)/(2\gamma)}} \left\{ \sqrt{1 + z_L(\gamma - 1)/(2\gamma)} - \sqrt{1 + z_L} \right\} > 0
\end{aligned}
$$

We can easily derive similar results when $\sigma > \mathbf{n} \cdot \mathbf{v}_R$. Our discussion concludes with the following two summaries.

Summary 4.4.6 *Suppose that the normal* $\mathbf{n}$ *to a shock satisfying* $\sigma < \mathbf{n} \cdot \mathbf{v}_L$ *is oriented toward the post-shock (right) state, and the Mach number is the ratio of the speed of the shock relative to the pre-shock velocity, divided by the pre-shock sound speed:*

$$
M_L = \frac{\mathbf{n} \cdot \mathbf{v}_L - \sigma}{c_L}.
$$

Given a state $(\rho_L, \mathbf{v}_L, p_L)$ *with* $\rho_L > 0$*, define* $\nu = \mathbf{n} \cdot \mathbf{v} - \sigma$*. Then states* $(\rho, \mathbf{v}, p)$ *on the Hugoniot locus for this shock satisfy*

$$\frac{\nu}{\nu_L} = \frac{\rho_L}{\rho} = r \equiv \frac{1 + \frac{1}{2}(\gamma - 1)M_L^2}{\frac{1}{2}(\gamma + 1)M_L^2}$$

$$\frac{[\mathbf{n} \cdot \mathbf{v}]}{c_L} = -\frac{2(M_L^2 - 1)}{(\gamma + 1)M_L}$$

$$\frac{p}{p_L} = 1 + \frac{2\gamma}{\gamma + 1}(M_L^2 - 1).$$

In order for the entropy to increase as the discontinuity passes a material particle, we require any of the following equivalent inequalities:

$$\rho > \rho_L, \ \mathbf{n} \cdot \mathbf{v} < \mathbf{n} \cdot \mathbf{v}_L, \ p > p_L, \ c > c_L \text{ or } M_L > 1.$$

In particular, the Lax admissibility condition

$$\mathbf{n} \cdot \mathbf{v}_L - c_L \geq \sigma \geq \mathbf{n} \cdot \mathbf{v} - c.$$

is satisfied by this shock.

Summary 4.4.7 *Suppose that the normal* $\mathbf{n}$ *to a shock satisfying* $\sigma > \mathbf{n} \cdot \mathbf{v}_R$ *is oriented toward the pre-shock (right) state, and the Mach number is the ratio of the speed of the shock relative to the pre-shock velocity, divided by the pre-shock sound speed:*

$$M_R = \frac{\sigma - \mathbf{n} \cdot \mathbf{v}_R}{c_R}.$$

Given a state (ρ_R, v_R, p_R) *with* $\rho_R \neq 0$*, define* $v = \mathbf{n} \cdot \mathbf{v} - \sigma$*. Then states* (ρ, v, p) *on the Hugoniot locus for this shock satisfy*

$$\frac{v}{v_R} = \frac{\rho_R}{\rho} = r \equiv \frac{1 + \frac{1}{2}(\gamma - 1)M_R^2}{\frac{1}{2}(\gamma + 1)M_R^2}$$

$$\frac{[\mathbf{n} \cdot \mathbf{v}]}{c_R} = -\frac{2(M_R^2 - 1)}{(\gamma + 1)M_R}$$

$$\frac{p}{p_R} = 1 + \frac{2\gamma}{\gamma + 1}(M_R^2 - 1).$$

In order for the entropy to increase as the discontinuity passes a material particle, we require any of the following equivalent inequalities:

$$\rho > \rho_R, \ \mathbf{n} \cdot \mathbf{v} > \mathbf{n} \cdot \mathbf{v}_R, \ p > p_R, \ c > c_R \text{ or } M_R > 1.$$

In particular, the Lax admissibility condition

$$\mathbf{n} \cdot \mathbf{v} + c \geq \sigma \geq \mathbf{n} \cdot \mathbf{v}_R + c_R.$$

is satisfied by this shock.

4.4.7 Riemann Problem

In sections 4.4.5 and 4.4.6 we developed the results we need to solve the Riemann problem for gas dynamics. Let us summarize the information we will need to describe the solution of this problem. Given a left state $(\rho_L, \mathbf{v}_L, p_L)$ and a right state $(\rho_R, \mathbf{v}_R, p_R)$, let $c_L \equiv \sqrt{\frac{\gamma p_L}{\rho_L}}$ and $c_R \equiv \sqrt{\frac{\gamma p_R}{\rho_R}}$ be the sound speeds at these two states. The solution of the Riemann problem for gas dynamics involves two intermediate states with the same normal velocity $\mathbf{n} \cdot \mathbf{v}_*$ and pressure p_* at the intersection of the slow wave curve

$$\mathbf{v}_-(p) \equiv \begin{cases} \mathbf{v}_L - \mathbf{n}\frac{p-p_L}{\gamma p_L} \frac{c_L}{\sqrt{1+\frac{1}{2}\frac{p-p_L}{p_L}\frac{1+\gamma}{\gamma}}}, & p > p_L \\ \mathbf{v}_L + \mathbf{n}\frac{2}{\gamma-1} c_L \left[1 - (\frac{p}{p_L})^{(\gamma-1)/(2\gamma)}\right], & p \leq p_L \end{cases}$$

and the fast wave curve

$$\mathbf{v}_+(p) \equiv \begin{cases} \mathbf{v}_R + \mathbf{n}\frac{p-p_R}{\gamma p_R} \frac{c_R}{\sqrt{1+\frac{1}{2}\frac{p-p_R}{p_R}\frac{1+\gamma}{\gamma}}}, & p > p_R \\ \mathbf{v}_R - \mathbf{n}\frac{2}{\gamma-1} c_R \left[1 - (\frac{p}{p_R})^{(\gamma-1)/(2\gamma)}\right], & p \leq p_R \end{cases}$$

In other words, p_* solves the nonlinear equation

$$\mathbf{n} \cdot \mathbf{v}_-(p_*) = \mathbf{n} \cdot \mathbf{v}_+(p_*).$$

Let

$$\rho_-(p) = \begin{cases} \rho_L \left[1 + \frac{p-p_L}{\gamma p_L} \frac{1}{1+\frac{\gamma-1}{2\gamma}\frac{p-p_L}{p_L}}\right], & p > p_L \\ \rho_L \left(\frac{p}{p_L}\right)^{1/\gamma}, & p \leq p_L \end{cases}$$

be the density as a function of pressure along the slow wave curve, and

$$\rho_+(p) = \begin{cases} \rho_R \left[1 + \frac{p - p_R}{\gamma p_R} \frac{1}{1 + \frac{\gamma - 1}{2\gamma} \frac{p - p_R}{p_R}} \right], & p > p_R \\ \rho_R \left(\frac{p}{p_R} \right)^{1/\gamma}, & p \le p_R \end{cases}$$

be the density as a function of pressure along the fast wave. Finally, let

$$\xi_L = \begin{cases} \mathbf{n} \cdot \mathbf{v}_L - c_L \sqrt{1 + \frac{1+\gamma}{2\gamma} \frac{p_* - p_L}{p_L}}, & p_* > p_L \\ \mathbf{n} \cdot \mathbf{v}_L - c_L, & p_* \le p_L \end{cases}$$

$$\xi_- = \begin{cases} \mathbf{n} \cdot \mathbf{v}_L - c_L \sqrt{1 + \frac{1+\gamma}{2\gamma} \frac{p_* - p_L}{p_L}}, & p_* > p_L \\ \mathbf{n} \cdot \mathbf{v}(p_*) - \sqrt{\frac{\gamma p_*}{\rho_-(p_*)}}, & p_* \le p_L \end{cases}$$

be the wave speeds at the beginning and the end of the slow wave, and

$$\xi_+ = \begin{cases} \mathbf{n} \cdot \mathbf{v}_R + c_R \sqrt{1 + \frac{1+\gamma}{2\gamma} \frac{p_* - p_R}{p_R}}, & p_* > p_R \\ \mathbf{n} \cdot \mathbf{v}_+(p_*) + \sqrt{\frac{\gamma p_*}{\rho_+(p_*)}}, & p_* \le p_R \end{cases}$$

$$\xi_R = \begin{cases} \mathbf{n} \cdot \mathbf{v}_R + c_R \sqrt{1 + \frac{1+\gamma}{2\gamma} \frac{p_* - p_R}{p_R}}, & p_* > p_R \\ \mathbf{n} \cdot \mathbf{v}_R + c_R, & p_* \le p_R \end{cases}$$

be the wave speeds at the beginning and the end of the fast wave.

Given a left state $\mathbf{w}_L = (\rho_L, \mathbf{v}_L, p_L)$, we construct the slow centered rarefaction curve in the direction of decreasing p, and the Hugoniot locus in the direction of increasing p. Similarly, given a right state $\mathbf{w}_R = (\rho_R, \mathbf{v}_R, p_R)$ we construct the fast centered rarefaction curve in the direction of decreasing p, and the Hugoniot locus in the direction of increasing p. We can draw these curves in the two-dimensional $\mathbf{v}$, p plane, and use the definitions of the centered rarefaction curves to determine the values of the density ρ along the curves. In this case, the slow rarefaction is the curve where $\mathbf{n} \cdot \mathbf{v} + 2c/(\gamma - 1)$ is constant, and the fast rarefaction is the curve where $\mathbf{n} \cdot \mathbf{v} - 2c/(\gamma - 1)$ is constant.

The remainder of the details in the solution of the Riemann problem can be found in the Summaries 4.4.1, 4.4.3, 4.4.5, 4.4.7, and 4.4.6.

Summary 4.4.8 *There are six structurally different states in the solution of the gas dynamics Riemann problem: either a slow shock or a slow rarefaction followed by a constant state, a contact discontinuity, another constant state, and then either a fast shock or a fast rarefaction. The state that moves with speed* ξ *in the Riemann* *problem is*

$$\begin{bmatrix} \rho_\xi \\ \mathbf{v}_\xi \\ p_\xi \end{bmatrix} = \begin{cases} \begin{bmatrix} \rho_L \\ \mathbf{v}_L \\ p_L \end{bmatrix}, & \xi \leq \xi_L \\ \begin{bmatrix} \rho_L \left(\frac{p_\xi}{p_L}\right)^{1/\gamma} \\ \mathbf{v}_L - \mathbf{n}\frac{2}{\gamma-1}(c_\xi - c_L) \\ p_L \left(\frac{c_\xi}{c_L}\right)^{2\gamma/(\gamma-1)} \end{bmatrix}, & \xi_L < \xi < \xi_- \text{ where } c_\xi = \frac{\gamma-1}{\gamma+1}[\mathbf{n}\cdot\mathbf{v}_L + \frac{2c_L}{\gamma-1} - \xi] \\ \begin{bmatrix} \rho_-(p_*) \\ \mathbf{v}_-(p_*) \\ p_* \end{bmatrix}, & \xi_- \leq \xi \leq \mathbf{n}\cdot\mathbf{v}_-(p_*) \\ \begin{bmatrix} \rho_+ \\ \mathbf{v}_+(p_*) \\ p_* \end{bmatrix}, & \mathbf{n}\cdot\mathbf{v}_+(p_*) < \xi < \xi_+ \\ \begin{bmatrix} \rho_R \left(\frac{p_\xi}{p_R}\right)^{1/\gamma} \\ \mathbf{v}_R + \mathbf{n}\frac{2}{\gamma-1}(c_\xi - c_R) \\ p_R \left(\frac{c_\xi}{c_R}\right)^{2\gamma/(\gamma-1)} \end{bmatrix}, & \xi_+ < \xi < \xi_R \text{ where } c_\xi = \frac{\gamma-1}{\gamma+1}[\xi - \mathbf{n}\cdot\mathbf{v}_R + \frac{2c_R}{\gamma-1}] \\ \begin{bmatrix} \rho_R \\ \mathbf{v}_R \\ p_R \end{bmatrix}, & \xi \geq \xi_R \end{cases}$$

In rarefactions, these equations can be used to compute the variables in the order c_ξ, $\mathbf{v}_\xi$, p_ξ and then ρ_ξ.

The file **Program 4.4-47: gas_dynamics.f** contains routines that solve the gas dynamics Riemann problem A simple program for visualizing the solution to the gas dynamics Riemann problem can be executed by clicking on **Executable 4.4-18: gas DynamicsRiemannProblem.** This program will open a window for users to select values of velocity and pressure for the left state in the Riemann problem by clicking with the left mouse button. The user can push and drag the left mouse button again to see how the solution of the Riemann problem would change as the right state is changed. Releasing the mouse button produces graphs of the various parameters in the Riemann problem versus self-similar coordinates x/t. A user can also execute a different program by clicking on the link **Executable 4.4-19: guiGasDynamics** The user can use the left mouse button to rotate the rectangular region of acceptable left and right states for the Riemann problem. Then the user can press the right mouse

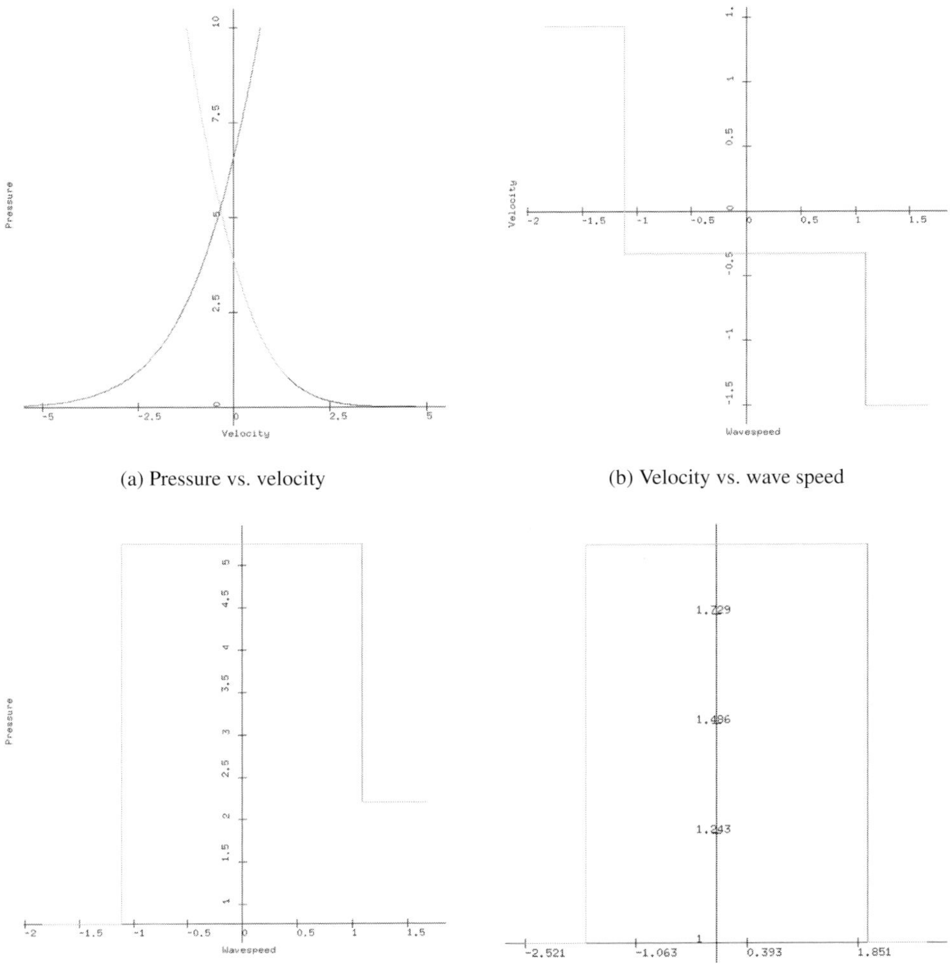

(a) Pressure vs. velocity

(b) Velocity vs. wave speed

(c) Pressure vs. wave speed

(d) Density vs. wave speed

Fig. 4.8 Shock–shock solution to gas dynamics Riemann problem

button and move in the general direction of the coordinate axes and then release the button to select the left state, consisting of the values for density, velocity and pressure. Afterward, the user can perform a similar process to select the right state. The program will show the wave curves involved in the solution of the Riemann problem, will display plots of velocity, pressure, density, characteristic speeds and entropy versus $\mathbf{x}/t$. The user can view the latter plots by bring each to the top of the view, and moving the windows around on the screen. When the user is finished viewing these plots, he find the window that asks "Are you finished viewing the results?" and click on "OK". At this point, it will be possible to interact with the original window where the left and right states were selected. When the user is

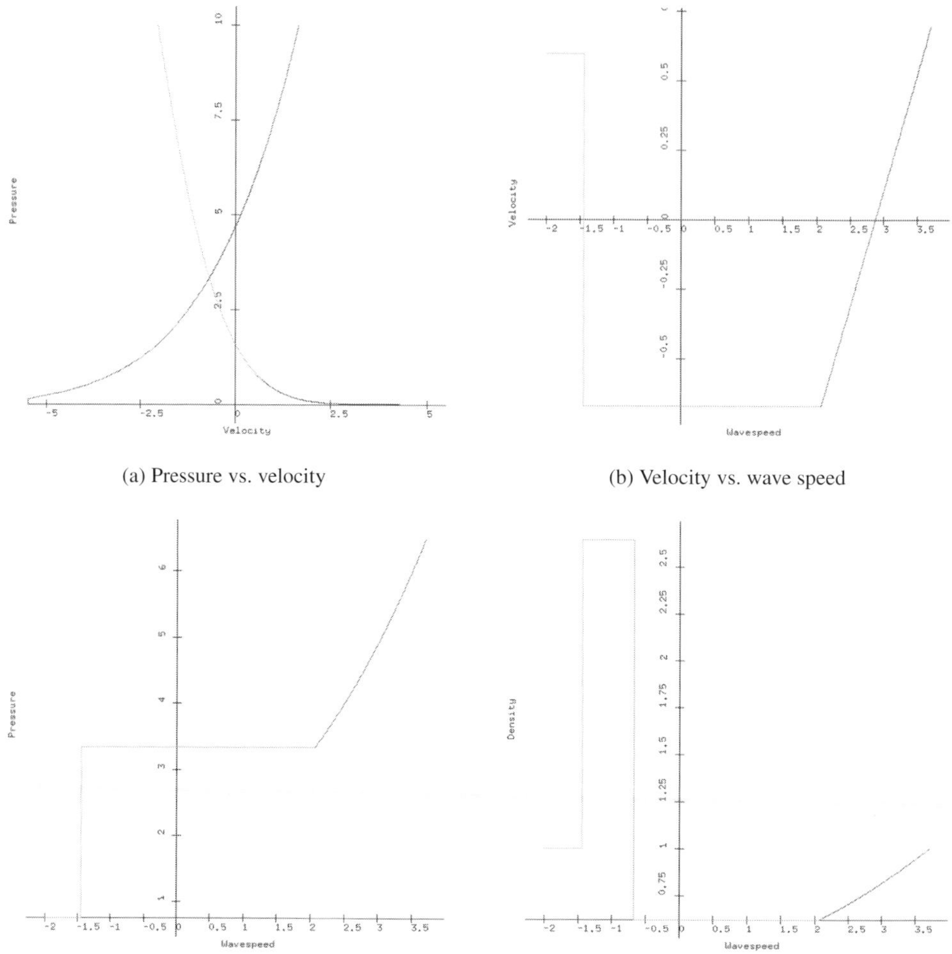

(a) Pressure vs. velocity

(b) Velocity vs. wave speed

(c) Pressure vs. wave speed

(d) Density vs. wave speed

Fig. 4.9 Shock–rarefaction solution to gas dynamics Riemann problem

finished with this window, he should use the mouse to pull down on the "File" label and release on "Quit". After answering "OK" to the question "Do you want to quit?", the user needs to respond to the window displaying "Run finished" by clicking on "Cleanup." The `guiGasDynamics` executable is a bit more difficult to use than the `gasDynamicsRiemannProblem` executable, so we recommend the former.

The solution of the gas dynamics Riemann problem is determined by finding the intersection of the two wave families. There are four cases. The first case involves shocks in both families; this is illustrated in Figure 4.8. The second case involves a slow shock and a fast rarefaction; this is illustrated in Figure 4.9. The third case

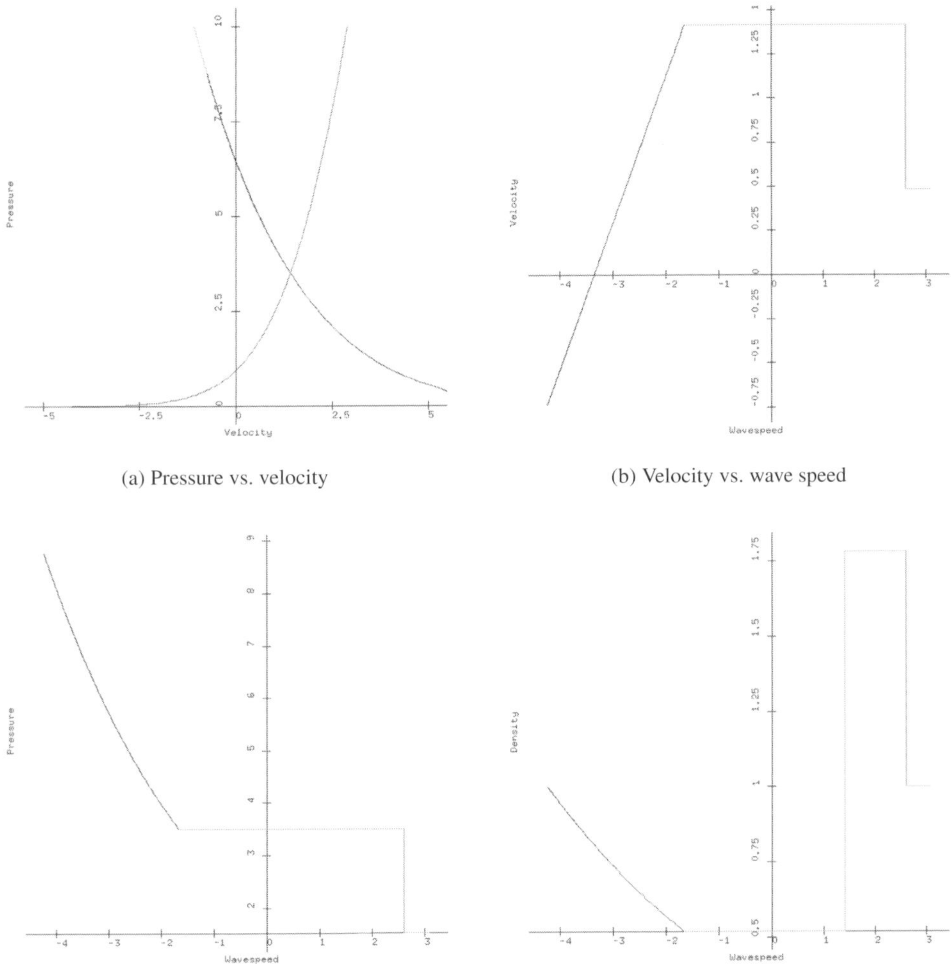

(a) Pressure vs. velocity

(b) Velocity vs. wave speed

(c) Pressure vs. wave speed

(d) Density vs. wave speed

Fig. 4.10 Rarefaction–shock solution to gas dynamics Riemann problem

involves a slow rarefaction and a fast shock; this is illustrated in Figure 4.10. The fourth case involves rarefactions in both families; this is illustrated in Figure 4.11.

4.4.8 Reflecting Walls

Let us comment on reflecting boundaries, which might represent a solid wall restricting the gas flow. At a reflecting boundary, the normal component of velocity is an odd function of distance from the wall, while ρ, p and transverse components of velocity are even functions of distance.

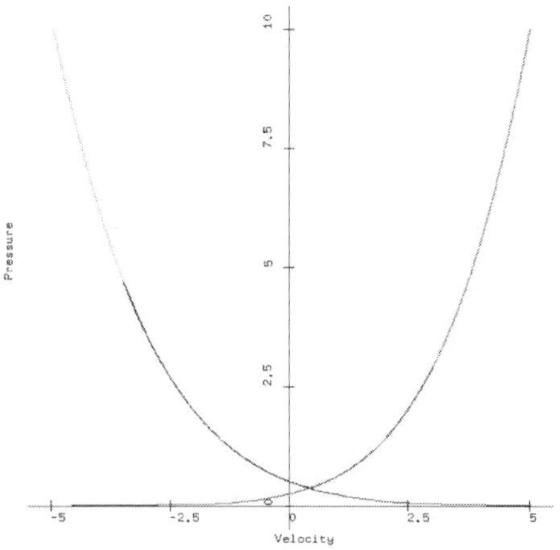

(a) Pressure vs. velocity

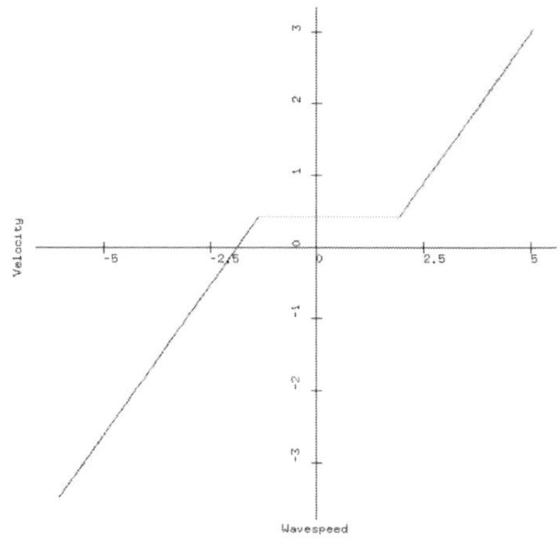

(b) Velocity vs. wave speed

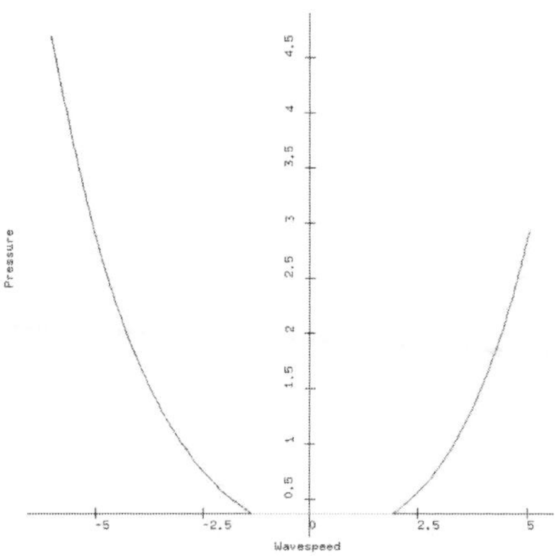

(c) Pressure vs. wave speed

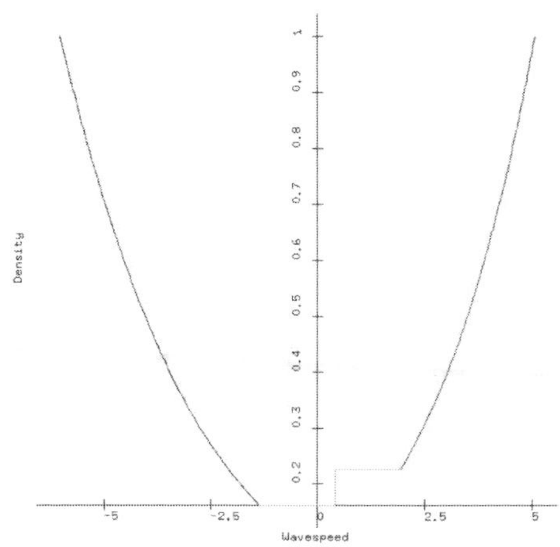

(d) Density vs. wave speed

Fig. 4.11 Rarefaction–rarefaction solution to gas dynamics Riemann problem

Exercises for 4.4

4.4.1 The Lagrangian form of the conservation laws for gas dynamics is a bit complicated. First, use Lemma 4.1.9 to convert the Eulerian conservation laws for gas dynamics to the Lagrangian frame of reference. This will give you

$$\mathbf{u}_L \equiv \begin{bmatrix} \rho_0 \\ \mathbf{v}\rho_0 \\ (e + \frac{1}{2}\mathbf{v}\cdot\mathbf{v})\rho_0 \end{bmatrix}, \quad \mathbf{F}_L \equiv \begin{bmatrix} 0 \\ \mathbf{J}^{-\top}|\mathbf{J}|p \\ p|\mathbf{J}|\mathbf{v}^\top\mathbf{J}^{-\top} \end{bmatrix}, \quad \mathbf{b}_L \equiv \begin{bmatrix} 0 \\ \mathbf{g}\rho_0 \\ \mathbf{g}\cdot\mathbf{v}\rho_0 \end{bmatrix}. \tag{4.42}$$

Note that the Lagrangian flux depends on flux variables ρ_0, $\mathbf{v}$, p, e and the deformation gradient $\mathbf{J}$. Use Equation (4.14) to expand the system of Lagrangian conservation laws for gas dynamics so that we have conservation laws for enough quantities to evolve the flux variables. Normally, gas dynamics problems are solved in the Eulerian frame of reference. Not only are the conservation laws simpler in the Eulerian frame, but also there is no problem with mesh tangling in displaying the Lagrangian results in the viewer's coordinate system $\mathbf{x}$.

4.4.2 Find the maximum value of the ratio of the post-shock density to the pre-shock density for a polytropic gas. Note that thermodynamic stability requires that the Mach number is greater than 1. What is this maximum value for air ($\gamma = 7/5$)?

4.4.3 Suppose that we have a discontinuity propagating with speed $\sigma > 0$ into a vacuum. In other words, the pre-shock state has $\rho_R = 0$, $p_R = 0$ and undefined $\mathbf{v}_R$. Describe the post-shock state determined by the Rankine–Hugoniot conditions. Does this state make sense physically ?

4.4.4 Given a left state and a discontinuity speed, solve the Rankine–Hugoniot jump conditions for Lagrangian gas dynamics to find the right state. Under what conditions is the discontinuity thermodynamically stable?

4.4.5 Find the characteristic speeds and directions for Lagrangian gas dynamics. Show that the Lagrangian acoustic tensor is a scalar multiple of the Eulerian acoustic tensor.

4.4.6 Describe how to solve the Riemann problem for Lagrangian gas dynamics. Write a computer program to implement your results.

4.4.7 Show that for a centered rarefaction wave in a polytropic gas, the velocity $\mathbf{v}$ and the sound speed c are linear functions of $\mathbf{x}/t$. (Hint: show that $\mathbf{x}/t = \mathbf{v} \pm c$, and that $\mathbf{v} \pm 2c/(\gamma - 1)$ is constant.)

4.4.8 Write a computer program to evaluate the vector $\mathbf{u}$ of conserved quantities and vector $\mathbf{g}$ of fluxes, given the flux variables $\mathbf{w}$ for Eulerian gas dynamics.

4.4.9 Write a computer program to evaluate the vector $\mathbf{w}$ of flux variables $\mathbf{w}$, given the vector $\mathbf{u}$ of conserved quantities for Eulerian gas dynamics.

4.4.10 Show that in the Lagrangian frame of reference away from discontinuities, conservation of entropy is given by

$$\frac{dS}{dt} = c_v \frac{\mathbf{g} \cdot \mathbf{v} \rho_0}{p|\mathbf{J}|} .$$

4.4.11 Given any left state $\rho_L, \mathbf{v}_L, p_L$, write a computer program to find those states $\rho_R, \mathbf{v}_R, p_R$ that can be connected to the left state by a slow shock of Mach number M_L. Given any right state $\rho_R, \mathbf{v}_R, p_R$, write a computer program to find those states $\rho_L, \mathbf{v}_L, p_L$ that can be connected to the right state state by a fast shock of Mach number M_R.

4.4.12 Given any left state $\rho_L, \mathbf{v}_L, p_L$, what conditions on the right state $\rho_R, \mathbf{v}_R, p_R$ will lead to a Riemann problem such that the slow and fast wave families do not intersect? Presumably, the solution of the Riemann problem would involve two centered rarefactions separated by a vacuum.

4.4.13 Suppose that we have a symmetric Riemann problem, in which $\rho_- = \rho_+, \mathbf{v}_- = -\mathbf{v}_+, p_- = p_+$. Show that the contact discontinuity moves with zero speed in this problem. Find the flux vector at the state that moves with zero speed. (Note that the flux vector will depend on the sign of $\mathbf{v}_-$.)

4.4.14 Suppose that we have a gas with arbitrarily large density on one side of a Riemann problem (possibly corresponding to a solid wall). To be specific, suppose that we have $\rho_+ \to \infty$, $\mathbf{v}_+ = 0$ and $p_+ = p_-$. If $\mathbf{v}_- > 0$, show that the contact discontinuity moves with zero speed, and find the flux at the state on the left side of the contact discontinuity. Perform similar calculations for $\mathbf{v}_- < 0$.

4.4.15 Program the Lax–Friedrichs scheme for polytropic gas dynamics. Plot the numerical solution (*i.e.*, ρ, $\mathbf{v}$, p and characteristic speeds $\mathbf{v} \pm c$ versus x/t) for the following Riemann problems:
   (a) a Mach 2 shock moving to the right into air with density 1, velocity 0 and pressure 1 (*i.e.*, $p_R = 1$, $\mathbf{v}_R = 0$, $\rho_R = 1$; $p_L = 9/2$, $\rho_L = 8/3$, $\mathbf{v}_L = \sqrt{35/16}$);
   (b) a stationary shock ($\sigma = 0$) in air with density 1, velocity -2 and pressure 1 on the right (*i.e.*, $p_R = 1$, $\mathbf{v}_R = -2$, $\rho_R = 1$; $p_L = 19/6$, $\rho_L = 24/11$, $\mathbf{v}_L = -11/12$);
   (c) a rarefaction moving to the right from air with density 1, pressure 1 and velocity 0 into a vacuum. (*i.e.*, $p_L = 1$, $\mathbf{v}_L = 0$, $\rho_L = 1$; $p_R = 0$, $\rho_R = 0$, $\mathbf{v}_R = \sqrt{35}$).

4.4.16 Program Rusanov's scheme for polytropic gas dynamics and apply it to the previous exercise.

4.4.17 Program Godunov's scheme for polytropic gas dynamics and apply it to the previous exercise.

4.4.18 Program Godunov's scheme for the **Sod shock tube problem** [131, page 116]. This is a Riemann problem for air ($\gamma = 1.4$) in which the left state is given by $\rho_L = 1$, $\mathbf{v}_L = 0$, $p_L = 1$ and the right state is $\rho_R = 0.125$, $\mathbf{v}_R = 0$, $p_R = 0.1$. Perform the calculation with 100 and 1000 cells. Plot ρ, $\mathbf{v}$, p and the characteristic speeds versus x/t at a time for which the fastest wave is near the boundary of the computational domain.

4.4.19 Program Godunov's scheme for the **Colella-Woodward interacting blast wave problem** [177]. The gas is assumed to be air ($\gamma = 1.4$) confined between two reflecting walls at $x = 0$ and $x = 1$. Initially, $\rho = 1$ and $\mathbf{v} = 0$ everywhere. The initial condition for pressure consists of three constant states:

$$p = \begin{cases} 1000., & 0 < x < 0.1 \\ 0.01, & 0.1 < x < 0.9 \\ 100., & 0.9 < x < 1.0. \end{cases}$$

Plot the numerical results for times 0.01, 0.016, 0.026, 0.028, 0.030, 0.032, 0.034 and 0.038. Plot ρ, $\mathbf{v}$, p and the temperature versus x. Try 100, 1000 and 10,000 cells.

4.5 Case Study: Magnetohydrodynamics (MHD)

4.5.1 Conservation Laws

An extension of gas dynamics to handle magnetic fields leads to **magnetohydrodynamics** (MHD). The system of equations has conserved quantities, fluxes and right-hand side given by

$$\mathbf{u} = \begin{bmatrix} \rho \\ \mathbf{v}\rho \\ \mathbf{B} \\ (e+\frac{1}{2}\mathbf{v}^\top\mathbf{v})\rho+\frac{1}{2}\mathbf{B}^\top\mathbf{B} \end{bmatrix}, \mathbf{F} = \begin{bmatrix} \rho\mathbf{v}^\top \\ \mathbf{v}\rho\mathbf{v}^\top + \mathbf{I}(p+\frac{1}{2}\mathbf{B}^\top\mathbf{B}) - \mathbf{B}\mathbf{B}^\top \\ \mathbf{B}\mathbf{v}^\top - \mathbf{v}\mathbf{B}^\top \\ \{e\rho+\frac{1}{2}\rho\mathbf{v}^\top\mathbf{v}+\mathbf{B}^\top\mathbf{B}+p\}\mathbf{v}^\top - \mathbf{v}^\top\mathbf{B}\mathbf{B}^\top \end{bmatrix}, \mathbf{r} = \begin{bmatrix} 0 \\ \mathbf{g}\rho \\ 0 \\ \mathbf{g}^\top\mathbf{v}\rho \end{bmatrix}.$$

Here **B** is the magnetic induction vector (see Maxwell's equations in section 4.3), and the other variables are the same as in polytropic gas dynamics (see section 4.4). As in Maxwell's equations, it is common to assume that the **B** field is divergence-free. Let us examine the implications of this assumption. Let k be the number of spatial dimensions in the problem. First, we expand the quasi-linear form for conservation of magnetic induction:

$$0 = \frac{\partial \mathbf{B}_i}{\partial t} + \sum_{j=1}^{k} \frac{\partial(\mathbf{B}_i \mathbf{v}_j - \mathbf{v}_i \mathbf{B}_j)}{\partial \mathbf{x}_j} = \frac{\partial \mathbf{B}_i}{\partial t} + \sum_{j=1}^{k} \left[\frac{\partial \mathbf{B}_i}{\partial \mathbf{x}_j} \mathbf{v}_j + \mathbf{B}_i \frac{\partial \mathbf{v}_j}{\partial \mathbf{x}_j} - \frac{\partial \mathbf{v}_i}{\partial \mathbf{x}_j} \mathbf{B}_j - \mathbf{v}_i \frac{\partial \mathbf{B}_j}{\partial \mathbf{x}_j} \right].$$

Next, we take the divergence of this equation to get

$$0 = \frac{\partial}{\partial t} \left(\sum_{i=1}^{k} \frac{\partial \mathbf{B}_i}{\partial \mathbf{x}_i} \right) + \sum_{i=1}^{k} \sum_{j=1}^{k} \left[\frac{\partial \mathbf{B}_i}{\partial \mathbf{x}_i \partial \mathbf{x}_j} \mathbf{v}_j + \frac{\partial \mathbf{B}_i}{\partial \mathbf{x}_j} \frac{\partial \mathbf{v}_j}{\partial \mathbf{x}_i} + \frac{\partial \mathbf{B}_i}{\partial \mathbf{x}_i} \frac{\partial \mathbf{v}_j}{\partial \mathbf{x}_j} + \mathbf{B}_i \frac{\partial^2 \mathbf{v}_j}{\partial \mathbf{x}_i \partial \mathbf{x}_j} \right.$$
$$\left. - \frac{\partial^2 \mathbf{v}_i}{\partial \mathbf{x}_i \partial \mathbf{x}_j} \mathbf{B}_j - \frac{\partial \mathbf{v}_i}{\partial \mathbf{x}_j} \frac{\partial \mathbf{B}_j}{\partial \mathbf{x}_i} - \frac{\partial \mathbf{v}_i}{\partial \mathbf{x}_i} \frac{\partial \mathbf{B}_j}{\partial \mathbf{x}_j} - \mathbf{v}_i \frac{\partial^2 \mathbf{B}_j}{\partial \mathbf{x}_i \partial \mathbf{x}_j} \right].$$

Inside the double sum, we can switch i and j to cancel the first term against the eighth, the second term against the sixth, the third term against the seventh, and the fourth term against the fifth.

Summary 4.5.1 *Suppose that the vector-valued functions* **B** *and* **v** *are twice continuously differentiable with respect to space and time. If for all* **x** *and all* $t > 0$ *we have*

$$\frac{\partial \mathbf{B}_i}{\partial t} + \sum_{j=1}^{k} \frac{\partial(\mathbf{B}_i \mathbf{v}_j - \mathbf{v}_i \mathbf{B}_j)}{\partial \mathbf{x}_j} = 0$$

then

$$\frac{\partial \nabla_{\mathbf{x}} \cdot \mathbf{B}}{\partial t} = 0.$$

4.5.2 Characteristic Analysis

In order to determine the characteristic speeds for MHD, we will first determine the quasi-linear form for the system. For smooth flow, the conservation laws imply

$$\begin{bmatrix} 0 \\ \mathbf{g}\rho \\ 0 \\ \mathbf{g}\cdot\mathbf{v}\rho \end{bmatrix} = \frac{\partial}{\partial t}\begin{bmatrix} \rho \\ \mathbf{v}\rho \\ \mathbf{B} \\ \frac{p}{\gamma-1}+\frac{\rho}{2}\mathbf{v}\cdot\mathbf{v}+\frac{1}{2}\mathbf{B}\cdot\mathbf{B} \end{bmatrix} + \sum_{i=1}^{k}\frac{\partial}{\partial \mathbf{x}_i}\begin{bmatrix} \rho\mathbf{v}_i \\ \mathbf{v}\rho\mathbf{v}_i + \mathbf{e}_i(p+\frac{1}{2}\mathbf{B}\cdot\mathbf{B}) - \mathbf{B}\mathbf{B}_i \\ \mathbf{B}\mathbf{v}_i - \mathbf{v}\mathbf{B}_i \\ \{p\frac{\gamma}{\gamma-1}+\frac{1}{2}\rho\mathbf{v}\cdot\mathbf{v}+\mathbf{B}\cdot\mathbf{B}\}\mathbf{v}_i - \mathbf{v}\cdot\mathbf{B}\mathbf{B}_i \end{bmatrix}$$

$$= \begin{bmatrix} 1 & 0 & 0 & 0 \\ \mathbf{v} & \mathbf{I}\rho & 0 & 0 \\ 0 & 0 & \mathbf{I} & 0 \\ \frac{1}{2}\mathbf{v}\cdot\mathbf{v} & \rho\mathbf{v}^\top & \mathbf{B}^\top & \frac{1}{\gamma-1} \end{bmatrix}\frac{\partial}{\partial t}\begin{bmatrix} \rho \\ \mathbf{v} \\ \mathbf{B} \\ p \end{bmatrix} + \sum_{i=1}^{k}\begin{bmatrix} \mathbf{v}_i \\ \mathbf{v}\mathbf{v}_i \\ 0 \\ \frac{1}{2}\mathbf{v}\cdot\mathbf{v}\mathbf{v}_i \end{bmatrix}\frac{\partial \rho}{\partial \mathbf{x}_i} + \sum_{i=1}^{k}\begin{bmatrix} 0 \\ \mathbf{e}_i \\ 0 \\ \frac{\gamma}{\gamma-1}\mathbf{v}_i \end{bmatrix}\frac{\partial p}{\partial \mathbf{x}_i}$$

$$+ \sum_{i=1}^{k}\begin{bmatrix} \rho\mathbf{e}_i^\top \\ \mathbf{v}\rho\mathbf{e}_i^\top + \mathbf{I}\rho\mathbf{v}_i \\ \mathbf{B}\mathbf{e}_i^\top - \mathbf{I}\mathbf{B}_i \\ (p\frac{\gamma}{\gamma-1}+\frac{1}{2}\rho\mathbf{v}\cdot\mathbf{v}+\mathbf{B}\cdot\mathbf{B})\mathbf{e}_i^\top + \rho\mathbf{v}_i\mathbf{v}^\top - \mathbf{B}_i\mathbf{B}^\top \end{bmatrix}\frac{\partial \mathbf{v}}{\partial \mathbf{x}_i} + \sum_{i=1}^{k}\begin{bmatrix} 0 \\ \mathbf{e}_i\mathbf{B}^\top - \mathbf{I}\mathbf{B}_i \\ \mathbf{I}\mathbf{v}_i \\ 2\mathbf{v}_i\mathbf{B}^\top - \mathbf{B}_i\mathbf{v}^\top \end{bmatrix}\frac{\partial \mathbf{B}}{\partial \mathbf{x}_i}.$$

Note that we used $0 = \nabla_\mathbf{x}\cdot\mathbf{B} = \sum_{i=1}^{k}\frac{\partial \mathbf{B}_i}{\partial \mathbf{x}_i}$ to eliminate the terms involving $\frac{\partial \mathbf{B}_i}{\partial \mathbf{x}_i}$.

Summary 4.5.2 *If the conserved quantities in MHD are continuously differentiable and $\nabla_\mathbf{x}\cdot\mathbf{B} = 0$, then the quasilinear form of MHD is*

$$\begin{bmatrix} 0 \\ \mathbf{g}\rho \\ 0 \\ \mathbf{g}\cdot\mathbf{v}\rho \end{bmatrix} = \begin{bmatrix} 1 & 0 & 0 & 0 \\ \mathbf{v} & \mathbf{I}\rho & 0 & 0 \\ 0 & 0 & \mathbf{I} & 0 \\ \frac{1}{2}\mathbf{v}\cdot\mathbf{v} & \rho\mathbf{v}^\top & \mathbf{B}^\top & \frac{1}{\gamma-1} \end{bmatrix}\frac{\partial}{\partial t}\begin{bmatrix} \rho \\ \mathbf{v} \\ \mathbf{B} \\ p \end{bmatrix}$$

$$+ \sum_{i=1}^{k}\begin{bmatrix} \mathbf{v}_i & \rho\mathbf{e}_i^\top & 0 & 0 \\ \mathbf{v}\mathbf{v}_i & \mathbf{I}\rho\mathbf{v}_i + \mathbf{v}\rho\mathbf{e}_i^\top & \mathbf{e}_i\mathbf{B}^\top - \mathbf{I}\mathbf{B}_i & \mathbf{e}_i \\ 0 & \mathbf{B}\mathbf{e}_i^\top - \mathbf{I}\mathbf{B}_i & \mathbf{I}\mathbf{v}_i & 0 \\ \frac{\mathbf{v}\cdot\mathbf{v}\mathbf{v}_i}{2} & [\frac{\gamma p}{\gamma-1}+\frac{\rho\mathbf{v}\cdot\mathbf{v}}{2}+\mathbf{B}\cdot\mathbf{B}]\mathbf{e}_i^\top + \rho\mathbf{v}_i\mathbf{v}^\top - \mathbf{B}_i\mathbf{B}^\top & 2\mathbf{v}_i\mathbf{B}^\top - \mathbf{B}_i\mathbf{v}^\top & \frac{\gamma\mathbf{v}_i}{\gamma-1} \end{bmatrix}\frac{\partial}{\partial \mathbf{x}_i}\begin{bmatrix} \rho \\ \mathbf{v} \\ \mathbf{B} \\ p \end{bmatrix}.$$

From the quasilinear form of MHD, we have

$$\frac{\partial \mathbf{u}}{\partial \mathbf{w}} = \begin{bmatrix} 1 & 0 & 0 & 0 \\ \mathbf{v} & \mathbf{I}\rho & 0 & 0 \\ 0 & 0 & \mathbf{I} & 0 \\ \frac{1}{2}\mathbf{v}\cdot\mathbf{v} & \rho\mathbf{v}^\top & \mathbf{B}^\top & \frac{1}{\gamma-1} \end{bmatrix}.$$

Let $[\mathbf{n}, \mathbf{N}]$ be an orthogonal matrix with first column equal to $\mathbf{n}$. With the orthogonal decompositions

$$\mathbf{v} \equiv \mathbf{n}\upsilon + \mathbf{N}\mathbf{v}^\perp, \qquad \mathbf{B} \equiv \mathbf{n}\beta + \mathbf{N}\mathbf{B}^\perp$$

we have

$$\frac{\partial \mathbf{Fn}}{\partial \mathbf{w}} = \begin{bmatrix} \nu & \rho \mathbf{n}^\top & 0 & 0 \\ \mathbf{v}\nu & \mathbf{I}\rho\nu + \mathbf{v}\rho\mathbf{n}^\top & \mathbf{n}\mathbf{B}^\top - \mathbf{I}\beta & \mathbf{n} \\ 0 & \mathbf{B}\mathbf{n}^\top - \mathbf{I}\beta & \mathbf{I}\nu & 0 \\ \frac{\nu \mathbf{v}\cdot\mathbf{v}}{2} & \rho\nu\mathbf{v}^\top + [\frac{p\gamma}{\gamma-1} + \frac{\rho\mathbf{v}\cdot\mathbf{v}}{2} + \mathbf{B}\cdot\mathbf{B}]\mathbf{n}^\top - \beta\mathbf{B}^\top & 2\nu\mathbf{B}^\top - \beta\mathbf{v}^\top & \frac{\gamma\nu}{\gamma-1} \end{bmatrix}.$$

The expressions for the flux derivatives can be obtained by applying a coordinate rotation to the result of Lemma 4.5.2 using Lemma 4.1.9. It follows from the discussion in Section 4.1.1 that we want to compute the eigenvalues Λ and eigenvectors $\mathbf{Y}$ of

$$\begin{aligned} \left(\frac{\partial \mathbf{u}}{\partial \mathbf{w}}\right)^{-1} \frac{\partial (\mathbf{Fn})}{\partial \mathbf{w}} &= \begin{bmatrix} \nu & \rho\mathbf{n}^\top & 0 & 0 \\ 0 & \mathbf{I}\nu & \mathbf{n}\frac{1}{\rho}\mathbf{B}^\top - \mathbf{I}\frac{\beta}{\rho} & \mathbf{n}\frac{1}{\rho} \\ 0 & \mathbf{B}\mathbf{n}^\top - \mathbf{I}\beta & \mathbf{I}\nu & 0 \\ 0 & \gamma p \mathbf{n}^\top & 0 & \nu \end{bmatrix} \\ &= \mathbf{I}\nu + \begin{bmatrix} 0 & \rho\mathbf{n}^\top & 0 & 0 \\ 0 & 0 & \mathbf{n}\frac{1}{\rho}\mathbf{B}^\top - \mathbf{I}\frac{\beta}{\rho} & \mathbf{n}\frac{1}{\rho} \\ 0 & \mathbf{B}\mathbf{n}^\top - \mathbf{I}\beta & 0 & 0 \\ 0 & \gamma p \mathbf{n}^\top & 0 & 0 \end{bmatrix} \equiv \mathbf{I}\nu + \mathbf{M}. \end{aligned}$$

If we find the eigenvalues of $\mathbf{M}$, then we can add the normal velocity ν to them to get the desired eigenvalues in Λ.

Summary 4.5.3 *Given a direction* $\mathbf{n}$, *and flux variables*

$$\mathbf{w} \equiv \begin{bmatrix} \rho \\ \mathbf{v} \\ \mathbf{B} \\ p \end{bmatrix}$$

define $\nu \equiv \mathbf{v}\cdot\mathbf{n}$ *and* $\beta \equiv \mathbf{B}\cdot\mathbf{n}$. *The characteristic speeds of MHD are of the form* $\nu + \lambda$, *where* λ *is an eigenvalue of*

$$\mathbf{M} \equiv \left(\frac{\partial \mathbf{u}}{\partial \mathbf{w}}\right)^{-1} \frac{\partial (\mathbf{Fn})}{\partial \mathbf{w}} - \mathbf{I}(\mathbf{v}\cdot\mathbf{n}) = \begin{bmatrix} 0 & \rho\mathbf{n}^\top & 0 & 0 \\ 0 & 0 & \mathbf{n}\frac{1}{\rho}\mathbf{B}^\top - \mathbf{I}\frac{\beta}{\rho} & \mathbf{n}\frac{1}{\rho} \\ 0 & \mathbf{B}\mathbf{n}^\top - \mathbf{I}\beta & 0 & 0 \\ 0 & \gamma p \mathbf{n}^\top & 0 & 0 \end{bmatrix}$$

We can write the eigenvector equation $\mathbf{Mz} = \mathbf{z}\lambda$ in the partitioned form

$$\begin{bmatrix} 0 & \rho\mathbf{n}^\top & 0 & 0 \\ 0 & 0 & \mathbf{n}\frac{1}{\rho}\mathbf{B}^\top - \mathbf{I}\frac{\beta}{\rho} & \mathbf{n}\frac{1}{\rho} \\ 0 & \mathbf{B}\mathbf{n}^\top - \mathbf{I}\beta & 0 & 0 \\ 0 & \gamma p \mathbf{n}^\top & 0 & 0 \end{bmatrix} \begin{bmatrix} \tau \\ \mathbf{x} \\ \mathbf{y} \\ \zeta \end{bmatrix} = \begin{bmatrix} \tau \\ \mathbf{x} \\ \mathbf{y} \\ \zeta \end{bmatrix} \lambda. \tag{4.43}$$

Note that $\mathbf{M}$ has a column of zeros, so it has a zero eigenvalue. If $\lambda = 0$ and $\beta \neq 0$, then the second equation in (4.43) implies that $\mathbf{y} = \mathbf{n}\eta$ for some scalar η. This same equation then implies that $\zeta = 0$. The third equation implies that $\mathbf{x} = \mathbf{B}\alpha$ for some scalar α, and the first or fourth equation implies that $\alpha = 0$. Thus we have discovered two eigenvectors of $\mathbf{M}$ with zero eigenvalue:

$$\begin{bmatrix} 0 & \rho\mathbf{n}^\top & 0 & 0 \\ 0 & 0 & \mathbf{n}\frac{1}{\rho}\mathbf{B}^\top - \mathbf{I}\frac{\beta}{\rho} & \mathbf{n}\frac{1}{\rho} \\ 0 & \mathbf{B}\mathbf{n}^\top - \mathbf{I}\beta & 0 & 0 \\ 0 & \gamma p\mathbf{n}^\top & 0 & 0 \end{bmatrix} \begin{bmatrix} 1 & 0 \\ 0 & 0 \\ 0 & \mathbf{n} \\ 0 & 0 \end{bmatrix} = 0.$$

The corresponding characteristic speed for MHD is ν. This characteristic speed is linearly degenerate:

$$\frac{\partial \mathbf{v}\cdot\mathbf{n}}{\partial \mathbf{w}} \begin{bmatrix} 1 & 0 \\ 0 & 0 \\ 0 & \mathbf{n} \\ 0 & 0 \end{bmatrix} = \begin{bmatrix} 0, & \mathbf{n}^\top, & 0, & 0 \end{bmatrix} \begin{bmatrix} 1 & 0 \\ 0 & 0 \\ 0 & \mathbf{n} \\ 0 & 0 \end{bmatrix} = \begin{bmatrix} 0 & 0 \end{bmatrix}.$$

On the other hand, if $\lambda = 0$ and $\beta = 0$, then the second equation in (4.43) implies that $\zeta = -\mathbf{B}\cdot\mathbf{y}$ and the first or fourth equation implies that $\mathbf{n}\cdot\mathbf{x} = 0$. Then if $\beta = 0$,

$$\begin{bmatrix} 0 & \rho\mathbf{n}^\top & 0 & 0 \\ 0 & 0 & \mathbf{n}\frac{1}{\rho}\mathbf{B}^\top - \mathbf{I}\frac{\beta}{\rho} & \mathbf{n}\frac{1}{\rho} \\ 0 & \mathbf{B}\mathbf{n}^\top - \mathbf{I}\beta & 0 & 0 \\ 0 & \gamma p\mathbf{n}^\top & 0 & 0 \end{bmatrix} \begin{bmatrix} 1 & 0 & 0 \\ 0 & \mathbf{N} & 0 \\ 0 & 0 & \mathbf{I} \\ 0 & 0 & -\mathbf{B}^\top \end{bmatrix} = \begin{bmatrix} 0 & 0 & 0 \\ 0 & 0 & -\mathbf{I}\frac{\beta}{\rho} \\ 0 & -\mathbf{N}\beta & 0 \\ 0 & 0 & 0 \end{bmatrix} = 0$$

shows that we have found $2k$ eigenvectors of $\mathbf{M}$ with zero eigenvalue, where k is the dimension of $\mathbf{B}$ and $\mathbf{v}$. Thus when $\beta \neq 0$, the eigenvalue 0 of $\mathbf{M}$ has multiplicity 2, and when $\mathbf{B}\cdot\mathbf{n} = 0$ it has multiplicity $2k$. This characteristic speed for MHD is still linearly degenerate:

$$\frac{\partial \nu}{\partial \mathbf{w}} \begin{bmatrix} 1 & 0 & 0 \\ 0 & \mathbf{N} & 0 \\ 0 & 0 & \mathbf{I} \\ 0 & 0 & -\mathbf{B}^\top \end{bmatrix} = \begin{bmatrix} 0, & \mathbf{n}^\top, & 0, & 0 \end{bmatrix} \begin{bmatrix} 1 & 0 & 0 \\ 0 & \mathbf{N} & 0 \\ 0 & 0 & \mathbf{I} \\ 0 & 0 & -\mathbf{B}^\top \end{bmatrix} = \begin{bmatrix} 0, & 0, & 0 \end{bmatrix}.$$

Summary 4.5.4 *Consider the matrix*

$$\mathbf{M} \equiv \begin{bmatrix} 0 & \rho\mathbf{n}^\top & 0 & 0 \\ 0 & 0 & \mathbf{n}\frac{1}{\rho}\mathbf{B}^\top - \mathbf{I}\frac{\beta}{\rho} & \mathbf{n}\frac{1}{\rho} \\ 0 & \mathbf{B}\mathbf{n}^\top - \mathbf{I}\beta & 0 & 0 \\ 0 & \gamma p\mathbf{n}^\top & 0 & 0 \end{bmatrix}.$$

If $\mathbf{B} \cdot \mathbf{n} \equiv \beta \neq 0$, then $\mathbf{M}$ has a zero eigenvalue of multiplicity 2:

$$\begin{bmatrix} 0 & \rho\mathbf{n}^\top & 0 & 0 \\ 0 & 0 & \mathbf{n}\frac{1}{\rho}\mathbf{B}^\top - \mathbf{I}\frac{\beta}{\rho} & \mathbf{n}\frac{1}{\rho} \\ 0 & \mathbf{B}\mathbf{n}^\top - \mathbf{I}\beta & 0 & 0 \\ 0 & \gamma p\mathbf{n}^\top & 0 & 0 \end{bmatrix} \begin{bmatrix} 1 & 0 & 0 \\ 0 & \mathbf{N} & 0 \\ 0 & 0 & \mathbf{I} \\ 0 & 0 & -\mathbf{B}^\top \end{bmatrix} = \begin{bmatrix} 0 & 0 & 0 \\ 0 & 0 & -\mathbf{I}\frac{\beta}{\rho} \\ 0 & -\mathbf{N}\beta & 0 \\ 0 & 0 & 0 \end{bmatrix} \begin{bmatrix} 0 & 0 \\ 0 & 0 \end{bmatrix}.$$

Otherwise, if $\mathbf{v}$ and $\mathbf{B}$ are k-vectors, then $\mathbf{M}$ has a zero eigenvalue of multiplicity 2k. The corresponding characteristic speed $v \equiv \mathbf{v} \cdot \mathbf{n}$ for MHD is linearly degenerate.

Next, suppose that the eigenvalue λ of $\mathbf{M}$ is nonzero. Since the third equation in (4.43) implies that we must have $\mathbf{n} \cdot \mathbf{y} = 0$, let us try $\mathbf{y} = \mathbf{B} \times \mathbf{n}$, provided that this vector is nonzero. (In two dimensions $\mathbf{n} \perp \mathbf{y} = 0 = \mathbf{N} \perp \mathbf{y}$ implies $\mathbf{y} = 0$.) Then the second equation in (4.43) implies that $\mathbf{x} = \mathbf{n}\alpha_1 + \mathbf{B} \times \mathbf{n}\alpha_2$ for some scalars α_1 and α_2. Since $\mathbf{B}$ and $\mathbf{B} \times \mathbf{n}$ are linearly independent when they are nonzero, the third equation implies that $\alpha_1 = 0$ and $-\alpha_2\beta = \lambda$. Since $\mathbf{x} = \mathbf{B} \times \mathbf{n}\alpha_2$ is orthogonal to $\mathbf{n}$, the first equation implies that $\tau = 0$ and the fourth equation implies that $\zeta = 0$. Also, the second equation implies that $-\beta = \alpha_2\rho\lambda$. We now see that $\alpha_2 = \mp \operatorname{sign}(\beta)/\sqrt{\rho}$, and $\lambda = \pm c_a$ where

$$c_a = |\mathbf{B} \cdot \mathbf{n}|/\sqrt{\rho}$$

is the the **Alfvén speed**. Note that if $\sigma \equiv \operatorname{sign}(\beta)$, then

$$\begin{bmatrix} 0 & \rho\mathbf{n}^\top & 0 & 0 \\ 0 & 0 & \mathbf{n}\frac{1}{\rho}\mathbf{B}^\top - \mathbf{I}\frac{\beta}{\rho} & \mathbf{n}\frac{1}{\rho} \\ 0 & \mathbf{B}\mathbf{n}^\top - \mathbf{I}\beta & 0 & 0 \\ 0 & \gamma p\mathbf{n}^\top & 0 & 0 \end{bmatrix} \begin{bmatrix} 0 & 0 \\ \mathbf{B}\times\mathbf{n}\sigma & -\mathbf{B}\times\mathbf{n}\sigma \\ \mathbf{B}\times\mathbf{n}\sqrt{\rho} & \mathbf{B}\times\mathbf{n}\sqrt{\rho} \\ 0 & 0 \end{bmatrix}$$
$$= \begin{bmatrix} 0 & 0 \\ \mathbf{B}\times\mathbf{n}\sigma & -\mathbf{B}\times\mathbf{n}\sigma \\ \mathbf{B}\times\mathbf{n}\sqrt{\rho} & \mathbf{B}\times\mathbf{n}\sqrt{\rho} \\ 0 & 0 \end{bmatrix} \begin{bmatrix} -c_a & 0 \\ 0 & c_a \end{bmatrix}.$$

so we have found two more eigenvalues and eigenvectors of $\mathbf{M}$.

Let us show that the Alfvén speed is linearly degenerate. Since $\rho c_a^2 = (\beta)^2$, we have

$$\frac{\partial \rho c_a^2}{\partial \mathbf{w}} = [0, 0, 2\beta\mathbf{n}^\top, 0],$$

and

$$\frac{\partial v \pm c_a}{\partial \mathbf{w}} = \frac{\partial v}{\partial \mathbf{w}} \pm \frac{1}{\rho c_a} \left\{ \frac{\partial \rho c_a^2}{\partial \mathbf{w}} - c_a^2 \frac{\partial \rho}{\partial \mathbf{w}} \right\} = \left[\mp\frac{c_a}{\rho}, \mathbf{n}^\top, \frac{2\beta}{\rho c_a}\mathbf{n}^\top, 0 \right].$$

Then we multiply the **w**-derivatives of the eigenvalue times the eigenvector to get

$$\frac{\partial v \pm c_a}{\partial \mathbf{w}} \begin{bmatrix} 0 \\ \mp \mathbf{B} \times \mathbf{n}\sigma \\ \mathbf{B} \times \mathbf{n}\sqrt{\rho} \\ 0 \end{bmatrix} = \left[\mp \frac{c_a}{\rho}, \mathbf{n}^\top, \frac{2\beta}{\rho c_a}\mathbf{n}^\top, 0 \right] \begin{bmatrix} 0 \\ \mp \mathbf{B} \times \mathbf{n}\sigma \\ \mathbf{B} \times \mathbf{n}\sqrt{\rho} \\ 0 \end{bmatrix} = 0.$$

Summary 4.5.5 *Given a direction* **n**, *define the* **Alfvén speed** c_a *by*

$$c_a \equiv |\mathbf{B} \cdot \mathbf{n}| / \sqrt{\rho} \tag{4.44}$$

and consider the matrix

$$\mathbf{M} \equiv \begin{bmatrix} 0 & \rho\mathbf{n}^\top & 0 & 0 \\ 0 & 0 & \mathbf{n}\frac{1}{\rho}\mathbf{B}^\top - \mathbf{I}\frac{\beta}{\rho} & \mathbf{n}\frac{1}{\rho} \\ 0 & \mathbf{B}\mathbf{n}^\top - \mathbf{I}\beta & 0 & 0 \\ 0 & \gamma p\mathbf{n}^\top & 0 & 0 \end{bmatrix}$$

where $\beta \equiv \mathbf{B} \cdot \mathbf{n}$. *Then* $\pm c_a$ *are two eigenvalues of* **M**:

$$\begin{bmatrix} 0 & \rho\mathbf{n}^\top & 0 & 0 \\ 0 & 0 & \mathbf{n}\frac{1}{\rho}\mathbf{B}^\top - \mathbf{I}\frac{\beta}{\rho} & \mathbf{n}\frac{1}{\rho} \\ 0 & \mathbf{B}\mathbf{n}^\top - \mathbf{I}\beta & 0 & 0 \\ 0 & \gamma p\mathbf{n}^\top & 0 & 0 \end{bmatrix} \begin{bmatrix} 0 & 0 \\ \mathbf{B}\times\mathbf{n}\sigma & -\mathbf{B}\times\mathbf{n}\sigma \\ \mathbf{B}\times\mathbf{n}\sqrt{\rho} & \mathbf{B}\times\mathbf{n}\sqrt{\rho} \\ 0 & 0 \end{bmatrix} = \begin{bmatrix} 0 & 0 \\ \mathbf{B}\times\mathbf{n}\sigma & -\mathbf{B}\times\mathbf{n}\sigma \\ \mathbf{B}\times\mathbf{n}\sqrt{\rho} & \mathbf{B}\times\mathbf{n}\sqrt{\rho} \\ 0 & 0 \end{bmatrix} \begin{bmatrix} -c_a & 0 \\ 0 & c_a \end{bmatrix}$$

where $\sigma \equiv sign(\beta)$. *The MHD characteristic speeds* $\mathbf{v} \cdot \mathbf{n} \pm c_a$ *are linearly degenerate. These characteristic directions are not possible for flow in fewer than three dimensions.*

Next, suppose that $\lambda \neq 0$, and that $\mathbf{y} = \mathbf{B} - \mathbf{n}\beta \neq 0$. Then the third equation in (4.43) implies that $\mathbf{x} = \mathbf{n}\alpha_1 + \mathbf{B}\alpha_2$ for some scalars α_1 and α_2. Since $\mathbf{y} \neq 0$ implies that **B** and **n** are linearly independent, the third equation implies that $\alpha_1 = \lambda$. The second equation implies that $-\beta = \alpha_2\rho\lambda$ and $\mathbf{B} \cdot \mathbf{B} + \zeta = \rho\lambda^2$. The fourth equation implies that $\rho c^2(\beta\alpha_2 + \lambda) = \zeta\lambda$. Thus the eigenvalue satisfies $c^2[\lambda^2\rho - \beta^2] = \zeta\lambda^2 = \rho\lambda^4 - \|\mathbf{B}\|^2\lambda^2$, which gives us a quadratic equation for λ^2:

$$\rho\lambda^4 - (\|\mathbf{B}\|^2 + \rho c^2)\lambda^2 + c^2\beta^2 = 0.$$

If we define $\rho c_*^2 = \frac{1}{2}(\|\mathbf{B}\|^2 + \rho c^2)$, then we can solve the quadratic equation for λ^2 to get

$$\lambda^2 = c_*^2 \pm \sqrt{c_*^4 - c^2 c_a^2}.$$

Let us prove that the eigenvalues λ given by this expression are real. Recall the Schwarz inequality in the form $|\mathbf{B}\cdot\mathbf{n}| \le \|\mathbf{B}\|\,\|\mathbf{n}\|$, and the inequality $ab \le \frac{1}{2}(a^2+b^2)$. It follows that

$$\rho c_a c = |\mathbf{B}\cdot\mathbf{n}|\sqrt{\rho}c \le \|\mathbf{B}\|\|\mathbf{n}\|\sqrt{\rho c^2} \le \frac{1}{2}(\|\mathbf{B}\|^2 + \rho c^2) = \rho c_*^2.$$

This inequality implies that λ^2 is real, no matter which sign is chosen. Since $\sqrt{c_*^4 - c_a^2 c^2} \le c_*^2$, λ is real.

Next, we will prove some inequalities regarding these eigenvalues. Since $\rho c_a^2 = (\mathbf{B}\cdot\mathbf{n})^2 \le \|\mathbf{B}\|^2$, it follows that

$$\rho(c_a^2 + c^2) \le \|\mathbf{B}\|^2 + \rho c^2 = 2\rho c_*^2.$$

This in turn implies that

$$(c_*^2 - c_a^2)^2 = c_*^4 - 2c_*^2 c_a^2 + c_a^4 \le c_*^4 - c_a^2 c^2.$$

Taking square roots, we obtain

$$-\sqrt{c_*^4 - c_a^2 c^2} \le c_a^2 - c_*^2 \le \sqrt{c_*^4 - c_a^2 c^2}.$$

It follows that

$$c_s^2 \equiv c_*^2 - \sqrt{c_*^4 - c_a^2 c^2} \le c_a^2 \le c_*^2 + \sqrt{c_*^4 - c_a^2 c^2} \equiv c_f^2.$$

Finally, we will prove (4.46). Note that the Schwarz inequality implies that $(\mathbf{B}\cdot\mathbf{n})^2 \le \|\mathbf{B}\|^2$. This implies that

$$\frac{1}{2}\gamma p\|\mathbf{B}\|^2 - \gamma p\beta^2 \ge -\frac{1}{2}\gamma p\|\mathbf{B}\|^2.$$

which in turn implies that

$$\frac{1}{4}(\gamma p + \|\mathbf{B}\|^2)^2 - \gamma p\beta^2 \ge \frac{1}{4}(\|\mathbf{B}\|^2 - \gamma p)^2.$$

It follows that

$$-\sqrt{\frac{1}{4}(\gamma p + \|\mathbf{B}\|^2)^2 - \gamma p\beta^2} \le \frac{1}{2}((\|\mathbf{B}\|^2 - \gamma p) \le \sqrt{\frac{1}{4}(\gamma p + \|\mathbf{B}\|^2)^2 - \gamma p\beta^2}.$$

Adding either of $\pm\frac{1}{2}(\gamma p + \|\mathbf{B}\|^2)$ to all terms in this inequality gives us

$$0 \le \rho c_s^2 \le \min\{\rho c^2, \beta^2\} \quad \text{and} \quad \max\{\rho c^2, \|\mathbf{B}\|^2\} \le \rho c_f^2.$$

We have shown that the corresponding eigenvector for a nonzero eigenvalue of **M** is

$$\begin{bmatrix}\tau\\ \mathbf{x}\\ \mathbf{y}\\ \zeta\end{bmatrix} = \begin{bmatrix}\rho-\rho\frac{c_a^2}{\lambda^2}\\ \mathbf{n}\lambda-\mathbf{B}\frac{\beta}{\rho\lambda}\\ \mathbf{B}-\mathbf{n}\beta\\ \rho\lambda^2-\|\mathbf{B}\|^2\end{bmatrix} = \begin{bmatrix}\frac{\rho\lambda^2-\|\mathbf{B}\|^2}{c^2}\\ \mathbf{n}\lambda-\mathbf{B}\frac{\beta}{\rho\lambda}\\ \mathbf{B}-\mathbf{n}\beta\\ \rho\lambda^2-\|\mathbf{B}\|^2\end{bmatrix}.$$

This gives us four more eigenvalues, for a total of eight in three dimensions.

Summary 4.5.6 *Given a direction* **n***, polytropic gas constant* γ *and flux variables* ρ, **v**, **B** *and* p, *define the speed of sound* c *by* $\rho c^2 = \gamma p$, *the* **Alfvén speed** c_a *by* $\rho c_a^2 = (\mathbf{B}\cdot\mathbf{n})^2 \equiv \beta^2$ *and the speed* c_* *by* $\rho c_*^2 = \frac{1}{2}(\mathbf{B}\cdot\mathbf{B}+\rho c^2)$. *Also define the slow speed* c_s *by* $c_s^2 = c_*^2 - \sqrt{c_*^4 - c^2 c_a^2}$ *and the fast speed* c_f *by* $c_f^2 = c_*^2 + \sqrt{c_*^4 - c^2 c_a^2}$. *Then*

$$0 \le c_s \le c_a \le c_f, \tag{4.45}$$

and

$$0 \le \rho c_s^2 \le \min\{\rho c^2, \beta^2\} \quad \text{and} \quad \max\{\rho c^2, \|\mathbf{B}\|^2\} \le \rho c_f^2. \tag{4.46}$$

Further, $\pm c_s$ *and* $\pm c_f$ *are four eigenvalues of*

$$\mathbf{M} \equiv \begin{bmatrix}0 & \rho\mathbf{n}^\top & 0 & 0\\ 0 & 0 & \mathbf{n}\frac{1}{\rho}\mathbf{B}^\top - \mathbf{I}\frac{\beta}{\rho} & \mathbf{n}\frac{1}{\rho}\\ 0 & \mathbf{B}\mathbf{n}^\top - \mathbf{I}\beta & 0 & 0\\ 0 & \gamma p\mathbf{n}^\top & 0 & 0\end{bmatrix}$$

since

$$\mathbf{M}\begin{bmatrix}\frac{\rho c_f^2-\|\mathbf{B}\|^2}{c^2} & \frac{\rho c_s^2-\|\mathbf{B}\|^2}{c^2} & \frac{\rho c_s^2-\|\mathbf{B}\|^2}{c^2} & \frac{\rho c_f^2-\|\mathbf{B}\|^2}{c^2}\\ -\mathbf{n}c_f+\mathbf{B}\frac{\beta}{\rho c_f} & -\mathbf{n}c_s+\mathbf{B}\frac{\beta}{\rho c_s} & \mathbf{n}c_s-\mathbf{B}\frac{\beta}{\rho c_s} & \mathbf{n}c_f-\mathbf{B}\frac{\beta}{\rho c_f}\\ \mathbf{B}-\mathbf{n}\beta & \mathbf{B}-\mathbf{n}\beta & \mathbf{B}-\mathbf{n}\beta & \mathbf{B}-\mathbf{n}\beta\\ \rho c_f^2-\|\mathbf{B}\|^2 & \rho c_s^2-\|\mathbf{B}\|^2 & \rho c_s^2-\|\mathbf{B}\|^2 & \rho c_f^2-\|\mathbf{B}\|^2\end{bmatrix}$$

$$= \begin{bmatrix}\frac{\rho c_f^2-\|\mathbf{B}\|^2}{c^2} & \frac{\rho c_s^2-\|\mathbf{B}\|^2}{c^2} & \frac{\rho c_s^2-\|\mathbf{B}\|^2}{c^2} & \frac{\rho c_f^2-\|\mathbf{B}\|^2}{c^2}\\ -\mathbf{n}c_f+\mathbf{B}\frac{\beta}{\rho c_f} & -\mathbf{n}c_s+\mathbf{B}\frac{\beta}{\rho c_s} & \mathbf{n}c_s-\mathbf{B}\frac{\beta}{\rho c_s} & \mathbf{n}c_f-\mathbf{B}\frac{\beta}{\rho c_f}\\ \mathbf{B}-\mathbf{n}\beta & \mathbf{B}-\mathbf{n}\beta & \mathbf{B}-\mathbf{n}\beta & \mathbf{B}-\mathbf{n}\beta\\ \rho c_f^2-\|\mathbf{B}\|^2 & \rho c_s^2-\|\mathbf{B}\|^2 & \rho c_s^2-\|\mathbf{B}\|^2 & \rho c_f^2-\|\mathbf{B}\|^2\end{bmatrix}$$

$$\times \begin{bmatrix}-c_f & 0 & 0 & 0\\ 0 & -c_s & 0 & 0\\ 0 & 0 & c_s & 0\\ 0 & 0 & 0 & c_f\end{bmatrix}.$$

The corresponding characteristic speeds for MHD are $\nu \pm c_s$ and $\nu \pm c_f$ where $\nu \equiv \mathbf{v} \cdot \mathbf{n}$; these are not necessarily genuinely nonlinear.

Summary 4.5.7 *Given a direction $\mathbf{n}$, polytropic gas constant γ and flux variables ρ, $\mathbf{v}$, $\mathbf{B}$ and p, let $[\mathbf{n}, \mathbf{N}]$ be an orthogonal matrix and define*

$$\mathbf{v} \equiv \mathbf{n}\nu + \mathbf{N}\mathbf{w}, \ \ \mathbf{B} \equiv \mathbf{n}\beta + \mathbf{N}\mathbf{B}^{\perp}.$$

Also define the speed of sound c by $\rho c^2 = \gamma p$, the **Alfvén speed** *c_a by $\rho c_a^2 = \beta^2$ and the speed c_* by $\rho c_*^2 = \frac{1}{2}(\mathbf{B} \cdot \mathbf{B} + \rho c^2)$. Also define the slow speed c_s by $c_s^2 = c_*^2 - \sqrt{c_*^4 - c^2 c_a^2}$ and the fast speed c_f by $c_f^2 = c_*^2 + \sqrt{c_*^4 - c^2 c_a^2}$. The equations of MHD are hyperbolic with characteristic speeds ν, $\nu \pm c_s$, $\nu \pm c_a$ and $\nu \pm c_f$. The characteristic speeds ν and $\nu \pm c_a$ are linearly degenerate. In fact,*

$$\left(\frac{\partial \mathbf{u}}{\partial \mathbf{w}}\right)^{-1} \frac{\partial (\mathbf{F}\mathbf{n})}{\partial \mathbf{w}} \mathbf{Y} = \mathbf{Y}\{\mathbf{I}\nu + \Lambda\}$$

where if $\beta = 0$ we have

$$\mathbf{Y} = \begin{bmatrix} \frac{\rho c_f^2 - \|\mathbf{B}\|^2}{c^2} & 1 & 0 & 0 & \frac{\rho c_f^2 - \|\mathbf{B}\|^2}{c^2} \\ -\mathbf{n}c_f & 0 & \mathbf{N}^{\perp} & 0 & \mathbf{n}c_f \\ \mathbf{B} & 0 & 0 & \mathbf{I} & \mathbf{B} \\ \rho c_f^2 - \|\mathbf{B}\|^2 & 0 & 0 & -\mathbf{B}^{\top} & \rho c_f^2 - \|\mathbf{B}\|^2 \end{bmatrix}, \ \Lambda = \begin{bmatrix} -c_f & 0 & 0 & 0 & 0 \\ 0 & 0 & 0 & 0 & 0 \\ 0 & 0 & 0 & 0 & 0 \\ 0 & 0 & 0 & 0 & 0 \\ 0 & 0 & 0 & 0 & c_f \end{bmatrix}$$

and if $\beta \neq 0$ we have $\sigma \equiv sign(\beta)$ and

$$\mathbf{Y} = \begin{bmatrix} \frac{\rho c_f^2 - \|\mathbf{B}\|^2}{c^2} & 0 & \frac{\rho c_s^2 - \|\mathbf{B}\|^2}{c^2} & 1 & 0 & \frac{\rho c_s^2 - \|\mathbf{B}\|^2}{c^2} & 0 & \frac{\rho c_f^2 - \|\mathbf{B}\|^2}{c^2} \\ -\mathbf{n}c_f + \mathbf{B}\frac{\beta}{\rho c_f} & \mathbf{B} \times \mathbf{n}\sigma & -\mathbf{n}c_s + \mathbf{B}\frac{\beta}{\rho c_s} & 0 & 0 & \mathbf{n}c_s - \mathbf{B}\frac{\beta}{\rho c_s} & -\mathbf{B} \times \mathbf{n}\sigma & \mathbf{n}c_f - \mathbf{B}\frac{\beta}{\rho c_f} \\ \mathbf{B} - \mathbf{n}\beta & \mathbf{B} \times \mathbf{n}\sqrt{\rho} & \mathbf{B} - \mathbf{n}\beta & 0 & \mathbf{n} & \mathbf{B} - \mathbf{n}\beta & \mathbf{B} \times \mathbf{n}\sqrt{\rho} & \mathbf{B} - \mathbf{n}\beta \\ \rho c_f^2 - \|\mathbf{B}\|^2 & 0 & \rho c_s^2 - \|\mathbf{B}\|^2 & 0 & 0 & \rho c_s^2 - \|\mathbf{B}\|^2 & 0 & \rho c_f^2 - \|\mathbf{B}\|^2 \end{bmatrix}$$

$$\Lambda = \begin{bmatrix} -c_f &&&&&&& \\ & -c_a &&&&&& \\ && -c_s &&&&& \\ &&& 0 &&&& \\ &&&& 0 &&& \\ &&&&& c_s && \\ &&&&&& c_a & \\ &&&&&&& c_f \end{bmatrix}.$$

The Alfvén speed does not occur if $\mathbf{v}$ and $\mathbf{B}$ are two-dimensional, and the only nonzero characteristic speeds are $\nu \pm c$ if $\mathbf{v}$ and $\mathbf{B}$ are one-dimensional.

4.5.3 Entropy Function

Recall that the specific entropy of a polytropic gas is $S = S_0 + c_v\{\ln \frac{p}{p_0} - \gamma \ln \frac{\rho}{\rho_0}\}$. It follows that the derivatives of the specific entropy with respect to the MHD flux variables are

$$\frac{\partial S}{\partial \mathbf{w}} = c_v \left[-\gamma/\rho, \ 0, \ 0, \ 1/p\right].$$

Similarly, the partial derivatives of the entropy per volume are

$$\frac{\partial S\rho}{\partial \mathbf{w}} = \left[S - c_v\gamma, \ 0, \ 0, \ c_v\rho/p\right].$$

We claim that the entropy flux for MHD is $S\rho\mathbf{v}^\top$, the same as in gas dynamics. To justify this claim, recall from Section 4.1.11 that we must show that

$$\frac{\partial S\mathbf{n}\cdot\mathbf{v}}{\partial \mathbf{w}} = \frac{\partial S\rho}{\partial \mathbf{w}} \left(\frac{\partial \mathbf{u}}{\partial \mathbf{w}}\right)^{-1} \frac{\partial \mathbf{Fn}}{\partial \mathbf{w}}.$$

To this end, we compute

$$\begin{aligned}
&\frac{\partial S\rho}{\partial \mathbf{w}} \left(\frac{\partial \mathbf{u}}{\partial \mathbf{w}}\right)^{-1} \frac{\partial \mathbf{Fn}}{\partial \mathbf{w}} \\
&= \left[S - c_v\gamma \ \ 0 \ \ 0 \ \ c_v\rho/p\right] \begin{bmatrix} (\mathbf{n}\cdot\mathbf{v}) & \rho\mathbf{n}^\top & 0 & 0 \\ 0 & \mathbf{I}(\mathbf{n}\cdot\mathbf{v}) & \mathbf{n}\frac{1}{\rho}\mathbf{B}^\top - \mathbf{I}\frac{\mathbf{B}\cdot\mathbf{n}}{\rho} & \mathbf{n}\frac{1}{\rho} \\ 0 & \mathbf{Bn}^\top - \mathbf{IB}\cdot\mathbf{n} & \mathbf{Iv}\cdot\mathbf{n} & 0 \\ 0 & \rho c^2\mathbf{n}^\top & 0 & \mathbf{n}\cdot\mathbf{v} \end{bmatrix} \\
&= \left[(S - c_v\gamma)\mathbf{n}\cdot\mathbf{v} \ \ S\rho\mathbf{n}^\top \ \ 0 \ \ c_v\frac{\rho}{p}\mathbf{n}\cdot\mathbf{v}\right] \\
&= \frac{\partial S}{\partial \mathbf{w}}\rho\mathbf{n}\cdot\mathbf{v} + S\mathbf{n}^\top\left[\mathbf{v} \ \ \mathbf{I}\rho \ \ 0 \ \ 0\right] = \frac{\partial S}{\partial \mathbf{w}}\rho\mathbf{n}\cdot\mathbf{v} + S\mathbf{n}^\top\frac{\partial \mathbf{v}\rho}{\partial \mathbf{w}} = \frac{\partial S\rho(\mathbf{n}\cdot\mathbf{v})}{\partial \mathbf{w}}.
\end{aligned}$$

The entropy function for MHD is a strictly concave function of ρ and p, and a concave function of $\mathbf{w}$.

4.5.4 Centered Rarefaction Curves

Recall Equation (4.28) for a centered rarefaction wave in one dimension:

$$\tilde{\mathbf{w}}' = \mathbf{Y}(\tilde{\mathbf{w}})\mathbf{e}_j\alpha, \ \ \tilde{\mathbf{w}}(0) = \mathbf{w}_L.$$

We will use this equation to determine the centered rarefaction waves for MHD.

The centered rarefaction corresponding to the characteristic speed $\mathbf{v} \cdot \mathbf{n} + \lambda$ is the solution of the ordinary differential equation

$$\frac{d}{d\mathbf{y}} \begin{bmatrix} \rho \\ \mathbf{v} \\ \mathbf{B} \\ p \end{bmatrix} = \begin{bmatrix} (\rho\lambda^2 - \|\mathbf{B}\|^2)/c^2 \\ \mathbf{n}\lambda - \mathbf{B}\beta/(\rho\lambda) \\ \mathbf{B} - \mathbf{n}\beta \\ \rho\lambda^2 - \|\mathbf{B}\|^2 \end{bmatrix} \alpha,$$

where $\mathbf{y}$ is some measure of distance along the rarefaction curve, and $\beta = \mathbf{B} \cdot \mathbf{n}$. Note that this system of ordinary differential equations says that

$$\frac{d\rho}{dp} = \frac{\rho}{\gamma p}.$$

We can solve this equation to get

$$\rho = \rho_L \left(\frac{p}{p_L} \right)^{1/\gamma}.$$

Note that this implies that

$$S = S_0 + c_v \ln \left\{ \frac{p}{\rho^\gamma} \right\}$$

is constant. Thus the specific entropy is constant along a centered rarefaction curve. Also note that we can multiply the third equation in the system by $\mathbf{n}^\top$ to get

$$\frac{d\beta}{dp} = 0.$$

Thus the normal component of $\mathbf{B}$ is constant as well. Since $\frac{d\mathbf{n}\beta}{dp} = 0$, the third equation in the system can be rewritten

$$\frac{d\mathbf{B}^\perp}{dp} = \mathbf{B}^\perp \frac{1}{\rho\lambda^2 - \|\mathbf{B}\|^2}.$$

Since $\mathbf{B}^\perp \perp \mathbf{n}$, the Pythagorean theorem implies that $\|\mathbf{B}\|^2 = \|\mathbf{B}^\perp\|^2 + \beta^2$. Further, the definitions of c_s, c_a and c_f imply that

$$\rho\lambda^2 - \|\mathbf{B}\|^2 = \frac{1}{2}(\gamma p - \|\mathbf{B}^\perp\|^2 - \beta^2) \pm \sqrt{\frac{1}{4}(\gamma p + \|\mathbf{B}^\perp\|^2 + \beta^2)^2 - \gamma p \beta^2}$$

Thus we have an ordinary differential equation for the vector $\mathbf{B}^\perp$.

The ordinary differential equation for $\mathbf{v}$ is

$$\frac{d\mathbf{v}}{dp} = (\mathbf{n}\rho\lambda^2 - \mathbf{B}\beta)\frac{1}{(\rho\lambda^2 - \|\mathbf{B}\|^2)\rho\lambda} = \{\mathbf{n}(\lambda^2 - c_a^2) - \mathbf{B}^\perp\beta/\rho\}\frac{\lambda}{(\rho\lambda^2 - \|\mathbf{B}\|^2)\lambda^2}$$
$$= \{\mathbf{n}(\lambda^2 - c_a^2) - \mathbf{B}^\perp\beta/\rho\}\frac{\lambda}{\gamma p(\lambda^2 - c_a^2)}.$$

If we take the inner product of this equation with $\mathbf{n}$, we obtain

$$\frac{d\mathbf{v}\cdot\mathbf{n}}{dp} = \{(\lambda^2 - c_a^2)\}\frac{\lambda}{\gamma p(\lambda^2 - c_a^2)} = \frac{\lambda}{\gamma p}.$$

Summary 4.5.8 *In magnetohydrodynamics, two Riemann invariants for a centered rarefaction, with characteristic speed $\mathbf{v}\cdot\mathbf{n} \pm c_s$ or $\mathbf{v}\cdot\mathbf{n} \pm c_f$, are $\beta \equiv \mathbf{B}\cdot\mathbf{n}$ and the specific entropy $S = c_v \ln(p/\rho^\gamma)$. The specific entropy can be used to determine the density as a function of pressure along a centered rarefaction; further, density is always an increasing function of pressure along a centered rarefaction. In a centered rarefaction, the perpendicular component of the **B** field, $\mathbf{B}^\perp \equiv \mathbf{B} - \mathbf{n}\beta$, satisfies the ordinary differential equation*

$$\frac{d\mathbf{B}^\perp}{dp} = \mathbf{B}^\perp \frac{1}{\rho\lambda^2 - \|\mathbf{B}\|^2}$$

where $\lambda = \pm c_s$ or $\lambda = \pm c_f$ are functions of p and $\mathbf{B}^\perp$. Further, the velocity satisfies the ordinary differential equation

$$\frac{d\mathbf{v}}{dp} = \{\mathbf{n}\rho(\lambda^2 - c_a^2) - \mathbf{B}^\perp\beta\}\frac{1}{(\rho\lambda^2 - \|\mathbf{B}\|^2)\rho\lambda}.$$

In particular, the normal velocity satisfies

$$\frac{d\mathbf{v}\cdot\mathbf{n}}{dp} = \frac{\lambda}{\gamma p}.$$

4.5.5 Jump Conditions

Recall from equation (4.22) that at a propagating discontinuity, the Rankine-Hugoniot conditions are $[\mathbf{F}_R - \mathbf{F}_L]\mathbf{n} = [\mathbf{u}_R - \mathbf{u}_L]\sigma$. Here, $\mathbf{n}$ is the normal to the discontinuity and σ is the normal velocity of the discontinuity. If we apply these jump conditions to MHD write the jump conditions separately, we obtain

$$\begin{aligned}
[\rho\mathbf{v}^\top\mathbf{n}] &= [\rho]\sigma, \\
\left[\mathbf{v}\rho\mathbf{v}^\top\mathbf{n} + \mathbf{n}p + \mathbf{n}\frac{\mathbf{B}^\top\mathbf{B}}{2} - \mathbf{B}\mathbf{B}^\top\mathbf{n}\right] &= [\mathbf{v}\rho]\sigma, \\
\left[\mathbf{B}\mathbf{v}^\top\mathbf{n} - \mathbf{v}\mathbf{B}^\top\mathbf{n}\right] &= [\mathbf{B}]\sigma, \\
\left[\left(p\frac{\gamma}{\gamma-1} + \frac{\rho}{2}\mathbf{v}^\top\mathbf{v} + \mathbf{B}^\top\mathbf{B}\right)\mathbf{v}^\top\mathbf{n} - \mathbf{v}^\top\mathbf{B}\mathbf{B}^\top\mathbf{n}\right] &= \left[p\frac{1}{\gamma-1} + \frac{\rho}{2}\mathbf{v}^\top\mathbf{v} + \frac{1}{2}\mathbf{B}^\top\mathbf{B}\right]\sigma.
\end{aligned}$$

While it is possible to solve these equations, the solution is pretty messy except in special circumstances. We will omit this discussion, because we do not plan to use the information to solve Riemann problems for MHD in numerical methods.

The first four exercises are intended to guide the student through the analysis of the jump conditions for MHD.

Exercises for 4.5

4.5.1 Suppose that $\rho_L v_L = \rho_R v_R \neq 0, \sigma \neq 0, \mathbf{B}_R^\perp = \mathbf{B}_L^\perp(1+\alpha)$ and $\mathbf{B}_R^\perp \neq 0$. Show that if $\alpha = 0$, then the Rankine–Hugoniot jump conditions imply that the left and right states are the same. Otherwise, show that the Rankine–Hugoniot jump conditions lead to a cubic equation for α, with coefficients that are functions of the left state and the discontinuity speed σ.

4.5.2 Suppose that $\rho_L v_L = \rho_R v_R \neq 0, \sigma \neq 0, \mathbf{B}_R^\perp = \mathbf{B}_L^\perp(1+\alpha)$ and $\mathbf{B}_R^\perp = 0$. If $\mathbf{B}_L^\perp = 0$, show that $[p] = 0$ implies that the left and right states are identical. If $\mathbf{B}_L^\perp = 0$ and $[p] \neq 0$, show that given the left state and p_R we can determine the discontinuity speed and the right state. If $\mathbf{B}_L^\perp \neq 0$, show that the discontinuity is an Alfvén wave.

4.5.3 Suppose that $\rho_L v_L = \rho_R v_R \neq 0, \sigma \neq 0$ and that $\mathbf{b}_R^\perp$ is not a scalar multiple of $\mathbf{B}_L^\perp$. Show that the Rankine–Hugoniot jump conditions imply that $\mathbf{B}_L^\perp = 0 \neq \mathbf{B}_R^\perp$. Then show that given the left state and $\mathbf{b}_R^\perp$, the discontinuity speed is determined by a quartic equation.

4.5.4 Suppose that $\rho_L v_L = \rho_R v_R \neq 0$ and $\sigma = 0$. If $\mathbf{B}_R^\perp$ is a linear combination of $\mathbf{B}_L$ and $\mathbf{v}_L^\perp$, show that the Rankine–Hugoniot jump conditions lead to a quartic equation for β_R, from which the remainder of the right state can be determined. Otherwise, show that the jump conditions lead to a quadratic equation for β_R^2, from which the remainder of the right state can be determined.

4.5.5 Suppose that $\rho_L v_L = \rho_R v_R = 0$ and $\sigma \neq 0$. Ignore zero density. Show that $[\beta] = 0$, and that if $\beta_L \neq 0$ then the left and right states are identical. On the other hand, if $\beta_L = 0$, then the discontinuity speed is equal to $\mathbf{v}_L \cdot \mathbf{n}$ and the only jump between the two states are in the transverse velocity and transverse $\mathbf{B}$ field.

4.5.6 Discuss the solution of the Rankine–Hugoniot jump conditions when $\rho_L v_L = \rho_R v_R = 0$ and $\sigma = 0$.

4.5.7 Determine the conditions to impose on the $\mathbf{B}$ field at a reflecting wall. See section 4.4.8 for the reflecting wall conditions with gas dynamics.

4.6 Case Study: Finite Deformation in Elastic Solids

Solid mechanics is a much more complicated subject than gas dynamics. The systems of conservation laws are larger, and the constitutive models are more complicated. Traditionally, these problems have been solved by finite element methods. The application of modern shock-capturing techniques to these problems is somewhat recent, and in need of additional development.

4.6.1 Eulerian Formulation of Equations of Motion for Solids

A description of the Eulerian forms of the conservation laws for finite deformation in solid mechanics can be found in [39]. Conservation of mass can be written either as the **continuity equation**

$$\frac{d\rho}{dt} + \rho \nabla_{\mathbf{x}} \cdot \mathbf{v} = 0, \tag{4.47}$$

or as the conservation law

$$\frac{\partial \rho}{\partial t} + \nabla_{\mathbf{x}} \cdot (\mathbf{v}\rho) = 0. \tag{4.48}$$

Conservation of momentum can be written either as **Newton's second law of motion**

$$\frac{d\mathbf{n} \cdot \mathbf{v}}{dt} - \frac{1}{\rho} \nabla_{\mathbf{x}} \cdot (\mathbf{S}\mathbf{n}) = \mathbf{n} \cdot \mathbf{g}, \tag{4.49}$$

or as a conservation law

$$\frac{\partial \mathbf{n} \cdot \mathbf{v}\rho}{\partial t} + \nabla_{\mathbf{x}} \cdot (\mathbf{v}\rho \mathbf{n} \cdot \mathbf{v} - \mathbf{S}\mathbf{n}) = \rho \mathbf{n} \cdot \mathbf{g}. \tag{4.50}$$

Here $\mathbf{n}$ is an arbitrary fixed direction, $\mathbf{g}$ is a body acceleration (such as gravity) and $\mathbf{S}$ is the **Cauchy stress tensor**. In some cases, $\mathbf{g}$ might include the effects of viscous forces. The Cauchy stress tensor $\mathbf{S}$ is symmetric in most practical problems, and requires a **constitutive law** to relate it to other variables. Finally, conservation of energy can be written either as the **first law of thermodynamics**

$$\frac{d\epsilon}{dt} - \frac{1}{\rho} \text{tr} \left(\mathbf{S} \frac{\partial \mathbf{v}}{\partial \mathbf{x}} \right) = \omega \tag{4.51}$$

or in conservation form

$$\frac{\partial \rho(\epsilon + \frac{1}{2}\mathbf{v} \cdot \mathbf{v})}{\partial t} + \nabla_{\mathbf{x}} \cdot \left\{ \mathbf{v}\rho \left(\epsilon + \frac{1}{2}\mathbf{v} \cdot \mathbf{v} \right) - \mathbf{S}\mathbf{v} \right\} = \rho(\omega + \mathbf{g} \cdot \mathbf{v}). \tag{4.52}$$

Here ϵ is the internal energy per mass, and ω is the radiative heat transfer per unit mass. In some cases, ω might include the effects of heat diffusion.

4.6.2 Lagrangian Formulation of Equations of Motion for Solids

We can use (4.20) to discover the equivalent conservation laws in the Lagrangian frame of reference. We will define the Lagrangian density $\rho_L = \rho|\mathbf{J}|$ and the **second Piola–Kirchhoff stress tensor** $\mathbf{S}_L = \mathbf{J}^{-1}\mathbf{S}\mathbf{J}^{-\top}|\mathbf{J}|$. Here $\mathbf{J} = \partial \mathbf{x}/\partial a$ is the deformation gradient, defined originally in equation (4.11).

Lagrangian conservation of mass can be written as

$$\frac{d\rho_L}{dt} = 0. \tag{4.53}$$

Conservation of momentum can be written either as Newton's second law of motion

$$\frac{d\mathbf{n}\cdot\mathbf{v}}{dt} - \frac{1}{\rho_L}\nabla_a \cdot (\mathbf{S}_L \mathbf{n}) = \mathbf{n}\cdot\mathbf{g}, \tag{4.54}$$

or as a conservation law

$$\frac{\partial \mathbf{n}\cdot\mathbf{v}\rho_L}{\partial t} - \nabla_a \cdot \left(\mathbf{S}_L \mathbf{J}^\top \mathbf{n}\right) = \rho_L \mathbf{n}\cdot\mathbf{g}. \tag{4.55}$$

Here $\mathbf{J}\mathbf{S}_L$ is the **first Piola–Kirchhoff stress tensor**. Note that the second Piola–Kirchhoff stress tensor $\mathbf{S}_L$ is symmetric whenever the Cauchy stress tensor $\mathbf{S}$ is symmetric; however, the first Piola–Kirchhoff stress tensor $\mathbf{J}\mathbf{S}_L$ is generally not symmetric. Finally, conservation of energy can be written either as the first law of thermodynamics

$$\frac{d\epsilon}{dt} - \frac{1}{\rho_L}\mathrm{tr}\left(\mathbf{S}_L \mathbf{J}^\top \frac{\partial \mathbf{v}}{\partial a}\right) = \omega \tag{4.56}$$

or in conservation form

$$\frac{\partial \rho_L(\epsilon + \frac{1}{2}\mathbf{v}\cdot\mathbf{v})}{\partial t} - \nabla_a \cdot (\mathbf{S}_L \mathbf{J}^\top \mathbf{v}) = \rho_L(\omega + \mathbf{g}\cdot\mathbf{v}). \tag{4.57}$$

Note that the Eulerian form of the conservation laws used equality of mixed partial derivatives in the form

$$\frac{\partial \mathbf{J}^{-1}}{\partial t} + \frac{\partial \mathbf{J}^{-1}\mathbf{v}}{\partial \mathbf{x}} = 0.$$

If we had transformed the Lagrangian equality of mixed partial derivatives to the Eulerian frame using (4.20), we would have obtained

$$\frac{\partial \mathbf{n}\cdot\mathbf{J}\mathbf{e}_i|\mathbf{J}^{-1}|}{\partial t} + \nabla_{\mathbf{x}} \cdot (\mathbf{v}\mathbf{n}\cdot\mathbf{J}\mathbf{e}_i|\mathbf{J}^{-1}| - \mathbf{J}\mathbf{e}_i|\mathbf{J}^{-1}|\mathbf{n}\cdot\mathbf{v}) = 0.$$

These two conservation laws are equivalent, provided that the deformation gradient satisfies $\nabla_a \times \mathbf{J} = 0$.

4.6.3 Constitutive Laws

In order to close the equations of motion, we need to provide a **constitutive law** for the stress. We can place the model on a firm thermodynamical foundation by assuming a **hyperelastic model**

$$\mathbf{S}_L = 2\frac{\partial \rho_L \psi}{\partial C}, \tag{4.58}$$

where $C = \mathbf{J}^\top\mathbf{J}$ is the **Green deformation tensor** and $\psi(C)$ is the **Helmholtz free energy** per unit mass. A particularly simple free energy function is given by the

Mooney–Rivlin model

$$\rho_L \psi(C) = \frac{\lambda}{8}[\ln \det(C)]^2 + \frac{\mu}{2}\mathrm{tr}(C) - \frac{\mu}{2}\ln \det(C). \tag{4.59}$$

The constants are the **shear modulus** μ and the **Lamé constant** $\lambda = \kappa - \frac{2}{3}\mu$, where κ is the **bulk modulus**. Thus a hyperelastic model considers the second Piola–Kirchhoff stress tensor to be a function of the deformation gradient $\mathbf{J}$; this naturally implies that the Cauchy stress tensor $\mathbf{S}$ is a function of $\mathbf{J}$.

In practice, it is more common that constitutive laws for solids are given as a system of ordinary differential equations. These **hypolelastic models** take the form

$$\frac{d\mathbf{J}\mathbf{S}_L}{dt} = \left(\frac{d\mathbf{J}}{dt} + \frac{d\mathbf{J}^\top}{dt}\right)\mu + \mathbf{I}\lambda \mathrm{tr}\frac{d\mathbf{J}}{dt}. \tag{4.60}$$

Again, these ordinary differential equations imply that the first Piola–Kirchhoff stress tensor is a function of the deformation gradient $\mathbf{J}$; as a result, the second Piola–Kirchhoff stress tensor $\mathbf{S}_L$ and the Cauchy stress tensor $\mathbf{S}$ are functions of the deformation gradient.

In general, the equation of state can either be differentiated in time or expressed directly in rate form

$$\frac{d\mathbf{J}\mathbf{S}_L\mathbf{e}_i}{dt} = -\sum_{j=1}^{3} \mathbf{H}_{ij}\frac{d\mathbf{J}\mathbf{e}_j}{dt} + \mathbf{h}_i\frac{d\theta}{dt}.$$

Here θ is the absolute temperature and

$$\mathbf{H}_{ij} = \frac{\partial \mathbf{J}\mathbf{S}_L\mathbf{e}_i}{\partial \mathbf{J}\mathbf{e}_j} \text{ and } \mathbf{h}_i = \frac{\partial \mathbf{J}\mathbf{S}_L\mathbf{e}_i}{\partial \theta}.$$

Here $\mathbf{H}_{ij}$ is a matrix, not the i, j entry of a matrix, and $\mathbf{h}_i$ is a vector. In many cases, the equations of motion for solids are assumed to be isothermal, and $\mathbf{h}_i = 0$. Similarly, the equation of state could either be differentiated in time or expressed directly in rate form as

$$\frac{d\mathbf{S}\mathbf{e}_i}{dt} = -\sum_{j=1}^{3} \tilde{\mathbf{H}}_{ij}\mathbf{J}\frac{d\mathbf{J}^{-1}\mathbf{e}_j}{dt} + \tilde{\mathbf{h}}_i\frac{d\theta}{dt}.$$

Here

$$\tilde{\mathbf{H}}_{ij} = -\frac{\partial \mathbf{S}\mathbf{e}_i}{\partial \mathbf{J}^{-1}\mathbf{e}_j}\mathbf{J}^{-1} \text{ and } \tilde{\mathbf{h}}_i = \frac{\partial \mathbf{S}\mathbf{e}_i}{\partial \theta}.$$

Typically, the internal energy is assumed to be a function of absolute temperature θ and the deformation gradient. Then the equation for the internal energy can either be differentiated in time or expressed directly in rate form as

$$\frac{d\epsilon}{dt} = c_j \cdot \frac{d\mathbf{J}\mathbf{e}_j}{dt} + \gamma \frac{d\theta}{dt}.$$

Here

$$c_j^\top = \frac{\partial \epsilon}{\partial \mathbf{J}\mathbf{e}_j} \text{ and } \gamma = \frac{\partial \epsilon}{\partial \theta}.$$

4.6.4 Conservation Form of the Equations of Motion for Solids

We can write our equations of motion in the Lagrangian frame as

$$\frac{d}{dt}\int_{\Omega_0} \mathbf{u}_L \, da + \int_{\partial\Omega_0} \mathbf{F}_L \mathbf{n} \, ds = \int_{\Omega_0} r_L \, da$$

where the vector of conserved quantities, array of fluxes and vector of body forces are

$$\mathbf{u}_L = \begin{bmatrix} \rho_L \\ \mathbf{v}\rho_L \\ (\epsilon + \frac{1}{2}\mathbf{v}\cdot\mathbf{v})\rho_L \\ \mathbf{J}\mathbf{e}_1 \\ \mathbf{J}\mathbf{e}_2 \\ \mathbf{J}\mathbf{e}_3 \end{bmatrix}, \quad \mathbf{F}_L = \begin{bmatrix} 0 \\ -\mathbf{J}\mathbf{S}_L \\ -\mathbf{v}^\top \mathbf{J}\mathbf{S}_L \\ -\mathbf{v}\mathbf{e}_1^\top \\ -\mathbf{v}\mathbf{e}_2^\top \\ -\mathbf{v}\mathbf{e}_3^\top \end{bmatrix}, \quad r_L = \begin{bmatrix} 0 \\ \mathbf{g}\rho_L \\ (\omega + \mathbf{g}\cdot\mathbf{v})\rho_L \\ 0 \\ 0 \\ 0 \end{bmatrix}.$$

Here we have considered the most general case of three dimensions. The first three entries in these arrays represent conservation of mass, momentum and energy, while the remaining entries represent the equality of mixed partial derivatives,

$$\frac{d\mathbf{J}}{dt} = \frac{\partial \mathbf{v}}{\partial a}. \tag{4.61}$$

In addition to these conservation laws, we also have the constitutive laws for stress and internal energy.

Also note that the deformation gradient must satisfy the constraint

$$\nabla_a \times \mathbf{J}^\top = 0.$$

This may be viewed as a constraint on the initial values for the differential equation; if this curl condition is satisfied initially, then the conservation law (4.61) guarantees that it is satisfied for all time.

In the Eulerian frame, the conservation laws can be written

$$\frac{d}{dt}\int_{\Omega} \mathbf{u}_E \, d\mathbf{x} + \int_{\partial\Omega} \mathbf{F}_E \mathbf{n} \, ds = \int_{\Omega} r_E \, d\mathbf{x}$$

where the vector of conserved quantities, array of fluxes and vector of body forces are

$$\mathbf{u}_E = \begin{bmatrix} \rho \\ \mathbf{v}\rho \\ (\epsilon + \frac{1}{2}\mathbf{v}\cdot\mathbf{v})\rho \\ \mathbf{J}^{-1}\mathbf{e}_1 \\ \mathbf{J}^{-1}\mathbf{e}_2 \\ \mathbf{J}^{-1}\mathbf{e}_3 \end{bmatrix}, \quad \mathbf{F}_E = \begin{bmatrix} \rho\mathbf{v}^\top \\ \mathbf{v}\rho\mathbf{v}^\top - \mathbf{S} \\ (\epsilon + \frac{1}{2}\mathbf{v}\cdot\mathbf{v})\rho\mathbf{v}^\top - \mathbf{v}^\top\mathbf{S} \\ \mathbf{J}^{-1}\mathbf{v}\mathbf{e}_1^\top \\ \mathbf{J}^{-1}\mathbf{v}\mathbf{e}_2^\top \\ \mathbf{J}^{-1}\mathbf{v}\mathbf{e}_3^\top \end{bmatrix}, \quad r_E = \begin{bmatrix} 0 \\ \mathbf{g}\rho \\ (\omega + \mathbf{g}\cdot\mathbf{v})\rho \\ 0 \\ 0 \\ 0 \end{bmatrix}.$$

The first three entries of these arrays represent conservation of mass, momentum and energy, while the remaining entries can be written as

$$\frac{\partial \mathbf{J}^{-1}}{\partial t} + \frac{\partial \mathbf{J}^{-1}\mathbf{v}}{\partial \mathbf{x}} = 0.$$

This equation can be derived from the equations

$$\frac{d\mathbf{J}^{-1}}{dt} = -\mathbf{J}^{-1}\frac{d\mathbf{J}}{dt}\mathbf{J}^{-1} = -\mathbf{J}^{-1}\frac{\partial \mathbf{v}}{\partial \mathbf{x}}, \quad \text{and} \quad \nabla_{\mathbf{x}} \times \mathbf{J}^{-\top} = 0.$$

4.6.5 Jump Conditions for Isothermal Solids

For isothermal solids, there is no dependence on the absolute temperature θ, and no need to consider conservation of energy. The Lagrangian form of the Rankine–Hugoniot jump conditions for isothermal solids can be written

$$\begin{aligned} 0 &= [\rho_L]\sigma_L, \\ [-\mathbf{J}\mathbf{S}_L\mathbf{n}_L] &= [\mathbf{v}\rho_L]\sigma_L, \\ [-\mathbf{v}]\mathbf{n}_L^\top &= [\mathbf{J}]\sigma_L. \end{aligned}$$

Note that we can multiply the last equation by the normal direction $\mathbf{n}_L$ and obtain

$$-[\mathbf{v}] = [\mathbf{J}]\mathbf{n}_L\sigma_L.$$

If the discontinuity speed σ_L is nonzero, then $[\rho_L] = 0$; it follows that traveling isothermal discontinuities satisfy

$$[\mathbf{J}\mathbf{S}_L]\mathbf{n}_L = [\mathbf{J}]\mathbf{n}_L \rho_L \sigma_L^2 .$$

The constitutive law can then be used to determine the nonzero discontinuity speeds. If the discontinuity speed $\sigma_L = 0$, then stationary discontinuities satisfy $[\mathbf{v}] = 0$ and $[\mathbf{J}\mathbf{S}_L]\mathbf{n}_L = 0$.

Similarly, the Eulerian form of the Rankine–Hugoniot jump conditions for isothermal solids can be written

$$\begin{aligned}
[\rho \mathbf{n}_E \cdot \mathbf{v}] &= [\rho]\sigma_E, \\
[\mathbf{v}\rho \mathbf{n}_E \cdot \mathbf{v} - \mathbf{S}\mathbf{n}_E] &= [\mathbf{v}\rho]\sigma_E, \\
[\mathbf{J}|\mathbf{J}^{-1}|\mathbf{n}_E \cdot \mathbf{v} - \mathbf{v}|\mathbf{J}^{-1}|\mathbf{n}_E^\top \mathbf{J}] &= [\mathbf{J}|\mathbf{J}^{-1}|]\sigma_E .
\end{aligned}$$

Recall from Section 4.1.3.3 that the discontinuity speeds are related by

$$\sigma_E = \|\mathbf{J}^\top \mathbf{n}_E\|\sigma_L + \mathbf{n}_E \cdot \mathbf{v},$$

and the normal directions to the discontinuity are related by

$$\mathbf{n}_E = \mathbf{J}^{-\top}\mathbf{n}_L \frac{1}{\|\mathbf{J}^{-\top}\mathbf{n}_L\|} .$$

Thus the jump conditions can be rewritten

$$\begin{aligned}
[\rho(\sigma_E - \mathbf{n}_E \cdot \mathbf{v})] &= 0, \\
-[\mathbf{S}]\mathbf{n}_E &= [\mathbf{v}\rho(\sigma_E - \mathbf{n}_E \cdot \mathbf{v})], \\
-\left[\mathbf{v}|\mathbf{J}^{-1}|\mathbf{n}_E^\top \mathbf{J}\right] &= \left[\mathbf{J}|\mathbf{J}^{-1}|(\sigma_E - \mathbf{n}_E \cdot \mathbf{v})\right] .
\end{aligned}$$

If $\sigma_E = \mathbf{n}_E \cdot \mathbf{v}$, then the normal component of velocity is continuous across the discontinuity, $[\mathbf{S}]\mathbf{n}_E = 0$ and $[\mathbf{v}|\mathbf{J}^{-1}|\mathbf{n}_E^\top \mathbf{J}] = 0$.

4.6.6 Characteristic Analysis for Solids

In regions of smooth motion, the quasilinear form for the Lagrangian equations of motion can be written in terms of Newton's second law, the first law of thermodynamics and the rate form of the constitutive equation for stress:

$$\begin{bmatrix} \mathbf{I}\rho_L & 0 & 0 & 0 & 0 \\ 0 & \rho_L\gamma & 0 & 0 & 0 \\ 0 & -\mathbf{h}_1 & \mathbf{I} & 0 & 0 \\ 0 & -\mathbf{h}_2 & 0 & \mathbf{I} & 0 \\ 0 & -\mathbf{h}_3 & 0 & 0 & \mathbf{I} \end{bmatrix} \frac{\partial}{\partial t} \begin{bmatrix} \mathbf{v} \\ \theta \\ \mathbf{J}\mathbf{S}_L\mathbf{e}_1 \\ \mathbf{J}\mathbf{S}_L\mathbf{e}_2 \\ \mathbf{J}\mathbf{S}_L\mathbf{e}_3 \end{bmatrix}$$

$$= \sum_{j=1}^{3} \begin{bmatrix} 0 & 0 & -\mathbf{I}\delta_{1j} & -\mathbf{I}\delta_{2j} & -\mathbf{I}\delta_{3j} \\ -\rho_L\gamma\mathbf{b}_j^\top & 0 & 0 & 0 & 0 \\ -\mathbf{H}_{1j} & 0 & 0 & 0 & 0 \\ -\mathbf{H}_{2j} & 0 & 0 & 0 & 0 \\ -\mathbf{H}_{3j} & 0 & 0 & 0 & 0 \end{bmatrix} \frac{\partial}{\partial a_j} \begin{bmatrix} \mathbf{v} \\ \theta \\ \mathbf{J}\mathbf{S}_L\mathbf{e}_1 \\ \mathbf{J}\mathbf{S}_L\mathbf{e}_2 \\ \mathbf{J}\mathbf{S}_L\mathbf{e}_3 \end{bmatrix} = \begin{bmatrix} \mathbf{g}\rho_L \\ \omega\rho_L \\ 0 \\ 0 \\ 0 \end{bmatrix}.$$

Here

$$\mathbf{b}_j^\top = \frac{1}{\gamma\rho_L}\left(\mathbf{e}_j^\top \mathbf{S}_L \mathbf{J}^\top - \rho_L c_j^\top\right),\ \gamma = \frac{\partial\epsilon}{\partial\theta},\ c_j^\top = \frac{\partial\epsilon}{\partial \mathbf{J}\mathbf{e}_j},$$

$$\mathbf{H}_{ij} = \frac{\partial \mathbf{J}\mathbf{S}_L\mathbf{e}_i}{\partial \mathbf{J}\mathbf{e}_j} \text{ and } \mathbf{h}_i = \frac{\partial \mathbf{J}\mathbf{S}_L\mathbf{e}_i}{\partial\theta}.$$

This is because the conserved quantities and flux are functions of the **flux variables**

$$\mathbf{w}_L = \begin{bmatrix} \mathbf{v} \\ \theta \\ \mathbf{J}\mathbf{S}_L\mathbf{e}_1 \\ \mathbf{J}\mathbf{S}_L\mathbf{e}_2 \\ \mathbf{J}\mathbf{S}_L\mathbf{e}_3 \end{bmatrix}.$$

We did not determine the quasilinear form directly from the conservation laws, although this could be done with careful additional use of the constraint that the deformation gradient is curl-free.

In order for these equations of motion to be hyperbolic, we require that

$$\mathbf{B} = \begin{bmatrix} \mathbf{I}\rho_L & 0 & 0 & 0 & 0 \\ 0 & \rho_L\gamma & 0 & 0 & 0 \\ 0 & -\mathbf{h}_1 & \mathbf{I} & 0 & 0 \\ 0 & -\mathbf{h}_2 & 0 & \mathbf{I} & 0 \\ 0 & -\mathbf{h}_3 & 0 & 0 & \mathbf{I} \end{bmatrix}^{-1} \begin{bmatrix} 0 & 0 & -\mathbf{I}n_1 & \mathbf{I}n_2 & \mathbf{I}n_3 \\ -\rho_L\gamma\sum_{j=1}^3 n_j\mathbf{b}_j^\top & 0 & 0 & 0 & 0 \\ -\sum_{j=1}^3 \mathbf{H}_{1j}n_j & 0 & 0 & 0 & 0 \\ -\sum_{j=1}^3 \mathbf{H}_{2j}n_j & 0 & 0 & 0 & 0 \\ -\sum_{j=1}^3 \mathbf{H}_{3j}n_j & 0 & 0 & 0 & 0 \end{bmatrix}$$

$$= \begin{bmatrix} 0 & 0 & -\mathbf{I}n_1/\rho_L & -\mathbf{I}n_2/\rho_L & -\mathbf{I}n_3/\rho_L \\ -\sum_{j=1}^3 n_j\mathbf{b}_j^\top & 0 & 0 & 0 & 0 \\ -\sum_{j=1}^3(\mathbf{H}_{1j}n_j - \mathbf{h}_1 n_j\mathbf{b}_j^\top) & 0 & 0 & 0 & 0 \\ -\sum_{j=1}^3(\mathbf{H}_{2j}n_j - \mathbf{h}_2 n_j\mathbf{b}_j^\top) & 0 & 0 & 0 & 0 \\ -\sum_{j=1}^3(\mathbf{H}_{3j}n_j - \mathbf{h}_3 n_j\mathbf{b}_j^\top) & 0 & 0 & 0 & 0 \end{bmatrix}$$

must have real eigenvalues.

Let us define the **acoustic tensor** for a given normal $\mathbf{n}$ by

$$\mathbf{A}_L = \sum_i n_i \sum_j (\mathbf{H}_{ij} + \mathbf{h}_i \mathbf{b}_j^\top) n_j = \frac{\partial \mathbf{J}\mathbf{S}_L \mathbf{n}}{\partial \mathbf{J}\mathbf{n}} + \frac{\partial \mathbf{J}\mathbf{S}_L \mathbf{n}}{\partial \theta} \frac{1}{\gamma \rho_L} \left(\mathbf{n}^\top \mathbf{S}_L \mathbf{J}^\top - \rho_L \frac{\partial \epsilon}{\partial \mathbf{J}\mathbf{n}} \right). \quad (4.62)$$

We suppose that we can find a nonsingular matrix $\mathbf{X}$ and a diagonal matrix Λ_L so that $\mathbf{A}_L \mathbf{X} = \mathbf{X} \Lambda_L^2 \rho_L$. Also, let $[\mathbf{n}, \mathbf{N}]$ be an orthogonal matrix, so that $\mathbf{n}^\top \mathbf{N} = 0$. Then it is easy to see that the characteristic speeds are real; in three dimensions we have

$$\begin{bmatrix} 0 & 0 & -\mathbf{I}n_1/\rho_L & -\mathbf{I}n_2/\rho_L & -\mathbf{I}n_3/\rho_L \\ -\sum_{j=1}^3 \mathbf{n}_j \mathbf{b}_j^\top & 0 & 0 & 0 & 0 \\ -\sum_{j=1}^3 (\mathbf{H}_{1j} + \mathbf{h}_1 \mathbf{b}_j^\top) \mathbf{n}_j & 0 & 0 & 0 & 0 \\ -\sum_{j=1}^3 (\mathbf{H}_{2j} + \mathbf{h}_2 \mathbf{b}_j^\top) \mathbf{n}_j & 0 & 0 & 0 & 0 \\ -\sum_{j=1}^3 (\mathbf{H}_{3j} + \mathbf{h}_3 \mathbf{b}_j^\top) \mathbf{n}_j & 0 & 0 & 0 & 0 \end{bmatrix}$$

$$\begin{bmatrix} \mathbf{X}\Lambda_L & 0 & 0 & 0 & -\mathbf{X}\Lambda_L \\ \sum_{j=1}^3 n_j \mathbf{b}_j^\top \mathbf{X} & 0 & 1 & 0 & \sum_{j=1}^3 n_j \mathbf{b}_j^\top \mathbf{X} \\ \sum_{j=1}^3 (\mathbf{H}_{1j} + \mathbf{h}_1 \mathbf{b}_j^\top) \mathbf{n}_j \mathbf{X} & \mathbf{I}\mathbf{N}_{11} & 0 & \mathbf{I}\mathbf{N}_{12} & \sum_{j=1}^3 (\mathbf{H}_{1j} + \mathbf{h}_1 \mathbf{b}_j^\top) \mathbf{n}_j \mathbf{X} \\ \sum_{j=1}^3 (\mathbf{H}_{2j} + \mathbf{h}_2 \mathbf{b}_j^\top) \mathbf{n}_j \mathbf{X} & \mathbf{I}\mathbf{N}_{21} & 0 & \mathbf{I}\mathbf{N}_{22} & \sum_{j=1}^3 (\mathbf{H}_{2j} + \mathbf{h}_2 \mathbf{b}_j^\top) \mathbf{n}_j \mathbf{X} \\ \sum_{j=1}^3 (\mathbf{H}_{3j} + \mathbf{h}_3 \mathbf{b}_j^\top) \mathbf{n}_j \mathbf{X} & \mathbf{I}\mathbf{N}_{31} & 0 & \mathbf{I}\mathbf{N}_{32} & \sum_{j=1}^3 (\mathbf{H}_{3j} + \mathbf{h}_3 \mathbf{b}_j^\top) \mathbf{n}_j \mathbf{X} \end{bmatrix}$$

$$= \begin{bmatrix} \mathbf{X}\Lambda_L & 0 & 0 & 0 & -\mathbf{X}\Lambda_L \\ \sum_{j=1}^3 \mathbf{n}_j \mathbf{b}_j^\top \mathbf{X} & 0 & 1 & 0 & \sum_{j=1}^3 \mathbf{n}_j \mathbf{b}_j^\top \mathbf{X} \\ \sum_{j=1}^3 (\mathbf{H}_{1j} + \mathbf{h}_1 \mathbf{b}_j^\top) \mathbf{n}_j \mathbf{X} & \mathbf{I}\mathbf{N}_{11} & 0 & \mathbf{I}\mathbf{N}_{12} & \sum_{j=1}^3 (\mathbf{H}_{1j} + \mathbf{h}_1 \mathbf{b}_j^\top) \mathbf{n}_j \mathbf{X} \\ \sum_{j=1}^3 (\mathbf{H}_{2j} + \mathbf{h}_2 \mathbf{b}_j^\top) \mathbf{n}_j \mathbf{X} & \mathbf{I}\mathbf{N}_{21} & 0 & \mathbf{I}\mathbf{N}_{22} & \sum_{j=1}^3 (\mathbf{H}_{2j} + \mathbf{h}_2 \mathbf{b}_j^\top) \mathbf{n}_j \mathbf{X} \\ \sum_{j=1}^3 (\mathbf{H}_{3j} + \mathbf{h}_3 \mathbf{b}_j^\top) \mathbf{n}_j \mathbf{X} & \mathbf{I}\mathbf{N}_{31} & 0 & \mathbf{I}\mathbf{N}_{32} & \sum_{j=1}^3 (\mathbf{H}_{3j} + \mathbf{h}_3 \mathbf{b}_j^\top) \mathbf{n}_j \mathbf{X} \end{bmatrix} \begin{bmatrix} -\Lambda_L & 0 & 0 & 0 & 0 \\ 0 & 0 & 0 & 0 & 0 \\ 0 & 0 & 0 & 0 & 0 \\ 0 & 0 & 0 & 0 & 0 \\ 0 & 0 & 0 & 0 & \Lambda_L \end{bmatrix}.$$

In the Eulerian frame of reference, the equations of motion for smooth flow can be written as the continuity equation, Newton's second law, the first law of thermodynamics, the rate equations representing the constitutive law, and

$$\frac{\partial \mathbf{J}^{-1}\mathbf{v}}{\partial t} - \mathbf{J}^{-1} \frac{\partial \mathbf{v}}{\partial t} + \frac{\partial \mathbf{J}^{-1}\mathbf{v}}{\partial \mathbf{x}} \mathbf{v} = 0.$$

This equation follows directly from the conservation law and product rule

$$\frac{\partial \mathbf{J}^{-1}}{\partial t} = -\frac{\partial \mathbf{J}^{-1}\mathbf{v}}{\partial \mathbf{x}} \quad \text{and} \quad \frac{\partial \mathbf{J}^{-1}\mathbf{v}}{\partial t} - \mathbf{J}^{-1} \frac{\partial \mathbf{v}}{\partial t} - \frac{\partial \mathbf{J}^{-1}}{\partial t} \mathbf{v} = 0.$$

Let us define

$$\tilde{\mathbf{h}}_i = \frac{\partial \mathbf{S}\mathbf{e}_i}{\partial \theta}, \quad \tilde{\mathbf{H}}_{ij} = -\frac{\partial \mathbf{S}\mathbf{e}_i}{\partial \mathbf{J}^{-1}\mathbf{e}_j}\mathbf{J}^{-1}, \quad \tilde{\mathbf{b}}_j^\top = -\frac{1}{\rho\gamma}\left(\mathbf{e}_j^\top \mathbf{S} + \rho \frac{\partial \epsilon}{\partial \mathbf{J}^{-1}\mathbf{e}_j}\mathbf{J}^{-1}\right).$$

Then we obtain the quasilinear form

$$\begin{bmatrix} \mathbf{I} & 0 & 0 & 0 & 0 & 0 & 0 \\ 0 & \mathbf{I}\rho & 0 & 0 & 0 & 0 & 0 \\ 0 & 0 & \rho\gamma & 0 & 0 & 0 & 0 \\ 0 & 0 & -\tilde{\mathbf{h}}_1 & \mathbf{I} & 0 & 0 & 0 \\ 0 & 0 & -\tilde{\mathbf{h}}_2 & 0 & \mathbf{I} & 0 & 0 \\ 0 & 0 & -\tilde{\mathbf{h}}_3 & 0 & 0 & \mathbf{I} & 0 \\ 0 & -\mathbf{J}^{-1} & 0 & 0 & 0 & 0 & \mathbf{I} \end{bmatrix} \frac{\partial}{\partial t} \begin{bmatrix} \rho \\ \mathbf{v} \\ \theta \\ \mathbf{S}\mathbf{e}_1 \\ \mathbf{S}\mathbf{e}_2 \\ \mathbf{S}\mathbf{e}_3 \\ \mathbf{J}^{-1}\mathbf{v} \end{bmatrix}$$

$$+ \sum_{j=1}^{3} \begin{bmatrix} \mathbf{v}_j & \rho\mathbf{e}_j^\top & 0 & 0 & 0 & 0 & 0 \\ 0 & \mathbf{I}\rho\mathbf{v}_j & 0 & -\mathbf{I}\delta_{1j} & -\mathbf{I}\delta_{2j} & -\mathbf{I}\delta_{3j} & 0 \\ 0 & \gamma\rho\tilde{\mathbf{b}}_j^\top & \gamma\rho\mathbf{v}_j & 0 & 0 & 0 & 0 \\ 0 & -\tilde{\mathbf{H}}_{1j} & \tilde{\mathbf{h}}_1\mathbf{v}_j & \mathbf{I}\mathbf{v}_j & 0 & 0 & 0 \\ 0 & -\tilde{\mathbf{H}}_{2j} & \tilde{\mathbf{h}}_2\mathbf{v}_j & 0 & \mathbf{I}\mathbf{v}_j & 0 & 0 \\ 0 & -\tilde{\mathbf{H}}_{3j} & \tilde{\mathbf{h}}_3\mathbf{v}_j & 0 & 0 & \mathbf{I}\mathbf{v}_j & 0 \\ 0 & 0 & 0 & 0 & 0 & 0 & \mathbf{I}\mathbf{v}_j \end{bmatrix} \frac{\partial}{\partial \mathbf{x}_j} \begin{bmatrix} \rho \\ \mathbf{v} \\ \theta \\ \mathbf{S}\mathbf{e}_1 \\ \mathbf{S}\mathbf{e}_2 \\ \mathbf{S}\mathbf{e}_3 \\ \mathbf{J}^{-1}\mathbf{v} \end{bmatrix} = \begin{bmatrix} 0 \\ \mathbf{g}\rho \\ \omega\rho \\ 0 \\ 0 \\ 0 \\ 0 \end{bmatrix}.$$

If we solve the equation for the time derivatives, we see that in order for the system to be hyperbolic, we require

$$\begin{bmatrix} \mathbf{v}\cdot\mathbf{n} & \rho\mathbf{n}^\top & 0 & 0 & 0 & 0 & 0 \\ 0 & \mathbf{I}\mathbf{v}\cdot\mathbf{n} & 0 & -\mathbf{I}\mathbf{n}_1/\rho & -\mathbf{I}\mathbf{n}_2/\rho & -\mathbf{I}\mathbf{n}_3/\rho & 0 \\ 0 & \sum_{j=1}^3 \mathbf{n}_j\tilde{\mathbf{b}}_j^\top & \mathbf{v}\cdot\mathbf{n} & 0 & 0 & 0 & 0 \\ 0 & -\sum_{j=1}^3(\tilde{\mathbf{H}}_{1j} - \mathbf{h}_1\mathbf{b}_j^\top)\mathbf{n}_j & 0 & \mathbf{I}\mathbf{v}\cdot\mathbf{n} & 0 & 0 & 0 \\ 0 & -\sum_{j=1}^3(\tilde{\mathbf{H}}_{2j} - \mathbf{h}_2\mathbf{b}_j^\top)\mathbf{n}_j & 0 & 0 & \mathbf{I}\mathbf{v}\cdot\mathbf{n} & 0 & 0 \\ 0 & -\sum_{j=1}^3(\tilde{\mathbf{H}}_{3j} - \mathbf{h}_3\mathbf{b}_j^\top)\mathbf{n}_j & 0 & 0 & 0 & \mathbf{I}\mathbf{v}\cdot\mathbf{n} & 0 \\ 0 & \mathbf{J}^{-1}\mathbf{v}\cdot\mathbf{n} & 0 & -\mathbf{J}^{-1}\mathbf{n}_1/\rho & -\mathbf{J}^{-1}\mathbf{n}_2/\rho & -\mathbf{J}^{-1}\mathbf{n}_3/\rho & \mathbf{I}\mathbf{v}\cdot\mathbf{n} \end{bmatrix}$$

to have real eigenvalues. We will assume that we can find a nonsingular matrix $\mathbf{X}$ and diagonal matrix Λ_E so that

$$\mathbf{A}_E\mathbf{X} \equiv \left\{ \sum_{i=1}^{3} \mathbf{n}_i \sum_{j=1}^{3} (\tilde{\mathbf{H}}_{ij} - \tilde{\mathbf{h}}_i\mathbf{b}_j^\top)\mathbf{n}_j \right\} \mathbf{X} = \mathbf{X}\Lambda_E^2\rho. \tag{4.63}$$

Then we see that

$$
\begin{bmatrix}
\mathbf{v}\cdot\mathbf{n} & \rho\mathbf{n}^\top & 0 & 0 & 0 & 0 & 0 \\
0 & \mathbf{I}\mathbf{v}\cdot\mathbf{n} & 0 & -\mathbf{I}\mathbf{n}_1/\rho & -\mathbf{I}\mathbf{n}_2/\rho & -\mathbf{I}\mathbf{n}_3/\rho & 0 \\
0 & \sum_{j=1}^3 \mathbf{n}_j\tilde{\mathbf{b}}_j^\top & \mathbf{v}\cdot\mathbf{n} & 0 & 0 & 0 & 0 \\
0 & -\sum_{j=1}^3(\tilde{\mathbf{H}}_{1j}-\mathbf{h}_1\mathbf{b}_j^\top)\mathbf{n}_j & 0 & \mathbf{I}\mathbf{v}\cdot\mathbf{n} & 0 & 0 & 0 \\
0 & -\sum_{j=1}^3(\tilde{\mathbf{H}}_{2j}-\mathbf{h}_2\mathbf{b}_j^\top)\mathbf{n}_j & 0 & 0 & \mathbf{I}\mathbf{v}\cdot\mathbf{n} & 0 & 0 \\
0 & -\sum_{j=1}^3(\tilde{\mathbf{H}}_{3j}-\mathbf{h}_3\mathbf{b}_j^\top)\mathbf{n}_j & 0 & 0 & 0 & \mathbf{I}\mathbf{v}\cdot\mathbf{n} & 0 \\
0 & \mathbf{J}^{-1}\mathbf{v}\cdot\mathbf{n} & 0 & -\mathbf{J}^{-1}\mathbf{n}_1/\rho & -\mathbf{J}^{-1}\mathbf{n}_2/\rho & -\mathbf{J}^{-1}\mathbf{n}_3/\rho & \mathbf{I}\mathbf{v}\cdot\mathbf{n}
\end{bmatrix}
\begin{bmatrix}
\rho\mathbf{n}^\top\mathbf{X} & 1 & 0 & 0 & 0 & 0 & -\rho\mathbf{n}^\top\mathbf{X} \\
\mathbf{X}\Lambda_E & 0 & 0 & 0 & 0 & 0 & -\mathbf{X}\Lambda_E \\
-\sum_{j=1}^3 \mathbf{n}_j\tilde{\mathbf{b}}_j^\top\mathbf{X} & 0 & 1 & 0 & 0 & 0 & -\sum_{j=1}^3 \mathbf{n}_j\tilde{\mathbf{b}}_j^\top\mathbf{X} \\
\sum_{j=1}^3(\tilde{\mathbf{H}}_{1j}-\tilde{\mathbf{h}}_1\tilde{\mathbf{b}}_j^\top)\mathbf{n}_j\mathbf{X} & 0 & 0 & \mathbf{I}\mathbf{N}_{11} & \mathbf{I}\mathbf{N}_{12} & 0 & \sum_{j=1}^3(\tilde{\mathbf{H}}_{1j}-\tilde{\mathbf{h}}_1\tilde{\mathbf{b}}_j^\top)\mathbf{n}_j\mathbf{X} \\
\sum_{j=1}^3(\tilde{\mathbf{H}}_{2j}-\tilde{\mathbf{h}}_2\tilde{\mathbf{b}}_j^\top)\mathbf{n}_j\mathbf{X} & 0 & 0 & \mathbf{I}\mathbf{N}_{21} & \mathbf{I}\mathbf{N}_{22} & 0 & \sum_{j=1}^3(\tilde{\mathbf{H}}_{2j}-\tilde{\mathbf{h}}_2\tilde{\mathbf{b}}_j^\top)\mathbf{n}_j\mathbf{X} \\
\sum_{j=1}^3(\tilde{\mathbf{H}}_{3j}-\tilde{\mathbf{h}}_3\tilde{\mathbf{b}}_j^\top)\mathbf{n}_j\mathbf{X} & 0 & 0 & \mathbf{I}\mathbf{N}_{31} & \mathbf{I}\mathbf{N}_{32} & 0 & \sum_{j=1}^3(\tilde{\mathbf{H}}_{3j}-\tilde{\mathbf{h}}_3\tilde{\mathbf{b}}_j^\top)\mathbf{n}_j\mathbf{X} \\
-\mathbf{J}^{-1}X(-\Lambda_E+\mathbf{I}\mathbf{v}\cdot\mathbf{n}) & 0 & 0 & 0 & 0 & \mathbf{I} & -\mathbf{J}^{-1}X(\Lambda_E+\mathbf{I}\mathbf{v}\cdot\mathbf{n})
\end{bmatrix}
$$

$$
=\begin{bmatrix}
\rho\mathbf{n}^\top\mathbf{X} & 1 & 0 & 0 & 0 & 0 & -\rho\mathbf{n}^\top\mathbf{X} \\
\mathbf{X}\Lambda_E & 0 & 0 & 0 & 0 & 0 & -\mathbf{X}\Lambda_E \\
-\sum_{j=1}^3 \mathbf{n}_j\tilde{\mathbf{b}}_j^\top\mathbf{X} & 0 & 1 & 0 & 0 & 0 & -\sum_{j=1}^3 \mathbf{n}_j\tilde{\mathbf{b}}_j^\top\mathbf{X} \\
\sum_{j=1}^3(\tilde{\mathbf{H}}_{1j}-\tilde{\mathbf{h}}_1\tilde{\mathbf{b}}_j^\top)\mathbf{n}_j\mathbf{X} & 0 & 0 & \mathbf{I}\mathbf{N}_{11} & \mathbf{I}\mathbf{N}_{12} & 0 & \sum_{j=1}^3(\tilde{\mathbf{H}}_{1j}-\tilde{\mathbf{h}}_1\tilde{\mathbf{b}}_j^\top)\mathbf{n}_j\mathbf{X} \\
\sum_{j=1}^3(\tilde{\mathbf{H}}_{2j}-\tilde{\mathbf{h}}_2\tilde{\mathbf{b}}_j^\top)\mathbf{n}_j\mathbf{X} & 0 & 0 & \mathbf{I}\mathbf{N}_{21} & \mathbf{I}\mathbf{N}_{22} & 0 & \sum_{j=1}^3(\tilde{\mathbf{H}}_{2j}-\tilde{\mathbf{h}}_2\tilde{\mathbf{b}}_j^\top)\mathbf{n}_j\mathbf{X} \\
\sum_{j=1}^3(\tilde{\mathbf{H}}_{3j}-\tilde{\mathbf{h}}_3\tilde{\mathbf{b}}_j^\top)\mathbf{n}_j\mathbf{X} & 0 & 0 & \mathbf{I}\mathbf{N}_{31} & \mathbf{I}\mathbf{N}_{32} & 0 & \sum_{j=1}^3(\tilde{\mathbf{H}}_{3j}-\tilde{\mathbf{h}}_3\tilde{\mathbf{b}}_j^\top)\mathbf{n}_j\mathbf{X} \\
-\mathbf{J}^{-1}X(-\Lambda_E+\mathbf{I}\mathbf{v}\cdot\mathbf{n}) & 0 & 0 & 0 & 0 & \mathbf{I} & -\mathbf{J}^{-1}X(\Lambda_E+\mathbf{I}\mathbf{v}\cdot\mathbf{n})
\end{bmatrix}
\begin{bmatrix}
-\Lambda_E+\mathbf{I}\mathbf{v}\cdot\mathbf{n} & 0 & 0 & 0 & 0 & 0 & 0 \\
0 & \mathbf{v}\cdot\mathbf{n} & 0 & 0 & 0 & 0 & 0 \\
0 & 0 & \mathbf{v}\cdot\mathbf{n} & 0 & 0 & 0 & 0 \\
0 & 0 & 0 & \mathbf{I}\mathbf{v}\cdot\mathbf{n} & 0 & 0 & 0 \\
0 & 0 & 0 & 0 & \mathbf{I}\mathbf{v}\cdot\mathbf{n} & 0 & 0 \\
0 & 0 & 0 & 0 & 0 & \mathbf{I}\mathbf{v}\cdot\mathbf{n} & 0 \\
0 & 0 & 0 & 0 & 0 & 0 & \Lambda_E+\mathbf{I}\mathbf{v}\cdot\mathbf{n}
\end{bmatrix}.
$$

In general, given any direction $\mathbf{n}$ we need to find a nonsingular matrix $\mathbf{X}$ and a diagonal matrix Λ_E so that

$$
\begin{aligned}
\sum_{i=1}^3 \left\{\tilde{\mathbf{H}}_{ii}\mathbf{n}_i^2 + \tilde{\mathbf{h}}_o\mathbf{n}_i\tilde{\mathbf{b}}_i^\top\right\}\mathbf{X} &= -\left\{\frac{\partial\mathbf{S}\mathbf{n}}{\partial\mathbf{J}^{-1}\mathbf{n}}\mathbf{J}^{-1} - \frac{\partial\mathbf{S}\mathbf{n}}{\partial\theta}\frac{1}{\gamma\rho}\left(\mathbf{n}^\top\mathbf{S} + \rho\frac{\partial\epsilon}{\partial\mathbf{J}^{-1}\mathbf{n}}\mathbf{J}^{-1}\right)\right\}\mathbf{X} \\
&= \mathbf{X}\Lambda_E^2\rho.
\end{aligned}
$$

Exercises for 4.6

4.6.1 Consider the Mooney–Rivlin model in one dimension.

(a) Compute $\mathbf{H} = \frac{\partial \mathbf{J}\mathbf{S}_L}{\partial \mathbf{J}}$ and $\tilde{\mathbf{H}} = -\frac{\partial \mathbf{S}}{\partial \mathbf{J}^{-1}}\mathbf{J}^{-1}$ for this model.

(b) Find the Eulerian and Lagrangian characteristic speeds.

(c) Describe the Rankine–Hugoniot jump conditions for this model in both the Eulerian and Lagrangian frames of reference.

(d) Program the Rusanov scheme for the Mooney–Rivlin model in the Lagrangian frame.

(e) Program the Rusanov scheme for the Mooney–Rivlin model in the Eulerian frame.

4.6.2 For the Mooney–Rivlin model in three dimensions, compute the derivatives

$$\mathbf{H} = \frac{\partial \mathbf{J}\mathbf{S}_L \mathbf{n}}{\partial \mathbf{J}\mathbf{m}}$$

for arbitrary fixed directions $\mathbf{n}$ and $\mathbf{m}$. Use your results to find the Lagrangian characteristic speeds.

4.6.3 Find an entropy function for the Mooney–Rivlin model in three dimensions.

4.6.4 Use the Mooney-Rivlin model to compute the derivatives

$$\tilde{\mathbf{H}} = -\frac{\partial \mathbf{S}\mathbf{n}}{\partial \mathbf{J}^{-1}\mathbf{m}}\mathbf{J}^{-1}$$

for arbitrary fixed directions $\mathbf{n}$ and $\mathbf{m}$. Use your results to find the Eulerian characteristic speeds. Relate the Eulerian characteristic speeds to the Lagrangian speeds found in the previous problem

4.6.5 Consider the one-dimensional hypoelastic model

$$\frac{\partial \mathbf{S}}{\partial t} = \frac{\partial \mathbf{v}}{\partial \mathbf{x}}\left(\kappa + \frac{4\mu}{3}\right).$$

(a) Compute $\mathbf{H} = \frac{\partial \mathbf{J}\mathbf{S}_L}{\partial \mathbf{J}}$ and $\tilde{\mathbf{H}} = -\frac{\partial \mathbf{S}}{\partial \mathbf{J}^{-1}}\mathbf{J}^{-1}$ for this model.

(b) Find the Eulerian and Lagrangian characteristic speeds.

(c) Describe the Rankine–Hugoniot jump conditions for this model in both the Eulerian and Lagrangian frames of reference.

(d) Program the Rusanov scheme for the hypoelastic model in the Lagrangian frame.

(e) Program the Rusanov scheme for the hypoelastic model in the Eulerian frame.

4.6.6 Consider the hypoelastic model

$$\dot{\mathbf{S}} \equiv \frac{d\mathbf{S}}{dt} + W^{\top}\mathbf{S} + \mathbf{S}W,$$

where W is the **spin tensor**

$$W = \frac{1}{2}\left[\frac{\partial \mathbf{v}}{\partial \mathbf{x}} - \left(\frac{\partial \mathbf{v}}{\partial \mathbf{x}}\right)^{\top}\right].$$

The derivative $\dot{\mathbf{S}}$ is called the **Jaumann stress rate**.

(a) Since W is antisymmetric, show that we can use it to generate an orthogonal matrix $\Omega(t)$ defined by the initial-value problem

$$\frac{d\Omega}{dt} = W\Omega, \ \Omega^\top \Omega|_{t=0} = \mathbf{I}.$$

(b) Show that the Jaumann stress rate satisfies

$$\dot{\mathbf{S}} = \Omega \frac{d\Omega^\top \mathbf{S}\Omega}{dt} \Omega^\top.$$

This shows that the Jaumann stress is determined by rotating the rate of unrotated stress $\Omega\mathbf{S}\Omega$.

(c) Suppose that the Jaumann stress rate is related to the **rate of deformation**

$$D = \frac{1}{2}\left\{\frac{\partial \mathbf{v}}{\partial \mathbf{x}} + \left(\frac{\partial \mathbf{v}}{\partial \mathbf{x}}\right)^\top\right\}$$

through **Hooke's law** in rate form:

$$\dot{\mathbf{S}} = D2\mu + \mathbf{I}\left(\kappa - \frac{2\mu}{3}\right)\mathrm{tr}D.$$

For arbitrary fixed directions $\mathbf{n}$ and $\mathbf{m}$, compute the partial derivatives

$$\tilde{\mathbf{H}} = -\frac{\partial \mathbf{S}\mathbf{n}}{\partial \mathbf{J}^{-1}\mathbf{m}}\mathbf{J}^{-1}$$

(d) Assuming that $0 < \mu < \kappa$, and that the absolute values of the eigenvalues of $\mathbf{S}$ are less that the shear modulus μ, show that the characteristic speeds are real in the Eulerian frame. (Hint: if $\mathbf{n} = \mathbf{m}$, one of the eigenvectors of $\tilde{\mathbf{H}}$ is $\mathbf{n}$, with eigenvalue $(\kappa + \mu 4/3)/\rho$).

(e) Program the Rusanov scheme for the hypoelastic model in the Eulerian frame.

4.7 Case Study: Linear Elasticity

To illustrate the ideas presented above for general finite deformation in solids, let us consider the simple case of linear elasticity. In this case, the deformation is assumed to be infinitesimal, and there is no distinction between Eulerian and Lagrangian frames of reference. The density is constant, and the material is isothermal. Our system of conservation laws is

$$\frac{\partial \mathbf{u}}{\partial t} + \sum_{k=1}^{3} \frac{\partial \mathbf{F}\mathbf{e}_k}{\partial \mathbf{x}_k} = \mathbf{r}$$

where the vector of conserved quantities **u**, array of fluxes **F** and vector of body forces **r** are

$$\mathbf{u} = \begin{bmatrix} \mathbf{v}\rho \\ \mathbf{J}\mathbf{e}_1 \\ \mathbf{J}\mathbf{e}_2 \\ \mathbf{J}\mathbf{e}_3 \end{bmatrix}, \quad \mathbf{F} = \begin{bmatrix} -\mathbf{S} \\ -\mathbf{v}\mathbf{e}_1^\top \\ -\mathbf{v}\mathbf{e}_2^\top \\ -\mathbf{v}\mathbf{e}_3^\top \end{bmatrix}, \quad \mathbf{r} = \begin{bmatrix} \mathbf{g}\rho \\ 0 \\ 0 \\ 0 \end{bmatrix}.$$

The constitutive law for linear elasticity says that

$$\mathbf{S} = (\mathbf{J} + \mathbf{J}^\top)\mu + \mathbf{I}\left(\kappa - \frac{2}{3}\mu\right) \operatorname{tr}(\mathbf{J}), \tag{4.64}$$

where κ is the **bulk modulus** and μ is the **shear modulus**. Examination of the flux **F** suggests that we choose our flux variables **w** to be

$$\mathbf{w} = \begin{bmatrix} \mathbf{v} \\ \mathbf{S}\mathbf{e}_1 \\ \mathbf{S}\mathbf{e}_2 \\ \mathbf{S}\mathbf{e}_3 \end{bmatrix}.$$

Lemma 4.7.1 *Linear elasticity with stress given by* (4.64) *is hyperbolic. The characteristic speeds are either zero,* $\pm\sqrt{(\kappa + 4\mu/3)/\rho}$ *or* $\pm\sqrt{\mu/\rho}$.

Proof Taking derivatives of (4.64) leads to

$$\frac{\partial \mathbf{S}_{ik}}{\partial \mathbf{J}_{j\ell}} = (\delta_{ij}\delta_{k\ell} + \delta_{jk}\delta_{i\ell})\mu + \delta_{ik}\delta_{j\ell}\left(\kappa - \frac{2}{3}\mu\right),$$

where δ_{ij} is the Kronecker delta; this can be rewritten in matrix form as

$$\frac{\partial \mathbf{S}\mathbf{e}_k}{\partial \mathbf{J}\mathbf{e}_\ell} = \left(\mathbf{I}\delta_{k\ell} + \mathbf{e}_\ell\mathbf{e}_k^\top\right)\mu + \mathbf{e}_k\left(\kappa - \frac{2}{3}\mu\right)\mathbf{e}_\ell^\top.$$

We can also write the constitutive law in the rate form

$$\frac{d\mathbf{S}}{dt} = \left[\frac{\partial \mathbf{v}}{\partial \mathbf{x}} + \left(\frac{\partial \mathbf{v}}{\partial \mathbf{x}}\right)^\top\right]\mu + \mathbf{I}\left(\kappa - \frac{2}{3}\mu\right) \operatorname{tr}\left(\frac{\partial \mathbf{v}}{\partial \mathbf{x}}\right).$$

The quasilinear form of the conservation law is

$$\frac{\partial}{\partial t}\begin{bmatrix} \mathbf{v} \\ \mathbf{S}\mathbf{e}_1 \\ \mathbf{S}\mathbf{e}_2 \\ \mathbf{S}\mathbf{e}_3 \end{bmatrix} + \sum_{j=1}^{3} \begin{bmatrix} 0 & -\mathbf{I}\delta_{1j}/\rho & -\mathbf{I}\delta_{2j}/\rho & -\mathbf{I}\delta_{3j}/\rho \\ -\partial\mathbf{S}\mathbf{e}_1/\partial\mathbf{J}\mathbf{e}_j & 0 & 0 & 0 \\ -\partial\mathbf{S}\mathbf{e}_2/\partial\mathbf{J}\mathbf{e}_j & 0 & 0 & 0 \\ -\partial\mathbf{S}\mathbf{e}_3/\partial\mathbf{J}\mathbf{e}_j & 0 & 0 & 0 \end{bmatrix} = \begin{bmatrix} \mathbf{g} \\ 0 \\ 0 \\ 0 \end{bmatrix}.$$

For a given direction **n**, let us define the linear elasticity acoustic tensor **A** by

$$\mathbf{A} = \sum_{i=1}^{3} \mathbf{n}_i \sum_{j=1}^{3} \frac{\partial \mathbf{Se}_i}{\partial \mathbf{Je}_j} \mathbf{n}_j = \mathbf{I}\mu + \mathbf{n}\left(\kappa + \frac{\mu}{3}\right)\mathbf{n}^\top .$$

If $\mathbf{X} = [\mathbf{n}, \mathbf{N}]$ is an orthogonal matrix, then the eigenvalues and eigenvectors of the acoustic tensor can be given by

$$\mathbf{AX} = \begin{bmatrix} n & \mathbf{N} \end{bmatrix} \begin{bmatrix} \kappa + \frac{4\mu}{3} & 0 \\ 0 & \mu \end{bmatrix} = \mathbf{X}\Lambda^2 \rho .$$

The eigenvalues and eigenvectors associated with the quasilinear form of the conservation law for direction **n** are given in the equation

$$\begin{bmatrix} 0 & -\mathbf{I}\frac{\mathbf{n}_1}{\rho} & \mathbf{I}\frac{\mathbf{n}_2}{\rho} & \mathbf{I}\frac{\mathbf{n}_3}{\rho} \\ -\sum_{j=1}^3 \frac{\partial \mathbf{Se}_1}{\partial \mathbf{Je}_j}\mathbf{n}_j & 0 & 0 & 0 \\ -\sum_{j=1}^3 \frac{\partial \mathbf{Se}_2}{\partial \mathbf{Je}_j}\mathbf{n}_j & 0 & 0 & 0 \\ -\sum_{j=1}^3 \frac{\partial \mathbf{Se}_3}{\partial \mathbf{Je}_j}\mathbf{n}_j & 0 & 0 & 0 \end{bmatrix} \begin{bmatrix} \mathbf{X}\Lambda & 0 & 0 & -\mathbf{X}\Lambda \\ \sum_{j=1}^3 \frac{\partial \mathbf{Se}_1}{\partial \mathbf{Je}_j}\mathbf{n}_j\mathbf{X} & 0 & 0 & \sum_{j=1}^3 \frac{\partial \mathbf{Se}_1}{\partial \mathbf{Je}_j}\mathbf{n}_j\mathbf{X} \\ \sum_{j=1}^3 \frac{\partial \mathbf{Se}_2}{\partial \mathbf{Je}_j}\mathbf{n}_j\mathbf{X} & 0 & 0 & \sum_{j=1}^3 \frac{\partial \mathbf{Se}_2}{\partial \mathbf{Je}_j}\mathbf{n}_j\mathbf{X} \\ \sum_{j=1}^3 \frac{\partial \mathbf{Se}_3}{\partial \mathbf{Je}_j}\mathbf{n}_j\mathbf{X} & 0 & 0 & \sum_{j=1}^3 \frac{\partial \mathbf{Se}_3}{\partial \mathbf{Je}_j}\mathbf{n}_j\mathbf{X} \end{bmatrix}$$

$$= \begin{bmatrix} \mathbf{X}\Lambda & 0 & 0 & -\mathbf{X}\Lambda \\ \sum_{j=1}^3 \frac{\partial \mathbf{Se}_1}{\partial \mathbf{Je}_j}\mathbf{n}_j\mathbf{X} & 0 & 0 & \sum_{j=1}^3 \frac{\partial \mathbf{Se}_1}{\partial \mathbf{Je}_j}\mathbf{n}_j\mathbf{X} \\ \sum_{j=1}^3 \frac{\partial \mathbf{Se}_2}{\partial \mathbf{Je}_j}\mathbf{n}_j\mathbf{X} & 0 & 0 & \sum_{j=1}^3 \frac{\partial \mathbf{Se}_2}{\partial \mathbf{Je}_j}\mathbf{n}_j\mathbf{X} \\ \sum_{j=1}^3 \frac{\partial \mathbf{Se}_3}{\partial \mathbf{Je}_j}\mathbf{n}_j\mathbf{X} & 0 & 0 & \sum_{j=1}^3 \frac{\partial \mathbf{Se}_3}{\partial \mathbf{Je}_j}\mathbf{n}_j\mathbf{X} \end{bmatrix} \begin{bmatrix} -\Lambda & 0 & 0 & 0 \\ 0 & 0 & 0 & 0 \\ 0 & 0 & 0 & 0 \\ 0 & 0 & 0 & \Lambda \end{bmatrix} .$$

□

The p-waves ($\lambda = \sqrt{(\kappa + \frac{4\mu}{3})\rho}$) travel fastest, and in them the velocity is normal to the propagating front; the s-waves ($\lambda = \sqrt{\mu/\rho}$) travel slower, with the velocity transverse to the propagating front.

4.8 Case Study: Vibrating String

In this section, we will consider the motion of a vibrating string that is constrained to move in a plane. This problem will be a simple example of the more general solid mechanics problem discussed in the previous section. We will see that the Riemann problem can be solved; its solution can involve states where two characteristic speeds are equal with no loss of characteristic directions.

4.8.1 Conservation Laws

Since a string is one-dimensional, we can consider its configuration at rest to be parameterized by a single Lagrangian coordinate a_1. Since the motion of the string is assumed to lie in a plane in the Eulerian frame of reference, the deformation gradient is

$$\mathbf{J} = \begin{bmatrix} \partial \mathbf{x}_1/\partial a_1 & 0 & 0 \\ \partial \mathbf{x}_2/\partial a_1 & 1 & 0 \\ 0 & 0 & 1 \end{bmatrix}.$$

Since the Green deformation tensor is

$$C = \mathbf{J}^\top \mathbf{J} = \begin{bmatrix} (\partial \mathbf{x}_1/\partial a_1)^2 + (\partial \mathbf{x}_2/\partial a_1)^2 & \partial \mathbf{x}_2/\partial a_1 & 0 \\ \partial \mathbf{x}_2/\partial a_1 & 1 & 0 \\ 0 & 0 & 1 \end{bmatrix},$$

its trace is

$$\operatorname{tr} C = \left(\frac{\partial \mathbf{x}_1}{\partial a_1}\right)^2 + \left(\frac{\partial \mathbf{x}_2}{\partial a_1}\right)^2 + 2 = \|\mathbf{J}\mathbf{e}_1\|^2 + 2.$$

Following Cristescu [39], we note that the second Piola–Kirchhoff stress $\mathbf{S}_{_L}$ is an isotropic function of $\|\mathbf{J}\mathbf{e}_1\|$, meaning that it has the functional form

$$\mathbf{S}_{_L} = \mathbf{I}\frac{\tau(\|\mathbf{J}\mathbf{e}_1\|)}{\|\mathbf{J}\mathbf{e}_1\|}.$$

Here τ is the tension. We assume that the tension is positive for all positive strain (*i.e.*, $\|\mathbf{J}\mathbf{e}_1\| > 1$ implies $\tau > 0$) and that zero strain implies zero tension ($\tau(1) = 0$). In order to guarantee real characteristic speeds, we also assume that the tension increases with strain.

If we write $\mathbf{f} \equiv \mathbf{J}\mathbf{e}_1$, then the Lagrangian equations of motion can be written

$$\rho\frac{\partial \mathbf{v}}{\partial t} - \frac{\partial}{\partial a_1}\left(\mathbf{f}\frac{\tau(\|\mathbf{f}\|)}{\|\mathbf{f}\|}\right) = 0,$$
$$\frac{\partial \mathbf{f}}{\partial t} - \frac{\partial \mathbf{v}}{\partial a_1} = 0.$$

Note that $\mathbf{f}$ is determined by its norm $\phi = \|\mathbf{f}\|$ and an angle θ:

$$\mathbf{f} = \begin{bmatrix} \cos\theta \\ \sin\theta \\ 0 \end{bmatrix}\phi.$$

This suggests that we choose our **flux variables** to be

$$\mathbf{w}^\top \equiv [\mathbf{v}_1, \quad \mathbf{v}_2, \quad \phi, \quad \theta].$$

In order to simplify the discussion, we will ignore the zero third components of $\mathbf{v}$ and $\mathbf{f}$. We will also write

$$\mathbf{n}(\theta) = \begin{bmatrix} \cos\theta \\ \sin\theta \end{bmatrix}.$$

Then the conservation laws can be rewritten in the form

$$\frac{\partial}{\partial t}\begin{bmatrix}\mathbf{v}\rho\\ \mathbf{n}(\theta)\phi\end{bmatrix}+\frac{\partial}{\partial a_1}\begin{bmatrix}-\mathbf{n}(\theta)\tau(\phi)\\ -\mathbf{v}\end{bmatrix}=0. \tag{4.65}$$

4.8.2 Characteristic Analysis

Let us examine the characteristic speeds for the vibrating string.

Lemma 4.8.1 *If $\tau(\phi) > 0$ and $\tau' > 0$ for all $\phi > 1$, then the system of conservation laws for the vibrating string (4.65) is hyperbolic, with characteristic speeds $\pm\sqrt{\tau'(\phi)/\rho}$ and $\pm\lambda_2 = \sqrt{\tau(\phi)/\rho\phi}$. The first two characteristic speeds are genuinely nonlinear and the second two are linearly degenerate.*

Proof We can easily compute the **quasilinear form** to be

$$\begin{bmatrix}\mathbf{I}\rho & 0 & 0\\ 0 & \mathbf{n} & \mathbf{n}'\phi\end{bmatrix}\frac{\partial}{\partial t}\begin{bmatrix}\mathbf{v}\\ \phi\\ \theta\end{bmatrix}+\begin{bmatrix}0 & -\mathbf{n}\tau' & -\mathbf{n}'\tau\\ -\mathbf{I} & 0 & 0\end{bmatrix}\frac{\partial}{\partial a_1}\begin{bmatrix}\mathbf{v}\\ \phi\\ \theta\end{bmatrix}=0.$$

If $\phi > 0$ then we can solve to get

$$\frac{\partial}{\partial t}\begin{bmatrix}\mathbf{v}\\ \phi\\ \theta\end{bmatrix}+\begin{bmatrix}0 & -\mathbf{n}\frac{\tau'}{\rho} & -\mathbf{n}'\frac{\tau}{\rho}\\ -\mathbf{n}^\top & 0 & 0\\ -\frac{1}{\phi}(\mathbf{n}')^\top & 0 & 0\end{bmatrix}\frac{\partial}{\partial a_1}\begin{bmatrix}\mathbf{v}\\ \phi\\ \theta\end{bmatrix}=0.$$

We see that the eigenvalues and eigenvectors of this matrix take the form

$$\begin{aligned}\mathbf{BY} &= \begin{bmatrix}0 & -\mathbf{n}\frac{\tau'}{\rho} & -\mathbf{n}'\frac{\tau}{\rho}\\ -\mathbf{n}^\top & 0 & 0\\ -\frac{1}{\phi}(\mathbf{n}')^\top & 0 & 0\end{bmatrix}\begin{bmatrix}-\mathbf{n}\lambda_1 & -\mathbf{n}'\phi\lambda_2 & \mathbf{n}'\phi\lambda_2 & \mathbf{n}\lambda_1\\ 1 & 0 & 0 & 1\\ 0 & 1 & 1 & 0\end{bmatrix}\\ &= \begin{bmatrix}-\mathbf{n}\lambda_1 & -\mathbf{n}'\phi\lambda_2 & \mathbf{n}'\phi\lambda_2 & \mathbf{n}\lambda_1\\ 1 & 0 & 0 & 1\\ 0 & 1 & 1 & 0\end{bmatrix}\begin{bmatrix}\lambda_1 & & & \\ & \lambda_2 & & \\ & & -\lambda_2 & \\ & & & -\lambda_1\end{bmatrix} \equiv \mathbf{Y}\Lambda\end{aligned}$$

where the (positive) characteristic speeds are

$$\lambda_1 = \sqrt{\frac{\tau'(\phi)}{\rho}}, \quad \lambda_2 = \sqrt{\frac{\tau(\phi)}{\rho\phi}}.$$

Since the tension τ is positive and increasing, the characteristic speeds are real and nonzero.

Let us examine the characteristic speeds to see if they are genuinely nonlinear or linearly degenerate. Note that both characteristic speeds are functions of ϕ alone. Thus for any of the characteristic speeds,

$$\frac{\partial \lambda}{\partial \mathbf{w}} = \begin{bmatrix} 0 & 0 & \frac{\partial \lambda}{\partial \phi} & 0 \end{bmatrix}.$$

Since this vector is orthogonal to the characteristic direction for $\pm\lambda_2$, this characteristic speed is linearly degenerate. The other characteristic speed is genuinely nonlinear, provided that

$$\frac{\partial \lambda}{\partial \phi} = \tau'' \neq 0.$$

□

Although the characteristic directions are always distinct, the characteristic speeds are not. Consider a nonlinear tension τ with $\tau(1) = 0$ and $\tau(\phi)$ concave for $\phi > 1$. Then $\lambda^2 \rho$ is either equal to the slope of the tension curve (if $\lambda = \lambda_1$), or the slope of the line from the origin to a point on the tension curve (if $\lambda = \lambda_2$). It is possible for the latter line to be tangent to the tension curve at some point; at this point, the two positive characteristic speeds are equal.

According to Cristescu, the linearly degenerate characteristic speeds are related to the propagation of transverse waves, or changes in the shape of the string without changes in tension. The genuinely nonlinear characteristic speeds are associated with longitudinal or tension waves, in which there is no change in the shape of the string.

4.8.3 Jump Conditions

Lemma 4.8.2 *In the vibrating string model* (4.65)*, suppose that* $\tau(\phi) > 0$ *and* $\tau' > 0$ *for all* $\phi > 1$*. Then there are three kinds of solutions to the Rankine-Hugoniot jump conditions. One is a contact discontinuity, in which*

$$\frac{\tau(\phi_R)}{\phi_R} = \frac{\tau(\phi_L)}{\phi_L}, \quad \sigma = \pm\sqrt{\frac{\tau(\phi_R)}{\rho\phi_R}}, \quad \mathbf{v}_R - \mathbf{v}_L = -\{\mathbf{n}(\theta_R)\phi_R - \mathbf{n}(\theta_L)\phi_L\}\sigma.$$

The second kind of solution has

$$\theta_L = \theta_R, \quad \sigma = \pm\sqrt{\frac{1}{\rho}\frac{\tau(\phi_R) - \tau(\phi_L)}{\phi_R - \phi_L}}, \quad \mathbf{v}_R - \mathbf{v}_L = -\mathbf{n}(\theta_R)(\phi_R - \phi_L)\sigma.$$

The third kind of solution, called an **anomalous** *discontinuity, satisfies*

$$\theta_R = \theta_L \pm \pi, \quad \sigma = \pm\sqrt{\frac{1}{\rho}\frac{\tau(\phi_R) + \tau(\phi_L)}{\phi_R + \phi_L}}, \quad \mathbf{v}_R - \mathbf{v}_L = \mathbf{n}(\theta_R)(\phi_R + \phi_L)\sigma.$$

Proof The Rankine–Hugoniot jump conditions require that a propagating discontinuity satisfy

$$\begin{bmatrix} -\mathbf{n}\tau \\ -\mathbf{v} \end{bmatrix} = \begin{bmatrix} \mathbf{v}\rho \\ \mathbf{n}\phi \end{bmatrix} \sigma.$$

This implies that the jumps satisfy

$$[\mathbf{n}\tau] = -[\mathbf{v}]\rho\sigma = [\mathbf{n}\phi]\rho\sigma^2.$$

We can rewrite the outer equation in the forms

$$(\tau_R - \phi_R\rho\sigma^2)\cos\theta_R = (\tau_L - \phi_L\rho\sigma^2)\cos\theta_L$$
$$(\tau_R - \phi_R\rho\sigma^2)\sin\theta_R = (\tau_L - \phi_L\rho\sigma^2)\sin\theta_L.$$

Thus either $\tau_R - \phi_R\rho\sigma^2 = 0 = \tau_L - \phi_L\rho\sigma^2$ or $\tan\theta_L = \tan\theta_R$. The former of these two cases corresponds to a contact discontinuity. In this case, we have

$$\frac{\tau(\phi_R)}{\phi_R} = \frac{\tau(\phi_L)}{\phi_L}, \quad \sigma = \pm\sqrt{\frac{\tau(\phi_R)}{\rho\phi_R}}, \quad \mathbf{v}_R - \mathbf{v}_L = -\{\mathbf{n}(\theta_R)\phi_R - \mathbf{n}(\theta_L)\phi_L\}\sigma.$$

In these equations, the jump in θ is arbitrary, and any jump in ϕ must be along the line from $(\phi = 1, \tau = 0)$ to points $(\phi, \tau(\phi))$ on the tension curve. Since $\tau(1) = 0$ and τ is increasing, these discontinuity speeds are real for $\phi \geq 1$. If we do not have a contact discontinuity, then we have $\tan\theta_L = \tan\theta_R$. An elastic discontinuity occurs if $\theta_R = \theta_L$ and $\phi_R \neq \phi_L$. In this case, we have

$$\sigma = \pm\sqrt{\frac{1}{\rho}\frac{\tau(\phi_R) - \tau(\phi_L)}{\phi_R - \phi_L}}$$
$$\mathbf{v}_R - \mathbf{v}_L = -\mathbf{n}(\theta_R)(\phi_R - \phi_L)\sigma.$$

Since τ is increasing, these discontinuity speeds are real. Otherwise, if $\theta_R = \theta_L \pm \pi$ then we have an **anomalous** discontinuity

$$\sigma = \pm\sqrt{\frac{1}{\rho}\frac{\tau(\phi_R) + \tau(\phi_L)}{\phi_R + \phi_L}}$$
$$\mathbf{v}_R - \mathbf{v}_L = \mathbf{n}(\theta_L)(\phi_R + \phi_L)\sigma.$$

Since τ and ϕ are nonnegative, these discontinuity speeds are real. □

The third kind of discontinuity is called "anomalous" because the left and right values of θ correspond to the string bending back on itself.

4.8.4 Lax Admissibility Conditions

In order for an elastic discontinuity to be a shock, we require it to satisfy the Lax admissibility conditions. For the vibrating string, these conditions require that the elastic characteristic speed on the left be greater than the shock speed, which must in turn be greater than the characteristic speed on the right. For a shock with positive speed, we require

$$\tau'(\phi_L) > \frac{\tau(\phi_R) - \tau(\phi_L)}{\phi_R - \phi_L} > \tau'(\phi_R)$$

which in turn implies that $\phi_L < \phi_R$ if τ is concave (and the reverse inequality if τ is convex). For an elastic shock with negative speed, we require

$$\tau'(\phi_L) < \frac{\tau(\phi_R) - \tau(\phi_L)}{\phi_R - \phi_L} < \tau'(\phi_R)$$

which in turn implies that $\phi_L > \phi_R$ if τ is concave (and the reverse inequality if τ is convex).

We do not consider the Lax admissibility conditions for a contact discontinuity. Also, we ignore the Lax admissibility conditions for the anomalous discontinuity, because we will not use it in our construction of the solution of the Riemann problem.

4.8.5 Entropy Function

The sum of the kinetic and strain energy $E \equiv \frac{1}{2}\rho \mathbf{v}^\top \mathbf{v} + \int^\phi \tau(\eta)\, d\eta$ is an entropy function for the vibrating string, with energy flux $\Psi \equiv -\tau(\phi)\mathbf{v}^\top \mathbf{n}(\theta)$. To show that this is so, we compute

$$\frac{\partial E}{\partial \mathbf{w}} = \begin{bmatrix} \rho \mathbf{v}^\top & \tau & 0 \end{bmatrix}$$

and

$$\frac{\partial \Psi}{\partial \mathbf{w}} = \begin{bmatrix} -\tau \mathbf{n}^\top & -\tau' \mathbf{v}^\top \mathbf{n} & -\tau \mathbf{v}^\top \mathbf{n}' \end{bmatrix}.$$

Then

$$\frac{\partial E}{\partial \mathbf{w}} \left(\frac{\partial \mathbf{u}}{\partial \mathbf{w}} \right)^{-1} \frac{\partial \mathbf{F}}{\partial \mathbf{w}} = \begin{bmatrix} \rho \mathbf{v}^\top & \tau & 0 \end{bmatrix} \begin{bmatrix} 0 & -\mathbf{n}\frac{\tau'}{\rho} & -\mathbf{n}'\frac{\tau}{\rho} \\ -\mathbf{n}^\top & 0 & 0 \\ -\frac{1}{\phi}(\mathbf{n}')^\top & 0 & 0 \end{bmatrix}$$
$$= \begin{bmatrix} -\tau \mathbf{n}^\top & -\tau' \mathbf{v}^\top \mathbf{n} & -\tau \mathbf{v}^\top \mathbf{n}' \end{bmatrix} = \frac{\partial \Psi}{\partial \mathbf{w}},$$

which verifies the equation (4.29) for an entropy function and flux.

4.8.6 Wave Families for Concave Tension

The case of concave tension is more interesting than convex tension. In this case, there exists an equivelocity value of ϕ:

$$\tau'(\phi_*) = \frac{\tau(\phi_*)}{\phi_*}.$$

Further, for any $\phi \in (1, \phi_*)$ there exist a reciprocal $\overline{\phi} \in (\phi_*, \infty)$ so that

$$\frac{\tau(\overline{\phi})}{\overline{\phi}} = \frac{\tau(\phi)}{\phi};$$

in other words, both values of ϕ correspond to the same contact discontinuity speed. Similarly, for any $\phi \in (\phi_*, \infty)$ there exist a reciprocal $\overline{\phi} \in (1, \phi_*)$ so that the same equation holds. In order to guarantee that the Riemann problem has a solution we assume that

$$\int^{\phi} \sqrt{\tau'(\eta)}\, d\eta \to \infty \text{ as } \phi \to \infty. \tag{4.66}$$

Given any left state $(\mathbf{v}_L, \phi_L, \theta_L)$ and any intermediate state values ϕ_0, θ_0, we will consider two cases. If $\phi_* < \phi_L$, then the negative contact discontinuity speed is less than the negative elastic speed at the left state. In this case, we have three possibilities:

SC: If $\phi_0 < \overline{\phi}_L$, then we have a shock moving left faster than a contact discontinuity, and the jump conditions imply that

$$\begin{aligned}\mathbf{v}_-(\phi_0, \theta_0) = \mathbf{v}_L &- \mathbf{n}(\theta_L)\sqrt{\frac{1}{\rho}\{\tau(\phi_L) - \tau(\phi_0)\}(\phi_L - \phi_0)}\\ &+ \{\mathbf{n}(\theta_0) - \mathbf{n}(\theta_L)\}\sqrt{\frac{1}{\rho}\tau(\phi_0)\phi_0}.\end{aligned}$$

CS: Else if $\overline{\phi}_L \le \phi_0 < \phi_L$, then we have a contact discontinuity moving left no slower than a shock, and the jump conditions imply that

$$\begin{aligned}\mathbf{v}_-(\phi_0, \theta_0) = \mathbf{v}_L &+ \{\mathbf{n}(\theta_0) - \mathbf{n}(\theta_L)\}\sqrt{\frac{1}{\rho}\tau(\phi_0)\phi_0}\\ &- \mathbf{n}(\theta_0)\sqrt{\frac{1}{\rho}\{\tau(\phi_L) - \tau(\phi_0)\}(\phi_L - \phi_0)}.\end{aligned}$$

CR: Else $\phi_L \le \phi_0$, so we have a contact discontinuity moving left no slower than a rarefaction, and

$$\mathbf{v}_-(\phi_0, \theta_0) = \mathbf{v}_L + \{\mathbf{n}(\theta_0) - \mathbf{n}(\theta_L)\}\sqrt{\frac{1}{\rho}\tau(\phi_0)\phi_0} + \mathbf{n}(\theta_0)\int_{\phi_L}^{\phi_0}\sqrt{\frac{1}{\rho}\tau'(\eta)}\,d\eta.$$

On the other hand, if $\phi_* \ge \phi_L > 1$ then the negative contact discontinuity is no less than the negative elastic speed at the left state. We again have three possibilities:

SC: If $1 < \phi_0 < \phi_L$, then we have a shock moving left faster than a contact discontinuity, and the jump conditions imply that

$$\mathbf{v}_-(\phi_0, \theta_0) = \mathbf{v}_L - \mathbf{n}(\theta_L)\sqrt{\frac{1}{\rho}\{\tau(\phi_L) - \tau(\phi_0)\}(\phi_L - \phi_0)} + \{\mathbf{n}(\theta_0) - \mathbf{n}(\theta_L)\}\sqrt{\frac{1}{\rho}\tau(\phi_0)\phi_0}.$$

RC: Else if $\phi_L \le \phi_0 < \phi_*$, then we have a rarefaction moving left no slower than a contact discontinuity, and the jump conditions imply that

$$\mathbf{v}_-(\phi_0, \theta_0) = \mathbf{v}_L + \mathbf{n}(\theta_L)\int_{\phi_L}^{\phi_0}\sqrt{\frac{1}{\rho}\tau'(\eta)}\,d\eta + \{\mathbf{n}(\theta_0) - \mathbf{n}(\theta_L)\}\sqrt{\frac{1}{\rho}\tau(\phi_0)\phi_0}.$$

RCR: Else $\phi_* \le \phi$, so we have a contact discontinuity moving left in the middle of rarefaction, and

$$\mathbf{v}_-(\phi_0, \theta_0) = \mathbf{v}_L + \mathbf{n}(\theta_L)\int_{\phi_L}^{\phi_*}\sqrt{\frac{1}{\rho}\tau'(\eta)}\,d\eta + \{\mathbf{n}(\theta_0) - \mathbf{n}(\theta_L)\}\sqrt{\frac{1}{\rho}\tau(\phi_*)\phi_*} + \mathbf{n}(\theta_0)\int_{\phi_*}^{\phi_0}\sqrt{\frac{1}{\rho}\tau'(\eta)}\,d\eta.$$

Similarly, given any right state $(\mathbf{v}_R, \phi_R, \theta_R)$ and any intermediate state values ϕ_0, θ_0, we will consider two cases. If $\phi_* < \phi_R$, then the positive contact discontinuity speed is greater than the positive elastic speed at the right state. In this case, we have three possibilities:

CS: If $\phi_0 < \overline{\phi_R}$, then we have a shock moving right faster than a contact discontinuity, and the jump conditions imply that

$$\mathbf{v}_+(\phi_0, \theta_0) = \mathbf{v}_R + \mathbf{n}(\theta_R)\sqrt{\frac{1}{\rho}\{\tau(\phi_R) - \tau(\phi_0)\}(\phi_R - \phi_0)} \\ + \{\mathbf{n}(\theta_R) - \mathbf{n}(\theta_0)\}\sqrt{\frac{1}{\rho}\tau(\phi_0)\phi_0}.$$

SC: Else if $\overline{\phi_R} \leq \phi_0 < \phi_R$, then we have a contact discontinuity moving right no slower than a shock, and the jump conditions imply that

$$\mathbf{v}_+(\phi_0, \theta_0) = \mathbf{v}_R + \{\mathbf{n}(\theta_R) - \mathbf{n}(\theta_0)\}\sqrt{\frac{1}{\rho}\tau(\phi_R)\phi_R} \\ + \mathbf{n}(\theta_0)\sqrt{\frac{1}{\rho}\{\tau(\phi_R) - \tau(\phi_0)\}(\phi_R - \phi_0)}.$$

RC: Else $\phi_R \leq \phi_0$, so we have a contact discontinuity moving left no slower than a rarefaction, and

$$\mathbf{v}_+(\phi_0, \theta_0) = \mathbf{v}_R + \{\mathbf{n}(\theta_R) - \mathbf{n}(\theta_0)\}\sqrt{\frac{1}{\rho}\tau(\phi_R)\phi_R} + \mathbf{n}(\theta_0)\int_{\phi_R}^{\phi_0}\sqrt{\frac{1}{\rho}\tau'(\eta)}\,d\eta.$$

On the other hand, if $\phi_* \geq \phi_R > 1$ then the positive contact discontinuity is no greater than the positive elastic speed at the left state. We again have three possibilities:

CS: If $1 < \phi_0 < \phi_R$, then we have a shock moving right faster than a contact discontinuity, and the jump conditions imply that

$$\mathbf{v}_+(\phi_0, \theta_0) = \mathbf{v}_R + \mathbf{n}(\theta_R)\sqrt{\frac{1}{\rho}\{\tau(\phi_R) - \tau(\phi_0)\}(\phi_R - \phi_0)} \\ + \{\mathbf{n}(\theta_R) - \mathbf{n}(\theta_0)\}\sqrt{\frac{1}{\rho}\tau(\phi_0)\phi_0}.$$

CR: Else if $\phi_R \leq \phi_0 < \phi_*$, then we have a rarefaction moving right no slower than a contact discontinuity, and

$$\mathbf{v}_+(\phi_0, \theta_0) = \mathbf{v}_R - \mathbf{n}(\theta_R)\int_{\phi_R}^{\phi_0}\sqrt{\frac{1}{\rho}\tau'(\eta)}\,d\eta + \{\mathbf{n}(\theta_R) - \mathbf{n}(\theta_0)\}\sqrt{\frac{1}{\rho}\tau(\phi_0)\phi_0}.$$

RCR: Else $\phi_* \le \phi_0$, so we have a contact discontinuity moving right in the middle of rarefaction, and

$$\mathbf{v}_+(\phi_0, \theta_0) = \mathbf{v}_R - \mathbf{n}(\theta_R) \int_{\phi_R}^{\phi_*} \sqrt{\frac{1}{\rho}\tau'(\eta)}\, d\eta + \{\mathbf{n}(\theta_R) - \mathbf{n}(\theta_0)\}\sqrt{\frac{1}{\rho}\tau(\phi_*)\phi_*}$$
$$- \mathbf{n}(\theta_0) \int_{\phi_*}^{\phi_0} \sqrt{\frac{1}{\rho}\tau'(\eta)}\, d\eta.$$

These exhaust all possible cases for values of ϕ and θ.

It is important to note that these wave families produce continuous functions $\mathbf{v}_-(\phi_0, \theta_0)$ and $\mathbf{v}_+(\phi_0, \theta_0)$. In fact, we found that

$$\mathbf{v}_-(\phi_0, \theta_0) = \mathbf{v}_L + \mathbf{n}(\theta_L)\nu_L(\phi_0) + \mathbf{n}(\theta_0)\nu_{L,0}(\phi_0) \tag{4.67a}$$
$$\mathbf{v}_+(\phi_0, \theta_0) = \mathbf{v}_R + \mathbf{n}(\theta_R)\nu_R(\phi_0) + \mathbf{n}(\theta_0)\nu_{R,0}(\phi_0) \tag{4.67b}$$

where

$$\nu_L(\phi_0) = \begin{cases} -\sigma(\phi_L - \phi_0) - \lambda_2(\phi_0)\phi_0, & \text{SC} \\ -\lambda_2(\phi_L)\phi_L, & \text{CS or CR} \\ \int_{\phi_L}^{\phi_0} \lambda_1(\eta)\, d\eta - \lambda_2(\phi_0)\phi_0, & \text{RC} \\ \int_{\phi_L}^{\phi_*} \lambda_1(\eta)\, d\eta - \lambda_2(\phi_*)\phi_*, & \text{RCR} \end{cases}$$

$$\nu_{L,0}(\phi_0) = \begin{cases} \lambda_2(\phi_0)\phi_0, & \text{SC or RC} \\ \lambda_2(\phi_L)\phi_L - \sigma(\phi_L - \phi_0), & \text{CS} \\ \lambda_2(\phi_L)\phi_L + \int_{\phi_L}^{\phi_0} \lambda_1(\eta)\, d\eta, & \text{CR} \\ \lambda_2(\phi_*)\phi_* + \int_{\phi_*}^{\phi_0} \lambda_1(\eta)\, d\eta, & \text{RCR} \end{cases}$$

$$\nu_R(\phi_0) = \begin{cases} \lambda_2(\phi_R)\phi_R, & \text{SC or RC} \\ \sigma(\phi_R - \phi_0) + \lambda_2(\phi_0)\phi_0, & \text{CS} \\ -\int_{\phi_R}^{\phi_0} \lambda_1(\eta)\, d\eta + \lambda_2(\phi_0)\phi_0, & \text{CR} \\ -\int_{\phi_R}^{\phi_*} \lambda_1(\eta)\, d\eta + \lambda_2(\phi_*)\phi_*, & \text{RCR} \end{cases}$$

and

$$\nu_{R,0}(\phi_0) = \begin{cases} -\lambda_2(\phi_R)\phi_R + \sigma(\phi_R - \phi_0), & \text{SC} \\ -\lambda_2(\phi_0)\phi_0, & \text{CS or CR} \\ -\lambda_2(\phi_R)\phi_R - \int_{\phi_R}^{\phi_0} \lambda_1(\eta)\, d\eta, & \text{RC} \\ -\lambda_2(\phi_*)\phi_* - \int_{\phi_*}^{\phi_0} \lambda_1(\eta)\, d\eta, & \text{RCR.} \end{cases}$$

Here λ_1 and λ_2 are the characteristic speeds we found above, and σ is the appropriate elastic shock speed. These equations have been written in a form that is independent of the convexity or concavity of τ, provided that τ is either always convex or always concave.

We can check the continuity of ν_L and $\nu_{L,0}$ as follows. We have

$$\nu_L(\phi_0) = -\lambda_2(\phi_L)\phi_L$$

for either a shock-contact transitioning to a rarefaction-contact (at $\phi_0 = \phi_L$), or a shock-contact transitioning to a contact-shock (at $\phi_0 = \overline{\phi}_L$), or a contact-shock transitioning to a contact-rarefaction (at $\phi_0 = \phi_L$). We also have

$$\nu_L(\phi_*) = -\lambda_2(\phi_*)\phi_* + \int_{\phi_L}^{\phi_*} \lambda_1(\eta)\, d\eta$$

for a rarefaction-contact transitioning to a rarefaction-contact-rarefaction (at $\phi_0 = \phi_*$). Note that

$$\begin{aligned} \nu_{L,0}(\phi_L) &= \lambda_2(\phi_L)\phi_L, && \text{SC to RC or SC to CR} \\ \nu_{L,0}(\phi_*) &= \lambda_2(\phi_*)\phi_*, && \text{RC to RCR} \\ \nu_{L,0}(\overline{\phi}_L) &= \lambda_2(\overline{\phi}_L)\overline{\phi}_L, && \text{SC to CS.} \end{aligned}$$

Similarly, we check the continuity of ν_R and $\nu_{R,0}$. We have

$$\nu_R(\phi_0) = \lambda_2(\phi_R)\phi_R$$

for either a contact-shock transitioning to a contact-rarefaction (at $\phi_0 = \phi_R$), or a contact-shock transitioning to a shock-contact (at $\phi_0 = \overline{\phi}_R$), or a shock-contact transitioning to a rarefaction-contact (at $\phi_0 = \phi_R$). We also have

$$\nu_R(\phi_*) = \lambda_2(\phi_*)\phi_* - \int_{\phi_R}^{\phi_*} \lambda_1(\eta)\, d\eta$$

for a contact-rarefaction transitioning to a rarefaction-contact-rarefaction (at $\phi_0 = \phi_*$). Note that

$$\begin{aligned} \nu_{R,0}(\phi_R) &= -\lambda_2(\phi_R)\phi_R, && \text{CS to CR or CS to RC} \\ \nu_{R,0}(\phi_*) &= -\lambda_2(\phi_*)\phi_*, && \text{CR to RCR} \\ \nu_{R,0}(\overline{\phi}_R) &= -\lambda_2(\overline{\phi}_R)\overline{\phi}_R, && \text{CS to SC} \end{aligned}$$

4.8.7 Wave Family Intersections

The discussion in this section has been adapted from work by Keyfitz and Kranzer [81], who (unfortunately) considered convex tension with $\tau(0) = 0$.

In order to solve the Riemann problem for the vibrating string, we need to show that for any values of $\mathbf{v}_L$ and $\mathbf{v}_R$ we can find values for ϕ_0 and θ_0 so that the negative and positive wave families intersect:

$$\mathbf{v}_-(\phi_0, \theta_0) = \mathbf{v}_+(\phi_0, \theta_0).$$

Using equations (4.67) we can rewrite this intersection condition in the form

$$\mathbf{v}_R - \mathbf{v}_L + \mathbf{n}(\theta_R)\nu_R(\phi_0) - \mathbf{n}(\theta_L)\nu_L(\phi_0) = \mathbf{n}(\theta_0)\{\nu_{L,0}(\phi_0) - \nu_{R,0}(\phi_0)\}$$

We can eliminate $\mathbf{n}(\theta_0)$ by taking the Euclidean norms of both sides of this equation. This leads to the nonlinear scalar equation

$$0 = g(\phi_0) \equiv -\|\mathbf{v}_R - \mathbf{v}_L + \mathbf{n}(\theta_R)\nu_R(\phi_0) - \mathbf{n}(\theta_L)\nu_L(\phi_0)\|^2 + \{\nu_{L,0}(\phi_0) - \nu_{R,0}(\phi_0)\}^2. \quad (4.68)$$

From the discussion in section 4.8.6 we know that $g(\phi_0)$ is continuous. In order to show that this function always has a unique zero, we will show that $g(1) < 0$, that $g(\phi_0) > 0$ for sufficiently large ϕ_0, and that $g'(\phi_0) > 0$ whenever $g(\phi_0) = 0$.

For concave tension τ, it is easy to compute

$$\nu_{L,0}(1) = 0 = \nu_{R,0}(1)$$

It follows that

$$g(1) = -\|\mathbf{v}_R - \mathbf{v}_L + \mathbf{n}(\theta_R)\nu_R(1) - \mathbf{n}(\theta_L)\nu_L(1)\|^2.$$

This is negative, except for one special circumstance in which it is zero. The special circumstance occurs precisely when

$$\mathbf{v}_R = \mathbf{v}_L - \mathbf{n}(\theta_R)\sqrt{\frac{1}{\rho}\tau(\phi_R)(\phi_R - 1)} + \mathbf{n}(\theta_L)\sqrt{\frac{1}{\rho}\tau(\phi_L)(\phi_L - 1)}.$$

For large values of ϕ_0 we have

$$\begin{aligned}
\nu_L(\phi_0) &= -\lambda_2(\max\{\phi_L, \phi_*\}) \\
\nu_{L,0}(\phi_0) &= \lambda_2(\max\{\phi_L, \phi_*\}) + \int_{\max\{\phi_L,\phi_*\}}^{\phi_0} \lambda_1(\eta)\, d\eta \\
\nu_R(\phi_0) &= \lambda_2(\max\{\phi_R, \phi_*\}) \\
\nu_{R,0}(\phi_0) &= -\lambda_2(\max\{\phi_R, \phi_*\}) - \int_{\max\{\phi_R,\phi_*\}}^{\phi_0} \lambda_1(\eta)\, d\eta.
\end{aligned}$$

Note that assumption (4.66) guarantees that both $\nu_{L,0}(\phi_0)$ and $\nu_{R,0}(\phi_0)$ become arbitrarily large as ϕ_0 approaches infinity. This shows that $g(\phi_0) > 0$ for sufficiently large ϕ_0. Since $g(\phi_0)$ is continuous and $g(1) \leq 0$, $g(\phi_0)$ must have at least one zero for $\phi_0 \geq 1$.

Next, we would like to show that g has a positive slope whenever it is zero. To show this, we will first prove that

$$\nu_{L,0}(\phi_0) \geq 0 \tag{4.69a}$$
$$\nu_{R,0}(\phi_0) \leq 0 \tag{4.69b}$$
$$0 \geq \nu_L'(\phi_0) \geq -\nu_{L,0}'(\phi) \tag{4.69c}$$
$$0 \leq \nu_R'(\phi_0) \leq -\nu_{R,0}'(\phi) \tag{4.69d}$$

for all $\phi_0 \geq 1$. This is simply a matter of checking the cases. First, we examine the cases to prove inequality (4.69a):

$$0 \leq \nu_{L,0}(\phi_0) = \begin{cases} \lambda_2(\phi_0)\phi_0, & \text{SC or RC} \\ (\lambda_2(\phi_L) - \sigma)\phi_L + \sigma\phi_0, & \text{CS} \\ \lambda_2(\phi_L)\phi_L + \int_{\phi_L}^{\phi_0} \lambda_1(\eta)\, d\eta, & \text{CR} \\ \lambda_2(\phi_*)\phi_* + \int_{\phi_*}^{\phi_0} \lambda_1(\eta)\, d\eta, & \text{RCR.} \end{cases}$$

Next, we examine the cases to prove inequality (4.69b):

$$0 \geq \nu_{R,0}(\phi_0) = \begin{cases} -(\lambda_2(\phi_R) - \sigma)\phi_R - \sigma\phi_0, & \text{SC} \\ -\lambda_2(\phi_0)\phi_0, & \text{SC or CS or CR} \\ -\lambda_2(\phi_R)\phi_R - \int_{\phi_R}^{\phi_0} \lambda_1(\eta)\, d\eta, & \text{RC} \\ -\lambda_2(\phi_*)\phi_* - \int_{\phi_*}^{\phi_0} \lambda_1(\eta)\, d\eta, & \text{RCR.} \end{cases}$$

In order to prove inequalities (4.69c) we will need to provide more detail.

SC: First, consider the case of a shock moving left faster than a contact discontinuity. In this case $\lambda_1(\phi_0) > \sigma > \lambda_1(\phi_L)$ and $\sigma \geq \lambda_2(\phi_0)$. Thus

$$\nu_L'(\phi_0) = -\frac{\{\sigma - \lambda_2(\phi_0)\}\{\lambda_1^2(\phi_0) - \sigma\lambda_2(\phi_0)\}}{2\sigma\lambda_2(\phi_0)} < 0$$
$$\nu_{L,0}'(\phi_0) = \frac{\lambda_1^2(\phi_0) + \lambda_2^2(\phi_0)}{2\lambda_2(\phi_0)} > 0$$
$$\nu_L'(\phi_0) + \nu_{L,0}'(\phi_0) = \frac{\sigma^2 + \lambda_1^2(\phi_0)}{2\sigma} > 0 > 0.$$

CS: Next, consider the case of a contact discontinuity moving left faster than a shock. In this case

$$\nu_L'(\phi_0) = 0$$
$$\nu_{L,0}'(\phi_0) = \frac{\lambda_1^2(\phi_0) + \sigma^2}{2\sigma} > 0.$$

RC: Next, consider the case of a rarefaction moving left faster than a contact discontinuity. In this case,

$$\nu'_L(\phi_0) = -\frac{\{\lambda_1(\phi_0) - \lambda_2(\phi_0)\}^2}{2\lambda_2(\phi_0)} < 0$$
$$\nu'_{L,0}(\phi_0) = \frac{\lambda_1^2(\phi_0) + \lambda_2^2(\phi_0)}{2\lambda_2(\phi_0)} > 0$$
$$\nu'_L(\phi_0) + \nu'_{L,0}(\phi_0) = \lambda_1(\phi_0) > 0 > 0.$$

CR or RCR: Finally, consider the case of a contact discontinuity moving left either faster than a rarefaction or in the middle of a rarefaction. In this case,

$$\nu'_L(\phi_0) = 0$$
$$\nu'_{L,0}(\phi_0) = \lambda_1(\phi_0) > 0.$$

Similarly, in order to prove inequalities (4.69d) we will examine the cases.

CS: First, consider the case of a shock moving right faster than a contact discontinuity. In this case $\lambda_1(\phi_0) > \sigma > \lambda_1(\phi_R)$ and $\sigma \geq \lambda_2(\phi_0)$. Thus

$$\nu'_R(\phi_0) = \frac{\{\sigma - \lambda_2(\phi_0)\}\{\lambda_1^2(\phi_0) - \sigma\lambda_2(\phi_0)\}}{2\sigma\lambda_2(\phi_0)} > 0$$
$$\nu'_{R,0}(\phi_0) = -\frac{\lambda_1^2(\phi_0) + \lambda_2^2(\phi_0)}{2\lambda_2(\phi_0)} < 0$$
$$\nu'_R(\phi_0) + \nu'_{R,0}(\phi_0) = -\frac{\sigma^2 + \lambda_1^2(\phi_0)}{2\sigma} < 0.$$

SC: Next, consider the case of a contact discontinuity moving right faster than a shock. In this case

$$\nu'_R(\phi_0) = 0$$
$$\nu'_{R,0}(\phi_0) = -\frac{\lambda_1^2(\phi_0) + \sigma^2}{2\sigma} < 0.$$

CR: Next, consider the case of a rarefaction moving right faster than a contact discontinuity. In this case,

$$\nu'_R(\phi_0) = \frac{\{\lambda_1(\phi_0) - \lambda_2(\phi_0)\}^2}{2\lambda_2(\phi_0)} > 0$$
$$\nu'_{R,0}(\phi_0) = -\frac{\lambda_1^2(\phi_0) + \lambda_2^2(\phi_0)}{2\lambda_2(\phi_0)} < 0$$
$$\nu'_R(\phi_0) + \nu'_{R,0}(\phi_0) = -\lambda_1(\phi_0) < 0.$$

RC or RCR: Finally, consider the case of a contact discontinuity moving right either faster than a rarefaction or in the middle of a rarefaction. In this case,

$$\nu'_R(\phi_0) = 0$$
$$\nu'_{R,0}(\phi_0) = -\lambda_1(\phi_0) < 0.$$

Now that we have verified the inequalities (4.69) we can examine the sign of g' at a zero of g, defined in Equation (4.68). Note that

$$\begin{aligned} g'(\phi_0) = & -2\{\mathbf{v}_R - \mathbf{v}_L + \mathbf{n}(\theta_R)\nu_R(\phi_0) - \mathbf{n}(\theta_L)\nu_L(\phi_0)\}^\top \{\mathbf{n}(\theta_R)\nu'_R(\phi_0) - \{\mathbf{n}(\theta_L)\nu'_L(\phi_0)\} \\ & + 2\left\{\nu_{L,0}(\phi_0) - \nu_{R,0}(\phi_0)\right\}\{\nu'_{L,0}(\phi_0) - \nu'_{R,0}(\phi_0)\} \\ \geq & -2\|\mathbf{v}_R - \mathbf{v}_L + \mathbf{n}(\theta_R)\nu_R(\phi_0) - \mathbf{n}(\theta_L)\nu_L(\phi_0)\|\{\nu'_{L,0}(\phi_0) - \nu'_{R,0}(\phi_0)\} \\ & + 2\left\{\nu_{L,0}(\phi_0) - \nu_{R,0}(\phi_0)\right\}\{\nu'_{L,0}(\phi_0) - \nu'_{R,0}(\phi_0)\}. \end{aligned}$$

The end of this inequality is zero at a zero of $g(\phi_0)$. Equality occurs only if $\mathbf{n}(\theta_R)$ and $\mathbf{n}(\theta_L)$ have the same direction as $\mathbf{v}_R - \mathbf{v}_L$. It follows that g has nonnegative slope at all of its zeros. Either g has a unique zero, or an interval in ϕ where it is zero.

4.8.8 Riemann Problem Solution

After finding ϕ_0 so that the wave families for the vibrating string intersect, we define Wavespeeds $\xi_1 \leq \xi_2 \leq \xi_3 \leq \xi_4 \leq 0$ as follows:

SC: If we have a shock moving left faster than a contact discontinuity, then we define

$$\xi_1 = \xi_2 = -\sqrt{\frac{1}{\rho}\frac{\tau(\phi_L) - \tau(\phi_0)}{\phi_L - \phi_0}}, \quad \xi_3 = \xi_4 = -\sqrt{\frac{\tau(\phi_0)}{\rho\phi_0}}$$
$$[\mathbf{v}_{2,3}, \phi_{2,3}, \theta_{2,3}] = [\mathbf{v}_L - \mathbf{n}(\theta_L)\sqrt{\{\tau(\phi_L) - \tau(\phi_0)\}(\phi_L - \phi_0)/\rho}, \phi_0, \theta_L].$$

CS: If we have a contact discontinuity moving left faster than a shock, then we define

$$\xi_1 = \xi_2 = -\sqrt{\frac{\tau(\phi_L)}{\rho\phi_L}}, \quad \xi_3 = \xi_4 = -\sqrt{\frac{1}{\rho}\frac{\tau(\phi_L) - \tau(\phi_0)}{\phi_L - \phi_0}}$$
$$[\mathbf{v}_{2,3}, \phi_{2,3}, \theta_{2,3}] = [\mathbf{v}_L + \{\mathbf{n}(\theta_0) - \mathbf{n}(\theta_L)\}\sqrt{\tau(\phi_L)\phi_L/\rho}, \phi_L, \theta_0].$$

RC: If we have a rarefaction moving left faster than a contact discontinuity, then we define

$$\xi_1 = -\sqrt{\frac{\tau'(\phi_L)}{\rho}}, \quad \xi_2 = -\sqrt{\frac{\tau'(\phi_0)}{\rho}}, \quad \xi_3 = \xi_4 = -\sqrt{\frac{\tau(\phi_0)}{\rho\phi_0}}$$

$$[\mathbf{v}_{2,3}, \phi_{2,3}, \theta_{2,3}] = [\mathbf{v}_L + \mathbf{n}(\theta_L)\int_{\phi_L}^{\phi_0} \sqrt{\tau'(\eta)/\rho}\, d\eta, \phi_0, \theta_L].$$

CR: If we have a contact discontinuity moving left faster than a rarefaction, then we define

$$\xi_1 = \xi_2 = -\sqrt{\frac{\tau(\phi_L)}{\rho\phi_L}}, \quad \xi_3 = -\sqrt{\frac{\tau'(\phi_L)}{\rho}}, \quad \xi_4 = -\sqrt{\frac{\tau'(\phi_0)}{\rho}}$$

$$[\mathbf{v}_{2,3}, \phi_{2,3}, \theta_{2,3}] = \left[\mathbf{v}_L + \{\mathbf{n}(\theta_0) - \mathbf{n}(\theta_L)\}\sqrt{\tau(\phi_L)\phi_L/\rho}, \phi_L, \theta_0\right].$$

RCR: If we have a contact discontinuity moving left in the middle of a rarefaction, then we define

$$\xi_1 = -\sqrt{\frac{\tau'(\phi_L)}{\rho}}, \quad \xi_2 = \xi_3 = -\sqrt{\frac{\tau'(\phi_*)}{\rho}}, \quad \xi_4 = -\sqrt{\frac{\tau'(\phi_0)}{\rho}}$$

$$[\mathbf{v}_{2,3}, \phi_{2,3}, \theta_{2,3}] = \left[\mathbf{v}_L + \mathbf{n}(\theta_L)\int_{\phi_L}^{\phi_*} \sqrt{\tau'(\eta)/\rho}\, d\eta, \phi_*, \theta_L\right].$$

We also define Wavespeeds $0 \leq \xi_5 \leq \xi_6 \leq \xi_7 \leq \xi_8$ as follows:

CS: If we have a shock moving right faster than a contact discontinuity, then we define

$$\xi_5 = \xi_6 = \sqrt{\frac{\tau(\phi_0)}{\rho\phi_0}}, \quad \xi_7 = \xi_8 = \sqrt{\frac{1}{\rho}\frac{\tau(\phi_R) - \tau(\phi_0)}{\phi_R - \phi_0}}$$

$$[\mathbf{v}_{6,7}, \phi_{6,7}, \theta_{6,7}] = [\mathbf{v}_R + \mathbf{n}(\theta_R)\sqrt{\{\tau(\phi_R) - \tau(\phi_0)\}(\phi_R - \phi_0)/\rho}, \phi_0, \theta_R].$$

SC: If we have a contact discontinuity moving right faster than a shock, then we define

$$\xi_5 = \xi_6 = \sqrt{\frac{1}{\rho}\frac{\tau(\phi_R) - \tau(\phi_0)}{\phi_R - \phi_0}}, \quad \xi_7 = \xi_8 = \sqrt{\frac{\tau(\phi_R)}{\rho\phi_R}}$$

$$[\mathbf{v}_{6,7}, \phi_{6,7}, \theta_{6,7}] = [\mathbf{v}_R - \{\mathbf{n}(\theta_0) - \mathbf{n}(\theta_R)\}\sqrt{\tau(\phi_R)\phi_R/\rho}, \phi_R, \theta_0].$$

CR: If we have a rarefaction moving right faster than a contact discontinuity, then

$$\xi_5 = \xi_6 = \sqrt{\frac{\tau(\phi_0)}{\rho\phi_0}}, \quad \xi_7 = \sqrt{\frac{\tau'(\phi_R)}{\rho}}, \quad \xi_8 = \sqrt{\frac{\tau'(\phi_0)}{\rho}}$$

$$[\mathbf{v}_{6,7}, \phi_{6,7}, \theta_{6,7}] = [\mathbf{v}_R - \mathbf{n}(\theta_R)\int_{\phi_R}^{\phi_0} \sqrt{\tau'(\eta)/\rho}\, d\eta, \phi_0, \theta_R].$$

RC: If we have a contact discontinuity moving right faster than a rarefaction, then

$$\xi_5 = \sqrt{\frac{\tau'(\phi_0)}{\rho}}, \quad \xi_6 = \sqrt{\frac{\tau'(\phi_R)}{\rho}}, \quad \xi_7 = \xi_8 = \sqrt{\frac{\tau(\phi_R)}{\rho\phi_R}}$$

$$[\mathbf{v}_{6,7}, \phi_{6,7}, \theta_{6,7}] = [\mathbf{v}_R - \{\mathbf{n}(\theta_0) - \mathbf{n}(\theta_R)\}\sqrt{\tau(\phi_R)\phi_R/\rho}, \phi_R, \theta_0].$$

RCR: If we have a contact discontinuity moving right in the middle of a rarefaction, then

$$\xi_5 = \sqrt{\frac{\tau'(\phi_0)}{\rho}}, \quad \xi_6 = \xi_7 = \sqrt{\frac{\tau'(\phi_*)}{\rho}}, \quad \xi_8 = \sqrt{\frac{\tau'(\phi_R)}{\rho}}$$

$$[\mathbf{v}_{6,7}, \phi_{6,7}, \theta_{6,7}] = \left[\mathbf{v}_R - \mathbf{n}(\theta_R)\int_{\phi_R}^{\phi_*} \sqrt{\tau'(\eta)/\rho}\, d\eta, \phi_*, \theta_R\right].$$

Given a wave speed ξ, our state $\mathbf{w}(\xi)$ in the solution of the Riemann problem for the vibrating string is given by

$$\begin{bmatrix} \mathbf{v} \\ \phi \\ \theta \end{bmatrix}(\xi) = \begin{cases} \begin{bmatrix} \mathbf{v}_L \\ \phi_L \\ \theta_L \end{bmatrix}, & \xi < \xi_1 \\ \begin{bmatrix} \mathbf{v}_L + \mathbf{n}(\theta_L) \int_{\phi_L}^{(\tau')^{-1}(\rho\xi^2)} \{\tau'(\eta)/\rho\}^{1/2}\, d\eta \\ (\tau')^{-1}(\rho\xi^2) \\ \theta_L \end{bmatrix}, & \xi_1 \le \xi \le \xi_2 \\ \begin{bmatrix} \mathbf{v}_{2,3} \\ \phi_{2,3} \\ \theta_{2,3} \end{bmatrix}, & \xi_2 < \xi < \xi_3 \\ \begin{bmatrix} \mathbf{v}_-(\phi_0, \theta_0) - \mathbf{n}(\theta_0) \int_{(\tau')^{-1}(\rho\xi^2)}^{\phi_0} \{\tau'(\eta)/\rho\}^{1/2}\, d\eta \\ (\tau')^{-1}(\rho\xi^2) \\ \theta_0 \end{bmatrix}, & \xi_3 \le \xi \le \xi_4 \\ \begin{bmatrix} \mathbf{v}_-(\phi_0, \theta_0) \\ \phi_0 \\ \theta_0 \end{bmatrix}, & \xi_4 < \xi < \xi_5 \\ \begin{bmatrix} \mathbf{v}_+(\phi_0, \theta_0) + \mathbf{n}(\theta_0) \int_{(\tau')^{-1}(\rho\xi^2)}^{\phi_0} \{\tau'(\eta)/\rho\}^{1/2}\, d\eta \\ (\tau')^{-1}(\rho\xi^2) \\ \theta_0 \end{bmatrix}, & \xi_5 \le \xi \le \xi_6 \\ \begin{bmatrix} \mathbf{v}_{6,7} \\ \phi_{6,7} \\ \theta_{6,7} \end{bmatrix}, & \xi_6 < \xi < \xi_7 \\ \begin{bmatrix} \mathbf{v}_R - \mathbf{n}(\theta_R) \int_{\phi_R}^{(\tau')^{-1}(\rho\xi^2)} \{\tau'(\eta)/\rho\}^{1/2}\, d\eta \\ (\tau')^{-1}(\rho\xi^2) \\ \theta_R \end{bmatrix}, & \xi_7 \le \xi \le \xi_8 \\ \begin{bmatrix} \mathbf{v}_R \\ \phi_R \\ \theta_R \end{bmatrix}, & \xi_8 < \xi \end{cases}$$

Programs to solve the Riemann problem for the vibrating string can be found in **Program 4.8-48: string.f.** You can execute this Riemann problem solver by clicking on the link **Executable 4.8-20: guiString** You can select input parameters for the Riemann problem by clicking on the arrow next to "String Parameters." When you have selected all of your input parameters, click on "Start Run Now" in the window labeled "guiString." Move the windows so that you can see them completely, then answer "OK" to the window that asks "Are you finished viewing the results?" Click

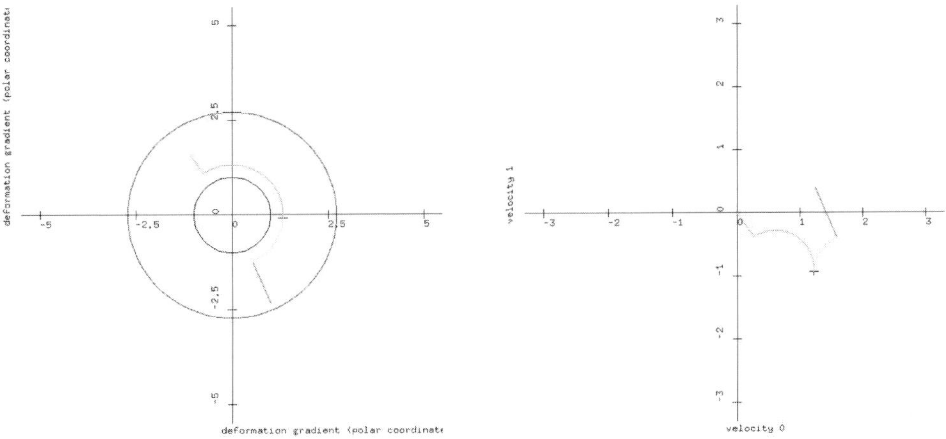

(a) Deformation gradient (polar coordinates) (b) Velocity

Fig. 4.12 Solution to vibrating string Riemann problem

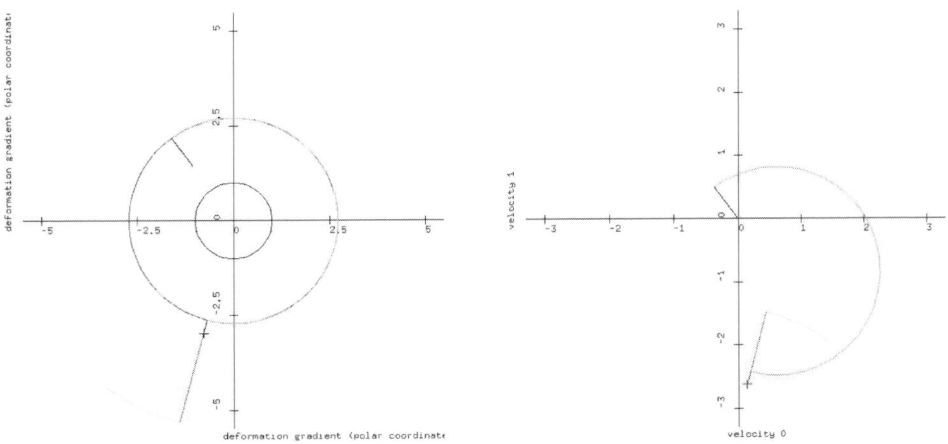

(a) Deformation gradient (polar coordinates) (b) Velocity

Fig. 4.13 Solution to vibrating string riemann problem

in the windows as requested to selected the left and right state for the Riemann problem. You may drag the mouse when selecting the right state to see how the solution of the Riemann problem changes with the right state. Figures 4.12 and 4.13 show two solutions to the vibrating string Riemann problem. In these figures, the equivelocity value of ϕ is drawn as a circle in the deformation gradient graph, and the velocity at the left state is always zero.

Exercises for 4.8

4.8.1 Show that θ and $\mathbf{v} \pm \mathbf{n}(\theta) \int^{\phi} \sqrt{\frac{\tau'(\eta)}{\rho}}\, d\eta$ are Riemann invariants for the vibrating string associated with the characteristic speed $\lambda = \pm\sqrt{\frac{\tau'(\phi)}{\rho}}$.

4.8.2 We would like to examine cases in which the function $g(\phi_0)$, defined in equation (4.68), has zero derivative at its zero. We expect that in such a case, we should have $\phi_0 = 1$. Thus, we want to examine two cases. First,

$$[\mathbf{v}_R, \phi_R, \phi_R] = [\mathbf{v}_L - \mathbf{n}(\theta_L)\{\sqrt{\tau(\phi_R)(\phi_R - 1)/\rho} + \{\sqrt{\tau(\phi_L)(\phi_L - 1)/\rho}, \phi_R, \theta_L]$$

and

$$[\mathbf{v}_R, \phi_R, \phi_R] = [\mathbf{v}_L + \mathbf{n}(\theta_L)\{\sqrt{\tau(\phi_R)(\phi_R - 1)/\rho} - \{\sqrt{\tau(\phi_L)(\phi_L - 1)/\rho}, \phi_R, \theta_L \pm \pi].$$

Do these Riemann problems have unique solutions? If you use numerical experiments to form your conclusion, then be careful with numerical oscillations that lead to values of $\phi < 1$, and therefore possibly negative tension.

4.8.3 Our construction of the solution of the Riemann problem omitted the anomalous shock

$$[\mathbf{v}_R, \phi_R, \phi_R] = [\mathbf{v}_L + \mathbf{n}(\theta_L)\{\sqrt{\tau(\phi_R + \tau(\phi_L)(\phi_R + \phi_L)/\rho}, \phi_R, \theta_L \pm \pi]$$

in the wave families. What waves does our Riemann problem solution produce in such a case?

4.8.4 Suppose that $\tau(\phi) = \log(\phi)$ and $\rho = 1$. Choose the left and right state for a Riemann problem so that the solution involves

- a shock moving left faster than a contact discontinuity, and a shock moving right faster than a contact discontinuity;
- a contact discontinuity moving left in the middle of a rarefaction, and a contact discontinuity moving right in the middle of a rarefaction

4.8.5 Formulate the Eulerian equations of motion for the vibrating string.

4.8.6 Suppose that the tension has the form $\tau(\phi) = \tau_0 + \tau_1\phi + \tau_2\phi^2$, where τ_0, τ_1 and $\tau_2 > 0$. Find the value of ϕ at which the two positive characteristic speeds are equal.

4.8.7 Program the Lax–Friedrichs scheme for the vibrating string and test it for problems involving shocks, rarefactions and contact discontinuities. How can you tell which waves are shocks, which are contact discontinuities, and which are rarefactions?

4.8.8 Repeat the previous exercise for the Rusanov scheme.

4.8.9 We would like to develop a technique to approximate the state that moves at zero speed in the solution of a Riemann problem for the vibrating string. For **weak waves** (in which the characteristic directions do not change much) we can approximate

$$\begin{bmatrix} (\mathbf{v}_1)_R - (\mathbf{v}_1)_L \\ (\mathbf{v}_2)_R - (\mathbf{v}_2)_L \\ \phi_R - \phi_L \\ \theta_R - \theta_L \end{bmatrix} = \begin{bmatrix} -\mathbf{X}\Lambda & \mathbf{X}\Lambda \\ \mathbf{I} & \mathbf{I} \end{bmatrix} \begin{bmatrix} c_L \\ c_R \end{bmatrix}.$$

Then the state that moves with zero speed can be approximated either by

$$\begin{bmatrix} (\mathbf{v}_1)_L \\ (\mathbf{v}_2)_L \\ \phi_L \\ \theta_L \end{bmatrix} + \begin{bmatrix} -\mathbf{X}\Lambda \\ \mathbf{I} \end{bmatrix} c_L$$

or

$$\begin{bmatrix} (\mathbf{v}_1)_R \\ (\mathbf{v}_2)_R \\ \phi_R \\ \theta_R \end{bmatrix} - \begin{bmatrix} \mathbf{X}\Lambda \\ \mathbf{I} \end{bmatrix} c_R .$$

Show that both of these approximations give the same value. Write a program to implement this weak wave approximation, by evaluating the characteristic directions at the average of the left and right states.

4.8.10 Program the Godunov scheme for the vibrating string, using the weak wave approximation from the previous problem.

4.9 Case Study: Plasticity

In this section we will consider another model for solid mechanics. This model will incorporate an important physical effect, known as **plasticity**. Solid materials that undergo finite deformation often suffer a realignment of the particles that make up the solid. If an applied force is sufficient, the material undergoes a permanent change in shape called a plastic deformation; if the applied force is removed, the material does not return to its original shape. Materials such as putty and clay can develop plastic deformation due to very small forces; materials such as steel require much larger applied forces to develop plastic deformation. The analysis of this model is interesting, because it necessarily involves more flux variables than conservation laws, in order to model hysteresis.

4.9.1 Lagrangian Equations of Motion

Following [4], we will consider one-dimension of motion in the Lagrangian frame. We will write conservation of momentum in the form

$$\frac{\partial v}{\partial t} - \frac{\partial s}{\partial a} = 0,$$

where s is the first Piola–Kirchhoff stress divided by the initial density. If the deformation gradient $\mathbf{J}$ has first entry $\mathbf{J}_{11} = 1 + \epsilon$, then equality of mixed partial derivatives for the particle position **x** can be written in the form

$$\frac{\partial \epsilon}{\partial t} - \frac{\partial v}{\partial a} = 0.$$

Since the deformation gradient must have positive determinant (the material can never be turned inside-out), we require $1 + \epsilon > 0$. For simplicity, we will refer to ϵ as the **strain**.

4.9.2 Constitutive Laws

For any physically realistic strain ϵ, there are upper and lower bounds on the stress s, given by functions $\gamma(\epsilon)$ and $\tau(\epsilon)$. These are called the plastic compression and tension curves, respectively. Once the stress reaches one of these bounds, the plastic strain π changes. Thus, between the bounds the stress is given by an elastic curve $s = e(\epsilon, \pi)$.

In order to construct a physically realistic model, an infinite force must be required to totally compress the material:

$$\epsilon \downarrow -1 \Longrightarrow \gamma(\epsilon) \downarrow -\infty. \tag{4.70}$$

Further, we assume that plastic compression occurs only for negative stress:

$$\text{for all } \epsilon > -1, \quad \gamma(\epsilon) < 0,$$

and that plastic tension occurs only for positive stress:

$$\text{for all } \epsilon > -1, \quad \tau(\epsilon) > 0.$$

We assume that there is some lower limit on the plastic strain $\pi_{\min} > -1$ corresponding to total compression:

$$e(\epsilon, \pi) \downarrow -\infty \implies \epsilon \downarrow -1 \quad \text{and} \quad \pi \downarrow \pi_{\min}.$$

We also assume that there are unique values $\epsilon_\gamma(\pi) < \epsilon_\tau(\pi)$ of the strain such that the elastic and plastic curves intersect:

$$\begin{aligned} &\text{for all } \pi > \pi_{\min} \quad \gamma(\epsilon_\gamma(\pi)) = e(\epsilon_\gamma(\pi), \pi), \\ &\text{for all } \pi > \pi_{\min} \quad \tau(\epsilon_\tau(\pi)) = e(\epsilon_\tau(\pi), \pi). \end{aligned}$$

We assume that for a given value of the plastic strain, the elastic stress lies between the values of the compression and tension yield stresses:

$$\text{for all } \epsilon_\gamma(\pi) < \epsilon < \epsilon_\tau(\pi) \quad \gamma(\epsilon_\gamma(\pi)) < e(\epsilon, \pi) < \tau(\epsilon_\tau(\pi)).$$

So that the characteristic speeds will be real, we assume that

$$\text{for all } \pi > \pi_{\min} \quad \frac{d\gamma}{d\epsilon}, \frac{d\tau}{d\epsilon} \text{ and } \frac{\partial e}{\partial \epsilon} > 0, \text{ for all } \epsilon > -1.$$

So that the characteristic speeds will be genuinely nonlinear, we will assume that

$$\text{for all } \epsilon > -1 \text{ for all } \pi > \pi_{\min} \quad \frac{d^2\gamma}{d\epsilon^2}, \frac{d^2\tau}{d\epsilon^2} \text{ and } \frac{\partial^2 e}{\partial \epsilon^2} > 0. \tag{4.71}$$

We also assume that the slopes of the plastic loading curves are less than the slopes of the elastic loading curves at the corresponding yield points:

$$\text{for all } \pi > \pi_{\min} \quad \frac{d\gamma}{d\epsilon}(\epsilon_\gamma(\pi)) < \frac{\partial e}{\partial \epsilon}(\epsilon_\gamma(\pi), \pi),$$
$$\text{for all } \pi > \pi_{\min} \quad \frac{d\tau}{d\epsilon}(\epsilon_\tau(\pi)) < \frac{\partial e}{\partial \epsilon}(\epsilon_\tau(\pi), \pi).$$

We assume that the elastic curves do not intersect for distinct values of π:

$$\text{for all } \pi > \pi_{\min} \quad \frac{\partial e}{\partial \pi}(\epsilon, \pi) < 0 \quad \text{for all } \epsilon_\gamma(\pi) < \epsilon < \epsilon_\tau(\pi).$$

Thus elastic curves move to the right (*i.e.*, toward increasing value of strain) as π increases. For example, we can choose

$$\begin{aligned} \gamma(\epsilon) &= -0.1 - (1+\epsilon)^{-2}, \\ \tau(\epsilon) &= 1.1 - (2+\epsilon)^{-0.5}, \\ e(\epsilon, \pi) &= -0.49(1+\epsilon+\pi_{\min}-\pi)^{-1} + (1+\epsilon+\pi_{\min}-\pi) \\ \pi_{\min} &= -0.3. \end{aligned}$$

During elastic response, π is fixed and $s = e(\epsilon, \pi)$. During plastic compression, $s = \gamma(\epsilon)$ and the plastic strain π varies so that $\gamma(\epsilon) = e(\epsilon, \pi)$. Similarly, during plastic tension $s = \tau(\epsilon) = e(\epsilon, \pi)$. The **hysteresis** rule says that plastic compression occurs if and only if the material is at yield and the time derivative of the strain is negative:

$$\text{during plastic compression } \epsilon = \epsilon_\gamma(\pi) \text{ and } \left.\frac{\partial \epsilon}{\partial t}\right|_a < 0.$$

Similarly,

$$\text{during plastic tension } \epsilon = \epsilon_\tau(\pi) \text{ and } \left.\frac{\partial \epsilon}{\partial t}\right|_a > 0.$$

Otherwise the material is elastic.

Finally, we assume that tensile failure is impossible, by requiring

$$\int_0^\infty \sqrt{\frac{d\tau}{d\epsilon}} d\epsilon < \infty. \tag{4.72}$$

This condition will guarantee that wave families in the solution of the Riemann problem will intersect.

4.9.3 Centered Rarefactions

During plastic tension or compression, the stress s is a function of the strain ϵ only, and the plastic strain π does not affect the equations of motion. (This statement is particular to our assumption that the motion is one-dimensional; in models allowing for plastic deformation in multiple dimensions, the hysteresis parameters usually affect the stress during plastic response.) On the other hand, during elastic response the plastic strain π is constant. Thus in either case, it is easy to see that the characteristic speeds are

$$\lambda = \pm\sqrt{\frac{\partial s}{\partial \epsilon}}.$$

Since the stress is always an increasing function of strain, the model is hyperbolic.

Centered rarefaction curves involve smooth motion in which the velocity and stress are function of a/t only. For centered rarefactions, we have the system of ordinary differential equations

$$0 = \frac{\partial v}{\partial t} - \frac{\partial s}{\partial a} = -\left\{ v' \frac{a}{t} + \frac{\partial s}{\partial \epsilon} \epsilon' \right\} \frac{1}{t},$$
$$0 = \frac{\partial \epsilon}{\partial t} - \frac{\partial v}{\partial a} = -\left\{ \epsilon' \frac{a}{t} + v' \right\} \frac{1}{t}.$$

We can use the latter equation to eliminate v' in the former equation, to get

$$0 = \left\{ \frac{\partial s}{\partial \epsilon} - \left(\frac{a}{t}\right)^2 \right\} \epsilon'.$$

This result determines the characteristic speeds, and shows that the centered rarefaction curves satisfy

$$v' = \pm\sqrt{\frac{\partial s}{\partial \epsilon}} \epsilon'.$$

During an admissible rarefaction, the characteristic speeds must increase from left to right. Thus inequality (4.71) implies that an elastic rarefaction moving to the left in physical space from a left state (v_L, ϵ_L, π_L) to a right state (v_R, ϵ_R, π_R) must satisfy

$$\epsilon_Y(\pi_L) < \epsilon_L < \epsilon_R \le \epsilon_\tau(\pi_L).$$

Similarly, an elastic rarefaction moving to the right satisfies

$$\epsilon_\gamma(\pi_L) \le \epsilon_R < \epsilon_L < \epsilon_\tau(\pi_L).$$

The determination of admissible centered rarefactions and shocks during plastic response is complicated by hysteresis. First, we will consider plastic compression; in other words, we assume that $\epsilon_L = \epsilon_\gamma(\pi_L)$. Recall that the stress is given by $s = \gamma(\epsilon)$. Also recall that during plastic response the time derivative of the displacement gradient must be negative; otherwise, the material relaxes elastically. Because $\epsilon_L < \epsilon_R$ corresponds to unloading during rarefactions moving to the left, and $\epsilon_L > \epsilon_R$ corresponds to unloading during rarefactions moving to the right, centered rarefactions are impossible during plastic compression. We are left only with shocks during plastic compression.

During plastic tension, shocks are unphysical and only rarefactions are possible; tension rarefactions with negative wave speed satisfy

$$\epsilon_L = \epsilon_\tau(\pi_L) < \epsilon_R,$$

and tension rarefactions with positive wave speed satisfy

$$-1 < \epsilon_R < \epsilon_L = \epsilon_\tau(\pi_L).$$

4.9.4 Hugoniot Loci

The Rankine–Hugoniot conditions show that the left and right states at traveling discontinuities are related by

$$-\begin{bmatrix} s_R - s_L \\ v_R - v_L \end{bmatrix} = \begin{bmatrix} v_R - v_L \\ \epsilon_R - \epsilon_L \end{bmatrix} \sigma,$$

where σ is the speed of the discontinuity. It is straightforward to solve these equations to get

$$\sigma = \pm\sqrt{\frac{s_R - s_L}{\epsilon_R - \epsilon_L}}.$$

It is fairly easy to see that our assumptions on the constitutive model guarantee that this discontinuity speed is real whenever the two states on either side of the discontinuity are both undergoing the same loading conditions (compression, tension or elastic response). The Rankine–Hugoniot conditions now show that traveling discontinuities satisfy

$$v_R - v_L = \pm\sqrt{\frac{s_R - s_L}{\epsilon_R - \epsilon_L}}(\epsilon_R - \epsilon_L).$$

During an admissible elastic shock with nonzero speed σ, we must have [40, 107, 122, 174]

$$\sigma \left[\frac{s_R - s_L}{\epsilon_R - \epsilon_L} - \frac{s(\epsilon) - s(\epsilon_L)}{\epsilon - \epsilon_L} \right] \leq 0 \text{ for all } \epsilon \text{ between } \epsilon_L \text{ and } \epsilon_R. \tag{4.73}$$

It is straightforward to see that the admissible elastic shocks moving to the left satisfy

$$\epsilon_\gamma(\pi_L) \leq \epsilon_R < \epsilon_L < \epsilon_\tau(\pi_L).$$

Similarly, admissible elastic shocks moving to the right satisfy

$$\epsilon_\gamma(\pi_L) < \epsilon_L < \epsilon_R \leq \epsilon_\tau(\pi_L).$$

During plastic compression, shocks with negative speed satisfy

$$\epsilon_s(\epsilon_L, \pi_L) < \epsilon_R < \epsilon_L = \epsilon_\gamma(\pi_L),$$

and shocks with positive speed satisfy

$$\epsilon_L = \epsilon_\gamma(\pi_L) < \epsilon_R.$$

Here, the transition point $\epsilon_s(\epsilon_L, \pi_L)$ is defined for $\epsilon_\tau(\pi_L) > \epsilon_L > \epsilon_\gamma(\pi_L)$ by

$$\epsilon_s(\epsilon_L, \pi_L) \equiv \sup \left\{ -1 < \epsilon < \epsilon_\gamma(\pi_L) : \frac{e(\epsilon_L, \pi_L) - \gamma(\epsilon_\gamma(\pi_L))}{\epsilon_L - \epsilon_\gamma(\pi_L)} \leq \frac{e(\epsilon_L, \pi_L) - \gamma(\epsilon)}{\epsilon_L - \epsilon} \right\}. \tag{4.74}$$

Note that condition (4.70) implies that $\epsilon_s > -1$. Also note that ϵ_s is a nondecreasing function of ϵ for fixed π; this can be seen by differentiating (4.74) and applying (4.71) and the mean value theorem.

It is possible to have a shock between a state undergoing elastic response and a state undergoing compression. If the shock speed is negative, we have

$$v_R = v_L - \sqrt{\{\gamma(\epsilon_R) - e(\epsilon_L, \pi_L)\}(\epsilon_R - \epsilon_L)}.$$

Admissibility for this shock requires that

$$-1 < \epsilon_R < \epsilon_s(\epsilon_L, \pi_L).$$

Similarly, an elastic–plastic shock with positive speed satisfies

$$v_L = v_R + \sqrt{\{\gamma(\epsilon_L) - e(\epsilon_R, \pi_R)\}(\epsilon_L - \epsilon_R)}$$

and

$$-1 < \epsilon_L < \epsilon_s(\epsilon_R, \pi_R).$$

4.9.5 Entropy Function

The sum of the kinetic and strain energy $E \equiv \frac{1}{2}v^2 + \int^{\epsilon} s(\theta)\,d\theta$ is an entropy function for the plasticity model, with energy flux $\Psi \equiv -s(\epsilon)v$. To show that this is so, we will verify Equation (4.29). We compute

$$\frac{\partial E}{\partial \mathbf{w}} = \begin{bmatrix} v & s \end{bmatrix}$$

and

$$\frac{\partial \Psi}{\partial \mathbf{w}} = \begin{bmatrix} -s & -v\frac{\partial s}{\partial \epsilon} \end{bmatrix}.$$

Then

$$\frac{\partial E}{\partial \mathbf{w}} \left(\frac{\partial \mathbf{u}}{\partial \mathbf{w}} \right)^{-1} \frac{\partial \mathbf{F}}{\partial \mathbf{w}} = \begin{bmatrix} v & s \end{bmatrix} \begin{bmatrix} 0 & \frac{\partial s}{\partial \epsilon} \\ -1 & 0 \end{bmatrix} = \begin{bmatrix} -s & -v\frac{\partial s}{\partial \epsilon} \end{bmatrix} = \frac{\partial \Psi}{\partial \mathbf{w}}.$$

It is easy to see that the entropy function is a convex function of the flux variables $\mathbf{w}$.

4.9.6 Riemann Problem

In all cases for waves with negative speed, given a left state the admissible right states involve stress increasing monotonically with strain, and velocity increasing monotonically with strain. Similarly, given a right state, admissible waves with positive speed have stress increasing with strain and velocity decreasing with strain. Given a left state (v_L, s_L) we construct the negative wave family by using rarefaction curves in the direction of increasing stress, and shock curves in the direction of decreasing stress. Given a right state (v_R, s_R) we construct the positive wave family by using rarefaction curves in the direction of increasing stress and shock curves in the direction of decreasing stress. The negative wave family involves all stresses between negative infinity and the maximum tension stress, and all real velocities because of inequality (4.72). Similarly, the positive wave family involves the same stresses and velocities, but with negative slope $\frac{dv}{ds}$. As a result, the two wave families must intersect. Let (v_0, s_0) be the velocity and stress at the intersection of the two wave families.

For waves with negative speed, we will identify four Wavespeeds, depending on a variety of cases. If the stress s_0 at the intersection of the wave families satisfies $s_0 \geq e(\epsilon_L, \pi_L)$, we define $\epsilon_{0,L}$ and $\pi_{0,L}$ as follows. If $e(\epsilon_L, \pi_L) \leq s_0 < \tau(\epsilon_\tau(\pi_L))$ then we take $\pi_{0,L} = \pi_L$ and solve $e(\epsilon_{0,L}, \pi_L) = s_0$ for $\epsilon_{0,L}$. On the other hand, if $\tau(\epsilon_\tau(\pi_L)) < s_0$ then we solve $\tau(\epsilon_{0,L}) = s_0$ for $\epsilon_{0,L}$ and then we solve $e(\epsilon_{0,L}, \pi_{0,L}) = s_0$

for $\pi_{0,L}$. We also define wavespeeds

$$\xi_1 = -\sqrt{\frac{\partial e}{\partial \epsilon}(\epsilon_L, \pi_L)}$$

$$\xi_2 = -\sqrt{\frac{\partial e}{\partial \epsilon}(\min\{\epsilon_{0,L}, \epsilon_\tau(\pi_L)\}, \pi_L)}$$

$$\xi_3 = \begin{cases} -\sqrt{\frac{d\tau}{d\epsilon}(\epsilon_\tau(\pi_L))}, & \epsilon_{0,L} > \epsilon_\tau(\pi_L) \\ \xi_2, & \epsilon_{0,L} \leq \epsilon_\tau(\pi_L) \end{cases}$$

$$\xi_4 = \begin{cases} -\sqrt{\frac{d\tau}{d\epsilon}(\epsilon_{0,L})}, & \epsilon_{0,L} > \epsilon_\tau(\pi_L) \\ \xi_2, & \epsilon_{0,L} \leq \epsilon_\tau(\pi_L) \end{cases}$$

and intermediate state

$$\begin{bmatrix} v_{23} \\ \epsilon_{23} \\ \pi_{23} \end{bmatrix} = \begin{cases} \begin{bmatrix} v_L + \int_{\epsilon_L}^{\epsilon_\tau(\pi_L)} \sqrt{\frac{\partial e}{\partial \epsilon}(\epsilon, \pi_L)}\, d\epsilon \\ \epsilon_\tau(\pi_L) \\ \pi_L \end{bmatrix}, & \tau(\epsilon_\tau(\pi_L)) < s_0 \\ \begin{bmatrix} v_L + \int_{\epsilon_L}^{\epsilon_{0,L}} \sqrt{\frac{\partial e}{\partial \epsilon}(\epsilon, \pi_L)}\, d\epsilon \\ \epsilon_{0,L} \\ \pi_L \end{bmatrix}, & e(\epsilon_L, \pi_L) \leq s_0 < \tau(\epsilon_\tau(\pi_L)). \end{cases}$$

For $s_0 < e(\epsilon_L, \pi_L)$ we define $\epsilon_{0,L}$ and $\pi_{0,L}$ as follows. If $\gamma(\epsilon_\gamma(\pi_L)) < s_0 < e(\epsilon_L, \pi_L)$ then we take $\pi_{0,L} = \pi_L$ and we solve $e(\epsilon_{0,L}, \pi_L) = s_0$ for $\epsilon_{0,L}$. On the other hand, if $s_0 < \gamma(\epsilon_\gamma(\pi_L))$ then we solve $\gamma(\epsilon_{0,L}) = s_0$ for $\epsilon_{0,L}$, then we solve $e(\epsilon_{0,L}, \pi_{0,L}) = s_0$ for $\pi_{0,L}$. If $\epsilon_s(\epsilon_L, \pi_L) \leq \epsilon_{0,L}$ we define wavespeeds

$$\xi_1 = -\sqrt{\frac{e(\epsilon_L, \pi_L) - e(\max\{\epsilon_{0,L}, \epsilon_\gamma(\pi_L)\}, \pi_L)}{\epsilon_L - \max\{\epsilon_{0,L}, e_\gamma(\pi_L)\}}}, \quad \xi_2 = \xi_1$$

$$\xi_3 = \begin{cases} -\sqrt{\frac{\gamma(\epsilon_\gamma(\pi_L)) - \gamma(\epsilon_{0,L})}{\epsilon_\gamma(\pi_L) - \epsilon_{0,L}}}, & \epsilon_{0,L} < \epsilon_\gamma(\pi_L) \\ \xi_2, & \epsilon_\gamma(\pi_L) \leq \epsilon_{0,L} \end{cases}, \quad \xi_4 = \xi_3$$

and intermediate state

$$\begin{bmatrix} v_{23} \\ \epsilon_{23} \\ \pi_{23} \end{bmatrix} = \begin{cases} \begin{bmatrix} v_L - \sqrt{\{e(\epsilon_L, \pi_L) - e(\epsilon_\gamma(\pi_L), \pi_L)\}(\epsilon_L - \epsilon_\gamma(\pi_L))} \\ \epsilon_\gamma(\pi_L) \\ \pi_L \end{bmatrix}, & \epsilon_\gamma(\pi_L)) > \epsilon_{0,L} \geq \epsilon_s(\epsilon_L, \pi_L) \\ \begin{bmatrix} v_L - \sqrt{\{e(\epsilon_L, \pi_L) - e(\epsilon_{0,L}, \pi_L)\}(\epsilon_L - \epsilon_{0,L})} \\ \epsilon_{0,L} \\ \pi_L \end{bmatrix}, & \epsilon_\gamma(\pi_L)) \leq \epsilon_{0,L} \end{cases}$$

On the other hand, if $\epsilon_{0,L} < \epsilon_s(\epsilon_L, \pi_L)$ we define wavespeeds

$$\xi_1 = \xi_2 = \xi_3 = \xi_4 = -\sqrt{\frac{e(\epsilon_L, \pi_L) - \gamma(\epsilon_{0,L})}{\epsilon_L - \epsilon_{0,L}}}$$

and intermediate state

$$\begin{bmatrix} v_{23} \\ \epsilon_{23} \\ \pi_{23} \end{bmatrix} = \begin{bmatrix} v_L - \sqrt{\{e(\epsilon_L, \pi_L) - \gamma(\epsilon_{0,L})\}(\epsilon_L - \epsilon_{0,L})} \\ \epsilon_{0,L} \\ \epsilon_\gamma^{-1}(\gamma^{-1}(s_0)) \end{bmatrix}.$$

Similarly, for waves with positive speed we will identify four additional wavespeeds. For $s_0 \geq e(\epsilon_R, \pi_R)$ we define $\epsilon_{0,R}$ and $\pi_{0,R}$ as follows. If $e(\epsilon_R, \pi_R) \leq s_0 < \tau(\epsilon_\tau(\pi_R))$ then we take $\pi_{0,R} = \pi_R$ and solve $e(\epsilon_{0,R}, \pi_R) = s_0$ for $\epsilon_{0,R}$. On the other hand, if $\tau(\epsilon_\tau(\pi_R)) < s_0$ then we solve $\tau(\epsilon_{0,R}) = s_0$ for $\epsilon_{0,R}$ and then we solve $e(\epsilon_{0,R}, \pi_{0,R}) = s_0$ for $\pi_{0,R}$. We also define wavespeeds

$$\xi_8 = \sqrt{\frac{\partial e}{\partial \epsilon}(\epsilon_R, \pi_R)}$$

$$\xi_7 = \sqrt{\frac{\partial e}{\partial \epsilon}(\min\{\epsilon_{0,R}, \epsilon_\tau(\pi_R)\}, \pi_R)}$$

$$\xi_6 = \begin{cases} \sqrt{\frac{d\tau}{d\epsilon}(\epsilon_\tau(\pi_R))}, & \epsilon_{0,R} > \epsilon_\tau(\pi_R) \\ \xi_7, & \epsilon_{0,R} \leq \epsilon_\tau(\pi_R) \end{cases}$$

$$\xi_5 = \begin{cases} \sqrt{\frac{d\tau}{d\epsilon}(\max\{\epsilon_{0,R}, \epsilon_\tau(\pi_R)\})}, & \epsilon_{0,R} > \epsilon_\tau(\pi_R) \\ \xi_7, & \epsilon_{0,R} \leq \epsilon_\tau(\pi_R) \end{cases}$$

and intermediate state

$$\begin{bmatrix} v_{67} \\ \epsilon_{67} \\ \pi_{67} \end{bmatrix} = \begin{cases} \begin{bmatrix} v_R - \int_{\epsilon_R}^{\epsilon_\tau(\pi_R)} \sqrt{\frac{\partial e}{\partial \epsilon}(\epsilon, \pi_R)}\, d\epsilon \\ \epsilon_\tau(\pi_R) \\ \pi_R \end{bmatrix}, & \tau(\epsilon_\tau(\pi_R)) < s_0 \\ \begin{bmatrix} v_R - \int_{\epsilon_R}^{\epsilon_{0,R}} \sqrt{\frac{\partial e}{\partial \epsilon}(\epsilon, \pi_R)}\, d\epsilon \\ \epsilon_{0,R} \\ \pi_R \end{bmatrix}, & e(\epsilon_R, \pi_R) \leq s_0 < \tau(\epsilon_\tau(\pi_R)). \end{cases}$$

For $s_0 < e(\epsilon_R, \pi_R)$ we define $\epsilon_{0,R}$ and $\pi_{0,R}$ as follows. If $\gamma(\epsilon_\gamma(\pi_R)) < s_0 < e(\epsilon_R, \pi_R)$ then we take $\pi_{0,R} = \pi_R$ and we solve $e(\epsilon_{0,R}, \pi_R) = s_0$ for $\epsilon_{0,R}$. On the other hand, if $s_0 < \gamma(\epsilon_\gamma(\pi_R))$ then we solve $\gamma(\epsilon_{0,R}) = s_0$ for $\epsilon_{0,R}$, then we solve $e(\epsilon_{0,R}, \pi_{0,R}) = s_0$ for $\pi_{0,R}$. If $\epsilon_s(\epsilon_R, \pi_R) \leq \epsilon_{0,R}$ we define wavespeeds

$$\xi_8 = \sqrt{\frac{e(\epsilon_R, \pi_R) - e(\max\{\epsilon_{0,R}, \epsilon_\gamma(\pi_R)\}, \pi_R)}{\epsilon_R - \max\{\epsilon_{0,R}, \epsilon_\gamma(\pi_R)\}}}, \qquad \xi_7 = \xi_8$$

$$\xi_6 = \begin{cases} \sqrt{\frac{\gamma(\epsilon_\gamma(\pi_R)) - \gamma(\epsilon_{0,R})}{\epsilon_\gamma(\pi_R) - \epsilon_{0,R}}}, & \epsilon_{0,R} < \epsilon_\gamma(\pi_R) \\ \xi_7, & \epsilon_\gamma(\pi_R) \leq \epsilon_{0,R} \end{cases}, \qquad \xi_5 = \xi_6$$

and intermediate state

$$\begin{bmatrix} v_{67} \\ \epsilon_{67} \\ \pi_{67} \end{bmatrix} = \begin{cases} \begin{bmatrix} v_R + \sqrt{\{e(\epsilon_R, \pi_R) - e(\epsilon_\gamma(\pi_R), \pi_R)\}(\epsilon_R - \epsilon_\gamma(\pi_R))} \\ \epsilon_\gamma(\pi_R) \\ \pi_R \end{bmatrix}, & \epsilon_\gamma(\pi_R)) > \epsilon_{0,R} \geq \epsilon_s(\epsilon_R, \pi_R) \\ \begin{bmatrix} v_R + \sqrt{\{e(\epsilon_R, \pi_R) - e(\epsilon_{0,R}, \pi_R)\}(\epsilon_R - \epsilon_{0,R})} \\ \epsilon_{0,R} \\ \pi_R \end{bmatrix}, & \epsilon_\gamma(\pi_R)) \leq \epsilon_{0,R} \end{cases}$$

On the other hand, if $\epsilon_{0,R} < \epsilon_s(\epsilon_R, \pi_R)$ we define wavespeeds

$$\xi_8 = \xi_7 = \xi_6 = \xi_5 = \sqrt{\frac{e(\epsilon_R, \pi_R) - \gamma(\epsilon_{0,R})}{\epsilon_R - \epsilon_{0,R}}}$$

and intermediate state

$$\begin{bmatrix} v_{67} \\ \epsilon_{67} \\ \pi_{67} \end{bmatrix} = \begin{bmatrix} v_R - \sqrt{\{e(\epsilon_R, \pi_R) - \gamma(\epsilon_{0,R})\}(\epsilon_R - \epsilon_{0,R})} \\ \epsilon_{0,R} \\ \epsilon_\gamma^{-1}(\gamma^{-1}(s_0)) \end{bmatrix}.$$

Given a wave speed ξ, the state in the solution of the Riemann problem that moves with speed ξ is given by

$$
\begin{bmatrix} v \\ \epsilon \\ \pi \end{bmatrix}(\xi) = \begin{cases}
\begin{bmatrix} v_L \\ \epsilon_L \\ \pi_L \end{bmatrix}, & \xi \le \xi_1 \\
\begin{bmatrix} v_L + \int_{\epsilon_L}^{\epsilon(\xi)} \sqrt{\frac{\partial e}{\partial \epsilon}} d\epsilon \\ \epsilon(\xi) \\ \pi_L \end{bmatrix}, & \xi_1 < \xi < \xi_2 \text{ with } \frac{\partial e}{\partial \epsilon}(\epsilon(\xi), \pi_L) = \xi^2 \\
\begin{bmatrix} v_{23} \\ \epsilon_{23} \\ \pi_{23} \end{bmatrix}, & \xi_2 \le \xi \le \xi_3 \\
\begin{bmatrix} v_L + \int_{\epsilon_L}^{\epsilon_\tau(\pi_L)} \sqrt{\frac{\partial e}{\partial \epsilon}} d\epsilon + \int_{\epsilon_\tau(\pi_L)}^{\epsilon(\xi)} \sqrt{\frac{d\tau}{d\epsilon}} d\epsilon \\ \epsilon(\xi) \\ \epsilon_\tau^{-1}(\epsilon(\xi)) \end{bmatrix}, & \xi_3 < \xi < \xi_4 \text{ with } \frac{d\tau}{d\epsilon}(\epsilon(\xi)) = \xi^2 \\
\begin{bmatrix} v_0 \\ \epsilon_{0,L} \\ \pi_{0,L} \end{bmatrix}, & \xi_4 < \xi < 0 \\
\begin{bmatrix} v_0 \\ \epsilon_{0,R} \\ \pi_{0,R} \end{bmatrix}, & 0 < \xi < \xi_5 \\
\begin{bmatrix} v_R - \int_{\epsilon_R}^{\epsilon_\tau(\pi_R)} \sqrt{\frac{\partial e}{\partial \epsilon}} d\epsilon - \int_{\epsilon_\tau(\pi_R)}^{\epsilon(\xi)} \sqrt{\frac{d\tau}{d\epsilon}} d\epsilon \\ \epsilon(\xi) \\ \epsilon_\tau^{-1}(\epsilon(\xi)) \end{bmatrix}, & \xi_5 < \xi < \xi_6 \text{ with } \frac{d\tau}{d\epsilon}(\epsilon(\xi)) = \xi^2 \\
\begin{bmatrix} v_{67} \\ \epsilon_{67} \\ \pi_{67} \end{bmatrix}, & \xi_6 \le \xi \le \xi_7 \\
\begin{bmatrix} v_R - \int_{\epsilon_R}^{\epsilon(\xi)} \sqrt{\frac{\partial e}{\partial \epsilon}} d\epsilon \\ \epsilon(\xi) \\ \pi_R \end{bmatrix}, & \xi_7 < \xi < \xi_8 \text{ with } \frac{\partial e}{\partial \epsilon}(\epsilon(\xi), \pi_R) = \xi^2 \\
\begin{bmatrix} v_R \\ \epsilon_R \\ \pi_R \end{bmatrix}, & \xi_8 \le \xi.
\end{cases}
$$

For more details concerning the solution of this Riemann problem, see [163].

Programs to solve the Riemann problem for the plasticity model can be found in **Program 4.9-49: plasticity.f.** You can execute this Riemann problem solver by clicking

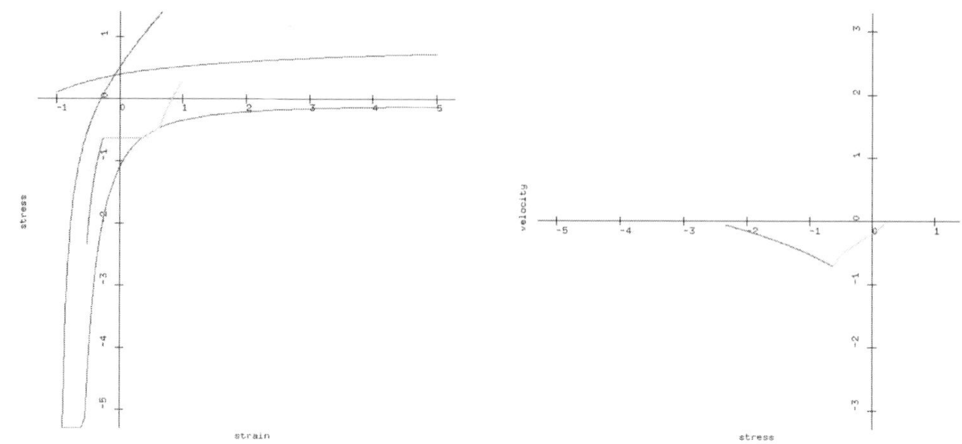

(a) Stress vs. strain

(b) Velocity vs. stress

Fig. 4.14 Solution to plasticity Riemann problem

on the link **Executable 4.9-21: guiPlasticity** You can select input parameters for the Riemann problem by pulling down on "View," releasing on "Main," and then clicking on the arrow next to "Plasticity Parameters." When you have selected all of your input parameters, click on "Start Run Now" in the window labeled "guiPlasticity." Move the windows so that you can see them completely, then answer "OK" to the window that asks "Are you finished viewing the results?" Click in the windows as requested to selected the left and right state for the Riemann problem. Note that the states you select must lie between the tension and compression plastic yield curves in brown. You may drag the mouse when selecting the right state to see how the solution of the Riemann problem changes with the right state. An example of the analytical solution to the plasticity Riemann problem is shown in Figure 4.14.

Exercises for 4.9

4.9.1 We would like to discover conditions under which centered rarefactions are admissible waves. Suppose that the left-hand state is given, and lies inside the elastic regime:

$$\epsilon_\gamma(\pi_L) < \epsilon_L < \epsilon_\tau(\pi_L).$$

Recall that in this regime the stress is given by $s = e(\epsilon, \pi_L)$, and $\pi = \pi_L$ is constant. Also recall that during a rarefaction, the characteristic speeds must increase from left to right. Show that an elastic rarefaction moving to the left satisfies

$$v_R = v_L + \int_{\epsilon_L}^{\epsilon_R} \sqrt{\frac{\partial e}{\partial \epsilon}}\, d\epsilon \text{ for all } \epsilon_\gamma(\pi_L) < \epsilon_L < \epsilon_R < \epsilon_\tau(\pi_L),$$

$s_R = e(\epsilon_R, \pi_R)$ and $\pi_R = \pi_L$. Similarly, show that an elastic rarefaction moving to the right satisfies

$$v_R = v_L + \int_{\epsilon_R}^{\epsilon_L} \sqrt{\frac{\partial e}{\partial \epsilon}}\, d\epsilon \text{ for all } \epsilon_\gamma(\pi_L) < \epsilon_R < \epsilon_L < \epsilon_\tau(\pi_L),$$

$s_R = e(\epsilon_R, \pi_R)$ and $\pi_R = \pi_L$.

4.9.2 Show that elastic shocks moving to the left satisfy

$$v_R = v_L - \sqrt{(\epsilon_R - \epsilon_L)(e(\epsilon_R, \pi_R) - e(\epsilon_L, \pi_L))} \text{ for all } \epsilon_\gamma(\pi_L) < \epsilon_R < \epsilon_L < \epsilon_\tau(\pi_L),$$

$s_R = e(\epsilon_R, \pi_R)$ and $\pi_R = \pi_L$. Similarly, show that elastic shocks moving to the right satisfy

$$v_R = v_L - \sqrt{(\epsilon_R - \epsilon_L)(e(\epsilon_R, \pi_R) - e(\epsilon_L, \pi_L))} \text{ for all } \epsilon_\gamma(\pi_L) < \epsilon_L < \epsilon_R < \epsilon_\tau(\pi_L),$$

$s_R = e(\epsilon_R, \pi_R)$ and $\pi_R = \pi_L$.

4.9.3 Since the time derivative of the strain must be negative during plastic response, show that centered rarefactions are impossible during plastic compression.

4.9.4 Show that during plastic tension, shocks are unphysical.

4.9.5 Since π is constant during elastic response, show that jumps in the plastic strain occur only when $s_R = s_L$ and $v_R = v_L$.

4.9.6 Program the Lax–Friedrichs scheme for this plasticity model. For suggestions of test problems, see [163].

4.9.7 Program Rusanov's scheme for this plasticity model. For suggestions of test problems, see [163].

4.10 Case Study: Polymer Model

We already saw several examples of nonlinear hyperbolic conservation laws for flow in porous media in Sections 3.2.2 and 3.2.3. However, these problems could be each formulated as a single conservation law. Beginning with this section, we will present some examples of problems for flow in porous media which involve systems of conservation laws.

A simple model of three-component two-phase flow can be found in [132]; the complete solution of the Riemann problem for this model in the absence of gravity can be found in [3, 81, 132]. In essentially all oil recovery processes, water is injected to maintain reservoir pressure and to force the oil toward production wells. Since water is less viscous than oil, the water front can develop viscous instabilities, which lead to the formation of "fingers." In some cases, these viscous instabilities are reduced by adding a polymer to the water. The polymer increases the viscosity of water, thereby reducing the viscous instability.

The Riemann problem for the polymer model is interesting because its solution involves states where the characteristic speeds are equal and there is a single characteristic direction. This could cause problems for numerical methods that depend too strongly on characteristic decompositions.

4.10.1 Constitutive Laws

As in the Buckley–Leverett model, we shall assume incompressibility, so that the mass densities of oil and water are constant and the porosity of the rock is constant. Thus the total mass per fluid volume in the oil phase is $\rho_o s_o$, and the total mass per fluid volume in the water phase is $\rho_w s_w$, where s_o and s_w are the saturations of the phases. Recall that saturation is phase volume per fluid volume, so $s_o + s_w = 1$.

In this model, polymer does not mix in the oil, but does mix in the water phase. Let us define the polymer concentration c to be the mass of polymer per water volume divided by the total mass per water volume. Then the mass of polymer per fluid volume is $c\rho_w s_w$. Then conservation of mass for the three chemical components (oil, polymer and water) can be written

$$\frac{d}{dt}\int_\Omega \begin{bmatrix} s_o\rho_o\phi \\ cs_w\rho_w\phi \\ (1-c)s_w\rho_w\phi \end{bmatrix} dx + \int_{\partial\Omega} \begin{bmatrix} \mathbf{n}\cdot\mathbf{v}_o\rho_o \\ \mathbf{n}\cdot\mathbf{v}_w c\rho_w \\ \mathbf{n}\cdot\mathbf{v}_w(1-c)\rho_w \end{bmatrix} ds = 0.$$

Here $\mathbf{v}_o$ and $\mathbf{v}_w$ are the Darcy phase velocities. Since the phase densities are constant, we can divide each equation in this system by the appropriate phase density to get

$$\frac{d}{dt}\int_\Omega \begin{bmatrix} s_o\phi \\ cs_w\phi \\ (1-c)s_w \end{bmatrix} dx + \int_{\partial\Omega} \begin{bmatrix} \mathbf{n}\cdot\mathbf{v}_o \\ \mathbf{n}\cdot\mathbf{v}_w c \\ \mathbf{n}\cdot\mathbf{v}_w(1-c) \end{bmatrix} ds = 0. \tag{4.75}$$

If we sum these equations, we obtain

$$\int_{\partial\Omega} \mathbf{n}\cdot(\mathbf{v}_o + \mathbf{v}_w)\, ds = 0.$$

This equation says that the total fluid velocity $\mathbf{v}_T \equiv \mathbf{v}_o + \mathbf{v}_w$ is divergence-free.

Darcy's law can be written

$$\mathbf{v}_o = \mathbf{K}[-\nabla p + \mathbf{g}\rho_o]\lambda_o(1-s_w),$$
$$\mathbf{v}_w = \mathbf{K}[-\nabla p + \mathbf{g}\rho_w]\lambda_w(s_w, c).$$

The matrix $\mathbf{K}$ represents the rock permeability, and the vector $\mathbf{g}$ represents the acceleration due to gravity. Here we have written the phase mobilities as the ratios of relative permeability divided by viscosity:

$$\lambda_o(1-s_w) = \frac{\kappa_{ro}(1-s_w)}{\mu_o}, \quad \lambda_w(s_w, c) = \frac{\kappa_{rw}(s_w)}{\mu_w(c)}.$$

In these expressions, we have ignored diffusive terms, such as capillary pressure, convective mixing and molecular diffusion.

If we sum these phase velocities and substitute into the sum of the conservation laws, we can obtain an elliptic equation for pressure:

$$\nabla \cdot \{\mathbf{K}(\lambda_o + \lambda_w)\nabla p\} = \nabla \cdot \{\mathbf{Kg}(\rho_o\lambda_o + \rho_w\lambda_w)\}.$$

With appropriate boundary conditions, this can be solved for fixed values of the water saturation s_o and the polymer concentration c to find the pressure p.

In one dimension, the divergence-free condition on the total fluid velocity implies that $\mathbf{v}_T$ is constant in space. This suggests that in general we might rewrite the pressure gradient in terms of the total fluid velocity as follows:

$$-\mathbf{K}\nabla p = \mathbf{v}_T \frac{1}{\lambda_o + \lambda_w} - \mathbf{Kg}\frac{\rho_o\lambda_o + \rho_w\lambda_w}{\lambda_o + \lambda_w}.$$

We can substitute this expression into the formulas for the phase velocities to get

$$\mathbf{v}_w = \mathbf{v}_T \frac{\lambda_w}{\lambda_o + \lambda_w} + \mathbf{Kg}(\rho_w - \rho_o)\frac{\lambda_o\lambda_w}{\lambda_o + \lambda_w}.$$

If we are given the total fluid velocity as a function of space, then one of the conservation laws (4.75) is redundant. We will ignore conservation of oil mass. The remaining system of conservation laws can be written

$$\frac{d}{dt}\int_\Omega \begin{bmatrix} s_w\phi \\ cs_w\phi \end{bmatrix} dx + \int_{\partial\Omega} \begin{bmatrix} \mathbf{n}\cdot\mathbf{v}_w \\ \mathbf{n}\cdot\mathbf{v}_w c \end{bmatrix} ds = 0.$$

In one dimension, this implies the partial differential equations

$$\frac{\partial}{\partial t}\begin{bmatrix} s_w\phi \\ cs_w\phi \end{bmatrix} + \frac{\partial}{\partial x}\begin{bmatrix} \mathbf{v}_w(s_w, c) \\ c\mathbf{v}_w(s_w, c) \end{bmatrix} = 0. \qquad (4.76)$$

4.10.2 Characteristic Analysis

Lemma 4.10.1 *Suppose that $\kappa_{rw}(s_w) \downarrow 0$ as $s_w \downarrow 0$, and $\kappa'_{rw} > 0$ is finite. Then the polymer model* (4.76) *is hyperbolic, with characteristic speeds $\frac{\partial \mathbf{v}_w}{\partial s_w}\frac{1}{\phi}$ and $\frac{\mathbf{v}_w}{s_w}\frac{1}{\phi}$. The latter characteristic speed is linearly degenerate. The two characteristic speeds are equal along a curve in state space, and there is only one characteristic direction at such points.*

Proof It is obvious that both the conserved quantities and the fluxes are functions of the flux variables

$$\mathbf{w} = \begin{bmatrix} s_w \\ c \end{bmatrix}.$$

Thus our quasilinear form of the conservation law is

$$\begin{bmatrix} 1 & 0 \\ c & s_w \end{bmatrix} \frac{\partial}{\partial t} \begin{bmatrix} s_w \\ c \end{bmatrix} + \begin{bmatrix} \phi \partial \mathbf{v}_w / \partial s_w & \partial \mathbf{v}_w / \partial c \\ c \partial \mathbf{v}_w / \partial s_w & c \partial \mathbf{v}_w / \partial c + \mathbf{v}_w \end{bmatrix} \frac{\partial}{\partial x} \begin{bmatrix} s_w \\ c \end{bmatrix} = 0.$$

Thus the system of conservation laws is hyperbolic if and only if the matrix

$$\mathbf{A} = \begin{bmatrix} 1 & 0 \\ c & s_w \end{bmatrix}^{-1} \begin{bmatrix} \partial \mathbf{v}_w / \partial s_w & \partial \mathbf{v}_w / \partial c \\ c \partial \mathbf{v}_w / \partial s_w & c \partial \mathbf{v}_w / \partial c + \mathbf{v}_w \end{bmatrix} \frac{1}{\phi} = \begin{bmatrix} \partial \mathbf{v}_w / \partial s_w & \partial \mathbf{v}_w / \partial c \\ 0 & \mathbf{v}_w / s_w \end{bmatrix} \frac{1}{\phi}$$

has real eigenvalues. Since this matrix is upper triangular, hyperbolicity is obvious.

Note that the second characteristic speed is $\mathbf{v}_w / (\phi s_w)$. At first glance, it would appear that this characteristic speed could be infinite. However, $\mathbf{v}_w$ is proportional to the mobility λ_w. Since $\lambda_w(s_w, c) = \kappa_{rw}(s_w)/\mu_w(c)$, where the relative permeability $\kappa_{rw}(s_w) \downarrow 0$ as $s_w \downarrow 0$ and $\kappa_{rw}' > 0$ is finite, this characteristic speed is finite.

The first characteristic speed, $\frac{\partial \mathbf{v}_w}{\partial s_w} \frac{1}{\phi}$, is called the **Buckley Leverett speed**. This is $1/\phi$ times the slope of the Darcy velocity of water with respect to saturation. The second characteristic speed, called the **particle velocity**, is $1/\phi$ times the Darcy velocity chord slope of the line through the origin. Since we typically have $\kappa_{rw}'(0) = 0$ and $\kappa_{ro}'(0) = 0$, this Darcy velocity curve is shaped like an "S". As a result, for any polymer concentration c there is at least one water saturation s_w so that the two characteristic speeds are equal. This defines a curve, called the **equivelocity curve**. Note that when the two characteristic speeds are equal, there is only one characteristic direction. □

4.10.3 Jump Conditions

Lemma 4.10.2 *Suppose that $\kappa_{rw}(s_w) \downarrow 0$ as $s_w \downarrow 0$, and $\kappa_{rw}' > 0$ is finite. Then the Rankine–Hugoniot jump conditions for the polymer model* (4.76) *have two kinds of solutions. Either*

$$c_{w,R} = c_{w,L}, \quad s_{w,R} \neq s_{w,L}, \quad \sigma = \frac{\mathbf{v}_{w,R} - \mathbf{v}_{w,L}}{s_{w,R} - s_{w,L}} \frac{1}{\phi}$$

or

$$c_{w,R} \neq c_{w,L}, \quad s_{w,R} = s_{w,L}, \quad \sigma = \frac{\mathbf{v}_{w,L}}{s_{w,L}\phi} = \frac{\mathbf{v}_{w,R}}{s_{w,R}\phi}.$$

Proof The Rankine–Hugoniot jump conditions for the polymer flooding model imply that

$$\begin{bmatrix} \mathbf{v}_{w,R} - \mathbf{v}_{w,L} \\ c_{w,R}\mathbf{v}_{w,R} - c_{w,L}\mathbf{v}_{w,L} \end{bmatrix} = \begin{bmatrix} s_{w,R} - s_{w,L} \\ c_{w,R}s_{w,R} - c_{w,L}s_{w,L} \end{bmatrix} \phi\sigma.$$

where σ is the discontinuity speed. The first of these jump conditions implies that

$$\mathbf{v}_{w,R} = \mathbf{v}_{w,L} + (s_{w,R} - s_{w,L})\phi\sigma.$$

Substituting this result into the second jump condition leads to

$$0 = (c_{w,R} - c_{w,L})(\mathbf{v}_{w,L} - s_{w,L}\phi\sigma)$$

This requires either that $c_{w,R} = c_{w,L}$ or that $\mathbf{v}_{w,L} = s_{w,L}\phi\sigma$.

If $c_{w,R} = c_{w,L}$, then we must have $s_{w,R} \neq s_{w,L}$ in order to have a jump. In this case, the first Rankine–Hugoniot jump condition implies that

$$\sigma = \frac{\mathbf{v}_{w,R} - \mathbf{v}_{w,L}}{s_{w,R} - s_{w,L}} \frac{1}{\phi}.$$

This is a Buckley–Leverett discontinuity, involving a jump in saturation but no jump in polymer concentration.

If $c_{w,R} \neq c_{w,L}$, then we must have $\mathbf{v}_{w,L} = s_{w,L}\phi\sigma$, which in turn implies that $\sigma = \mathbf{v}_{w,L}/(s_{w,L}\phi)$. The first jump condition now implies that

$$\mathbf{v}_{w,R} = \mathbf{v}_{w,L} + (s_{w,R} - s_{w,L})\phi\sigma = s_{w,L}\phi\sigma + (s_{w,R} - s_{w,L})\phi\sigma = s_{w,R}\phi\sigma.$$

It follows that $\sigma = \mathbf{v}_{w,R}/s_{w,R}\phi$ as well. □

This discontinuity speed is the particle velocity discussed in the characteristic analysis above, and is a contact discontinuity.

4.10.4 Riemann Problem Solution

The waves in this problem can be explained graphically in terms of the velocity curves as follows. Buckley–Leverett waves correspond to changes in saturation for a fixed polymer concentration; these follow the convex or concave hull of the velocity function, depending on the ordering of the saturations in the Riemann problem. The particle velocity waves correspond to changes in concentration and saturation; these follow line through the origin, intersecting the velocity functions for different polymer concentrations.

The solution of the Riemann problem is complicated to describe, due to the non-convexity of the velocity function. An interactive program to solve the Riemann problem for polymer flooding can be executed by clicking on the link **Executable 4.10-22: guiPolymer** You can select input parameters for the Riemann problem by pulling down on "View," releasing on "Main," and then clicking on the arrow next to "Polymer Parameters." When you have selected all of your input parameters, click on "Start Run Now" in the window labeled "1d/polymer_riemann." Move the windows so that you can see them completely, then answer "OK" to the window that asks "Are you finished viewing the results?" Click in the windows as requested

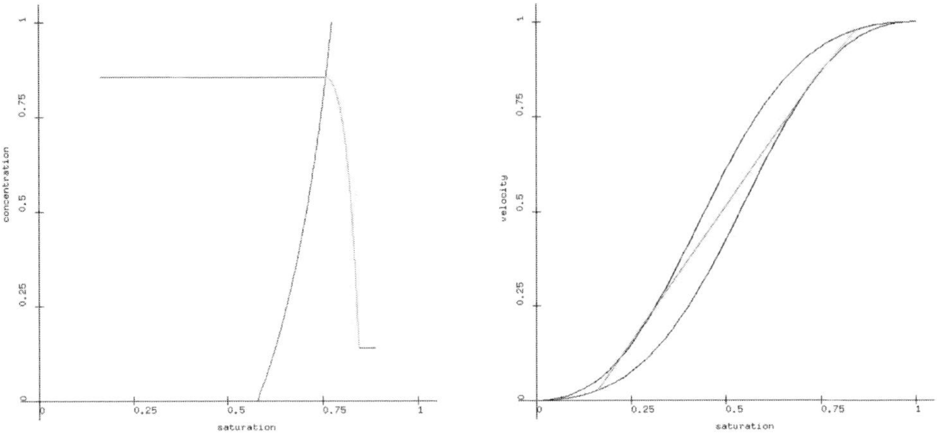

(a) Concentration vs. saturation

(b) Velocity vs. saturation

Fig. 4.15 Solution to polymer Riemann problem

to selected the left and right state for the Riemann problem. You may drag the mouse when selecting the right state to see how the solution changes. Students interested in the details of the solution of the Riemann problem can examine the code in **Program 4.10-50: polymer.f.** Figure 4.15 shows one solution of the polymer flooding Riemann problem.

The Riemann problem for the polymer flooding model was solved by Isaacson, but never published. A number of petroleum engineers [71, 132] published particular solutions to polymer flooding and set forth the general principles for the solution of the Riemann problem. A paper by Keyfitz and Kranzer [81] on the Riemann problem for the vibrating string included a discussion of a general system of two conservation laws that encompasses the polymer model, under the assumption that the velocity function is convex. See also [53, 80] for more general solutions involving adsorption and hysteresis.

Exercises for 4.10

4.10.1 Show that the particle velocity is linearly degenerate. Also show that since the water velocity is "S"-shaped, the Buckley–Leverett speed is neither genuinely nonlinear nor linearly degenerate.

4.10.2 Determine equations for a centered rarefaction in the Buckley–Leverett wave family. Under what conditions is the centered rarefaction admissible?

4.10.3 Determine the equations for the Hugoniot loci in either wave family. Under what conditions are they admissible?

4.10.4 Show that the polymer concentration is the Riemann invariant for the Buckley–Leverett characteristic speed, and the particle velocity is the Riemann invariant for the other wave family.

4.10.5 Show that $S(s_w, c) = s_w \beta(c)$ is an entropy function for polymer model, with entropy flux $\mathbf{v}\beta(c)/\phi$, where β is any function of polymer concentration. Also show that this entropy function is neither convex nor concave for any choice of β.

4.10.6 Describe the solution of the Riemann problem in the absence of gravity. For help, see [3].

4.10.7 Program Rusanov's scheme for the polymer flooding problem.

4.10.8 Program the Lax–Friedrichs scheme for three-phase Buckley-Leverett flow.

4.11 Case Study: Three-Phase Buckley–Leverett Flow

4.11.1 Constitutive Models

Another interesting model for flow in porous media concerns the flow of three immiscible and incompressible phases. We shall call these phases oil, gas and water, and ignore the obvious fact that a gas phase should be considered to be compressible, especially at reservoir pressures. However, nearly incompressible three-phase flow does occur in other circumstances, such as chemically-enhanced recovery. (For example, surfactant flooding leads to greatest mobilization of the oil when the fluid forms water, oil and microemulsion phases; further, it is possible for the injection of carbon dioxide to involve the formation of two separate hydrocarbon phases, in addition to water and gas.)

Since the saturations are the volume fractions of the phases, they sum to one: $s_w + s_o + s_g = 1$. Since the phases are incompressible, their mass densities (*i.e.*, mass of phase per phase volume) are constant; it follows that the masses per fluid volume of the phases are $\rho_w s_w$, $\rho_o s_o$ and $\rho_g s_g$. We will assume that Darcy's law is valid for each phase:

$$\mathbf{v}_w = -\mathbf{K}(\nabla p - \mathbf{g}\rho_w)\lambda_w(s_w),$$

$$\mathbf{v}_o = -\mathbf{K}(\nabla p - \mathbf{g}\rho_o)\lambda_o(s_o),$$

$$\mathbf{v}_g = -\mathbf{K}(\nabla p - \mathbf{g}\rho_g)\lambda_g(s_g).$$

Here we have ignored diffusive forces, and recalled that

$$\lambda_w(s_w) = \frac{\kappa_{rw}(s_w)}{\mu_w}, \quad \lambda_w(s_o) = \frac{\kappa_{ro}(s_o)}{\mu_o} \quad \text{and} \quad \lambda_w(s_g) = \frac{\kappa_{rg}(s_g)}{\mu_g}$$

are the phase mobilities, namely the ratios of phase relative permeability to phase viscosity. Then conservation of mass takes the form

$$\frac{d}{dt}\int_{\Omega}\begin{bmatrix}\rho_w s_w\\ \rho_o s_o\\ \rho_g s_g\end{bmatrix}\phi\,dt+\int_{\partial\Omega}\begin{bmatrix}\mathbf{n}\cdot\mathbf{v}_w\rho_w\\ \mathbf{n}\cdot\mathbf{v}_o\rho_o\\ \mathbf{n}\cdot\mathbf{v}_g\rho_g\end{bmatrix}ds=0.$$

As before, we can divide each of the mass conservations by the corresponding phase density to obtain

$$\frac{d}{dt}\int_{\Omega}\begin{bmatrix}s_w\\ s_o\\ s_g\end{bmatrix}\phi\,dt+\int_{\partial\Omega}\begin{bmatrix}\mathbf{n}\cdot\mathbf{v}_w\\ \mathbf{n}\cdot\mathbf{v}_o\\ \mathbf{n}\cdot\mathbf{v}_g\end{bmatrix}ds=0.$$

If we sum these equations, we obtain

$$\int_{\partial\Omega}\mathbf{n}\cdot(\mathbf{v}_w+\mathbf{v}_o+\mathbf{v}_g)\,ds=0.$$

This equation says that the sum of the phase velocities is divergence-free. If we substitute Darcy's law for the phase velocities, we obtain

$$\nabla\cdot[\mathbf{K}(\lambda_w+\lambda_o+\lambda_g)\nabla p]=\nabla\cdot[\mathbf{K}\mathbf{g}(\lambda_w\rho_w+\lambda_o\rho_o+\lambda_g\rho_g)].$$

This gives us an elliptic equation for the fluid pressure, given the phase saturations (and therefore the phase mobilities).

We can also write the phase velocities in terms of the total fluid velocity $\mathbf{v}_T = \mathbf{v}_w + \mathbf{v}_o + \mathbf{v}_g$:

$$\begin{bmatrix}\mathbf{v}_w\\ \mathbf{v}_o\\ \mathbf{v}_g\end{bmatrix}=\begin{bmatrix}\lambda_w\\ \lambda_o\\ \lambda_g\end{bmatrix}\frac{\mathbf{v}_T}{\lambda_w+\lambda_o+\lambda_g}+\begin{bmatrix}\lambda_w\{\lambda_o(\rho_w-\rho_o)+\lambda_g(\rho_w-\rho_g)\}\\ \lambda_o\{\lambda_w(\rho_o-\rho_w)+\lambda_g(\rho_o-\rho_g)\}\\ \lambda_g\{\lambda_w(\rho_g-\rho_w)+\lambda_o(\rho_g-\rho_o)\}\end{bmatrix}\mathbf{K}\mathbf{g}\frac{1}{\lambda_w+\lambda_o+\lambda_g}.$$

If the total fluid velocity $\mathbf{v}_T$ is given, then the phase velocities can be considered to be functions of s_w and s_o. In particular, in one dimension the divergence-free condition on the total fluid velocity implies that it is constant in space, and this assumption is reasonable. In particular, in one dimension, we can simplify the conservation laws to

$$\frac{\partial}{\partial t}\begin{bmatrix}s_w\phi\\ s_o\phi\end{bmatrix}+\frac{\partial}{\partial x}\begin{bmatrix}\mathbf{v}_w\\ \mathbf{v}_o\end{bmatrix}=0.$$

4.11.2 Characteristic Analysis

It is easy to see that the system of conservation laws is hyperbolic if and only if the matrix

$$\mathbf{A} = \begin{bmatrix} \partial \mathbf{v}_w / s_w & \partial \mathbf{v}_w / s_o \\ \partial \mathbf{v}_o / s_w & \partial \mathbf{v}_o / s_o \end{bmatrix} \frac{1}{\phi}$$

has real eigenvalues. The quadratic formula for the eigenvalues of $\mathbf{A}\phi$ shows that the eigenvalues will be real if and only if the discriminant is positive:

$$0 \leq \left(\frac{\partial \mathbf{v}_w}{\partial s_w} + \frac{\partial \mathbf{v}_o}{s_o} \right)^2 - \left(\frac{\partial \mathbf{v}_w}{\partial s_w} \frac{\partial \mathbf{v}_o}{s_o} - \frac{\partial \mathbf{v}_w}{s_o} \frac{\partial \mathbf{v}_o}{s_w} \right) = \left(\frac{\partial \mathbf{v}_w}{\partial s_w} - \frac{\partial \mathbf{v}_o}{s_o} \right)^2 + \frac{\partial \mathbf{v}_w}{s_o} \frac{\partial \mathbf{v}_o}{s_w}.$$

To simplify the discussion, let us ignore gravity: $\mathbf{g} = 0$. In this case,

$$\left(\frac{\partial \mathbf{v}_w}{\partial s_w} - \frac{\partial \mathbf{v}_o}{s_o} \right)^2 + \frac{\partial \mathbf{v}_w}{s_o} \frac{\partial \mathbf{v}_o}{s_w} = m^\top \mathbf{P} m.$$

Here we have used the notation

$$m \equiv \begin{bmatrix} \mu_w \\ \mu_o \\ \mu_g \end{bmatrix} \frac{1}{\lambda_w + \lambda_o + \lambda_g} \frac{1}{\mu_w \mu_o \mu_g}.$$

We have also used the matrix

$$\mathbf{P} = \begin{bmatrix} \kappa_{ww} & \kappa_{wo} & \kappa_{wg} \\ \kappa_{ow} & \kappa_{oo} & \kappa_{og} \\ \kappa_{gw} & \kappa_{go} & \kappa_{gg} \end{bmatrix}$$

where the entries of **P** are functions of the relative permeabilities:

$$\begin{aligned}
\kappa_{ww} &= (\kappa'_{ro}\kappa_{rg} + \kappa'_{rg}\kappa_{ro})^2, \\
\kappa_{oo} &= (\kappa'_{rw}\kappa_{rg} + \kappa'_{rg}\kappa_{rw})^2, \\
\kappa_{gg} &= (\kappa'_{ro}\kappa_{rw} + \kappa'_{rw}\kappa_{ro})^2, \\
\kappa_{wo} &= -(\kappa'_{rg}\kappa_{rw} + \kappa'_{rw}\kappa_{rg})(\kappa'_{rg}\kappa_{ro} + \kappa'_{ro}\kappa_{rg}) + 2\kappa_{rw}\kappa_{ro}(\kappa'_{rg})^2, \\
\kappa_{wg} &= (\kappa'_{rw}\kappa_{ro} - \kappa'_{ro}\kappa_{rw})(\kappa'_{rg}\kappa_{ro} + \kappa'_{ro}\kappa_{rg}) - 2\kappa_{rw}\kappa_{ro}\kappa'_{rg}\kappa'_{ro}, \\
\kappa_{og} &= -(\kappa'_{ro}\kappa_{rw} - \kappa'_{rw}\kappa_{ro})(\kappa'_{rg}\kappa_{rw} + \kappa'_{rw}\kappa_{ro}) - 2\kappa_{rw}\kappa_{ro}\kappa'_{rg}\kappa'_{rw}.
\end{aligned}$$

This matrix is singular, but it does have a factorization $\mathbf{P} = LDL^\top$ where

$$L = \begin{bmatrix} \kappa_{ww} & 0 \\ -\kappa_{gw} & \kappa'_{rg} \\ -\kappa_{go} & -\kappa'_{ro} \end{bmatrix}$$

and

$$D = \begin{bmatrix} 1 & 0 \\ 0 & 4\kappa_{rw}\kappa_{ro}\kappa_{rg}(\kappa'_{rg}\kappa'_{ro}\kappa_{rw} + \kappa_{rg}\kappa'_{ro}\kappa'_{rw} + \kappa'_{rg}\kappa_{ro}\kappa'_{rw}) \end{bmatrix} \frac{1}{\kappa_{ww}}.$$

It is reasonable to assume that $\kappa_{rw}(s_w)$, $\kappa_{ro}(s_o)$ and $\kappa_{rg}(s_g)$ are all increasing functions, taking values between zero and one, and zero at zero. It follows that if at least two of the saturations are positive, then $\mathbf{K}$ is nonnegative and the system of conservation laws is hyperbolic. It is possible to show that the three-phase Buckley–Leverett model is hyperbolic even when gravity is included, and that the relative permeabilities described here are the only functional forms for which this is true for all values of gravity and the viscosities [161].

4.11.3 Umbilic Point

Note that there is a special point, called the **umbilic point** where the eigenvalues of $\mathbf{A}$ are equal. This occurs when $m^\top \mathbf{K} m = 0$. If at least two saturations are nonzero, this occurs where $L^\top m = 0$. An examination of the nullspace of $L^\top$ shows that this equation is equivalent to

$$\frac{\mu_w}{\kappa'_w} = \frac{\mu_o}{\kappa'_o} = \frac{\mu_g}{\kappa'_g} = \alpha.$$

In other words, the saturations are defined by

$$s_w = (\kappa'_{rw})^{-1}\left(\frac{\mu_w}{\alpha}\right), \quad s_o = (\kappa'_{ro})^{-1}\left(\frac{\mu_o}{\alpha}\right), \quad s_g = (\kappa'_{rg})^{-1}\left(\frac{\mu_g}{\alpha}\right).$$

Since the saturations are constrained to sum to one, α is defined by

$$1 = (\kappa'_{rw})^{-1}\left(\frac{\mu_w}{\alpha}\right) + (\kappa'_{ro})^{-1}\left(\frac{\mu_o}{\alpha}\right) + (\kappa'_{rg})^{-1}\left(\frac{\mu_g}{\alpha}\right).$$

In the special case where

$$\kappa_{rw}(s_w) = s_w^2, \quad \kappa_{ro}(s_o) = s_o^2, \quad \kappa_{rg}(s_g) = s_g^2,$$

the umbilic point occurs at

$$s_w = \frac{\mu_w}{\mu_w + \mu_o + \mu_g}, \quad s_o = \frac{\mu_o}{\mu_w + \mu_o + \mu_g}, \quad s_g = \frac{\mu_g}{\mu_w + \mu_o + \mu_g}.$$

4.11.4 Elliptic Regions

Other choices of the relative permeability functions can lead to **elliptic regions**. These are regions in the saturation triangle where the characteristic speeds are not real. Problems of this sort lead to all kinds of interesting mathematical analysis, but are of little practical interest. Rather, we should view the difficulty as a modeling problem: people who develop models need to take care that their models are well-posed (in this case, for initial-value problems). For more information regarding three-phase flow, see [5, 8, 15, 21, 36, 74, 147, 148, 161].

Exercise for 4.11

4.11.1 Program the Rusanov scheme for three-phase Buckley–Leverett flow. Experiment with numerical solution of Riemann problems with states chosen near the umbilic point.

4.12 Case Study: Schaeffer–Schechter–Shearer System

An example of a system of conservation laws with no known physical application has been suggested by Schaeffer, Schecter and Shearer [143]. This model is interesting because the Hugoniot loci for the wave families terminate in the middle of state space. As a result, it is possible for the two wave families associated with the left and right states in a Riemann problem to have no intersection. Such a problem could cause difficulties for a numerical computation, appearing to the unwise as a problem with the numerical method rather than with the problem.

The system of conservation laws for this model is

$$\frac{\partial}{\partial t}\begin{bmatrix} p \\ q \end{bmatrix} + \frac{\partial}{\partial \mathbf{x}}\begin{bmatrix} p^2 - q \\ \frac{1}{3}p^3 - p \end{bmatrix} = 0. \tag{4.77}$$

In this case, we take the flux variables to be $\mathbf{w} = \mathbf{u} = \begin{bmatrix} p \\ q \end{bmatrix}$.

Lemma 4.12.1 *The system of conservation laws in Equation* (4.77) *is hyperbolic, with genuinely nonlinear characteristic speeds* $p \pm 1$.

Proof The matrix of flux derivatives

$$\frac{\partial \mathbf{f}}{\partial \mathbf{u}} = \begin{bmatrix} 2p & -1 \\ p^2 - 1 & 0 \end{bmatrix}$$

has eigenvalues and eigenvectors given by the expression

$$\frac{\partial \mathbf{f}}{\partial \mathbf{u}}\mathbf{X} = \begin{bmatrix} 2p & -1 \\ p^2 - 1 & 0 \end{bmatrix}\begin{bmatrix} 1 & 1 \\ p+1 & p-1 \end{bmatrix} = \begin{bmatrix} 1 & 1 \\ p+1 & p-1 \end{bmatrix}\begin{bmatrix} p-1 & 0 \\ 0 & p+1 \end{bmatrix} = \mathbf{X}\Lambda.$$

Thus the characteristic speeds are $p \mp 1$. Both characteristic speeds are genuinely nonlinear, since

$$\frac{\partial \lambda_i}{\partial \mathbf{u}}\mathbf{X}\mathbf{e}_i = \frac{\partial (p \pm 1)}{\partial \mathbf{u}}\begin{bmatrix} 1 \\ p \pm 1 \end{bmatrix} = \begin{bmatrix} 1 & 0 \end{bmatrix}\begin{bmatrix} 1 \\ p \pm 1 \end{bmatrix} = 1.$$

□

Lemma 4.12.2 *Given a left state (p_L, q_L), the points (p, q) on the slow Rankine–Hugoniot locus for the Schaeffer–Schechter–Shearer system (4.77) satisfy*

$$q(p) = q_L + (p - p_L)\left[\frac{p + p_L}{2} + \sqrt{1 - \frac{(p - p_L)^2}{12}}\right] \text{ where } p_L > p > p_L - \sqrt{12}$$

with shock speed

$$\sigma_-(p) = \frac{p + p_L}{2} - \sqrt{1 - \frac{(p - p_L)^2}{12}}.$$

Similarly, given a right state (p_R, q_R), the points (p, q) on the fast Rankine–Hugoniot locus satisfy

$$q(p) = q_R + (p - p_R)\left[\frac{p + p_R}{2} - \sqrt{1 - \frac{(p - p_R)^2}{12}}\right] \text{ where } p_R < p < p_R + \sqrt{12}$$

with shock speed

$$\sigma_+(p) = \frac{p + p_R}{2} + \sqrt{1 - \frac{(p - p_R)^2}{12}}.$$

Proof The jump conditions for this hyperbolic system are

$$[p^2 - q] = [p]\sigma, \quad \left[\frac{1}{3}p^3 - p\right] = [q]\sigma.$$

There are five variables (namely p_L, p_R, q_L, q_R and σ) and two equations. In order to solve the jump conditions, we will assume that we know the three variables p_L, p_R and q_L. Then the jump conditions imply

$$\left[\frac{1}{3}p^3\right] - [p] = [p^2]\sigma - [p]\sigma^2.$$

We can divide this equation by $[p]$ to get

$$\frac{1}{3}(p_R^2 + p_L p_R + p_L^2) - 1 = (p_R + p_L)\sigma - \sigma^2.$$

Now let $z = (p_R - p_L)/p_L$. Then we can write $p_R = p_L(1 + z)$, and

$$\frac{1}{3}p_L^2\left\{(1 + z)^2 + (1 + z) + 1\right\} - 1 = p_L(2 + z)\sigma - \sigma^2.$$

If we solve this equation for σ, we get

$$\sigma = \frac{1}{2}\left[p_L(2+z) \pm \sqrt{p_L^2(2+z)^2 - 4\left(1+z+\frac{1}{3}z^2\right)p_L^2 + 4}\right]$$
$$= \frac{1}{2}(p_L + p_R) \pm \sqrt{1 - \frac{(p_R - p_L)^2}{12}}.$$

We can use the jump relation $[q] = [p^2] - [p]\sigma$ to obtain

$$q_R = q_L + (p_R^2 - p_L^2) - (p_R - p_L)\left[\frac{p_R + p_L}{2} \pm \sqrt{1 - \frac{(p_R - p_L)^2}{12}}\right]$$
$$= q_L + (p_R - p_L)\left[\frac{p_R + p_L}{2} \mp \sqrt{1 - \frac{(p_R - p_L)^2}{12}}\right].$$

Note that as the jump in p tends to zero, the discontinuity speed tends to one of the characteristic speeds $p \pm 1$. On the other hand, for a given value of p_L there is a real solution for σ only for $|p_R - p_L| \leq \sqrt{12}$.

The slow discontinuity is admissible when $p_L - 1 > \sigma > p_R - 1$. Using our solution to the jump conditions in that example, we see that these inequalities imply that

$$0 > \frac{[p]}{2} - \sqrt{1 - \frac{[p]^2}{12}} + 1 > [p].$$

This can be rewritten

$$1 + \frac{[p]}{2} < \sqrt{1 - \frac{[p]^2}{12}} < 1 - \frac{[p]}{2}. \tag{4.78}$$

The outermost inequalities imply that $[p] < 0$. Thus it is permissible to square both sides of the right-hand inequality in (4.78) to obtain

$$1 - \frac{[p]^2}{12} < 1 - [p] + \frac{[p]^2}{4};$$

this can be rewritten

$$0 < [p]\left(\frac{[p]}{3} - 1\right).$$

This inequality is satisfied whenever $[p] < 0$. If $-2 < [p] < 0$, then we can square both sides of the left inequality in (4.78) to obtain

$$1 - [p] + \frac{[p]^2}{4} < 1 - \frac{[p]^2}{12};$$

this can be rewritten

$$0 < -[p]\left(\frac{[p]}{3}+1\right)$$

and places no further restrictions on $[p]$. If $[p] < -2$, then we notice that the discontinuity speed is real provided that $[p] > -\sqrt{12}$. Thus the slow discontinuity in this example exists and is admissible if and only if $0 > [p] > -\sqrt{12}$. A similar analysis for the fast discontinuity arrives at the same admissibility condition. □

Lemma 4.12.3 *The Riemann invariants for the Schaeffer–Schechter–Shearer system* (4.77) *are* $q - p - \frac{1}{2}p^2$ *for the slow rarefaction, and* $q + p - \frac{1}{2}p^2$ *for the fast rarefaction.*

Proof Centered rarefactions for this system satisfy

$$\frac{d}{d\mathbf{y}}\begin{bmatrix} p \\ q \end{bmatrix} = \begin{bmatrix} 1 \\ p \pm 1 \end{bmatrix}.$$

It follows that along centered rarefactions $\frac{dq}{dp} = p \pm 1$. If we integrate this ordinary differential equation, we get

$$q_R - q_L = \frac{1}{2}(p_R^2 - p_L^2) \pm (p_R - p_L).$$

The Riemann invariant for the slow rarefaction is $q - p - \frac{1}{2}p^2$, and the Riemann invariant for the fast rarefaction is $q + p - \frac{1}{2}p^2$. □

Theorem 4.12.4 *(Schaeffer–Schechter–Shearer Riemann Problem) Suppose that we are given a left state* (p_L, q_L) *and a right state* (p_R, q_R) *in a Riemann problem for the Schaeffer–Schechter–Shearer model* (4.77). *If a solution exists, it involves the intermediate state* (p_*, q_*) *at the intersection of the slow wave curve*

$$q_-(p) \equiv \begin{cases} q_L + (p - p_L)\left[\frac{p+p_L}{2} + 1.\right], & p > p_L \\ q_L + (p - p_L)\left[\frac{p+p_L}{2} + \sqrt{1 - \frac{(p-p_L)^2}{12}}\right], & p_L - \sqrt{12} \le p \le p_L \end{cases}$$

and the fast wave curve

$$q_+(p) \equiv \begin{cases} q_R - (p_R - p)\left[\frac{p+p_R}{2} - 1.\right], & p < p_R \\ q_R - (p_R - p)\left[\frac{p+p_R}{2} - \sqrt{1 - \frac{(p_R-p)^2}{12}}\right], & p_R + \sqrt{12} \ge p \ge p_R. \end{cases}$$

Thus there are four different structural forms for the solution of the Riemann problem: either a slow shock or a slow rarefaction followed by either a fast shock or a fast rarefaction. Let

$$\xi_L = \begin{cases} p_L - 1, & p_* > p_L \\ \frac{p_*+p_L}{2} - \sqrt{1 - \frac{(p_L-p_*)^2}{12}}, & p_L - \sqrt{12} \leq p_* \leq p_L \end{cases}$$

$$\xi_- = \begin{cases} p_* - 1, & p_* > p_L \\ \frac{p_*+p_L}{2} - \sqrt{1 - \frac{(p_L-p_*)^2}{12}}, & p_L - \sqrt{12} \leq p_* \leq p_L \end{cases}$$

be the wave speeds at the beginning and the end of the slow wave, and

$$\xi_+ = \begin{cases} p_R + 1, & p_* < P_R \\ \frac{p_*+p_R}{2} + \sqrt{1 - \frac{(p_R-p_*)^2}{12}}, & p_R + \sqrt{12} \geq p_* \geq p_R \end{cases}$$

$$\xi_R = \begin{cases} p_R + 1, & p_* < h_R \\ \frac{p_*+p_R}{2} + \sqrt{1 - \frac{(p_R-p_*)^2}{12}}, & p_R + \sqrt{12} \geq p_* \geq p_R \end{cases}$$

be the wave speeds at the beginning and the end of the fast wave. Then the state that moves with speed ξ in the Riemann problem is

$$(p_\xi, q_\xi) = \begin{cases} (p_L, q_L), & \xi < \xi_L \\ \left(\xi + 1, q_L + (p_\xi - p_L)\left[1 + \frac{p_\xi+p_L}{2}\right]\right), & \xi_L < \xi < \xi_- \\ (p_*, q_*), & \xi_- < \xi < \xi_+ \\ \left(\xi - 1, q_R + (p_R - p_\xi)\left[1 - \frac{p_\xi+p_R}{2}\right]\right), & \xi_+ < \xi < \xi_R \\ (p_R, q_R), & \xi_R < \xi. \end{cases}$$

Proof This follows from the results in Lemmas 4.12.1, 4.12.2 and 4.12.3. □

Programs to solve the Riemann problem for the Schaeffer–Schechter–Shearer model can be found in **Program 4.12-51: schaeffer_shearer.f.** You can execute this Riemann problem solver by clicking on the link **Executable 4.12-23: guiSchaefferShearer.** You can select input parameters for the Riemann problem by pulling down on "View," releasing on "Main," and then clicking on arrows as needed. When you have selected all of your input parameters, click on "Start Run Now" in the window labeled "1d/schaeffer_shearer_riemann." Click in the window to select the left and right state for the Riemann problem. You may drag the mouse when selecting the right state to see how the solution changes. Some experimentation easily shows that there are choices of the left and right states for which the two wave families do not intersect, and the Riemann problem has no solution. We would expect numerical methods to have trouble if they were given such a Riemann problem to integrate. For example, the Riemann problem in Figure 4.16 involves wave families that come close to intersecting, but do not.

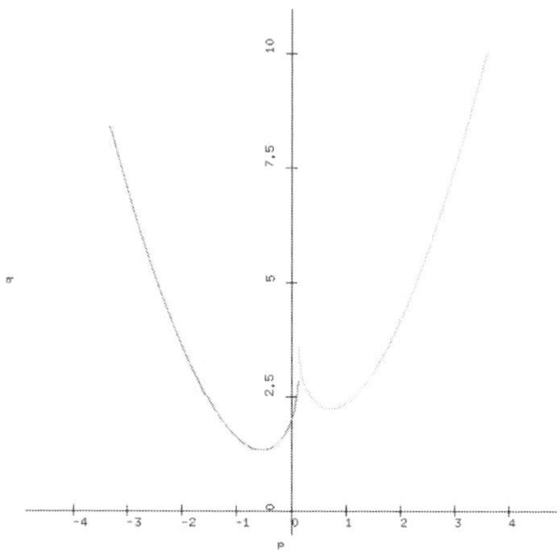

p vs. q

Fig. 4.16 Solution to Schaeffer–Schechter–Shearer Riemann problem

The next lemma discusses entropy functions for this model.

Lemma 4.12.5 *For all constants* α, β *and* γ, $S(p, q) = \alpha[q^2 + p^2(1 - p^2/6)] + \beta p + \gamma q$ *is an entropy function for* (4.77) *with entropy flux* $\Psi(p, q) = \alpha p[q(p^2/3 - 1) + 2p^2/3(1 - p^2/5)] + \beta(p^2 - q) + \gamma p(p^2/3 - 1)$.

Proof We compute

$$\frac{\partial S}{\partial \mathbf{u}} = \left[\alpha p(1 - p^2/3) + \beta, \quad \alpha q + \gamma\right]$$

and

$$\frac{\partial \Psi}{\partial \mathbf{u}} = \left[\alpha q(p^2 - 1) + 2\alpha p^2(1 - p^2/3) + 2\beta p + \gamma(p^2 - 1), \quad -\alpha p - \beta\right]$$

Since

$$\frac{\partial \mathbf{f}}{\partial \mathbf{u}} = \begin{bmatrix} 2p & -1 \\ p^2 - 1 & 0 \end{bmatrix}$$

we have that

$$\begin{aligned} \frac{\partial S}{\partial \mathbf{u}} \frac{\partial \mathbf{f}}{\partial \mathbf{u}} &= \left[\alpha p(1 - p^2/3) + \beta, \quad \alpha q + \gamma\right] \begin{bmatrix} 2p & -1 \\ p^2 - 1 & 0 \end{bmatrix} \\ &= \left[\alpha q(p^2 - 1) + 2\alpha p^2(1 - p^2/3) + 2\beta p + \gamma(p^2 - 1), \quad -\alpha p - \beta\right] = \frac{\partial \Psi}{\partial \mathbf{u}}. \end{aligned}$$

□

This form was discovered by assuming that $S(p,q) = P(p) + Q(q)$ and using the equation (4.29) to determine the possible forms of P and Q.

Exercises for 4.12

4.12.1 Given an arbitrary state $\mathbf{w} = \begin{bmatrix} p \\ q \end{bmatrix}$, plot the fast and slow rarefaction curves going out of $\mathbf{w}$ in the directions of increasing characteristic speed. Also plot the fast and slow Hugoniot loci going out of $\mathbf{w}$ in the directions of decreasing characteristic speed.

4.12.2 Give an example of a choice of left and right states for which the Riemann problem does not have a finite intersection of the wave families.

4.12.3 How would you expect the analytical solution to this example Riemann problem to behave? Program Rusanov's scheme for your Riemann problem and verify your expectations.

4.13 Approximate Riemann Solvers

The solution of Riemann problems seems to make the use of Godunov's method more daunting than other first-order schemes. The analytical solutions of the vibrating string and polymer flooding Riemann problems are pretty difficult to program, while the analytical solution of the three-phase Buckley–Leverett model exists only as a proprietary software program (not ours). Fortunately, there are several good techniques for approximating the solution of Riemann problems.

4.13.1 Design of Approximate Riemann Solvers

The discussion in this section has been adapted from the review paper by Harten *et al.* [68]. Recall that the solution to the Riemann problem $\mathcal{R}(\mathbf{u}_L, \mathbf{u}_R, x/t)$ is the exact solution to the initial value problem in one spatial dimension

$$\frac{\partial \mathbf{u}}{\partial t} + \frac{\partial \mathbf{f}(\mathbf{u})}{\partial x} = 0, \quad \mathbf{u}(x,0) = \begin{cases} \mathbf{u}_L, & x < 0 \\ \mathbf{u}_R, & x > 0 \end{cases}.$$

Let $\lambda_{\max}$ be the largest absolute value of any characteristic speed in any part of the solution of this Riemann problem. If we integrate the conservation law in space over the interval $(-\triangle x_L/2, \triangle x_R/2)$, and in time over the interval $(0, \triangle t)$ where $\lambda_{\max}\triangle t < \min\{\triangle x_L/2, \triangle x_R/2\}$, then the divergence theorem implies that

$$\int_{-\triangle x_L/2}^{\triangle x_R/2} \mathcal{R}\left(\mathbf{u}_L, \mathbf{u}_R, \frac{x}{\triangle t}\right) dx = \mathbf{u}_L \frac{\triangle x_L}{2} + \mathbf{u}_R \frac{\triangle x_R}{2} - \triangle t[\mathbf{f}(\mathbf{u}_R) - \mathbf{f}(\mathbf{u}_L)].$$

We can also use the divergence theorem over quadrilaterals in space and time. For any $\xi \in (-\triangle x_L/2\triangle t, \triangle x_R/2\triangle t)$ we integrate over the quadrilateral with corners

$(-\triangle x_L/2, 0)$, $(0, 0)$, $(\xi\triangle t, \triangle t)$ and $(-\triangle x_L/2, \triangle t)$ to get

$$0 = \int_{-\Delta x_L/2}^{\xi\Delta t} \mathcal{R}\left(\mathbf{u}_L, \mathbf{u}_R, \frac{x}{\Delta t}\right) dx - \mathbf{u}_L \frac{\Delta x_L}{2} - \triangle t \mathbf{f}(\mathbf{u}_L)$$
$$+ \int_0^{\Delta t} \frac{1}{\sqrt{1+\xi^2}} \begin{bmatrix} -\xi & 1 \end{bmatrix} \begin{bmatrix} \mathbf{u} \\ \mathbf{f} \end{bmatrix} \sqrt{1+\xi^2}\, dt.$$

Because of the self-similarity of the solution of Riemann problems, u and $\mathbf{f}(u)$ are constant along the line $x = \xi t$, so the integral at the far right of this equation has the value $\triangle t \lim_{x\uparrow\xi t}(\mathbf{f} - \mathbf{u}\xi)$. We conclude that the flux along the curve $x = \xi t$ is given by

$$\lim_{x\uparrow\xi t}(\mathbf{f} - \mathbf{u}\xi) = \mathbf{f}(\mathbf{u}_L) + \mathbf{u}_L \frac{\Delta x_L}{2\triangle t} - \frac{1}{\triangle t} \int_{-\Delta x_L/2}^{\xi\Delta t} \mathcal{R}\left(\mathbf{u}_L, \mathbf{u}_R, \frac{x}{\Delta t}\right) dx \tag{4.79a}$$

Alternatively, we can apply the divergence theorem over the quadrilateral with corners $(0, 0)$, $(\triangle x_R/2, 0)$, $(\triangle x_R/2, \triangle t)$ and $(\xi\triangle t, \triangle t)$ to get

$$\int_{\xi\Delta t}^{\Delta x_R/2} \mathcal{R}\left(\mathbf{u}_L, \mathbf{u}_R, \frac{x}{\Delta t}\right) dx - \mathbf{u}_R \frac{\Delta x_R}{2}$$
$$+ \int_0^{\Delta t} \frac{1}{\sqrt{1+\xi^2}} \begin{bmatrix} \xi & -1 \end{bmatrix} \begin{bmatrix} \mathbf{u} \\ \mathbf{f} \end{bmatrix} \sqrt{1+\xi^2}\, dt - \triangle t \mathbf{f}(\mathbf{u}_R) = 0,$$

so we conclude that the flux along the curve $x = \xi t$ is given by

$$\lim_{x\downarrow\xi t}(\mathbf{f} - \mathbf{u}\xi) = \mathbf{f}(\mathbf{u}_R) - \mathbf{u}_R \frac{\Delta x_R}{2\triangle t} + \frac{1}{\triangle t} \int_{\xi\Delta t}^{\Delta x_R/2} \mathcal{R}\left(\mathbf{u}_L, \mathbf{u}_R, \frac{x}{\Delta t}\right) dx. \tag{4.79b}$$

Now suppose that we have an approximate Riemann solver of the form

$$\tilde{\mathcal{R}}(\mathbf{u}_L, \mathbf{u}_R, \xi) = \mathbf{u}_L + \sum_{j:\lambda_j<\xi} \Delta\mathbf{u}_j$$

where $\sum_j \Delta\mathbf{u}_j = \mathbf{u}_R - \mathbf{u}_L$. The constraint guarantees that the approximate Riemann solver reproduces the left and right states for sufficiently large (either positive or negative) ξ. Then we can integrate the approximate Riemann solver to get

$$\int_{-\Delta x_L/2}^{\xi\Delta t} \tilde{\mathcal{R}}\left(\mathbf{u}_L, \mathbf{u}_R, \frac{x}{\Delta t}\right) dx = \mathbf{u}_L\left(\xi\triangle t + \frac{\Delta x_L}{2}\right) + \triangle t \sum_j \max\{\xi - \lambda_j, 0\}\Delta\mathbf{u}_j$$

and

$$\int_{\xi\Delta t}^{\Delta x_R/2} \tilde{\mathcal{R}}\left(\mathbf{u}_L, \mathbf{u}_R, \frac{x}{\Delta t}\right) dx = \mathbf{u}_R\left(-\xi\triangle t + \frac{\Delta x_R}{2}\right) - \triangle t \sum_j \max\{\lambda_j - \xi, 0\}\Delta\mathbf{u}_j.$$

Combining these two equations leads to

$$\int_{-\Delta x_L/2}^{\Delta x_R/2} \tilde{\mathcal{R}}\left(\mathbf{u}_L, \mathbf{u}_R, \frac{x}{\Delta t}\right) dx = \frac{1}{2}(\mathbf{u}_L\Delta x_L + \mathbf{u}_R\Delta x_R) - \triangle t \sum_j \lambda_j\Delta\mathbf{u}_j.$$

We will typically approximate the flux along the curve $x = \xi t$ by averaging the expressions for the flux on either side of this curve:

$$\begin{aligned}
[\mathbf{f}(\mathbf{u}) - \mathbf{u}\xi](\xi t, t) &\approx \tilde{\mathbf{f}}_\xi(\mathbf{u}_L, \mathbf{u}_R) \\
&= \frac{1}{2}\left\{\mathbf{f}(\mathbf{u}_L) + \mathbf{u}_L \frac{\triangle x_L}{2\triangle t} - \frac{1}{\triangle t}\int_{-\triangle x_L/2}^{\xi\triangle t} \tilde{\mathcal{R}}\left(\mathbf{u}_L, \mathbf{u}_R, \frac{x}{\triangle t}\right) dx \right. \\
&\quad \left. + \mathbf{f}(\mathbf{u}_R) - \mathbf{u}_R \frac{\triangle x_R}{2\triangle t} + \frac{1}{\triangle t}\int_{\xi\triangle t}^{\triangle x_R/2} \tilde{\mathcal{R}}\left(\mathbf{u}_L, \mathbf{u}_R, \frac{x}{\triangle t}\right) dx \right\} \\
&= \frac{1}{2}\left\{\mathbf{f}(\mathbf{u}_L) + \mathbf{f}(\mathbf{u}_R) - (\mathbf{u}_L + \mathbf{u}_R)\xi - \sum_j |\lambda_j - \xi| \Delta \mathbf{u}_j \right\}. \qquad (4.80)
\end{aligned}$$

Note that Godunov's method does not use the full information from the Riemann problem solution; it averages that function on the grid at the new time. Thus, it should be possible to approximate the solution to the Riemann problem in order to reduce the computational work. When we do so, we need to make sure that the resulting scheme is consistent with the original conservation law, and satisfies the entropy condition.

Lemma 4.13.1 *Suppose that* $\tilde{\mathcal{R}}(\mathbf{u}_L, \mathbf{u}_R, \xi)$ *is* **consistent**, *meaning that for all states* $\mathbf{u}$ *and all speeds* ξ *we have* $\tilde{\mathcal{R}}(\mathbf{u}, \mathbf{u}, \xi) = \mathbf{u}$. *Also assume that* $\tilde{\mathcal{R}}(\mathbf{u}_L, \mathbf{u}_R, \xi)$ *is* **conservative**, *meaning that for all states* $\mathbf{u}_L$ *and* $\mathbf{u}_R$ *and for all timesteps satisfying* $\triangle t < \frac{1}{2}\max_j\{|\lambda_j|\}$,

$$\int_{-\triangle x_L/2}^{\triangle x_R/2} \tilde{\mathcal{R}}\left(\mathbf{u}_L, \mathbf{u}_R, \frac{x}{\triangle t}\right) dx = \frac{\mathbf{u}_L \triangle x_L + \mathbf{u}_R \triangle x_R}{2} - \triangle t[\mathbf{f}(\mathbf{u}_R) - \mathbf{f}(\mathbf{u}_L)].$$

Then with $x_i = \frac{1}{2}(x_{i+1/2} + x_{i-1/2})$ *and* $\triangle x_i = x_{i+1/2} - x_{i-1/2}$, *the scheme*

$$\begin{aligned}
\tilde{\mathbf{u}}_i^{n+1} \triangle x_i &= \int_{x_i}^{x_{i+1/2}} \tilde{\mathcal{R}}\left(\tilde{\mathbf{u}}_i^n, \tilde{\mathbf{u}}_{i+1}^n, \frac{x - x_{i+1/2}}{\triangle t^{n+\frac{1}{2}}}\right) dx \\
&\quad + \int_{x_{i-1/2}}^{x_i} \tilde{\mathcal{R}}\left(\tilde{\mathbf{u}}_{i-1}^n, \tilde{\mathbf{u}}_i^n, \frac{x - x_{i-1/2}}{\triangle t^{n+\frac{1}{2}}}\right) dx \qquad (4.81)
\end{aligned}$$

can be rewritten as a conservative difference where the numerical flux

$$\begin{aligned}
\tilde{\mathbf{f}}_{i+\frac{1}{2}}(\tilde{\mathbf{u}}_i^n, \tilde{\mathbf{u}}_{i+1}^n) &= \mathbf{f}(\tilde{\mathbf{u}}_i^n) - \frac{\triangle x_i}{2\triangle t^{n+\frac{1}{2}}}\tilde{\mathbf{u}}_i^n - \frac{1}{\triangle t^{n+\frac{1}{2}}}\int_{-\triangle x_i/2}^{0} \tilde{\mathcal{R}}\left(\tilde{\mathbf{u}}_i^n, \tilde{\mathbf{u}}_{i+1}^n, \frac{x}{\triangle t^{n+\frac{1}{2}}}\right) dx \\
&= \mathbf{f}(\tilde{\mathbf{u}}_{i+1}^n) - \frac{\triangle x_{i+1}}{2\triangle t^{n+\frac{1}{2}}}\tilde{\mathbf{u}}_{i+1}^n + \frac{1}{\triangle t^{n+\frac{1}{2}}}\int_0^{\triangle x_{i+1}/2} \tilde{\mathcal{R}}\left(\tilde{\mathbf{u}}_i^n, \tilde{\mathbf{u}}_{i+1}^n, \frac{x}{\triangle t^{n+\frac{1}{2}}}\right) dx
\end{aligned}$$

is **consistent** *with the original flux, meaning that for all* $\mathbf{u}$, $\tilde{\mathbf{f}}_{i+\frac{1}{2}}(\mathbf{u}, \mathbf{u}) = \mathbf{f}(\mathbf{u})$.

Proof Using the fact that $\tilde{\mathcal{R}}$ is conservative, we can rewrite

$$
\begin{aligned}
\tilde{\mathbf{u}}_i^{n+1} &= \frac{1}{\triangle x_i}\int_{x_i}^{x_{i+1/2}} \tilde{\mathcal{R}}\left(\tilde{\mathbf{u}}_i^n, \tilde{\mathbf{u}}_{i+1}^n, \frac{x - x_{i+1/2}}{\triangle t^{n+\frac{1}{2}}}\right) dx + \frac{1}{\triangle x_i}\int_{x_{i-1/2}}^{x_i} \tilde{\mathcal{R}}\left(\tilde{\mathbf{u}}_{i-1}^n, \tilde{\mathbf{u}}_i^n, \frac{x - x_{i-1/2}}{\triangle t^{n+\frac{1}{2}}}\right) dx \\
&= \frac{\tilde{\mathbf{u}}_i \triangle x_i + \tilde{\mathbf{u}}_{i+1}^n \triangle x_{i+1}}{2\triangle x_i} - \frac{\triangle t^{n+\frac{1}{2}}}{\triangle x_i}[\mathbf{f}(\tilde{\mathbf{u}}_{i+1}^n) - \mathbf{f}(\tilde{\mathbf{u}}_i^n)] - \frac{1}{\triangle x_i}\int_0^{\triangle x_{i+1}/2} \tilde{\mathcal{R}}\left(\tilde{\mathbf{u}}_i^n, \tilde{\mathbf{u}}_{i+1}^n, \frac{x}{\triangle t^{n+\frac{1}{2}}}\right) dx \\
&\quad + \frac{\tilde{\mathbf{u}}_{i-1}\triangle x_{i-1} + \tilde{\mathbf{u}}_i^n \triangle x_i}{2\triangle x_i} - \frac{\triangle t^{n+\frac{1}{2}}}{\triangle x_i}[\mathbf{f}(\tilde{\mathbf{u}}_i^n) - \mathbf{f}(\tilde{\mathbf{u}}_{i-1}^n)] - \frac{1}{\triangle x_i}\int_{-\triangle x_{i-1}/2}^{0} \tilde{\mathcal{R}}\left(\tilde{\mathbf{u}}_{i-1}^n, \tilde{\mathbf{u}}_i^n, \frac{x}{\triangle t^{n+\frac{1}{2}}}\right) dx \\
&= \tilde{\mathbf{u}}_i^n - \frac{\triangle t^{n+\frac{1}{2}}}{\triangle x_i}\left\{\mathbf{f}(\tilde{\mathbf{u}}_{i+1}^n) - \frac{\triangle x_{i+1}}{2\triangle t^{n+\frac{1}{2}}}\tilde{\mathbf{u}}_{i+1}^n + \frac{1}{\triangle t^{n+\frac{1}{2}}}\int_0^{\triangle x_{i+1}/2} \tilde{\mathcal{R}}\left(\tilde{\mathbf{u}}_i^n, \tilde{\mathbf{u}}_{i+1}^n, \frac{x}{\triangle t^{n+\frac{1}{2}}}\right) dx\right\} \\
&\quad + \frac{\triangle t^{n+\frac{1}{2}}}{\triangle x_i}\left\{\mathbf{f}(\tilde{\mathbf{u}}_{i-1}^n) - \frac{\triangle x_{i-1}}{2\triangle t^{n+\frac{1}{2}}}\tilde{\mathbf{u}}_{i-1}^n - \frac{1}{\triangle t^{n+\frac{1}{2}}}\int_{-\triangle x_{i-1}/2}^{0} \tilde{\mathcal{R}}\left(\tilde{\mathbf{u}}_i^n, \tilde{\mathbf{u}}_{i+1}^n, \frac{x}{\triangle t^{n+\frac{1}{2}}}\right) dx\right\} \\
&\equiv \tilde{\mathbf{u}}_i^n - \frac{\triangle t^{n+\frac{1}{2}}}{\triangle x_i}[\mathbf{f}_{i+\frac{1}{2}}^{+}(\tilde{\mathbf{u}}_i^n, \tilde{\mathbf{u}}_{i+1}^n) - \mathbf{f}_{i-\frac{1}{2}}^{-}(\tilde{\mathbf{u}}_{i-1}^n, \tilde{\mathbf{u}}_i^n)]
\end{aligned}
$$

Since $\tilde{\mathcal{R}}$ is consistent,

$$
\mathbf{f}_{i+\frac{1}{2}}^{+}(\mathbf{u}, \mathbf{u}) = \mathbf{f}(\mathbf{u}) - \frac{\triangle x_{i+1}}{2\triangle t^{n+\frac{1}{2}}}\mathbf{u} + \frac{1}{\triangle t^{n+\frac{1}{2}}}\int_0^{\triangle x_{i+1}/2} \tilde{\mathcal{R}}\left(\mathbf{u}, \mathbf{u}, \frac{x}{\triangle t^{n+\frac{1}{2}}}\right) dx = \mathbf{f}(\mathbf{u})
$$

$$
\mathbf{f}_{i-\frac{1}{2}}^{-}(\mathbf{u}, \mathbf{u}) = \mathbf{f}(\mathbf{u}) + \frac{\triangle x_{i-1}}{2\triangle t^{n+\frac{1}{2}}}\mathbf{u} - \frac{1}{\triangle t^{n+\frac{1}{2}}}\int_{-\triangle x_{i-1}/2}^{0} \tilde{\mathcal{R}}\left(\mathbf{u}, \mathbf{u}, \frac{x}{\triangle t^{n+\frac{1}{2}}}\right) dx = \mathbf{f}(\mathbf{u}).
$$

Further, since $\tilde{\mathcal{R}}$ is conservative

$$
\begin{aligned}
&\mathbf{f}_{i+\frac{1}{2}}^{+}(\mathbf{u}_L, \mathbf{u}_R) - \mathbf{f}_{i+\frac{1}{2}}^{-}(\mathbf{u}_L, \mathbf{u}_R) \\
&= \mathbf{f}(\mathbf{u}_R) - \frac{\triangle x_{i+1}}{2\triangle t^{n+\frac{1}{2}}}\mathbf{u}_R + \frac{1}{\triangle t^{n+\frac{1}{2}}}\int_0^{\triangle x_{i+1}/2} \tilde{\mathcal{R}}\left(\mathbf{u}_L, \mathbf{u}_R, \frac{x}{\triangle t^{n+\frac{1}{2}}}\right) dx \\
&\quad - \mathbf{f}(\mathbf{u}_L) - \frac{\triangle x_i}{2\triangle t^{n+\frac{1}{2}}}\mathbf{u}_L + \frac{1}{\triangle t^{n+\frac{1}{2}}}\int_{-\triangle x_i/2}^{0} \tilde{\mathcal{R}}\left(\mathbf{u}_L, \mathbf{u}_R, \frac{x}{\triangle t^{n+\frac{1}{2}}}\right) dx = 0.
\end{aligned}
$$

As a result, we can drop the $\pm$ superscripts from these numerical fluxes. Thus the scheme (4.81) has been rewritten as a conservative difference. □

Note that the notions of conservative difference scheme and conservative approximate Riemann solvers are distinct. It is possible to have a conservative difference that leads to a convergent finite difference approximation, without using a conservative approximate Riemann problem solver.

Let us return to the exact solution of the Riemann problem. If there is a convex entropy function $S(\mathbf{u})$ with entropy flux $\Psi(\mathbf{u})$, then in the limit of vanishing diffusion the inequality

$$\frac{\partial S(\mathbf{u})}{\partial t} + \frac{\partial \Psi(\mathbf{u})}{\partial x} \leq 0$$

holds weakly. Integrating in space and time and applying the divergence theorem leads to

$$\int_{-\triangle x_L/2}^{\triangle x_R/2} S\left(\mathcal{R}\left(\mathbf{u}_L, \mathbf{u}_R, \frac{x}{\triangle t}\right)\right) dx \leq S(\mathbf{u}_L)\frac{\triangle x_L}{2} + S(\mathbf{u}_R)\frac{\triangle x_R}{2} - \triangle t[\Psi(\mathbf{u}_R) - \Psi(\mathbf{u}_L)]. \quad (4.82)$$

Compare this result to the next lemma.

Lemma 4.13.2 *Suppose that* $\tilde{\mathcal{R}}(\mathbf{u}_L, \mathbf{u}_R, \xi)$ *is* consistent *and* conservative. *With ;* $x_i = \frac{1}{2}(x_{i+1/2} + x_{i-1/2})$ *and* $\triangle x_i = x_{i+1/2} - x_{i-1/2}$, *consider the scheme*

$$\tilde{\mathbf{u}}_i^{n+1} = \frac{1}{\triangle x_i}\int_{x_i}^{x_{i+1/2}} \tilde{\mathcal{R}}\left(\tilde{\mathbf{u}}_i^n, \tilde{\mathbf{u}}_{i+1}^n, \frac{x - x_{i+1/2}}{\triangle t^{n+\frac{1}{2}}}\right) dx + \frac{1}{\triangle x_i}\int_{x_{i-1/2}}^{x_i} \tilde{\mathcal{R}}\left(\tilde{\mathbf{u}}_{i-1}^n, \tilde{\mathbf{u}}_i^n, \frac{x - x_{i-1/2}}{\triangle t^{n+\frac{1}{2}}}\right) dx. \quad (4.83)$$

If $\tilde{\mathcal{R}}$ *satisfies the* **entropy inequality**

for all $\mathbf{u}_L, \mathbf{u}_R$ *for all* $\triangle t < \dfrac{\min\{\triangle x_L, \triangle x_R\}}{2\max_j\{|\lambda_j|\}}$

$$\int_{-\triangle x_L/2}^{\triangle x_R/2} S\left(\tilde{\mathcal{R}}\left(\mathbf{u}_L, \mathbf{u}_R, \frac{x}{\triangle t}\right)\right) dx \geq \frac{S(\mathbf{u}_L)\triangle x_L + S(\mathbf{u}_R)\triangle x_R}{2} - \triangle t[\Psi(\mathbf{u}_R) - \Psi(\mathbf{u}_L)], \quad (4.84)$$

then the scheme satisfies the entropy inequality

$$S(\tilde{\mathbf{u}}_i^{n+1}) \geq S(\tilde{\mathbf{u}}_i^n) - \frac{\triangle t^{n+1/2}}{\triangle x_i}\left[\tilde{\Psi}_{i+\frac{1}{2}}(\tilde{\mathbf{u}}_i^n, \tilde{\mathbf{u}}_{i+1}^n) - \tilde{\Psi}_{i-\frac{1}{2}}(\tilde{\mathbf{u}}_{i-1}^n, \tilde{\mathbf{u}}_i^n)\right]$$

where the numerical entropy flux

$$\begin{aligned}\tilde{\Psi}_{i+\frac{1}{2}}(\mathbf{u}_i^n, \mathbf{u}_{i+1}^n) &= \Psi(\mathbf{u}_i^n) + \frac{\triangle x_i}{2\triangle t^{n+\frac{1}{2}}} S(\mathbf{u}_i^n) + \frac{1}{\triangle t^{n+\frac{1}{2}}}\int_{-\triangle x_i/2}^{0} S\left(\tilde{\mathcal{R}}\left(\mathbf{u}_i^n, \mathbf{u}_{i+1}^n, \frac{x}{\triangle t^{n+\frac{1}{2}}}\right)\right) dx\\ &= \Psi(\mathbf{u}_{i+1}^n) - \frac{\triangle x_{i+1}}{2\triangle t^{n+\frac{1}{2}}} S(\mathbf{u}_{i+1}^n) + \frac{1}{\triangle t^{n+\frac{1}{2}}}\int_{0}^{-\triangle x_{i+1}/2} S\left(\tilde{\mathcal{R}}\left(\mathbf{u}_i^n, \mathbf{u}_{i+1}^n, \frac{x}{\triangle t^{n+\frac{1}{2}}}\right)\right) dx\end{aligned}$$

is **consistent** *with the original entropy flux, meaning that for all* $\mathbf{u}$ *we have* $\tilde{\Psi}_{i+\frac{1}{2}}(\mathbf{u}, \mathbf{u}) = \Psi(\mathbf{u})$.

Proof Let us examine the entropy inequality for the scheme (4.83). Since $\tilde{\mathcal{R}}$ satisfies the entropy inequality (4.84),

$$\int_{x_i}^{x_{i+1/2}} S\left(\tilde{\mathcal{R}}\left(\tilde{\mathbf{u}}_i^n, \tilde{\mathbf{u}}_{i+1}^n, \frac{x - x_{i+1/2}}{\triangle t^{n+\frac{1}{2}}}\right)\right) \triangle x_i + \int_{x_{i-1/2}}^{x_i} S\left(\tilde{\mathcal{R}}\left(\tilde{\mathbf{u}}_{i-1}^n, \tilde{\mathbf{u}}_i^n, \frac{x - x_{i-1/2}}{\triangle t^{n+\frac{1}{2}}}\right)\right) \triangle x_i$$

$$\geq \frac{\triangle x_i}{2} S(\tilde{\mathbf{u}}_i^n) + \frac{\triangle x_{i+1}}{2} S(\tilde{\mathbf{u}}_{i+1}^n) - \frac{\triangle t^{n+\frac{1}{2}}}{2} [\Psi(\tilde{\mathbf{u}}_{i+1}^n) - \Psi(\tilde{\mathbf{u}}_i^n)]$$

$$- \int_0^{\triangle x_{i+1}/2} S\left(\tilde{\mathcal{R}}\left(\tilde{\mathbf{u}}_i^n, \tilde{\mathbf{u}}_{i+1}^n, \frac{x}{\triangle t^{n+\frac{1}{2}}}\right)\right) dx$$

$$+ \frac{\triangle x_{i-1}}{2} S(\tilde{\mathbf{u}}_{i-1}^n) + \frac{\triangle x_i}{2} S(\tilde{\mathbf{u}}_i^n) - \frac{\triangle t^{n+\frac{1}{2}}}{2} [\Psi(\tilde{\mathbf{u}}_i^n) - \Psi(\tilde{\mathbf{u}}_{i-1}^n)]$$

$$- \int_{-\triangle x_{i+1}/2}^{0} S\left(\tilde{\mathcal{R}}\left(\tilde{\mathbf{u}}_{i-1}^n, \tilde{\mathbf{u}}_i^n, \frac{x}{\triangle t^{n+\frac{1}{2}}}\right)\right) dx$$

$$= S(\tilde{\mathbf{u}}_i^n) \triangle x_i - \triangle t^{n+\frac{1}{2}} \left\{ \Psi(\tilde{\mathbf{u}}_{i+1}^n) - \frac{\triangle x_{i+1}}{2 \triangle t^{n+\frac{1}{2}}} S(\tilde{\mathbf{u}}_{i+1}^n) \right.$$

$$\left. + \frac{1}{\triangle t^{n+\frac{1}{2}}} \int_0^{\triangle x_{i+1}/2} S\left(\tilde{\mathcal{R}}\left(\tilde{\mathbf{u}}_i^n, \tilde{\mathbf{u}}_{i+1}^n, \frac{x}{\triangle t^{n+\frac{1}{2}}}\right)\right) dx \right\}$$

$$+ \triangle t^{n+\frac{1}{2}} \left\{ \Psi(\tilde{\mathbf{u}}_{i-1}^n) + \frac{\triangle x_{i-1}}{2 \triangle t^{n+\frac{1}{2}}} S(\tilde{\mathbf{u}}_{i-1}^n) \right.$$

$$\left. + \frac{1}{\triangle t^{n+\frac{1}{2}}} \int_{-\triangle x_{i-1}/2}^{0} S\left(\tilde{\mathcal{R}}\left(\tilde{\mathbf{u}}_{i-1}^n, \tilde{\mathbf{u}}_i^n, \frac{x}{\triangle t^{n+\frac{1}{2}}}\right)\right) dx \right\}$$

$$\equiv S(\tilde{\mathbf{u}}_i^n) \triangle x_i - \triangle t^{n+\frac{1}{2}} \left[\Psi_{i+\frac{1}{2}}^{+}(\tilde{\mathbf{u}}_i^n, \tilde{\mathbf{u}}_{i+1}^n) - \Psi_{i-\frac{1}{2}}^{-}(\tilde{\mathbf{u}}_{i-1}^n, \tilde{\mathbf{u}}_i^n) \right].$$

Since $\tilde{\mathcal{R}}$ is consistent,

$$\Psi_{i+\frac{1}{2}}^{+}(\mathbf{u}, \mathbf{u}) = \Psi(\mathbf{u}) - \frac{\triangle x_{i+1}}{2 \triangle t^{n+\frac{1}{2}}} S(\mathbf{u}) + \frac{1}{\triangle t^{n+\frac{1}{2}}} \int_0^{\triangle x_{i+1}/2} S\left(\tilde{\mathcal{R}}\left(\mathbf{u}, \mathbf{u}, \frac{x}{\triangle t^{n+\frac{1}{2}}}\right)\right) dx = \Psi(\mathbf{u})$$

$$\Psi_{i-\frac{1}{2}}^{-}(\mathbf{u}, \mathbf{u}) = \Psi(\mathbf{u}) + \frac{\triangle x_{i-1}}{2 \triangle t^{n+\frac{1}{2}}} S(\mathbf{u}) + \frac{1}{\triangle t^{n+\frac{1}{2}}} \int_{-\triangle x_{i+1}/2}^{0} S\left(\tilde{\mathcal{R}}\left(\mathbf{u}, \mathbf{u}, \frac{x}{\triangle t^{n+\frac{1}{2}}}\right)\right) dx = \Psi(\mathbf{u}).$$

Thus both $\Psi_{i+\frac{1}{2}}^{+}$ and $\Psi_{i-\frac{1}{2}}^{-}$ are consistent with the original entropy flux. We can use either one, or any average of these, to define a conservative difference for the entropy. Suppose that we use $\Psi^+_{i+\frac{1}{2}}$. Then

$$\begin{aligned}\sum_i S(\tilde{\mathbf{u}}_i^{n+1}) &\geq \sum_i S(\tilde{\mathbf{u}}_i^n) - \frac{\triangle t^{n+\frac{1}{2}}}{\triangle x_i}\sum_i [\Psi^+_{i+1/2}(\tilde{\mathbf{u}}_i^n, \tilde{\mathbf{u}}_{i+1}^n) - \Psi^+_{i-1/2}(\tilde{\mathbf{u}}_{i-1}^n, \tilde{\mathbf{u}}_i^n)] \\ &= \sum_i S(\tilde{\mathbf{u}}_i^n) - \frac{\triangle t^{n+\frac{1}{2}}}{\triangle x_i}\sum_i [\Psi^+_{i+1/2}(\tilde{\mathbf{u}}_i^n, \tilde{\mathbf{u}}_{i+1}^n) - \Psi^+_{i+1/2}(\tilde{\mathbf{u}}_i^n, \tilde{\mathbf{u}}_{i+1}^n)] \\ &= \sum_i S(\tilde{\mathbf{u}}_i^n).\end{aligned}$$

□

Definition 4.13.3 The conservative difference scheme

$$\mathbf{u}_i^{n+1} = \mathbf{u}_i^n - \frac{\triangle t^{n+\frac{1}{2}}}{\triangle x_i}[\tilde{\mathbf{f}}_{i+\frac{1}{2}}(\mathbf{u}_i^n, \mathbf{u}_{i+1}^n) - \tilde{\mathbf{f}}_{i-\frac{1}{2}}(\mathbf{u}_{i-1}^n, \mathbf{u}_i^n)]$$

is an **upstream scheme** if and only if whenever all signal speeds are positive we have $\tilde{\mathbf{f}}_{i+\frac{1}{2}}(\mathbf{u}_L, \mathbf{u}_R) = \mathbf{f}(\mathbf{u}_L)$ for all $\mathbf{u}_L$ and all $\mathbf{u}_R$, and whenever all signal speeds are negative we have $\tilde{\mathbf{f}}_{i+\frac{1}{2}}(\mathbf{u}_L, \mathbf{u}_R) = \mathbf{f}(\mathbf{u}_R)$ for all $\mathbf{u}_L$ and all $\mathbf{u}_R$.

Lemma 4.13.4 *Suppose that* $\tilde{\mathcal{R}}(\mathbf{u}_L, \mathbf{u}_R, \xi)$ *is* consistent *and* conservative. *Also suppose that for all speeds* ξ *and all states* $\mathbf{u}_L, \mathbf{u}_R$, *whenever all signal speeds are positive we have* $\tilde{\mathcal{R}}(\mathbf{u}_L, \mathbf{u}_R, \xi) = \mathbf{u}_L$ *and whenever all signal speeds are negative we have* $\tilde{\mathcal{R}}(\mathbf{u}_L, \mathbf{u}_R, \xi) = \mathbf{u}_R$. *Then the scheme*

$$\begin{aligned}\tilde{\mathbf{u}}_i^{n+1} &= \frac{1}{\triangle x_i}\int_{x_i}^{x_{i+1/2}} \tilde{\mathcal{R}}\left(\tilde{\mathbf{u}}_i^n, \tilde{\mathbf{u}}_{i+1}^n, \frac{x - x_{i+1/2}}{\triangle t^{n+\frac{1}{2}}}\right) dx \\ &+ \frac{1}{\triangle x_i}\int_{x_{i-1/2}}^{x_i} \tilde{\mathcal{R}}\left(\tilde{\mathbf{u}}_{i-1}^n, \tilde{\mathbf{u}}_i^n, \frac{x - x_{i-1/2}}{\triangle t^{n+\frac{1}{2}}}\right) dx \end{aligned} \tag{4.85}$$

is an upstream scheme.

Proof The proof of Lemma 4.13.1 shows that the scheme (4.85) can be rewritten as a conservative difference via the numerical flux

$$\begin{aligned}\tilde{\mathbf{f}}_{i+\frac{1}{2}}(\mathbf{u}_L, \mathbf{u}_R) &= \mathbf{f}(\mathbf{u}_R) - \frac{\triangle x_{i+1}}{2\triangle t^{n+\frac{1}{2}}}\mathbf{u}_R + \frac{1}{\triangle x_i}\int_0^{\triangle x_{i+1}/2} \tilde{\mathcal{R}}\left(\mathbf{u}_L, \mathbf{u}_R, \frac{x}{\triangle t^{n+\frac{1}{2}}}\right) dx \\ &= \mathbf{f}(\mathbf{u}_L) + \frac{\triangle x_i}{2\triangle t^{n+\frac{1}{2}}}\mathbf{u}_L - \frac{1}{\triangle x_i}\int_{-\triangle x_i/2}^0 \tilde{\mathcal{R}}\left(\mathbf{u}_L, \mathbf{u}_R, \frac{x}{\triangle t^{n+\frac{1}{2}}}\right) dx.\end{aligned}$$

The lemma now follows immediately from the definition of an upstream scheme. □

4.13.2 Artificial Diffusion

We can write any numerical flux in the form

$$
\begin{aligned}
(\mathbf{f}-\mathbf{u}\xi)(\mathbf{u}_L,\mathbf{u}_R) &= \frac{1}{2}[\mathbf{f}(\mathbf{u}_L)+\mathbf{f}(\mathbf{u}_R)] - \frac{\varepsilon}{2}[\mathbf{u}_L+\mathbf{u}_R] - \frac{1}{2}\mathrm{A}\ \mathrm{sign}(\mathbf{A}-\mathbf{I}\xi)[\mathbf{u}_R-\mathbf{u}_L] \quad (4.86)\\
d(\mathbf{u}_L,\mathbf{u}_R) &= \mathrm{A}\ \mathrm{sign}(\mathbf{A}-\mathbf{I}\xi)(\mathbf{u}_R-\mathbf{u}_L)
\end{aligned}
$$

for some function d. The function d acts as a numerical diffusion. Let us discuss the typical situation, in which we seek the state moving at zero speed in the solution of the Riemann problem ($\xi = 0$). For convex entropy, in order to achieve perfect resolution of stationary shocks we want

$$\mathbf{f}(\mathbf{u}_L)=\mathbf{f}(\mathbf{u}_R) \quad \text{and} \quad \Psi(\mathbf{u}_R) > \Psi(\mathbf{u}_L) \Longrightarrow d(\mathbf{u}_L,\mathbf{u}_R)=0.$$

If we can identify a particular wave family in which the shock belongs, then we can replace the test on the entropy flux with a test on characteristic speeds: $\lambda_j(\mathbf{u}_L) > \lambda_j(\mathbf{u}_R)$. However, for transonic rarefactions, in which for the relevant wave family we have $\lambda_j(\mathbf{u}_L) < 0 < \lambda_j(\mathbf{u}_R)$, we want

$$\mathbf{f}(\mathbf{u}_L)=\mathbf{f}(\mathbf{u}_R) \quad \text{and} \quad \Psi(\mathbf{u}_R) < \Psi(\mathbf{u}_L) \Longrightarrow d(\mathbf{u}_L,\mathbf{u}_R)\neq 0.$$

Thus, the choice of d is subtle.

Example 4.13.5 If $\mathbf{f}(\mathbf{u}) = \mathbf{A}\mathbf{u}$, then the flux in Godunov's scheme is

$$(\mathbf{f}-\mathbf{u}\xi)(\mathbf{u}_L,\mathbf{u}_R) = \frac{1}{2}[\mathbf{A}\mathbf{u}_L+\mathbf{A}\mathbf{u}_R] - \frac{\varepsilon}{2}[\mathbf{u}_L+\mathbf{u}_R] - \frac{1}{2}\mathrm{A}\ \mathrm{sign}(\mathbf{A}-\mathbf{I}\xi)[\mathbf{u}_R-\mathbf{u}_L].$$

In this case, $d(\mathbf{u}_L,\mathbf{u}_R) = \mathbf{A}\ \mathrm{sign}(\mathbf{A}-\mathbf{I}\xi)(\mathbf{u}_R-\mathbf{u}_L)$. Since there are neither shocks nor rarefactions for this problem, there are no constraints on d to check.

Example 4.13.6 In Burgers' equation we have $\mathbf{f}(\mathbf{u}) = \frac{1}{2}\mathbf{u}^2$, and the flux in Godunov's scheme is

$$
\begin{aligned}
(\mathbf{f}-\mathbf{u}\xi)(\mathbf{u}_L,\mathbf{u}_R) &= \begin{cases} \frac{1}{2}\max\{\mathbf{u}_L,\min\{\mathbf{u}_R,\xi\}\}^2 - \xi\max\{\mathbf{u}_L,\min\{\mathbf{u}_R,\xi\}\}, & \mathbf{u}_L < \mathbf{u}_R \\ \max\{\frac{1}{2}\mathbf{u}_R^2-\mathbf{u}_R\xi,\ \frac{1}{2}\mathbf{u}_L^2-\mathbf{u}_L\xi\}, & \mathbf{u}_L \geq \mathbf{u}_R \end{cases}\\
&= \begin{cases} \frac{1}{2}\mathbf{u}_L^2-\mathbf{u}_L\xi,\ \xi < \mathbf{u}_L < \mathbf{u}_R \text{ or } 2\xi < \mathbf{u}_L+\mathbf{u}_R < 2\mathbf{u}_L \\ -\frac{1}{2}\xi^2,\ \mathbf{u}_L < \xi < \mathbf{u}_R \\ \frac{1}{2}\mathbf{u}_R^2-\mathbf{u}_R\xi,\ \mathbf{u}_L < \mathbf{u}_R < \xi \text{ or } 2\mathbf{u}_R < \mathbf{u}_L+\mathbf{u}_R < 2\xi. \end{cases}
\end{aligned}
$$

In this case,

$$
d(\mathbf{u}_L,\mathbf{u}_R) = \begin{cases} \frac{1}{2}(\mathbf{u}_R-\xi)^2 - \frac{1}{2}(\mathbf{u}_L-\xi)^2, & \xi < \mathbf{u}_L < \mathbf{u}_R \text{ or } 2\xi < \mathbf{u}_L+\mathbf{u}_R < 2\mathbf{u}_L \\ \frac{1}{2}(\xi-\mathbf{u}_L)^2 + \frac{1}{2}(\mathbf{u}_R-\xi)^2, & \mathbf{u}_L < \xi < \mathbf{u}_R \\ \frac{1}{2}(\xi-\mathbf{u}_L)^2 - \frac{1}{2}(\xi-\mathbf{u}_R)^2, & \mathbf{u}_L < \mathbf{u}_R < \xi \text{ or } 2\mathbf{u}_R < \mathbf{u}_L+\mathbf{u}_R < 2\xi. \end{cases}
$$

Note that Godunov adds numerical diffusion in each case, whether the wave involves a shock or rarefaction.

Example 4.13.7 The flux in Rusanov's scheme is

$$\mathbf{f}(\mathbf{u}_L, \mathbf{u}_R) = \frac{1}{2}\left[\mathbf{f}(\mathbf{u}_L) + \mathbf{f}(\mathbf{u}_R)\right] - \frac{1}{2}\max_j \left\{|\lambda_j|\right\}(\mathbf{u}_R - \mathbf{u}_L)$$

where λ_j are the characteristic speeds of the conservation law. This numerical diffusion $d(\mathbf{u}_L, \mathbf{u}_R) = \frac{1}{2}\max_j\{|\lambda_j|\}(\mathbf{u}_R - \mathbf{u}_L)$ is active for all jumps, even stationary shocks and transonic rarefactions.

Example 4.13.8 One form of the Lax–Wendroff scheme takes

$$\mathbf{f}(\mathbf{u}_L, \mathbf{u}_R) = \frac{1}{2}[\mathbf{f}(\mathbf{u}_L) + \mathbf{f}(\mathbf{u}_R)] - \frac{1}{2}\frac{\partial \mathbf{f}}{\partial \mathbf{u}}\left(\frac{\mathbf{u}_L + \mathbf{u}_R}{2}\right)[\mathbf{f}(\mathbf{u}_R) - \mathbf{f}(\mathbf{u}_L)]\frac{\triangle x}{\triangle t}.$$

A similar Lax–Wendroff flux is

$$\mathbf{f}(\mathbf{u}_L, \mathbf{u}_R) = \frac{1}{2}[\mathbf{f}(\mathbf{u}_L) + \mathbf{f}(\mathbf{u}_R)] - \frac{1}{2}\left|\frac{\partial \mathbf{f}}{\partial \mathbf{u}}\left(\frac{\mathbf{u}_L + \mathbf{u}_R}{2}\right)\right|[\mathbf{u}_R - \mathbf{u}_L].$$

Both have trouble with transonic rarefactions in Burgers' equation, because $\mathbf{f}(\mathbf{u}_L) = \mathbf{f}(\mathbf{u}_R)$ implies that

$$\frac{\partial \mathbf{f}}{\partial \mathbf{u}}\left(\frac{\mathbf{u}_L + \mathbf{u}_R}{2}\right) = 0,$$

and thus that $d(\mathbf{u}_L, \mathbf{u}_R) = 0$. In order to avoid this problem, van Leer has suggested the following modification of the Lax–Wendroff scheme:

$$\mathbf{f}(\mathbf{u}_L, \mathbf{u}_R) = \frac{1}{2}[\mathbf{f}(\mathbf{u}_L) + \mathbf{f}(\mathbf{u}_R)] - \frac{1}{4}\left[\left|\frac{\partial \mathbf{f}}{\partial \mathbf{u}}(\mathbf{u}_L)\right| + \left|\frac{\partial \mathbf{f}}{\partial \mathbf{u}}(\mathbf{u}_R)\right|\right][\mathbf{u}_R - \mathbf{u}_L].$$

Many schemes incorporate a **numerical diffusion** to help spread discontinuities and to help entropy production. One approach is to replace

$$\mathbf{f} \leftarrow \mathbf{f} + (\mathbf{u}_R - \mathbf{u}_L)\alpha\psi_2(a) \tag{4.87}$$

where a is some velocity associate with the problem. For example, in gas dynamics we might choose

$$\psi_2(a) = \min\{\mathbf{v}_R - \mathbf{v}_L, 0\}$$

so that diffusion is added only in compressions (*i.e.*, shocks). Such an numerical diffusion makes sense for gas dynamics, but not for all systems. More generally, one might try

$$\psi_2(a) = \max_i\{\max\{\lambda_{i,L} - \lambda_{i,R}, 0\}\},$$

as suggested in [13].

The parameter α in (4.87) is user-adjustable, but typically $\alpha = 0.1$ works well. Note that both of the numerical viscosities in the previous paragraph are such that $\psi(a) = O(\triangle x)$, so the modification of the flux is $O(\triangle x^2)$. Such numerical viscosities are called **quadratic viscosities**. This means that the use of these viscosities for methods with order greater than 2 would reduce the order of the scheme.

Sometimes a linear diffusion is used, of the form

$$\mathbf{f} \leftarrow \mathbf{f} + (\mathbf{u}_R - \mathbf{u}_L)\psi(a), \tag{4.88}$$

where a is some velocity associated with the problem, and $\psi(a) = O(1)$. For example, a **Rusanov numerical diffusion** would choose $\psi(a) = \max_i |\lambda_i|$, where λ_i are the characteristic speeds. A **Lax–Friedrichs numerical diffusion** would choose $\psi(a) = \frac{\triangle x}{\triangle t}$. Another choice, due to Osher and Solomon [123] computes a weighted average of the jumps by choosing

$$\psi(a) = \frac{|[\mathbf{f}]^\top [\mathbf{u}]|}{[\mathbf{u}]^\top [\mathbf{u}]}.$$

If a convex or concave entropy function S is available, we can also use

$$\psi(a) = \frac{|[\frac{\partial S}{\partial \mathbf{u}}][\mathbf{f}]|}{[\frac{\partial S}{\partial \mathbf{u}}][\mathbf{u}]},$$

as suggested in [68].

Linear viscosities destroy the order of schemes of order greater than one, but quadratic viscosities are too big for large jumps. Often, some hybrid of the two numerical viscosities is used, perhaps by taking the minimum of the quadratic and linear numerical diffusion coefficients, and then multiplying times the jump $\mathbf{u}_R - \mathbf{u}_L$.

4.13.3 Rusanov Solver

The Rusanov scheme described in Section 4.2.2 can be interpreted as coming from an approximate Riemann solver. Let λ be an upper bound on the characteristic speeds in all waves associated with the Riemann problem arising from $\mathbf{u}_L$ and $\mathbf{u}_R$. The Rusanov approximate Riemann solver is

$$\tilde{\mathcal{R}}(\mathbf{u}_L, \mathbf{u}_R, \xi) = \begin{cases} \mathbf{u}_L, & \xi < -\lambda \\ \mathbf{u}_{LR}, & -\lambda < \xi < \lambda \\ \mathbf{u}_R, & \lambda < \xi, \end{cases}$$

where the intermediate state $\mathbf{u}_{LR}$ is chosen so that the the approximate Riemann solver is conservative:

$$\begin{aligned} \int_{-\triangle x_L/2}^{\triangle x_R/2} \tilde{\mathcal{R}}\left(\mathbf{u}_L, \mathbf{u}_R, \frac{x}{\triangle t}\right) dx &= \mathbf{u}_L\left(-\lambda\triangle t + \frac{\triangle x_L}{2}\right) + \mathbf{u}_{LR} 2\lambda\triangle t + \mathbf{u}_R\left(-\lambda\triangle t + \frac{\triangle x_R}{2}\right) \\ &= (\mathbf{u}_L \triangle x_L + \mathbf{u}_R \triangle x_R)\frac{1}{2} - [\mathbf{f}_R - \mathbf{f}_L]\triangle t. \end{aligned}$$

This equation implies that

$$\mathbf{u}_{LR} = (\mathbf{u}_L + \mathbf{u}_R)\frac{1}{2} - [\mathbf{f}_R - \mathbf{f}_L]\frac{1}{2\lambda}.$$

The flux associated with the state moving at speed ξ is given by Equation (4.80):

$$\begin{aligned}
&(\mathbf{f} - \mathbf{u}\xi)_{\text{Rusanov}}(\mathbf{u}_L, \mathbf{u}_R) \\
&= \frac{1}{2}\{\mathbf{f}_L + \mathbf{f}_R - (\mathbf{u}_L + \mathbf{u}_R)\xi - |-\lambda - \xi|(\mathbf{u}_{LR} - \mathbf{u}_L) - |\lambda - \xi|(\mathbf{u}_R - \mathbf{u}_{LR})\} \\
&= \frac{1}{2}\left\{\mathbf{f}_L + \mathbf{f}_R - (\mathbf{u}_L + \mathbf{u}_R)\xi - (\mathbf{u}_R - \mathbf{u}_L)\frac{|\lambda + \xi| + |\lambda - \xi|}{2} + (\mathbf{f}_R - \mathbf{f}_L)\frac{|\lambda + \xi| - |\lambda - \xi|}{2\lambda}\right\}
\end{aligned} \tag{4.89}$$

The flux associated with speed $\xi = 0$ for this approximate Riemann solver is the usual Rusanov flux.

We can interpret this approximate Riemann solver as decomposing the jump into two waves associated with the relative wavespeeds $-\lambda - \xi$ and $\lambda - \xi$:

$$\begin{aligned}
(\mathbf{f}_R - \mathbf{u}_R\xi) - (\mathbf{f}_L - \mathbf{u}_L\xi) &= \left\{[\mathbf{u}_R - \mathbf{u}_L] - [\mathbf{f}_R - \mathbf{f}_L]\frac{1}{\lambda}\right\}\frac{1}{2}(-\lambda - \xi) \\
&+ \left\{[\mathbf{u}_R - \mathbf{u}_L] + [\mathbf{f}_R - \mathbf{f}_L]\frac{1}{\lambda}\right\}\frac{1}{2}(\lambda - \xi).
\end{aligned} \tag{4.90}$$

This formulation will be useful for the wave propagation scheme in Section 6.2.6 below. If the jump between the states is nonzero, we can also write this flux difference as a sum of fluctuations

$$\begin{aligned}
(\mathbf{f}_R - \mathbf{u}_R\xi) - (\mathbf{f}_L - \mathbf{u}_L\xi) &= \left[\left\{\left(\triangle\mathbf{u} - \triangle\mathbf{f}\frac{1}{\lambda}\right)\frac{1}{\|\triangle\mathbf{u}\|}\right\}\frac{1}{2}(-\lambda - \xi)\left\{\frac{1}{\|\triangle\mathbf{u}\|}\triangle\mathbf{u}^\top\right\}\right. \\
&\left. + \left\{\left(\triangle\mathbf{u} + \triangle\mathbf{f}\frac{1}{\lambda}\right)\frac{1}{\|\triangle\mathbf{u}\|}\right\}\frac{1}{2}(\lambda - \xi)\left\{\frac{1}{\|\triangle\mathbf{u}\|}\triangle\mathbf{u}^\top\right\}\right]\triangle\mathbf{u} \\
&= \mathbf{A}\triangle\mathbf{u}
\end{aligned} \tag{4.91}$$

where $\triangle\mathbf{u} = \mathbf{u}_R - \mathbf{u}_L$ and $\triangle\mathbf{f} = \mathbf{f}_R - \mathbf{f}_L$. This formulation will be useful for the 2D wave propagation scheme in Section 7.1.3.

4.13.4 Weak Wave Riemann Solver

Given any linearly independent set of characteristic directions $\mathbf{X}$ with corresponding speeds λ_j, we might approximate the solution of the Riemann problem by

$$\tilde{\mathcal{R}}(\mathbf{u}_L, \mathbf{u}_R, \xi) = \mathbf{u}_L + \sum_{j:\lambda_j < \xi} \mathbf{X}\mathbf{e}_j\alpha_j,$$

where the scalars α_j are determined so that the approximate Riemann solver produces the right state at large wave speed:

$$\sum_j \mathbf{X}\mathbf{e}_j \alpha_j = \mathbf{u}_R - \mathbf{u}_L .$$

In order for this approximate Riemann solver to be conservative, it would be necessary that

$$\sum_j \mathbf{X}\mathbf{e}_j \alpha_j \lambda_j = \mathbf{f}(\mathbf{u}_R) - \mathbf{f}(\mathbf{u}_L).$$

If we solve the linear systems

$$\mathbf{X}\mathbf{y} = \mathbf{u}_R - \mathbf{u}_L \quad \text{and} \quad \mathbf{X}\mathbf{z} = \mathbf{f}(\mathbf{u}_R) - \mathbf{f}(\mathbf{u}_L)$$

then we can see that conservation would require that we take the wave speed λ_j to be

$$\lambda_j = \frac{\mathbf{e}_j \cdot \mathbf{z}}{\mathbf{e}_j \cdot \mathbf{y}} = \frac{\mathbf{e}_j^\top \mathbf{X}^{-1}[\mathbf{f}(\mathbf{u}_R) - \mathbf{f}(\mathbf{u}_L)]}{\mathbf{e}_j^\top \mathbf{X}^{-1}[\mathbf{u}_R - \mathbf{u}_L]} . \tag{4.92}$$

The numerical flux associated with zero wave speed would be evaluated by

$$\mathbf{f}(\mathbf{u}_L, \mathbf{u}_R) = \begin{cases} \mathbf{f}(\mathbf{u}_L) + \sum_{j:\lambda_j<0} \mathbf{X}\mathbf{e}_j \mathbf{e}_j \cdot \mathbf{z}, & \text{most wavespeeds positive} \\ \mathbf{f}(\mathbf{u}_R) - \sum_{j:\lambda_j>0} \mathbf{X}\mathbf{e}_j \mathbf{e}_j \cdot \mathbf{z}, & \text{most wavespeeds negative.} \end{cases}$$

Although Equation (4.92) for λ_j can involve significant rounding errors and possibly division by zero, all that is really needed is the sign of λ_j; this can be determined without dividing by zero.

A non-conservative form of the weak-wave Riemann solver is also sometimes used. This approach takes the λ_j to be the characteristic speeds associated with the same state as the characteristic directions $\mathbf{X}$. After solving $\mathbf{X}\mathbf{y} = \mathbf{u}_R - \mathbf{u}_L$, the numerical flux associated with zero wave speed would be evaluated by

$$\mathbf{f}(\mathbf{u}_L, \mathbf{u}_R) = \begin{cases} \mathbf{f}(\mathbf{u}_L) + \sum_{j:\lambda_j<0} \mathbf{X}\mathbf{e}_j \lambda_j \mathbf{e}_j \cdot \mathbf{y}, & \text{most wavespeeds positive} \\ \mathbf{f}(\mathbf{u}_R) - \sum_{j:\lambda_j>0} \mathbf{X}\mathbf{e}_j \lambda_j \mathbf{e}_j \cdot \mathbf{y}, & \text{most wavespeeds negative.} \end{cases}$$

One common way to estimate the sign of “most of the wavespeeds” is to examine the sign of

$$\lambda = \frac{[\mathbf{u}_R - \mathbf{u}_L] \cdot [\mathbf{f}(\mathbf{u}_R) - \mathbf{f}(\mathbf{u}_L)]}{[\mathbf{u}_R - \mathbf{u}_L] \cdot [\mathbf{u}_R - \mathbf{u}_L]} .$$

Example 4.13.9 Let us determine a weak wave Riemann solver flux for Burgers' equation. The conservative form of the weak wave Riemann solver takes

$$\lambda = \frac{\mathbf{f}(\mathbf{u}_R) - \mathbf{f}(\mathbf{u}_L)}{\mathbf{u}_R - \mathbf{u}_L} = \frac{\mathbf{u}_R + \mathbf{u}_L}{2}.$$

Then

$$\mathbf{f}(\mathbf{u}_L, \mathbf{u}_R) = \begin{cases} \mathbf{f}(\mathbf{u}_L), & \lambda \geq 0 \\ \mathbf{f}(\mathbf{u}_R), & \lambda < 0. \end{cases}$$

(Note that when $\lambda = 0$ it does not matter whether we use the left or right flux.) In practice, this approximate Riemann solver does not perform well for transonic rarefactions: see Example 4.13.15 and Section 4.13.9 below. However, with the addition of numerical diffusion it can be reliable for small disturbances.

4.13.5 Colella–Glaz Riemann Solver

Colella and Glaz [34] suggested replacing the rarefaction curves by shock curves in an approximate solution of the Riemann problem. This approximation is acceptable for weak rarefactions, or for rarefactions that have spread out over the grid. Their original purpose was to avoid a numerical integrator for the ordinary differential equations associated with a "real gas" equation of state, namely one that includes such effects as disassociation and recombination of air molecules. However, this approach is still based on finding a state in the solution of the Riemann problem at which to evaluate the flux, and depends on some knowledge of the ordering of the waves and the determination of the state associated with propagation at a given speed. Such information is difficult to obtain for nonstrictly hyperbolic conservation laws, or conservation laws with local linear degeneracies. Of all our case studies, it is reasonably easy to program this approximate Riemann solver only for shallow water and gas dynamics.

Example 4.13.10 The Colella–Glaz approximate Riemann problem solver for a polytropic gas is initialized by computing

$$C_L^2 = \gamma p_L \rho_L, \; W_L = \sqrt{C_L^2},$$

$$C_R^2 = \gamma p_R \rho_R, \; W_R = \sqrt{C_R^2},$$

$$p^* = [W_R p_R + W_L p_L - W_L W_R (\mathbf{v}_R - \mathbf{v}_L)]/(W_R + W_L).$$

Afterward, a Newton iteration is used to solve $\mathbf{v}_R(p^*) = \mathbf{v}_L(p^*)$ given by the Rankine–Hugoniot jump conditions. This amounts to performing the following steps:

$$W_L = \{C_L^2(1 + [p^*/p_L - 1](\gamma + 1)/(2\gamma)\}^{1/2},$$

$$W_R = \{C_R^2(1 + [p^*/p_R - 1](\gamma + 1)/(2\gamma)\}^{1/2},$$

$$\zeta_R = \frac{2W_R^3}{W_R^2 + C_R^2}, \qquad \zeta_L = \frac{2W_L^3}{W_L^2 + C_L^2},$$

$$\mathbf{v}_R^* = \mathbf{v}_R + (p_R - p^*)/W_R, \ \mathbf{v}_L^* = \mathbf{v}_L - (p_L - p^*)/W_L,$$

$$p^* := p^* - (\mathbf{v}_R^* - \mathbf{v}_L^*)\zeta_R\zeta_L/(\zeta_R + \zeta_L).$$

Once the pressure p^* at the contact discontinuity has been found, the velocity of the contact discontinuity is computed by

$$\mathbf{v}^* = (\zeta_R\mathbf{v}_R^* + \zeta_R\mathbf{v}_L^*)/(\zeta_R + \zeta_L).$$

If $\mathbf{v}^* \le \xi$, then the state moving with speed ξ is between the contact discontinuity and the right state; in this case, the velocity is initialized by

$$\mathbf{v}_\pm = \mathbf{v}_R, \quad p_\pm = p_R, \quad \rho_\pm = \rho_R.$$

Otherwise, the state moving with speed ξ is between the contact discontinuity and the left state; the velocity is initialized by

$$\mathbf{v}_\pm = \mathbf{v}_L, \quad p_\pm = p_L, \quad \rho_\pm = \rho_L.$$

Afterword, calculations serve to identify the state on the Hugoniot locus:

$$c_\pm = \sqrt{\gamma p_\pm/\rho_\pm}, \quad W_\pm^2 = \gamma\rho_\pm p_\pm\{1 + (p^*/p_\pm - 1)(\gamma + 1)/(2\gamma)\},$$

$$\rho^* = \rho_\pm \bigg/ \left(1 - \frac{\gamma\rho_\pm(p^* - p_\pm)}{W_\pm^2}\right), \qquad c^* = \sqrt{\gamma p^*/\rho^*}.$$

If $p^* \ge p_\pm$, then the wave is a shock, and the shock speed is

$$\lambda_o = \lambda_i = \begin{cases} \mathbf{v}^* - W_\pm/\rho^*, \ \mathbf{v}^* > \xi \\ \mathbf{v}^* + W_\pm/\rho^*, \ \mathbf{v}^* \le \xi. \end{cases}$$

Otherwise the wave is a rarefaction and the characteristic speeds at the end of the wave curve are

$$\lambda_i = \begin{cases} \mathbf{v}^* - c^*, & \mathbf{v}^* > \xi \\ \mathbf{v}^* + c^*, & \mathbf{v}^* \le \xi \end{cases}$$

$$\lambda_0 = \begin{cases} \mathbf{v}_\pm - c_\pm, & \mathbf{v}^* > \xi \\ \mathbf{v}_\pm + c_\pm, & \mathbf{v}^* \le \xi. \end{cases}$$

If $(\mathbf{v}^* - \xi)(\xi - \mathbf{v}_i) \geq 0$, then the solution of the Riemann problem is taken to be the state on this wave family at the contact discontinuity:

$$\rho = \rho^*, \quad p = p^*, \quad \mathbf{v} = \mathbf{v}^*.$$

If $(\mathbf{v}^* - \xi)(\mathbf{v}_0 - \xi) \geq 0$, the solution of the Riemann problem is taken to be the outer state (i.e., the appropriate left or right state in the description of the Riemann problem):

$$\rho = \rho_0, \quad p = p_0, \quad \mathbf{v} = \mathbf{v}_0.$$

Otherwise, the state is interpolated within the rarefaction:

$$\begin{aligned}
&\alpha = \frac{\lambda_0 - \xi}{\lambda_0 - \lambda_i}, \\
&\rho = \alpha\rho^* + (1-\alpha)\rho_\pm, \quad \mathbf{v} = \alpha\mathbf{v}^* + (1-\alpha)\mathbf{v}_\pm, \\
&c = \alpha c^* + (1-\alpha)c_\pm, \quad p = c^2\rho/\gamma.
\end{aligned}$$

A Fortran program implementing this approximate Riemann solver for $\xi = 0$ is available in **Program 4.13-52: riemnv.f.**

4.13.6 Osher–Solomon Riemann Solver

Another approach to the solution of the Riemann problem for gas dynamics is described in [123]. In order to avoid the nonlinear iteration required in the analytical solution of the Riemann problem (or in the Colella–Glaz Riemann solver [34]), we could assume that the rarefaction curves are determined from Riemann invariants, and approximate all waves by rarefactions or contact discontinuities. Along a transonic rarefaction, we could use the characteristic speeds at the left and right states as if they were associated with the respective ends of the waves families, and use these speeds to estimate the location of internal sonic points. The **Engquist–Osher flux** could be used to approximate the path integrals. The resulting numerical flux is a sum of fluxes at the initial left or right states, and interior (approximate) sonic points.

Example 4.13.11 For shallow water, the rarefaction curves satisfy

$$\mathbf{n}\cdot\mathbf{v}_-(h) = \mathbf{n}\cdot\mathbf{v}_L + (h_L - h)\frac{2g}{\sqrt{gh_L} + \sqrt{gh}} = \mathbf{n}\cdot\mathbf{v}_L + 2(c_L - c)$$
$$\mathbf{n}\cdot\mathbf{v}_+(h) = \mathbf{n}\cdot\mathbf{v}_R - (h_R - h)\frac{2g}{\sqrt{gh_R} + \sqrt{gh}} = \mathbf{n}\cdot\mathbf{v}_R - 2(c_R - c)$$

where $c = \sqrt{gh}$. At the intersection of these curves we have

$$\sqrt{gh_*} = c_* = \frac{c_R + c_L}{2} - \frac{\mathbf{n} \cdot (\mathbf{v}_R - \mathbf{v}_L)}{4}$$
$$\mathbf{n} \cdot \mathbf{v}_* = \frac{\mathbf{n} \cdot (\mathbf{v}_R + \mathbf{v}_L)}{2} + c_L - c_R.$$

One difficulty is that we could have $c_* < 0$. Suppose that we are given a speed ξ. If $h_L > h_ast$ and $\mathbf{n} \cdot \mathbf{v}_L - c_L < \xi < \mathbf{n} \cdot \mathbf{v}_* - c_*$ then we have a transonic rarefaction; the state that moves with speed ξ satisfies

$$\sqrt{gh} = c = (\mathbf{n} \cdot \mathbf{v}_L + 2c_L - \xi)/3$$
$$\mathbf{n} \cdot \mathbf{v} = \{\mathbf{n} \cdot \mathbf{v}_L + 2(c_L + \xi)\}/3.$$

There is no need for an Engquist–Osher approximation here. Similarly, if $h_R > h_*$ and $\mathbf{n} \cdot \mathbf{v}_R + c_R > \xi > \mathbf{n} \cdot \mathbf{v}_* + c_*$ then we have a transonic rarefaction; the state that moves with speed ξ satisfies

$$\sqrt{gh} = c = (\xi - \mathbf{n} \cdot \mathbf{v}_R + 2c_R)/3$$
$$\mathbf{n} \cdot \mathbf{v} = \{\mathbf{n} \cdot \mathbf{v}_R + 2(\xi - c_R)\}/3.$$

The Osher–Solomon Riemann problem solver for gas dynamics requires an iteration to determine the intersection of the rarefaction curves, if it exists. For non-strictly hyperbolic problems or problems with local linear degeneracies, the Osher-Solomon Riemann problem solver is even more difficult to implement.

4.13.7 Bell–Colella–Trangenstein Approximate Riemann Problem Solver

The previous approximate Riemann problem solvers assumed either that the Riemann invariants are known, or that wave curves are approximated well by Hugoniot loci. For problems such as gas dynamics, these techniques are very effective. However, these approaches do not apply to many important problems, such as solid mechanics or flow in porous media. We will describe another approach [13], which only requires that the characteristic speeds and directions can be computed.

Suppose that we are given flux variables $\mathbf{w}_L$ and $\mathbf{w}_R$. If all of the characteristic speeds at both $\mathbf{w}_L$ and W_r are positive (as well as at all intermediate states in the solution of the Riemann problem), then we will evaluate

$$\mathbf{f}(\mathcal{R}(\mathbf{w}_L, \mathbf{w}_R; 0)) = \mathbf{f}(\mathbf{w}_L).$$

Similarly, if all of the characteristic speeds are negative, then we will compute

$$\mathbf{f}(\mathcal{R}(\mathbf{w}_L, \mathbf{w}_R; 0)) = \mathbf{f}(\mathbf{w}_R).$$

The discussion that follows concerns the case when at least one of the characteristic speeds changes sign between $\mathbf{w}_L$ and $\mathbf{w}_R$.

First, we determine some average state $\overline{\mathbf{w}}$ and find the generalized eigenvectors $\mathbf{X}$ and eigenvalues Λ so that

$$\left.\frac{\partial \mathbf{f}}{\partial \mathbf{w}}\right|_{\overline{\mathbf{w}}} \mathbf{X} = \left.\frac{\partial \mathbf{u}}{\partial \mathbf{w}}\right|_{\overline{\mathbf{w}}} \mathbf{X}\Lambda.$$

For example, we might choose $\overline{\mathbf{w}} = \frac{1}{2}(\mathbf{w}_L + \mathbf{w}_R)$. Afterward, we solve

$$\mathbf{X}\mathbf{y} = \mathbf{w}_R - \mathbf{w}_L$$

for the characteristic expansion coefficients $\mathbf{y}$. We can approximate the path from the left state to the right state in the solution of the Riemann problem by

$$\mathbf{w}_R = \mathbf{w}_L + \sum_i \mathbf{X}\mathbf{e}_i \mathbf{e}_i \cdot \mathbf{y}.$$

If most of the characteristic speeds at $\mathbf{w}_L$ and $\mathbf{w}_R$ are positive, then we will traverse the path from $\mathbf{w}_L$ to $\mathbf{w}_R$; otherwise, we go from right to left. Let us suppose that we will traverse from left to right. Arrange the wave families i in order of increasing characteristic speed, and let $\boldsymbol{I}$ be the largest wave family index that involves a negative characteristic speed. For all wave families $k \leq \boldsymbol{I}$ we compute the intermediate states

$$\mathbf{w}_k = \mathbf{w}_L + \sum_{i \leq k} \mathbf{X}\mathbf{e}_i \mathbf{e}_i \cdot \mathbf{y}.$$

We check that each of these states is physically realistic; if not, we will abandon the approximate Riemann solver and use a low-order diffusive flux, such as Rusanov's. The final step is to determine an approximate path integral, mimicking the Engquist–Osher flux:

$$\mathbf{f} = \mathbf{f}(\mathbf{w}_L) + \sum_{k \leq \mathbf{I}} \mathbf{X}\mathbf{e}_i \int_0^{\mathbf{e}_i \cdot \mathbf{y}} \min\{\lambda_i(\eta), 0\}\, d\eta.$$

In this expression, we will use some approximate model $\lambda_i(\eta)$ for the characteristic speed along the path. A common model is linear interpolation:

$$\lambda_i(\eta) = \lambda_i|_{\mathbf{w}_L} + \frac{\lambda_i|_{\mathbf{w}_R} - \lambda_i|_{\mathbf{w}_L}}{\mathbf{e}_i \cdot \mathbf{y}} \eta.$$

If most of the characteristic speeds are negative, then we will traverse the path from right to left. Again, arrange the wave families i in order of increasing characteristic speed. Let I be the smallest wave family index that involves a negative characteristic speed. For all wave families $k \geq I$ we compute the intermediate states

$$\mathbf{w}_k = \mathbf{w}_R - \sum_{i \geq k} \mathbf{X}\mathbf{e}_i \mathbf{e}_i \cdot \mathbf{y}.$$

We check that each of these states is physically realistic; if not, we will abandon the approximate Riemann solver and use a low-order diffusive flux, such as Rusanov's. The final step is to determine an approximate path integral, mimicking the Engquist-Osher flux:

$$\mathbf{f} = \mathbf{f}(\mathbf{w}_R) - \sum_{k \geq I} \mathbf{X}\mathbf{e}_i \int_0^{\mathbf{e}_i \cdot \mathbf{y}} \max\{\lambda_i(\eta), 0\}\, d\eta.$$

Example 4.13.12 Let us discuss the application of the Bell–Colella–Trangenstein approximate Riemann solver to gas dynamics. Notice that for gas dynamics

$$\left(\frac{\partial \mathbf{u}}{\partial \mathbf{w}}\right)^{-1} \frac{\partial \mathbf{f}}{\partial \mathbf{w}} = \begin{bmatrix} 1 & 0 & 0 \\ \mathbf{v} & \mathbf{v} & 0 \\ \frac{1}{2}\mathbf{v}^2 & \mathbf{v}\rho & \frac{1}{\gamma-1} \end{bmatrix}^{-1} \begin{bmatrix} \mathbf{v} & \rho & 0 \\ \mathbf{v}^2 & 2\mathbf{v}\rho & 1 \\ \frac{1}{2}\mathbf{v}^3 & \rho(e + \frac{1}{2}\mathbf{v}^2) + p + \mathbf{v}^2\rho & \frac{\gamma}{\gamma-1}\mathbf{v} \end{bmatrix}$$

$$= \begin{bmatrix} \mathbf{v} & \rho & 0 \\ 0 & \mathbf{v} & \frac{1}{\rho} \\ 0 & \gamma p & \mathbf{v} \end{bmatrix} \equiv \mathbf{A} + \mathbf{I}\mathrm{v}.$$

We compute the average state $\overline{\mathbf{w}} = \frac{1}{2}(\mathbf{w}_L + \mathbf{w}_R)$, and find the eigenvectors $\mathbf{X}$ and eigenvalues Λ of $\left(\frac{\partial \mathbf{u}}{\partial \mathbf{w}}\right)^{-1} \frac{\partial \mathbf{f}}{\partial \mathbf{w}}$ at this state:

$$\mathbf{X} = \begin{bmatrix} \rho & 1 & \rho \\ -c & 0 & c \\ \rho c^2 & 0 & \rho c^2 \end{bmatrix}, \quad \Lambda = \begin{bmatrix} -c & 0 & 0 \\ 0 & 0 & 0 \\ 0 & 0 & c \end{bmatrix}.$$

Afterward, we solve $\mathbf{X}\mathbf{y} = \mathbf{w}_R - \mathbf{w}_L$ for the characteristic expansion coefficients $\mathbf{y}$. This gives us

$$\mathbf{y} = \begin{bmatrix} 0 & -1/(2c) & 1/(2\rho c^2) \\ 1 & 0 & -1/c^2 \\ 0 & 1/(2c) & 1/(2\rho c^2) \end{bmatrix} \begin{bmatrix} \rho_R - \rho_L \\ \mathbf{v}_R - \mathbf{v}_L \\ p_R - p_L \end{bmatrix} = \begin{bmatrix} \frac{1}{2}(\alpha - \beta) \\ \rho - \overline{\rho}\alpha \\ \frac{1}{2}(\alpha + \beta) \end{bmatrix}$$

where

$$\overline{\rho} = \frac{1}{2}(\rho_L + \rho_R), \quad \overline{p} = \frac{1}{2}(p_L + p_R), \quad \overline{\mathbf{v}} = \frac{1}{2}(\mathbf{v}_L + \mathbf{v}_R), \quad \overline{c} = \sqrt{\gamma \overline{p}/\overline{\rho}},$$

$$\alpha = \frac{[p]}{\gamma \overline{p}}, \quad \beta = \frac{[\mathbf{v}]}{\overline{c}}.$$

We now have a path from $\mathbf{w}_L$ to $\mathbf{w}_R$, using the intermediate states

$$\mathbf{w}_1 = \mathbf{w}_L + \mathbf{X}\mathbf{e}_1\mathbf{e}_1 \cdot \mathbf{y} = \begin{bmatrix} \rho_L \\ \mathbf{v}_L \\ p_L \end{bmatrix} + \begin{bmatrix} \overline{\rho} \\ -\overline{c} \\ \gamma \overline{p} \end{bmatrix} \frac{1}{2}(\alpha - \beta),$$

$$\mathbf{w}_2 = \mathbf{w}_R - \mathbf{X}\mathbf{e}_3\mathbf{e}_3 \cdot \mathbf{y} = \begin{bmatrix} \rho_R \\ \mathbf{v}_R \\ p_R \end{bmatrix} - \begin{bmatrix} \overline{\rho} \\ \overline{c} \\ \gamma \overline{p} \end{bmatrix} \frac{1}{2}(\alpha + \beta).$$

at either end of the contact discontinuity. If there are no changes in sign of the characteristic speed along any of the path segments (*i.e.*, in the genuinely nonlinear wave from $\mathbf{w}_L$ to $\mathbf{w}_1$, in the contact discontinuity from $\mathbf{w}_1$ to $\mathbf{w}_2$, or in the genuinely nonlinear wave from $\mathbf{w}_2$ to $\mathbf{w}_R$), then we evaluate the flux at whichever of these states in the path corresponds to zero characteristic speed. For example, if all of the characteristic speeds are positive, we evaluate the flux at the left state. If $(\mathbf{v} - c)|_{\mathbf{w}_L} < 0$ and $(\mathbf{v} - c)|_{\mathbf{w}_R} < 0$ but all other characteristic speeds are positive at $\mathbf{w}_L$ and $\mathbf{w}_R$, then we will evaluate the flux at the left side of the contact discontinuity, namely $\mathbf{w}_1$. If we need to evaluate the flux at either $\mathbf{w}_1$ or $\mathbf{w}_2$, we require that both the density ρ and the pressure p are positive at this state; otherwise, we will resort to a Rusanov flux.

In the remainder of the discussion, we will assume that at least one of the wave families involves a change in sign of the characteristic speed between $\mathbf{w}_L$ and $\mathbf{w}_R$. Because the contact discontinuity is linearly degenerate, and because the Engquist–Osher flux is diffusive, we want to avoid using an Engquist–Osher flux contribution for the contact discontinuity. In order to remove ambiguity, we will say that the speed of the contact discontinuity is the velocity $\mathbf{v}$ at the average state $\frac{1}{2}(\mathbf{w}_L + \mathbf{w}_R)$. If the speed of the contact discontinuity is positive, we will begin our Engquist–Osher path integral at $\mathbf{w}_L$; otherwise, we will begin at $\mathbf{w}_R$. Note that for gas dynamics

$$\mathbf{w}_2 - \mathbf{w}_1 = \mathbf{X}\mathbf{e}_2\mathbf{e}_2 \cdot \mathbf{y} = \begin{bmatrix} 1 \\ 0 \\ 0 \end{bmatrix} \mathbf{e}_2 \cdot \mathbf{y}$$

so $\mathbf{w}_1$ and $\mathbf{w}_2$ have the same entries for $\mathbf{v}$ and p.

If we have a transonic wave, then we must determine an approximate path integral, mimicking the Engquist–Osher flux. If the speed of the contact discontinuity is positive, we begin the path integral at the left state

$$\mathbf{f} = \mathbf{f}(\mathbf{w}_L) + \sum_i \mathbf{X}\mathbf{e}_i \int_0^{\mathbf{e}_i \cdot \mathbf{y}} \min\{\lambda_i, 0\} \, d\eta;$$

otherwise, we begin at the right state:

$$\mathbf{f} = \mathbf{f}(\mathbf{w}_R) - \sum_i \mathbf{X}\mathbf{e}_i \int_0^{\mathbf{e}_i \cdot \mathbf{y}} \max\{\lambda_i, 0\}\, d\eta.$$

Since the waves other than the contact discontinuity are genuinely nonlinear, there are no local extrema of the characteristic speeds. Thus in these Engquist–Osher path integrals, it is sufficient to use a piecewise-linear approximation to $\lambda_i(\eta)$:

$$\lambda_i(\eta) = \lambda_i|_{\mathbf{w}_L} + \frac{\lambda_i|_{\mathbf{w}_R} - \lambda_i|_{\mathbf{w}_L}}{\mathbf{e}_i \cdot \mathbf{y}} \eta.$$

Note that the zero of this linear model occurs at

$$\eta_{i,s} = \frac{\lambda_i|_{\mathbf{w}_L}}{\lambda_i|_{\mathbf{w}_L} - \lambda_i|_{\mathbf{w}_R}} \mathbf{e}_i \cdot \mathbf{y}.$$

Then we have four cases:

++: if $\lambda_i|_{\mathbf{w}_L} > 0$ and $\lambda_i|_{\mathbf{w}_R} > 0$ then

$$\int_0^{\mathbf{e}_i \cdot \mathbf{y}} \min\{\lambda_i(\eta), 0\} d\eta = 0,$$
$$\int_0^{\mathbf{e}_i \cdot \mathbf{y}} \max\{\lambda_i(\eta), 0\} d\eta = \frac{1}{2}(\lambda_i|_{\mathbf{w}_L} + \lambda_i|_{\mathbf{w}_R})\mathbf{e}_i \cdot \mathbf{y};$$

−+: if $\lambda_i|_{\mathbf{w}_L} < 0$ and $\lambda_i|_{\mathbf{w}_R} > 0$ then the flux integrals are approximated by

$$\int_0^{\mathbf{e}_i \cdot \mathbf{y}} \min\{\lambda_i(\eta), 0\}\, d\eta = \frac{1}{2}\lambda_i|_{\mathbf{w}_L} \eta_{i,s} = -\frac{1}{2} \frac{\lambda_i^2|_{\mathbf{w}_L}}{\lambda_i|_{\mathbf{w}_R} - \lambda_i|_{\mathbf{w}_L}} \mathbf{e}_i \cdot \mathbf{y},$$
$$\int_0^{\mathbf{e}_i \cdot \mathbf{y}} \max\{\lambda_i(\eta), 0\}\, d\eta = \frac{1}{2}\lambda_i|_{\mathbf{w}_R}(\mathbf{e}_i \cdot \mathbf{y} - \eta_{i,s}) = \frac{1}{2} \frac{\lambda_i^2|_{\mathbf{w}_R}}{\lambda_i|_{\mathbf{w}_R} - \lambda_i|_{\mathbf{w}_L}} \mathbf{e}_i \cdot \mathbf{y};$$

+−: if $\lambda_i|_{\mathbf{w}_L} > 0$ and $\lambda_i|_{\mathbf{w}_R} < 0$ then the flux integrals are approximated by

$$\int_0^{\mathbf{e}_i \cdot \mathbf{y}} \min\{\lambda_i(\eta), 0\}\, d\eta = \frac{1}{2}\lambda_i|_{\mathbf{w}_R}(\mathbf{e}_i \cdot \mathbf{y} - \eta_{i,s}) = -\frac{1}{2} \frac{\lambda_i^2|_{\mathbf{w}_R}}{\lambda_i|_{\mathbf{w}_L} - \lambda_i|_{\mathbf{w}_R}} \mathbf{e}_i \cdot \mathbf{y},$$
$$\int_0^{\mathbf{e}_i \cdot \mathbf{y}} \max\{\lambda_i(\eta), 0\}\, d\eta = \frac{1}{2}\lambda_i|_{\mathbf{w}_L} \eta_{i,s} = \frac{1}{2} \frac{\lambda_i^2|_{\mathbf{w}_L}}{\lambda_i|_{\mathbf{w}_L} - \lambda_i|_{\mathbf{w}_R}} \mathbf{e}_i \cdot \mathbf{y};$$

−−: if $\lambda_{L,i} < 0$ and $\lambda_{R,i} < 0$ then the flux integral would be approximated by

$$\int_0^{\mathbf{e}_i \cdot \mathbf{y}} \min\{\lambda_i(\eta), 0\}\, d\eta = \frac{1}{2}(\lambda_i|_{\mathbf{w}_L} + \lambda_i|_{\mathbf{w}_R})\mathbf{e}_i \cdot \mathbf{y},$$
$$\int_0^{\mathbf{e}_i \cdot \mathbf{y}} \max\{\lambda_i(\eta), 0\}\, d\eta = 0.$$

The contributions to the path integral are summed over all waves, as necessary. Note that for gas dynamics, the choice of the initial point for the path integral implies that contact discontinuity makes no contribution to the sum.

4.13.8 Roe Riemann Solver

Roe [138] suggested an even simpler approximation to the solution of the Riemann problem for gas dynamics. He showed that for gas dynamics it is possible to find a matrix $\mathbf{A}(\mathbf{w}_L, \mathbf{w}_R)$ that represents an appropriate average of the flux derivative. He then approximated the solution of the nonlinear Riemann problem by the solution of the Riemann problem with linear flux $\mathbf{Au}$.

It is possible to prove the existence of a Roe matrix for any hyperbolic system of conservation laws with a convex or concave entropy function.

Theorem 4.13.13 *[68] For any conservation law*

$$\frac{\partial \mathbf{u}}{\partial t} + \sum_{i=1}^{k} \frac{\partial \mathbf{F}\mathbf{e}_i}{\partial x_i} = 0,$$

with convex entropy function there is a Roe matrix $\bar{\mathbf{A}}(\mathbf{w}_L, \mathbf{w}_R)$ such that

(i) $[\mathbf{F}(\mathbf{w}_R) - \mathbf{F}(\mathbf{w}_L)]\mathbf{n} = \bar{\mathbf{A}}(\mathbf{w}_L, W_R)[\mathbf{u}(\mathbf{w}_R) - \mathbf{u}(\mathbf{w}_L)]$
(ii) $\bar{\mathbf{A}}(\mathbf{w}_L, W_R)\bar{\mathbf{X}} = \bar{\mathbf{X}}\bar{\Lambda}$ *where* $\bar{\mathbf{X}}$ *is nonsingular and* $\bar{\Lambda}$ *is real and diagonal*
(iii) $\bar{\mathbf{A}}(\mathbf{w}, \mathbf{w}) = \frac{\partial \mathbf{F}\mathbf{n}}{\partial \mathbf{w}} \left(\frac{\partial \mathbf{u}}{\partial \mathbf{w}}\right)^{-1}$.

Proof Suppose that S is an entropy function with entropy flux Ψ:

$$\frac{\partial \Psi \mathbf{n}}{\partial \mathbf{w}} = \frac{\partial S}{\partial \mathbf{w}} \left(\frac{\partial \mathbf{u}}{\partial \mathbf{w}}\right)^{-1} \frac{\partial \mathbf{F}\mathbf{n}}{\partial \mathbf{w}}.$$

Let

$$\mathbf{z}^\top = \frac{\partial S}{\partial \mathbf{w}} \left(\frac{\partial \mathbf{u}}{\partial \mathbf{w}}\right)^{-1}.$$

Suppose that we differentiate the entropy flux equation

$$\frac{\partial \Psi \mathbf{n}}{\partial \mathbf{u}_j} = \sum_{i=1}^{m} \frac{\partial S}{\partial \mathbf{u}_i} \frac{\partial \mathbf{e}_i^\top \mathbf{F}\mathbf{n}}{\partial \mathbf{u}_j}$$

with respect to $\mathbf{u}_k$:

$$\frac{\partial \Psi \mathbf{n}}{\partial \mathbf{u}_j \partial \mathbf{u}_k} = \sum_{i=1}^{m} \frac{\partial^2 S}{\partial \mathbf{u}_i \partial \mathbf{u}_k} \frac{\partial \mathbf{e}_i^\top \mathbf{F}\mathbf{n}}{\partial \mathbf{u}_j} + \sum_{i=1}^{m} \frac{\partial S}{\partial \mathbf{u}_i} \frac{\partial^2 \mathbf{e}_i^\top \mathbf{F}\mathbf{n}}{\partial \mathbf{u}_j \partial \mathbf{u}_k}.$$

From the definition of $\mathbf{z}$,

$$\frac{\partial \mathbf{z}}{\partial \mathbf{u}} = \frac{\partial^2 S}{\partial \mathbf{u} \partial \mathbf{u}}.$$

It follows that

$$\frac{\partial \mathbf{Fn}}{\partial \mathbf{z}} = \frac{\partial \mathbf{Fn}}{\partial \mathbf{u}} \left(\frac{\partial \mathbf{z}}{\partial \mathbf{u}} \right)^{-1} = \frac{\partial \mathbf{Fn}}{\partial \mathbf{u}} \left(\frac{\partial^2 S}{\partial \mathbf{u} \partial \mathbf{u}} \right)^{-1} = \left[\frac{\partial \Psi \mathbf{n}}{\partial \mathbf{u}_j \partial \mathbf{u}_k} - \sum_{i=1}^{m} \frac{\partial S}{\partial \mathbf{u}_i} \frac{\partial^2 \mathbf{e}_i^\top \mathbf{Fn}}{\partial \mathbf{u}_j \partial \mathbf{u}_k} \right].$$

The first term in the right-hand side in this equation is symmetric, and the second term on the right is a linear combination of symmetric matrices. It follows that $\frac{\partial \mathbf{Fn}}{\partial \mathbf{z}}$ is symmetric.

Next, we write

$$\begin{aligned}
[\mathbf{F}(\mathbf{w}_R) - \mathbf{F}(\mathbf{w}_L)]\mathbf{n} &= \int_0^1 \frac{d}{d\theta} \mathbf{F}(\mathbf{z}(\mathbf{w}_R)\theta + \mathbf{z}(\mathbf{w}_L)[1-\theta]) \, d\theta \\
&= \int_0^1 \frac{\partial \mathbf{Fn}}{\partial \mathbf{z}} (\mathbf{z}(\mathbf{w}_R)\theta + \mathbf{z}(\mathbf{w}_L)[1-\theta]) \, d\theta [\mathbf{z}(\mathbf{w}_R) - \mathbf{z}(\mathbf{w}_L)] \\
&\equiv \mathbf{B}[\mathbf{z}(\mathbf{w}_R) - \mathbf{z}(\mathbf{w}_L)],
\end{aligned}$$

and

$$\begin{aligned}
[\mathbf{z}(\mathbf{w}_R) - \mathbf{z}(\mathbf{w}_L)] &= \int_0^1 \frac{d}{d\theta} \mathbf{z}(\mathbf{u}_R\theta + \mathbf{u}_L[1-\theta]) \, d\theta \\
&= \int_0^1 \frac{\partial \mathbf{z}}{\partial \mathbf{u}} (\mathbf{u}_R\theta + \mathbf{u}_L[1-\theta]) \, d\theta [\mathbf{u}_R - \mathbf{u}_L] \equiv \mathbf{P}[\mathbf{u}_R - \mathbf{u}_L].
\end{aligned}$$

From the discussion in the previous paragraph, it is obvious that $\mathbf{B}$ is symmetric and $\mathbf{P}$ is positive-definite. Then $\bar{\mathbf{A}} = \mathbf{BP}$ is similar to a symmetric matrix:

$$(\mathbf{P})^{-1/2}\mathbf{BP}(\mathbf{P})^{-1/2} = (\mathbf{P})^{1/2}\mathbf{B}(\mathbf{P})^{1/2}.$$

It follows that $\bar{\mathbf{A}}$ is diagonalizable with real eigenvalues. □

Lemma 4.13.14 *Suppose that $\bar{\mathbf{A}}(\mathbf{u}_L, \mathbf{u}_R)$ is a Roe matrix for some flux $\mathbf{f}$; in other words,*

$$\text{for all } \mathbf{u}_L \text{ for all } \mathbf{u}_R \quad \mathbf{f}(\mathbf{u}_R) - \mathbf{f}(\mathbf{u}_L) = \bar{\mathbf{A}}(\mathbf{u}_L, \mathbf{u}_R)[\mathbf{u}_R - \mathbf{u}_L]$$

where $\bar{\mathbf{A}}$ has real eigenvalues $\bar{\Lambda}$ and real eigenvectors $\bar{\mathbf{X}}$. Then the approximate Riemann solver

$$\tilde{\mathcal{R}}(\mathbf{u}_L, \mathbf{u}_R, \xi) = \mathbf{u}_L + \sum_{j:\bar{\lambda}_j < \xi} \bar{\mathbf{X}}\mathbf{e}_j\mathbf{e}_j^\top \bar{\mathbf{X}}^{-1}[\mathbf{u}_R - \mathbf{u}_L]$$

is consistent and conservative.

Proof The definition of $\tilde{\mathcal{R}}$ shows that it is consistent:

$$\text{for all } \mathbf{u} \text{ for all } \xi \; \tilde{\mathcal{R}}(\mathbf{u},\mathbf{u},\xi) = \mathbf{u} + \sum_{j:\bar{\lambda}_j<\xi} \bar{\mathbf{X}}\mathbf{e}_j\mathbf{e}_j^\top\bar{\mathbf{X}}^{-1}[\mathbf{u}-\mathbf{u}] = \mathbf{u}.$$

To show that $\tilde{\mathcal{R}}$ is conservative, we compute

$$\begin{aligned}
\int_{-\Delta x_L/2}^{\Delta x_R/2} \tilde{\mathcal{R}}\left(\mathbf{u}_L,\mathbf{u}_R,\frac{x}{\Delta t}\right) dx &= \int_{-\Delta x_L/2}^{\Delta x_R/2} \mathbf{u}_L + \sum_{j:\bar{\lambda}_j\Delta t<x} \bar{\mathbf{X}}\mathbf{e}_j\mathbf{e}_j^\top\bar{\mathbf{X}}^{-1}[\mathbf{u}_R-\mathbf{u}_L]\,dx \\
&= \mathbf{u}_L\frac{\Delta x_R+\Delta x_L}{2} + \sum_{k=1}^{m-1}\int_{\bar{\lambda}_k\Delta t}^{\bar{\lambda}_{k+1}\Delta t}\sum_{j=1}^{k}\bar{\mathbf{X}}\mathbf{e}_j\mathbf{e}_j^\top\bar{\mathbf{X}}^{-1}[\mathbf{u}_R-\mathbf{u}_L]\,dx \\
&\quad + \int_{\bar{\lambda}_m\Delta t}^{\Delta x_R/2}\sum_{j=1}^{m}\bar{\mathbf{X}}\mathbf{e}_j\mathbf{e}_j^\top\bar{\mathbf{X}}^{-1}[\mathbf{u}_R-\mathbf{u}_L]\,dx \\
&= \mathbf{u}_L\frac{\Delta x_R+\Delta x_L}{2} + \sum_{k=1}^{m-1}\sum_{j=1}^{k}\bar{\mathbf{X}}\mathbf{e}_j(\bar{\lambda}_{k+1}-\bar{\lambda}_k)\Delta t\mathbf{e}_j^\top\bar{\mathbf{X}}^{-1}[\mathbf{u}_R-\mathbf{u}_L] \\
&\quad + \sum_{j=1}^{m}\bar{\mathbf{X}}\mathbf{e}_j\left(\frac{\Delta x_R}{2}-\bar{\lambda}_m\Delta t\right)\mathbf{e}_j^\top\bar{\mathbf{X}}^{-1}[\mathbf{u}_R-\mathbf{u}_L] \\
&= \mathbf{u}_L\frac{\Delta x_R+\Delta x_L}{2} + \sum_{j=1}^{m}\bar{\mathbf{X}}\mathbf{e}_j\left(\frac{\Delta x_R}{2}-\bar{\lambda}_j\Delta t\right)\mathbf{e}_j^\top\bar{\mathbf{X}}^{-1}[\mathbf{u}_R-\mathbf{u}_L] \\
&= \mathbf{u}_L\frac{\Delta x_R+\Delta x_L}{2} + \frac{\Delta x_R}{2}[\mathbf{u}_R-\mathbf{u}_L] - \Delta t\bar{\mathbf{A}}(\mathbf{u}_L,\mathbf{u}_R)[\mathbf{u}_R-\mathbf{u}_L] \\
&= \mathbf{u}_L\frac{\Delta x_L}{2} + \mathbf{u}_R\frac{\Delta x_R}{2} - \Delta t[\mathbf{f}(\mathbf{u}_R)-\mathbf{f}(\mathbf{u}_L)]
\end{aligned}$$

□

Our results in Lemma 4.13.14 show us that the Roe flux associated with speed ξ is

$$\begin{aligned}
(\mathbf{f}-\mathbf{u}\xi)(\mathbf{u}_L,\mathbf{u}_R) = \tfrac{1}{2}\Big\{&\mathbf{f}(\mathbf{u}_L)+\mathbf{f}(\mathbf{u}_R)-(\mathbf{u}_L+\mathbf{u}_R)\xi \\
&-\textstyle\sum_{j=1}^m \bar{\mathbf{X}}\mathbf{e}_j|\bar{\lambda}_j-\xi|\mathbf{e}_j^\top\bar{\mathbf{X}}^{-1}[\mathbf{u}_R-\mathbf{u}_L]\Big\}.
\end{aligned} \tag{4.93}$$

If we need to express the flux jump for wave propagation (Section 6.2.6 below), then we write

$$(\mathbf{f}_R-\mathbf{u}_R\xi)-(\mathbf{f}_L-\mathbf{u}_L\xi) = \bar{\mathbf{X}}(\bar{\Lambda}-\mathbf{I}\xi)(\bar{\mathbf{X}})^{-1}(\mathbf{u}_R-\mathbf{u}_L).$$

The problem with the definition of the Roe matrix in Theorem 4.13.13 is that the matrix $\bar{\mathbf{A}}$ is difficult to compute. We will usually choose different approaches that depend on the problem.

Example 4.13.15 Consider the scalar conservation law

$$\frac{\partial \mathbf{u}}{\partial t} + \frac{\partial \mathbf{f}(\mathbf{u})}{\partial x} = 0.$$

The Roe "matrix" for this problem must be $\bar{\lambda} = [\mathbf{f}]/[\mathbf{u}]$. The Roe approximate Riemann solver is easily seen to be

$$\tilde{\mathcal{R}}(\mathbf{u}_L, \mathbf{u}_R, \xi) = \begin{cases} \mathbf{u}_L, & \bar{\lambda} > \xi \\ \mathbf{u}_R, & \bar{\lambda} < \xi \end{cases}$$

and the flux associated with speed ξ in this approximate Riemann solver is

$$\begin{aligned}(\mathbf{f} - \mathbf{u}\xi)(\mathbf{u}_L, \mathbf{u}_R) &= \frac{1}{2}\left\{\mathbf{f}(\mathbf{u}_L) + \mathbf{f}(\mathbf{u}_R) - (\mathbf{u}_L + \mathbf{u}_R)\xi - |\bar{\lambda} - \xi|[\mathbf{u}_R - \mathbf{u}_L]\right\} \\ &= \begin{cases} \mathbf{f}(\mathbf{u}_L) - \mathbf{u}_L\xi, & \bar{\lambda} > \xi \\ \mathbf{f}(\mathbf{u}_R) - \mathbf{u}_R\xi, & \bar{\lambda} < \xi. \end{cases}\end{aligned}$$

Example 4.13.16 We will develop a Roe matrix for the shallow water equations, following the discussion in LeVeque [97, page 320ff]. Instead of using $\mathbf{z} = \nabla_{\mathbf{u}} E$ as the intermediate variable to use in computing the Roe matrix, we will use

$$\mathbf{z} = \begin{bmatrix} c \\ \mathbf{v}c \end{bmatrix} \equiv \begin{bmatrix} \zeta_1 \\ \mathbf{z}_2 \end{bmatrix},$$

where $c = \sqrt{gh}$. Then the vector of conserved quantities can be written

$$\mathbf{u} = \begin{bmatrix} \zeta_1^2 \\ \mathbf{z}_2\zeta_1 \end{bmatrix} \frac{1}{g},$$

so

$$\frac{\partial \mathbf{u}}{\partial \mathbf{z}} = \begin{bmatrix} 2\zeta_1 & 0 \\ \mathbf{z}_2 & \mathbf{I}\zeta_1 \end{bmatrix} \frac{1}{g}$$

is linear in the entries of $\mathbf{z}$. Similarly,

$$\mathbf{F} = \begin{bmatrix} \zeta_1 \mathbf{z}_2^\top \\ \mathbf{z}_2\mathbf{z}_2^\top + \mathbf{I}\frac{1}{2}\zeta_1^4 \end{bmatrix} \frac{1}{g}$$

so

$$\frac{\partial \mathbf{F}\mathbf{n}}{\partial \mathbf{z}} = \begin{bmatrix} \mathbf{z}_2^\top \mathbf{n} & \zeta_1 \mathbf{n}^\top \\ \mathbf{n}2\zeta_1^3 & \mathbf{I}\mathbf{z}_2^\top \mathbf{n} + \mathbf{z}_2\mathbf{n}^\top \end{bmatrix} \frac{1}{g}$$

involves polynomials in the entries of $\mathbf{z}$ as well.

Next, we define matrices $\mathbf{B}$ and $\mathbf{P}^{-1}$ by

$$\mathbf{B} \equiv g \int_0^1 \frac{\partial \mathbf{f}}{\partial \mathbf{z}}(\mathbf{z}_R\theta + \mathbf{z}_L[1-\theta])\, d\theta, \quad \mathbf{P}^{-1} \equiv g \int_0^1 \frac{\partial \mathbf{u}}{\partial \mathbf{z}}(\mathbf{z}_R\theta + \mathbf{z}_L[1-\theta])\, d\theta.$$

This would suggest that we compute

$$\bar{\zeta}_1 \equiv \frac{\int_{(\zeta_1)_L}^{(\zeta_1)_R} \zeta_1 \, d\zeta_1}{(\zeta_1)_R - (\zeta_1)_L} = \frac{1}{2}\{(\zeta_1)_R + (\zeta_1)_L\}$$

$$\bar{\mathbf{z}}_2 \equiv \frac{\int_{(\mathbf{z}_2)_L}^{(\mathbf{z}_2)_R} \mathbf{z}_2 \, d\mathbf{z}_2}{(\mathbf{z}_2)_R - (\mathbf{z}_2)_L} = \frac{1}{2}\{(\mathbf{z}_2)_R + (\mathbf{z}_2)_L\}$$

$$\overline{\zeta_1^3} \equiv \frac{\int_{(\zeta_1)_L}^{(\zeta_1)_R} \zeta_1^3 \, d\zeta_1}{(\zeta_1)_R - (\zeta_1)_L} = \frac{(\zeta_1)_R^4 - (\zeta_1)_L^4}{4[(\zeta_1)_R - (\zeta_1)_L]}$$

$$= \frac{(\zeta_1)_R + (\zeta_1)_L}{2} \frac{(\zeta_1)_R^2 + (\zeta_1)_L^2}{2} \equiv \bar{\zeta}_1 \frac{(\zeta_1)_R^2 + (\zeta_1)_L^2}{2}$$

and use the polynomials for the partial derivatives of **u** and **f** to compute

$$\mathbf{B} = \begin{bmatrix} \bar{\mathbf{z}}_2^\top \mathbf{n} & \bar{\zeta}_1 \mathbf{n}^\top \\ \mathbf{n} 2\overline{\zeta_1^3} & \mathbf{I}\bar{\mathbf{z}}_2^\top \mathbf{n} + \bar{\mathbf{z}}_2 \mathbf{n}^\top \end{bmatrix} \quad \text{and} \quad \mathbf{P}^{-1} = \begin{bmatrix} 2\bar{\zeta}_1 & 0 \\ \bar{\mathbf{z}}_2 & \mathbf{I}\bar{\zeta}_1 \end{bmatrix}.$$

Then

$$\mathbf{f}(\mathbf{z}_R) - \mathbf{f}(\mathbf{z}_L) = \mathbf{BP}[\mathbf{u}(\mathbf{z}_R) - \mathbf{u}(z_L)]$$

where the Roe matrix is

$$\mathbf{BP} = \begin{bmatrix} 0 & \mathbf{n}^\top \\ \mathbf{n}\frac{\overline{\zeta_1^3}}{\bar{\zeta}_1} - (\bar{\mathbf{z}}_2/\bar{\zeta}_1)(\bar{\mathbf{z}}_2/\bar{\zeta}_1)^\top \mathbf{n} & \mathbf{I}(\bar{\mathbf{z}}_2/\bar{\zeta}_1)^\top \mathbf{n} + (\bar{\mathbf{z}}_2/\bar{\zeta}_1)\mathbf{n}^\top \end{bmatrix}.$$

If we define

$$\bar{\mathbf{v}} \equiv \frac{\bar{\mathbf{z}}_2}{\bar{\zeta}_1} = \frac{\mathbf{v}_R c_R + \mathbf{v}_L c_L}{c_R + c_L}, \quad \bar{h} \equiv \frac{\overline{\zeta_1^3}}{\bar{\zeta}_1} = \frac{h_R + h_L}{2}, \quad \bar{c} \equiv \sqrt{g\bar{h}}$$

then we can write

$$\mathbf{BP} = \begin{bmatrix} 0 & \mathbf{n}^\top \\ \mathbf{n}g\bar{h} - \bar{\mathbf{v}}\bar{\mathbf{v}}^\top \mathbf{n} & \mathbf{I}\bar{\mathbf{v}}^\top \mathbf{n} + \bar{\mathbf{v}}\mathbf{n}^\top \end{bmatrix}.$$

The eigenvectors of the Roe matrix are

$$\bar{\mathbf{X}} = \begin{bmatrix} 1 & 0 & 1 \\ \bar{\mathbf{v}} - \mathbf{n}\bar{c} & \mathbf{N} & \bar{\mathbf{v}} + \mathbf{n}\bar{c} \end{bmatrix}$$

where $[\mathbf{n}, \mathbf{N}]$ is an orthogonal matrix, and the eigenvalues are

$$\bar{\Lambda} = \begin{bmatrix} \bar{\mathbf{v}}^\top \mathbf{n} - \bar{c} & 0 & 0 \\ 0 & \mathbf{I}\bar{\mathbf{v}}^\top \mathbf{n} & 0 \\ 0 & 0 & \bar{\mathbf{v}}^\top \mathbf{n} + \bar{c} \end{bmatrix}.$$

In order to solve the Riemann problem, Roe solves the linear system $\mathbf{u}_R - \mathbf{u}_L = \bar{\mathbf{X}}\mathbf{y}$ for the expansion coefficients $\mathbf{y}$ of the waves:

$$\mathbf{y} = \bar{\mathbf{X}}^{-1}(\mathbf{u}_R - \mathbf{u}_L)] = \begin{bmatrix} \bar{\mathbf{v}}^\top\mathbf{n} + \bar{c} & -\mathbf{n}^\top \\ -\mathbf{N}^\top\bar{\mathbf{v}}2\bar{c} & \mathbf{N}^\top 2\bar{c} \\ -\bar{\mathbf{v}}^\top\mathbf{n} + \bar{c} & \mathbf{n}^\top \end{bmatrix} \begin{bmatrix} h_R - h_L \\ \mathbf{v}_R h_R - \mathbf{v}_L h_L \end{bmatrix} \frac{1}{2\bar{c}}$$

$$= \begin{bmatrix} (\bar{\mathbf{v}}^\top\mathbf{n} + \bar{c})(h_R - h_L) - \mathbf{n}^\top(\mathbf{v}_R h_R - \mathbf{v}_L h_L) \\ \mathbf{N}^\top\{-\bar{\mathbf{v}}(h_R - h_L) + (\mathbf{v}_R h_R - \mathbf{v}_L h_L)\}2\bar{c} \\ \mathbf{n}^\top(\mathbf{v}_R h_R - \mathbf{v}_L h_L) - (\bar{\mathbf{v}}^\top\mathbf{n} - \bar{c})(h_R - h_L) \end{bmatrix} \frac{1}{2\bar{c}}.$$

These expansion coefficients give us an approximate path from left state to right state. Roe approximated the flux at the state moving at speed ξ by

$$(\mathbf{f} - \mathbf{u}\xi)_{\text{Roe}} = \frac{1}{2}\{\mathbf{f}_R + \mathbf{f}_L - (\mathbf{u}_R + \mathbf{u}_L)\xi - \sum_i \bar{\mathbf{X}}\mathbf{e}_i|\bar{\lambda}_i - \xi|\mathbf{y}_i\}$$

$$= \frac{1}{2}\{\mathbf{f}_L + \mathbf{f}_R - (\mathbf{u}_R + \mathbf{u}_L)\xi - \bar{\mathbf{X}}|\bar{\Lambda} - \mathbf{I}\xi|\bar{\mathbf{X}}^{-1}(\mathbf{u}_R - \mathbf{u}_L)\}.$$

Example 4.13.17 The following discussion of a Roe solver for gas dynamics follows that in Hirsch [73, page 463ff]. Instead of using $\mathbf{z} = \nabla_{\mathbf{u}} S$ as the intermediate variable to use in computing the Roe matrix for gas dynamics, we will use

$$\mathbf{z} = \begin{bmatrix} 1 \\ \mathbf{v} \\ H \end{bmatrix} \sqrt{\rho} \equiv \begin{bmatrix} \zeta_1 \\ \mathbf{z}_2 \\ \zeta_3 \end{bmatrix},$$

where $H \equiv e + \frac{1}{2}\mathbf{v}\cdot\mathbf{v} + p/\rho$ is the total specific enthalpy. Then the vector of conserved quantities can be written

$$\mathbf{u} = \begin{bmatrix} \zeta_1^2 \\ \mathbf{z}_2\zeta_1 \\ \frac{\zeta_3\zeta_1}{\gamma} + \frac{\gamma-1}{2\gamma}\mathbf{z}_2\cdot\mathbf{z}_2 \end{bmatrix}$$

so

$$\frac{\partial\mathbf{u}}{\partial\mathbf{z}} = \begin{bmatrix} 2\zeta_1 & 0 & 0 \\ \mathbf{z}_2 & \mathbf{I}\zeta_1 & 0 \\ \frac{\zeta_3}{\gamma} & \frac{\gamma-1}{\gamma}\mathbf{z}_2^\top & \frac{\zeta_1}{\gamma} \end{bmatrix}$$

is linear in the entries of $\mathbf{z}$. Similarly,

$$\mathbf{F}\mathbf{n} = \begin{bmatrix} \zeta_1\mathbf{z}_2\cdot\mathbf{n} \\ \mathbf{z}_2\mathbf{z}_2\cdot\mathbf{n} + \mathbf{n}(\zeta_1\zeta_3 - \frac{\mathbf{z}_2\cdot\mathbf{z}_2}{2})\frac{\gamma-1}{\gamma} \\ \zeta_3\mathbf{z}_2\cdot\mathbf{n} \end{bmatrix}$$

so

$$\frac{\partial \mathbf{F}\mathbf{n}}{\partial \mathbf{z}} = \begin{bmatrix} \mathbf{z}_2 \cdot \mathbf{n} & \zeta_1 \mathbf{n}^\top & 0 \\ \mathbf{n}\zeta_3 \frac{\gamma-1}{\gamma} & \mathbf{I}\mathbf{z}_2 \cdot \mathbf{n} + \mathbf{z}_2 \mathbf{n}^\top - \mathbf{n}\frac{\gamma-1}{\gamma}\mathbf{z}_2^\top & \mathbf{n}\frac{\gamma-1}{\gamma}\zeta_1 \\ 0 & \zeta_3 \mathbf{n}^\top & \mathbf{z}_2 \cdot \mathbf{n} \end{bmatrix}$$

is linear in the entries of $\mathbf{z}$ as well.

Next, we define matrices $\mathbf{B}$ and $\mathbf{P}^{-1}$ by

$$[\mathbf{F}(\mathbf{z}_R) - \mathbf{F}(\mathbf{z}_L)]\mathbf{n} = \int_0^1 \frac{\partial \mathbf{F}\mathbf{n}}{\partial \mathbf{z}}(\mathbf{z}_R\theta + \mathbf{z}_L[1-\theta])\, d\theta(\mathbf{z}_R - \mathbf{z}_L) \equiv \mathbf{B}(\mathbf{z}_R - \mathbf{z}_L)$$

and

$$\mathbf{u}(\mathbf{z}_R) - \mathbf{u}(\mathbf{z}_L) = \int_0^1 \frac{\partial \mathbf{u}}{\partial \mathbf{z}}(\mathbf{z}_R\theta + \mathbf{z}_L[1-\theta])\, d\theta(\mathbf{z}_R - \mathbf{z}_L) \equiv \mathbf{P}^{-1}(\mathbf{z}_R - \mathbf{z}_L).$$

This would suggest that we compute

$$\bar{\mathbf{z}} \equiv \begin{bmatrix} \sqrt{\rho_R} + \sqrt{\rho_L} \\ \mathbf{v}_R\sqrt{\rho_R} + \mathbf{v}_L\sqrt{\rho_L} \\ H_R\sqrt{\rho_R} + H_L\sqrt{\rho_L} \end{bmatrix} \frac{1}{2}$$

and use the linearity of the partial derivatives of $\mathbf{u}$ and $\mathbf{F}\mathbf{n}$ to compute

$$\mathbf{P}^{-1} = \begin{bmatrix} 2\bar{\zeta}_1 & 0 & 0 \\ \bar{\mathbf{z}}_2 & \mathbf{I}\bar{\zeta}_1 & 0 \\ \frac{\bar{\zeta}_3}{\gamma} & \frac{\gamma-1}{\gamma}\bar{\mathbf{z}}_2^\top & \frac{\bar{\zeta}_1}{\gamma} \end{bmatrix}$$

and

$$\mathbf{B} = \begin{bmatrix} \bar{\mathbf{z}}_2 \cdot \mathbf{n} & \bar{\zeta}_1 \mathbf{n}^\top & 0 \\ \mathbf{n}\bar{\zeta}_3 \frac{\gamma-1}{\gamma} & \mathbf{I}\bar{\mathbf{z}}_2 \cdot \mathbf{n} + \bar{\mathbf{z}}_2 \mathbf{n}^\top - \mathbf{n}\frac{\gamma-1}{\gamma}\bar{\mathbf{z}}_2^\top & \mathbf{n}\frac{\gamma-1}{\gamma}\bar{\zeta}_1 \\ 0 & \bar{\zeta}_3 \mathbf{n}^\top & \bar{\mathbf{z}}_2 \cdot \mathbf{n} \end{bmatrix}.$$

Then

$$[\mathbf{F}(\mathbf{z}_R) - \mathbf{F}(\mathbf{z}_L)]\mathbf{n} = \mathbf{B}\mathbf{P}[\mathbf{u}(\mathbf{z}_R) - \mathbf{u}(\mathbf{z}_L)]$$

where the Roe matrix is

$$
\mathbf{BP} = \begin{bmatrix} \bar{\mathbf{z}}_2 \cdot \mathbf{n} & \bar{\zeta}_1 \mathbf{n}^\top & 0 \\ \mathbf{n}\bar{\zeta}_3 \frac{\gamma-1}{\gamma} & \mathbf{I}\bar{\mathbf{z}}_2 \cdot \mathbf{n} + \bar{\mathbf{z}}_2 \mathbf{n}^\top - \mathbf{n}\frac{\gamma-1}{\gamma}\bar{\mathbf{z}}_2^\top & \mathbf{n}\frac{\gamma-1}{\gamma}\bar{\zeta}_1 \\ 0 & \bar{\zeta}_3 \mathbf{n}^\top & \bar{\mathbf{z}}_2 \cdot \mathbf{n} \end{bmatrix} \begin{bmatrix} \frac{1}{2\bar{\zeta}_1} & 0 & 0 \\ -\bar{\mathbf{z}}_2 \frac{1}{2\bar{\zeta}_1^2} & \mathbf{I}\frac{1}{\bar{\zeta}_1} & 0 \\ (\gamma-1)\frac{\bar{\mathbf{z}}_2 \cdot \bar{\mathbf{z}}_2}{2\bar{\zeta}_1^3} - \frac{\bar{\zeta}_3}{2\bar{\zeta}_1^2} & -\frac{\gamma-1}{\bar{\zeta}_1^2}\bar{\mathbf{z}}_2^\top & \frac{\gamma}{\bar{\zeta}_1} \end{bmatrix}
$$

$$
= \begin{bmatrix} 0 & \mathbf{n}^\top & 0 \\ \mathbf{n}\frac{\gamma-1}{2}\frac{\bar{\mathbf{z}}_2 \cdot \bar{\mathbf{z}}_2}{\bar{\zeta}_1^2} - \bar{\mathbf{z}}_2 \frac{\bar{\mathbf{z}}_2 \cdot \mathbf{n}}{\bar{\zeta}_1^2} & \mathbf{I}\frac{\bar{\mathbf{z}}_2 \cdot \mathbf{n}}{\bar{\zeta}_1} + \bar{\mathbf{z}}_2 \frac{1}{\bar{\zeta}_1}\mathbf{n}^\top - \mathbf{n}\frac{\gamma-1}{\bar{\zeta}_1}\bar{\mathbf{z}}_2^\top & \mathbf{n}(\gamma-1) \\ -\bar{\zeta}_3 \frac{\bar{\mathbf{z}}_2 \cdot \mathbf{n}}{\bar{\zeta}_1^2} + (\gamma-1)\frac{\bar{\mathbf{z}}_2 \cdot \bar{\mathbf{z}}_2}{2\bar{\zeta}_1^3}\bar{\mathbf{z}}_2 \cdot \mathbf{n} & \frac{\bar{\zeta}_3}{\bar{\zeta}_1}\mathbf{n}^\top - (\gamma-1)\frac{\bar{\mathbf{z}}_2 \cdot \mathbf{n}}{\bar{\zeta}_1^2}\bar{\mathbf{z}}_2^\top & \gamma \frac{\bar{\mathbf{z}}_2 \cdot \mathbf{n}}{\bar{\zeta}_1} \end{bmatrix}.
$$

Next, we note that $\bar{\mathbf{A}} = \mathbf{BP}$ depends only on the average velocity

$$
\bar{\mathbf{v}} \equiv \bar{\mathbf{z}}_2 / \bar{\zeta}_1 = \frac{\mathbf{v}_R \sqrt{\rho_R} + \mathbf{v}_L \sqrt{\rho_L}}{\sqrt{\rho_R} + \sqrt{\rho_L}}
$$

and the average total specific enthalpy

$$
\bar{H} \equiv \bar{\zeta}_3 / \bar{\zeta}_1 = \frac{H_R \sqrt{\rho_R} + H_L \sqrt{\rho_L}}{\sqrt{\rho_R} + \sqrt{\rho_L}}.
$$

Thus we can write

$$
\bar{\mathbf{A}} = \begin{bmatrix} 0 & \mathbf{n}^\top & 0 \\ \mathbf{n}\frac{\gamma-1}{2}\bar{\mathbf{v}} \cdot \bar{\mathbf{v}} - \bar{\mathbf{v}}\bar{\mathbf{v}} \cdot \mathbf{n} & \mathbf{I}\bar{\mathbf{v}} \cdot \mathbf{n} + \bar{\mathbf{v}}\mathbf{n}^\top - \mathbf{n}(\gamma-1)\bar{\mathbf{v}}^\top & \mathbf{n}(\gamma-1) \\ (\frac{\gamma-1}{2}\bar{\mathbf{v}} \cdot \bar{\mathbf{v}} - \bar{H})\bar{\mathbf{v}} \cdot \mathbf{n} & \bar{H}\mathbf{n}^\top - (\gamma-1)\bar{\mathbf{v}} \cdot \mathbf{n}\bar{\mathbf{v}}^\top & \gamma \bar{\mathbf{v}} \cdot \mathbf{n} \end{bmatrix}.
$$

For simplicity, we will define the average sound speed by

$$
\bar{c} \equiv \sqrt{(\gamma-1)\left(\bar{H} - \frac{1}{2}\bar{\mathbf{v}} \cdot \bar{\mathbf{v}}\right)}.
$$

Then we can see that the eigenvectors of the Roe matrix are

$$
\bar{\mathbf{X}} = \begin{bmatrix} 1 & 0 & 1 & 1 \\ \bar{\mathbf{v}} - \mathbf{n}\bar{c} & \mathbf{N} & \bar{\mathbf{v}} & \bar{\mathbf{v}} + \mathbf{n}\bar{c} \\ \bar{H} - \bar{\mathbf{v}} \cdot \mathbf{n}\bar{c} & \bar{\mathbf{v}}^\top \mathbf{N} & \frac{1}{2}\bar{\mathbf{v}} \cdot \bar{\mathbf{v}} & \bar{H} + \bar{\mathbf{v}} \cdot \mathbf{n}\bar{c} \end{bmatrix}
$$

and the eigenvalues are

$$
\bar{\Lambda} = \begin{bmatrix} \bar{\mathbf{v}} \cdot \mathbf{n} - \bar{c} & 0 & 0 & 0 \\ 0 & \mathbf{I}\bar{\mathbf{v}} \cdot \mathbf{n} & 0 & 0 \\ 0 & 0 & \bar{\mathbf{v}} \cdot \mathbf{n} & 0 \\ 0 & 0 & 0 & \bar{\mathbf{v}} \cdot \mathbf{n} + \bar{c} \end{bmatrix}.
$$

Here $[\mathbf{n}, \mathbf{N}]$ is an orthogonal matrix.

In order to solve the Riemann problem, Roe solves the linear system $\mathbf{u}_R - \mathbf{u}_L = \bar{\mathbf{X}}\mathbf{y}$ for the expansion coefficients $\mathbf{y}$ of the waves by computing

$$\mathbf{y} = \bar{\mathbf{X}}^{-1}[\mathbf{u}] = \begin{bmatrix} \frac{1}{2}(\beta_1 + \mathbf{b}\cdot\mathbf{n}) & -\frac{1}{2}(\beta_2\bar{\mathbf{v}}^\top + \frac{1}{\bar{c}}\mathbf{n}^\top) & \frac{1}{2}\beta_2 \\ -\mathbf{N}^\top\bar{\mathbf{v}} & \mathbf{N}^\top & 0 \\ 1-\beta_1 & \beta_2\bar{\mathbf{v}}^\top & -\beta_2 \\ \frac{1}{2}(\beta_1 - \mathbf{b}\cdot\mathbf{n}) & -\frac{1}{2}(\beta_2\bar{\mathbf{v}}^\top - \frac{1}{\bar{c}}\mathbf{n}^\top) & \frac{1}{2}\beta_2 \end{bmatrix} \begin{bmatrix} \rho_R - \rho_L \\ \mathbf{v}_R\rho_R - \mathbf{v}_L\rho_L \\ \rho_R(e_R + \frac{1}{2}\mathbf{v}_R^2) - \rho_L(e_L + \frac{1}{2}\mathbf{v}_L^2) \end{bmatrix}.$$

Here

$$\beta_2 = (\gamma - 1)/\bar{c}^2, \quad \beta_1 = \frac{1}{2}\beta_2\bar{\mathbf{v}}\cdot\bar{\mathbf{v}}, \quad \mathbf{b} = \bar{\mathbf{v}}/\bar{c}.$$

These expansion coefficients give us an approximate path from left state to right state. Roe approximated the flux at the state moving with speed ξ by

$$(\mathbf{f} - \mathbf{u}\xi)_{\text{Roe}} = \frac{1}{2}\{\mathbf{f}_L + \mathbf{f}_R - (\mathbf{u}_R + \mathbf{u}_L)\xi - \bar{\mathbf{X}}|\bar{\Lambda} - \mathbf{I}\xi|\bar{\mathbf{X}}^{-1}(\mathbf{u}_R - \mathbf{u}_L)\}.$$

Example 4.13.18 Let us see if we can develop a Roe solver for the Schaeffer–Schechter–Shearer model. Recall from Equation (4.77) that this model has

$$\mathbf{u} = \begin{bmatrix} p \\ q \end{bmatrix} \text{ and } \mathbf{f} = \begin{bmatrix} p^2 - q \\ \frac{1}{3}p^3 - p \end{bmatrix}.$$

Notice that $\partial\mathbf{f}/\partial\mathbf{u}$ is a function of p only. We will take $\mathbf{z} = \mathbf{u}$, and compute

$$\mathbf{B} = \int_0^1 \frac{\partial\mathbf{f}}{\partial\mathbf{u}}(p_R\theta + p_L[1-\theta])\,d\theta = \begin{bmatrix} \frac{1}{p_R - p_L}\int_{p_L}^{p_R} 2p\,dp & -1 \\ \frac{1}{p_R - p_L}\int_{p_L}^{p_R} p^2 - 1\,dp & 0 \end{bmatrix} = \begin{bmatrix} 2\bar{p} & -1 \\ \overline{p^2} & 0 \end{bmatrix}$$

where

$$\bar{p} = \frac{p_R + p_L}{2} \quad \text{and} \quad \overline{p^2} = \frac{p_R^2 + p_R p_L p_L^2}{3}.$$

Since $\mathbf{z} = \mathbf{u}$, we have $\mathbf{P} = \mathbf{I}$. Now $\mathbf{B}\bar{\mathbf{X}} = \bar{\mathbf{X}}\bar{\Lambda}$ where

$$\bar{\mathbf{X}} = \begin{bmatrix} 1 & 1 \\ \bar{p} + r & \bar{p} - r \end{bmatrix} \text{ and } \bar{\Lambda} = \begin{bmatrix} \bar{p} - r & 0 \\ 0 & \bar{p} + r \end{bmatrix}$$

and

$$r = \sqrt{1 + \bar{p}^2 - \overline{p^2}} = \sqrt{1 - \frac{(p_R - p_L)^2}{12}}.$$

Thus the Roe matrix does not have real eigenvalues if the jump between the states is too large.

It is, in general, very difficult to construct Roe solvers for the conservation laws in our other case studies. The vibrating string, plasticity, polymer and Buckley–Leverett models all involve nonlinear functions that greatly complicate the construction of the Roe average matrices **B** and **P**. Roe solvers have been discussed for magnetohydrodynamics [23], but only for special cases ($\gamma = 2$). In general, MHD computations approximate a Roe matrix by flux derivatives at some intermediate state; this is more like a weak-wave Riemann solver.

Note that if the solution to the Riemann problem consists of a single discontinuity, then the Roe solver gets the exact solution. On the other hand, if the solution to the Riemann problem involves a transonic rarefaction, then Roe's solver typically produces an entropy-violating discontinuity. Also recall that a Roe solver leads to a numerical flux of the form (4.93). By examining the definition of the numerical diffusion relative to the centered differences flux in Equation (4.86), it is easy to see that the numerical diffusion associated with a Roe solver has the form

$$d(\mathbf{u}_L, \mathbf{u}_R) = |\bar{\mathbf{A}}(\mathbf{u}_L, \mathbf{u}_R) - \mathbf{I}\xi|[\mathbf{u}_R - \mathbf{u}_L]. \tag{4.94}$$

Typically, we are interested in those states that move with zero speed in the solution of the Riemann problem ($\xi = 0$). Across a transonic rarefaction, it is common for the Roe matrix to have a zero eigenvalue, resulting in zero diffusion and an entropy violation. One solution to this problem is to add an numerical diffusion, as discussed in Section 4.13.2; however, this approach typically results in a nonconservative approximate Riemann solver. We will discuss some other alternatives in the following sections.

4.13.9 Harten–Hyman Modification of the Roe Solver

Suppose that we use a Roe solver for some flux **f**, and we compute the characteristic speeds λ_{jL} and λ_{jR} at the left and right states. The Roe approximate Riemann solver takes

$$\tilde{\mathcal{R}}_{Roe}(\mathbf{u}_L, \mathbf{u}_R, \xi) = \mathbf{u}_L + \sum_{\lambda_j < \xi} \bar{\mathbf{X}}\mathbf{e}_j\mathbf{e}_j^\top\bar{\mathbf{X}}^{-1}[\mathbf{u}_R - \mathbf{u}_L]\} = \mathbf{u}_L + \sum_{\lambda_j < \xi} \triangle\mathbf{u}_j$$

where $\triangle\mathbf{u}_j = \bar{\mathbf{X}}\mathbf{e}_j\mathbf{e}_j^\top\bar{\mathbf{X}}^{-1}[\mathbf{u}_R - \mathbf{u}_L]$. If j is such that $\lambda_{jL} < \xi < \lambda_{jR}$, Harten and Hyman [66] suggest that the jth jump be subdivided in the form $\triangle\mathbf{u}_j = \triangle\mathbf{u}_{jL} + \triangle\mathbf{u}_{jR}$ where $\triangle\mathbf{u}_{jL} = \triangle\mathbf{u}_j\beta_j$ is associated with speed λ_{jL}, and $\triangle\mathbf{u}_{jR} = \triangle\mathbf{u}_j(1 - \beta_j)$ is associated with speed λ_{jR}. The associated approximate Riemann solver will be

$$\tilde{\mathcal{R}}_{HH}(\mathbf{u}_L, \mathbf{u}_R, \xi) = \mathbf{u}_L + \sum_{k<j} \triangle\mathbf{u}_j + \beta_j\triangle\mathbf{u}_j.$$

The resulting approximate Riemann solver will still be consistent, because the change to the solution of the Riemann problem is zero when $\mathbf{u}_L = \mathbf{u}_R$. The resulting numerical flux can be evaluated as

$$\begin{aligned}(\mathbf{f} - \mathbf{u}\xi)_{HH}(\mathbf{u}_L, \mathbf{u}_R) = &\frac{1}{2}[\mathbf{f}(\mathbf{u}_L) + \mathbf{f}(\mathbf{u}_R)] - [\mathbf{u}_L + \mathbf{u}_R]\frac{\xi}{2} \\ &-\frac{1}{2}\sum_{k \neq j} |\lambda_k - \xi| \triangle \mathbf{u}_k \\ &-\frac{1}{2}\triangle \mathbf{u}_j\{(\xi - \lambda_{jL})\beta_j + (\lambda_{j,R} - \xi)(1 - \beta_j)\}.\end{aligned}$$

From Lemma 4.13.14, we see that in order for the resulting approximate Riemann solver to be conservative, it is sufficient to require that

$$\lambda_{jL}\beta_j + \lambda_{jR}(1 - \beta_j) = \bar{\lambda}_j.$$

It is easy to see that this implies $\beta_j = (\lambda_{jR} - \bar{\lambda}_j)/(\lambda_{jR} - \lambda_{jL})$. We also require $0 \leq \beta_j \leq 1$, so we actually compute

$$\beta_j = \max\left\{0, \min\left\{1, \frac{\lambda_{jR} - \bar{\lambda}_j}{\lambda_{jR} - \lambda_{jL}}\right\}\right\}.$$

Recall that Equation (4.94) showed that the Roe solver numerical diffusion has the form $\overline{\mathbf{X}}|\overline{\Lambda} - \mathbf{I}\xi|\overline{\mathbf{X}}^{-1}\triangle\mathbf{u}$. In this context, the next lemma shows that the Harten–Hyman modification to the Roe approximate Riemann solver will involve additional numerical diffusion.

Lemma 4.13.19 *If* $\lambda_{jL} < \xi < \lambda_{jR}$ *and* $\lambda_{jL} \leq \bar{\lambda}_j \leq \lambda_{jR}$*, then*

$$(\lambda_{jR} - \xi)\left[1 - \frac{\lambda_{jR} - \bar{\lambda}_j}{\lambda_{jR} - \lambda_{jL}}\right] - (\lambda_{jL} - \xi)\left[\frac{\lambda_{jR} - \bar{\lambda}_j}{\lambda_{jR} - \lambda_{jL}}\right] \geq |\bar{\lambda}_j - \xi|.$$

Proof Note that

$$\begin{aligned}0 &\leq 2(\xi - \lambda_{jL})(\lambda_{jR} - \bar{\lambda}_j) \\ &= (\lambda_{jR} - \xi)\left[\bar{\lambda}_j - \lambda_{jL} - (\bar{\lambda}_j - \xi)\right] - (\lambda_{jL} - \xi)\left[\lambda_{jR} - \bar{\lambda}_j - (\bar{\lambda}_j - \xi)\right]\end{aligned}$$

so if $\bar{\lambda}_j \geq \xi$ we have

$$0 \leq (\lambda_{jR} - \xi)(\bar{\lambda}_j - \lambda_{jL} - |\bar{\lambda}_j - \xi|) - (\lambda_{jL} - \xi)(\lambda_{jR} - \bar{\lambda}_j - |\bar{\lambda}_j - \xi|). \tag{4.95}$$

On the other hand,

$$\begin{aligned}0 &\leq 2(\lambda_{jR} - \xi)(\bar{\lambda}_j - \lambda_{jL}) \\ &= (\lambda_{jR} - \xi)\left[\bar{\lambda}_j - \lambda_{jL} + (\bar{\lambda}_j - \xi)\right] - (\lambda_{jL} - \xi)\left[\lambda_{jR} - \bar{\lambda}_j + (\bar{\lambda}_j - \xi)\right],\end{aligned}$$

so if $\bar{\lambda}_j < \xi$ we see that inequality (4.95) is still true. This inequality implies that

$$(\lambda_{jR} - \xi)(\bar{\lambda}_j - \lambda_{jL}) - (\lambda_{jL} - \xi)(\lambda_{jR} - \bar{\lambda}_j) \geq |\bar{\lambda}_j - \xi|(\lambda_{jR} - \lambda_{jL}),$$

from which it follows that

$$(\lambda_{jR} - \xi)\left[1 - \frac{\lambda_{jR} - \bar{\lambda}_j}{\lambda_{jR} - \lambda_{jL}}\right] - (\lambda_{jL} - \xi)\left[\frac{\lambda_{jR} - \bar{\lambda}_j}{\lambda_{jR} - \lambda_{jL}}\right] \geq |\bar{\lambda}_j - \xi|.$$

□

Thus this modification of the Roe solver should never be performed for linearly degenerate waves, such as the contact discontinuity in gas dynamics. This is because the Roe solver captures discontinuities correctly, and because we do not want to add additional numerical diffusion to contact discontinuities.

4.13.10 Harten–Lax–van Leer Scheme

A vastly simpler approximate Riemann solver is due to Harten, Lax and van Leer [68]. They assume that we can find lower and upper bounds $\underline{\lambda}$ and $\overline{\lambda}$ on the characteristic speeds in the solution of the Riemann problem involving states $\mathbf{u}_L$ and $\mathbf{u}_R$. In practice [48], these bounds are often approximated by

$$\underline{\lambda} = \min_j \left\{\min\{\lambda_{jL}, \bar{\lambda}_j\}\right\} \quad \text{and} \quad \overline{\lambda} = \max_j \left\{\max\{\lambda_{jR}, \bar{\lambda}_j\}\right\}$$

where λ_{jL} are the characteristic speeds at $\mathbf{u}_L$, λ_{jR} are the characteristic speeds at $\mathbf{u}_R$, and $\bar{\lambda}_j$ are the eigenvalues of the Roe matrix. The **HLL** approximate Riemann solver takes the form

$$\tilde{\mathcal{R}}(\mathbf{u}_L, \mathbf{u}_R, \xi) = \begin{cases} \mathbf{u}_L, & \xi < \underline{\lambda} \\ \mathbf{u}_{LR}, & \underline{\lambda} < \xi < \overline{\lambda} \\ \mathbf{u}_R, & \overline{\lambda} < \xi. \end{cases}$$

Here the intermediate state $\mathbf{u}_{LR}$ is chosen so that $\tilde{\mathcal{R}}$ is conservative. (Note that this approach is very similar to the Rusanov solver in section 4.13.3.) Conservation requires

$$\int_{-\triangle x_L/2}^{\triangle x_R/2} \tilde{\mathcal{R}}\left(\mathbf{u}_L, \mathbf{u}_R, \frac{x}{\triangle t}\right) dx = \mathbf{u}_L\left(\underline{\lambda}\triangle t + \frac{\triangle x_L}{2}\right) + \mathbf{u}_{LR}(\overline{\lambda} - \underline{\lambda})\triangle t + \mathbf{u}_R\left(\frac{\triangle x_R}{2} - \overline{\lambda}\triangle t\right)$$
$$= \frac{\triangle x_L \mathbf{u}_L + \triangle x_R \mathbf{u}_R}{2} - \triangle t[\mathbf{f}(\mathbf{u}_R) - \mathbf{f}(\mathbf{u}_L)].$$

We can solve for $\mathbf{u}_{LR}$ to get

$$\mathbf{u}_{LR} = \frac{\overline{\lambda}\mathbf{u}_R - \underline{\lambda}\mathbf{u}_L}{\overline{\lambda} - \underline{\lambda}} - \frac{\mathbf{f}(\mathbf{u}_R) - \mathbf{f}(\mathbf{u}_L)}{\overline{\lambda} - \underline{\lambda}}.$$

The flux at the state that moves with speed ξ in this approximate Riemann problem solution can be computed by applying Equation (4.80):

$$(\mathbf{f} - \mathbf{u}\xi)_{\text{HLL}}(\mathbf{u}_L, \mathbf{u}_R) = \frac{1}{2}\{\mathbf{f}(\mathbf{u}_L) + \mathbf{f}(\mathbf{u}_R) - (\mathbf{u}_L + \mathbf{u}_R)\xi - |\underline{\lambda} - \xi|(\mathbf{u}_{LR} - \mathbf{u}_L) - |\overline{\lambda} - \xi|(\mathbf{u}_R - \mathbf{u}_{LR})\}. \tag{4.96}$$

If $\xi < \underline{\lambda}$ then

$$\begin{aligned}
&(\mathbf{f} - \mathbf{u}\xi)_{\text{HLL}}\ (\mathbf{u}_L, \mathbf{u}_R) \\
&= \frac{1}{2}\{\mathbf{f}(\mathbf{u}_L) + \mathbf{f}(\mathbf{u}_R) - (\mathbf{u}_L + \mathbf{u}_R)\xi - (\underline{\lambda} - \xi)(\mathbf{u}_{LR} - \mathbf{u}_L) - (\overline{\lambda} - \xi)(\mathbf{u}_R - \mathbf{u}_{LR})\} \\
&= \frac{1}{2}[\mathbf{f}(\mathbf{u}_L) + \mathbf{f}(\mathbf{u}_R) - 2\mathbf{u}_L\xi + \mathbf{u}_L\underline{\lambda} - \mathbf{u}_R\overline{\lambda} + \mathbf{u}_{LR}(\overline{\lambda} - \underline{\lambda})] \\
&= \frac{1}{2}[\mathbf{f}(\mathbf{u}_L) + \mathbf{f}(\mathbf{u}_R) - 2\mathbf{u}_L\xi + \mathbf{u}_L\underline{\lambda} - \mathbf{u}_R\overline{\lambda}\mathbf{u}_R\overline{\lambda} - \mathbf{u}_L\underline{\lambda} - \mathbf{f}(\mathbf{u}_L) + \mathbf{f}(\mathbf{u}_R)] \\
&= \mathbf{f}(\mathbf{u}_L) - \mathbf{u}_L\xi.
\end{aligned}$$

Similarly, if $\xi > \overline{\lambda}$ then $(\mathbf{f} - \mathbf{u}\xi)_{\text{HLL}}(\mathbf{u}_L, \mathbf{u}_R) = \mathbf{f}(\mathbf{u}_R) - \mathbf{u}_R\xi$. Otherwise $\underline{\lambda} < \xi < \overline{\lambda}$ and

$$\begin{aligned}
&(\mathbf{f} - \mathbf{u}\xi)_{\text{HLL}}(\mathbf{u}_L, \mathbf{u}_R) \\
&= \frac{1}{2}\{\mathbf{f}(\mathbf{u}_L) + \mathbf{f}(\mathbf{u}_R) - (\mathbf{u}_L + \mathbf{u}_R)\xi + (\underline{\lambda} - \xi)(\mathbf{u}_{LR} - \mathbf{u}_L) - (\overline{\lambda} - \xi)(\mathbf{u}_R - \mathbf{u}_{LR})\} \\
&= \frac{1}{2}[\mathbf{f}(\mathbf{u}_L) + \mathbf{f}(\mathbf{u}_R) - \mathbf{u}_L\underline{\lambda} - \mathbf{u}_R\overline{\lambda} + \mathbf{u}_{LR}(\overline{\lambda} + \underline{\lambda})] - \mathbf{u}_{LR}\xi \\
&= [\mathbf{f}(\mathbf{u}_L)\overline{\lambda} - \mathbf{f}(\mathbf{u}_R)\underline{\lambda} + (\mathbf{u}_R - \mathbf{u}_L)\overline{\lambda}\underline{\lambda}]\frac{1}{\overline{\lambda} - \underline{\lambda}} - \mathbf{u}_{LR}\xi.
\end{aligned}$$

It is interesting to note that this corresponds to decomposing the jump in the flux as follows:

$$\begin{aligned}
&(\mathbf{f}(\mathbf{u}_R) - \mathbf{u}_R\xi) - (\mathbf{f}(\mathbf{u}_L) - \mathbf{u}_L\xi) \\
&\quad= (\mathbf{u}_R - \mathbf{u}_{LR})(\overline{\lambda} - \xi) + (\mathbf{u}_{LR} - \mathbf{u}_L)(\underline{\lambda} - \xi) \\
&\quad= \left\{[\mathbf{f}(\mathbf{u}_R) - \mathbf{f}(\mathbf{u}_L)] - [\mathbf{u}_R - \mathbf{u}_L]\underline{\lambda}\right\}\frac{\overline{\lambda} - \xi}{\overline{\lambda} - \underline{\lambda}} \\
&\qquad+ \left\{[\mathbf{u}_R - \mathbf{u}_L]\overline{\lambda} - [\mathbf{f}(\mathbf{u}_R) - \mathbf{f}(\mathbf{u}_L)]\right\}\frac{\underline{\lambda} - \xi}{\overline{\lambda} - \underline{\lambda}}.
\end{aligned} \tag{4.97}$$

If we need to write the flux difference as a sum of waves for wave propagation (see section 6.2.6), then we can use equation (4.97). We can also write this equation in the form

$$\triangle\mathbf{f} - \triangle\mathbf{u}\xi = \left\{ \left[\triangle\mathbf{f} - \triangle\mathbf{u}\underline{\lambda}\right] \frac{1}{\|\triangle\mathbf{u}\|} \frac{\overline{\lambda} - \xi}{\overline{\lambda} - \underline{\lambda}} \frac{1}{\|\triangle\mathbf{u}\|} \triangle\mathbf{u}^\top + \left[\triangle\mathbf{u}\overline{\lambda} - \triangle\mathbf{f}\right] \frac{1}{\|\triangle\mathbf{u}\|} \frac{\underline{\lambda} - \xi}{\overline{\lambda} - \underline{\lambda}} \frac{1}{\|\triangle\mathbf{u}\|} \triangle\mathbf{u}^\top \right\} \triangle\mathbf{u}$$

$$\equiv \mathbf{A}_{\mathrm{HLL}} \triangle\mathbf{u} \tag{4.98}$$

for use with wave propagation methods in multiple dimensions (section 7.1.3).

By comparing the formula (4.96) for the Harten-Lax-vanLeer flux with the definition (4.86) of numerical diffusion relative to centered differences, we see that the HLL numerical diffusion is

$$\begin{aligned} d(\mathbf{u}_L, \mathbf{u}_R) &= |\underline{\lambda} - \xi|(\mathbf{u}_{LR} - \mathbf{u}_L) + |\overline{\lambda} - \xi|(\mathbf{u}_R - \mathbf{u}_{LR}) \\ &= \frac{|\overline{\lambda} - \xi| - |\underline{\lambda} - \xi|}{\overline{\lambda} - \underline{\lambda}}[\mathbf{f}(\mathbf{u}_R) - \mathbf{f}(\mathbf{u}_L)] + \frac{\overline{\lambda}|\underline{\lambda} - \xi| - \underline{\lambda}|\overline{\lambda} - \xi|}{\overline{\lambda} - \underline{\lambda}}[\mathbf{u}_R - \mathbf{u}_L]. \end{aligned}$$

If we have a jump in a single wave family, so that $\mathbf{f}(\mathbf{u}_R) - \mathbf{f}(\mathbf{u}_L) = [\mathbf{u}_R - \mathbf{u}_L]\sigma$, where $\underline{\lambda} \le \sigma \le \overline{\lambda}$, then

$$d(\mathbf{u}_L, \mathbf{u}_R) = \frac{|\overline{\lambda} - \xi|(\sigma - \underline{\lambda}) + |\underline{\lambda} - \xi|(\overline{\lambda} - \sigma)}{\overline{\lambda} - \underline{\lambda}}[\mathbf{u}_R - \mathbf{u}_L].$$

Thus the HLL approximate Riemann solver adds numerical diffusion in any wave family that has a jump, even a contact discontinuity. In general, the HLL solver adds too much numerical diffusion to linearly degenerate waves, though not as much as the Rusanov flux.

However, the HLL solver does satisfy an entropy inequality. Recall that the intermediate state $\mathbf{u}_{LR}$ was chosen to make $\tilde{\mathcal{R}}$ conservative, Whenever S is convex, Jensen's inequality and inequality (4.82) show us that

$$\begin{aligned} \int_{-\Delta x_L/2}^{\Delta x_R/2} S\left(\tilde{\mathcal{R}}\left(\mathbf{u}_L, \mathbf{u}_R, \frac{x}{\Delta t}\right)\right) dx &\le S\left(\int_{-\Delta x_L/2}^{\Delta x_R/2} \tilde{\mathcal{R}}\left(\mathbf{u}_L, \mathbf{u}_R, \frac{x}{\Delta t}\right) dx\right) \\ &= S\left(\int_{-\Delta x_L/2}^{\Delta x_R/2} \mathcal{R}\left(\mathbf{u}_L, \mathbf{u}_R, \frac{x}{\Delta t}\right) dx\right) \\ &\le \frac{\triangle x_L S(\mathbf{u}_L) + \triangle x_R S(\mathbf{u}_R)}{2} - \triangle t[\Psi(\mathbf{u}_R) - \Psi(\mathbf{u}_L)]. \end{aligned}$$

4.13.11 HLL Solvers with Two Intermediate States

The HLL solver described in section 4.13.10 does a good job in resolving the slowest and fastest waves, but not so good a job in resolving waves associated with intermediate characteristic speeds. This is because the HLL solver is designed to use information from those two waves only.

A modification of the HLL solver has been developed by Linde [103], based on extending ideas presented in [68]. The approximate Riemann solver has the form

$$\tilde{\mathcal{R}}_{\text{HLLM}}(\mathbf{u}_L, \mathbf{u}_R, \xi) = \begin{cases} \mathbf{u}_L, & \xi < \underline{\lambda} \\ \mathbf{u}_{L*}, & \underline{\lambda} < \xi < \lambda_* \\ \mathbf{u}_{R*}, & \lambda_* < \xi < \overline{\lambda} \\ \mathbf{u}_R, & \overline{\lambda} < \xi \end{cases} \tag{4.99}$$

where $\underline{\lambda} \le \lambda_* \le \overline{\lambda}$ are wavespeeds to be specified below. In order for this Riemann solver to be conservative, we require

$$\begin{aligned} &\frac{\triangle x_L \mathbf{u}_L + \triangle x_R \mathbf{u}_R}{2} - \triangle t[\mathbf{f}(\mathbf{u}_R) - \mathbf{f}(\mathbf{u}_L)] \\ &= \int_{-\Delta x_L/2}^{\Delta x_R/2} \tilde{\mathcal{R}}\left(\mathbf{u}_L, \mathbf{u}_R, \frac{x}{\Delta t}\right) dx \\ &= \mathbf{u}_L \left(\underline{\lambda}\triangle t + \frac{\Delta x_L}{2}\right) + \mathbf{u}_{L*}(\lambda_* - \underline{\lambda})\triangle t + \mathbf{u}_{R*}(\overline{\lambda} - \lambda_*)\triangle t + \mathbf{u}_R \left(\frac{\Delta x_R}{2} - \overline{\lambda}\triangle t\right). \end{aligned}$$

This condition can be simplified to

$$\mathbf{f}(\mathbf{u}_R) - \mathbf{f}(\mathbf{u}_L) = (\mathbf{u}_R - \mathbf{u}_{R*})\overline{\lambda} + (\mathbf{u}_{R*} - \mathbf{u}_{R*})\lambda_* + (\mathbf{u}_{L*} - \mathbf{u}_L)\underline{\lambda}. \tag{4.100}$$

The flux at the state that moves with speed ξ in the Riemann problem is approximated by (4.80). This produces the following formula for the modified HLL flux:

$$(\mathbf{f} - \mathbf{u}\xi)_{\text{HLLM}}(\mathbf{u}_L, \mathbf{u}_R) = \begin{cases} \mathbf{f}(\mathbf{u}_L) - \mathbf{u}_L\xi, & 0 < \underline{\lambda} \\ \mathbf{f}(\mathbf{u}_L) + (\mathbf{u}_{L*} - \mathbf{u}_L)\underline{\lambda} - \mathbf{u}_{L*}\xi, & \underline{\lambda} < 0 < \lambda_* \\ \mathbf{f}(\mathbf{u}_R) - (\mathbf{u}_R - \mathbf{u}_{R*})\overline{\lambda} - \mathbf{u}_{R*}\xi, & \lambda_* < 0 < \overline{\lambda} \\ \mathbf{f}(\mathbf{u}_R), & \overline{\lambda} < 0. \end{cases}$$

We will require $\mathbf{u}_{L*}$ and $\mathbf{u}_{R*}$ to satisfy $\mathbf{u}_{R*} - \mathbf{u}_{L*} = (\mathbf{u}_R - \mathbf{u}_L)\alpha$ for some scalar α that will be specified below. This equation says that the jump in the intermediate states lies in the same direction as the jump between the left and right states. If we were willing to use information from characteristic directions, we might hope to do better, but one purpose of this approximate Riemann solver is to avoid using information from characteristic directions. The scalar α will be determined in such a way that $\alpha = 1$ for isolated discontinuities, and $\alpha = 0$ if there is no discontinuity.

We now have a linear system for $\mathbf{u}_{L*}$ and $\mathbf{u}_{R*}$:

$$\begin{bmatrix} \mathbf{I}(\lambda_* - \underline{\lambda}) & \mathbf{I}(\overline{\lambda} - \lambda_*) \\ -\mathbf{I} & \mathbf{I} \end{bmatrix} \begin{bmatrix} \mathbf{u}_{L*} \\ \mathbf{u}_{R*} \end{bmatrix} = \begin{bmatrix} \mathbf{u}_R\overline{\lambda} - \mathbf{u}_L\underline{\lambda} - \mathbf{f}(\mathbf{u}_R) + \mathbf{f}(\mathbf{u}_L) \\ (\mathbf{u}_R - \mathbf{u}_L)\alpha \end{bmatrix}.$$

We can solve this linear system to get

$$\begin{bmatrix} \mathbf{u}_{L*} \\ \mathbf{u}_{R*} \end{bmatrix} = \begin{bmatrix} \mathbf{I} & \mathbf{I}(\lambda_* - \overline{\lambda}) \\ \mathbf{I} & \mathbf{I}(\lambda_* - \underline{\lambda}) \end{bmatrix} \begin{bmatrix} \mathbf{u}_R\overline{\lambda} - \mathbf{u}_L\underline{\lambda} - \mathbf{f}(\mathbf{u}_R) + \mathbf{f}(\mathbf{u}_L) \\ (\mathbf{u}_R - \mathbf{u}_L)\alpha \end{bmatrix} \frac{1}{\overline{\lambda} - \underline{\lambda}}$$
$$= \begin{bmatrix} \mathbf{u}_R\overline{\lambda} - \mathbf{u}_L\underline{\lambda} - \mathbf{f}(\mathbf{u}_R) + \mathbf{f}(\mathbf{u}_L) - (\mathbf{u}_R - \mathbf{u}_L)(\overline{\lambda} - \lambda_*)\alpha \\ \mathbf{u}_R\overline{\lambda} - \mathbf{u}_L\underline{\lambda} - \mathbf{f}(\mathbf{u}_R) + \mathbf{f}(\mathbf{u}_L) + (\mathbf{u}_R - \mathbf{u}_L)(\lambda_* - \underline{\lambda})\alpha \end{bmatrix} \frac{1}{\overline{\lambda} - \underline{\lambda}}.$$

In order to evaluate the flux, we will need to compute

$$\mathbf{u}_{L*} - \mathbf{u}_L = \left\{[\mathbf{u}_R - \mathbf{u}_L](\overline{\lambda}(1-\alpha) + \lambda_*\alpha) - [\mathbf{f}(\mathbf{u}_R) - \mathbf{f}(\mathbf{u}_L)]\right\} / (\overline{\lambda} - \underline{\lambda}) \tag{4.101a}$$
$$\mathbf{u}_R - \mathbf{u}_{R*} = \left\{-[\mathbf{u}_R - \mathbf{u}_L](\underline{\lambda}(1-\alpha) + \lambda_*\alpha) + [\mathbf{f}(\mathbf{u}_R) - \mathbf{f}(\mathbf{u}_L)]\right\} / (\overline{\lambda} - \underline{\lambda}). \tag{4.101b}$$

Then equation (4.100) implies that

$$(\mathbf{f}(\mathbf{u}_R) - \mathbf{u}_R\xi) - (\mathbf{f}(\mathbf{u}_L) - \mathbf{u}_L\xi) = [\mathbf{f}(\mathbf{u}_R) - \mathbf{f}(\mathbf{u}_L)] - [\mathbf{u}_R - \mathbf{u}_L]\xi$$
$$= [(\mathbf{u}_R - \mathbf{u}_{R*}](\overline{\lambda} - \xi) + [(\mathbf{u}_{R*} - \mathbf{u}_{L*}](\lambda_* - \xi) + [(\mathbf{u}_{L*} - \mathbf{u}_L](\underline{\lambda} - \xi). \tag{4.102}$$

This equation can be used with wave propagation methods (see Section 6.2.6 below.)

In order to complete the evaluation of the flux, we need expressions for $\underline{\lambda}$, $\overline{\lambda}$, λ_* and α. Linde assumes that a strictly convex entropy $S(\mathbf{u})$ is available, and computes

$$\lambda_* = \frac{\left[\frac{\partial S}{\partial \mathbf{u}}(\mathbf{u}_R) - \frac{\partial S}{\partial \mathbf{u}}(\mathbf{u}_L)\right][\mathbf{f}(\mathbf{u}_R) - \mathbf{f}(\mathbf{u}_L)]}{\left[\frac{\partial S}{\partial \mathbf{u}}(\mathbf{u}_R) - \frac{\partial S}{\partial \mathbf{u}}(\mathbf{u}_L)\right][\mathbf{u}_R - \mathbf{u}_L]}.$$

Next, he computes

$$\underline{\lambda} = \min\{\lambda_*, \lambda_{1L}, \lambda_{1R}\} \quad \text{and} \quad \overline{\lambda} = \max\{\lambda_*, \lambda_{mL}, \lambda_{mR}\},$$

where the eigenvalues at the left and right states have been assumed to be ordered from smallest to largest.

The definition of λ_* implies that

$$\left[\frac{\partial S}{\partial \mathbf{u}}(\mathbf{u}_R) - \frac{\partial S}{\partial \mathbf{u}}(\mathbf{u}_L)\right]([\mathbf{f}(\mathbf{u}_R) - \mathbf{f}(\mathbf{u}_L)] - [\mathbf{u}_R - \mathbf{u}_L]\lambda_*) = 0.$$

Since S is strictly convex, we must have that $\frac{\partial^2 S}{\partial \mathbf{u} \partial \mathbf{u}}$ is positive definite, and that

$$\left[\frac{\partial S}{\partial \mathbf{u}}(\mathbf{u}_R) - \frac{\partial S}{\partial \mathbf{u}}(\mathbf{u}_L)\right] = \mathbf{P}[\mathbf{u}_R - \mathbf{u}_L]$$

for some positive-definite matrix $\mathbf{P}$, as discussed in the proof of theorem 4.13.13. The Pythagorean theorem implies that

$$1 = \frac{[\mathbf{u}_R - \mathbf{u}_L]^\top \mathbf{P}[\mathbf{u}_R - \mathbf{u}_L]}{[\mathbf{f}(\mathbf{u}_R) - \mathbf{f}(\mathbf{u}_L)]^\top \mathbf{P}[\mathbf{f}(\mathbf{u}_R) - \mathbf{f}(\mathbf{u}_L)]}\lambda_*^2$$
$$+ \frac{[\mathbf{f}(\mathbf{u}_R) - \mathbf{f}(\mathbf{u}_L) - (\mathbf{u}_R - \mathbf{u}_L)\lambda_*]^\top \mathbf{P}[\mathbf{f}(\mathbf{u}_R) - \mathbf{f}(\mathbf{u}_L) - (\mathbf{u}_R - \mathbf{u}_L)\lambda_*]}{[\mathbf{f}(\mathbf{u}_R) - \mathbf{f}(\mathbf{u}_L)]^\top \mathbf{P}[\mathbf{f}(\mathbf{u}_R) - \mathbf{f}(\mathbf{u}_L)]}.$$

The first term on the right represents the relative strength of waves moving with speed λ_*, while the second term represents the relative strength of waves moving with different speeds. We will choose α^2 to be an approximation to the first term.

The difficulty is that the matrix $\mathbf{P}$ is not available. Linde [103] suggests approximating $\mathbf{P}$ by the diagonal part $\tilde{\mathbf{P}}$ of $\frac{\partial^2 S}{\partial \mathbf{u} \partial \mathbf{u}}$ evaluated at $\frac{1}{2}(\mathbf{u}_L + \mathbf{u}_R)$. His final expression is

$$\alpha = \begin{cases} \frac{([\mathbf{u}]^\top \tilde{\mathbf{P}}[\mathbf{f}])^2}{[\mathbf{f}]^\top \tilde{\mathbf{P}}[\mathbf{f}][\mathbf{u}]^\top \tilde{\mathbf{P}}[\mathbf{u}]}, & [S]\lambda_* - [\Psi] \geq 0 \text{ (shock)} \\ 0, & [S]\lambda_* - [\Psi] < 0 \text{ (rarefaction)} \end{cases}$$

where $[\mathbf{u}] = \mathbf{u}_R - \mathbf{u}_L$, and so on.

Let us use equations (4.80) and (4.99) to summarize our results:

$$(\mathbf{f} - \mathbf{u}\xi)_{\text{HLLM}}(\mathbf{u}_L, \mathbf{u}_R) = \begin{cases} \mathbf{f}(\mathbf{u}_L) - \mathbf{u}_L\xi, & \xi < \underline{\lambda} \\ \mathbf{f}(\mathbf{u}_L) - \mathbf{u}_L\xi + (\mathbf{u}_{L*} - \mathbf{u}_L)(\underline{\lambda} - \xi), & \underline{\lambda} < \xi < \lambda_* \\ \mathbf{f}(\mathbf{u}_R) - \mathbf{u}_R\xi - (\mathbf{u}_R - \mathbf{u}_{R*})(\overline{\lambda} - \lambda_*), & \lambda_* < 0 < \overline{\lambda} \\ \mathbf{f}(\mathbf{u}_R) - \mathbf{u}_R\xi, & \overline{\lambda} < \xi \end{cases}$$

In these expressions, the differences involving the intermediate states can be computed by equations (4.101).

An alternative when an entropy function is not available would be to take

$$\lambda_* = \frac{[\mathbf{u}_R - \mathbf{u}_L]^\top [\mathbf{f}(\mathbf{u}_R) - \mathbf{f}(\mathbf{u}_L)]}{[\mathbf{u}_R - \mathbf{u}_L]^\top [\mathbf{u}_R - \mathbf{u}_L]}.$$

This corresponds to taking $\mathbf{P} = \mathbf{I}$. Then the definition of α would become

$$\alpha(\mathbf{u}_L, \mathbf{u}_R) = \begin{cases} \frac{([\mathbf{u}]^\top [\mathbf{f}])^2}{\|[\mathbf{f}]\|^2 \|[\mathbf{u}]\|^2}, & \text{if discontinuity would be admissible} \\ 0, & \text{otherwise.} \end{cases}$$

4.13.12 Approximate Riemann Solver Recommendations

Our discussion of approximate Riemann solvers has been somewhat long. In order to give the student some guidance, we will make the following recommendations. For problems in which the exact Riemann problem solution is inexpensive, Godunov's method should use the exact Riemann solver. Burgers' equation is an example of such a case. For those problems in which the exact Riemann problem solution is expensive but a Roe solver is available, the Roe solver should be accompanied with one of the modifications of the Roe solver or a numerical diffusion. Shallow water, gas dynamics, plasticity and the Schaeffer–Schechter–Shearer model are examples of conservation laws for which Roe solvers are available. For Lagrangian solid mechanics, the characteristic speeds come in plus/minus pairs, and a Roe solver is almost never available. Here the Harten–Lax–vanLeer solver or Rusanov solver

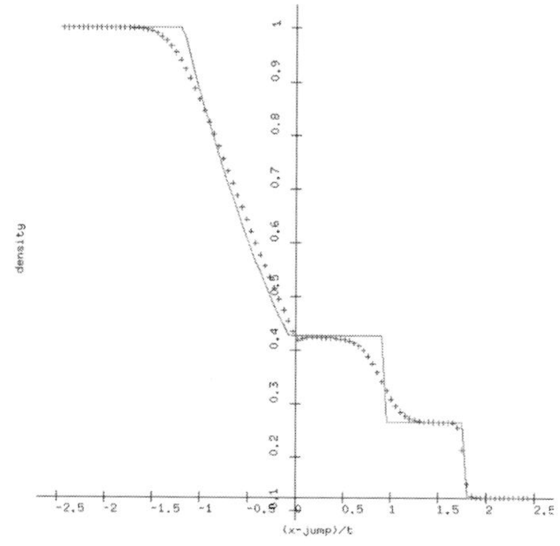

(a) Exact

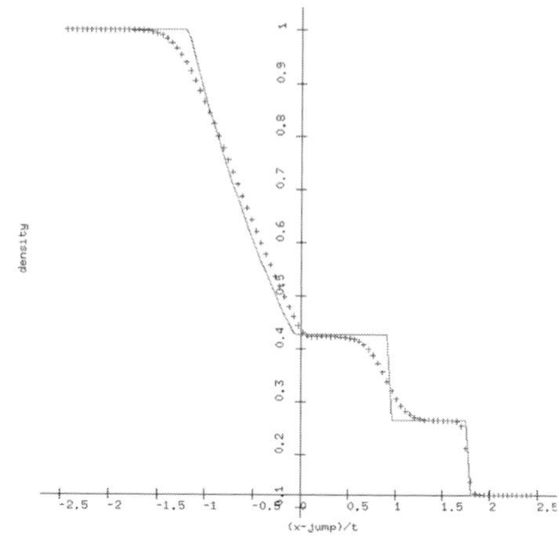

(b) Harten–Hyman

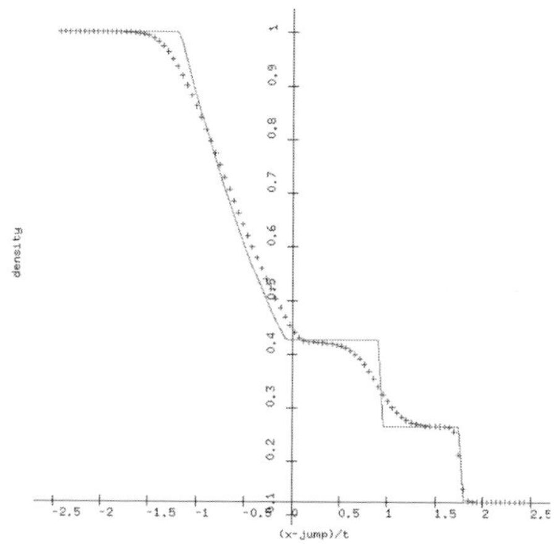

(c) Harten–Lax–vanLeer

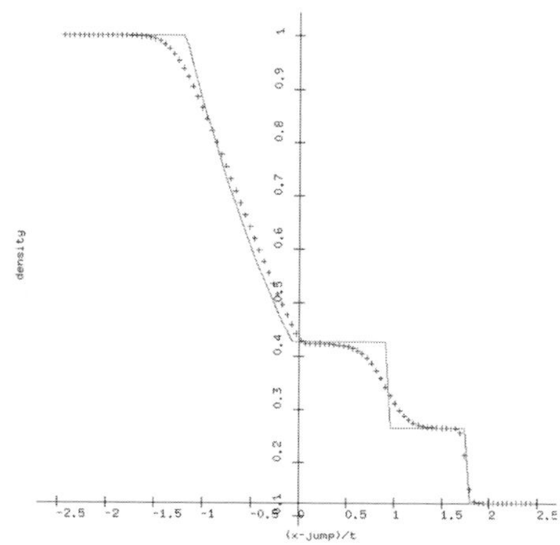

(d) Linde

Fig. 4.17 Various Riemann solvers for sod shock tube problem: Density vs. x/t

could be used. Another useful option is the weak wave solver, with characteristic directions and speeds taken from the left state for positive speeds and from the right state for negative wave speeds. The vibrating string is an example of this situation.

The polymer model and the three-phase Buckley–Leverett model are especially challenging. In both cases, characteristic speeds can coalesce with a loss of a linearly independent characteristic direction. This makes approximate Riemann solvers especially difficult to construct. We have used the Harten–Lax–vanLeer scheme for such problems with some success. An attempt to use characteristic directions

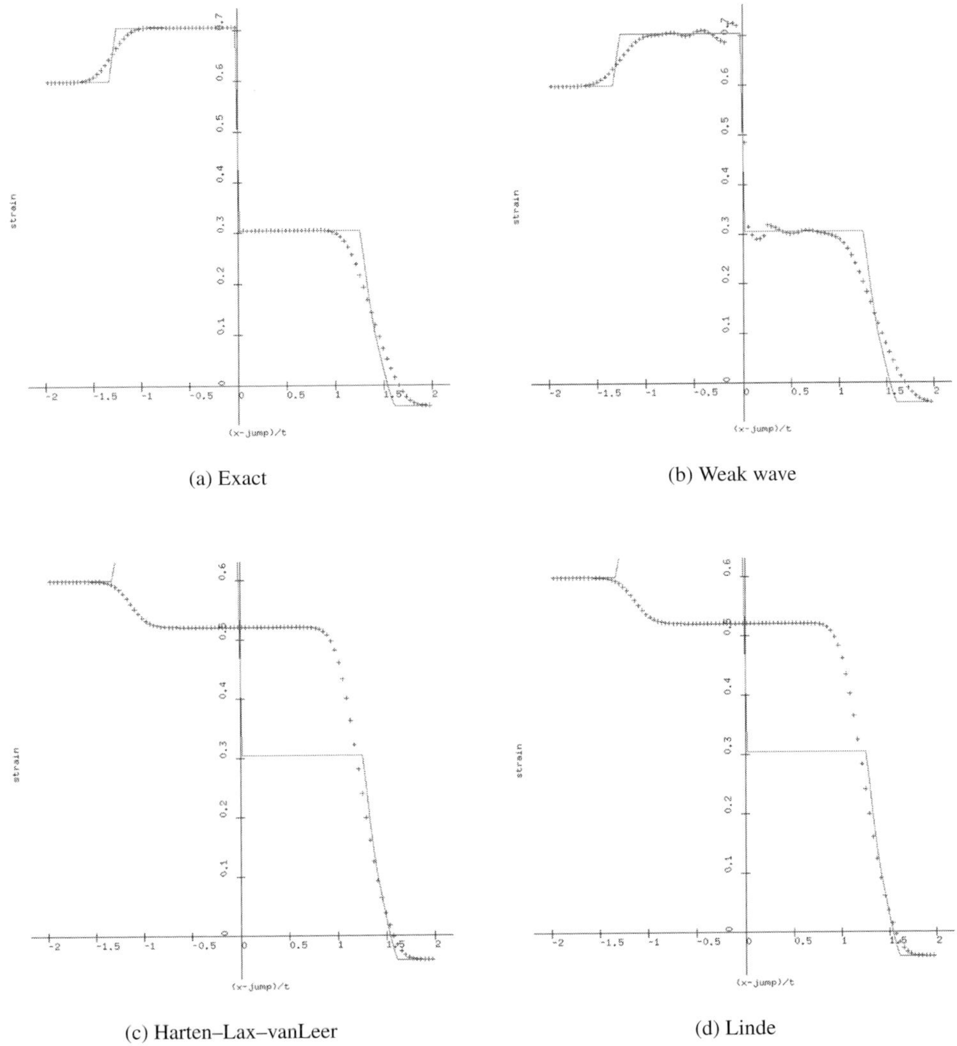

(a) Exact (b) Weak wave

(c) Harten–Lax–vanLeer (d) Linde

Fig. 4.18 Various Riemann solvers for plasticity: elastic strain vs. x/t

for such problems was made in [13], but this approach has not proved to be robust enough for general use.

Executable 4.13-24: guiGodunov has been provided for testing Riemann solvers. The user can select input parameters by dragging down on "View" and releasing on "Main." The physical model can be selected under "Riemann Problem Parameters." In particular, the user can choose to work with shallow water, gas dynamics, magnetohydrodynamics, vibrating string, plasticity, polymer flooding, three-phase Buckley-Leverett flow or the Schaeffer-Schechter-Shearer model. Specific input parameters for the individual models can also be selected. The user can select the approximate Riemann solver under "Numerical Method Parameters." These options include Godunov (if an exact Riemann problem solution is available), weak wave, Colella-Glaz, Osher-Solomon, Roe, Harten-Hyman, Harten-Lax-vanLeer and Linde. Some options are unavailable for certain models.

Figures 4.17 and 4.18 provide evidence that most of the approximate Riemann solvers perform fairly much the same for most of our case studies. Of course, the Roe solvers fail for transonic rarefactions without the Harten–Hyman modification. An exception is the plasticity model, for which only the exact Riemann problem solver and the weak wave approximation seem to produce the correct results. Here the failure may be due to hysteresis effects.

Details of how the Riemann solvers are implemented can be found by examining the code. The code **Program 4.13-53: shallow_water.f** contains routines to compute conserved quantities from the flux variables and *vice versa*, to compute the fluxes from the flux variables, to compute characteristic speeds, to solve Riemann problems, to perform various approximate Riemann problem solvers, and to select initial data for Riemann problems. Students can also view code for **Program 4.13-54: gas dynamics** or **Program 4.13-55: magnetohydrodynamics** or **Program 4.13-56: vibrating string** or **Program 4.13-57: plasticity** or **Program 4.13-58: polymer** or **Program 4.13-59: three-phase Buckley-Leverett flow** or the **Program 4.13-60: Schaeffer-Schechter-Shearer model**. The main program for this executable is **Program 4.13-61: GUIGodunov.C**. As always, a list of all of the files in the directory can be obtained by deleting the file name from the web link (*i.e.*, backing up to the first "/") in the browser window.

Exercises for 4.13

4.13.1 Describe the analogue of the weak wave Riemann solver for the shallow water equations. Compare it to Rusanov's scheme by programming both schemes and testing them for a problem involving a transonic rarefaction and a problem involving a transonic shock at various values of CFL. Use 100 grid cells in your calculations. Plot the numerical solution versus versus x/t in all cases, at a time at which the wave has traveled across 80% of the computational domain. Plot h, v and the characteristic speeds versus x/t. Also plot h versus v at the final timestep.

4.13.2 Trangenstein and Colella [162] suggested using the following weak wave solver for solid mechanics. If $\mathbf{Y}$ is any nonsingular matrix with corresponding eigenvalues Λ, then solve $\mathbf{Yd} = \mathbf{w}_R - \mathbf{w}_L$ for the characteristic expansion coefficients $\mathbf{d}$ of the jump in the flux variables. The state in the solution of the Riemann problem moving with zero speed was approximated either by $\mathbf{w}_L + \sum_{i:\lambda_i<0} \mathbf{Y}\mathbf{e}_i\mathbf{e}_i \cdot \mathbf{d}$ or by $\mathbf{w}_R - \sum_{i:\lambda_i>0} \mathbf{Y}\mathbf{e}_i\mathbf{e}_i \cdot c$.

(a) Describe the general approximate Riemann solver $\tilde{\mathcal{R}}(\mathbf{w}_L, \mathbf{w}_R, \xi)$ being proposed here.

(b) Are the two values for the solution of the Riemann problem the same?

(c) Under what circumstances is this approximate Riemann solver conservative?

4.13.3 LeVeque [97, p. 333f] suggested using the following weak wave solver. If $\mathbf{X}$ is any nonsingular matrix with corresponding real eigenvalues Λ, then solve $\mathbf{Xz} = \mathbf{f}(\mathbf{u}_R) - \mathbf{f}(\mathbf{u}_L)$ for the characteristic expansion coefficients $\mathbf{z}$ of the jump in the flux. The flux at the solution of the Riemann problem moving with zero speed was approximated either by $\mathbf{f}(\mathbf{u}_L) + \sum_{i:\lambda_i<0} \mathbf{X}\mathbf{e}_i\mathbf{e}_i \cdot \mathbf{z}_i$ or by $\mathbf{f}(\mathbf{u}_R) - \sum_{i:\lambda_i>0} \mathbf{X}\mathbf{e}_i\mathbf{e}_i \cdot \mathbf{z}_i$.

(a) Describe the general approximate Riemann solver $\tilde{\mathcal{R}}(\mathbf{u}_L, \mathbf{u}_R, \xi)$ being proposed here.

(b) Are the two values for the flux at the solution of the Riemann problem the same?

(c) Is is possible to decide if this approximate Riemann solver is conservative?

4.13.4 The initial guess in the Colella–Glaz Riemann solver is based on a **weak wave approximation**. The idea is the following. Using the results from section 4.4.3, we will decompose the jump between the states in terms of the characteristic directions from left and right:

$$\begin{bmatrix} \rho_L & 1 & \rho_R \\ -c_L & 0 & c_R \\ \rho_L c_L^2 & 0 & \rho_R c_R^2 \end{bmatrix} \begin{bmatrix} \mathbf{y}_1 \\ \mathbf{y}_2 \\ \mathbf{y}_3 \end{bmatrix} = \begin{bmatrix} \rho_R - \rho_L \\ \mathbf{v}_R - \mathbf{v}_L \\ p_R - p_L \end{bmatrix}.$$

Show that the initial guess in the Colella–Glaz Riemann solver takes $\mathbf{v}$ and p from the weak wave approximation to the state at the contact discontinuity, given by

$$\begin{bmatrix} \rho \\ \mathbf{v} \\ p \end{bmatrix} = \begin{bmatrix} \rho_L \\ \mathbf{v}_L \\ p_L \end{bmatrix} + \begin{bmatrix} \rho_L \\ -c_L \\ \rho_L c_L^2 \end{bmatrix} \mathbf{y}_1.$$

4.13.5 Show that in the Colella–Glaz Riemann solver, $\zeta_L = -\frac{d\mathbf{v}_L(p)}{dp}$ and $\zeta_R = -\frac{d\mathbf{v}_R(p)}{dp}$. Then show that the iteration in their Riemann solver is a Newton iteration.

4.13.6 Compare the Colella–Glaz Riemann solver to the exact solution to the Riemann problem for the following test problems:

(a) a Mach 2 shock moving to the right into air with density 1, velocity 0 and pressure 1 (*i.e.*, $p_R = 1$, $\mathbf{v}_R = 0$, $\rho_R = 1$; $p_L = 9/2$, $\rho_L = 8/3$, $\mathbf{v}_L = \sqrt{35/16}$);

(b) a stationary shock ($\sigma = 0$) in air with density 1, velocity -2 and pressure 1 on the right (*i.e.*, $p_R = 1$, $\mathbf{v}_R = -2$, $\rho_R = 1$; $p_L = 19/6$, $\rho_L = 24/11$, $\mathbf{v}_L = -11/12$);

(c) a rarefaction moving to the right from air with density 1, pressure 1 and velocity 0 into a vacuum. (*i.e.*, $p_L = 1$, $\mathbf{v}_L = 0$, $\rho_L = 1$; $p_R = 0$, $\rho_R = 0$, $\mathbf{v}_R = \sqrt{35}$).

(d) the **Colella–Woodward interacting blast wave problem** [177]. The gas is assumed to be air ($\gamma = 1.4$) confined between two reflecting walls at $x = 0$ and $x = 1$. Initially, $\rho = 1$ and $\mathbf{v} = 0$ everywhere. The initial condition for pressure consists of three constant states:

$$p = \begin{cases} 1000., & 0 < x < 0.1 \\ 0.01, & 0.1 < x < 0.9 \\ 100., & 0.9 < x < 1.0. \end{cases}$$

Plot the numerical results for times 0.01, 0.016, 0.026, 0.028, 0.030, 0.032, 0.034 and 0.038. Plot ρ, $\mathbf{v}$, p and the temperature versus x. Try 100, 1000 and 10,000 cells.

4.13.7 Program the Roe solver for gas dynamics, and compare it to the exact solution to the Riemann problem. Plot the numerical solution (*i.e.*, ρ, $\mathbf{v}$, p and characteristic speeds $\mathbf{v} \pm c$ versus x/t), at a time for which the fastest wave is near the boundary of the computational domain, for the following problems:

(a) a Mach 2 shock moving to the right into air with density 1, velocity 0 and pressure 1 (*i.e.*, $p_R = 1$, $\mathbf{v}_R = 0$, $\rho_R = 1$; $p_L = 9/2$, $\rho_L = 8/3$, $\mathbf{v}_L = \sqrt{35/16}$);
(b) a stationary shock ($\sigma = 0$) in air with density 1, velocity -2 and pressure 1 on the right (*i.e.*, $p_R = 1$, $\mathbf{v}_R = -2$, $\rho_R = 1$; $p_L = 19/6$, $\rho_L = 24/11$, $\mathbf{v}_L = -11/12$);
(c) a rarefaction moving to the right from air with density 1, pressure 1 and velocity 0 into a vacuum. (*i.e.*, $p_L = 1$, $\mathbf{v}_L = 0$, $\rho_L = 1$; $p_R = 0$, $\rho_R = 0$, $\mathbf{v}_R = \sqrt{35}$);
(d) the **Sod shock tube problem** [131, page 116]. This is a Riemann problem for air ($\gamma = 1.4$) in which the left state is given by $\rho_L = 1$, $\mathbf{v}_L = 0$, $p_L = 1$ and the right state is $\rho_R = 0.125$, $\mathbf{v}_R = 0$, $p_R = 0.1$. Perform the calculation with 100 and 1000 cells. Plot ρ, $\mathbf{v}$, p and the characteristic speeds versus x/t at a time for which the fastest wave is near the boundary of the computational domain;
(e) the **Colella-Woodward interacting blast wave problem** [177]. The gas is assumed to be air ($\gamma = 1.4$) confined between two reflecting walls at $x = 0$ and $x = 1$. Initially, $\rho = 1$ and $\mathbf{v} = 0$ everywhere. The initial condition for pressure consists of three constant states:

$$p = \begin{cases} 1000., & 0 < x < 0.1 \\ -0.01, & 0.1 < x < 0.9 \\ -100., & 0.9 < x < 1.0. \end{cases}$$

Plot the numerical results for times 0.01, 0.016, 0.026, 0.028, 0.030, 0.032, 0.034 and 0.038. Plot ρ, $\mathbf{v}$, p and the temperature versus x. Try 100, 1000 and 10,000 cells.

4.13.8 Suppose that we want to develop a Roe solver for the vibrating string, described in Section 4.8. Show that the Roe matrix should take the form

$$\bar{\mathbf{A}} = \begin{bmatrix} 0 & -\mathbf{A}_{12} \\ -\mathbf{I}\frac{1}{\rho} & 0 \end{bmatrix}.$$

If the deformation gradients at the left and right states are $\mathbf{f}_L$ and $\mathbf{f}_R$, and if

$$\mathbf{f}(\alpha) = \mathbf{f}_R\alpha + \mathbf{f}_L(1-\alpha)$$

then show that $\mathbf{A}_{12}$ should be given by

$$\mathbf{A}_{12} = \int_0^1 \left[\mathbf{I}\frac{\tau(\|\mathbf{f}(\alpha)\|}{\|\mathbf{f}(\alpha)\|} + \frac{\mathbf{f}(\alpha)}{\|\mathbf{f}(\alpha)\|} \left\{ \tau'(\|\mathbf{f}(\alpha)\|) - \frac{\tau(\|\mathbf{f}(\alpha)\|}{\|\mathbf{f}(\alpha)\|} \right\} \frac{\mathbf{f}(\alpha)^\top}{\|\mathbf{f}(\alpha)\|} \right] d\alpha.$$

How hard is it to compute α?

4.13.9 Show that the plasticity model, described in section 4.9, has Roe matrix

$$\bar{\mathbf{A}} = \begin{bmatrix} 0 & -[s]/[\epsilon] \\ -1 & 0 \end{bmatrix}.$$

Here s is the stress, ϵ is the strain, and $[u] = u_R - u_L$ is the jump in some state u.

4.13.10 Choose one of the models in the case studies, and examine the available approximate Riemann solvers in combination with Godunov's scheme. Test your Riemann solvers for problems involving both shocks and rarefactions. If possible, test your solvers for a problem involving a transonic rarefaction. Which solvers are most accurate in general? Which are most efficient?

4.13.11 Modify the Godunov scheme to solve Riemann problems on a moving mesh, designed to expand in a self-similar fashion. To do this, generate a uniform mesh in *wave speeds* covering the range of wave speeds in your Riemann problem. Next, set the initial mesh locations to be equal to the mesh speeds (*i.e.*, the mesh is initialized at time $t = 1$). Now, run Godunov's method with mesh expanding at the given mesh speeds; in other words, solve the Riemann problems to find the flux $\mathbf{f} - \mathbf{u}\xi$ associated with the state that moves with speed ξ, which is the speed associated with each mesh point. In this way, the numerical method should reach a steady state on the self-similar grid, and the numerical solution should never reach the numerical boundary.

5

Methods for Scalar Laws

In Chapters 3 and 4 we examined hyperbolic partial differential equations and some basic numerical methods to solve these problems. In this chapter we will develop theory to help us understand the convergence of these methods, and use the theory to help us develop more accurate methods.

5.1 Convergence

It is important to determine analytically whether our numerical methods should converge. This question is reasonably easy to answer for linear methods applied to linear advection, which will be the topic for this section. Later, we will examine whether nonlinear schemes converge, and whether they converge to the correct solution.

5.1.1 Consistency and Order

Suppose that we want to approximate the solution of the conservation law

$$\frac{\partial u}{\partial t} + \frac{\partial f(u)}{\partial x} = 0$$

by means of an explicit numerical method

$$u_i^{n+1} = H(u_{i-k}^n, \ldots, u_{i+k}^n; \triangle t^{n+1/2})$$

on a mesh with cell widths, cell centers and timesteps given by (respectively)

$$\triangle x_i = x_{i+1/2} - x_{i-1/2}, \; x_i = \frac{1}{2}(x_{i-1/2} + x_{i+1/2}), \; \triangle t^{n+1/2} = t^{n+1} - t^n.$$

We shall say that the **pointwise error** in the method is $e_i^n = u_i^n - u(x_i, t^n)$. We should not expect the pointwise error to go to zero uniformly as the mesh is refined in the neighborhood of a discontinuity. Typically, the numerical solution selects some intermediate state as its approximate value for the solution at the discontinuity, and the error in the solution at such points does not go to zero. Instead, we will consider other measures of the error in the solution.

Definition 5.1.1 For suitable functions $w(x)$, the 1-norm is

$$\|w\|_1 \equiv \int_{-\infty}^{\infty} |w(x)|\, dx,$$

and for suitable functions $v(x,t)$ the 1-norm is

$$\|v\|_{1,T} \equiv \int_0^T \int_{-\infty}^{\infty} |v(x,t)|\, dx\, dt.$$

This suggests that we define the spaces of functions $\mathcal{L}_1 \equiv \{w(x) : \|w\|_1 < \infty\}$ and $\mathcal{L}_{1,T} \equiv \{v(x,t) : \|v\|_{1,T} < \infty\}$. These spaces can be taken to be the completions of C_0^∞ functions with respect to the appropriate norms.

For piecewise constant functions, such as we typically see in finite difference numerical methods for conservation laws, the 1-norm of the pointwise error simplifies to

$$\|e^n\|_1 \equiv \sum_i |e_i^n| \triangle x_i .$$

In order to estimate the error in the numerical solution before it is computed, we will investigate the error in the numerical approximation to the partial differential equation.

Definition 5.1.2 If the solution $u(x,t)$ of the conservation law

$$\frac{\partial u}{\partial t} + \frac{\partial f(u)}{\partial x} = 0$$

is approximated by the explicit numerical method

$$u_i^{n+1} = H(u_{i-k}^n, \ldots, u_{i+k}^n; \triangle t^{n+1/2}),$$

then **local truncation error** in the method is

$$\text{for all } x \in (x_{i-1/2}, x_{i+1/2})$$
$$L(x,t;\triangle t) = \frac{1}{\triangle t}[u(x,t+\triangle t) - H\left(u(x_{i-k},t), \ldots, u(x_{i+k},t); \triangle t\right)].$$

A numerical method with local truncation error L is **consistent** if and only if for all x and t in the problem, the local truncation error $L(x,t;\triangle t)$ satisfies $\|L(x,t;\triangle t)\|_1 \to 0$ as $\triangle t \to 0$.

A numerical method with local truncation error L has **order** p if and only if for all sufficiently smooth initial data with compact support (meaning that the initial data is identically zero for $|x|$ sufficiently large), we have that

$$\text{there exists } T > 0, \ \text{there exists } \triangle t_0 > 0, \ \text{there exists } C > 0, \ \text{such that}$$
$$\text{for all } \triangle t < \triangle t_0 \ \text{for all } 0 < t < T,$$
$$\|L(x,t;\triangle t)\|_1 \le C \triangle t^p.$$

5.1.2 Linear Methods and Stability

Linear methods allow us to study the local truncation error more easily than others. By definition, these methods satisfy

$$\begin{aligned}\text{for all } u^n \ \text{for all } v^n,\ & H\left(u^n_{i-k}+v^n_{i-k},\ldots,u^n_{i+k}+v^n_{i+k},\triangle t\right)\\ &= H\left(u^n_{i-k},\ldots,u^n_{i+k},\triangle t\right) + H\left(v^n_{i-k},\ldots,v^n_{i+k},\triangle t\right).\end{aligned}$$

Since we can write the exact solution in terms of the method and the truncation error, *i.e.*

$$\text{for all } x_{i-1/2} < x < x_{i+1/2},\, u(x,t+\triangle t) = H\left(u(x_{i-k},t),\ldots,u(x_{i+k},t)\,\triangle t\right) + L(x,t;\triangle t)\triangle t,$$

the pointwise error at the cell center x_i for a linear method is

$$\begin{aligned}e^{n+1}_i &\equiv u^{n+1}_i - u(x_i, t^n + \triangle t^{n+1/2})\\ &= H\left(u^n_{i-k} - u(x_{i-k},t^n),\ldots,u^n_{i+k} - u(x_{i+k},t^n);\triangle t^{n+1/2}\right)\\ &\quad - \triangle t L(x_i,t^n;\ \triangle t^{n+1/2}).\end{aligned}$$

The solution of this linear recurrence leads to the formula

$$\begin{aligned}e^n_i &= \left\{\prod_{\ell=0}^{n-1} H(\cdot,\ldots,\cdot;\triangle t^{\ell+1/2})\right\}(e^0)\\ &\quad - \sum_{i=0}^{n-1} \triangle t^{i+1/2}\left\{\prod_{\ell=i}^{n-1} H(\cdot,\ldots,\cdot;\triangle t^{\ell+1/2})\right\}(L(\cdot,t^i;\triangle t^{i+1/2})).\end{aligned} \tag{5.1}$$

In this formula, the products indicate composition of the linear operators $H(\cdot,\ldots,\cdot;\triangle t^{\ell+1/2})$; most of the arguments for these linear operators have been omitted to simplify the expressions. We will denote the norms of these linear operators by

$$\|H(\cdot,\ldots,\cdot;\triangle t^{\ell+1/2})\|_1 \equiv \sup_{u^\ell} \frac{\|H(u^\ell_{i-k},\ldots,u^\ell_{i+k};\triangle t^{\ell+1/2})\|_1}{\|u^\ell\|_1}.$$

The following lemma is similar to Theorem 2.4.5.

Lemma 5.1.3 *Suppose that the scheme* $u_i^{n+1} = H(u_{i-k}^n, \ldots, u_{i+k}^n; \triangle t^{n+1/2})$ *involves a linear method* H*. Further, suppose that* H *is bounded close to 1, in the sense*

$$\text{there exists } \triangle t_0 > 0 \text{ there exists } \alpha > 0 \text{ for all } \triangle t < \triangle t_0 \; \|H(\cdot, \ldots, \cdot; \triangle t)\|_1 \leq 1 + \alpha \triangle t. \tag{5.2}$$

Then the linear method H *is* **Lax–Richtmyer stable** *meaning that*

for all $T > 0$ *there exists* $C_T > 0$ *there exists* $\triangle t_0 > 0$ *there exists* n *such that*

if for all $0 \leq \ell < n$ *and* $\triangle t^{\ell+1/2} < \triangle t_0$ *and* $t^n = \sum_{\ell=0}^{n-1} \triangle t^{\ell+1/2} \leq T$

$$\text{then } \left\| \prod_{\ell=0}^{n-1} H(\cdot, \ldots, \cdot; \triangle t^{\ell+1/2}) \right\|_1 \leq C_T,$$

and if L *is the local truncation error (see Definition 5.1.2), then the pointwise error satisfies*

for all $T > 0$ *there exists* $\triangle t_0 > 0$ *there exists* n *such that*

if for all $0 \leq \ell < n$ *we have* $\triangle t^{\ell+1/2} < \triangle t_0$ *and* $t^n \equiv \sum_{i=0}^{n-1} \triangle t^{i+1/2} \leq T$

then $\|e^n\|_1 \leq e^{\alpha T} \|e_0\|_1 + T e^{\alpha T} \max_\ell \{\|L(\cdot, t^\ell, \triangle t^{\ell+1/2})\|_1$.

Proof Assumption (5.2) implies that

$$\left\| \prod_{\ell=0}^{n-1} H(\cdot, \ldots, \cdot; \triangle t^{\ell+1/2}) \right\|_1 \leq \prod_{\ell=0}^{n-1} (1 + \alpha \triangle t^{\ell+1/2}) \leq e^{\alpha \sum_{\ell=0}^{n-1} \triangle t^{\ell+1/2}} \leq e^{\alpha T}.$$

From this and the solution (5.1) of the linear recurrence for the scheme, it follows that

$$\begin{aligned}
\|e^n\|_1 &\leq e^{\alpha T} \|e^0\|_1 + e^{\alpha T} \sum_{i=0}^{n-1} \triangle t^{i+1/2} \|L(\cdot, t^i; \triangle t^{i+1/2})\|_1 \\
&\leq e^{\alpha T} \|e^0\|_1 + T e^{\alpha T} \max_\ell \left\{ \|L(\cdot, t^\ell; \triangle t^{\ell+1/2})\|_1 \right\}.
\end{aligned}$$

□

Convergence up to a given time T will follow if the initial error and the maximum truncation error tend to zero, as the mesh width and timestep tends to zero. Further, the order of the scheme (see Definition 5.1.2) will determine the rate of convergence.

5.1.3 Convergence of Linear Methods

Recall that in Theorems 2.7.14 and 2.7.15 we proved that stability is necessary and sufficient for convergence. This statement is limited to consistent linear schemes (see Definition 2.7.5) applied to scalar partial differential equations involving a single derivative in time (and subject to modest assumptions on the symbols of the partial differential equation and the scheme).

Example 5.1.4 Consider the explicit upwind scheme for linear advection. We approximate the solution of the linear advection equation

$$\frac{\partial u}{\partial t} + \lambda \frac{\partial u}{\partial x} = 0$$

by computing the explicit upwind solution

$$u_i^{n+1} = u_i^n - \frac{\lambda \triangle t}{\triangle x}[u_i^n - u_{i-1}^n] = \frac{\lambda \triangle t}{\triangle x} u_{i-1}^n + \left(1 - \frac{\lambda \triangle t}{\triangle x}\right) u_i^n \equiv H(u_{i-1}^n, u_i^n; \triangle t).$$

The local truncation error (see Definition 5.1.2) is

$$\begin{aligned}
L(x, t; \triangle t) &= \frac{1}{\triangle t}\left[u(x, t + \triangle t) - \frac{\lambda \triangle t}{\triangle x} u(x - \triangle x, t) - \left(1 - \frac{\lambda \triangle t}{\triangle x}\right) u(x, t)\right] \\
&\approx \frac{1}{\triangle t}\left[\left\{u(x, t) + \frac{\partial u}{\partial t}\triangle t + \frac{\partial^2 u}{\partial t^2}\frac{\triangle t^2}{2}\right\} - \frac{\lambda \triangle t}{\triangle x}\left\{u(x, t) - \frac{\partial u}{\partial x}\triangle x + \frac{\partial^2 u}{\partial x^2}\frac{\triangle x^2}{2}\right\}\right. \\
&\quad \left. - \left(1 - \frac{\lambda \triangle t}{\triangle x}\right) u(x, t)\right] \\
&= \left(\frac{\partial u}{\partial t} + \lambda \frac{\partial u}{\partial x}\right) + \left(\frac{\partial^2 u}{\partial t^2}\frac{\triangle t}{2} - \frac{\partial^2 u}{\partial x^2}\frac{\lambda \triangle x}{2}\right) = \left(\frac{\partial u}{\partial t} + \lambda \frac{\partial u}{\partial x}\right) + \frac{\lambda}{2}(\lambda \triangle t - \triangle x)\frac{\partial^2 u}{\partial x^2}.
\end{aligned}$$

If $\| \frac{\partial^2 u}{\partial x^2} \|$ is bounded, and if $\triangle t \to 0$ as $\triangle x \to 0$, then it follows that the method is consistent.

Since the method is linear, we can easily investigate its stability. Note that if $1 > \lambda \triangle t / \triangle x$ then $\|H\|_1 \leq 1$:

$$\begin{aligned}
\|u^{n+1}\|_1 &\equiv \sum_i |u_i^{n+1}|\triangle x = \sum_i \left| \frac{\lambda \triangle t}{\triangle x} u_{i-1}^n + \left(1 - \frac{\lambda \triangle t}{\triangle x}\right) u_i^n \right| \\
&\leq \frac{\lambda \triangle t}{\triangle x} \sum_i |u_{i-1}^n|\triangle x + \left(1 - \frac{\lambda \triangle t}{\triangle x}\right) \sum_i |u_i^n|\triangle x = \sum_i |u_i^n|\triangle x \equiv \|u^n\|_1.
\end{aligned}$$

Since $\|H^n\|_1 \leq 1$, this shows that the method is stable if $\lambda \triangle t / \triangle x < 1$. Since the explicit upwind scheme is consistent and stable for $\lambda \triangle t / \triangle x < 1$, the Lax equivalence theorem 2.7.14 show that it is convergent under these restrictions on $\triangle t$ and $\triangle x$. In particular, note that $\triangle t$ must approach zero at the same rate as $\triangle x$; in other words, $\triangle t < \triangle x / \lambda$. It is not hard to see either from the proof of the Lax Equivalence Theorem 2.7.14 or from Lemma 5.1.3 that the scheme will have order one under these circumstances.

Exercises for 5.1

5.1.1 Consider the scheme $u_i^{n+1} = u_i^n$ for the linear advection equation.
  (a) Is this a linear method?
  (b) Under what conditions is this method stable?
  (c) Under what conditions is this method consistent?
  (d) Under what conditions is this method convergent?
  (e) What is the order of this scheme?

5.1.2 Answer the questions of the first exercise for the explicit centered difference scheme applied to the linear advection equation.

5.1.3 Answer the questions of the first exercise for the Lax–Friedrichs scheme applied to linear advection.

5.1.4 Answer the questions of the first exercise for the Lax–Wendroff scheme applied to linear advection.

5.2 Entropy Conditions and Difference Approximations

In Section 5.1 we studied linear finite difference methods. This study of linear methods necessarily restricted our attention to linear conservation laws. In this section, we will begin to develop the tools we need to understand methods for nonlinear conservation laws.

5.2.1 Bounded Convergence

Before stating our next result, we need a couple of definitions.

Definition 5.2.1 A numerical flux function $F(u_{i-k+1}^n, \ldots, u_{i+k}^n)$ is **consistent** with a physical flux $f(u)$ if and only if for all w we have $F(w, \ldots, w) = f(w)$.

The piecewise constant function $U(x,t)$ **converges to** $u(x,t)$ **in** $\mathcal{L}^1$ if and only if over every bounded set $\Omega = [a,b] \times [0,T]$ we have that whenever the mesh is sufficiently small then the error in the numerical solution is small:

for all $T > 0$ for all $-\infty < a < b < \infty$ for all $\epsilon > 0$

there exists $h > 0$ there exists $\delta > 0$ there exists $J \geq \frac{b-a}{h}$ there exists $N \geq \frac{T}{\delta}$

so that $a = x_{-1/2} < x_{1/2} < \ldots < x_{J+1/2} = b$ with for all $0 \leq i \leq J$ $\triangle x_i \equiv x_{i+1/2} - x_{i-1/2} \leq h$

and $0 = t^0 < t^1 < \ldots < t^N = T$ with for all $0 \leq n < N$ $\triangle t^{n+1/2} \equiv t^{n+1} - t^n \leq \delta$

implies that $\int_0^T \int_a^b |U(x,t) - u(x,t)| \, dx \, dt < \epsilon$.

If so, then we write $\|U - u\|_{1,\Omega} \to 0$.

Theorem 5.2.2 *(Lax–Wendroff) [90] Consider the conservative difference scheme*

$$u_i^{n+1} = u_i^n - \frac{\triangle t^{n+1/2}}{\triangle x_i} \left[\tilde{f}(u_i^n, u_{i+1}^n) - \tilde{f}(u_{i-1}^n, u_i^n) \right]$$

approximating the conservation law

$$\frac{\partial u}{\partial t} + \frac{\partial f(u)}{\partial x} = 0.$$

Assume that both f and $\tilde{f}$ are continuous, and that the numerical flux $\tilde{f}$ is consistent (Definition 5.1.2). Given any $\epsilon > 0$, suppose that we can choose a mesh in space and initial data for the scheme so that

$$\sum_{i=-\infty}^{\infty} \int_{x_{i-1/2}}^{x_{i+1/2}} \|u_i^0 - u(x,0)\| \, dx < \epsilon.$$

Further, suppose that the numerical solution u_i^n converges to the function u in the following sense (c.f. *Definition 5.2.1): given any $\epsilon > 0$ we can choose a mesh in space and time, apply the numerical scheme and find that*

$$\sum_{n=1}^{\infty} \sum_{i=-\infty}^{\infty} \int_{t^{n-1}}^{t^n} \int_{x_{i-1/2}}^{x_{i+1/2}} \|u_i^n - u(x,t)\| \, dx \, dt < \epsilon. \tag{5.3}$$

Finally, suppose that u is locally bounded, meaning that

for all compact $K \subset \mathbf{R} \times (0,\infty)$ there exists $C_{u,K} > 0$ so that $\max_{(x,t) \in K} \|u(x,t)\| \leq C_{u,K}$,

and that the numerical solution is locally bounded, meaning that

for all compact $K \subset \mathbf{R} \times (0,\infty)$ there exists $C_{\tilde{u},K} > 0$ so that $\max_{(x_i,t^n) \in K} \|u_i^n\| \leq C_{\tilde{u},K}$,

Then $u(x,t)$ is a weak solution of the conservation law.

Proof First, we will perform a computation to set up the integrals we need to examine for limits. Let $\phi(x,t) \in C_0^\infty(\mathbf{R} \times (0,\infty))$. Define $\phi_i^n = \phi(x_i, t^n)$, and assume that the support of ϕ is contained in the compact set $K \subset \mathbf{R}$. Then

$$\begin{aligned}
0 &= \sum_{n=0}^{\infty} \sum_{i=-\infty}^{\infty} \phi_i^n [u_i^{n+1} - u_i^n] \triangle x_i \\
&\quad + \sum_{n=0}^{\infty} \sum_{i=-\infty}^{\infty} \frac{\triangle t^{n+1/2}}{\triangle x_i} \phi_i^n [\tilde{f}(u_i^n, u_{i+1}^n) - \tilde{f}(u_{i-1}^n, u_i^n)] \triangle x_i \\
&= \sum_{i=-\infty}^{\infty} \left[\sum_{n=1}^{\infty} \phi_i^{n-1} u_i^n - \sum_{n=0}^{\infty} \phi_i^n u_i^n \right] \triangle x_i \\
&\quad + \sum_{n=0}^{\infty} \triangle t^{n+1/2} \left[\sum_{i=-\infty}^{\infty} \phi_{i-1}^n \tilde{f}(u_{i-1}^n, u_i^n) - \sum_{i=-\infty}^{\infty} \phi_i^n \tilde{f}(u_{i-1}^n, u_i^n) \right] \\
&= - \sum_{i=-\infty}^{\infty} \sum_{n=1}^{\infty} \frac{\phi_i^n - \phi_i^{n-1}}{\triangle t^{n+1/2}} u_i^n \triangle x_i \triangle t^{n+1/2} - \sum_{i=-\infty}^{\infty} \phi_i^0 u_i^0 \triangle x_i \\
&\quad - \sum_{n=0}^{\infty} \sum_{i=-\infty}^{\infty} \frac{\phi_i^n - \phi_{i-1}^n}{\frac{1}{2}(\triangle x_{i-1} + \triangle x_i)} \tilde{f}(u_{i-1}^n, u_i^n) \frac{\triangle x_{i-1} + \triangle x_i}{2} \triangle t^{n+1/2} \\
&\equiv -S_u - S_0 - S_f.
\end{aligned} \tag{5.4}$$

Let us discuss the convergence of each of these sums. First, we will deal with the initial conditions in S_0. We note that

$$\begin{aligned}
&\sum_{i=-\infty}^{\infty} \phi_i^0 u_i^0 \triangle x_i - \int_{-\infty}^{\infty} \phi(x,0) u(x,0)\, dx \\
&\quad = \sum_{i=-\infty}^{\infty} \int_{x_{i-1/2}}^{x_{i+1/2}} \phi(x_i,0) u_i^0 - \phi(x,0) u(x,0)\, dx \\
&\quad = \sum_{i=-\infty}^{\infty} \int_{x_{i-1/2}}^{x_{i+1/2}} \phi(x_i,0)[u_i^0 - u(x,0)] + [\phi(x_i,0) - \phi(x,0)] u(x,0)\, dx.
\end{aligned}$$

Given $\epsilon > 0$, we can choose the mesh and initial data so that

$$\begin{aligned}
\left\| \sum_{i=-\infty}^{\infty} \int_{x_{i-1/2}}^{x_{i+1/2}} \phi(x_i,0)[u_i^0 - u(x,0)]\, dx \right\| &\le \sum_{i=-\infty}^{\infty} \int_{x_{i-1/2}}^{x_{i+1/2}} \phi(x_i,0) \| u_i^0 - u(x,0) \|\, dx \\
&\le \|\phi\|_\infty \sum_{i=-\infty}^{\infty} \int_{x_{i-1/2}}^{x_{i+1/2}} \| u_i^0 - u(x,0) \|\, dx < \frac{\epsilon}{6}.
\end{aligned}$$

We can further restrict the mesh, if necessary, so that

$$
\begin{aligned}
&\left\| \sum_{i=-\infty}^{\infty} \int_{x_{i-1/2}}^{x_{i+1/2}} [\phi(x_i,0) - \phi(x,0)] u(x,0)\, dx \right\| \\
&\le \sum_{i=-\infty}^{\infty} \int_{x_{i-1/2}}^{x_{i+1/2}} |\phi(x_i,0) - \phi(x,0)| \, \|u(x,0)\| \, dx \\
&= \sum_{i=-\infty}^{\infty} \int_{x_{i-1/2}}^{x_{i+1/2}} \left| \int_x^{x_i} \frac{\partial \phi}{\partial x}(\xi,0)\, d\xi \right| \, \|u(x,0)\| \, dx \\
&\le C_{u,K} \sum_{i=-\infty}^{\infty} \int_{x_{i-1/2}}^{x_i} \int_x^{x_i} \left| \frac{\partial \phi}{\partial x}(\xi,0) \right| d\xi\, dx \\
&\quad + C_{u,K} \sum_{i=-\infty}^{\infty} \int_{x_i}^{x_{i+1/2}} \int_{x_i}^{x} \left| \frac{\partial \phi}{\partial x}(\xi,0) \right| d\xi\, dx \\
&= C_{u,K} \left[\sum_{i=-\infty}^{\infty} \int_{x_{i-1/2}}^{x_i} \left| \frac{\partial \phi}{\partial x}(\xi,0) \right| \int_{x_{i-1/2}}^{\xi} dx\, d\xi \right. \\
&\quad \left. + \sum_{i=-\infty}^{\infty} \int_{x_i}^{x_{i+1/2}} \left| \frac{\partial \phi}{\partial x}(\xi,0) \right| \int_{\xi}^{x_{i+1/2}} dx\, d\xi \right] \\
&\le C_{u,K} \sum_{i=-\infty}^{\infty} \int_{x_{i-1/2}}^{x_{i+1/2}} \left| \frac{\partial \phi}{\partial x}(\xi,0) \right| \triangle x_i\, d\xi \le C_{u,K} \max_i\{\triangle x_i\} \left\| \frac{\partial \phi}{\partial x}(\cdot,0) \right\|_1 < \frac{\epsilon}{6}.
\end{aligned}
$$

Thus

$$
\begin{aligned}
\left\| S_0 - \int_{-\infty}^{\infty} \phi(x,0) u(x,0) dx \right\| &\le \left\| \sum_{i=-\infty}^{\infty} \int_{x_{i-1/2}}^{x_{i+1/2}} \phi(x_i,0)[u_i^0 - u(x,0)]\, dx \right\| \\
&\quad + \left\| \sum_{i=-\infty}^{\infty} \int_{x_{i-1/2}}^{x_{i+1/2}} [\phi(x_i,0) - \phi(x,0)] u(x,0)\, dx \right\| < \frac{\epsilon}{3}. \quad (5.5)
\end{aligned}
$$

Let us examine the terms related to time derivatives of ϕ, namely the sum S_u in (5.4) Note that

$$
\begin{aligned}
&\sum_{n=1}^{\infty} \sum_{i=-\infty}^{\infty} \frac{\phi_i^n - \phi_i^{n-1}}{\triangle t^{n-1/2}} u_i^n \triangle x_i \triangle t^{n+1/2} - \int_0^{\infty} \int_{-\infty}^{\infty} \frac{\partial \phi}{\partial t}(x,t) u(x,t)\, dx\, dt \\
&= \sum_{n=1}^{\infty} \sum_{i=-\infty}^{\infty} \int_{t^{n-1}}^{t^n} \int_{x_{i-1/2}}^{x_{i+1/2}} \frac{\phi_i^n - \phi_i^{n-1}}{\triangle t^{n-1/2}} u_i^n - \frac{\partial \phi}{\partial t}(x,t) u(x,t)\, dx\, dt \\
&= \sum_{n=1}^{\infty} \sum_{i=-\infty}^{\infty} \int_{t^{n-1}}^{t^n} \int_{x_{i-1/2}}^{x_{i+1/2}} \frac{\phi_i^n - \phi_i^{n-1}}{\triangle t^{n-1/2}} [u_i^n - u(x,t)]\, dx\, dt \\
&\quad + \sum_{n=1}^{\infty} \sum_{i=-\infty}^{\infty} \int_{t^{n-1}}^{t^n} \int_{x_{i-1/2}}^{x_{i+1/2}} \left[\frac{\phi_i^n - \phi_i^{n-1}}{\triangle t^{n-1/2}} - \frac{\partial \phi}{\partial t}(x,t) \right] u(x,t)\, dx\, dt \\
&\equiv S_{u1} + S_{u2}. \qquad (5.6)
\end{aligned}
$$

Let us examine $S_{u,1}$ first. Given $\epsilon > 0$, we can further restrict the mesh so that

$$\begin{aligned}
\|S_{u,1}\| &= \left\| \sum_{n=1}^{\infty} \sum_{i=-\infty}^{\infty} \int_{t^{n-1}}^{t^n} \int_{x_{i-1/2}}^{x_{i+1/2}} \frac{\phi_i^n - \phi_i^{n-1}}{\triangle t^{n-1/2}} [u_i^n - u(x,t)] \, dx \, dt \right\| \\
&\leq \sum_{n=1}^{\infty} \sum_{i=-\infty}^{\infty} \int_{t^{n-1}}^{t^n} \int_{x_{i-1/2}}^{x_{i+1/2}} \left| \frac{\phi_i^n - \phi_i^{n-1}}{\triangle t^{n-1/2}} \right| \|u_i^n - u(x,t)\| \, dx \, dt \\
&= \sum_{n=1}^{\infty} \sum_{i=-\infty}^{\infty} \int_{t^{n-1}}^{t^n} \int_{x_{i-1/2}}^{x_{i+1/2}} \left| \frac{1}{\triangle t^{n-1/2}} \int_{t^{n-1}}^{t^n} \frac{\partial \phi}{\partial t}(x_i, s) \, ds \right| \|u_i^n - u(x,t)\| \, dx \, dt \\
&\leq \left\| \frac{\partial \phi}{\partial t} \right\|_\infty \sum_{n=1}^{\infty} \sum_{i=-\infty}^{\infty} \int_{t^{n-1}}^{t^n} \int_{x_{i-1/2}}^{x_{i+1/2}} \|u_i^n - u(x,t)\| \, dx \, dt < \frac{\epsilon}{6}.
\end{aligned} \tag{5.7}$$

Next, let us bound $S_{u,2}$. Since u is locally bounded, we can further restrict the mesh so that

$$\begin{aligned}
\|S_{u,2}\| &= \left\| \sum_{n=1}^{\infty} \sum_{i=-\infty}^{\infty} \int_{t^{n-1}}^{t^n} \int_{x_{i-1/2}}^{x_{i+1/2}} \left[\frac{\phi_i^n - \phi_i^{n-1}}{\triangle t^{n-1/2}} - \frac{\partial \phi}{\partial t} \right] u(x,t) \, dx \, dt \right\| \\
&\leq C_{u,K} \sum_{n=1}^{\infty} \sum_{i=-\infty}^{\infty} \int_{t^{n-1}}^{t^n} \int_{x_{i-1/2}}^{x_{i+1/2}} \left| \frac{\phi_i^n - \phi_i^{n-1}}{\triangle t^{n-1/2}} - \frac{\partial \phi}{\partial t} \right| dx \, dt \\
&= C_{u,K} \sum_{n=1}^{\infty} \sum_{i=-\infty}^{\infty} \int_{t^{n-1}}^{t^n} \int_{x_{i-1/2}}^{x_{i+1/2}} \left| \frac{1}{\triangle t^{n-1/2}} \int_{t^{n-1}}^{t^n} \frac{\partial \phi}{\partial t}(x_i, s) - \frac{\partial \phi}{\partial t}(x,t) \, ds \right| dx \, dt \\
&\leq C_{u,K} \sum_{n=1}^{\infty} \sum_{i=-\infty}^{\infty} \int_{t^{n-1}}^{t^n} \int_{x_{i-1/2}}^{x_{i+1/2}} \frac{1}{\triangle t^{n-1/2}} \int_{t^{n-1}}^{t^n} \left| \frac{\partial \phi}{\partial t}(x_i, s) - \frac{\partial \phi}{\partial t}(x,s) \right| ds \, dx \, dt \\
&\quad + C_{u,K} \sum_{n=1}^{\infty} \sum_{i=-\infty}^{\infty} \int_{t^{n-1}}^{t^n} \int_{x_{i-1/2}}^{x_{i+1/2}} \frac{1}{\triangle t^{n-1/2}} \int_{t^{n-1}}^{t^n} \left| \frac{\partial \phi}{\partial t}(x, s) - \frac{\partial \phi}{\partial t}(x,t) \right| ds \, dx \, dt \\
&= C_{u,K} \sum_{n=1}^{\infty} \sum_{i=-\infty}^{\infty} \int_{t^{n-1}}^{t^n} \int_{x_{i-1/2}}^{x_{i+1/2}} \frac{1}{\triangle t^{n-1/2}} \int_{t^{n-1}}^{t^n} \left| \int_{x}^{x_i} \frac{\partial^2 \phi}{\partial t \partial x}(\xi, s) \, d\xi \right| ds \, dx \, dt \\
&\quad + C_{u,K} \sum_{n=1}^{\infty} \sum_{i=-\infty}^{\infty} \int_{t^{n-1}}^{t^n} \int_{x_{i-1/2}}^{x_{i+1/2}} \frac{1}{\triangle t^{n-1/2}} \int_{t^{n-1}}^{t^n} \left| \int_{t}^{s} \frac{\partial^2 \phi}{\partial t^2}(x, \tau) \, d\tau \right| ds \, dx \, dt \\
&= B_{u2a} + B_{u2b}.
\end{aligned} \tag{5.8}$$

We will change order of integration to estimate both of these bounds. First

$$
\begin{aligned}
B_{u2a} &= \sum_{n=1}^{\infty}\sum_{i=-\infty}^{\infty}\int_{t^{n-1}}^{t^n}\int_{x_{i-1/2}}^{x_{i+1/2}} \frac{1}{\triangle t^{n-1/2}}\int_{t^{n-1}}^{t^n}\left|\int_x^{x_i}\frac{\partial^2\phi}{\partial t\partial x}(\xi,s)\,d\xi\right|\,ds\,dx\,dt\\
&= \sum_{n=1}^{\infty}\sum_{i=-\infty}^{\infty}\int_{t^{n-1}}^{t^n}\int_{x_{i-1/2}}^{x_i}\left|\frac{\partial^2\phi}{\partial t\partial x}(\xi,s)\right|\int_{t^{n-1}}^{t^n}\int_{x_{i-1/2}}^{\xi}\frac{1}{\triangle t^{n-1/2}}\,dx\,dt\,d\xi\,ds\\
&\quad+\sum_{n=1}^{\infty}\sum_{i=-\infty}^{\infty}\int_{t^{n-1}}^{t^n}\int_{x_i}^{x_{i+1/2}}\left|\frac{\partial^2\phi}{\partial t\partial x}(\xi,s)\right|\int_{t^{n-1}}^{t^n}\int_{\xi}^{x_{i+1/2}}\frac{1}{\triangle t^{n-1/2}}\,dx\,dt\,d\xi\,ds\\
&= \sum_{n=1}^{\infty}\sum_{i=-\infty}^{\infty}\int_{t^{n-1}}^{t^n}\int_{x_{i-1/2}}^{x_i}\left|\frac{\partial^2\phi}{\partial t\partial x}(\xi,s)\right|(\xi-x_{i-1/2})\,d\xi\,ds\\
&\quad+\sum_{n=1}^{\infty}\sum_{i=-\infty}^{\infty}\int_{t^{n-1}}^{t^n}\int_{x_i}^{x_{i+1/2}}\left|\frac{\partial^2\phi}{\partial t\partial x}(\xi,s)\right|(x_{i+1/2}-\xi)\,d\xi\,ds\\
&\le \left\|\frac{\partial^2\phi}{\partial t\partial x}\right\|_1 \max_i \triangle x_i .
\end{aligned}
$$

Next,

$$
\begin{aligned}
B_{u2b} &= \sum_{n=1}^{\infty}\sum_{i=-\infty}^{\infty}\int_{t^{n-1}}^{t^n}\int_{x_{i-1/2}}^{x_{i+1/2}} \frac{1}{\triangle t^{n-1/2}}\int_{t^{n-1}}^{t^n}\left|\int_t^{s}\frac{\partial^2\phi}{\partial t^2}(x,\tau)\,d\tau\right|\,ds\,dx\,dt\\
&= \sum_{n=1}^{\infty}\sum_{i=-\infty}^{\infty}\int_{t^{n-1}}^{t^n}\int_{x_{i-1/2}}^{x_{i+1/2}}\left|\frac{\partial^2\phi}{\partial t^2}(x,\tau)\right|\int_{t^{n-1}}^{s}\int_{t^{n-1}}^{\tau}\frac{1}{\triangle t^{n-1/2}}\,dt\,d\tau\,dx\,ds\\
&\quad+\sum_{n=1}^{\infty}\sum_{i=-\infty}^{\infty}\int_{t^{n-1}}^{t^n}\int_{x_{i-1/2}}^{x_{i+1/2}}\left|\frac{\partial^2\phi}{\partial t^2}(x,\tau)\right|\int_{s}^{t^n}\int_{\tau}^{t^n}\frac{1}{\triangle t^{n-1/2}}\,dt\,d\tau\,dx\,ds\\
&= \sum_{n=1}^{\infty}\sum_{i=-\infty}^{\infty}\int_{t^{n-1}}^{t^n}\int_{x_{i-1/2}}^{x_{i+1/2}}\left|\frac{\partial^2\phi}{\partial t^2}(x,\tau)\right|\frac{(\tau-t^{n-1})(t^n-\tau)}{\triangle t^{n-1/2}}i\,dx\,d\tau\\
&\quad+\sum_{n=1}^{\infty}\sum_{i=-\infty}^{\infty}\int_{t^{n-1}}^{t^n}\int_{x_{i-1/2}}^{x_{i+1/2}}\left|\frac{\partial^2\phi}{\partial t^2}(x,\tau)\right|\frac{(\tau-t^{n-1})(t^n-\tau)}{\triangle t^{n-1/2}}i\,dx\,d\tau\\
&\le 2\left\|\frac{\partial^2\phi}{\partial t^2}\right\|_1 \max_n \triangle t^{n-1/2}.
\end{aligned}
$$

Putting the inequality (5.8) together with the bounds on B_{u2a} and B_{u2b}, we see that we can restrict the mesh and timesteps so that

$$
\begin{aligned}
\|S_{u,2}\| &= \left\|\sum_{n=1}^{\infty}\sum_{i=-\infty}^{\infty}\int_{t^{n-1}}^{t^n}\int_{x_{i-1/2}}^{x_{i+1/2}}\left[\frac{\phi_i^n-\phi_i^{n-1}}{\triangle t^{n-1/2}}-\frac{\partial\phi}{\partial t}\right]u(x,t)\,dx\,dt\right\| \le B_{u2a}+B_{u2b}\\
&\le C_{u,K}\left[\left\|\frac{\partial^2\phi}{\partial t\partial x}\right\|_1 \max_i \triangle x_i + 2\left\|\frac{\partial^2\phi}{\partial t^2}\right\|_1 \max_n \triangle t^{n-1/2}\right] < \frac{\epsilon}{6}. \qquad (5.9)
\end{aligned}
$$

Finally, let us examine the terms related to S_f in (5.4). Note that

$$\begin{aligned}
&\sum_{n=0}^{\infty}\sum_{i=-\infty}^{\infty}\frac{\phi_i^n-\phi_{i-1}^n}{\frac{1}{2}(\triangle x_{i-1}+\triangle x_i)}\tilde{f}(u_{i-1}^n,u_i^n)\frac{\triangle x_{i-1}+\triangle x_i}{2}\triangle t^{n-1/2} \\
&\quad-\int_0^\infty\int_{-\infty}^\infty\frac{\partial\phi}{\partial x}(x,t)f(u(x,t))\,dx\,dt \\
&=\sum_{n=1}^{\infty}\sum_{i=-\infty}^{\infty}\int_{t^{n-1}}^{t^n}\int_{x_{i-1}}^{x_i}\frac{\phi_i^n-\phi_{i-1}^n}{\frac{1}{2}(\triangle x_{i-1}+\triangle x_i)}\tilde{f}(u_{i-1}^n,u_i^n)-\frac{\partial\phi}{\partial x}(x,t)f(u(x,t))\,dx\,dt \\
&=\sum_{n=1}^{\infty}\sum_{i=-\infty}^{\infty}\int_{t^{n-1}}^{t^n}\int_{x_{i-1}}^{x_i}\frac{\phi_i^n-\phi_{i-1}^n}{\frac{1}{2}(\triangle x_{i-1}+\triangle x_i)}[\tilde{f}(u_{i-1}^n,u_i^n)-f(u(x,t))] \\
&\quad+\sum_{n=1}^{\infty}\sum_{i=-\infty}^{\infty}\int_{t^{n-1}}^{t^n}\int_{x_{i-1}}^{x_i}+\left[\frac{\phi_i^n-\phi_{i-1}^n}{\frac{1}{2}(\triangle x_{i-1}+\triangle x_i)}-\frac{\partial\phi}{\partial x}\right]f(u(x,t))\,dx\,dt \\
&=S_{f1}+S_{f2}.
\end{aligned}\tag{5.10}$$

We can extend the numerical solution to a piecewise-constant function on the mesh in space and time. Since the numerical method satisfies the convergence assumption (5.3), the numerical solution converges to the true solution almost everywhere. Since the numerical flux is consistent (see definition 5.1.2) and continuous, it converges to the true flux almost everywhere. Since the flux is locally bounded, it is locally integrable, especially on compact sets containing the support of ϕ. Since the numerical flux is continuous and the numerical solution is locally bounded, the numerical flux is locally integrable. As a result, we can bound the norm of the difference between the numerical flux and the true flux by a locally integrable function. Then Lebesgue's dominated convergence theorem [139, page 26], shows that the integral of the norm of the error in the numerical flux tends to zero. In other words, given $\delta > 0$, we can choose the mesh so that if $\chi_K(x,t)$ is the indicator function for the compact set K containing the support of ϕ, then

$$\sum_{n=1}^{\infty}\sum_{i=-\infty}^{\infty}\int_{t^{n-1}}^{t^n}\int_{x_{i-1}}^{x_i}\chi_K(x,t)\|\tilde{f}(u_{i-1}^n,u_i^n)-f(u(x,t))\|\,dx\,dt<\delta.$$

Thus we can further restrict the mesh, if necessary, so that

$$\begin{aligned}
\|S_{f1}\| &= \left\|\sum_{n=1}^{\infty}\sum_{i=-\infty}^{\infty}\int_{t^{n-1}}^{t^n}\int_{x_{i-1}}^{x_i}\frac{\phi_i^n-\phi_{i-1}^n}{\frac{1}{2}(\triangle x_{i-1}+\triangle x_i)}[\tilde{f}(u_{i-1}^n,u_i^n)-f(u(x,t))]\,dx\,dt\right\| \\
&\le\sum_{n=1}^{\infty}\sum_{i=-\infty}^{\infty}\int_{t^{n-1}}^{t^n}\int_{x_{i-1}}^{x_i}\left|\frac{2}{\triangle x_{i-1}+\triangle x_i}\int_{x_{i-1}}^{x_i}\frac{\partial\phi}{\partial x}(\xi,t^n)d\xi\right|\|\tilde{f}(u_{i-1}^n,u_i^n)-f(u(x,t))\|\,dx\,dt \\
&\le\left\|\frac{\partial\phi}{\partial x}\right\|_\infty\sum_{n=1}^{\infty}\sum_{i=-\infty}^{\infty}\int_{t^{n-1}}^{t^n}\int_{x_{i-1}}^{x_i}\chi_K(x,t)\|\tilde{f}(u_{i-1}^n,u_i^n)-f(u(x,t))\|\,dx\,dt\|<\frac{\epsilon}{6}.
\end{aligned}$$

Now, let us turn our attention to S_{f2} in (5.10) Since u is locally bounded and the flux f is continuous, the flux is locally bounded. The fundamental theorem of calculus, followed by the triangle inequality, then imply that

$$
\begin{aligned}
&\|S_{f2}\| \\
&= \left\| \sum_{n=1}^{\infty} \sum_{i=-\infty}^{\infty} \int_{t^{n-1}}^{t^n} \int_{x_{i-1}}^{x_i} \left[\frac{\phi_i^n - \phi_{i-1}^n}{\frac{1}{2}(\triangle x_{i-1} + \triangle x_i)} - \frac{\partial \phi}{\partial x}(x,t) \right] f(u(x,t)) \, dx \, dt \right\| \\
&\leq C_{f,K} \sum_{n=1}^{\infty} \sum_{i=-\infty}^{\infty} \int_{t^{n-1}}^{t^n} \int_{x_{i-1}}^{x_i} \frac{2}{\triangle x_{i-1} + \triangle x_i} \int_{x_{i-1}}^{x_i} \left| \frac{\partial \phi}{\partial x}(\xi, t^n) - \frac{\partial \phi}{\partial x}(x,t) \right| \, d\xi \, dx \, dt \\
&\leq C_{f,K} \sum_{n=1}^{\infty} \sum_{i=-\infty}^{\infty} \int_{t^{n-1}}^{t^n} \int_{x_{i-1}}^{x_i} \frac{2}{\triangle x_{i-1} + \triangle x_i} \int_{x_{i-1}}^{x_i} \left| \frac{\partial \phi}{\partial x}(\xi, t^n) - \frac{\partial \phi}{\partial x}(x,t^n) \right| \, d\xi \, dx \, dt \\
&\quad + C_{f,K} \sum_{n=1}^{\infty} \sum_{i=-\infty}^{\infty} \int_{t^{n-1}}^{t^n} \int_{x_{i-1}}^{x_i} \frac{2}{\triangle x_{i-1} + \triangle x_i} \int_{x_{i-1}}^{x_i} \left| \frac{\partial \phi}{\partial x}(x, t^n) - \frac{\partial \phi}{\partial x}(x,t) \right| \, d\xi \, dx \, dt \\
&= C_{f,K}(S_{f2a} + S_{f2b}).
\end{aligned}
$$

Then changing the order of integration implies that

$$
\begin{aligned}
S_{f2a} &= \sum_{n=1}^{\infty} \sum_{i=-\infty}^{\infty} \int_{t^{n-1}}^{t^n} \int_{x_{i-1}}^{x_i} \frac{2}{\triangle x_{i-1} + \triangle x_i} \int_{x_{i-1}}^{x_i} \left| \frac{\partial \phi}{\partial x}(\xi, t^n) - \frac{\partial \phi}{\partial x}(x,t^n) \right| \, d\xi \, dx \, dt \\
&= \sum_{n=1}^{\infty} \sum_{i=-\infty}^{\infty} \int_{t^{n-1}}^{t^n} \int_{x_{i-1}}^{x_i} \frac{2}{\triangle x_{i-1} + \triangle x_i} \int_{x_{i-1}}^{x} \left| \int_{\xi}^{x} \frac{\partial^2 \phi}{\partial x^2}(\eta, t^n) \, d\eta \right| \, d\xi \, dx \, dt \\
&\quad + \sum_{n=1}^{\infty} \sum_{i=-\infty}^{\infty} \int_{t^{n-1}}^{t^n} \int_{x_{i-1}}^{x_i} \frac{2}{\triangle x_{i-1} + \triangle x_i} \int_{x}^{x_i} \left| \int_{x}^{\xi} \frac{\partial^2 \phi}{\partial x^2}(\eta, t^n) \, d\eta \right| \, d\xi \, dx \, dt \\
&\leq \sum_{n=1}^{\infty} \sum_{i=-\infty}^{\infty} \int_{t^{n-1}}^{t^n} \int_{x_{i-1}}^{x_i} \frac{2}{\triangle x_{i-1} + \triangle x_i} \left| \frac{\partial^2 \phi}{\partial x^2}(\eta, t^n) \right| \int_{\eta}^{x_i} \int_{x_{i-1}}^{\eta} \, d\xi \, dx \, d\eta \, dt \\
&\quad + \sum_{n=1}^{\infty} \sum_{i=-\infty}^{\infty} \int_{t^{n-1}}^{t^n} \int_{x_{i-1}}^{x_i} \frac{2}{\triangle x_{i-1} + \triangle x_i} \left| \frac{\partial^2 \phi}{\partial x^2}(\eta, t^n) \right| \int_{x_{i-1}}^{\eta} \int_{\eta}^{x_i} \, d\xi \, dx \, d\eta \, dt \\
&\leq 2 \max_i \frac{\triangle x_{i-1} + \triangle x_i}{2} \left\| \frac{\partial^2 \phi}{\partial x^2} \right\|_1
\end{aligned}
$$

and

$$
\begin{aligned}
S_{f2b} &= \sum_{n=1}^{\infty}\sum_{i=-\infty}^{\infty}\int_{t^{n-1}}^{t^n}\int_{x_{i-1}}^{x_i}\frac{2}{\triangle x_{i-1}+\triangle x_i}\int_{x_{i-1}}^{x_i}\left\|\frac{\partial\phi}{\partial x}(x,t^n)-\frac{\partial\phi}{\partial x}(x,t)\right\|\,d\xi\,dx\,dt\\
&=\sum_{n=1}^{\infty}\sum_{i=-\infty}^{\infty}\int_{t^{n-1}}^{t^n}\int_{x_{i-1}}^{x_i}\int_{t}^{t^n}\left|\frac{\partial^2\phi}{\partial x\partial t}(x,s)\right|\,ds\,dx\,dt\\
&=\sum_{n=1}^{\infty}\sum_{i=-\infty}^{\infty}\int_{t^{n-1}}^{t^n}\int_{x_{i-1}}^{x_i}\left|\frac{\partial^2\phi}{\partial x\partial t}(x,s)\right|\int_{t^{n-1}}^{s}\,dt\,dx\,ds\\
&\le \max_n \triangle t^{n-1/2}\left\|\frac{\partial^2\phi}{\partial x\partial t}\right\|_1 .
\end{aligned}
$$

It follows that we can choose the mesh so that

$$
\begin{aligned}
\|S_{f2}\| &= \left\|\sum_{n=1}^{\infty}\sum_{i=-\infty}^{\infty}\int_{t^{n-1}}^{t^n}\int_{x_{i-1}}^{x_i}\left[\frac{\phi_i^n-\phi_{i-1}^n}{\frac{1}{2}(\triangle x_{i-1}+\triangle x_i)}-\frac{\partial\phi}{\partial x}(x,t)\right]f(u(x,t))\,dx\,dt\right\|\\
&\le C_{f,K}(S_{f2a}+S_{f2b})\\
&\le C_{f,K}\max_i\{\triangle x_{i-1}+\triangle x_i\}\left\|\frac{\partial^2\phi}{\partial x^2}\right\|_1 + C_{f,K}\max_n \triangle t^{n-1/2}\left\|\frac{\partial^2\phi}{\partial x\partial t}\right\|_1 < \frac{\epsilon}{6}.
\end{aligned}
$$

Putting all our results together, we have

$$
\begin{aligned}
&\left|-\int_0^{\infty}\int_{-\infty}^{\infty}\left[\frac{\partial\phi}{\partial t}(x,t)u(x,t)+\frac{\partial\phi}{\partial x}(x,t)f(u(x,t))\right]dx\,dt+\int_{-\infty}^{\infty}\phi(x,0)u(x,0)\,dx\right|\\
&\le\left|\sum_{i=-\infty}^{\infty}\sum_{n=1}^{\infty}\frac{\phi_i^n-\phi_i^{n-1}}{\triangle t^{n+1/2}}u_i^n\triangle x_i\triangle t^{n+1/2}-\int_0^{\infty}\int_{-\infty}^{\infty}\frac{\partial\phi}{\partial t}(x,t)u(x,t)\,dx\,dt\right|\\
&+\left|\sum_{i=-\infty}^{\infty}\sum_{n=1}^{\infty}[\phi_i^n-\phi_{i-1}^n]\tilde{f}(u_{i-1}^n,u_i^n)\triangle t^{n+1/2}-\int_0^{\infty}\int_{-\infty}^{\infty}\frac{\partial\phi}{\partial x}(x,t)f(u(x,t))\,dx\,dt\right|\\
&+\left|\sum_{i=-\infty}^{\infty}\phi_i^0u_i^0\triangle x_i-\int_{-\infty}^{\infty}\phi(x,0)u(x,0)dx\right|<\frac{\epsilon}{3}+\frac{\epsilon}{3}+\frac{\epsilon}{3}=\epsilon.
\end{aligned}
$$

Since ϵ is arbitrary, u is a weak solution of the conservation law. □

It is useful to note that the Lax–Wendroff Theorem 5.2.2 can be extended to treat numerical fluxes of the form $\tilde{f}(u_{i-k+1}^n,\ldots,u_{i+k}^n)$. Furthermore, it is possible to extend the theorem to treat partial differential inequalities, such as those that might arise from numerical approximations to entropy inequalities. However, we should note that this theorem does not guarantee that the limit $u(x,t)$ satisfies an entropy condition. In other words, it is possible that the scheme could converge to a solution of the conservation law that does not result from vanishing diffusion.

Example 5.2.3 This example concerns the development of a Roe solver for a scalar conservation law; see Section 4.13.8 for the corresponding development in hyperbolic systems. For a general nonlinear scalar conservation law we can define

$$A(u_-, u_+) = \frac{f(u_+) - f(u_-)}{u_+ - u_-}.$$

Note that if the Riemann problem for states u_- and u_+ involves only a discontinuity, then $[f] = A(u_-, u_+)[u]$. The Roe flux associated with a state moving at zero speed is defined to be

$$\begin{aligned}
f_{j+1/2}^{n+1/2} &= \frac{1}{2}\left\{f\left(u_j^n\right) + f\left(u_{j+1}^n\right) - \left|A\left(u_j^n, u_{j+1}^n\right)\right|\left(u_{j+1}^n - u_j^n\right)\right\} \\
&= \frac{1}{2}\left\{f\left(u_j^n\right) + f\left(u_{j+1}^n\right) - \left|\frac{f\left(u_{j+1}^n\right) - f\left(u_j^n\right)}{u_{j+1}^n - u_j^n}\right|\left(u_{j+1}^n - u_j^n\right)\right\} \\
&= \begin{cases} \frac{1}{2}f(u_j^n) + \frac{1}{2}f(u_{j+1}^n) - \frac{1}{2}\{f(u_{j+1}^n) - f(u_j^n)\}, & A(u_j^n, u_{j+1}^n) \geq 0 \\ \frac{1}{2}f(u_j^n) + \frac{1}{2}f(u_{j+1}^n) + \frac{1}{2}\{f(u_{j+1}^n) - f(u_{j+1}^n)\}, & A(u_j^n, u_{j+1}^n) < 0 \end{cases} \\
&= \begin{cases} f(u_j^n), & A(u_j^n, u_{j+1}^n) \geq 0 \\ f(u_{j+1}^n), & A(u_j^n, u_{j+1}^n) < 0. \end{cases}
\end{aligned}$$

Note that this numerical flux function is consistent (see definition 5.1.2).

Figure 5.1 shows some numerical results obtained from **Program 5.2-62: roe.f**. Note that the numerical scheme does a good job on the problem involving a shock, but computes a combination of a stationary shock and a rarefaction in place of a transonic rarefaction. This is an example of a problem with a scheme that converges, but does not converge to the analytical solution in the limit of vanishing diffusion.

There are other examples of failure to converge to the correct solution of a conservation law. For the Buckley–Leverett model, the approximate Riemann solvers due to Roe and Harten–Hyman both fail to compute the correct solution in cases involving two shocks and an intermediate transonic rarefaction. For this model, the Harten–Lax–vanLeer and Linde approximate Riemann solvers work if modified to account for the larger characteristic speeds at inflections points.

Students can experiment with different numerical schemes for Riemann problems in a variety of models (Linear Advection, Burgers', Traffic and Buckley–Leverett) by clicking on **Executable 5.2-25: guiriemann** In the "View" menu, the user can select input parameters for the Riemann problem, numerical method, Buckley–Leverett model and graphics. The Riemann problem input parameters include the left and right states, the mesh bounds and the initial jump location, and the choice of

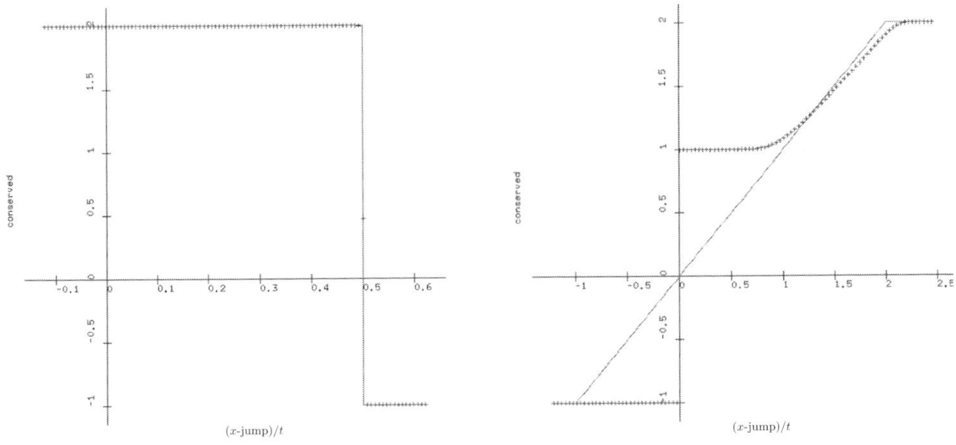

(a) Shock (b) Rarefaction

Fig. 5.1 Roe solver for Burgers' equation (solid = exact, plus = numerical solution)

the model. The numerical method parameters include the number of grid cells and timesteps, the max simulation time and the CFL number for selecting the timestep. Users can also select which scheme to use. If the number of grid cells is positive, then the program will use the first scheme selected; if the number of grid cells is zero, then the program will perform a mesh refinement study with all selected schemes to compare accuracy and efficiency. Errors in the schemes are computed as the average of the absolute values in the errors in the cell averages, and the slopes of the log error versus log mesh width are printed to indicate the rate of convergence.

5.2.2 Monotone Schemes

In this section, we will present a condition on schemes for scalar conservation laws in one dimension that will guarantee that when the scheme converges (see definition 5.2.1), it converges to the entropy-satisfying solution of the conservation law.

Definition 5.2.4 An explicit numerical method $u_j^{n+1} = H(u_{j-k}^n, \ldots, u_{j+k}^n)$ is **monotone** if and only if it preserves inequalities between sets of numerical results:

$$
\begin{aligned}
&\text{for all } u_i^n \text{ for all } v_i^n \\
&\text{if for all } i, \quad u_i^n \leq v_i^n \\
&\text{then for all } i \quad u_i^{n+1} = H(u_{i-k}^n, \ldots, u_{i+k}^n; \triangle x_i, \triangle t^{n+1/2}) \\
&\qquad\qquad\qquad \leq H(v_{i-k}^n, \ldots, v_{i+k}^n; \triangle x_i, \triangle t^{n+1/2}) \\
&\qquad\qquad\qquad = v_i^{n+1} .
\end{aligned}
$$

Here is common way to determine if a scheme is monotone.

Lemma 5.2.5 *Suppose that* $u_j^{n+1} = H(u_{j-k}^n, \ldots, u_{j+k}^n; \triangle x_i, \triangle t^{n+1/2})$ *is a monotone scheme, and that* $H(w_{-k}, \ldots, w_k; \triangle x, \triangle t)$ *is differentiable in each of its* w_ℓ *arguments for* $-k \leq \ell \leq k$. *Then for all* $-k \leq \ell \leq k$, $\partial H / \partial w_\ell \geq 0$. *Conversely, if* $\partial H / \partial w_\ell \geq 0$ *for all* $-k \leq \ell \leq k$, *then* $u_j^{n+1} = H(u_{j-k}^n, \ldots, u_{j+k}^n; \triangle x_i, \triangle t^{n+1/2})$ *is a monotone scheme.*

Proof Given any initial data u_i^0 and $\epsilon > 0$, let $v_i^0 = u_i^0 + \delta_{ij}\epsilon$. Then since the scheme is monotone and for all i it is true that $v_i^0 \geq u_i^0$, we have

$$H(u_{i-k}^n, \ldots, u_{i+k}^n; \triangle x_i, \triangle t^{n+1/2}) = u_i^1 \leq v_i^1 = H(v_{i-k}^n, \ldots, v_{i+k}^n; \triangle x_i, \triangle t^{n+1/2}).$$

It follows that with $i + \ell = j$ for $-k \leq \ell \leq k$,

$$\begin{aligned} 0 &\leq \frac{1}{\epsilon}[H(v_{i-k}^n, \ldots, v_{i+k}^n; \triangle x_i, \triangle t^{n+1/2}) - H(u_{i-k}^n, \ldots, u_{i+k}^n; \triangle x_i, \triangle t^{n+1/2})] \\ &= \frac{1}{\epsilon}[H(u_{i-k}^n, \ldots, u_j^n + \epsilon, \ldots, u_{i+k}^n; \triangle x_i, \triangle t^{n+1/2}) \\ &\quad - H(u_{i-k}^n, \ldots, u_j^n, \ldots, u_{i+k}^n; \triangle x_i, \triangle t^{n+1/2})] \\ &\to \frac{\partial H}{\partial w_\ell}(u_{i-k}^n, \ldots, \ldots, u_{i+k}^n; \triangle x_i, \triangle t^{n+1/2}) \quad \text{as } \epsilon \to 0. \end{aligned}$$

To prove the converse, we note that the intermediate value theorem (see, for example, [44, pages 70–71]) implies that there is some vector w on the line segment between u and v such that

$$\begin{aligned} H(v_{i-k}^n, \ldots, v_{i+k}^n; \triangle x_i, \triangle t^{n+1/2}) &= H(u_{i-k}^n, \ldots, u_{i+k}^n; \triangle x_i, \triangle t^{n+1/2}) \\ &\quad + \sum_{\ell=-k}^{k} \frac{\partial H}{\partial w_{i+\ell}}(w_{i-k}, \ldots, w_{i+k}; \triangle x_i, \triangle t^{n+1/2})(v_{i+\ell}^n - u_{i+\ell}^n) \end{aligned}$$

Since the partial derivatives of H are all nonnegative, this equation easily shows that the scheme is monotone. □

Monotone schemes are useful because they converge to an entropy-satisfying solution of the conservation law, as the next lemma shows.

Theorem 5.2.6 *(Harten–Hyman–Lax) [67] Consider the numerical scheme*

$$\begin{aligned} u_i^{n+1} &= u_i^n - \frac{\triangle t^{n+1/2}}{\triangle x_i}[\tilde{f}(u_{i-k+1}^n, \ldots, u_{i+k}^n) - \tilde{f}(u_{i-k}^n, \ldots, u_{i+k-1}^n)] \\ &\equiv H(u_{i-k}^n, \ldots, u_{i+k}^n; \triangle x_i, \triangle t^{n+1/2}) \end{aligned} \tag{5.11}$$

approximating the scalar conservation law

$$\frac{\partial u}{\partial t} + \frac{\partial f(u)}{\partial x} = 0$$

in one dimension. Assume that both f and $\tilde{f}$ are continuous, and that the numerical flux $\tilde{f}$ is consistent (Definition 5.1.2). Also suppose that the scheme is monotone (Definition 5.2.4). Finally, suppose that the numerical solution is locally bounded, and that it converges (Definition 5.2.1) to a locally bounded function u. Then u is a weak solution of the conservation law, and satisfies the weak entropy condition

$$\begin{aligned}
&\textit{for all } \phi \in C_0^\infty(\mathbf{R} \times [0,\infty)) \textit{ for all } z \in \mathbf{R} \\
&-\int_0^\infty \int_{-\infty}^\infty \frac{\partial \phi}{\partial t}(x,t)|u(x,t)-z|\,dx\,dt \\
&-\int_0^\infty \int_{-\infty}^\infty \frac{\partial \phi}{\partial x}(x,t)[f(u(x,t))-f(z)]\,\mathrm{sign}\{u(x,t)-z\}\,dx\,dt \\
&-\int_{-\infty}^\infty \phi(x,0)|u(x,0)-z|\,dx \le 0.
\end{aligned}$$

Proof From Lemma 3.1.18, we note that for any $z \in \mathbf{R}$ $s(u) \equiv |u-z|$ is an entropy function for the scalar conservation law, with corresponding entropy flux $\psi(u) \equiv [f(u)-f(z)]\,\mathrm{sign}(u-z)$. For fixed mesh cell index i, timestep n, scalar $\theta \in (0,1]$ and entropy function offset z, define

$$\begin{aligned}
w(\theta) &\equiv u\theta + z(1-\theta) \\
w_\ell(\theta) &\equiv u_{i+\ell}^n\theta + z(1-\theta) \\
v_j^{n+1}(\theta) &\equiv H\left(w_{-k}(\theta),\ldots,w_k(\theta);\triangle x_i,\triangle t^{n+1/2}\right).
\end{aligned}$$

Because the numerical flux $\tilde{f}$ is consistent,

$$v_j^{n+1}(0) = z - \frac{\triangle t^{n+1/2}}{\triangle x_i}[\tilde{f}(z,\ldots,z)-\tilde{f}(z,\ldots,z)] = z.$$

Note that the definitions of the scheme (5.11) and of the interpolation function $v_j^{n+1}(\theta)$ imply that

$$v_j^{n+1}(1) = u_j^n - \frac{\triangle t^{n+1/2}}{\triangle x_i}[\tilde{f}(u_{i-k+1}^n,\ldots,u_{i+k}^n)-\tilde{f}(u_{i-k}^n,\ldots,u_{i+k-1}^n)] = u_j^{n+1}.$$

Define the numerical entropy flux by

$$\tilde{\psi}(u_{i-k+1}^n,\ldots,u_{i+\ell}^n) \equiv \sum_{\ell=-k+1}^{k} |u_{i+\ell}^n - z| \int_0^1 \frac{\partial \tilde{f}}{\partial w_\ell}(w_{i-k+1}(\theta),\ldots,w_{i+k}(\theta))\,d\theta. \qquad (5.12)$$

Since the numerical flux $\tilde{f}$ is consistent with f (see definition 5.1.2), we can easily show that $\tilde{\psi}$ is continuous and consistent with the true entropy flux ψ:

$$\begin{aligned}\tilde{\psi}(u, \ldots, u) &= \operatorname{sign}(u-z) \int_0^1 \sum_{\ell=-k+1}^{k} \frac{\partial \tilde{f}}{\partial w_\ell}(w(\theta), \ldots, w(\theta))\,(u-z)\, d\theta \\ &= \operatorname{sign}(u-z) \int_0^1 \frac{d\tilde{f}}{d\theta}(w(\theta), \ldots, w(\theta))\, d\theta \\ &= \operatorname{sign}(u-z)[\tilde{f}(u, \ldots, u) - \tilde{f}(z, \ldots, z)] = \psi(u).\end{aligned}$$

Since sign $(u_j^{n+1} - z) = \operatorname{sign}(v_j^{n+1}(\theta) - z)$ for all $0 < \theta \leq 1$, we have that

$$\begin{aligned}s(u_j^{n+1}) &\equiv (u_j^{n+1} - z) \operatorname{sign}(u_j^{n+1} - z) \\ &= \int_0^1 \operatorname{sign}(v_j^{n+1}(\theta) - z) \frac{dv_j^{n+1}}{d\theta}\, d\theta \\ &= \sum_{\ell=-k}^{k} \int_0^1 \operatorname{sign}(v_j^{n+1}(\theta) - z) \\ &\quad \times \frac{\partial H}{\partial w_\ell}(w_{-k}(\theta), \ldots, w_k(\theta); \triangle x_i, \triangle t^{n+1/2})(u_{i+\ell}^n - z) d\theta.\end{aligned}$$

Since sign $(w_\ell(\theta) - z) = \operatorname{sign}(u_{i+\ell}^n - z)$ for all $0 < \theta \leq 1$, the definition (5.12) of the numerical entropy flux $\tilde{\psi}$ implies that

$$\begin{aligned}&-s(u_i^n) + \frac{\triangle t^{n+1/2}}{\triangle x_i}\left[\tilde{\psi}(u_{i-k+1}^n, \ldots, u_{i+k}^n) - \tilde{\psi}(u_{i-k}^n, \ldots, u_{i+k-1}^n)\right] \\ &= -\int_0^1 \operatorname{sign}(w_0(\theta) - z) \frac{dw_0}{d\theta}\, d\theta \\ &\quad + \frac{\triangle t^{n+1/2}}{\triangle x_i} \sum_{\ell=-k+1}^{k} \int_0^1 \operatorname{sign}(w_\ell(\theta) - z) \frac{\partial \tilde{f}}{\partial w_\ell}(w_{-k+1}(\theta), \ldots, w_k(\theta))\, d\theta \\ &\quad - \frac{\triangle t^{n+1/2}}{\triangle x_i} \sum_{\ell=-k}^{k-1} \int_0^1 \operatorname{sign}(w_\ell(\theta) - z) \frac{\partial \tilde{f}}{\partial w_\ell}(w_{-k}(\theta), \ldots, w_{k-1}(\theta))\, d\theta \\ &= -\int_0^1 \sum_{\ell=-k}^{k} \operatorname{sign}(w_\ell(\theta) - z) \frac{\partial H}{\partial w_\ell}\left(w_{-k}(\theta), \ldots, w_k(\theta); \triangle x_i, \triangle t^{n+1/2}\right)(u_{i+\ell}^n - z)\, d\theta.\end{aligned}$$

It follows that

$$s(u_j^{n+1}) - s(u_i^n) + \frac{\triangle t^{n+1/2}}{\triangle x_i}\left[\tilde{\psi}(u_{i-k+1}^n, \ldots, u_{i+k}^n) - \tilde{\psi}(u_{i-k}^n, \ldots, u_{i+k-1}^n)\right]$$
$$= \sum_{\ell=-k}^{k} (u_{i+\ell}^n - z) \int_0^1 \frac{\partial H}{\partial w_\ell}\left(w_{-k}(\theta), \ldots, w_k(\theta); \triangle x_i, \triangle t^{n+1/2}\right)$$
$$\times \left[\,\mathrm{sign}\left(v_i^{n+1}(\theta) - z\right) - \mathrm{sign}\,(w_\ell(\theta) - z)\right] d\theta .$$

Note that for all $0 < \theta \leq 1$,

$$(u_{i+\ell}^n - z)[\mathrm{sign}\left(v_i^{n+1}(\theta) - z\right) - \mathrm{sign}\,(w_\ell(\theta) - z)]$$
$$= (u_{i+\ell}^n - z)\,\mathrm{sign}(u_i^{n+1} - z) - |u_{i+\ell}^n - z|$$
$$= \begin{cases} 0, & \mathrm{sign}(u_i^{n+1} - z) = \mathrm{sign}(u_{i+\ell}^n - z) \\ -2|u_{i+\ell}^n - z|, & \mathrm{sign}(u_i^{n+1} - z) = -\mathrm{sign}(u_{i+\ell}^n - z) \end{cases} \leq 0.$$

Thus the monotone scheme satisfies the numerical entropy inequality

$$s(u_j^{n+1}) - s(u_i^n) + \frac{\triangle t^{n+1/2}}{\triangle x_i}[\tilde{\psi}(u_{i-k+1}^n, \ldots, u_{i+k}^n) - \tilde{\psi}(u_{i-k}^n, \ldots, u_{i+k-1}^n)] \leq 0.$$

The Lax–Wendroff theorem 5.2.2 now implies that if the scheme converges, it converges to a solution of the weak conservation law, satisfying the weak form of the entropy condition. □

Example 5.2.7 We will show that Godunov's method is a monotone scheme for Burgers' equation. Godunov's method is a conservative difference scheme, taking the form

$$u_j^{n+1} = u_j^n - \frac{\Delta t}{\Delta x}[f_{j+1/2}^{n+1/2} - f_{j-1/2}^{n+1/2}],$$

where the numerical fluxes are given by

$$f_{j+1/2}^{n+1/2} = F(u_j^n, u_{j+1}^n) \equiv \begin{cases} f(\max\{u_j^n,\ \min\{u_{j+1}^n,\ 0\}\}), & u_j^n \leq u_{j+1}^n \\ f(\max\{|u_j^n|,\ |u_{j+1}^n|\}), & u_j^n > u_{j+1}^n \end{cases}$$
$$= \begin{cases} 0, & u_j^n < 0 < u_{j+1}^n \\ f(u_j^n), & u_j^n \geq u_{j+1}^n \geq -u_j^n \text{ or } 0 \leq u_j^n \leq u_{j+1}^n \\ f(u_{j+1}^n), & -|u_j^n| \geq u_{j+1}^n \text{ or } u_j^n \leq u_{j+1}^n \leq 0. \end{cases}$$

It is easy to see that this flux is consistent (see definition 5.1.2). It is also easy to see that

$$\frac{\partial F}{\partial u_j^n} = \begin{cases} u_j^n, & u_j^n \geq u_{j+1}^n \geq -u_j^n \text{ or } 0 \leq u_j^n \leq u_{j+1}^n \\ 0, & \text{otherwise,} \end{cases}$$

and

$$\frac{\partial F}{\partial u_{j+1}^n} = \begin{cases} u_{j+1}^n, & -|u_j^n| \geq u_{j+1}^n \text{ or } u_j^n \leq u_{j+1}^n \leq 0 \\ 0, & \text{otherwise.} \end{cases}$$

Next, we will show that the Godunov flux is monotone provided that $\triangle t \max |u_j^n| \leq \triangle x$ for all j. The method

$$H(u_{j-1}^n, u_j^n, u_{j+1}^n) = u_j^n - \frac{\triangle t}{\triangle x}[F(u_j^n, u_{j+1}^n) - F(u_{j-1}^n, u_j^n)]$$

depends on the three values u_{j-1}^n, u_j^n and u_{j+1}^n. Using the partial derivatives of the fluxes $f_{j\pm 1/2}^{n+1/2}$ above, we see that

$$\begin{aligned} \frac{\partial H}{\partial u_j^n} &= 1 - \frac{\triangle t}{\triangle x}\left[\begin{cases} u_j^n, & u_j^n \geq u_{j+1}^n \geq -u_j^n \text{ or } 0 \leq u_j^n \leq u_{j+1}^n \\ 0, & \text{otherwise} \end{cases}\right. \\ &\quad \left. - \begin{cases} u_j^n, & -|u_{j-1}^n| \geq u_j^n \text{ or } u_{j-1}^n \leq u_j^n \leq 0 \\ 0, & \text{otherwise} \end{cases}\right] \\ &= \begin{cases} 1 - \frac{\triangle t}{\triangle x}u_j^n, & (u_j^n \geq u_{j+1}^n \geq -u_j^n \text{ or } 0 \leq u_j^n \leq u_{j+1}^n) \text{ and } (u_{j-1}^n > u_j^n \text{ and } u_j^n > 0) \\ 1 + \frac{\triangle t}{\triangle x}u_j^n, & (-|u_{j-1}^n| \geq u_j^n \text{ or } u_{j-1}^n \leq u_j^n \leq 0) \text{ and} \\ & (u_j^n < u_{j+1}^n \text{ or } u_{j+1}^n < -u_j^n) \text{ and } (u_j^n < 0 \text{ or } u_j^n > u_{j+1}^n) \\ 0, & \text{otherwise} \end{cases} \\ &\geq 0, \end{aligned}$$

and

$$\frac{\partial H}{\partial u_{j-1}^n} = \frac{\triangle t}{\triangle x}\begin{cases} u_{j-1}^n, & u_{j-1}^n \geq u_j^n \geq -u_{j-1}^n \text{ or } 0 \leq u_{j-1}^n \leq u_j^n \\ 0, & \text{otherwise} \end{cases} \geq 0,$$

and

$$\frac{\partial H}{\partial u_{j+1}^n} = \frac{\triangle t}{\triangle x}\begin{cases} u_{j+1}^n, & -|u_j^n| \geq u_{j+1}^n \text{ or } u_j^n \leq u_{j+1}^n \leq 0 \\ 0, & \text{otherwise} \end{cases} \geq 0.$$

Lemma 5.2.5 now shows that the method is monotone. Since Godunov's method is consistent (see Definition 5.1.2) and monotone for Burgers' equation, it follows from Lemma 5.2.6 that Godunov's method converges to the entropy-satisfying solution. This is not surprising, since we previously showed that Godunov's method converges to the correct solution in Lemma 4.2.2.

Example 5.2.8 Given the results of Theorem 5.2.6 and Example 5.2.7 we expect that the Roe solver does not produce a monotone flux for Godunov's method applied to Burgers' equation. To verify this, we compute

$$\frac{\partial H}{\partial u_j^n} = 1 - \frac{\triangle t}{\triangle x}\left[\begin{cases} u_j^n, & u_{j+1}^n + u_j^n \geq 0 \\ 0, & \text{otherwise} \end{cases} - \begin{cases} u_j^n, & u_j^n + u_{j-1}^n < 0 \\ 0, & \text{otherwise} \end{cases}\right]$$

$$= \begin{cases} 1 - \frac{\Delta t}{\Delta x} u_j^n, & u_{j+1}^n + u_j^n \geq 0 \text{ and } u_j^n + u_{j-1}^n \geq 0 \\ 1 + \frac{\Delta t}{\Delta x} u_j^n, & u_{j+1}^n + u_j^n < 0 \text{ and } u_j^n + u_{j-1}^n < 0 \geq 0 \\ 1, & \text{otherwise.} \end{cases}$$

This partial derivative does not pose a problem for the method. However,

$$\frac{\partial H}{\partial u_{j-1}^n} = \frac{\triangle t}{\triangle x}\begin{cases} u_{j-1}^n, & u_j^n + u_{j-1}^n \geq 0 \\ 0, & \text{otherwise} \end{cases}$$

and

$$\frac{\partial H}{\partial u_{j+1}^n} = -\frac{\triangle t}{\triangle x}\begin{cases} u_{j+1}^n, & u_{j+1}^n + u_j^n < 0 \\ 0, & \text{otherwise.} \end{cases}$$

The former of these two partial derivatives fails to be nonnegative when $u_j^n \geq -u_{j-1}^n > 0$, and the latter fails to be nonnegative when $-u_j^n \geq u_{j+1}^n > 0$. The former failure corresponds to a transonic rarefaction between u_{j-1}^n and u_j^n, while the latter failure corresponds to a transonic rarefaction between u_j^n and u_{j+1}^n.

Although monotone schemes have the advantages of convergence to the correct solution of their respective conservation laws, there is the following disadvantage.

Lemma 5.2.9 *Consider the numerical scheme*

$$u_i^{n+1} = u_i^n - \frac{\Delta t}{\Delta x}\left[\tilde{f}(u_{i-k+1}^n, \ldots, u_{i+k}^n) - \tilde{f}(u_{i-k}^n, \ldots, u_{i+k-1}^n)\right]$$
$$\equiv H(u_{i-k}^n, \ldots, u_{i+k}^n; \triangle x, \triangle t^{n+1/2})$$

approximating the scalar conservation law

$$\frac{\partial u}{\partial t} + \frac{\partial f(u)}{\partial x} = 0$$

in one dimension on a uniform mesh in space and time. Assume that both f and $\tilde{f}$ are twice continuously differentiable, and that the numerical flux $\tilde{f}$ is consistent (Definition 5.2.1). Further, suppose that the scheme is monotone (Definition 5.2.4) and that $\triangle t/\triangle x \equiv \lambda$ is fixed as the mesh is refined. If H depends on more than one of its arguments, then the scheme is at most first-order accurate (Definition 5.1.2).

Proof We will determine the modified equation for a monotone scheme. We assume that $u_i^n \approx \tilde{u}(i\triangle x, n\triangle t)$. A Taylor expansion leads to

$$\tilde{u}(x, t + \triangle t) = \tilde{u} + \triangle t\frac{\partial \tilde{u}}{\partial t} + \frac{\triangle t^2}{2}\frac{\partial^2 \tilde{u}}{\partial t^2} + o(\triangle t^2).$$

We assume that $\tilde{u}$ satisfies the modified equation

$$\frac{\partial \tilde{u}}{\partial t} + \frac{\partial f(\tilde{u})}{\partial x} = e = O(\triangle t) + O(\triangle x).$$

Then

$$\begin{aligned}
\frac{\partial^2 \tilde{u}}{\partial t^2} &= \frac{\partial}{\partial t}\left[e - \frac{\partial f(\tilde{u})}{\partial x}\right] = \frac{\partial e}{\partial t} - \frac{\partial}{\partial x}\left[\frac{\partial f(\tilde{u})}{\partial t}\right] \\
&= \frac{\partial e}{\partial t} - \frac{\partial}{\partial x}\left[\frac{\partial f}{\partial u}\frac{\partial \tilde{u}}{\partial t}\right] = \frac{\partial e}{\partial t} + \frac{\partial}{\partial x}\left[\frac{\partial f}{\partial u}\left\{-e + \frac{\partial f(\tilde{u})}{\partial x}\right\}\right] \\
&= \frac{\partial}{\partial x}\left[\left(\frac{\partial f}{\partial u}\right)^2 \frac{\partial \tilde{u}}{\partial x} - \frac{\partial f}{\partial u} e\right] + \frac{\partial e}{\partial t}.
\end{aligned}$$

Thus

$$\tilde{u}(x, t + \triangle t) = \tilde{u} + \triangle t \frac{\partial \tilde{u}}{\partial t} + \frac{\triangle t^2}{2}\frac{\partial}{\partial x}\left[\left(\frac{\partial f}{\partial u}\right)^2 \frac{\partial \tilde{u}}{\partial x}\right] + o(\triangle t^2)$$

It is a bit more intricate to treat the Taylor expansions for the spatial differences. Note that

$$\begin{aligned}
&H\left(\tilde{u}(x_{i-k}, t), \ldots, \tilde{u}(x_{i+k}, t); \triangle x, \triangle t\right) \\
&\quad = H\left(\tilde{u}(x_i, t), \ldots, \tilde{u}(x_i, t); \triangle x, \triangle t\right) \\
&\qquad + \sum_{\ell=-k}^{k} \frac{\partial H}{\partial w_\ell}\left(\tilde{u}(x_i, t), \ldots, \tilde{u}(x_i, t); \triangle x, \triangle t\right)\left[\tilde{u}(x_{i+\ell}, t) - \tilde{u}(x_i, t)\right] \\
&\qquad + \frac{1}{2}\sum_{\ell=-k}^{k}\sum_{m=-k}^{k} \frac{\partial^2 H}{\partial w_\ell \partial w_m}\left(\tilde{u}(x_i, t), \ldots, \tilde{u}(x_i, t); \triangle x, \triangle t\right) \\
&\qquad\qquad \left[\tilde{u}(x_{i+\ell}, t) - \tilde{u}(x_i, t)\right]\left[\tilde{u}(x_{i+m}, t) - \tilde{u}(x_i, t)\right] + o(\triangle x^2).
\end{aligned}$$

Now

$$\begin{aligned}
&H\left(\tilde{u}(x_i, t), \ldots, \tilde{u}(x_i, t); \triangle x, \triangle t\right) \\
&\quad = \tilde{u}(x_i, t) - \frac{\triangle t}{\triangle x}\left[\tilde{f}\left(\tilde{u}(x_i, t), \ldots, \tilde{u}(x_i, t)\right) - \tilde{f}\left(\tilde{u}(x_i, t), \ldots, \tilde{u}(x_i, t)\right)\right] = \tilde{u}(x_i, t)
\end{aligned}$$

and

$$\begin{aligned}
&\sum_{\ell=-k}^{k} \frac{\partial H}{\partial w_\ell}\left(\tilde{u}(x_i, t), \ldots, \tilde{u}(x_i, t); \triangle x, \triangle t\right)\left[\tilde{u}(x_{i+\ell}, t) - \tilde{u}(x_i, t)\right] \\
&\quad = \sum_{\ell=-k}^{k} \frac{\partial H}{\partial w_\ell}\left(\tilde{u}(x_i, t), \ldots, \tilde{u}(x_i, t); \triangle x, \triangle t\right)\left[\ell \triangle x \frac{\partial \tilde{u}}{\partial x} + \frac{\ell^2 \triangle x^2}{2}\frac{\partial^2 \tilde{u}}{\partial x^2} + o(\triangle x^2)\right] \\
&\quad = \triangle x \frac{\partial \tilde{u}}{\partial x}\sum_{\ell=-k}^{k} \ell \frac{\partial H}{\partial w_\ell}\left(\tilde{u}(x_i, t), \ldots, \tilde{u}(x_i, t); \triangle x, \triangle t\right) \\
&\qquad + \frac{\triangle x^2}{2}\frac{\partial^2 \tilde{u}}{\partial x^2}\sum_{\ell=-k}^{k} \ell^2 \frac{\partial H}{\partial w_\ell}\left(\tilde{u}(x_i, t), \ldots, \tilde{u}(x_i, t); \triangle x, \triangle t\right) + o(\triangle x^2)
\end{aligned}$$

and

$$\frac{1}{2}\sum_{\ell=-k}^{k}\sum_{m=-k}^{k}\frac{\partial^2 H}{\partial w_\ell \partial w_m}(\tilde{u}(x_i,t),\ldots,\tilde{u}(x_i,t);\triangle x,\triangle t)$$
$$\times[\tilde{u}(x_{i+\ell},t)-\tilde{u}(x_i,t)][\tilde{u}(x_{i+m},t)-\tilde{u}(x_i,t)]$$
$$=\frac{1}{2}\sum_{\ell=-k}^{k}\sum_{m=-k}^{k}\frac{\partial^2 H}{\partial w_\ell \partial w_m}(\tilde{u}(x_i,t),\ldots,\tilde{u}(x_i,t);\triangle x,\triangle t)$$
$$\times\left[\ell\triangle x\frac{\partial\tilde{u}}{\partial x}+o(\triangle x)\right]\left[m\triangle x\frac{\partial\tilde{u}}{\partial x}+o(\triangle x)\right]$$
$$=\frac{\triangle x^2}{2}\left(\frac{\partial\tilde{u}}{\partial x}\right)^2\sum_{\ell=-k}^{k}\sum_{m=-k}^{k}\ell m\frac{\partial^2 H}{\partial w_\ell \partial w_m}(\tilde{u}(x_i,t),\ldots,\tilde{u}(x_i,t);\triangle x,\triangle t)+o(\triangle x^2).$$

Thus

$$H\,(\tilde{u}(x_{i-k},t),\ldots,\tilde{u}(x_{i+k},t);\triangle x,\triangle t)$$
$$=\tilde{u}(x_i,t)+\triangle x\frac{\partial\tilde{u}}{\partial x}\sum_{\ell=-k}^{k}\ell\frac{\partial H}{\partial w_\ell}(\tilde{u}(x_i,t),\ldots,\tilde{u}(x_i,t);\triangle x,\triangle t)$$
$$+\frac{\triangle x^2}{2}\left[\frac{\partial^2\tilde{u}}{\partial x^2}\sum_{\ell=-k}^{k}\ell^2\frac{\partial H}{\partial w_\ell}(\tilde{u}(x_i,t),\ldots,\tilde{u}(x_i,t);\triangle x,\triangle t)\right.$$
$$\left.+\left(\frac{\partial\tilde{u}}{\partial x}\right)^2\sum_{\ell=-k}^{k}\sum_{m=-k}^{k}\ell m\frac{\partial^2 H}{\partial w_\ell \partial w_m}(\tilde{u}(x_i,t),\ldots,\tilde{u}(x_i,t);\triangle x,\triangle t)\right]+o(\triangle x^2)$$
$$=\tilde{u}(x_i,t)+\triangle x\frac{\partial\tilde{u}}{\partial x}\sum_{\ell=-k}^{k}\ell\frac{\partial H}{\partial w_\ell}(\tilde{u}(x_i,t),\ldots,\tilde{u}(x_i,t);\triangle x,\triangle t)$$
$$+\frac{\triangle x^2}{2}\left[\frac{\partial}{\partial x}\left(\frac{\partial\tilde{u}}{\partial x}\sum_{\ell=-k}^{k}\ell^2\frac{\partial H}{\partial w_\ell}(\tilde{u}(x_i,t),\ldots,\tilde{u}(x_i,t);\triangle x,\triangle t)\right)\right.$$
$$\left.+\left(\frac{\partial\tilde{u}}{\partial x}\right)^2\sum_{\ell=-k}^{k}\sum_{m=-k}^{k}(\ell m-\ell^2)\frac{\partial^2 H}{\partial w_\ell \partial w_m}(\tilde{u}(x_i,t),\ldots,\tilde{u}(x_i,t);\triangle x,\triangle t)\right]+o(\triangle x^2).$$

Next, we note that

$$\sum_{\ell=-k}^{k}\ell\frac{\partial H}{\partial w_\ell}(\tilde{u}(x_i,t),\ldots,\tilde{u}(x_i,t);\triangle x,\triangle t)=-\frac{\triangle t}{\triangle x}\sum_{\ell=-k}^{k}(\ell+1)\frac{\partial\tilde{f}}{\partial w_\ell}+\frac{\triangle t}{\triangle x}\sum_{\ell=-k}^{k}\ell\frac{\partial\tilde{f}}{\partial w_\ell}$$
$$=-\frac{\triangle t}{\triangle x}\sum_{\ell=-k}^{k}\frac{\partial\tilde{f}}{\partial w_\ell}=-\frac{\triangle t}{\triangle x}\frac{\partial f}{\partial u}.$$

We can also compute

$$\sum_{\ell=-k}^{k}\sum_{m=-k}^{k}(\ell m-\ell^2)\frac{\partial^2 H}{\partial w_\ell \partial w_m}(\tilde{u}(x_i,t),\ldots,\tilde{u}(x_i,t);\triangle x,\triangle t)$$
$$=\frac{1}{2}\sum_{\ell=-k}^{k}\sum_{m=-k}^{k}(\ell m-\ell^2)\frac{\partial^2 H}{\partial w_\ell \partial w_m}(\tilde{u}(x_i,t),\ldots,\tilde{u}(x_i,t);\triangle x,\triangle t)$$
$$+\frac{1}{2}\sum_{\ell=-k}^{k}\sum_{m=-k}^{k}(\ell m-\ell^2)\frac{\partial^2 H}{\partial w_m \partial w_\ell}(\tilde{u}(x_i,t),\ldots,\tilde{u}(x_i,t);\triangle x,\triangle t)$$
$$=\sum_{\ell=-k}^{k}\sum_{m=-k}^{k}\frac{\ell m-\ell^2+\ell m-m^2}{2}\frac{\partial^2 H}{\partial w_\ell \partial w_m}(\tilde{u}(x_i,t),\ldots,\tilde{u}(x_i,t);\triangle x,\triangle t)$$
$$=\frac{1}{2}\sum_{\ell=-k}^{k}\sum_{m=-k}^{k}(\ell-m)^2\frac{\partial^2 H}{\partial w_\ell \partial w_m}(\tilde{u}(x_i,t),\ldots,\tilde{u}(x_i,t);\triangle x,\triangle t)$$
$$=-\frac{\triangle t}{2\triangle x}\sum_{\ell=-k}^{k}\sum_{m=-k}^{k}(\ell+1-m-1)^2\frac{\partial^2 \tilde{f}}{\partial w_\ell \partial w_m}(\tilde{u}(x_i,t),\ldots,\tilde{u}(x_i,t);\triangle x,\triangle t)$$
$$+\frac{\triangle t}{2\triangle x}\sum_{\ell=-k}^{k}\sum_{m=-k}^{k}(\ell-m)^2\frac{\partial^2 \tilde{f}}{\partial w_\ell \partial w_m}(\tilde{u}(x_i,t),\ldots,\tilde{u}(x_i,t);\triangle x,\triangle t)=0.$$

It follows that

$$H(\tilde{u}(x_{i-k},t),\ldots,\tilde{u}(x_{i+k},t);\triangle x,\triangle t)$$
$$=\tilde{u}(x_i,t)-\triangle t\frac{\partial f(\tilde{u})}{\partial x}$$
$$+\frac{\triangle x^2}{2}\frac{\partial}{\partial x}\left(\frac{\partial \tilde{u}}{\partial x}\sum_{\ell=-k}^{k}\ell^2\frac{\partial H}{\partial w_\ell}(\tilde{u}(x_i,t),\ldots,\tilde{u}(x_i,t);\triangle x,\triangle t)\right)+o(\triangle x^2).$$

Thus the modified equation is

$$0=\frac{\tilde{u}(x_i,t+\triangle t)-H(\tilde{u}(x_{i-k},t),\ldots,\tilde{u}(x_{i+k},t);\triangle x,\triangle t)}{\triangle t}$$
$$=\frac{\partial \tilde{u}}{\partial t}+\frac{\partial f(\tilde{u})}{\partial x}$$
$$+\frac{\triangle t}{2}\frac{\partial}{\partial x}\left[\left(\frac{\partial f}{\partial u}\right)^2\frac{\partial \tilde{u}}{\partial x}\right]-\left(\frac{\triangle x}{\triangle t}\right)^2\frac{\partial}{\partial x}\left(\frac{\partial \tilde{u}}{\partial x}\sum_{\ell=-k}^{k}\ell^2\frac{\partial H}{\partial w_\ell}(\tilde{u}(x_i,t),\ldots,\tilde{u}(x_i,t);\triangle x,\triangle t)\right)$$
$$+o(\triangle t)+o(\triangle x^2/\triangle t)$$

which implies

$$\begin{aligned}
&\frac{\partial \tilde{u}}{\partial t} + \frac{\partial f(\tilde{u})}{\partial x} \\
&= \frac{\triangle x^2}{2\triangle t}\frac{\partial}{\partial x}\left[\left\{\sum_{\ell=-k}^{k} \ell^2 \frac{\partial H}{\partial w_\ell}(\tilde{u}(x_i,t),\ldots,\tilde{u}(x_i,t);\triangle x,\triangle t) - \left(\frac{\triangle t}{\triangle x}\frac{\partial f}{\partial u}\right)^2\right\}\frac{\partial \tilde{u}}{\partial x}\right] \\
&\quad + o(\triangle t) + o(\triangle x^2/\triangle t)
\end{aligned}$$

Note that since $\partial H/\partial w_\ell \geq 0$ for all ℓ, the Schwarz inequality implies that

$$\begin{aligned}
\left(\frac{\triangle t}{\triangle x}\frac{\partial f}{\partial u}\right)^2 &= \left(\sum_{\ell=-k}^{k} \ell \frac{\partial H}{\partial w_\ell}\right)^2 = \left[\sum_{\ell=-k}^{k}\left(\ell\sqrt{\frac{\partial H}{\partial w_\ell}}\right)\sqrt{\frac{\partial H}{\partial w_\ell}}\right]^2 \\
&\leq \left[\sum_{\ell=-k}^{k} \ell^2 \frac{\partial H}{\partial w_\ell}\right]\left[\sum_{\ell=-k}^{k}\frac{\partial H}{\partial w_\ell}\right] = \sum_{\ell=-k}^{k} \ell^2 \frac{\partial H}{\partial w_\ell}
\end{aligned}$$

since $\sum_{\ell=-k}^{k} \frac{\partial H}{\partial w_\ell} = 1$. Thus the right-hand side in the modified equation is diffusive and first-order in $\triangle t$ or $\triangle x$, assuming that these two mesh increments are held in strict proportion. The diffusion vanishes if equality holds in the Schwarz inequality; this occurs if and only if

$$\text{there exists } \alpha \text{ so that for all } \ell, \quad \sqrt{\frac{\partial H}{\partial w_\ell}}\alpha = \ell\sqrt{\frac{\partial H}{\partial w_\ell}}$$

which implies that $\partial H/\partial w_\ell \neq 0$ for at most one value of ℓ. □

Monotone schemes are also $\mathcal{L}^1$ contractive, as the next lemma shows.

Lemma 5.2.10 *[38, 67, 76]. Suppose that*

$$\begin{aligned}
u_i^{n+1} &= u_i^n - \frac{\triangle t}{\triangle x}[\tilde{f}(u_{i-k+1}^n,\ldots,u_{i+k}^n) - \tilde{f}(u_{i-k}^n,\ldots,u_{i+k-1}^n)] \\
&\equiv H(u_{i-k}^n,\ldots,u_{i+k}^n;\triangle x_i,\triangle t^{n+1/2})
\end{aligned}$$

is a monotone scheme (see Definition 5.2.4). If u_i^n and v_i^n are generated by this scheme using initial data u_i^0 and v_i^0, respectively, then

$$\text{for all } n \geq 0 \text{ for all } m > n,\ \|u^m - v^m\|_1 \leq \|u^n - v^n\|_1.$$

Proof It suffices to prove the result for $m = n + 1$. We define the functions

$$w_j(\theta) = u_j^n \theta + v_j^n (1 - \theta).$$

Then

$$\begin{aligned}
&H(u_{i-k}^n, \ldots, u_{i+k}^n; \triangle x_i, \triangle t^{n+1/2}) - H(v_{i-k}^n, \ldots, v_{i+k}^n; \triangle x_i, \triangle t^{n+1/2}) \\
&\quad = \int_0^1 \frac{d}{d\theta} H(w_{i-k}(\theta), \ldots, w_{i+k}(\theta), \triangle x_i, \triangle t^{n+1/2})\, d\theta \\
&\quad = \int_0^1 \sum_{\ell=-k}^{k} \frac{\partial H}{\partial w_\ell}(w_{i-k}(\theta), \ldots, w_{i+k}(\theta); \triangle x_i, \triangle t^{n+1/2})(u_{i+\ell}^n - v_{i+\ell}^n)\, d\theta.
\end{aligned}$$

Since the partial derivatives of H are nonnegative,

$$\begin{aligned}
&\|u^{n+1} - v^{n+1}\|_1 \\
&\quad = \sum_i \operatorname{sign}(u_i^{n+1} - v_i^{n+1})[H(u_{i-k}^n, \ldots, u_{i+k}^n; \triangle x_i, \triangle t^{n+1/2}) \\
&\quad\quad - H(v_{i-k}^n, \ldots, v_{i+k}^n; \triangle x_i, \triangle t^{n+1/2})]\triangle x_i \\
&\quad \leq \sum_i \sum_{\ell=-k}^{k} \int_0^1 \frac{\partial H}{\partial w_\ell}(w_{i-k}(\theta), \ldots, w_{i+k}(\theta); \triangle x_i, \triangle t^{n+1/2})|u_{i+\ell}^n - v_{i+\ell}^n|\, d\theta \triangle x_i \\
&\quad = \sum_j |u_j^n - v_j^n| \int_0^1 \sum_{\ell=-k}^{k} \frac{\partial H}{\partial w_\ell}(w_{j-\ell-k}(\theta), \ldots, w_{j-\ell+k}(\theta); \triangle x_i, \triangle t^{n+1/2}) d\theta \triangle x_{j-\ell}.
\end{aligned}$$

We compute

$$\begin{aligned}
&\sum_{\ell=-k}^{k} \frac{\partial H}{\partial w_\ell}(w_{j-\ell-k}(\theta), \ldots, w_{j-\ell+k}(\theta); \triangle x_i, \triangle t^{n+1/2})\triangle x_{j-\ell} \\
&\quad = \sum_{\ell=-k}^{k} \delta_{\ell,0} \triangle x_{j-\ell} - \triangle t^{n+1/2} \sum_{\ell=-k+1}^{k} \frac{\partial \tilde{f}}{\partial w_{\ell-1}}(w_{j-\ell-k+1}, \ldots, w_{j-\ell+k}) \\
&\quad\quad + \triangle t^{n+1/2} \sum_{\ell=-k}^{k-1} \frac{\partial \tilde{f}}{\partial w_\ell}(w_{j-\ell-k}, \ldots, w_{j-\ell+k-1}) = \triangle x_j.
\end{aligned}$$

Thus

$$\|u^{n+1} - v^{n+1}\|_1 \leq \sum_j |u_j^n - v_j^n| \int_0^1 \triangle x_j d\theta = \|u^n - v^n\|_1$$

□

Exercises for 5.2

5.2.1 Determine circumstances under which Rusanov's method is monotone for Burgers' equation.

5.2.2 Determine circumstances under which Rusanov's method is monotone for the traffic flow problem with logarithmic flux.

5.2.3 Determine circumstances under which the Lax–Friedrichs scheme is monotone for Burgers' equation.

5.2.4 When is Rusanov's method monotone for the Buckley–Leverett problem?

5.2.5 When is Godunov's method monotone for the Buckley–Leverett problem?

5.3 Nonlinear Stability

The discussion in this section follows that in LeVeque [96]. So far, we have a few results to guide our selection of numerical methods for scalar hyperbolic conservation laws. First, the Lax equivalence theorem 2.7.14 shows that for a consistent linear scheme, stability is equivalent to convergence. However, this result says nothing about nonlinear schemes (such as Godunov's method), and does not guarantee convergence to the entropy-satisfying solution. Next, the Lax–Wendroff theorem 5.2.2 shows that if the solution to a conservative difference scheme converges in $\mathcal{L}_1$, then it converges to a weak solution of the conservation law. However, this theorem does not guarantee that the scheme converges, and does not guarantee that the limit is an entropy-satisfying solution. Finally, the Harten–Hyman–Lax theorem 5.2.6 shows that an explicit scheme that is monotone, conservative and consistent has an entropy-satisfying limit, but unfortunately is at most first-order accurate.

In general, we will want to understand the circumstances under which we can guarantee convergence, in particular to the entropy-satisfying solution. We will also want to understand conditions under which we can obtain better than first-order accuracy. In order to understand convergence, we will need to develop the correct notion of compactness. We will see that one way to guarantee convergence to the entropy-satisfying solution of the differential equation will be to build more of the solution of the differential equation into the method, typically through approximate Riemann problem solvers. Finally, in order to obtain better than first-order convergence, we will have to forego monotonicity.

5.3.1 Total Variation

Definition 5.3.1 A set K in a normed linear space is **bounded** if and only if there is some radius R such for all $v \in K$, $\|v\| \leq R$. A set K is **closed** if and only if $\{v_k\}_{k=1}^{\infty} \subset K$ and $v_k \rightarrow v$ implies that $v \in K$. A set K in a normed linear space is **compact** if and only if for any sequence $\{v_k\}_{k=1}^{\infty} \subset K$ there is a subsequence $\{v_{k_j}\} \subset \{v_k\}$ and a limit $v \in K$ so that $\lim_{j \rightarrow \infty} \|v_{k_j} - v\| = 0$. The **support** of a function $w(x)$ is the set

$$\mathrm{supp}(w) = \{x : w(x) \neq 0\}.$$

It is well-known (see, for example, [83, p. 77]) that sets in $\mathbf{R}^n$ are compact if and only if they are closed and bounded.

Definition 5.3.2 The **total variation** of a function $w(x)$ is

$$TV(w) \equiv \limsup_{\epsilon \to 0} \frac{1}{\epsilon} \int_{-\infty}^{\infty} |w(x+\epsilon) - w(x)| \, dx,$$

and the total variation of a function $v(x, t)$ is

$$\begin{aligned} TV_T(v) \equiv & \limsup_{\epsilon \to 0} \frac{1}{\epsilon} \int_0^T \int_{-\infty}^{\infty} |v(x+\epsilon, t) - v(x, t)| \, dx \, dt \\ & + \limsup_{\epsilon \to 0} \frac{1}{\epsilon} \int_0^T \int_{-\infty}^{\infty} |v(x, t+\epsilon) - v(x, t)| \, dx \, dt. \end{aligned}$$

Note that

$$\text{for all } w \in \mathcal{C}^1(-\infty, \infty), \; TV(w) = \int_{-\infty}^{\infty} |w'(x)| \, dx \quad \text{and}$$

$$\text{for all } v \in \mathcal{C}^1\left((-\infty, \infty) \times (0, T)\right), \; TV_T(v) = \int_0^T \int_{-\infty}^{\infty} \left|\frac{\partial v}{\partial x}\right| + \left|\frac{\partial v}{\partial t}\right| dx \, dt.$$

We can extend mesh-values w_j to piecewise-constant functions on the mesh, and find that the total variation in space is

$$TV(w) \equiv \sum_{j=-\infty}^{\infty} |w_{j+1} - w_j|,$$

or the total variation in space and time is

$$TV(v) \equiv \sum_{n=0}^{N-1} \sum_{j=-\infty}^{\infty} \left[|v_{j+1}^n - v_j^n| \triangle t^{n+1/2} + |v_j^{n+1} - v_j^n| \triangle x_j \right],$$

where $\sum_{n=0}^{N-1} \triangle t^{n+1/2} = T$. Note that if $w(x, t)$ has bounded total variation for $x \in (-\infty, \infty)$, then $w(x, t)$ must approach constant values as $x \to \pm\infty$.

5.3.2 Total Variation Stability

The next definition help us extend the ideas of section 5.1.2 to nonlinear schemes.

Definition 5.3.3 Given initial data $u_0(x)$ with compact support and flux function $f(u)$, a numerical approximation u_j^n for the solution of the scalar law $\partial u/\partial t + \partial f(u)/\partial x = 0$ with initial data $u(x, 0) = u_0(x)$ is **TV stable** on the closed time interval $[0, T]$ if and only if there exist $R > 0$, $M > 0$, $\triangle t > 0$ and $N > 0$ so that for all j we have $\infty < x_j < x_{j+1} < \infty$ and $\triangle x_{j+1/2} \equiv x_{j+1} - x_j$ and so that for all $0 \leq n < N$ we have $0 \leq t^n < t^{n+1} \leq T$ and $\triangle t^{n+1/2} \equiv t^{n+1} - t^n$ so that $TV(u) \leq R$ and $|x_j| > M \Longrightarrow u_j^n = 0$.

Lemma 5.3.4 *[96, page 163] Suppose that*

(i) *u_j^n is a numerical solution generated by a conservative difference scheme $u_j^{n+1} = u_j^n - \frac{\Delta t^{n+1/2}}{\Delta x_j}[f_{j+1/2}^{n+1/2} - f_{j-1/2}^{n+1/2}]$;*
(ii) *the numerical flux function is given by $f_{j+1/2}^{n+1/2} = F(u_{j-k+1}^n, \ldots, u_{j+k}^n)$, where F is Lipschitz continuous in each of its arguments with Lipschitz constant K;*
(iii) *the initial data has compact support: there exist $M > 0$ and $J > 0$ such that for all $|j| > J$ we have $|x_j| > M_0$ and $u_j^0 = 0$;*
(iv) *the cell widths are bounded above proportional to the timesteps: there exists $C > 0$ such that for all $0 \leq n < N$ we have $\triangle x_j \leq \triangle t^{n+1/2}$;*
(v) *given $t^N = T > 0$ and initial data $u_0(x)$ there exist constants $\triangle t_0 > 0$ and $R > 0$ so that for all $n < N$ we have $\triangle t^{n+1/2} < \triangle t_0$ and $TV(u^n) \leq R$.*

Then

(i) *for all $n < N - 1$, $\|u^{n+1} - u^n\|_1 \leq 2kKR \triangle t^{n+1/2}$;*
(ii) *the method is TV-stable (Definition 5.3.3) on $[0, T]$;*
(iii) *if F is consistent with with a continuous function f (Definition 5.1.2), then u_j^n converges to a weak solution of $\partial u/\partial t + \partial f(u)/\partial x = 0$.*

Proof To prove the first result, we note that the conservative difference, Lipschitz continuity of the numerical flux function and bound on the total variation of u^n imply

$$\begin{aligned}\|u^{n+1} - u^n\|_1 &= \sum_j \left|u_j^{n+1} - u_j^n\right| \triangle x_j = \sum_j \left|f_{j+1/2}^{n+1/2} - f_{j-1/2}^{n+1/2}\right| \triangle t^{n+1/2} \\ &\leq \sum_j K \max_{-k \leq \ell < k} |u_{j+\ell+1}^n - u_{j+\ell}^n| \triangle t^{n+1/2} \\ &\leq K \triangle t^{n+1/2} \sum_{\ell=-k}^{k} \sum_j K |u_{j+1}^n - u_j^n| \\ &= 2kK\ TV(u^n) \leq 2kKR \triangle t^{n+1/2}.\end{aligned}$$

Next, let us prove that the method is TV-stable. Since the assumptions imply that the effect of the nonzero initial data can spread at most k cells per timestep, it is easy to see that $|j| > J + kn$ implies that $u_j^n = 0$. The assumption that the cell widths are bounded above implies that

$$x_{J+kn} = x_J + \sum_{m=0}^{n-1} \sum_{\ell=J+km+1}^{J+k(m+1)} (\triangle x_{\ell-1/2} + \triangle x_{\ell+1/2})\frac{1}{2} \leq x_J + \sum_{m=0}^{n-1} kC\triangle t^{n+1/2}$$

$$\leq x_J + kC \sum_{m=0}^{N-1} \triangle t^{n+1/2} = x_J + kCT$$

$$x_{-J-kn} = x_{-J} - \sum_{m=0}^{n-1} \sum_{\ell=-J-k(m+1)}^{-J-km-1} (\triangle x_{\ell-1/2} + \triangle x_{\ell+1/2})\frac{1}{2}$$

$$\geq x_{-J} - \sum_{m=0}^{n-1} kC\triangle t^{n+1/2} \geq x_{-J} - kC \sum_{m=0}^{N-1} \triangle t^{n+1/2} = x_{-J} - kCT.$$

Also, the total variation of u_j^n is bounded above because

$$TV_T(u) = \sum_{n=0}^{N-1} \sum_j \left[|u_{j+1}^n - u_j^n|\triangle t^{n+1/2} + |u_j^{n+1} - u_j^n|\triangle x_j \right]$$

$$\leq \sum_{n=0}^{N-1} [R\triangle t^{n+1/2} + 2kKR\triangle t^{n+1/2}] = (2kK+1)RT.$$

Thus the numerical solution is TV-stable.

The proof of the third result will be by contradiction. Suppose that for any weak solution w of the conservation law, there exists $\epsilon > 0$ such that for all discretizations $\triangle x_j$ and $\triangle t^{n+1/2}$ satisfying the hypotheses of the lemma we have $\|u_j^n - w\| > \epsilon$. Since the set of all $\mathcal{L}_1$ functions with bounded total variation and compact support is compact [50], there is a convergent subsequence of numerical approximations satisfying the hypotheses of the lemma. Let v be the limit of this convergent subsequence. For all sufficiently fine mesh in this subsequence, $\|u - v\|_1 < \epsilon$. Since the method is consistent and conservative, and the flux functions are continuous and consistent, the Lax–Wendroff Theorem 5.2.2 implies that u_j^n converges to a weak solution of the conservation law; this is a contradiction. □

The useful feature of this lemma is that it presents a list of circumstances under which nonlinear stability implies convergence. The difficulty with using this lemma is the need to verify that the third assumption is valid for all timesteps. One way to satisfy the third assumption is to require that the total variation in the scheme not increase from one timestep to the next; this will be the approach in Section 5.8. Another way to satisfy the third assumption of this lemma is to design the scheme so that the total variation has bounded growth; this approach is used by Osher and Chakravarthy in [126] and by the ENO schemes described in section 5.13.

5.3.3 Other Stability Notions

Note that weak solutions to scalar conservation laws are $\mathcal{L}_1$**-contracting**.

Theorem 5.3.5 *($\mathcal{L}_1$ Contraction) [85, 133] Suppose that*

(i) *u_1 and u_2 are piecewise-continuously differentiable solutions of the scalar conservation law $\partial u/\partial t + \partial f(u)/\partial x = 0$, and*
(ii) *the initial data $u_1(x,0)$ and $u_2(x,0)$ are piecewise-continuously differentiable and $\mathcal{L}_1$-integrable in x.*

Then two solutions u_1 and u_2 are $\mathcal{L}_1$-contracting, meaning that

$$\textit{for all } t > 0 \ \|u_2(\cdot,t) - u_1(\cdot,t)\|_1 \le \|u_2(\cdot,0) - u_1(\cdot,0)\|_1,$$

if and only if u_1 and u_2 satisfy the Oleinik chord condition (3.17) *at all shocks.*

It is tempting to require numerical solutions to scalar conservation laws to have similar stability properties.

Definition 5.3.6 Two mesh functions u_i^n and v_i^n are $\mathcal{L}_1$**-contracting** if and only if

$$\text{for all } m \ge n \ge 0 \ \|u^m - v^m\|_1 \le \|u^n - v^n\|_1.$$

These mesh functions are **total variation diminishing** if and only if

$$\text{for all } m \ge n \ge 0 \ TV(u^m) \le TV(u^n).$$

Finally, these mesh functions are **monotonicity-preserving** if and only if

$$\begin{aligned}
&\text{for all } i \ u_i^n \le u_{i+1}^n \Longrightarrow \text{ for all } m > n \ \text{ for all } i \ u_i^m \le u_{i+1}^m \text{ and} \\
&\text{for all } i \ u_i^n \ge u_{i+1}^n \Longrightarrow \text{ for all } m > n \ \text{ for all } i \ u_i^m \ge u_{i+1}^m.
\end{aligned}$$

For numerical schemes, some of these nonlinear stability notions are stronger than others.

Lemma 5.3.7 *[96, page 167]. Suppose that the numerical scheme*

$$u_i^{n+1} = H(u_{i-k}^n, \ldots, u_{i+k}^n; \triangle x, \triangle t^{n+1/2})$$

on a uniform mesh is such that H does not depend explicitly on x or t. If the numerical solution is $\mathcal{L}_1$**-contracting***, then the scheme is* **total variation diminishing**.

Proof It suffices to prove the conclusion for $m = n+1$. Given any $u_i^n = H(u_{i-k}^n, \ldots, u_{i+k}^n; \triangle x, \triangle t^{n+1/2})$ generated by the scheme, note that $v_i^n = u_{i-1}^n = H(u_{i-k-1}^n, \ldots, u_{i+k-1}^n; \triangle x, \triangle t^{n+1/2})$ is also generated by the same scheme. Since the scheme is $\mathcal{L}_1$-contracting,

$$\begin{aligned}
TV(u^{n+1}) &= \sum_{i=-\infty}^{\infty} |u_i^{n+1} - u_{i-1}^{n+1}| = \sum_{i=-\infty}^{\infty} |u_i^{n+1} - v_i^{n+1}| \\
&= \frac{1}{\Delta x}\|u^{n+1} - v^{n+1}\|_1 \leq \frac{1}{\Delta x}\|u^n - v^n\|_1 \\
&= \sum_{i=-\infty}^{\infty} |u_i^n - v_i^n| = \sum_{i=-\infty}^{\infty} |u_i^n - u_{i-1}^n| = TV(u^n).
\end{aligned}$$

□

Lemma 5.3.8 *[96, page 166]. Consider the numerical scheme*

$$u_i^{n+1} = u_i^n - \frac{\Delta t^{n+1/2}}{\Delta x_i}\left[\tilde{f}(u_{i-k+1}^n, \ldots, u_{i+k}^n) - \tilde{f}(u_{i-k}^n, \ldots, u_{i+k-1}^n)\right].$$

If the numerical solution is **total variation diminishing***, and if the numerical solution is constant for large spatial indices, meaning that*

$$\begin{aligned}
&\textit{there exists } u_-, u_+ \textit{ so that for all } n > 0 \textit{ there exists } M > 0 \\
&\textit{so that for all } i \leq -M,\ u_i^n = u_- \\
&\textit{and for all } i \geq M,\ u_i^n = u_+
\end{aligned}$$

then the scheme is **monotonicity-preserving**.

Proof It suffices to prove the conclusion for $m = n + 1$ and non-decreasing u_i^n. Note that for $i \geq M + k$,

$$\begin{aligned}
u_i^{n+1} &= u_i^n - \frac{\Delta t^{n+1/2}}{\Delta x_i}\left[\tilde{f}(u_{i-k+1}^n, \ldots, u_{i+k}^n) - \tilde{f}(u_{i-k}^n, \ldots, u_{i+k-1}^n)\right] \\
&= u_+ - \frac{\Delta t^{n+1/2}}{\Delta x_i}[\tilde{f}(u_+, \ldots, u_+) - \tilde{f}(u_+, \ldots, u_+)] = u_+.
\end{aligned}$$

Similarly, for $i \leq -M - k$, $u_i^{n+1} = u_-$. Next, we note that since u_i^n is non-decreasing and $TV(u^{n+1}) \leq TV(u^n)$,

$$\begin{aligned}
u_+ - u_- &= \sum_{i=-\infty}^{\infty} \left[u_{i+1}^{n+1} - u_i^{n+1}\right] \leq \sum_{i=-\infty}^{\infty} \left|u_{i+1}^{n+1} - u_i^{n+1}\right| = TV(u^{n+1}) \leq TV(u^n) \\
&= \sum_{i=-\infty}^{\infty} |u_{i+1}^n - u_i^n| = \sum_{i=-\infty}^{\infty} [u_{i+1}^n - u_i^n] = u_+ - u_-.
\end{aligned}$$

It follows that

$$\sum_{i=-\infty}^{\infty} \left[u_{i+1}^{n+1} - u_i^{n+1}\right] = \sum_{i=-\infty}^{\infty} \left|u_{i+1}^{n+1} - u_i^{n+1}\right|,$$

which implies that $u_{i+1}^{n+1} - u_i^{n+1} \geq 0$ for all i. □

Now that we have examined the inter-relationships of some of these nonlinear stability concepts, let us review their implications. First, a theorem due to Godunov [57, 91] shows that any linear monotonicity-preserving scheme is at best first-order accurate. Lemma 5.3.8 shows that total variation diminishing schemes are monotonicity-preserving. Total variation diminishing schemes are convergent [96, page 166], however, convergence to the entropy-satisfying solution is not guaranteed. Lemma 5.3.7 shows that an $\mathcal{L}_1$-contracting scheme is total variation diminishing. Lemma 5.2.10. shows that monotone schemes are $\mathcal{L}_1$-contracting, and the Harten–Hyman–Lax theorem 5.2.6 showed that monotone schemes converge to the entropy-satisfying solution, and are at best first-order accurate.

In summary, if we want to guarantee convergence to the entropy-satisfying solution then we can use a monotone scheme, provided that we are satisfied with first-order accuracy. If we want to guarantee convergence then we can use a total variation diminishing scheme, but such a scheme will be monotonicity-preserving and therefore first-order if it is linear. Thankfully, these are not our only options. Our results so far indicate that if we want to achieve better than first-order accuracy and simultaneously preserve monotonicity, then we cannot use a linear scheme.

5.4 Propagation of Numerical Discontinuities

Our goal in this section is to examine how numerical discontinuities propagate. Since almost all numerical schemes involve numerical diffusion in order to guarantee convergence to the desired physical solution, we will study the solution of a modified equation. The simplest of these is the convection–diffusion equation. For more discussion related to the ideas in this section, see [87, 113, 142].

We will consider the convection–diffusion equation with Riemann-problem initial data

$$\frac{\partial u_\epsilon}{\partial t} + \lambda \frac{\partial u_\epsilon}{\partial x} = \epsilon \frac{\partial^2 u_\epsilon}{\partial x^2}, \quad u_\epsilon(x,0) = \begin{cases} 1, & x < 0 \\ 0, & x > 0. \end{cases}$$

The analytical solution for $\epsilon = 0$ is

$$u_0(x,t) = \begin{cases} 1, & x < \lambda t \\ 0, & x > \lambda t, \end{cases}$$

and the analytical solution for $\epsilon > 0$ and $t > 0$ is

$$u_\epsilon(x,t) = \operatorname{erf}\left(\frac{x - \lambda t}{\sqrt{4\epsilon t}}\right).$$

The difference between the analytical solutions with and without diffusion is

$$\begin{aligned}\|u_0(\cdot,t)-u_\epsilon(\cdot,t)\|_1 &= \int_{-\infty}^{\lambda t}\left|1-\operatorname{erf}\left(\frac{x-\lambda t}{\sqrt{4\epsilon t}}\right)\right| dx+\int_{\lambda t}^{\infty}\left|\operatorname{erf}\left(\frac{x-\lambda t}{\sqrt{4\epsilon t}}\right)\right| dx\\ &= \int_{-\infty}^{0}\left|1-\operatorname{erf}\left(\frac{z}{\sqrt{4\epsilon t}}\right)\right| dz+\int_{0}^{\infty}\left|\operatorname{erf}\left(\frac{z}{\sqrt{4\epsilon t}}\right)\right| dz\\ &= 2\int_{-\infty}^{0}\operatorname{erf}\left(\frac{z}{\sqrt{4\epsilon t}}\right) dz = 2\sqrt{4\epsilon t}\int_{-\infty}^{0}\operatorname{erf}(z)\,dz=\sqrt{4\epsilon t}.\end{aligned}$$

Thus if a physical problem involves diffusion, we expect the initial discontinuity to spread a distance proportional to the square root of the product of the diffusion constant and time.

Numerical schemes typically involve numerical diffusion. For example, recall that in example 3.3.1 we saw that the modified equation analysis for the Lax-Friedrichs scheme produces

$$\epsilon \approx \frac{\triangle x^2}{4\triangle t}\left[1-\left(\frac{\lambda\triangle t}{\triangle x}\right)^2\right].$$

Decreasing the timestep or computing to large time will increase the spreading of the discontinuity.

No scheme is perfect. Suppose that we have a numerical scheme that spreads a discontinuity over a fixed number k of cells, and that the greatest contribution to the error is due to the resolution of the discontinuity. Then the error in the numerical solution is

$$\|u(x_j,t^n)-u_j^n\|_1 \approx \sum_{j\in\text{ discontinuity}} \triangle x_j|u(x_j,t^n)-u_j^n| \le kM\triangle x,$$

where M is the size of the jump in u and $\triangle x$ is an upper bound for the cell width. This suggests that the error should be no better than first-order (see definition 5.1.2) accurate at discontinuities.

Order p accuracy in the $\mathcal{L}_1$ norm at a discontinuity would require that the numerical width of the discontinuity be $O(\triangle x^p)$. This would in turn require that the position of the discontinuity within the cell be accurately determined. In particular, this would require a model for the variation of the solution within a grid cell, such as in [65]. Currently, no known scheme can do this for general problems.

Exercises for 5.4

5.4.1 Consider the linear advection problem with the initial data described in this section. Define the front width to be the distance (in x) between the points where $u = 0.95$ and $u = 0.05$. Run the explicit upwind, Lax-Friedrichs and Lax–Wendroff schemes for various values of CFL and $\triangle x$. Plot front width versus time and explain your results.

5.4.2 Consider the linear advection problem with initial data described in this section. For each of the explicit upwind, Lax-Friedrichs and Lax–Wendroff schemes, plot the logarithm of the $\mathcal{L}_1$ error versus the logarithm of the mesh width $\triangle x$. The results should be computed at CFL = 0.9 for the time at which the discontinuity has crossed 90% of the grid. Plot the results for $\triangle x = 2^{-5}, 2^{-6} \ldots 2^{-10}$. What rate of convergence do you observe from this plot?

5.5 Monotonic Schemes

In this section, we will follow a line of development due to van Leer [168, 169, 170, 171]. Since a linear monotonicity-preserving scheme is at best first-order accurate, a monotonicity-preserving higher-order scheme must be nonlinear. We will develop a *nonlinear* monotonicity-preserving scheme that is designed to obtain second-order accuracy as much as possible.

5.5.1 Smoothness Monitor

We begin with a definition.

Definition 5.5.1 If $\lambda > 0$, a scheme for linear advection $\partial u/\partial t + \lambda \partial u/\partial x = 0$ is **monotonic** if and only if u_j^{n+1} lies between u_{j-1}^n and u_j^n. An equivalent requirement is that $u_j^{n+1} - u_j^n$ lies between 0 and $u_{j-1}^n - u_j^n$. This can be rewritten in the form

$$u_j^n - u_{j-1}^n \neq 0 \Longrightarrow 0 \leq -\frac{u_j^{n+1} - u_j^n}{u_j^n - u_{j-1}^n} \leq 1.$$

Thus the **van Leer smoothness monitor** for the scheme is defined to be

$$r_j^n = \frac{u_j^n - u_{j-1}^n}{u_{j+1}^n - u_j^n}. \tag{5.13}$$

Note that a monotonic scheme is monotonicity-preserving.

Example 5.5.2 The explicit upwind difference $u_j^{n+1} - u_j^n = -\frac{\lambda \Delta t^{n+1/2}}{\Delta x_j}[u_j^n - j_{j-1}^n]$ can be rewritten

$$-\frac{u_j^{n+1} - u_j^n}{u_j^n - u_{j-1}^n} = \frac{\lambda \triangle t^{n+1/2}}{\triangle x_j}.$$

It follows that explicit upwind differencing for linear advection is monotonic if and only the Courant numbers satisfy

$$\frac{\lambda \triangle t^{n+1/2}}{\triangle x_j} \leq 1$$

for all timesteps.

There are several reasons why the smoothness monitor is useful. Note that $r_j^n < 0$ implies a local extremum in the numerical solution, possibly due to a numerical oscillation. On the other hand, $r_j^n > 0$ implies monotonic behavior in the numerical solution, either monotonically increasing or decreasing. Furthermore, $r_j^n \approx 1$ implies smooth behavior in the numerical solution, and $r_j^n \approx 0$ or ∞ indicates a numerical discontinuity.

5.5.2 Monotonizations

Consider the Lax-Wendroff scheme for linear advection:

$$0 = \frac{u_j^{n+1} - u_j^n}{\triangle t} + \lambda \frac{u_{j+1}^n - u_{j-1}^n}{2\triangle x} - \frac{\lambda \triangle t}{\triangle x} \frac{\lambda}{2} \left[\frac{u_{j+1}^n - u_j^n}{\triangle x} - \frac{u_j^n - u_{j-1}^n}{\triangle x} \right].$$

If we multiply this equation by $\triangle t/(u_j^n - u_{j-1}^n)$, we obtain

$$-\frac{u_j^{n+1} - u_j^n}{u_j^n - u_{j-1}^n} = \gamma \left\{ 1 + \frac{1}{2}(1-\gamma)\left(\frac{1}{r_j^n} - 1\right) \right\}$$

where $\gamma = \lambda \Delta T / \Delta X$. Recall from section 3.5 that this scheme has second-order local truncation error. Since this scheme is linear and second-order, Godunov's theorem [96, p. 174] shows that it cannot be monotonic. However, since this scheme is a function of the smoothness monitor r_j^n and the Courant number γ, we can consider monotonizing the Lax-Wendroff scheme as follows:

$$-\frac{u_j^{n+1} - u_j^n}{u_j^n - u_{j-1}^n} = \gamma \left\{ 1 + \frac{1}{2}(1-\gamma)\left(\frac{1}{r_j^n} - 1\right)\left(1 - Q\left(\frac{1}{r_j^n}\right)\right) \right\} \equiv \phi_{LW}(\gamma, r_j^n). \quad (5.14)$$

Thus, in order for this scheme to be monotonic we want

$$\text{for all } 0 \le \gamma \le 1 \ \text{ for all } r \in \mathbf{R}, 0 \le \phi_{LW}(\gamma, r) \le 1.$$

Lemma 5.5.3 *If* $0 \le \gamma \le 1$ *and the function* $Q(r)$ *satisfies the inequality*

$$\textit{for all } r \left| \left(\frac{1}{r} - 1\right)\left(1 - Q\left(\frac{1}{r}\right)\right) \right| \le 2,$$

which is equivalent to the inequalities

$$\textit{for all } r \ \min\left\{\frac{r+1}{1-r}, \frac{1-3r}{1-r}\right\} \le Q\left(\frac{1}{r}\right) \le \max\left\{\frac{r+1}{1-r}, \frac{1-3r}{1-r}\right\}, \quad (5.15)$$

then the function ϕ_{LW} *defined by* (5.14) *satisfies for all* $0 \le \gamma \le 1$ *for all* $r \in \mathbf{R}$ $0 \le \phi_{LW}(\gamma, r) \le 1$.

Proof Note that $\phi_{LW}(0,r)=0$ and $\phi_{LW}(1,r)=1$, so ϕ_{LW} is at the required bounds when γ is at either of its extreme values. It follows that we must have

$$\text{for } \gamma=0 \text{ or } 1 \ \text{ for all } r,\ 0\le \frac{\partial\phi_{LW}}{\partial\gamma}=1+\left[\frac{1}{2}-\gamma\right]\left[\frac{1}{r}-1\right]\left[1-Q\left(\frac{1}{r}\right)\right].$$

In particular,

$$\text{for all } r\ 0\le\frac{\partial\phi_{LW}}{\partial\gamma}(0,r)=1+\frac{1}{2}\left(\frac{1}{r}-1\right)\left(1-Q\left(\frac{1}{r}\right)\right)\ \text{ and} \tag{5.16a}$$

$$\text{for all } r\ 0\le\frac{\partial\phi_{LW}}{\partial\gamma}(1,r)=1-\frac{1}{2}\left(\frac{1}{r}-1\right)\left(1-Q\left(\frac{1}{r}\right)\right). \tag{5.16b}$$

Note that

$$\frac{\partial\phi_{LW}}{\partial\gamma}=(1-\gamma)\frac{\partial\phi_{LW}}{\partial\gamma}(0,r)+\gamma\frac{\partial\phi_{LW}}{\partial\gamma}(1,r).$$

It follows that if both of the constraints (5.16) are satisfied, then for all $0\le\gamma\le 1$ for all $r\in\mathbf{R}$ $0\le\phi(\gamma,r)\le 1$. The requirements (5.16) can be simplified to $|(\frac{1}{r}-1)(1-Q(\frac{1}{r}))|\le 2$, which can be rewritten in the form (5.15). □

Note that $Q(1/r)\equiv 1$ produces the upwind scheme. Also, note that the Lax–Wendroff scheme is itself monotonic precisely when $Q(1/r)=0$ satisfies the inequalities (5.15), that is, when $-1\le 1/r\le 3$. Unfortunately, even if we monotonize the Lax-Wendroff method by means of $Q(\frac{1}{r})$, the resulting scheme is not conservative.

Next, consider the Beam–Warming scheme

$$0=\frac{u_j^{n+1}-u_j^n}{\triangle t}+\lambda\frac{(3u_j^n-u_{j-1}^n)-(3u_{j-1}^n-u_{j-2}^n)}{2\triangle x}-\frac{\lambda\triangle t}{\triangle x}\frac{\lambda}{2}\left[\frac{u_j^n-u_{j-1}^n}{\triangle x}-\frac{u_{j-1}^n-u_{j-2}^n}{\triangle x}\right],$$

which can be rewritten in the form

$$-\frac{u_j^{n+1}-u_j^n}{u_j^n-u_{j-1}^n}=\gamma\left\{1+\frac{1}{2}(1-\gamma)(1-r_{j-1}^n)\right\}.$$

Again, since this scheme is linear and second-order, it cannot be monotonic. However, since this scheme is a function of the smoothness monitor r_{j-1}^n and the Courant number γ, consider monotonizing the Beam–Warming scheme as follows:

$$-\frac{u_j^{n+1}-u_j^n}{u_j^n-u_{j-1}^n}=\gamma\left\{1+\frac{1}{2}(1-\gamma)(1-r_{j-1}^n)\left(1-R(r_{j-1}^n)\right)\right\}\equiv\phi_{BW}(\gamma,r_{j-1}^n). \tag{5.17}$$

Following the same approach as in Lemma 5.5.3, we can easily prove the following lemma.

Lemma 5.5.4 *If* $0 \leq \gamma \leq 1$ *and the function* $R(r)$ *satisfies the inequalities*

$$\text{for all } r \ |(1-r)(1-R(r))| \leq 2,$$

which can be rewritten in the form

$$\text{for all } r \ \min\left\{\frac{r+1}{r-1}, \frac{r-3}{r-1}\right\} \leq R(r) \leq \max\left\{\frac{r+1}{r-1}, \frac{r-3}{r-1}\right\}, \tag{5.18}$$

then $\phi_{BW}(\gamma, r)$ *defined in* (5.17) *satisfies* *for all* $0 \leq \gamma \leq 1$ *for all* $r \in \mathbf{R}$ $0 \leq \phi_{BW}(\gamma, r) \leq 1$.

Note that $R(r) \equiv 1$ produces the upwind scheme. Also, note that the Beam–Warming scheme is itself monotonic precisely when $R(r) = 0$ satisfies the inequalities (5.15), that is, when $-1 \leq r \leq 3$. Unfortunately, even if we monotonize the Beam–Warming method by means of $R(r)$, the resulting scheme is not conservative.

The **Fromm scheme** [51] is the average of the Lax–Wendroff and Beam–Warming schemes, when applied to linear advection. If we average the monotonized Lax–Wendroff and monotonized Beam–Warming schemes, the result will be monotonic:

$$\begin{aligned}
-\frac{u_j^{n+1} - u_j^n}{u_j^n - u_{j-1}^n} &= \gamma \left\{1 + \frac{1-\gamma}{4}\left[\left(\frac{1}{r_j^n} - 1\right)\left(1 - Q\left(\frac{1}{r_j^n}\right)\right) + (1 - r_{j-1}^n)\left(1 - R(r_{j-1}^n)\right)\right]\right\}. \\
&= \frac{\gamma}{u_j^n - u_{j-1}^n}\left\{u_j^n + \frac{1-\gamma}{4}\left[(u_{j+1}^n - u_{j-1}^n) - (u_{j+1}^n - 2u_j^n + u_{j-1}^n) Q\left(\frac{1}{r_j^n}\right)\right]\right. \\
&\quad \left. - u_{j-1}^n + \frac{1-\gamma}{4}[-(u_j^n - u_{j-2}^n) - (u_j^n - 2u_{j-1}^n + u_{j-2}^n) R(r_{j-1}^n)]\right\}.
\end{aligned}$$

This expression shows that in order for this scheme to be conservative, it suffices to find a function S so that $S(1/r) = Q(1/r)$ and $S(r) = -R(r)$.

5.5.3 MUSCL Scheme

With this function $S(r)$ in the Fromm monotonization, Lemma 5.5.3shows that the Lax–Wendroff scheme will be monotonic if

$$\left|\left(\frac{1}{r} - 1\right)\left(1 - S\left(\frac{1}{r}\right)\right)\right| \leq 2$$

and Lemma 5.5.4 shows that the Beam–Warming scheme will be monotonic if

$$|(1-r)(1+S(r))| \leq 2.$$

By solving these inequalities, we see that for the average scheme to be monotonic

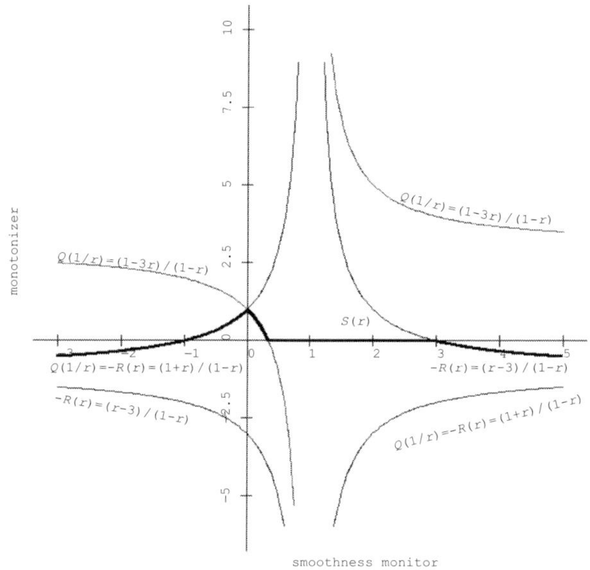

Fig. 5.2 MUSCL curves: Monotonizers vs. smoothness monitor; $S(r)$ is thick curve

we must require

$$\begin{aligned} r \leq 0 &\Longrightarrow S(r) = (1+r)/(1-r) \\ 0 \leq r \leq 1 &\Longrightarrow (1-3r)/(1-r) \leq S(r) \leq (1+r)/(1-r) \\ 1 \leq r &\Longrightarrow -(1+r)/(r-1) \leq S(r) \leq -(r-3)/(r-1). \end{aligned}$$

Next, because Q is a function of $1/r$ and R is a function of r, we require that $S(1/r) = -S(r)$. This is condition is satisfied for $r < 0$, where S has already been defined. As a consequence of this choice, $S(1) = 0$. Finally, we would like to minimize the value of the monotonizer, so that we differ from the original Fromm scheme as little as possible. As a result, we take $S(r) = (1-3r)/(1-r)$ for $0 \leq r \leq 1/3$ and $S(r) = -(r-3)/(r-1)$ for $3 \leq r$. Similarly, we take $S(r) = 0$ for $1/3 \leq r \leq 3$. In summary, our monotonizer is

$$S(r) = \begin{cases} \frac{1+r}{1-r}, & r \leq 0 & \Longrightarrow \text{upwind} \\ \frac{1-3r}{1-r}, & 0 \leq r \leq 1/3 & \Longrightarrow \text{Beam–Warming} \\ 0, & \frac{1}{3} \leq r \leq 3 & \Longrightarrow \text{Fromm} \\ -\frac{r-3}{r-1}, & 3 \leq r & \Longrightarrow \text{Lax–Wendroff.} \end{cases}$$

This function is illustrated in Figure 5.2.

An equivalent formulation of this method has numerical flux

$$f_{j+1/2}^{n+1/2} = \lambda \left[u_j^n + \frac{1}{2}(1-\gamma) s_j^n \right],$$

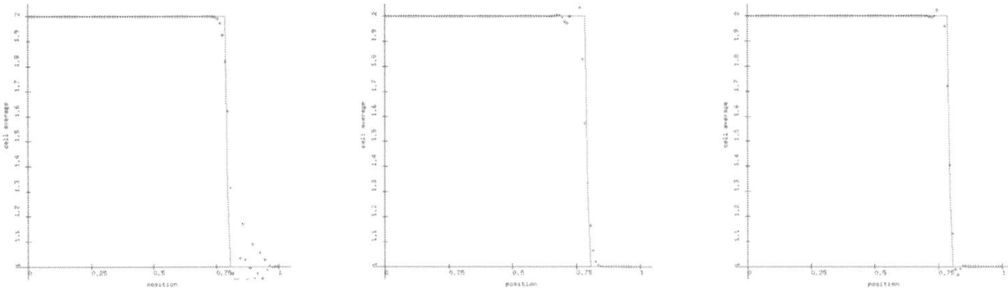

(a) Beam-Warming (b) Lax-Wendroff (c) Fromm

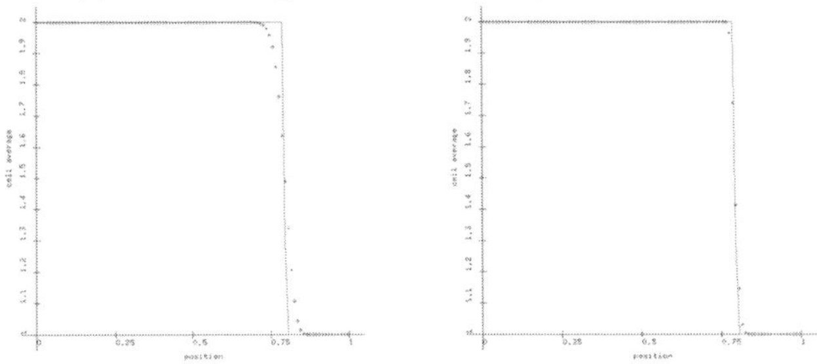

(d) Upwind (e) MUSCL

Fig. 5.3 Schemes for Linear Advection (CFL = 0.9) (red = exact, blue = numerical solution)

where $\gamma = \lambda \triangle t / \triangle x$ is the Courant number and

$$s_j^n = \frac{1}{2}\delta_{j+1/2}^n [1 + r_j^n - (1 - r_j^n) S(r_j^n)]$$

is the **MUSCL slope**. If we define the side-centered slope

$$\delta_{j+1/2}^n = u_{j+1}^n - u_j^n,$$

then we can write

$$s_j^n = \begin{cases} \text{sign}(\delta_{j+1/2}^n + \delta_{j-1/2}^n) \min\{2|\delta_{j+1/2}^n|, 2|\delta_{j-1/2}^n|, \frac{1}{2}(|\delta_{j+1/2}^n + \delta_{j-1/2}^n|)\}, & \delta_{j+1/2}^n \delta_{j-1/2}^n \geq 0 \\ 0, & \text{otherwise.} \end{cases}$$

Here "MUSCL" is an acronym for "Monotone Upstream-centered Scheme for Conservation Laws." Note that we can interpret $f_{j+1/2}^{n+1/2}$ as the flux function evaluated at the state $u_j^n + \frac{1}{2}(1 - \gamma) s_j^n$. In Section 5.9, we will generalize the MUSCL scheme to nonlinear scalar conservation laws.

Figure 5.3 shows the computational results with various first-order and second-order schemes. This figure was generated by **Program 5.5-63: GUILinearAdvectionMain.C.** The MUSCL scheme has been implemented in **Program 5.5-64: muscl.f**. Students can test and compare the upwind, Beam–Warming, Lax–Wendroff, Fromm and MUSCL schemes for the Zalesak linear advection test problems (see exercise 2.2.5 of section 2.2) by clicking on **Executable 5.5-26: guilinearad**. Selecting a positive number of grid cells causes the program to run the first method selected on the chosen initial data, while selecting zero grid cells causes the program to perform mesh refinement and efficiency studies on all selected programs.

Exercises for 5.5

5.5.1 Program the MUSCL scheme for linear advection. Compare it to Lax–Wendroff, Beam–Warming and Fromm for the Zalesak initial data in Exercise 2.2.5 of Section 2.2. Which scheme produces the smallest error for a given mesh width? Which scheme is most efficient? What convergence rates do the schemes produce? For which initial data are any of the schemes second-order accurate?

5.5.2 VanLeer also suggested

$$S(r) = \frac{1 - |r|}{1 + |r|}.$$

Show that this corresponds to using harmonically averaged slopes when the numerical solution has no local extremum.

5.6 Discrete Entropy Conditions

In section 3.1 we discussed the usefulness of an entropy function for scalar conservation laws. In this section we would like to study the usefulness of an entropy function for understanding the behavior of numerical methods.

Suppose that we have an explicit conservative numerical method

$$u_j^{n+1} = u_j^n - \frac{\Delta t^{n+1/2}}{\Delta x_j} \left[f_{j+1/2}^{n+1/2} - f_{j-1/2}^{n+1/2} \right]$$

that approximates the solution of the system of conservation laws

$$\frac{\partial u}{\partial t} + \frac{\partial f(u)}{\partial x} = 0.$$

Also suppose that η is a convex entropy function (definition 3.1.14) for the conservation law, with entropy flux ψ; in other words,

$$\frac{\partial \psi}{\partial u} = \frac{\partial \eta}{\partial u} \frac{\partial f}{\partial u}.$$

Suppose that η is concave, and we can find a numerical entropy flux Ψ that is consistent with ψ (definition 5.1.2) and satisfies

$$\eta(u_j^{n+1}) \geq \eta(u_j^n) - \frac{\Delta t^{n+1/2}}{\Delta x_j} [\Psi(u_{j-k+1}^n, \ldots, u_{j+k}^n) - \Psi(u_{j-k}^n, \ldots, u_{j+k+1}^n)].$$

Then it is possible to modify the proof of the Lax–Wendroff Theorem 5.2.2 (see [96, section 12.5]) to show that if $\{u_j^n\} \to u$, then the total entropy $\int \eta(u)\, dx$ increases in time. Similarly, if η is concave and the inequality on Ψ is reversed, then the total entropy decreases in time.

In Lemma 4.2.2, we proved that if Godunov's method converges in one dimension, then it converges to an entropy-satisfying solution of the conservation law. Thus exact Riemann solvers are useful in guaranteeing convergence to the entropy-satisfying solution of initial value problems for conservation laws. The problems with their use are their programming complexity and computational cost. It would be useful to establish a general principle for numerical schemes which would guarantee convergence to the correct solution at less expense than with the true Godunov method.

5.7 E-Schemes

Definition 5.7.1 [124]. An **E-scheme** is a consistent (Definition 5.1.2) conservative (Equation (2.14)) scheme for a scalar conservation law $\partial u/\partial t + \partial f(u)/\partial x = 0$, with numerical flux $f^{n+1/2}_{j+1/2}$ satisfying

$$\text{for all } u \text{ between } u^n_j,\ u^n_{j+1},\ \text{sign}(u^n_{j+1} - u^n_j)\left[f^{n+1/2}_{j+1/2} - f(u)\right] \leq 0.$$

Another way to state this definition is that for an E-scheme,

$$u^n_j \leq u^n_{j+1} \Longrightarrow f^{n+1/2}_{j+1/2} \leq \min_{u^n_j \leq u \leq u^n_{j+1}} f(u),$$

$$u^n_j \geq u^n_{j+1} \Longrightarrow f^{n+1/2}_{j+1/2} \geq \max_{u^n_j \leq u \leq u^n_{j+1}} f(u).$$

Osher [124] proved the following theorem for continuous-time schemes of the form

$$\frac{du_j}{dt} = -\frac{f_{j+1/2} - f_{j-1/2}}{\triangle x_j},$$

and Tadmor [156] proved the following theorem for the discrete-time scheme

$$\frac{u^{n+1}_j - u^n_j}{\triangle t^{n+1/2}} = -\frac{f_{j+1/2} - f_{j-1/2}}{\triangle x_j}.$$

Theorem 5.7.2 *If they converge, E-schemes (see Definition 5.7.1) converge to the entropy-satisfying solution, and are at most first-order accurate.*

Example 5.7.3 Godunov's method is an E-scheme, since the Oleinik chord condition (Lemma 3.1.19) shows that the Godunov flux is

$$F^G_{j+1/2} = f(\mathcal{R}(u_j, u_{j+1}; 0)) = \begin{cases} \min_{u_j \leq u \leq u_{j+1}} f(u), & u_j \leq u_{j+1} \\ \max_{u_j \geq u \geq u_{j+1}} f(u), & u_j \geq u_{j+1}. \end{cases}$$

Note that any flux $f_{j+1/2}^{n+1/2}$ of the form $f_{j+1/2}^{n+1/2} = f_{j+1/2}^{G} - \lambda(u_{j+1} - u_j)$, where $\lambda \geq 0$, generates an E-scheme, and all other E-scheme fluxes are bounded by the Godunov flux. Thus the flux for an E-scheme is equal to the Godunov flux plus a numerical diffusion. In other words, Godunov's method is the least diffusive E-scheme.

Example 5.7.4 The **Engquist–Osher flux** is

$$f_{j+1/2}^{EO} = \frac{1}{2}\left[f(u_{j+1}) + f(u_j) - \int_{u_j}^{u_{j+1}} |f'(v)|\, dv\right]. \tag{5.19}$$

This can be rewritten either as

$$\begin{aligned} f_{j+1/2}^{EO} &= f(u_j) + \frac{1}{2}\left\{f(u_{j+1}) - f(u_j) - \int_{u_j}^{u_{j+1}} |f'(v)|\, dv\right\} \\ &= f(u_j) + \int_{u_j}^{u_{j+1}} \min\{f'(v), 0\}\, dv, \end{aligned}$$

or

$$f_{j+1/2}^{EO} = f(u_{j+1}) - \int_{u_j}^{u_{j+1}} \max\{f'(v), 0\}\, dv.$$

Both of these forms can be rewritten as a sum or difference of fluxes at the end-points and intermediate sonic points. When viewed in these forms, it is easy to see that the Engquist–Osher scheme is an E-scheme. The Engquist-Osher scheme has been implemented in **Program 5.7-65: Schemes.C.** Students can experiment with the Engquist–Osher scheme for Riemann problems in a variety of models (Linear Advection, Burgers', Traffic and Buckley–Leverett) by clicking on **Executable 5.7-27: guiriemann.**

Exercises for 5.7

5.7.1 Consider a scheme with Rusanov-type flux

$$f_{j+1/2}^{n+1/2} = \frac{1}{2}\left[f(u_j^n) + f(u_{j+1}^n)\right] - \lambda_{j+1/2}^n(u_{j+1}^n - u_j^n).$$

Show that this flux generates an E-scheme if and only if

for all $u \in int[u_j^n, u_{j+1}^n]$ $\lambda_{j+\frac{1}{2}}^n \geq \dfrac{\frac{1}{2}\left[f(u_j^n) + f(u_{j+1}^n)\right] - f(u)}{u_{j+1}^n - u_j^n}$.

5.7.2 Under what circumstances does Marquina's flux (see also exercise 3.4.3), given by

$$f_{i+1/2}^{n+1/2} = \begin{cases} f(u_L),\ f'(u) > 0 & \text{for all } u \in int[u_L, u_R] \\ f(u_R),\ f'(u) < 0 & \text{for all } u \in int[u_L, u_R] \\ \frac{1}{2}[f(u_L) + f(u_R)] - (u_R - u_L)\max_{u \in int[u_L, u_r]} |f'(u)|, & \text{otherwise} \end{cases} .$$

generate an E-scheme? (Here $int[a, b]$ is the closed interval bounded by a and b.)

5.7.3 Under which circumstances is Murman's scheme [120], in which

$$f_{j+1/2}^{n+1/2} = \frac{1}{2}[f(u_j) + f(u_{j+1})] - \lambda_{j+1/2}(u_{j+1} - u_j),\ \lambda_{j+1/2} = \frac{f(u_{j+1}) - f(u_j)}{u_{j+1} - u_j},$$

an E-scheme? (You may want to compare this problem to Example 5.2.3.)

5.7.4 Show that for any continuously differentiable flux function f, the Engquist–Osher flux is a sum of values of f at the states u_j and/or u_{j+1} and critical points of f between these two states.

5.7.5 Consider the Engquist–Osher flux for Burgers' equation.
   (a) Show that for this problem, the Engquist–Osher flux is equal to the Godunov flux, except in the case of a transonic shock (*i.e.*, $u_j^n > 0 > u_{j+1}^n$).
   (b) Show that the Engquist–Osher flux has Lipschitz continuous partial derivatives with respect to the states u_j^n and u_{j+1}^n, while the Godunov flux does not.

5.7.6 Describe the Engquist–Osher flux for the traffic flow problem with flux $f(\rho) = -\rho\log(\rho)$. (Note that $u_{\text{sonic}} = 1/e$ is important because $f'(1/e) = 0$.)

5.7.7 Describe the Engquist–Osher flux for the Buckley–Leverett problem.

5.8 Total Variation Diminishing Schemes

5.8.1 Sufficient Conditions for Diminishing Total Variation

Recall from Section 5.3.3 that solutions of scalar conservation laws are **total variation diminishing** (TVD). This means that the total variation $TV(u)$ (Definition 5.3.2) is non-increasing in time. In this section, we will determine conditions under which a conservative difference scheme $u_j^{n+1} = u_j^n - \frac{\Delta t^{n+1/2}}{\Delta x_j}\left[f_{j+1/2}^{n+1/2} - f_{j-1/2}^{n+1/2}\right]$ is total variation diminishing.

The following lemma is crucial to our development of total variation diminishing schemes.

Lemma 5.8.1 *(Harten [64]) If $\triangle u_{j+1/2}^n \equiv u_{j+1}^n - u_j^n$, and the difference scheme*

$$u_j^{n+1} = u_j^n + D_{j+1/2}\triangle u_{j+1/2}^n - C_{j-1/2}\triangle u_{j-1/2}^n$$

is such that

$$\text{for all } j \quad 0 \le C_{j+1/2} \quad \text{and} \quad 0 \le D_{j+1/2} \quad \text{and} \quad C_{j+1/2} + D_{j+1/2} \le 1,$$

then the scheme is TVD.

Proof The total variation at the new time is

$$\begin{aligned}
&TV(u^{n+1})\\
&=\sum_j |u^{n+1}_{j+1}-u^{n+1}_j|\\
&=\sum_j |u^n_{j+1}+D_{j+\frac{3}{2}}\triangle u^n_{j+\frac{3}{2}}-C_{j+1/2}\triangle u^n_{j+1/2}-u^n_j-D_{j+1/2}\triangle u^n_{j+1/2}+C_{j-1/2}\triangle u^n_{j-1/2}|\\
&=\sum_j |D_{j+\frac{3}{2}}\triangle u^n_{j+\frac{3}{2}}+\{1-C_{j+1/2}-D_{j+1/2}\}\triangle u^n_{j+1/2}+C_{j-1/2}\triangle u^n_{j-1/2}|\\
&\le\sum_j D_{j+\frac{3}{2}}|\triangle u^n_{j+\frac{3}{2}}|+\sum_j[1-C_{j+1/2}-D_{j+1/2}]|\triangle u^n_{j+1/2}|+\sum_j C_{j-1/2}|\triangle u^n_{j-1/2}|\\
&=\sum_j[D_{j+1/2}+\{1-C_{j+1/2}-D_{j+1/2}\}+C_{j+1/2}]|\triangle u^n_{j+1/2}|\\
&=\sum_j|\triangle u^n_{j+1/2}|=\sum_j|u^n_{j+1}-u^n_j|=TV(u^n).
\end{aligned}$$

□

Our discussion in the remainder of this section follows the original development due to Sweby [155]. Let us write

$$\triangle f^n_{j+1/2}\equiv f(u^n_{j+1})-f(u^n_j) \text{ and } \triangle u^n_{j+1/2}\equiv u^n_{j+1}-u^n_j.$$

Note that we can perform the **flux splitting**

$$\triangle f^n_{j+1/2}=\triangle f^+_{j+1/2}+\triangle f^-_{j+1/2}$$

where

$$\triangle f^+_{j+1/2}\equiv-\left[f^{n+1/2}_{j+1/2}-f(u^n_{j+1})\right],\ \triangle f^-_{j+1/2}\equiv f^{n+1/2}_{j+1/2}-f(u^n_j).$$

Lemma 5.8.2 *If $f^{n+1/2}_{j+1/2}$ is generated by an E-scheme (Definition 5.7.1), then the split CFL numbers*

$$\gamma^+_{j+1/2}\equiv\frac{2\triangle t^{n+1/2}}{\triangle x_j+\triangle x_{j+1}}\frac{\triangle f^+_{j+1/2}}{\triangle u^n_{j+1/2}} \quad \text{and} \quad \gamma^-_{j+1/2}\equiv-\frac{2\triangle t^{n+1/2}}{\triangle x_j+\triangle x_{j+1}}\frac{\triangle f^-_{j+1/2}}{\triangle u^n_{j+1/2}} \tag{5.20}$$

are nonnegative.

Proof Using the definitions of the split flux differences and the definition of an E-scheme, we compute

$$\gamma^+_{j+1/2}=-\frac{2\triangle t^{n+1/2}}{\triangle x_j+\triangle x_{j+1}}\frac{\text{sign}(\triangle u^n_{j+1/2})\left[f^{n+1/2}_{j+1/2}-f(u^n_{j+1})\right]}{|\triangle u^n_{j+1/2}|}\ge 0,$$

$$\gamma^-_{j+1/2}=-\frac{2\triangle t^{n+1/2}}{\triangle x_j+\triangle x_{j+1}}\frac{\text{sign}(\triangle u^n_{j+1})\left[f^{n+1/2}_{j+1/2}-f(u^n_j)\right]}{|\triangle u^n_{j+1}|}\ge 0.$$

□

In order to simplify the expressions, we will define the dimensionless mesh factors

$$\alpha_{j+1/2} = \frac{2\triangle x_j}{\triangle x_j + \triangle x_{j+1}} \text{ and } \beta_{j+1/2} = \frac{2\triangle x_{j+1}}{\triangle x_{j+1} + \triangle x_j}. \tag{5.21}$$

Note that $\alpha_{j+1/2} + \beta_{j+1/2} = 2$. In fact, on a uniform grid, $\alpha_{j+1/2} = 1 = \beta_{j+1/2}$. Also note that we can relate the flux difference to the solution difference by

$$\frac{2\triangle t^{n+1/2}}{\triangle x_j + \triangle x_{j+1}} \triangle f^n_{j+1/2} = (\gamma^+_{j+1/2} - \gamma^-_{j+1/2})\triangle u^n_{j+1/2}.$$

Lemma 5.8.3 *Let the mesh factors* $\alpha_{j+1/2}$ *and* $\beta_{j+1/2}$ *be given by* (5.21) *and the split CFL numbers* $\gamma^+_{j+1/2}$ *and* $\gamma^-_{j+1/2}$ *be given by (5.20). Then E-scheme fluxes (Definition 5.7.1) generate a TVD conservative difference scheme if the timestep* $\triangle t^{n+1/2}$ *is chosen so that*

$$\frac{\gamma^+_{j+1/2}}{\beta_{j+1/2}} + \frac{\gamma^-_{j+1/2}}{\alpha_{j+1/2}} \leq 1.$$

Proof We can write the numerical scheme in the form

$$\begin{aligned}
u^{n+1}_j &= u^n_j - \frac{\triangle t^{n+1/2}}{\triangle x_j}\left[\left\{f^{n+1/2}_{j+1/2} - f(u^n_j)\right\} - \left\{f^{n+1/2}_{j-1/2} - f(u^n_j)\right\}\right] \\
&= u^n_j - \frac{\triangle t^{n+1/2}}{\triangle x_j}\left[\triangle f^-_{j+1/2} + \triangle f^+_{j-1/2}\right] \\
&= u^n_j + \frac{\triangle x_j + \triangle x_{j+1}}{2\triangle x_j}\gamma^-_{j+1/2}\triangle u^n_{j+1/2} - \frac{\triangle x_j + \triangle x_{j-1}}{2\triangle x_j}\gamma^+_{j-1/2}\triangle u^n_{j-1/2} \\
&= u^n_j + \frac{\gamma^-_{j+1/2}}{\alpha_{j+1/2}}\triangle u^n_{j+1/2} - \frac{\gamma^+_{j-1/2}}{\beta_{j-1/2}}\triangle u^n_{j-1/2}.
\end{aligned}$$

The result follows from Harten's Lemma 5.8.1. □

The next lemma describes a more clearly understandable strategy for selecting a timestep for the Engquist–Osher scheme.

Lemma 5.8.4 *The Engquist–Osher scheme* (5.19) *is TVD if*

$$\text{for all } j,\ \triangle t^{n+1/2} \max_{u \text{ between } u^n_j,\, u^n_{j+1}} \left|\frac{df(u)}{du}\right| \leq \min\{\triangle x_j, \triangle x_{j+1}\}.$$

Proof Since the Engquist-Osher scheme is an E-scheme, $\gamma^{\pm}_{j+1/2} \geq 0$. According to Lemma 5.8.3, we only need to prove that

$$\frac{\gamma^+_{j+1/2}}{\beta_{j+1/2}} + \frac{\gamma^-_{j+1/2}}{\alpha_{j+1/2}} \leq 1.$$

For any u, the definition (5.19) of the Engquist–Osher flux can be rewritten

$$f_{j+1/2}^{n+1/2} = f(u) + \int_u^{u_j^n} \max\{f'(v), 0\}\, dv + \int_u^{u_{j+1}^n} \min\{f'(v), 0\}\, dv.$$

By choosing $u = u_{j+1}^n$ we obtain

$$\triangle f_{j+1/2}^{+} \equiv f(u_{j+1}^n) - f_{j+1/2}^{n+1/2} = \int_{u_j^n}^{u_{j+1}^n} \max\{f'(v), 0\}\, dv.$$

By choosing $u = u_j^n$ we obtain

$$\triangle f_{j+1/2}^{-} \equiv f_{j+1/2}^{n+1/2} - f(u_j^n) = \int_{u_j^n}^{u_{j+1}^n} \min\{f'(v), 0\}\, dv.$$

It follows that

$$\begin{aligned}
\frac{\gamma_{j+1/2}^{+}}{\beta_{j+1/2}} + \frac{\gamma_{j+1/2}^{-}}{\alpha_{j+1/2}} &= \frac{\triangle t^{n+1/2}}{\triangle x_{j+1}} \frac{\triangle f_{j+1/2}^{+}}{\triangle u_{j+1/2}^{n}} - \frac{\triangle t^{n+1/2}}{\triangle x_j} \frac{\triangle f_{j+1/2}^{-}}{\triangle u_{j+1/2}^{n}} \\
&= \frac{\triangle t^{n+1/2}}{\triangle x_{j+1}} \frac{1}{u_{j+1}^n - u_j^n} \int_{u_j^n}^{u_{j+1}^n} \max\{f'(v), 0\}\, dv \\
&\quad - \frac{\triangle t^{n+1/2}}{\triangle x_j} \frac{1}{u_{j+1}^n - u_j^n} \int_{u_j^n}^{u_{j+1}^n} \min\{f'(v), 0\}\, dv \\
&\leq \frac{\triangle t^{n+1/2}}{\min\{\triangle x_j, \triangle x_{j+1}\}} \frac{1}{u_{j+1}^n - u_j^n} \int_{u_j^n}^{u_{j+1}^n} |f'(v)|\, dv \\
&\leq \frac{\triangle t^{n+1/2}}{\min\{\triangle x_j, \triangle x_{j+1}\}} \max_{u \text{ between } u_j^n,\, u_{j+1}^n} \left|\frac{df}{du}\right|.
\end{aligned}$$

□

Note that the proof of this lemma actually says that the Engquist–Osher scheme will be TVD if the timestep is chosen to be less than the min of the cell widths divided by the *average of the absolute value of the flux derivative*. If the flux derivative has constant sign on the interval between u_j^n and u_{j+1}^n, then the average of the absolute value of the flux derivative is the absolute value of the chord slope $\triangle f_{j+1/2}^n / \triangle u_{j+1/2}^n$. Sometimes, developers will try to take a larger timestep based on this chord slope. However, it is possible that numerical diffusion will cause the evolution of the scheme to generate intermediate points, some of which may come close to a local extremum of the flux derivative. In order to keep the timestep from varying abruptly when this happens, the timestep selection in Lemma 5.8.4 is suggested.

5.8.2 Higher-Order TVD Schemes for Linear Advection

We would like to extend the TVD analysis to higher-order schemes. As in the flux-corrected transport scheme [22], we will view the higher-order flux as the sum of a lower-order flux and a correction:

$$f_{j+1/2}^{n+1/2} = f_{j+1/2}^{L} + \phi_j \left[f_{j+1/2}^{H} - f_{j+1/2}^{L} \right].$$

Here $f_{j+1/2}^{L}$ and $f_{j+1/2}^{H}$ are the lower-order and higher-order fluxes (respectively), and ϕ_j is a limiter that is yet to be determined.

Consider the linear advection problem, in which $f(u) = \lambda u$ with $\lambda > 0$. Suppose that we use upwind differences for our low-order scheme, and the **Lax-Wendroff method** for the higher-order scheme. Then

$$\begin{aligned} f_{j+1/2}^{L} &= \lambda u_j^n \text{ and} \\ f_{j+1/2}^{H} - f_{j+1/2}^{L} &= \lambda \left(1 - \lambda \frac{\triangle t^{n+1/2}}{\triangle x_j}\right)(u_{j+1}^n - u_j^n) \frac{\triangle x_j}{\triangle x_{j+1} + \triangle x_j} \\ &= \frac{1}{2}(\alpha_{j+1/2} - \gamma_{j+1/2}^{+}) \lambda \triangle u_{j+1/2}^n, \end{aligned}$$

where we previously defined the mesh factor $\alpha_{j+1/2}$ in Equation (5.21) and the split Courant number $\gamma_{j+1/2}^{+}$ in equation (5.20). The limited scheme is

$$\begin{aligned} u_j^{n+1} &= u_j^n - \frac{\lambda \triangle t^{n+1/2}}{\triangle x_j} \left[u_j^n - u_{j-1}^n \right] \\ &\quad - \frac{1}{2} \frac{\lambda \triangle t^{n+1/2}}{\triangle x_j} \left\{ \phi_j (\alpha_{j+1/2} - \gamma_{j+1/2}^{+}) \triangle u_{j+1/2}^n - \phi_{j-1} (\alpha_{j-1/2} - \gamma_{j-1/2}^{+}) \triangle u_{j-1/2}^n \right\} \\ &= u_j^n - \frac{\gamma_{j+\frac{1}{2}}^{+}}{\alpha_{j+1/2}} \left[1 + \frac{1}{2}(\alpha_{j-1/2} - \gamma_{j-1/2}^{+}) \left\{ \phi_j \frac{(\alpha_{j+1/2} - \gamma_{j+1/2}^{+}) \triangle u_{j+1/2}^n}{(\alpha_{j-1/2} - \gamma_{j-1/2}^{+}) \triangle u_{j-1/2}^n} - \phi_{j-1} \right\} \right] \triangle u_{j-1/2}^n. \end{aligned}$$

Note that since the characteristic speed is always positive, we wrote the scheme so that it involves a factor times the difference $\triangle u_{j-1/2}^n$. In the notation of Harten's lemma 5.8.1, we have $D_{j+1/2} = 0$ and

$$C_{j-1/2} = \frac{\gamma_{j+1/2}^{+}}{\alpha_{j+1/2}} \left[1 + \frac{1}{2}(\alpha_{j-1/2} - \gamma_{j-1/2}^{+}) \left\{ \phi_j \frac{(\alpha_{j+1/2} - \gamma_{j+1/2}^{+}) \triangle u_{j+1/2}^n}{(\alpha_{j-1/2} - \gamma_{j-1/2}^{+}) \triangle u_{j-1/2}^n} - \phi_{j-1} \right\} \right] \tag{5.22}$$

So that the limited scheme is TVD, we want $0 \le C_{j-1/2} \le 1$.

In order to develop a general-purpose approach to the limiter, we will choose $\phi_j = \phi(r_j^n)$ where

$$r_j^n \equiv \frac{\left(\alpha_{j-1/2} - \gamma_{j-1/2}^{+}\right) \triangle u_{j-1/2}^n}{\left(\alpha_{j+1/2} - \gamma_{j+1/2}^{+}\right) \triangle u_{j+1/2}^n} \tag{5.23}$$

is the **smoothness monitor**, which on a uniform mesh is the same as that defined previously in Equation (5.13). The limiter function ϕ should have the following properties. First $\phi \geq 0$, so that the limited scheme involves a nonnegative contribution from the higher-order method. Second $\phi(r) = 0$ for $r < 0$, so that we use the low-order scheme near local extrema. Third $\phi(1) = 1$, so that we use the higher-order scheme in regions where the solution is smooth. Finally, ϕ should be as large as possible while remaining TVD, so that the scheme has as little numerical diffusion as possible.

Lemma 5.8.5 *Consider the scheme*

$$u_j^{n+1} = u_j^n - \frac{\gamma_{j+1/2}^{+}}{\alpha_{j+1/2}} \left[1 + \frac{1}{2}(\alpha_{j-1/2} - \gamma_{j-1/2}^{+}) \left\{ \frac{\phi(r_j^n)}{r_j^n} - \phi(r_{j-1}^n) \right\} \right] \triangle u_{j-1/2}^n \tag{5.24}$$

for linear advection, where the mesh factors $\alpha_{j+1/2}$ are given by (5.21)*, the split Courant numbers $\gamma_{j+1/2}^{+}$ are given by* (5.20)*, and the smoothness monitor r_j^n is given by* (5.23)*. If the timestep $\triangle t^{n+1/2}$ is chosen so that*

$$\text{for all } j,\ \frac{1}{2}(\triangle x_j - \triangle x_{j+1}) \leq \lambda \triangle t^{n+1/2} \leq \min\left\{ \triangle x_j, \frac{1}{2}(\triangle x_j + \triangle x_{j+1}) \right\}$$

and the limiter $\phi(r)$ is chosen so that

$$r \leq 0 \Longrightarrow \phi(r) = 0 \tag{5.25a}$$

$$r > 0 \Longrightarrow 0 \leq \phi(r) \leq \min\{2, 2r\} \tag{5.25b}$$

then the scheme (5.24) *is TVD.*

Proof Recall from Equation (5.22) that

$$C_{j-1/2} = \frac{\gamma_{j+1/2}^{+}}{\alpha_{j+1/2}} \left[1 + \frac{1}{2}(\alpha_{j-1/2} - \gamma_{j-1/2}^{+}) \left\{ \frac{\phi(r_j^n)}{r_j^n} - \phi(r_{j-1}^n) \right\} \right].$$

The assumed timestep restriction $\lambda \triangle t^{n+1/2} \leq \triangle x_j$ is equivalent to $\gamma^+_{j+1/2} \leq \alpha_{j+1/2}$, and the assumed bounds on the limiter ϕ imply that

$$C_{j-1/2} \geq \frac{\gamma^+_{j+1/2}}{\alpha_{j+1/2}}\{1 - \alpha_{j-1/2} + \gamma^+_{j-1/2}\}$$
$$C_{j-1/2} \leq \frac{\gamma^+_{j+1/2}}{\alpha_{j+1/2}}\{1 + \alpha_{j+1/2} - \gamma^+_{j+1/2}\}.$$

Note that the timestep restriction $2\lambda \triangle t^{n+1/2} \geq \triangle x_j - \triangle x_{j+1}$ implies that $\gamma^+_{j-1/2} \geq \alpha_{j+1/2} - 1$, which in turn implies that $C_{j-1/2} \geq 0$. Also, the timestep restriction $\lambda \triangle t^{n+1/2} \leq \min\{\triangle x_j,\ \frac{1}{2}(\triangle x_j + \triangle x_{j+1})\}$ implies that $\gamma^+_{j+1/2} \leq \min\{1, \alpha_{j+1/2}\}$, which in turn implies that

$$1 - \frac{\gamma^+_{j+1/2}}{\alpha_{j+1/2}}(1 + \alpha_{j+1/2} - \gamma^+_{j+1/2}) = (1 - \gamma^+_{j+1/2})(\alpha_{j+1/2} - \gamma^+_{j+1/2})/\alpha_{j+1/2} \geq 0.$$

As a result, $C_{j-1/2} \leq 1$. Harten's Lemma 5.8.1 now completes the proof. □

Lemma 5.8.6 *In addition to the hypotheses of Lemma 5.8.5 assume that the limiter $\phi(r)$ is chosen so that*

$$r \leq 0 \Longrightarrow \phi(r) = 0 \tag{5.26a}$$
$$0 < r \leq 1 \Longrightarrow r \leq \phi(r) \leq \min\{1, 2r\} \tag{5.26b}$$
$$1 \leq r \Longrightarrow 1 \leq \phi(r) \leq \min\{r, 2\}. \tag{5.26c}$$

Then the scheme (5.24) *is TVD, has a three-point stencil, and is second-order accurate away from local extrema in the numerical solution.*

Proof Since the limited scheme is supposed to be second-order accurate and have a three-point stencil, it follows that the limited scheme must be a weighted average of the Lax–Wendroff and Beam–Warming schemes. Since the Lax–Wendroff scheme corresponds to choosing $\phi(r) = 1$, and the Beam–Warming scheme corresponds to choosing $\phi(r) = r$, we require that

$$\phi(r) = (1 - \theta(r)) + \theta(r)r = 1 + \theta(r)(r - 1)$$

for some weighting factor $\theta(r)$ satisfying $0 \leq \theta(r) \leq 1$. For $0 \leq r \leq 1$, the TVD conditions (5.25) require that $0 \leq \phi(r) \leq 2r$, and the assumption that $0 \leq \theta(r) \leq 1$ imposes the additional constraint that $r \leq \phi(r) = 1 - \theta(r)(1 - r) \leq 1$. For $1 \leq r$, the TVD conditions (5.25) require that $0 \leq \phi(r) \leq 2$, and the assumption that $0 \leq \theta(r) \leq 1$ imposes the additional constraint that $1 \leq \phi(r) = 1 + \theta(r)(r - 1) \leq r$. This gives us the conditions (5.26). □

There are several examples of second-order limiters, corresponding to different schemes. All of the schemes will work with the traced flux increments

$$\triangle f_{j+1/2}^{n+1/2} = (\alpha_{j+1/2} - \gamma_{j+1/2}^{+})\lambda \triangle u_{j+1/2}^{n}.$$

The **van Leer limiter** chooses $\phi(r) = (r + |r|)/(1 + |r|)$. This corresponds to choosing the conservative flux to be

$$f_{j+1/2}^{n+1/2} = \lambda u_j^n + \frac{1}{2} \triangle f_j^n, \tag{5.27}$$

where the monotonized harmonic slope is defined to be

$$\triangle f_j^n = \begin{cases} \frac{2\triangle f_{j-1/2}^n \triangle f_{j+1/2}^n}{\triangle f_{j-1/2}^n + \triangle f_{j+1/2}^n}, & \triangle f_{j-1/2}^n \triangle f_{j+1/2}^n > 0 \\ 0, & \text{otherwise.} \end{cases}$$

In fact, all of the TVD limiters necessarily choose the monotonized slope $\triangle f_j^n$ to be zero when the side slopes $\triangle f_{j-1/2}^n$ and $\triangle f_{j+1/2}^n$ have opposite signs. The **minmod** limiter chooses $\phi(r) = \max\{0, \min\{1, r\}\}$. For this limiter, the conservative flux (5.27) chooses the monotonized slope to be

$$\triangle f_j^n = \text{sign}(\triangle f_{j+1/2}^n) \min\{|\triangle f_{j+1/2}^n|, |\triangle f_{j-1/2}^n|\},$$

provided that the side slopes have the same sign. This limiter corresponds to using the Beam–Warming scheme for $r \leq 1$ and the Lax–Wendroff scheme for $r \geq 1$. The **MUSCL** limiter chooses $\phi(r) = \max\{0, \min\{2, 2r, 1/2(1 + r)\}\}$. For this limiter, the monotonized slope for the conservative flux (5.27) is given by

$$\triangle f_j^n = \text{sign}(\triangle f_{j+1/2}^n) \min\left\{2|\triangle f_{j+1/2}^n|, 2|\triangle f_{j-1/2}^n|, \left|\frac{1}{2}\left(\triangle f_{j+1/2}^n + \triangle f_{j-1/2}^n\right)\right|\right\},$$

provided that the side slopes have the same sign. Finally, the **superbee** limiter chooses $\phi(r) = \max\{0, \min\{1, 2r\}, \min\{r, 2\}\}$. Here the monotonized slope is defined to be

$$\triangle f_j^n = \text{sign}(\triangle f_{j+1/2}^n) \max\left\{\min\{|\triangle f_{j+1/2}^n|, 2|\triangle f_{j-1/2}^n|\}, \min\{|\triangle f_{j-1/2}^n|, 2|\triangle f_{j+1/2}^n|\}\right\}$$

provided that the side slopes have the same sign. Figure 5.4 shows the graph of ϕ versus r, and several of the limiters.

A C^{++}program to implement the TVD scheme for linear advection can be found in **Program 5.8-66: LinearAdvectionSchemes.C** The main program is in **Program 5.8-67: GUILinearAdvectionMain2.C** Students can exercise this program by clicking on **Executable 5.8-28: guilinearad2** The user can select a variety of initial values from the Zalesak test problems in exercise 2.2.5 of section 2.2. In addition, the user can select from a variety of limiters. By setting the number of cells to 0, the user will cause the program to perform a mesh refinement study.

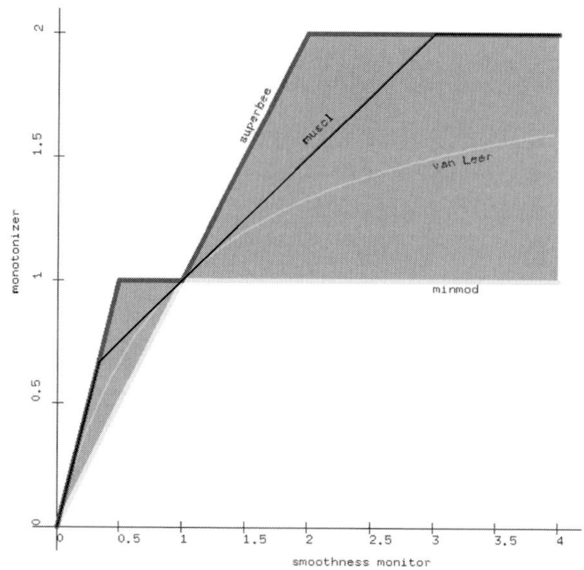

Fig. 5.4 Second-order TVD region and TVD limiters: ϕ versus r

5.8.3 Extension to Nonlinear Scalar Conservation Laws

Next, let us extend our development of TVD schemes to general nonlinear scalar conservation laws. Suppose that $f^L_{j+1/2}(u^n_j, u^n_{j+1})$ is a low-order diffusive flux, such as an E-scheme. As before, we will define the flux differences $\triangle f^+_{j+1/2} = f(u^n_{j+1}) - f^L_{j+1/2}$ and $\triangle f^-_{j+1/2} = f^L_{j+1/2} - f(u^n_j)$. Recall that the split Courant numbers were defined by Equation (5.20). Recall from Equation (3.28) that the Lax–Wendroff flux is

$$f^{LW}_{j+1/2} = \frac{f(u^n_j)\triangle x_{j+1} + f(u^n_{j+1})\triangle x_j - \triangle t^{n+1/2}\frac{df}{du}[f(u^n_{j+1}) - f(u^n_j)]}{\triangle x_j + \triangle x_{j+1}}$$

$$= \frac{1}{2}\left[\alpha_{j+1/2} f(u^n_{j+1}) + \beta_{j+1/2} f(u^n_j)\right] - \frac{\triangle t^{n+1/2}\frac{df}{du}}{\triangle x_j + \triangle x_{j+1}}\left[\triangle f^+_{j+\frac{1}{2}} + \triangle f^-_{j+\frac{1}{2}}\right]$$

$$= f^L_{j+\frac{1}{2}} + \frac{1}{2}\left[\alpha_{j+1/2} - \frac{2\triangle t^{n+1/2}\frac{df}{du}}{\triangle x_j + \triangle x_{j+1}}\right]\triangle f^+_{j+1/2}$$

$$+ \frac{1}{2}\left[\beta_{j+1/2} - \frac{\triangle t^{n+1/2}\frac{df}{du}}{\triangle x_j + \triangle x_{j+1}}\right]\triangle f^-_{j+\frac{1}{2}}.$$

Here the mesh factors $\alpha_{j+1/2}$ and $\beta_{j+1/2}$ are defined in (5.21), and the flux derivative is approximated by difference quotients. For our purposes in this section, the flux derivative approximations will give us the following form for the Lax–Wendroff flux:

$$f^{LW}_{j+1/2} = f^{L}_{j+1/2} + \frac{1}{2}(\alpha_{j+1/2} - \gamma^{+}_{j+1/2}) \triangle f^{+}_{j+1/2} - \frac{1}{2}(\beta_{j+1/2} - \gamma^{-}_{j+1/2}) \triangle f^{-}_{j+1/2}.$$

Recall that Lemma 5.8.2 showed that if $f^{L}_{j+1/2}$ is generated by an E-scheme, then $\gamma^{+}_{j+1/2}$ and $\gamma^{-}_{j+1/2}$ are nonnegative.

Next, we will modify the second-order terms by limiters. The revised numerical flux is described in the following lemma.

Lemma 5.8.7 *Suppose that we have a mesh* $\ldots < x_{j-1/2} < x_{j+1/2} < \ldots$ *and mesh values* u^n_j *approximating cell averages of the solution* u *to the conservation law* $\partial u/\partial t + \partial f(u)/\partial x = 0$. *Define the solution increments,* $\triangle u^n_{j+1/2} = u^n_{j+1} - u^n_j$. *Suppose that* $f^{L}_{j+1/2}(u^n_j, u^n_{j+1})$ *is generated by an E-scheme (definition 5.7.1). Define the flux increments* $\triangle f^{+}_{j+1/2} = f(u^n_{j+1}) - f^{L}_{j+1/2}$ *and* $\triangle f^{-}_{j+1/2} = f^{L}_{j+1/2} - f(u^n_j)$, *the mesh factors* $\alpha_{j+1/2}$ *and* $\beta_{j+1/2}$ *as in equation (5.21), and the split Courant numbers as in equation (5.20). Define the smoothness monitors by*

$$r^{+}_j = \frac{(\alpha_{j-1/2} - \gamma^{+}_{j-1/2}) \triangle f^{+}_{j-1/2}}{(\alpha_{j+1/2} - \gamma^{+}_{j+1/2}) \triangle f^{+}_{j+1/2}} \quad \text{and} \quad r^{-}_j = \frac{(\beta_{j-1/2} - \gamma^{-}_{j-1/2}) \triangle f^{-}_{j-1/2}}{(\beta_{j+1/2} - \gamma^{-}_{j+1/2}) \triangle f^{-}_{j+1/2}}.$$

Suppose that $1 \leq \Phi \leq 2$ *and the limiter* ϕ *satisfies*

$$\text{for all } r,\ 0 \leq \phi(r) \text{ and for all } r \neq 0,\ \max\left\{\phi(r), \frac{\phi(r)}{r}\right\} \leq \Phi.$$

If the timestep $\triangle t^{n+1/2}$ *is chosen so that*

$$\gamma^{+}_{j+1/2} > 0 \Longrightarrow |\alpha_{j+1/2} - \gamma^{+}_{j+1/2}| \leq \frac{2}{\Phi}$$

$$\gamma^{-}_{j+1/2} > 0 \Longrightarrow |\beta_{j+1/2} - \gamma^{-}_{j+1/2}| \leq \frac{2}{\Phi}$$

$$\frac{\gamma^{+}_{j+1/2}}{\beta_{j+1/2}} + \frac{\gamma^{-}_{j+1/2}}{\alpha_{j+1/2}} \leq \frac{1}{2}$$

and the flux is given by

$$f^{n+1/2}_{j+1/2} = f^{L}_{j+1/2} + \phi(r^{+}_j)\frac{1}{2}(\alpha_{j+1/2} - \gamma^{+}_{j+1/2}) \triangle f^{+}_{j+1/2} - \phi(r^{-}_{j+1})\frac{1}{2}(\beta_{j+1/2} - \gamma^{-}_{j+1/2}) \triangle f^{-}_{j+1/2} \tag{5.28}$$

then the conservative difference scheme

$$u^{n+1}_j = u^n_j - \triangle t^{n+1/2}\left[f^{n+1/2}_{j+1/2} - f^{n+1/2}_{j-1/2}\right]/\triangle x_j$$

is TVD.

Proof To check that the scheme is TVD, we write

$$
\begin{aligned}
u_j^{n+1} &= u_j^n \\
&\quad - \frac{\triangle t^{n+1/2}}{\triangle x_j}\left[\triangle f_{j+1/2}^- + \frac{1}{2}\phi(r_j^+)(\alpha_{j+1/2} - \gamma_{j+1/2}^+)\triangle f_{j+1/2}^+ \right.\\
&\quad - \frac{1}{2}\phi(r_{j+1}^-)(\beta_{j+1/2} - \gamma_{j+1/2}^-)\triangle f_{j+1/2}^- \\
&\quad + \triangle f_{j-1/2}^+ - \frac{1}{2}\phi(r_{j-1}^+)(\alpha_{j-1/2} - \gamma_{j-1/2}^+)\triangle f_{j-1/2}^+ \\
&\quad \left. + \frac{1}{2}\phi(r_j^-)(\beta_{j-1/2} - \gamma_{j-1/2}^-)\triangle f_{j-1/2}^-\right] \\
&= u_j^n - \frac{\triangle u_{j+1/2}^n}{\alpha_{j+1/2}}\left[-\gamma_{j+1/2}^- + \frac{1}{2}\phi(r_j^+)(\alpha_{j+1/2} - \gamma_{j+1/2}^+)\gamma_{j+1/2}^+ \right.\\
&\quad \left. - \frac{1}{2}\phi(r_{j+1}^-)(\beta_{j+1/2} - \gamma_{j+1/2}^-)\gamma_{j+1/2}^-\right] \\
&\quad - \frac{\triangle u_{j-1/2}^n}{\beta_{j-1/2}}\left[-\gamma_{j-1/2}^+ + \frac{1}{2}\phi(r_{j-1}^+)(\alpha_{j-1/2} - \gamma_{j-1/2}^+)\gamma_{j-1/2}^+ \right.\\
&\quad \left. + \frac{1}{2}\phi(r_j^-)(\beta_{j-1/2} - \gamma_{j-1/2}^-)\gamma_{j-1/2}^-\right] \\
&= u_j^n - \frac{\gamma_{j-1/2}^+\triangle u_{j-1/2}^n}{\beta_{j-1/2}}\left\{1 - \frac{1}{2}(\alpha_{j-1/2} - \gamma_{j-1/2}^+)\left[\phi(r_{j-1}^+) - \frac{\phi(r_j^+)}{r_j^+}\right]\right\} \\
&\quad + \frac{\gamma_{j+1/2}^-\triangle u_{j+1/2}^n}{\alpha_{j+1/2}}\left\{1 - \frac{1}{2}(\beta_{j+1/2} - \gamma_{j+1/2}^-)\left[\phi(r_{j+1}^-) - \frac{\phi(r_j^-)}{r_j^-}\right]\right\}.
\end{aligned}
$$

This suggests that we define

$$C_{j-1/2} = \frac{\gamma_{j-1/2}^+}{\beta_{j-1/2}}\left\{1 - \frac{1}{2}(\alpha_{j-1/2} - \gamma_{j-1/2}^+)\left[\phi(r_{j-1}^+) - \frac{\phi(r_j^+)}{r_j^+}\right]\right\} \tag{5.29a}$$

$$D_{j+1/2} = \frac{\gamma_{j+1/2}^-}{\alpha_{j+1/2}}\left\{1 - \frac{1}{2}(\beta_{j+1/2} - \gamma_{j+1/2}^-)\left[\phi(r_{j+1}^-) - \frac{\phi(r_j^-)}{r_j^-}\right]\right\}. \tag{5.29b}$$

Since $f_{j+1/2}^L$ is generated by an E-scheme, lemma 5.8.2 shows that the split Courant numbers $\gamma_{j-1/2}^\pm$ are nonnegative. It is easy to see that the timestep restriction $|\alpha_{j+1/2} - \gamma_{j+1/2}^+| \le 2/\Phi$ implies that $C_{j-1/2} \ge 0$, and the restriction $|\beta_{j+1/2} - \gamma_{j+1/2}^-| \le 2/\Phi$ implies that $D_{j+1/2} \ge 0$. In order for Harten's lemma 5.8.1 to guarantee that the scheme is TVD, we want $C_{j+1/2} + D_{j+1/2} \le 1$. Note that the timestep restrictions imply that

$$
\begin{aligned}
&C_{j+1/2} + D_{j+1/2} \\
&\le \frac{\gamma^+_{j+1/2}}{\beta_{j+1/2}} \left\{ 1 + \frac{1}{2} |\alpha_{j+1/2} - \gamma^+_{j+1/2}| \Phi \right\} + \frac{\gamma^-_{j+1/2}}{\alpha_{j+1/2}} \left\{ 1 + \frac{1}{2} |\beta_{j+1/2} - \gamma^-_{j+1/2}| \Phi \right\} \\
&\le \left[\frac{\gamma^+_{j+1/2}}{\beta_{j+1/2}} + \frac{\gamma^-_{j+1/2}}{\alpha_{j+1/2}} \right] 2 \le 1.
\end{aligned}
$$

□

Lemma 5.8.8 *Suppose that the hypotheses of Lemma 5.8.7 are satisfied. In addition, suppose that $f^L_{j+\frac{1}{2}}(u^n_j, j^n_{j+1})$ is given by the Engquist–Osher flux and the timestep $\triangle t^{n+1/2}$ is chosen so that*

$$
\triangle x_j - \frac{\triangle x_j + \triangle x_{j+1}}{\Phi} \le \triangle t^{n+1/2} \frac{\int_{u^n_j}^{u^n_{j+1}} \max\{f'(u), 0\}\, du}{u^n_{j+1} - u^n_j} \le \triangle x_j \tag{5.30a}
$$

$$
\triangle x_{j+1} - \frac{\triangle x_j + \triangle x_{j+1}}{\Phi} \le \triangle t^{n+1/2} \frac{\int_{u^n_j}^{u^n_{j+1}} \max\{-f'(u), 0\}\, du}{u^n_{j+1} - u^n_j} \le \triangle x_{j+1} \tag{5.30b}
$$

$$
\triangle t^{n+1/2} \frac{\int_{u^n_j}^{u^n_{j+1}} |f'(u)|\, du}{u^n_{j+1} - u^n_j} \max\left\{ \frac{1 + \alpha_{j+1/2}\Phi/2}{\triangle x_{j+1}}, \frac{1 + \beta_{j+1/2}\Phi/2}{\triangle x_j} \right\} \le 1. \tag{5.30c}
$$

Then the conservative difference scheme with fluxes given by (5.28) *is TVD.*

Proof Since the lower-order flux is given by the Engquist-Osher flux,

$$
\begin{aligned}
\frac{\gamma^+_{j+1/2}}{\beta_{j+1/2}} &= \frac{\triangle t^{n+1/2}}{\triangle x_{j+1}} \frac{\int_{u^n_j}^{u^n_{j+1}} \max\{f', 0\} du}{u^n_{j+1} - u^n_j} \\
\frac{\gamma^-_{j+1/2}}{\alpha_{j+1/2}} &= \frac{\triangle t^{n+1/2}}{\triangle x_j} \frac{\int_{u^n_j}^{u^n_{j+1}} \max\{-f', 0\} du}{u^n_{j+1} - u^n_j}.
\end{aligned}
$$

From the definitions of $C_{j+1/2}$ and $D_{j+1/2}$ in Equations (5.29), it is easy to see that the timestep restriction (5.30a) implies that $C_{j+1/2} \ge 0$, and the timestep restriction (5.30b) implies that $D_{j+1/2} \ge 0$. Finally, the timestep restriction (5.30c) implies that

$$
\begin{aligned}
C_{j+1/2} + D_{j+1/2} \le\ & \frac{\int_{u^n_j}^{u^n_{j+1}} \max\{f', 0\} du + \int_{u^n_j}^{u^n_{j+1}} \max\{-f', 0\} du}{u^n_{j+1} - u^n_j} \\
& \times \triangle t^{n+1/2} \max\left\{ \frac{1 + \frac{1}{2}(\alpha_{j+1/2} - \gamma^+_{j+1/2})\Phi}{\triangle x_{j+1}}, \frac{1 + \frac{1}{2}(\beta_{j+1/2} - \gamma^-_{j+1/2})\Phi}{\triangle x_j} \right\} \\
\le\ & \frac{\int_{u^n_j}^{u^n_{j+1}} |f'| du}{u^n_{j+1} - u^n_j} \triangle t^{n+1/2} \max\left\{ \frac{1 + \alpha_{j+1/2}\Phi/2}{\triangle x_{j+1}}, \frac{1 + \beta_{j+1/2}\Phi/2}{\triangle x_j} \right\} \le 1.
\end{aligned}
$$

The result follows from Harten's lemma 5.8.1. □

On a uniform grid, these requirements are satisfied whenever

$$\triangle t^{n+1/2} \max_{u \text{ between } u_j^n, u_{j+1}^n} \max\{f'(u), 0\} \leq \triangle x$$

$$\triangle t^{n+1/2} \max_{u \text{ between } u_j^n, u_{j+1}^n} \max\{-f'(u), 0\} \leq \triangle x$$

$$\triangle t^{n+1/2} \max_{u \text{ between } u_j^n, u_{j+1}^n} |f'(u)| \leq \frac{\triangle x}{1 + \Phi/2}.$$

For the minmod limiter we have $\Phi = 1$ and $\frac{1}{1+\Phi/2} = 2/3$; for the van Leer, MUSCL or superbee limiters we have $\Phi = 2$ and $\frac{1}{1+\Phi/2} = 1/2$.

A C^{++}program to implement the TVD scheme for a variety of nonlinear scalar conservation laws can be found in **Program 5.8-68: Schemes2.C**. The main program is in **Program 5.8-69: GUIRiemannProblem2.C**. Students can exercise this program by clicking on **Executable 5.8-29: guiriemann2** The user can select Riemann problem initial data for linear advection, Burgers' equation, traffic models and the Buckley-Leverett model. In addition, the user can select from a variety of limiters. By setting the number of cells to 0, the user will cause the program to perform a mesh refinement study.

Exercises for 5.8

5.8.1 Under what conditions are explicit upwind differences TVD for linear advection?

5.8.2 Which of the example limiters are such that $0 \leq r \leq 1$ implies that $0 \leq \theta(r) \leq 1$?

5.8.3 Program the TVD scheme for linear advection. Compare the minmod, van Leer, MUSCL and superbee limiters at CFL = 0.1, 0.5 and 0.9 for the Zalesak test problems of Exercise 2.2.5 of Section 2.2. In each case, plot the analytical solution with a continuous curve, and the numerical solution with discrete markers.

5.8.4 Program the TVD scheme for the Burgers equation. Compare the minmod, van Leer, MUSCL and superbee limiters at various values of CFL for calculations using at most 100 cells for a transonic rarefaction, a non-stationary shock and a stationary shock.

5.9 Slope-Limiter Schemes

The flux-limiter approach in Section 5.8 required two kinds of fluxes. One was a low-order entropy-satisfying flux $f^L_{j+1/2}$ evaluated at cell sides. The other was the flux $f(u_j^n)$, which was used to compute the anti-diffusive flux corrections. Because of the flux splitting, the resulting scheme had a restricted timestep, depending on the choice of the limiter (*i.e.*, $\lambda \triangle t/\triangle x \leq 2/3$ for minmod, $\lambda \triangle t/\triangle x \leq 1/2$ for MUSCL and superbee). The flux-splitting approach did avoid the solution of Riemann problems. Further, since the resulting higher-order scheme was still TVD it was necessarily convergent, but some choices of the limiter may allow the limited scheme to converge to an entropy-violating solution.

In this section, we will develop a second-order scheme that can take a larger timestep than the TVD schemes. This method will consist of four basic steps. First, **piecewise polynomial reconstruction** (Section 5.9.2) will be used to construct an approximate solution function $u^n(x)$ from the cell averages u^n_j. Second, **characteristic tracing** (Section 5.9.4) will produce values of the solution at cell sides and half-time by tracing characteristics back to data provided by the piecewise polynomial reconstruction $u^n(x)$. Third, **numerical quadrature** (Section 5.9.3) will approximate time integrals of the flux at cell sides by an appropriate quadrature rule. Finally, a **conservative difference** will compute the new solution by applying the divergence theorem to the conservation law. In this scheme, higher-order accuracy will be the result of using a higher-order reconstruction of the solution than the piecewise constant function used by Godunov's method, and by using a sufficiently accurate quadrature rule for the temporal integrals. The resulting scheme is convergent, as shown in [125].

5.9.1 Exact Integration for Constant Velocity

We will motivate the development of the slope-limiter scheme by considering the linear advection problem with piecewise constant initial data. We would like to develop a conservative difference scheme that is exact for this problem, given piecewise linear initial data.

Lemma 5.9.1 *Suppose that we want to solve the linear advection problem $\partial u/\partial t + v \partial u/\partial x = 0$ where v is constant. Let $v^+ = \max\{v, 0\}$ and $v^- = \min\{v, 0\}$ be the positive and negative parts of the advection speed. Suppose that in each grid cell we are given piecewise linear initial data at time t^n:*

$$u(x, t^n) = u^n_i + s^n_i (x - x_i), \quad x_{i-1/2} < x < x_{i+1/2}.$$

Suppose that the fluxes at the cell sides are given by

$$f^{n+1/2}_{i+1/2} = v^+ \left[u^n_i + \frac{s^n_i \triangle x_i}{2} \left(1 - \frac{v^+ \triangle t^{n+1/2}}{\triangle x_i} \right) \right] + v^- \left[u^n_{i+1} - \frac{s^n_{i+1} \triangle x_{i+1}}{2} \left(1 + \frac{v^- \triangle t^{n+1/2}}{\triangle x_{i+1}} \right) \right].$$

Then the conservative difference

$$u^{n+1}_i \triangle x_i = u^n_i \triangle x_i - \triangle t^{n+1/2} \left[f^{n+1/2}_{i+1/2} - f^{n+1/2}_{i-1/2} \right]$$

produces exact cell averages at time t^{n+1}.

Proof Since the initial data is piecewise linear, the initial cell average is

$$\int_{x_{i-1/2}}^{x_{i+1/2}} u(x, t^n)\, dx = u_i^n \triangle x_i .$$

The exact solution of this problem is $u(x,t) = u\left(x - v(t - t^n), t^n\right)$ for $t \geq t^n$. For $v \geq 0$ the cell average at $t^{n+1} = t^n + \triangle t^{n+1/2}$ is

$$\begin{aligned}
&\int_{x_{i-1/2}}^{x_{i+1/2}} u(x, t^{n+1})\, dx \\
&= \int_{x_{i-1/2}}^{x_{i+1/2}} u(x - v\triangle t^{n+1/2}, t^n)\, dx \\
&= \int_{x_{i-1/2}-v\triangle t^{n+1/2}}^{x_{i-1/2}} u_{i-1}^n + s_{i-1}^n (x - x_{i-1})\, dx + \int_{x_{i-1/2}}^{x_{i+1/2}-v\triangle t^{n+1/2}} u_i^n + s_i^n (x - x_i)\, dx \\
&= u_{i-1}^n v\triangle t^{n+1/2} + \frac{1}{2} s_{i-1}^n \triangle x_{i-1}^2 \left[\frac{1}{4} - \left(\frac{1}{2} - \frac{v\triangle t^{n+1/2}}{\triangle x_{i-1}}\right)^2\right] \\
&\quad + u_i^n(\triangle x_i - v\triangle t^{n+1/2}) + \frac{1}{2} s_i^n \triangle x_i^2 \left[\left(\frac{1}{2} - \frac{v\triangle t^{n+1/2}}{\triangle x_i}\right)^2 - \frac{1}{4}\right] \\
&= u_i^n \triangle x_i - v\triangle t^{n+1/2}[u_i^n - u_{i-1}^n] \\
&\quad + \frac{s_{i-1}^n \triangle x_{i-1}}{2} v\triangle t^{n+1/2}\left[1 - \frac{v\triangle t^{n+1/2}}{\triangle x_{i-1}}\right] - \frac{s_i^n \triangle x_i}{2} v\triangle t^{n+1/2}\left[1 - \frac{v\triangle t^{n+1/2}}{\triangle x_i}\right] \\
&= u_i^n \triangle x_i - v\triangle t^{n+1/2}\left\{\left[u_i^n + \frac{s_i^n \triangle x_i}{2}\left(1 - \frac{v\triangle t^{n+1/2}}{\triangle x_i}\right)\right] - \left[u_{i-1}^n + \frac{s_{i-1}^n \triangle x_{i-1}}{2}\left(1 - \frac{v\triangle t^{n+1/2}}{\triangle x_{i-1}}\right)\right]\right\}.
\end{aligned}$$

Similarly, for $v \leq 0$ we have

$$\begin{aligned}
&\int_{x_{i-1/2}}^{x_{i+1/2}} u(x, t^{n+1})\, dx \\
&= u_i^n \triangle x_i - v\triangle t^{n+1/2}\left\{\left[u_{i+1}^n - \frac{s_{i+1}^n \triangle x_{i+1}}{2}\left(1 + \frac{v\triangle t^{n+1/2}}{\triangle x_{i+1}}\right)\right] - \left[u_i^n - \frac{s_i^n \triangle x_i}{2}\left(1 + \frac{v\triangle t^{n+1/2}}{\triangle x_i}\right)\right]\right\}.
\end{aligned}$$

We can combine these two results in the form

$$\begin{aligned}
&\int_{x_{i-1/2}}^{x_{i+1/2}} u(x, t^{n+1})\, dx = u_i^n \triangle x_i \\
&\quad - v^+\triangle t^{n+1/2}\left\{\left[u_i^n + \frac{s_i^n \triangle x_i}{2}\left(1 - \frac{v^+\triangle t^{n+1/2}}{\triangle x_i}\right)\right] - \left[u_{i-1}^n + \frac{s_{i-1}^n \triangle x_{i-1}}{2}\left(1 - \frac{v^+\triangle t^{n+1/2}}{\triangle x_{i-1}}\right)\right]\right\} \\
&\quad - v^-\triangle t^{n+1/2}\left\{\left[u_{i+1}^n - \frac{s_{i+1}^n \triangle x_{i+1}}{2}\left(1 + \frac{v^-\triangle t^{n+1/2}}{\triangle x_{i+1}}\right)\right] - \left[u_i^n - \frac{s_i^n \triangle x_i}{2}\left(1 + \frac{v^-\triangle t^{n+1/2}}{\triangle x_i}\right)\right]\right\}.
\end{aligned}$$

This equation has the form of a conservative difference using the fluxes given in the lemma. □

In order to apply these ideas to general nonlinear conservation laws, we need to construct a piecewise linear profile from specified cell averages, and generalize the computation of the fluxes.

5.9.2 Piecewise Linear Reconstruction

In one dimension, the piecewise linear reconstruction step in the **MUSCL** (Monotone Upwind Scheme for Conservation Laws) takes the form

$$u^n(x) = u_j^n + s_j^n(x - x_j) \quad \text{for all} \quad x \in (x_{j-1/2}, x_{j+1/2}).$$

As we have already seen, the cell average of this function is the same for any choice of the slope s_j^n. The choice $s_j^n = 0$ will produce Godunov's method.

In order to determine a value for s_j^n, we will construct a cubic polynomial interpolant to the integral of u (because we want to respect the cell averages), and take the average over $[x_{j-1/2}, x_{j+1/2}]$ of its second derivative. The cubic polynomial can be constructed by using Newton interpolation. If we want slopes on cells $0, \ldots, J-1$, then we compute the first-order divided differences

$$u^n[x_{k+1}, x_k] \equiv \frac{u_{k+1}^n - u_k^n}{\triangle x_{k+1} + \triangle x_k}, \; -1 \leq k \leq J$$

and the second-order divided differences

$$u^n[x_{k+1}, x_k, x_{k-1}] \equiv \frac{u^n[x_{k+1}, x_k] - u^n[x_k, x_{k-1}]}{\triangle x_{k+1} + \triangle x_k + \triangle x_{k-1}}, \; 0 \leq k < J.$$

Then the cubic polynomial interpolating the integral of $u^n(x)$ at $x_{j-3/2}$, $x_{j-1/2}$, $x_{j+1/2}$ and $x_{j+3/2}$ is

$$\begin{aligned} c_j^n(x) \equiv & \, u_{j-1}^n(x - x_{j-3/2}) + u^n[x_j, x_{j-1}](x - x_{j-1/2})(x - x_{j+3/2}) \\ & + u^n[x_{j+1}, x_j, x_{j-1}](x - x_{j+1/2})(x - x_{j-1/2})(x - x_{j-3/2}). \end{aligned}$$

The derivative of this Newton interpolating polynomial at $x_{j+1/2}$ is

$$\frac{dc_j^n}{dx}(x_{j+1/2}) = u_{j-1}^n + u[x_j, x_{j-1}](2\triangle x_j + \triangle x_{j-1}) + u^n[x_{j+1}, x_j, x_{j-1}]\triangle x_j(\triangle x_j + \triangle x_{j-1}),$$

and the derivative at $x_{j-1/2}$ is

$$\frac{dc_j^n}{dx}(x_{j-1/2}) = u_{j-1}^n + u[x_j, x_{j-1}]\triangle x_{j-1} - u^n[x_{j+1}, x_j, x_{j-1}]\triangle x_j \triangle x_{j-1}.$$

It follows that the average of the second derivative of the polynomial is

$$\begin{aligned}
\tilde{s}_j^n &= \frac{1}{\triangle x_j}\int_{x_{j-1/2}}^{x_{j+1/2}} \frac{d^2 c_j^n}{dx^2}\,dx = \frac{1}{\triangle x_j}\left\{\frac{dc_j^n}{dx}(x_{j+1/2}) - \frac{dc_j^n}{dx}(x_{j-1/2})\right\} \\
&= \frac{1}{\triangle x_j}\{u[x_j, x_{j-1}]2\triangle x_j + u^n[x_{j+1}, x_j, x_{j-1}]\triangle x_j(\triangle x_j + 2\triangle x_{j-1})\} \\
&= 2u[x_j, x_{j-1}] + \frac{u^n[x_{j+1}, x_j] - u^n[x_j, x_{j-1}]}{\triangle x_{j+1} + \triangle x_j + \triangle x_{j-1}}(\triangle x_j + 2\triangle x_{j-1}) \\
&= u[x_j, x_{j-1}]\frac{\triangle x_j + 2\triangle x_{j+1}}{\triangle x_{j-1} + \triangle x_j + \triangle x_{j+1}} + u[x_{j+1}, x_j]\frac{\triangle x_j + 2\triangle x_{j-1}}{\triangle x_{j+1} + \triangle x_j + \triangle x_{j-1}}.
\end{aligned} \tag{5.31}$$

On a uniform grid, this simplifies to

$$\tilde{s}_j^n = u[x_{j-1}, x_j] + u[x_j, x_{j+1}] = \frac{u_{j+1}^n - u_{j-1}^n}{2\triangle x}.$$

The slopes s_j^n used in the flux computation are found by applying a limiter to the cell average slopes $\tilde{s}_j^n$. The purpose of the limiter is to prevent new extrema in the piecewise linear reconstruction. Specifically, if the cell averages are monotonically increasing (*i.e.*, $u_{j-1}^n \le u_j^n \le u_{j+1}^n$), then we want

$$u_{j-1}^n \le u^n(x_{j-1/2}) = u_j^n - s_j^n \triangle x_j/2 \quad \text{and} \quad u_{j+1}^n \ge u^n(x_{j+1/2}) = u_j^n + s_j^n \triangle x_j/2.$$

These inequalities imply that

$$s_j \triangle x_j \le 2\min\{u_{j+1}^n - u_j^n, u_j^n - u_{j-1}^n\}.$$

Similarly, if the cell averages are monotonically decreasing then we want

$$s_j \triangle x_j \ge 2\max\{u_{j+1}^n - u_j^n, u_j^n - u_{j-1}^n\}.$$

If the cell averages are not monotonic, then we choose $s_j^n = 0$ so that the piecewise linear reconstruction does not produce any new extrema. In the MUSCL scheme, we define the side differences

$$\triangle u_{j+1/2}^n = u_{j+1}^n - u_j^n,$$

and compute the limited slope by the formula

$$s_j^n \triangle x_j = \begin{cases} \operatorname{sign}(\tilde{s}_j^n \triangle x_j)\min\{2|\triangle u_{j-1/2}^n|, 2|\triangle u_{j+1/2}^n|, |\tilde{s}_j^n \triangle x_j|\}, & \triangle u_{j-1/2}^n \triangle u_{j+1/2}^n > 0 \\ 0, & \text{otherwise.} \end{cases} \tag{5.32}$$

Note that the MUSCL slopes actually allow the piecewise linear reconstruction $u^n(x)$ to have greater total variation than the discrete data u_j^n. In practice, it is said that the MUSCL reconstruction can develop a "sawtooth" profile. Sometimes, people will use the **minmod** slopes to prevent growth in the total variation:

$$s_j^n \triangle x_j = \begin{cases} \text{sign}(\triangle u_j^n) \min\{|\triangle u_{j-1/2}^n|, |\triangle u_{j+1/2}^n|\}, & \triangle u_{j-1/2}^n \triangle u_{j+1/2}^n > 0 \\ 0, & \text{otherwise .} \end{cases}$$

This choice typically leads to greater smearing of discontinuities. On the other hand, some people have suggested the use of the **superbee** slopes:

$$s_j^n \triangle x_j = \text{sign}(\triangle u_j^n) \max\left\{\min\left\{2|\triangle u_{j-1/2}^n|, |\triangle u_{j+1/2}^n|\right\}, \min\left\{|\triangle u_{j-1/2}^n|, 2|\triangle u_{j+1/2}^n|\right\}\right\}$$

if $\triangle u_{j-1/2}^n \triangle u_{j+1/2}^n > 0$, and $s_j^n \triangle x_j = 0$ otherwise. This choice has an even greater tendency to develop sawtooth profiles than MUSCL, and occasionally converges to inappropriate discontinuities.

Note that it is not necessary that $TV(u^n(x)) \leq TV(u_j^n)$ for the scheme to be TVD. This is because there is extra diffusion in computing the new cell averages in the conservative difference. In fact, the analysis by van Leer (see section 5.5) shows that MUSCL, minmod and superbee slopes are all TVD for linear advection.

5.9.3 Temporal Quadrature for Flux Integrals

Recall that we can integrate the conservation law over a space-time rectangle and apply the divergence theorem to obtain

$$\int_{x_{j-1/2}}^{x_{j+1/2}} u(x, t^{n+1})\, dx = \int_{x_{j-1/2}}^{x_{j+1/2}} u(x, t^n)\, dx - \int_{t^n}^{t^{n+1}} f(u(x_{j+1/2}, t))\, dt + \int_{t^n}^{t^{n+1}} f(u(x_{j-1/2}, t))\, dt.$$

In order to compute with this integral form of the conservation law, we need to approximate the time integrals of the flux. In the van Leer MUSCL scheme, we will use the midpoint rule and the solution of a Riemann problem to approximate the integrals:

$$\int_{t^n}^{t^{n+1}} f(u(x_{j+1/2}, t))\, dt \approx f_{j+1/2}^{n+1/2} \triangle t^{n+1/2}.$$

Here $f_{j+1/2}^{n+1/2} \approx f(u(x_{j+1/2}, t^n + \frac{1}{2}\triangle t^{n+1/2}))$ where $u(x, t)$ represents the exact solution of the conservation law with initial data given by the reconstruction $u^n(x)$. Because of the discontinuity in the piecewise linear reconstruction at the cell side $x_{j+1/2}$, this function value will require additional approximation.

5.9.4 Characteristic Tracing

In regions of smooth behavior, we can approximate

$$
\begin{aligned}
u&\left(x_j \pm \frac{1}{2}\triangle x_j, t^n + \frac{\triangle t^{n+1/2}}{2}\right)\\
&\approx u(x_j,t^n) \pm \frac{\partial u}{\partial x}(x_j,t^n)\frac{\triangle x_j}{2} + \frac{\partial u}{\partial t}(x_j,t^n)\frac{\triangle t^{n+1/2}}{2}\\
&= u(x_j,t^n) \pm \frac{\partial u}{\partial x}(x_j,t^n)\frac{\triangle x_j}{2} - \frac{\partial f}{\partial u}(u(x_j,t^n))\frac{\partial u}{\partial x}(x_j,t^n)\frac{\triangle t^{n+1/2}}{2}\\
&= u_j^n \pm s_j^n\frac{\triangle x_j}{2} - \lambda_j^n s_j^n \frac{\triangle t^{n+1/2}}{2} = u^n\left(x_j \pm \frac{\triangle x_j}{2}\left[1 \mp \frac{\lambda_j \triangle t^{n+1/2}}{\triangle x_j}\right]\right).
\end{aligned}
$$

Thus for arbitrarily small $\epsilon > 0$ we approximate

$$
\begin{aligned}
u\left(x_{j+1/2} - \epsilon, t^n + \frac{1}{2}\triangle t^{n+1/2}\right) &\approx u_j^n + \frac{1}{2}\left(1 - \frac{\lambda_j \triangle t^{n+1/2}}{\triangle x_j}\right) s_j^n \triangle x_j\\
u\left(x_{j+1/2} + \epsilon, t^n + \frac{1}{2}\triangle t^{n+1/2}\right) &\approx u_{j+1}^n - \frac{1}{2}\left(1 + \frac{\lambda_{j+1} \triangle t^{n+1/2}}{\triangle x_{j+1}}\right) s_{j+1}^n \triangle x_{j+1}.
\end{aligned}
$$

This corresponds to tracing the solution backward along characteristics to the initial data.

Note that, depending on the sign of λ_j, the characteristics could trace backward into the wrong cell. To avoid this problem, it is common to use the **characteristic projection**

$$
\begin{aligned}
u_{j+1/2}^L &= \begin{cases} u_j^n + \frac{1}{2}\left(1 - \frac{\lambda_j \triangle t^{n+1/2}}{\triangle x_j}\right) s_j^n \triangle x_j, & \lambda_j^n > 0\\ u_j^n, & \lambda_j^n \le 0 \end{cases}\\
u_{j+1/2}^R &= \begin{cases} u_{j+1}^n - \frac{1}{2}\left(1 + \frac{\lambda_{j+1} \triangle t^{n+1/2}}{\triangle x_{j+1}}\right) s_{j+1}^n \triangle x_{j+1}, & \lambda_{j+1}^n < 0\\ u_{j+1}^n, & \lambda_{j+1}^n \ge 0. \end{cases}
\end{aligned}
$$

In other words, the characteristic projection discards slope information coming from characteristics going in the wrong direction.

When a characteristic traces the wrong way, the state determined by characteristic projection is only first-order accurate. For example, if $\lambda_j^n \le 0$, then $u_{j+1/2}^L = u_j^n$ is a first-order approximation to $w^n(x_{j+1/2}, t^n + \frac{1}{2}\triangle t^{n+1/2})$. If $\lambda_j^n \le 0$ and $\lambda_{j+1}^n \ge 0$ (corresponding to a transonic rarefaction), then both $u_{j+1/2}^L$ and $u_{j+1/2}^R$ will be first-order accurate. This can significantly degrade the accuracy of the MUSCL scheme. We have chosen not to use the characteristic projection in our numerical experiments.

5.9.5 Flux Evaluation

The flux in the MUSCL scheme is computed at the solution of the Riemann problem with left and right states given by $u^L_{j+1/2}$ and $u^R_{j+1/2}$:

$$f^{n+1/2}_{j+1/2} = f(\mathcal{R}(u^L_{j+1/2}, u^R_{j+1/2}; 0)).$$

Note that if $\partial f/\partial u > 0$ for all u, then

$$\mathcal{R}(u^L_{j+1/2}, u^R_{j+1/2}; 0) = u^L_{j+1/2} = u^n_j + \frac{1}{2}\left(1 - \frac{\lambda_j \triangle t^{n+1/2}}{\triangle x_j}\right) s^n_j \triangle x_j$$

is second-order accurate in regions of smooth behavior; the first-order evaluation $u^R_{j+1/2} = u^n_{j+1}$ had no effect on the flux. A similar statement holds in the case when $\partial f/\partial u < 0$ for all u. For transonic expansions (*i.e.*, when $(\partial f/\partial u)_j < 0 < (\partial f)(\partial u)_{j+1},$) the characteristic projection reduces the method to Godunov's (first-order) method. For transonic compressions (*i.e.*, when $(\partial f/\partial u)_j > 0 > (\partial f)(\partial u)_{j+1},$) the characteristic tracing is used on both sides of $x_{j+1/2}$, but the limited slopes should be zero.

Example 5.9.2 Let us describe the MUSCL scheme for Burgers' equation on a uniform grid. Given the values u^n_j, we compute $\lambda^n_j = \frac{\partial f}{\partial u}(u^n_j) = u^n_j$, and choose $\triangle t^{n+1/2}$ so that

$$\triangle t^{n+1/2} = \gamma \triangle x \min_j \left\{\frac{1}{|\lambda_j|}\right\}.$$

Here $0 < \gamma \le 1$ is the *CFL* factor, and is chosen by the user. Next, we compute the side increments

$$\triangle u^n_{j+1/2} = u^n_{j+1} - u^n_j,$$

the centered increments

$$\triangle u^n_j = \frac{1}{2}(\triangle u^n_{j+1/2} + \triangle u^n_{j-1/2})$$

and the MUSCL slopes

$$s^n_j \triangle x = \begin{cases} \operatorname{sign}(\triangle u^n_j) \min\{2|\triangle u^n_{j-1/2}|, 2|\triangle u^n_{j+1/2}|, |\triangle u^n_j|\}, & \triangle u^n_{j+1/2}\triangle u^n_{j-1/2} > 0 \\ 0, & \text{otherwise.} \end{cases}$$

Afterward, compute the left and right states:

$$u^L_{j+1/2} = u^n_j + \frac{1}{2}\left(1 - \frac{\lambda^n_j \triangle t^{n+1/2}}{\triangle x}\right) s^n_j \triangle x$$

$$u^R_{j+1/2} = u^n_{j+1} - \frac{1}{2}\left(1 + \frac{\lambda^n_{j+1} \triangle t^{n+1/2}}{\triangle x}\right) s^n_{j+1} \triangle x.$$

These states are used to evaluate the flux at the solution of the Riemann problem for Burgers' equation:

$$f_{j+1/2}^{n+1/2} = \begin{cases} \frac{1}{2}\max\{|u_{j+1/2}^L|, |u_{j+1/2}^R|\}^2, & u_{j+1/2}^L > u_{j+1/2}^R \\ \frac{1}{2}\max\{u_{j+1/2}^L, \min\{u_{j+1/2}^R, 0\}\}^2, & u_{j+1/2}^L \le u_{j+1/2}^R. \end{cases}$$

Finally, we use a conservative difference to compute the new solution:

$$u_j^{n+1} = u_j^n - \frac{\triangle t^{n+1/2}}{\triangle x}[f_{j+1/2}^{n+1/2} - f_{j-1/2}^{n+1/2}].$$

A C^{++} program to implement the MUSCL scheme for nonlinear scalar conservation laws can be found in **Program 5.9-70: Schemes2.C** Students can exercise this program by clicking on **Executable 5.9-30: guiriemann2** The user can select Riemann problem initial data for linear advection, Burgers' equation, traffic models and the Buckley-Leverett model. In addition, the user can select from a variety of limiters and Riemann solvers. By setting the number of cells to 0, the user will cause the program to perform a mesh refinement study, and the user can select several different schemes for comparison.

5.9.6 Non-Reflecting Boundaries with the MUSCL Scheme

Finally, let us discuss the treatment of a non-reflecting boundary with the MUSCL scheme. For simplicity, let us assume that the non-reflecting boundary is at the right-hand side of the domain. Suppose that we use ghost cells and set the solution in the ghost cells to be equal to u_j^n, the solution in the last cell within the domain. Then we will have $\triangle u_{j+1/2}^n = 0$, and the slope in the last cell will be $s_j^n \triangle x_j = 0$. Characteristic tracing will approximate $w^n(x_{j+1/2}, t^n + \triangle t^{n+1/2}/2) \approx u_j^n$ on either side of the right-hand boundary. This gives us a first-order treatment of the nonreflecting boundary. Since the waves should be outgoing at the nonreflecting boundary, the first-order treatment should not significantly degrade the quality of the solution in the interior of the domain.

Exercises for 5.9

5.9.1 Program the slope-limiter scheme for linear advection, and test it on the problems in Exercise 2.2.5 of Section 2.2.

5.9.2 Program van Leer's MUSCL scheme for Burgers' equation on the domain $-1 < x < 1$. For each of the following test problems, plot the logarithm of the $\mathcal{L}_1$ norm of the error in the solution versus $\log \triangle x$ at fixed time $t = 0.4$:

(a) $u(x,0) = 1$ for $x < 0$ and $u(x,0) = 2$ for $x > 0$

(b) $u(x,0) = 2$ for $x < 0$ and $u(x,0) = 1$ for $x > 0$

(c) $u(x,0) = 2$ for $x < 0$ and $u(x,0) = -1$ for $x > 0$
(d) $u(x,0) = -2$ for $x < 0$ and $u(x,0) = 1$ for $x > 0$.
Describe how you evaluated the $\mathcal{L}_1$ norm. If it is the case, describe why you did not see second-order convergence. Also plot the error for Godunov's method for these problems.

5.9.3 Repeat the previous exercise on a non-uniform mesh. Let $\triangle x_j = 2/(3N)$ for j even and $\triangle x_j = 1/(3N)$ for j odd.

5.9.4 In Exercise 4.13.11 of Section 4.13 we suggested the use of a moving mesh to capture the solutions of Riemann problems. Describe how you could modify the MUSCL scheme to operate on a moving mesh. Program your scheme and test it on the problems in Exercise 4.13.11

5.10 Wave Propagation Slope Limiter Schemes

Goodman and LeVeque [59] and LeVeque [97] have suggested alternative forms of the slope limiter scheme that contains features of Sweby's TVD scheme. These are intended to be used for convex flux functions only, although they work well for the Buckley–Leverett problem. There are some distinct differences in the two approaches, so we will present them separately.

5.10.1 Cell-Centered Wave Propagation

Suppose that we want to solve $\partial u/\partial t + \partial f(u)/\partial x = 0$ where $f(u)$ is convex. Following [59] we will use a piecewise linear reconstruction

$$u(x,t^n) \approx \overline{u}^n_j(x) \equiv u^n_j + s^n_j(x - x_j),\, x \in (x_{j-1/2}, x_{j+1/2})$$

where $s^n_j \triangle x_j \equiv \text{minmod}(\triangle u^n_{j-1/2}, \triangle u^n_{j+1/2})$. Recall that the **minmod limiter** is defined by

$$\text{minmod}(a,b) = \begin{cases} \min\{|a|,|b|\}, & ab > 0 \\ 0, & ab \le 0. \end{cases}$$

It is important that the minmod limiter be used in this scheme, so that the resulting method is TVD.

Define the values of the reconstruction at the cell sides by

$$u^L_{j+1/2} = \overline{u}^n_j(x_{j+1/2}) = u^n_j + \frac{1}{2} s^n_j \triangle x_j$$

$$u^R_{j+1/2} = \overline{u}^n_{j+1}(x_{j+1/2}) = u^n_{j+1} - \frac{1}{2} s^n_{j+1} \triangle x_{j+1}.$$

We will replace $f(u)$ by a piecewise linear interpolant $\tilde{f}$; in regions of monotonicity for $u^R_{j-1/2}, u^L_{j+1/2}, u^R_{j+1/2}$ and $u^L_{j+3/2}$ this piecewise linear interpolant will be a well-defined function described below. We will approximate the solution of the original problem by the solution of

$$\frac{\partial \tilde{u}}{\partial t} + \frac{\partial \tilde{f}(u)}{\partial x} = 0$$
$$\tilde{u}(x, t^n) = \overline{u}_j^n(x), \; x \in (x_{j-1/2}, x_{j+1/2}).$$

Define the slopes

$$\tilde{f}_j' = \begin{cases} [f(u_{j+1/2}^L) - f(u_{j-1/2}^R)]/[u_{j+1/2}^L - u_{j-1/2}^R], & u_{j+1/2}^L - u_{j-1/2}^R \neq 0 \\ f'(u_j^n), & u_{j+1/2}^L - u_{j-1/2}^R = 0 \end{cases}$$

$$\tilde{f}_{j+1/2}' = \begin{cases} [f(u_{j+1/2}^R) - f(u_{j+1/2}^L)]/[u_{j+1/2}^R - u_{j+1/2}^L], & u_{j+1/2}^R - u_{j+1/2}^L \neq 0 \\ f'(u_{j+1/2}^L), & u_{j+1/2}^R - u_{j+1/2}^L = 0. \end{cases}$$

Then for $u \in \text{int}(u_{j-1/2}^R, u_{j+1/2}^L)$,

$$\tilde{f}(u) = f(u_{j+1/2}^L) + \tilde{f}_j'(u - u_{j+1/2}^L) = f(u_{j-1/2}^R) + \tilde{f}_j'(u - u_{j-1/2}^R)$$

and for $u \in \text{int}(u_{j+1/2}^L, u_{j+1/2}^R)$,

$$\tilde{f}(u) = f(u_{j+1/2}^L) + \tilde{f}_{j+1/2}'(u - u_{j+1/2}^L) = f(u_{j+1/2}^R) + \tilde{f}_{j+1/2}'(u - u_{j+1/2}^R).$$

In order to compute

$$\frac{1}{\triangle x_j} \int_{x_{j-1/2}}^{x_{j+1/2}} \tilde{u}(x, t^{n+1}) \, dx = \frac{1}{\triangle x_j} \int_{x_{j-1/2}}^{x_{j+1/2}} \tilde{u}(x, t^n) \, dx$$
$$- \frac{\triangle t^{n+1/2}}{\triangle x_j} \left[\frac{1}{\triangle t^{n+1/2}} \int_{t^n}^{t^{n+1}} \tilde{f}(u(x_{j+1/2}, t) \, dt - \frac{1}{\triangle t^{n+1/2}} \int_{t^n}^{t^{n+1}} \tilde{f}(u(x_{j-1/2}, t) \, dt \right]$$

we need to determine the temporal integrals of the piecewise linear flux interpolant. First, let us consider the case in which $\tilde{f}_j' > 0$ and $\tilde{f}_{j+1}' > 0$. Convexity of f will imply that $\tilde{f}_{j+1/2}' > 0$ as well. Then Equation (3.5) for the solution of a scalar nonlinear hyperbolic conservation law gives us

$$\frac{1}{\triangle t^{n+1/2}} \int_{t^n}^{t^{n+1}} \tilde{f}(u(x_{j+1/2}, t) \, dt$$
$$= \frac{1}{\triangle t^{n+1/2}} \int_{t^n}^{t^{n+1}} \tilde{f}(\overline{u}^n(x_{j+1/2} - [t - t^n]\tilde{f}'(u(x_{j+1/2}, t)))) \, dt$$
$$= \frac{1}{\triangle t^{n+1/2}} \int_{t^n}^{t^{n+1}} f(u_{j+1/2}^L) + \tilde{f}_j' \left[\overline{u}_j(x_{j+1/2} - [t - t^n]\tilde{f}_j') - u_{j+1/2}^L \right] \, dt$$
$$= f(u_{j+1/2}^L) + \frac{\tilde{f}_j'}{\triangle t^{n+1/2}} \int_{t^n}^{t^{n+1}} \left\{ u_j^n + s_j^n \left[\frac{1}{2}\triangle x_j - [t - t^n]\tilde{f}_j' \right] - u_j^n - s_j^n \frac{\triangle x_j}{2} \right\} \, dt$$
$$= f(u_{j+1/2}^L) - \frac{(\tilde{f}_j')^2}{\triangle t^{n+1/2}} s_j^n \int_{t^n}^{t^{n+1}} t - t^n \, dt = f(u_{j+1/2}^L) - (\tilde{f}_j')^2 s_j^n \triangle t^{n+1/2} \equiv \tilde{f}_{j+1/2}^L$$

provided that $\tilde{f}'_j \triangle t^{n+1/2} \leq \triangle x_j$. On the other hand, if $\tilde{f}'_j < 0$ and $\tilde{f}'_{j+1} < 0$, then

$$\begin{aligned}
&\frac{1}{\triangle t^{n+1/2}} \int_{t^n}^{t^{n+1}} \tilde{f}(u(x_{j+1/2}, t)\, dt \\
&= \frac{1}{\triangle t^{n+1/2}} \int_{t^n}^{t^{n+1}} f(u^R_{j+1/2}) + \tilde{f}'_{j+1} \left[\overline{u}_{j+1}(x_{j+1/2} - [t - t^n]\tilde{f}'_{j+1}) - u^R_{j+1/2} \right] dt \\
&= f(u^R_{j+1/2}) + \frac{\tilde{f}'_{j+1}}{\triangle t^{n+1/2}} \int_{t^n}^{t^{n+1}} \left\{ u^n_{j+1} + s^n_{j+1} \left[-\frac{1}{2}\triangle x_{j+1} - [t - t^n]\tilde{f}'_{j+1} \right] - u^n_{j+1} + s^n_{j+1}\frac{\triangle x_{j+1}}{2} \right\} dt \\
&= f(u^R_{j+1/2}) - \frac{(\tilde{f}'_{j+1})^2}{\triangle t^{n+1/2}} s^n_{j+1} \int_{t^n}^{t^{n+1}} t - t^n \, dt = f(u^R_{j+1/2}) - (\tilde{f}'_{j+1})^2 s^n_{j+1} \triangle t^{n+1/2} \equiv \tilde{f}^R_{j+1/2}
\end{aligned}$$

provided that $-\tilde{f}'_{j+1} \triangle t^{n+1/2} \leq \triangle x_{j+1}$.

Next, let us consider cases in which the slopes change sign. If $\tilde{f}'_j < 0 < \tilde{f}'_{j+1}$, then we have a rarefaction at $x_{j+1/2}$. If

$$u_* = \text{argmin}(f)$$

then we can add $(u_*, f(u_*))$ as a interpolation point for $\tilde{f}$, and get

$$\frac{1}{\triangle t^{n+1/2}} \int_{t^n}^{t^{n+1}} \tilde{f}(u(x_{j+1/2}, t)\, dt = \frac{1}{\triangle t^{n+1/2}} \int_{t^n}^{t^{n+1}} \tilde{f}(u_*)\, dt = f(u_*).$$

On the other hand, if $\tilde{f}'_j > 0 > \tilde{f}'_{j+1}$, then we have a shock at $x_{j+1/2}$. Because the initial data is not constant, the speed of this shock is not constant at $x_{j+1/2}$ and $t > t^n$. If $\tilde{f}'_{j+1/2} \neq 0$ and $u^L_{j+1/2} \neq u^R_{j+1/2}$, then we will use the initial motion of the shock to determine an approximation to the flux integral:

$$\frac{1}{\triangle t^{n+1/2}} \int_{t^n}^{t^{n+1}} \tilde{f}(u(x_{j+1/2}, t)\, dt \approx \begin{cases} \tilde{f}^L_{j+1/2}, & \tilde{f}'_{j+1/2} > 0 \\ \tilde{f}^R_{j+1/2}, & \tilde{f}'_{j+1/2} < 0. \end{cases}$$

If either $\tilde{f}'_{j+1/2} = 0$ or $u^L_{j+1/2} = u^R_{j+1/2}$, then we will approximate

$$\frac{1}{\triangle t^{n+1/2}} \int_{t^n}^{t^{n+1}} \tilde{f}(u(x_{j+1/2}, t)\, dt \approx \begin{cases} \tilde{f}^L_{j+1/2}, & (\tilde{f}'_j)^2 s^n_j > (\tilde{f}'_{j+1})^2 s^n_{j+1} \\ \tilde{f}^R_{j+1/2}, & \text{otherwise.} \end{cases}$$

We take the new solution to be

$$\begin{aligned}
u^{n+1}_j &= \frac{1}{\triangle x_j} \int_{x_{j-1/2}}^{x_{j+1/2}} \tilde{u}(x, t^n)\, dx \\
&\quad - \frac{\triangle t^{n+1/2}}{\triangle x_j} \left[\frac{1}{\triangle t^{n+1/2}} \int_{t^n}^{t^{n+1}} \tilde{f}(u(x_{j+1/2}, t)\, dt - \frac{1}{\triangle t^{n+1/2}} \int_{t^n}^{t^{n+1}} \tilde{f}(u(x_{j-1/2}, t)\, dt \right].
\end{aligned}$$

The resulting algorithm is second-order accurate, because we used a second-order accurate (piecewise-linear) approximation to the flux function. The timestep can be chosen so that

$$\triangle t^{n+1/2} \max \left\{ \left| \frac{df}{du} \right| \right\} \leq \triangle x_j .$$

5.10.2 Side-Centered Wave Propagation

The next algorithm is described in even greater detail in [97, p. 118ff], and is the basis for the CLAWPACK software. We are given the current cell averages u^n_j, mesh widths $\triangle x_j$ and a timestep $\triangle t^{n+1/2}$ satisfying

$$\triangle t^{n+1/2} \max \left\{ \left| \frac{df}{du} \right| \right\} \leq \triangle x_j .$$

For each cell side $j+\frac{1}{2}$ we compute the solution increment $\triangle u^n_{j+1/2} = u^n_{j+1} - u^n_j$ and the average speed

$$\lambda^n_{j+1/2} = \frac{f(u^n_{j+1}) - f(u^n_j)}{u^n_{j+1} - u^n_j} .$$

Next, for each cell side $j+\frac{1}{2}$ we compute the limited slope

$$\overline{\triangle u}^n_{j+1/2} = \begin{cases} \text{limiter}(\triangle u_{j-1/2}, \triangle u_{j+1/2}), & \lambda^n_{j+1/2} \geq 0 \\ \text{limiter}(\triangle u_{j+1/2}, \triangle u_{j+3/2}), & \lambda^n_{j+1/2} < 0. \end{cases}$$

Then the flux at side $j+\frac{1}{2}$ is

$$f^{n+1/2}_{j+1/2} = f(\mathcal{R}(u^n_j, u^n_{j+1}; 0)) + \frac{1}{2} |\lambda^n_{j+1/2}| \left(1 - \frac{2\triangle t^{n+1/2}}{\triangle x_j + \triangle x_{j+1}} |\lambda^n_{j+1/2}| \right) \overline{\triangle u}^n_{j+1/2} .$$

Here $f(\mathcal{R}(u^n_j, u^n_{j+1}; 0))$ is the flux at the state that moves with zero speed in the solution of the Riemann problem, or any numerical flux that approximates this value. The solution is updated by a conservative difference.

LeVeque prefers to implement the scheme by decomposing the flux jump into waves (CLAWPACK subroutine `rp1`),

$$\begin{aligned} f(u^n_{j+1}) - f(u^n_j) &= [f(u^n_{j+1}) - f(\mathcal{R}(u^n_j, u^n_{j+1}; 0))] - [f(u^n_j) - f(\mathcal{R}(u^n_j, u^n_{j+1}; 0))] \\ &\equiv \alpha^+_{j+1/2} \triangle u^n_{j+1/2} + \alpha^-_{j+1/2} \triangle u^n_{j+1/2} , \end{aligned}$$

defining the second-order corrections to the flux integrals (loop 120 in CLAWPACK subroutine `step1`)

$$\triangle \tilde{f}_{j+1/2} = |\lambda^n_{j+1/2}| \left(1 - \frac{2\triangle t^{n+1/2}}{\triangle x_j + \triangle x_{j+1}} |\lambda^n_{j+1/2}| \right) \overline{\triangle u}^n_{j+1/2}$$

and writing the conservative difference in the form

$$u_j^{n+1} = u_j^n - \frac{\triangle t^{n+1/2}}{\triangle x_j}\left[\alpha_{j-1/2}^{+}\triangle u_{j-1/2}^n + \alpha_{j+1/2}^{-}\triangle u_{j+1/2}^n + \frac{1}{2}(\triangle \tilde{f}_{j+1/2} - \triangle \tilde{f}_{j-1/2})\right].$$

Some care must be taken in the evaluation of $\lambda_{j+1/2}^n$ in constant states within the numerical solution. LeVeque also avoids computing the characteristic speeds at the cell-centered states u_j^n in order to compute a stable $\triangle t^{n+1/2}$; instead, he uses the average speeds $\lambda_{j+1/2}^n$ and repeats a timestep if the CFL stability restriction was violated.

A C^{++} program to implement the wave propagation schemes for nonlinear scalar conservation laws can be found in **Program 5.10-71: Schemes2.C** Simplified versions of these schemes for linear advection on uniform grids can be found in **Program 5.10-72: LinearAdvectionSchemes.C**. Students can exercise this program by clicking on **Executable 5.10-31: guiriemann2** The user can select Riemann problem initial data for linear advection, Burgers' equation, traffic models and the Buckley-Leverett model. In addition, the user can select from a variety of limiters and Riemann solvers. By setting the number of cells to 0, the user will cause the program to perform a mesh refinement study, and the user can select several different schemes for comparison. Students can also exercise this program for linear advection by clicking on **Executable 5.10-32: guilinearad2** The user can select a variety of initial values from the Zalesak test problems in exercise 2.2.5 of section 2.2. In addition, the user can select from a variety of limiters. By setting the number of cells to 0, the user will cause the program to perform a mesh refinement study.

5.11 Higher-Order Extensions of the Lax–Friedrichs Scheme

Nessyahu and Tadmor [121] have suggested a second-order version of the Lax–Friedrichs scheme (see Section 3.3.2) which involves no Riemann solvers, but requires a staggered grid. The essential differences between the Nessyahu–Tadmor scheme and classical Lax–Friedrichs are the use of a piecewise linear reconstruction of the solution in space, and the use of second-order quadratures in time.

The method begins by computing slopes $\triangle u_{j+1/2}^n = u_{j+1}^n - u_j^n$. A limiter is then applied to obtain cell-centered slopes $\triangle u_j^n = \text{limiter}(\triangle u_{j-1/2}^n, \triangle u_{j+1/2}^n)$. Any of the limiters in Section 5.8.2 could be used. The cell-centered slope provides a piecewise-linear reconstruction $u_j^n(x) = u_j^n + \triangle u_j^n(x - x_j)/(x_{j+1/2} - x_{j-1/2})$ as in the slope-limiter scheme of Section 5.9.2. We will assume that the timestep is chosen by the CFL condition

$$\triangle t^{n+1/2} \max_u \left|\frac{\partial f}{\partial u}\right| \leq \min_j \triangle x_j .$$

Integration of the conservation law over the two half-cells $x \in (x_j, x_{j+1})$ and a half-timestep $t \in (t^n, t^n + \triangle t^{n+1/2}/2)$ with this piecewise linear initial data leads to

$$\begin{aligned}\int_{x_j}^{x_{j+1}} u(x, t^{n+1/2})\, dx &= \int_{x_j}^{x_{j+1/2}} u_j^n + \triangle u_j^n \frac{x - x_j}{\triangle x_j}\, dx + \int_{x_{j+1/2}}^{x_{j+1}} u_{j+1}^n + \triangle u_{j+1}^n \frac{x - x_{j+1}}{\triangle x_{j+1}}\, dx \\ &\quad - \int_{t^n}^{t^n+\triangle t^{n+1/2}/2} f(u(x_{j+1}, t))\, dt + \int_{t^n}^{t^n+\triangle t^{n+1/2}/2} f(u(x_j, t))\, dt \\ &= \left[u_j^n + \frac{\triangle u_j^n}{4}\right]\frac{\triangle x_j}{2} + \left[u_{j+1}^n - \frac{\triangle u_{j+1}^n}{4}\right]\frac{\triangle x_{j+1}}{2} \\ &\quad - \int_{t^n}^{t^n+\triangle t^{n+1/2}/2} f(u(x_{j+1}, t))\, dt + \int_{t^n}^{t^n+\triangle t^{n+1/2}/2} f(u(x_j, t))\, dt.\end{aligned}$$

Here $u(x, t)$ refers to the exact solution of the conservation law with the piecewise linear initial data. In order to approximate the flux time integrals, we will use the midpoint rule for time integration, and a Taylor approximation for the function value:

$$\begin{aligned}\int_{t^n}^{t^n+\triangle t^{n+1/2}/2} f(u(x_j, t))\, dt &\approx \frac{\triangle t^{n+1/2}}{2} f\left(u\left(x_j, \frac{\triangle t^{n+1/2}}{4}\right)\right) \\ &\approx \frac{\triangle t^{n+1/2}}{2} f\left(u(x_j, t^n) + \frac{\partial u}{\partial t}(x_j, t^n)\frac{\triangle t^{n+1/2}}{4}\right) \\ &= \frac{\triangle t^{n+1/2}}{2} f(u_j^n) - \frac{\partial f}{\partial u}(u(x_j, t^n))\frac{\partial u}{\partial x}(x_j, t^n)\frac{\triangle t^{n+1/2}}{4}.\end{aligned}$$

Note that the use of a Taylor approximation is justified because $u(x_j, t)$ is smooth for $t^n \le t < t^n + \triangle t^{n+1/2}/2$. These results suggest the following computations for the first half-step of the scheme. At each cell center we compute the conserved quantity and flux

$$u_j^{n+1/4} = u_j^n - \frac{\partial f}{\partial u}(u_j^n)\triangle u_j^n \frac{\triangle t^{n+1/2}}{4\triangle x_j} \quad \text{and} \quad f_j^{n+1/4} = f(u_j^{n+1/4}).$$

and then at each cell side we compute the conserved quantity

$$\begin{aligned}u_{j+1/2}^{n+1/2} = &\left\{\left[u_j^n + \frac{\triangle u_j^n}{4}\right]\triangle x_j + \left[u_{j+1}^n - \frac{\triangle u_{j+1}^n}{4}\right]\triangle x_{j+1}\right. \\ &\left. - \left[f_{j+1}^{n+1/4} - f_j^{n+1/4}\right]\triangle t^{n+1/2}\right\}\frac{1}{\triangle x_j + \triangle x_{j+1}}.\end{aligned}$$

The second half-step is similar. First, we compute $\triangle u_j^{n+1/2} = u_{j+1/2}^{n+1/2} - u_{j-1/2}^{n+1/2}$. At the cell sides, we apply the limiter to get

$$\triangle u_{j+1/2}^{n+1/2} = \text{limiter}(\triangle u_j^{n+1/2}, \triangle u_{j+1}^{n+1/2}),$$

and compute the conserved quantity and flux by

$$u_{j+1/2}^{n+3/4} = u_{j+1/2}^{n} - \frac{\partial f}{\partial u}(u_{j+1/2}^{n+1/2}) \triangle u_{j+1/2}^{n+1/2} \frac{\triangle t^{n+1/2}}{2(\triangle x_j + \triangle x_{j+1})} \quad \text{and} \quad f_{j+1/2}^{n+3/4} = f(u_{j+1/2}^{n+3/4}).$$

Then at each cell center we compute

$$\begin{aligned} u_j^{n+1} = \frac{1}{2} \Bigg\{ & \left[u_{j-1/2}^{n+1/2} + \frac{\triangle u_{j-1/2}^{n+1/2}}{4} \right] + \left[u_{j+1/2}^{n+1/2} - \frac{\triangle u_{j+1/2}^{n+1/2}}{4} \right] \\ & - \left[f_{j+1/2}^{n+3/4} - f_{j-1/2}^{n+3/4} \right] \frac{\triangle t^{n+1/2}}{\triangle x_j} \Bigg\}. \end{aligned}$$

In a later paper [110], Liu and Tadmor describe a third-order version of the Lax–Friedrichs scheme. This scheme requires higher-order piecewise polynomial reconstruction in space, and higher-order quadratures in time. Their algorithm is designed for uniform grids only, but the algorithm described below is a natural extension of their ideas.

Let $p_j(x)$ interpolate $\int_{x_{j-3/2}}^{x} u(s, t^n)\, ds$ at $x_{j\pm 3/2}$ and $x_{j\pm 1/2}$. Then the Newton form of this interpolating polynomial is

$$\begin{aligned} p_j(x) = & u_{j-1}^n (x - x_{j-3/2}) + u^n[x_{j-1}, x_j](x - x_{j-3/2})(x - x_{j-1/2}) \\ & + u^n[x_{j-1}, x_j, x_{j+1}](x - x_{j-3/2})(x - x_{j-1/2})(x - x_{j+1/2}). \end{aligned}$$

where the divided differences are

$$u^n[x_{j-1}, x_j] = \frac{u_j^n - u_{j-1}^n}{\triangle x_j + \triangle x_{j-1}} \quad \text{and} \quad u^n[x_{j-1}, x_j, x_{j+1}] = \frac{u^n[x_j, x_{j+1}] - u^n[x_{j-1}, x_j]}{\triangle x_{j+1} = \triangle x_j + \triangle x_{j-1}}.$$

The derivative of this interpolating polynomial is

$$q_j(x) = \frac{dp_j}{dx} = u_j^n + \alpha_j + \xi_j(x)\beta_j + \frac{1}{2}\xi_j(x)^2 \gamma_j \text{ where } \xi_j(x) \equiv \frac{x - x_j}{\triangle x_j}.$$

It is easy to see that

$$\begin{aligned} \alpha_j = q_j(x_j) - u_j^n = & -\triangle u_{j-1/2}^n + \frac{\triangle u_{j-1/2}^n}{\triangle x_j + \triangle x_{j-1}} \left\{ \left(\frac{\triangle x_j}{2} + \triangle x_{j-1} \right) + \frac{\triangle x_j}{2} \right\} \\ & + \left[\frac{\triangle u_{j+1/2}^n}{\triangle x_{j+1} + \triangle x_j} - \frac{\triangle u_{j-1/2}^n}{\triangle x_j + \triangle x_{j-1}} \right] \frac{(\frac{\triangle x_j}{2} + \triangle x_{j-1})\frac{\triangle x_j}{2} - (\frac{\triangle x_j}{2}\frac{\triangle x_j}{2}) - \frac{\triangle x_j}{2}(\frac{\triangle x_j}{2} + \triangle x_{j-1})}{\triangle x_{j+1} + \triangle x_j + \triangle x_{j-1}} \\ = & \frac{1}{4} \left[\frac{\triangle u_{j+1/2}^n}{\triangle x_{j+1} + \triangle x_j} - \frac{\triangle u_{j-1/2}^n}{\triangle x_j + \triangle x_{j-1}} \right] \frac{\triangle x_j^2}{\triangle x_{j+1} + \triangle x_j + \triangle x_{j-1}}. \end{aligned}$$

The derivative of q_j is

$$q'_j(x) = 2u^n[x_{j-1}, x_j] + 2u^n[x_{j-1}, x_j, x_{j+1}]\{(x - x_{j-3/2}) + (x - x_{j-1/2}) + (x - x_{j+1/2})\}$$
$$= \frac{\beta_j + \xi_j(x)\gamma_j}{\triangle x_j}$$

From this it is easy to compute

$$\beta_j = \triangle x_j q'_j(x_j) = 2\triangle x_j \frac{\triangle u^n_{j-1/2}}{\triangle x_j + \triangle x_{j-1}}$$
$$+ 2\triangle x_j \left[\frac{\triangle u^n_{j+1/2}}{\triangle x_{j+1} + \triangle x_j} - \frac{\triangle u^n_{j-1/2}}{\triangle x_j + \triangle x_{j-1}}\right] \frac{(\frac{\triangle x_j}{2} + \triangle x_{j-1}) + \frac{\triangle x_j}{2} - \frac{\triangle x_j}{2}}{\triangle x_{j+1} + \triangle x_j + \triangle x_{j-1}}$$
$$= \triangle x_j \left\{2\frac{\triangle u^n_{j-1/2}}{\triangle x_j + \triangle x_{j-1}} + \left[\frac{\triangle u^n_{j+1/2}}{\triangle x_{j+1} + \triangle x_j} - \frac{\triangle u^n_{j-1/2}}{\triangle x_j + \triangle x_{j-1}}\right] \frac{\triangle x_j + 2\triangle x_{j-1}}{\triangle x_{j+1} + \triangle x_j + \triangle x_{j-1}}\right\}$$
$$= \left\{\frac{\triangle u^n_{j-1/2}}{\triangle x_j + \triangle x_{j-1}}(2\triangle x_{j+1} + \triangle x_j) + \frac{\triangle u^n_{j+1/2}}{\triangle x_{j+1} + \triangle x_j}(\triangle x_j + 2\triangle x_{j-1})\right\} \frac{\triangle x_j}{\triangle x_{j+1}+\triangle x_j+\triangle x_{j-1}}.$$

The second derivative of q_j is

$$q''_j(x_j) = 6u^n[x_{j-1}, x_j, x_{j+1}] = \frac{\gamma_j}{\triangle x_j^2}$$

so

$$\gamma_j = 6\left[\frac{\triangle u^n_{j+1/2}}{\triangle x_{j+1} + \triangle x_j} - \frac{\triangle u^n_{j-1/2}}{\triangle x_j + \triangle x_{j-1}}\right] \frac{\triangle x_j^2}{\triangle x_{j+1} + \triangle x_j + \triangle x_{j-1}} = -24\alpha_j.$$

On a uniform grid, these simplify to

$$\alpha_j = \frac{\triangle u^n_{j-1/2} - \triangle u^n_{j+1/2}}{24},\ \beta_j = \frac{\triangle u^n_{j+1/2} + \triangle u^n_{j-1/2}}{2} \quad \text{and} \quad \gamma_j = \triangle u^n_{j+1/2} - \triangle u^n_{j-1/2}.$$

We will also need the values

$$q_j(x_{j-1/2}) = u^n_j + \alpha_j - \frac{1}{2}\beta_j + \frac{1}{8}\gamma_j = u^n_j - 2\alpha_j - \frac{1}{2}\beta_j$$
$$q_j(x_{j+1/2}) = u^n_j + \alpha_j + \frac{1}{2}\beta_j + \frac{1}{8}\gamma_j = u^n_j - 2\alpha_j + \frac{1}{2}\beta_j.$$

The extreme point of $q_j(x)$ occurs at x_j^* where $\xi_j(x_j^*) = -\beta_j/\gamma_j$. Note that $x_j^* \in (x_{j-1/2}, x_{j+1/2})$ if and only if $|\xi_j(x_j^*)| \le \frac{1}{2}$; equivalently, $2|\beta_j| \le |\gamma_j|$. Using the equations for β_j and γ_j, we see that $x_j^* \in (x_{j-1/2}, x_{j+1/2})$ if and only if either

$$\frac{\triangle u^n_{j+1/2}}{\triangle x_{j+1} + \triangle x_j} > \frac{\triangle u^n_{j-1/2}}{\triangle x_j + \triangle x_{j-1}} \text{ and}$$

$$\frac{\triangle u^n_{j+1/2}}{\triangle x_{j+1} + \triangle x_j} > \frac{\triangle u^n_{j-1/2}}{\triangle x_j + \triangle x_{j-1}} \frac{\triangle x_j - \triangle x_{j+1}}{2\triangle x_j + \triangle x_{j-1}} \text{ and}$$

$$\frac{\triangle u^n_{j-1/2}}{\triangle x_j + \triangle x_{j-1}} < \frac{\triangle u^n_{j+1/2}}{\triangle x_{j+1} + \triangle x_j} \frac{\triangle x_j - \triangle x_{j-1}}{2\triangle x_j + \triangle x_{j+1}}$$

or

$$\frac{\triangle u^n_{j+1/2}}{\triangle x_{j+1} + \triangle x_j} < \frac{\triangle u^n_{j-1/2}}{\triangle x_j + \triangle x_{j-1}} \text{ and}$$

$$\frac{\triangle u^n_{j+1/2}}{\triangle x_{j+1} + \triangle x_j} < \frac{\triangle u^n_{j-1/2}}{\triangle x_j + \triangle x_{j-1}} \frac{\triangle x_j - \triangle x_{j+1}}{2\triangle x_j + \triangle x_{j-1}} \text{ and}$$

$$\frac{\triangle u^n_{j-1/2}}{\triangle x_j + \triangle x_{j-1}} < \frac{\triangle u^n_{j+1/2}}{\triangle x_{j+1} + \triangle x_j} \frac{\triangle x_j - \triangle x_{j-1}}{2\triangle x_j + \triangle x_{j+1}}.$$

On a uniform grid, these conditions are equivalent to either $\triangle u^n_{j+1/2} > 0 > \triangle u^n_{j-1/2}$ or $\triangle u^n_{j+1/2} < 0 < \triangle u^n_{j-1/2}$. Thus on a uniform grid, the quadratic reconstruction has no extrema interior to a grid cell unless the cell averages have a local extremum there as well.

In order to avoid introducing spurious new extrema, the scheme will work with an average of the original quadratic and a constant:

$$\overline{q}_j(x) \equiv u^n_j + \theta_j[q_j(x) - u^n_j].$$

Here $\theta_j \in [0, 1]$ is chosen so that if $0 \le \min\{\triangle u^n_{j-1/2}, \triangle u^n_{j+1/2}\}$ then

$$\min\left\{\frac{u^n_j + u^n_{j-1}}{2}, q_{j-1}(x_{j-1/2})\right\} \le \overline{q}_j(x_{j-1/2})$$

and

$$\overline{q}_j(x_{j+1/2}) \le \max\left\{\frac{u^n_j + u^n_{j+1}}{2}, q_{j+1}(x_{j+1/2})\right\},$$

and if $0 \ge \max\{\triangle u^n_{j-1/2}, \triangle u^n_{j+1/2}\}$ then

$$\max\left\{\frac{u^n_j + u^n_{j-1}}{2}, q_{j-1}(x_{j-1/2})\right\} \ge \overline{q}_j(x_{j-1/2})$$

and

$$\overline{q}_j(x_{j+1/2}) \geq \min\left\{\frac{u_j^n + u_{j+1}^n}{2}, q_{j+1}(x_{j+1/2})\right\}.$$

Otherwise, we choose $\theta_j = 1$. These conditions suggest that we compute

$$M_j = \max_{x \in [x_{j-1/2}, x_{j+1/2}]} q_j(x) \quad \text{and} \quad m_j = \min_{x \in [x_{j-1/2}, x_{j+1/2}]} q_j(x).$$

If $0 \leq \min\{\triangle u_{j-1/2}^n, \triangle u_{j+1/2}^n\}$ then

$$\theta_j = \min\left\{\frac{u_j^n - \min\{\frac{u_j + u_{j-1}}{2}, q_{j-1}(x_{j-1/2})\}}{u_j^n - m_j}, \frac{\max\{\frac{u_j + u_{j+1}}{2}, q_{j+1}(x_{j+1/2}) - u_j^n\}}{M_j - u_j^n}, 1\right\}$$

and if $0 \geq \max\{\triangle u_{j-1/2}^n, \triangle u_{j+1/2}^n\}$ then

$$\theta_j = \min\left\{\frac{\max\{\frac{u_j + u_{j-1}}{2}, q_{j-1}(x_{j-1/2})\} - u_j^n}{M_j - u_j^n}, \frac{u_j^n - \min\{\frac{u_j + u_{j+1}}{2}, q_{j+1}(x_{j+1/2})\}}{u_j^n - m_j}, 1\right\}.$$

Conservation requires that

$$\int_{x_j}^{x_{j+1}} u\left(x, t^n + \frac{\triangle t^{n+1/2}}{2}\right) dx = \int_{x_j}^{x_{j+1/2}} \overline{q}_j(x, t^n)\, dx + \int_{x_{j+1/2}}^{x_{j+1}} \overline{q}_{j+1}(x, t^n)\, dx$$
$$- \int_{t^n}^{t^n + \triangle t^{n+1/2}/2} f(u(x_{j+1}, t))\, dt + \int_{t^n}^{t^n + \triangle t^{n+1/2}/2} f(u(x_j, t))\, dt.$$

The integrals of the limited quadratic reconstruction are

$$\int_{x_j}^{x_{j+1/2}} \overline{q}_j(x)\, dx = \int_0^{1/2} u_j^n + \theta_j \left\{\alpha_j + \xi\beta_j + \frac{1}{2}\xi^2\gamma_j\right\} d\xi \triangle x_j$$
$$= \frac{\triangle x_j}{2}\left\{u_j^n + \theta_j\alpha_j + \theta_j\beta_j\frac{1}{4} + \theta_j\gamma_j\frac{1}{24}\right\} = \frac{\triangle x_j}{2}\left[u_j^n + \frac{1}{4}\theta_j\beta_j\right] \quad (5.33a)$$
$$\int_{x_{j+1/2}}^{x_{j+1}} \overline{q}_{j+1}(x)\, dx = \int_{-1/2}^{0} u_{j+1}^n + \theta_{j+1}\left\{\alpha_{j+1} + \xi\beta_{j+1} + \frac{1}{2}\xi^2\gamma_{j+1}\right\} d\xi \triangle x_{j+1}$$
$$= \frac{\triangle x_{j+1}}{2}\left[u_{j+1}^n - \frac{1}{4}\theta_{j+1}\beta_{j+1}\right] \quad (5.33b)$$

The time integrals are approximated by Simpson's rule

$$\int_{t^n}^{t^n+\triangle t^{n+1/2}/2} f(u(x_j,t))\,dt$$
$$\approx \frac{\triangle t^{n+1/2}}{12}[f(u(x_j,t^n))+4f(u(x_j,t^n+\triangle t^{n+1/2}/4))+f(u(x_j,t^n+\triangle t^{n+1/2}/2))].$$

The values of $u(x,t+\tau)$ can be provided by a Taylor expansion of the form

$$\begin{aligned}u(x,t+\tau) &\approx u+\tau\frac{\partial u}{\partial t}+\frac{\tau^2}{2}\frac{\partial^2 u}{\partial t^2} = u-\tau\frac{\partial f}{\partial x}-\frac{\tau^2}{2}\frac{\partial}{\partial x}\left(\frac{\partial f}{\partial t}\right)\\ &= u-\tau\frac{\partial f}{\partial u}\frac{\partial u}{\partial x}+\frac{\tau^2}{2}\frac{\partial}{\partial x}\left(\left[\frac{\partial f}{\partial u}\right]^2\frac{\partial u}{\partial x}\right)\\ &= u-\tau\frac{\partial f}{\partial u}\frac{\partial u}{\partial x}+\frac{\tau^2}{2}\left(\left[\frac{\partial f}{\partial u}\right]^2\frac{\partial^2 u}{\partial x^2}+2\frac{\partial f}{\partial u}\frac{\partial^2 f}{\partial u^2}\left[\frac{\partial u}{\partial x}\right]^2\right).\end{aligned}$$

The values of u and its derivatives are provided by the limited quadratic reconstruction $\overline{q}_j(x)$; in particular, the function value used in the Taylor series is $\overline{q}_j(x_j)=u_j^n+\theta\alpha_j$, the slope is $\overline{q}_j'(x_j)=\theta\beta_j/\triangle x_j$, and the second derivative is $\overline{q}_j''(x_j)=\theta\gamma_j/\triangle x_j^2$.

Recently, the piecewise quadratic reconstruction and limiting of Liu and Tadmor has been deprecated in favor of the central WENO reconstruction in [99].

The second-order and third-order versions of the Lax–Friedrichs scheme produce accurate solutions for the problems we have tested. However, the use of the staggered grid requires twice as many applications of the algorithm as for our previous schemes that use unstaggered grids. Some boundary conditions require that the scheme be modified in a non-staggered manner at the boundaries. A C^{++} program to implement the higher-order versions of the Lax-Friedrichs scheme for nonlinear scalar conservation laws can be found in **Program 5.11-73: Schemes2.C** Simplified versions of these schemes for linear advection on uniform grids can be found in **Program 5.11-74: LinearAdvectionSchemes.C**. Students can exercise this program by clicking on **Executable 5.11-33: guiriemann2** The user can select Riemann problem initial data for linear advection, Burgers' equation, traffic models and the Buckley-Leverett model. In addition, the user can select from a variety of limiters and Riemann solvers. By setting the number of cells to 0, the user will cause the program to perform a mesh refinement study, and the user can select several different schemes for comparison. Students can also exercise these schemes for linear advection by clicking on **Executable 5.11-34: guilinearad2** The user can select a variety of initial values from the Zalesak test problems in exercise 2.2.5 of section 2.2. In addition, the user can select from a variety of limiters. By setting the number of cells to 0, the user will cause the program to perform a mesh refinement study.

Exercises for 5.11

5.11.1 Determine how to implement the Lax–Friedrichs scheme for a boundary condition on the left with specified flux. Describe how this idea could be implemented in the higher-order versions of the Lax–Friedrichs scheme.

5.11.2 The flux integrals in the Nessyahu–Tadmor scheme can be approximated in other ways. For example, we could approximate

$$u(x, t+\tau) \approx u - \frac{\partial f}{\partial x}\tau .$$

Here $\frac{\partial f}{\partial x}$ could be approximated as a limited slope $\triangle f_j^n$ in cell-centered fluxes divided by the mesh width $\triangle x_j$. Program this version of the Nessyahu–Tadmor scheme and compare it to the version presented above.

5.11.3 The flux integrals in the Liu–Tadmor scheme can also be approximated in another way, corresponding to a Runge–Kutta integration. Starting with the third-order approximation

$$u(x, t+\tau) \approx u_j^n - \frac{\partial f}{\partial x}\left(x, t+\frac{\tau}{2}\right)\tau ,$$

we compute fluxes at $(x, t+\tau/2)$ as in the previous exercise, limit slopes in these fluxes, and then compute $u(x, t+\tau)$ from the slopes in the fluxes at $(x, t+\tau/2)$. Program this version of the Liu–Tadmor scheme and compare it to the version presented above.

5.11.4 At the right-hand side of a zero state, we expect $|u_{j+1/2}^n| \gg |u_{j-1/2}^n|$. In fact, $u_{j-1/2}^n$ could be either positive or negative, depending on the effects of rounding error. Show that the Liu–Tadmor limiter θ is not a continuous function of $u_{j-1/2}^n$ and $u_{j+1/2}^n$. Experiment with ways to evaluate θ when $u_{j-1/2}^n$ and $u_{j+1/2}^n$ have opposite signs so that θ remains continuous, without destroying the order of the scheme.

5.12 Piecewise Parabolic Method

In this section, we will develop a version of Godunov's method that formally has even higher order than MUSCL. The ideas are due to Colella and Woodward [35], and will lead to other higher-order extensions in Section 5.13. The new approach will involve a piecewise quadratic reconstruction $u^n(x)$, using the cell averages u_j^n. The new reconstruction will be mass-conserving and monotonicity-preserving. The scheme will also use a more accurate quadrature for the temporal flux integrals.

In each cell we will have

$$u^n(x) = \alpha_j + \xi(x)\{\beta_j + [1 - \xi(x)]\gamma_j\}, \quad \text{where} \quad \xi(x) \equiv \frac{x - x_{j-1/2}}{x_{j+1/2} - x_{j-1/2}}. \tag{5.34}$$

Note that $u^n(x_{j-1/2}) = \alpha_j = u^R_{j-1/2}$, and $u^n(x_{j+1/2}) = \alpha_j + \beta_j = u^L_{j+1/2}$, which in turn implies that $\beta_j = u^L_{j+1/2} - u^R_{j-1/2}$. Also

$$u^n_j = \frac{1}{\triangle x_j}\int_{x_{j-1/2}}^{x_{j+1/2}} u^n(x)\, dx = \int_0^1 \alpha_j + \xi[\beta_j + (1-\xi)\gamma_j]\, d\xi = \alpha_j + \frac{1}{2}\beta_j + \frac{1}{6}\gamma_j,$$

so we require $\gamma_j = 6[u^n_j - \alpha_j - \frac{1}{2}\beta_j] = 6[u^n_j - \frac{1}{2}(u^L_{j+1/2} + u^R_{j-1/2})]$. These calculations show that the piecewise quadratic reconstruction $u^n(x)$ is completely determined by the cell averages u^n_j and the values at the cell sides.

In order to determine a value for $u_{j+1/2}$, we will construct a quartic interpolation to $\int_{x_j-3/2}^{x} u(s, t^n)\, ds$, and evaluate the derivative of this quartic at $x_{j+1/2}$ to get $u^n_{j+1/2}$. The quartic polynomial can be constructed by using **Newton interpolation** as in Section 5.9. The piecewise quartic reconstruction will require the first-order divided differences

$$u^n[x_{k+1}, x_k] \equiv \frac{u^n_{k+1} - u^n_k}{\triangle x_{k+1} + \triangle x_k}, \; j-1 \le k \le j+1$$

the second-order divided differences

$$u^n[x_{k+1}, x_k, x_{k-1}] \equiv \frac{u^n[x_{k+1}, x_k] - u^n[x_k, x_{k-1}]}{\triangle x_{k+1} + \triangle x_k + \triangle x_{k-1}}, \; j \le k \le j+1$$

and the third-order divided difference

$$u^n[x_{j+2}, x_{j+1}, x_j, x_{j-1}] \equiv \frac{u^n[x_{j+2}, x_{j+1}, x_j] - u^n[x_{j+1}, x_j, x_{j-1}]}{\triangle x_{j+2} + \triangle x_{j+1} + \triangle x_j + \triangle x_{j-1}}.$$

Then the quartic Newton interpolating polynomial is

$$\begin{aligned}
\int_{x_{j-\frac{3}{2}}}^{x} & u(s, t^n)\, ds \approx Q^n_j(x) \\
&\equiv u^n_{j-1}(1 - x_{j-3/2}) + u^n[x_j, x_{j-1}](x - x_{j-1/2})(x - x_{j-3/2}) \\
&\quad + u^n[x_{j+1}, x_j, x_{j-1}](x - x_{j+1/2})(x - x_{j-1/2})(x - x_{j-3/2}) \\
&\quad + u^n[x_{j+2}, x_{j+1}, x_j, x_{j-1}](x - x_{j+3/2})(x - x_{j+1/2})(x - x_{j-1/2})(x - x_{j-3/2}).
\end{aligned}$$

The derivative of this interpolating polynomial at $x_{j+1/2}$ is

$$\begin{aligned}
\frac{dQ_j^n}{dx}(x_{j+1/2}) &= u_{j+1/2}^n \\
&\equiv u_{j-1}^n + u^n[x_j, x_{j-1}](2\triangle x_j + \triangle x_{j-1}) \\
&\quad + u^n[x_{j+1}, x_j, x_{j-1}]\triangle x_j(\triangle x_j + \triangle x_{j-1}) - u^n[x_{j+2}, x_{j+1}, x_j, x_{j-1}](\triangle x_j + \triangle x_{j-1})\triangle x_{j+1}\triangle x_j \\
&= u_j^n + u[x_j, x_{j-1}]\triangle x_j + \{u^n[x_{j+1}, x_j] - u^n[x_j, x_{j-1}]\}\frac{\triangle x_j(\triangle x_j + \triangle x_{j-1})}{\triangle x_{j+1} + \triangle x_j + \triangle x_{j-1}} \\
&\quad - u^n[x_{j+2}, x_{j+1}, x_j, x_{j-1}](\triangle x_j + \triangle x_{j-1})\triangle x_j\triangle x_{j+1} \\
&= u_j^n + u[x_{j+1}, x_j]\triangle x_j + \{u^n[x_{j+1}, x_j] - u^n[x_j, x_{j-1}]\}\triangle x_j\left\{1 - \frac{\triangle x_j + \triangle x_{j-1}}{\triangle x_{j+1} + \triangle x_j + \triangle x_{j-1}}\right\} \\
&\quad - u^n[x_{j+2}, x_{j+1}, x_j, x_{j-1}](\triangle x_j + \triangle x_{j-1})\triangle x_j\triangle x_{j+1} \\
&= u_j^n + u[x_{j+1}, x_j]\triangle x_j + u^n[x_{j+1}, x_j, x_{j-1}]\triangle x_j\triangle x_{j+1} \\
&\quad - \{u^n[x_{j+2}, x_{j+1}, x_j] - u^n[x_{j+1}, x_j, x_{j-1}]\}\frac{(\triangle x_j + \triangle x_{j-1})\triangle x_j\triangle x_{j+1}}{\triangle x_{j+2} + \triangle x_{j+1} + \triangle x_j + \triangle x_{j-1}} \\
&= u_j^n + u[x_{j+1}, x_j]\triangle x_j \\
&\quad - \left\{u^n[x_{j+1}, x_j, x_{j-1}](\triangle x_{j+2} + \triangle x_{j+1}) + u^n[x_{j+2}, x_{j+1}, x_j](\triangle x_j + \triangle x_{j-1})\right\} \\
&\quad \times \frac{\triangle x_j\triangle x_{j+1}}{\triangle x_{j+2} + \triangle x_{j+1} + \triangle x_j + \triangle x_{j-1}} \\
&= u_j^n + u[x_{j+1}, x_j]\triangle x_j \\
&\quad - \frac{\triangle x_j\triangle x_{j+1}}{\triangle x_{j+2} + \triangle x_{j+1} + \triangle x_j + \triangle x_{j-1}}\left\{u^n[x_{j+1}, x_j]\frac{\triangle x_{j+2} + \triangle x_{j+1}}{\triangle x_{j+1} + \triangle x_j + \triangle x_{j-1}}\right. \\
&\quad - u^n[x_j, x_{j-1}]\frac{\triangle x_{j+2} + \triangle x_{j+1}}{\triangle x_{j+1} + \triangle x_j + \triangle x_{j-1}} + u^n[x_{j+2}, x_{j+1}]\frac{\triangle x_j + \triangle x_{j-1}}{\triangle x_{j+2} + \triangle x_{j+1} + \triangle x_j} \\
&\quad \left. - u^n[x_{j+1}, x_j]\frac{\triangle x_j + \triangle x_{j-1}}{\triangle x_{j+2} + \triangle x_{j+1} + \triangle x_j}\right\}.
\end{aligned}$$

Using the formula (5.31) for the average $\tilde{s}_j^n$ of the second derivative of the cubic interpolant to $\int_{x_{j-3/2}}^{x} u^n(s)\,ds$ at $x_{j-3/2}, x_{j-1/2}, x_{j+1/2}$ and $x_{j+3/2}$, we obtain

$$u^n[x_{j+2}, x_{j+1}] = \left\{\tilde{s}_{j+1}^n - u^n[x_{j+1}, x_j]\frac{2\triangle x_{j+2} + \triangle x_{j+1}}{\triangle x_{j+2} + \triangle x_{j+1} + \triangle x_j}\right\}\frac{\triangle x_{j+2} + \triangle x_{j+1} + \triangle x_j}{\triangle x_{j+1} + 2\triangle x_j}$$

$$u^n[x_j, x_{j-1}] = \left\{\tilde{s}_j^n - u^n[x_{j+1}, x_j]\frac{\triangle x_j + 2\triangle x_{j-1}}{\triangle x_{j+1} + \triangle x_j + \triangle x_{j-1}}\right\}\frac{\triangle x_{j+1} + \triangle x_j + \triangle x_{j-1}}{2\triangle x_{j+1} + \triangle x_j}.$$

As a result, we can write

$$
\begin{aligned}
u^n_{j+1/2} &= u^n_j + u[x_{j+1}, x_j]\triangle x_j \\
&- \frac{\triangle x_j \triangle x_{j+1}}{\triangle x_{j+2} + \triangle x_{j+1} + \triangle x_j + \triangle x_{j-1}} \left\{ u^n[x_{j+1}, x_j] \frac{\triangle x_{j+2} + \triangle x_{j+1}}{\triangle x_{j+1} + \triangle x_j + \triangle x_{j-1}} \right. \\
&- \left(\tilde{s}^n_j - u^n[x_{j+1}, x_j] \frac{\triangle x_j + 2\triangle x_{j-1}}{\triangle x_{j+1} + \triangle x_j + \triangle x_{j-1}} \right) \frac{\triangle x_{j+2} + \triangle x_{j+1}}{2\triangle x_{j+1} + \triangle x_j} \\
&+ \left(\tilde{s}^n_{j+1} - u[x_{j+1}, x_j] \frac{2\triangle x_{j+2} + \triangle x_{j+1}}{\triangle x_{j+2} + \triangle x_{j+1} + \triangle x_j} \right) \frac{\triangle x_j + \triangle x_{j-1}}{\triangle x_{j+1} + 2\triangle x_j} \\
&\left. - u^n[x_{j+1}, x_j] \frac{\triangle x_j + \triangle x_{j-1}}{\triangle x_{j+2} + \triangle x_{j+1} + \triangle x_j} \right\} \\
&= u^n_j + u[x_{j+1}, x_j]\triangle x_j \\
&- u[x_{j+1}, x_j] \frac{\triangle x_{j+1} \triangle x_j}{\triangle x_{j+2} + \triangle x_{j+1} + \triangle x_j + \triangle x_{j-1}} \\
&\times \left\{ \frac{\triangle x_{j+2} + \triangle x_{j+1}}{\triangle x_{j+1} + \triangle x_j + \triangle x_{j-1}} + \frac{\triangle x_j + 2\triangle x_{j-1}}{2\triangle x_{j+1} + \triangle x_j} \frac{\triangle x_{j+2} + \triangle x_{j+1}}{\triangle x_{j+1} + \triangle x_j + \triangle x_{j-1}} \right. \\
&\left. - \frac{2\triangle x_{j+2} + \triangle x_{j+1}}{\triangle x_{j+1} + 2\triangle x_j} \frac{\triangle x_j + \triangle x_{j-1}}{\triangle x_{j+2} + \triangle x_{j+1} + \triangle x_j} - \frac{\triangle x_j + \triangle x_{j-1}}{\triangle x_{j+2} + \triangle x_{j+1} + \triangle x_j} \right\} \\
&+ \left\{ \tilde{s}^n_j \frac{\triangle x_{j+2} + \triangle x_{j+1}}{2\triangle x_{j+1} + \triangle x_j} - \tilde{s}^n_{j+1} \frac{\triangle x_j + \triangle x_{j-1}}{\triangle x_{j+1} + 2\triangle x_j} \right\} \frac{\triangle x_{j+1} \triangle x_j}{\triangle x_{j+2} + \triangle x_{j+1} + \triangle x_j + \triangle x_{j-1}} \\
&= u^n_j + u[x_{j+1}, x_j]\triangle x_j \\
&- u[x_{j+1}, x_j] \frac{2\triangle x_{j+1} \triangle x_j}{\triangle x_{j+2} + \triangle x_{j+1} + \triangle x_j + \triangle x_{j-1}} \left\{ \frac{\triangle x_{j+2} + \triangle x_{j+1}}{2\triangle x_{j+1} + \triangle x_j} - \frac{\triangle x_j + \triangle x_{j-1}}{\triangle x_{j+1} + 2\triangle x_j} \right\} \\
&+ \left\{ \tilde{s}^n_j \frac{\triangle x_{j+2} + \triangle x_{j+1}}{2\triangle x_{j+1} + \triangle x_j} - \tilde{s}^n_{j+1} \frac{\triangle x_j + \triangle x_{j-1}}{\triangle x_{j+1} + 2\triangle x_j} \right\} \frac{\triangle x_{j+1} \triangle x_j}{\triangle x_{j+2} + \triangle x_{j+1} + \triangle x_j + \triangle x_{j-1}} .
\end{aligned}
$$

On a uniform mesh, these simplify to

$$
u^n_{j+1/2} = \frac{1}{2}(u^n_j + u^n_{j+1}) - \frac{\triangle x}{6}[\tilde{s}^n_{j+1} - \tilde{s}^n_j].
$$

In practice, Colella and Woodward replace the slopes $\tilde{s}^n_j$ and $\tilde{s}^n_{j+1}$ with their limited values s^n_j and s^n_{j+1}, given by formula (5.32).

These values at the cell sides must be adjusted further to produce the values $u^L_{j+1/2}$ and $u^R_{j-1/2}$ used in the piecewise quadratic reconstruction. The extremum of $u^n_j(x)$ occurs at the point x^*_j where

$$
0 = \frac{du^n_j}{dx}(x^*_j) = \frac{1}{\triangle x_j}\left(\beta_j + \frac{x_{j+1/2} - x^*_j}{\triangle x_j} \gamma_j \right) - \frac{x^*_j - x_{j-1/2}}{\triangle x_j} \frac{\gamma_j}{\triangle x_j},
$$

which implies that

$$x_j^* = \frac{x_{j-1/2} + x_{j+1/2}}{2} + \frac{\beta_j \triangle x_j}{2\gamma_j}.$$

If $\beta_j^2 \geq |\beta_j \gamma_j|$, then $x_j^* \notin (x_{j-1/2}, x_{j+1/2})$, and we have the rule

$$\beta_j^2 \geq |\beta_j \gamma_j| \Longrightarrow u_{j-1/2}^R = u_{j-1/2}^n,\ u_{j+1/2}^L = u_{j+1/2}^n.$$

If $\beta_j^2 < -\beta_j \gamma_j$, then $(x_{j-1/2} + x_{j+1/2})/2 \geq x_j^* > x_{j-1/2}$, so we modify $u_{j+1/2}^L$ so that we have an extremum at $x_{j-1/2}$. This is equivalent to taking $\beta_j = -\gamma_j$, producing the rule

$$\beta_j^2 < -\beta_j \gamma_j \Longrightarrow u_{j-1/2}^R = u_{j-1/2}^n, \quad u_{j+1/2}^L = 3u_j^n - 2u_{j-1/2}^R.$$

If $\beta_j^2 < \beta_j \gamma_j$, then $(x_{j-1/2} + x_{j+1/2})/2 \leq x_j^* < x_{j+1/2}$, so we modify $u_{j-1/2}^R$ so that we have an extremum at $x_{j+1/2}$. This is equivalent to taking $\beta_j = \gamma_j$, producing the rule

$$\beta_j^2 < \beta_j \gamma_j \Longrightarrow u_{j+1/2}^L = u_{j+1/2}^n, \quad u_{j-1/2}^R = 3u_j^n - 2u_{j+1/2}^L.$$

It remains for us to compute approximations to the flux time integrals for the conservative difference. Colella and Woodward [35] describe their algorithm for linear advection and gas dynamics only. The following approach appears to follow their ideas. Recall that in smooth flow, the solution of the conservation law is constant along characteristics; as a result,

$$u(x_{j+1/2}, t^n + \tau) = u^n(x_{j+1/2} - \lambda\tau)$$

where the characteristic speed λ solves

$$\lambda = \frac{df}{du}(u(x_{j+1/2} - \lambda\tau)).$$

Here u^n refers to the piecewise quadratic reconstruction, either in cell j or $j+1$. The flux at the cell side is computed by

$$\begin{aligned}
&\frac{1}{\triangle t^{n+1/2}} \int_0^{\triangle t^{n+1/2}} f(\mathcal{R}(u(x_{j+1/2} - 0, t^n + \tau),\ u(x_{j+1/2} + 0, t^n + \tau);\ 0))\, d\tau \\
&= \frac{1}{\triangle t^{n+1/2}} \int_0^{\triangle t^{n+1/2}} f(\mathcal{R}(u_j^n(x_{j+1/2} - \lambda\tau),\ u_{j+1}^n(x_{j+1/2} - \lambda\tau);\ 0))\, d\tau \\
&\approx f\left(\mathcal{R}\left(\frac{1}{\triangle t^{n+1/2}} \int_0^{\triangle t^{n+1/2}} u_j^n(x_{j+1/2} - \lambda\tau)\, d\tau, \frac{1}{\triangle t^{n+1/2}} \int_0^{\triangle t^{n+1/2}} u_{j+1}^n(x_{j+1/2} - \lambda\tau)\, d\tau; 0\right)\right).
\end{aligned}$$

Colella and Woodward apparently keep the characteristic speed constant in these integrals. Provided that $\lambda > 0$ they compute

$$\begin{aligned}
\overline{u}^L_{j+1/2} &= \frac{1}{\Delta t^{n+1/2}} \int_0^{\Delta t^{n+1/2}} u^n_j(x_{j+1/2} - \lambda^n_j \tau)\, d\tau \\
&= \frac{1}{\Delta t^{n+1/2}} \int_0^{\Delta t^{n+1/2}} \alpha_j + \left(1 - \frac{\lambda\tau}{\Delta x_j}\right)\left(\beta_j + \frac{\lambda\tau}{\Delta x_j}\gamma_j\right) d\tau \\
&= \frac{\Delta x_j}{\lambda^n_j \Delta t^{n+1/2}} \int_0^{\lambda^n_j \Delta t^{n+1/2}/dx_j} \alpha_j + (1-\sigma)(\beta_j + \sigma\gamma_j)\, d\sigma \\
&= u^L_{j+1/2} - \frac{1}{2}\frac{\lambda^n_j \Delta t^{n+1/2}}{\Delta x_j}\left[\beta_j - \gamma_j\left(1 - \frac{2}{3}\frac{\lambda^n_j \Delta t^{n+1/2}}{\Delta x_j}\right)\right];
\end{aligned}$$

otherwise they take $\overline{u}^L_{j+1/2} = u^L_{j+1/2}$. Provided that $\lambda < 0$ they compute

$$\begin{aligned}
\overline{u}^R_{j+1/2} &= \frac{1}{\Delta t^{n+1/2}} \int_0^{\Delta t^{n+1/2}} u^n_{j+1}(x_{j+1/2} - \lambda^n_{j+1} \tau)\, d\tau \\
&= \frac{1}{\Delta t^{n+1/2}} \int_0^{\Delta t^{n+1/2}} \alpha_{j+1} - \frac{\lambda\tau}{\Delta x_{j+1}}\left(\beta_{j+1} + [1 + \frac{\lambda\tau}{\Delta x_j}]\gamma_{j+1}\right) d\tau \\
&= \frac{\Delta x_{j+1}}{\lambda^n_j \Delta t^{n+1/2}} \int_0^{\lambda^n_{j+1} \Delta t^{n+1/2}/\Delta x_{j+1}} \alpha_{j+1} - \sigma\left(\beta_{j+1} + [1+\sigma]\gamma_{j+1}\right) d\sigma \\
&= u^R_{j+1/2} - \frac{1}{2}\frac{\lambda^n_{j+1} \Delta t^{n+1/2}}{\Delta x_{j+1}}\left[\beta_{j+1} + \gamma_{j+1}\left(1 + \frac{2}{3}\frac{\lambda^n_{j+1} \Delta t^{n+1/2}}{\Delta x_{j+1}}\right)\right];
\end{aligned}$$

otherwise they take $\overline{u}^R_{j+1/2} = u^R_{j+1/2}$. The two states $\overline{u}^L_{j+1/2}$ and $\overline{u}^R_{j+1/2}$ are used in a Riemann problem to compute the flux for the conservative difference.

A C++ program to implement the piecewise parabolic method for nonlinear scalar conservation laws can be found in **Program 5.12-75: Schemes2.C** A simplified version of this scheme for linear advection on uniform grids can be found in **Program 5.12-76: LinearAdvectionSchemes.C.** Students can exercise this program by clicking on **Executable 5.12-35: guiriemann2** The user can select Riemann problem initial data for linear advection, Burgers' equation, traffic models and the Buckley-Leverett model. In addition, the user can select from a variety of limiters and Riemann solvers. By setting the number of cells to 0, the user will cause the program to perform a mesh refinement study, and the user can select several different schemes for comparison. Students can also exercise the piecewise parabolic method for linear advection by clicking on **Executable 5.12-36: guilinearad2** The user can select a variety of initial values from the Zalesak test problems in exercise 2.2.5 of section 2.2. In addition, the user can select from a variety of limiters. By setting the number of cells to 0, the user will cause the program to perform a mesh refinement study.

5.13 Essentially Non-Oscillatory Schemes

In this section we will present a scheme that is designed to reach an arbitrarily high order of accuracy. In order to do so, it will assume that the flux function in the conservation law has as many continuous derivatives as needed. The development here is due to Harten, Osher and Shu [65, 69, 149].

Let w be an arbitrary function of x and define the moving cell average

$$\overline{w}(x) = \frac{1}{\triangle x} \int_{-\triangle x/2}^{\triangle x/2} w(x+y)\, dy \equiv (A_{\triangle x} w)(x).$$

Note that $\overline{w}$ is smoother than w by one derivative. At points x where w is smooth,

$$\overline{w}(x) = w(x) + O(\triangle x^2).$$

Next, consider an initial value problem for a nonlinear scalar conservation law $\partial u/\partial t + \partial f/\partial x = 0$ with initial data $u(x,0) = u_0(x)$, and solution

$$u(x,t) = (E(t)u_0)(x).$$

Integrating the conservation law in space leads to the ordinary differential equations

$$\frac{\partial \overline{u}}{\partial t} + \frac{1}{\triangle x}\left[f\left(u\left(x + \frac{\triangle x}{2}, t\right)\right) - f\left(u\left(x - \frac{\triangle x}{2}, t\right)\right)\right] = 0.$$

The basic approach in approximating the solution to the conservation law will be to construct a higher-order interpolant to $\tilde{f}(x,t;u)$, and then apply the **method of lines** to solve

$$\frac{\partial \overline{u}}{\partial t} = \mathcal{L}(u) \equiv -\frac{1}{\triangle x}\left[f\left(u\left(x + \frac{\triangle x}{2}, t\right)\right) - f\left(u\left(x - \frac{\triangle x}{2}, t\right)\right)\right].$$

The analytical form of the **ENO scheme** is $u_j^{n+1} = AE_{\triangle x}(\triangle t)R(\cdot, u_j^n)$, where R constructs a piecewise polynomial interpolant to the cell averages u_j^n, $E_{\triangle x}$ evolves the polynomial interpolant through a time increment $\triangle t$, and A averages the result onto the mesh. For a scalar law, the exact evolution operator is monotone. For any conservation law, the cell average is always monotone. For a scalar conservation law, the evolution is total variation diminishing. It follows that for a scalar conservation law

$$TV(u_j^{n+1}) = TV(AE_{\triangle x}(\triangle t)R(\cdot, u_j^n)) \le TV(R(\cdot, u_j^n)) \le TV(u_j^n) + O(\triangle x^r).$$

Note that if the numerical flux satisfies

$$f_{j+1/2} = \tilde{f}(x_{j+1/2}, t^n) + d(x_{j+1/2})\triangle x^r + O(\triangle x^{r+1}),$$

where $d(x)$ is Lipschitz continuous, and if the numerical solution is given by the conservative difference

$$u_j^{n+1} = \overline{u}(x_j, t^n) - \frac{\triangle t}{\triangle x}[f_{j+1/2} - f_{j-1/2}] \equiv (E_{\triangle x}(\triangle t)\overline{u}(\cdot, t^n))_j,$$

then the local truncation error in the cell averages satisfies

$$\begin{aligned}\overline{u}(x_j, t^{n+1}) - (E_{\triangle x}(\triangle t)\overline{u}(\cdot, t^n))_j &= \frac{\triangle t}{\triangle x}[d(x_{j+1/2}) - d(x_{j-1/2})]\triangle x^r + O(\triangle x^{r+1}) \\ &= O(\triangle x^{r+1}).\end{aligned}$$

By extending the idea in the PPM reconstruction, we will show how to construct a piecewise polynomial function $R(x; w)$ so that the reconstruction has order r (*i.e.*, if w is smooth at x then $R(x, \overline{w}) = w(x) + e(x)\triangle x^r + O(\triangle x^{r+1})$, the reconstruction preserves cell averages (*i.e.* $\overline{R}(x_j; \overline{w}) = \overline{w}(x_j)$) and the reconstruction is **essentially non-oscillatory (ENO)** (*i.e.*, $TV(R(\cdot, \overline{w})) \le TV(w) + O(\triangle x^r)$).

When the error coefficient $e(x) = [R(x, \overline{w}) - w(x)]/\triangle x^r$ fails to be Lipschitz continuous at a point, then the local truncation error of ENO is only $O(\triangle x^r)$. For MUSCL, this happens at local extrema of the interpolant; for ENO, this may occur at zeros of higher derivatives of w. Due to local accumulation, the pointwise error at t^N with $N = O(1/\triangle x)$ is $O(\triangle x^{r-1})$. Away from these points, the global error is $O(\triangle x^r)$. Shu and Osher claim that the scheme has order $r - 1$ in $\mathcal{L}_\infty$, and order r in $\mathcal{L}_1$, where r is the order of accuracy of the reconstruction function. In practice, we have not observed this order of convergence for Riemann problems. Rather, we observe first-order convergence at propagating discontinuities and second-order convergence for rarefactions surrounded by constant states for all versions of the ENO scheme, other than the first-order scheme.

There are two major pieces to the ENO scheme. One is an appropriate ordinary differential equation solver to integrate $\partial u/\partial t(t) = L(u(t))$. Let $u^{(0)}$ approximate $u(x, t^n)$. For first-order ENO, the suggested ordinary differential equation solver is the forward Euler method. For second-order ENO, the suggested ordinary differential equation solver is Heun's method,

$$\begin{aligned}u^{(1)} &= u^{(0)} + \triangle t L(u^{(0)}) \\ u^{(2)} &= \frac{1}{2}[u^{(0)} + u^{(1)} + \triangle t L(u^{(1)})],\end{aligned}$$

producing $u^{(2)}$ approximating $u(x, t^n + \triangle t)$. The resulting scheme is stable for CFL ≤ 1. For third-order ENO, the suggested ordinary differential equation solver is

$$\begin{aligned}
u^{(1)} &= u^{(0)} + \triangle t L(u^{(0)}) \\
u^{(2)} &= \frac{1}{4}[3u^{(0)} + u^{(1)} + \triangle t L(u^{(1)})] \\
u^{(3)} &= \frac{1}{3}[u^{(0)} + 2u^{(2)} + 2\triangle t L(u^{(2)})].
\end{aligned}$$

The resulting scheme is also stable for CFL ≤ 1. For fourth-order ENO, the suggested ordinary differential equation solver is

$$\begin{aligned}
u^{(1)} &= u^{(0)} + \frac{1}{2}\triangle t L(u^{(0)}) \\
u^{(2)} &= \frac{1}{2}[u^{(0)} + u^{(1)}] + \frac{\triangle t}{4}[-\tilde{L}(u^{(0)}) + 2L(u^{(1)})] \\
u^{(3)} &= \frac{1}{9}[u^{(0)} + 2u^{(1)} + 6u^{(2)}] + \frac{\triangle t}{9}[-\tilde{L}(u^{(0)}) - 3\tilde{L}(u^{(1)}) + 9L(u^{(2)})] \\
u^{(4)} &= \frac{1}{3}[u^{(1)} + u^{(2)} + u^{(3)}] + \frac{\triangle t}{6}[L(u^{(1)}) + L(u^{(3)})].
\end{aligned}$$

The resulting scheme is stable for CFL $\leq 2/3$, and the reason for the notation $\tilde{L}$ will be explained at the end of the next paragraph.

The second major piece of the ENO scheme is the piecewise polynomial reconstruction. Here the goal is to construct a polynomial of degree $r + 1$ approximating the integral of the flux, and then evaluate the derivative of this polynomial at each cell side $x_{j+1/2}$. We also want to choose the index $i_m(j)$ for the start of the interpolation stencil so that the $m + 1$ points are chosen from the smoothest region for $\overline{w}$. In more recent versions of ENO, priority has been given to making the stencil vary more smoothly with j [77, 109]. If the characteristic speeds are all positive between $u_j^{n+\alpha}$ and $u_{j+1}^{n+\alpha}$, then we construct a divided difference table for values of the flux integral at $x_{j-r+1/2}, \ldots, x_{j+r-1/2}$, computing divided differences up to order r. Similarly, if the characteristic speeds are all negative between $u_j^{n+\alpha}$ and $u_{j+1}^{n+\alpha}$, then we construct a divided difference table at $x_{j-r+3/2}, \ldots, x_{j+r+1/2}$. If the characteristic speeds change sign, then we construct the former divided difference table for $\frac{1}{2}[f(u) + \lambda u]$, and the latter divided difference table for $\frac{1}{2}[f(u) - \lambda u]$. The actual piecewise polynomial interpolation within these stencils is selected by choosing the smaller divided difference of two alternatives as the order of the divided difference is increased. The resulting algorithm takes the form

$$
\begin{aligned}
&\text{for all } -r \leq j \leq r \;\; \delta_{j,0} = f_j \\
&\text{for all } 1 \leq k \leq r \\
&\quad \text{for all } -r \leq j \leq r-k \;\; \delta_{j,k} = \frac{\delta_{j+1,k-1} - \delta_{j,k-1}}{x_{j+k+1/2} - x_{j-1/2}} \\
&\ell_0 = 0 \\
&p'_{j+1/2}(\xi) = \delta_{0,0} \\
&\text{for all } 1 \leq k \leq r \\
&\quad D = \frac{d}{dx} \prod_{j=\ell_{k-1}}^{\ell_{k-1}+k} (x - x_{j-1/2})|_{x=\xi} \\
&\quad \text{if } |\delta_{\ell_{k-1},k}| \geq |\delta_{\ell_{k-1},k}| \text{ then } \ell_k = \ell_{k-1} - 1 \text{ else } \ell_k = \ell_{k-1} \\
&\quad p'_{j+1/2}(\xi) += D\delta_{\ell_k,k}.
\end{aligned}
$$

In the fourth-order version of ENO, two stages of the Runge–Kutta scheme involve negative coefficients, so the corresponding interpolation of the flux gradient $\tilde{L}$ must be computed with the stencil for $-f$ rather than f.

A C^{++} program to implement the ENO schemes for a variety of nonlinear scalar conservation laws can be found in **Program 5.13-77: Schemes2.C**, which calls several Fortran subroutines in **Program 5.13-78: eno.f.** A simplified version of these schemes for linear advection on uniform grids can be found in **Program 5.13-79: LinearAdvectionSchemes.C**. Students can exercise the former program by clicking on **Executable 5.13-37: guiriemann2** Students can also exercise the ENO scheme for linear advection by clicking on **Executable 5.13-38: guilinearad2** By setting the number of cells to 0, the user will cause the program to perform a mesh refinement study.

Exercises for 5.13

5.13.1 Show that the first-order version of the ENO scheme is the same as Godunov's method using Marquina's flux (see exercise 3.4.3).

5.13.2 Assuming that the characteristic speeds are all positive, show that the numerical flux in the second-order version of the ENO scheme is $f_{j+1/2}$ where

$$
\frac{f_{j+1/2} - f_j}{\triangle x_j} = \begin{cases} \frac{f_{j+1}-f_j}{\Delta x_{j+1}+\Delta x_j}, & |\frac{f_{j+1}-f_j}{\Delta x_{j+1}+\Delta x_j}| \leq |\frac{f_j - f_{j-1}}{\Delta x_j + \Delta x_{j-1}}| \\ \frac{f_j-f_{j-1}}{\Delta x_j+\Delta x_{j-1}}, & |\frac{f_{j+1}-f_j}{\Delta x_{j+1}+\Delta x_j}| > |\frac{f_j - f_{j-1}}{\Delta x_j + \Delta x_{j-1}}|. \end{cases}
$$

On a uniform grid, this is similar to using a minmod limiter, except that the limited slope is not set to zero when the left and right slopes have opposite signs.

5.14 Discontinuous Galerkin Methods

5.14.1 Weak Formulation

The discontinuous Galerkin method [29] is another technique for generating arbitrarily high order schemes for hyperbolic conservation laws. This approach rewrites the conservation law $\partial u/\partial t + \partial f(u)/\partial x = 0$ in a weak form. For any $w(x) \in C^\infty(\Omega)$ we have

$$0 = \int_\Omega \left[\frac{\partial u}{\partial t} + \frac{\partial f}{\partial x} \right] w \, dx = \sum_i \int_{x_{i-1/2}}^{x_{i+1/2}} \left[\frac{\partial u}{\partial t} w + \frac{\partial f w}{\partial x} - f \frac{\partial w}{\partial x} \right] dx$$
$$= \sum_i \left[\frac{d}{dt} \int_{x_{i-1/2}}^{x_{i+1/2}} u w \, dx + (f w)|_{x_{i-1/2}}^{x_{i+1/2}} - \int_{x_{i-1/2}}^{x_{i+1/2}} f \frac{\partial w}{\partial x} \, dx \right].$$

Several difficulties arise. One is that $C^\infty(\Omega)$ is infinite-dimensional. In practice, we will replace the set of functions $C^\infty(\Omega)$ with a finite-dimensional space of piecewise polynomials. Another difficulty is that the fluxes at the cell sides are difficult to determine. In practice, we will require that the flux at the cell side $x_{i+1/2}$ be evaluated at the solution of the Riemann problem taking left and right state from the values of $u(x_{i+1/2}, t)$ from the neighboring cells. This will be acceptable for a numerical timestep small enough that there are no interactions from Riemann problems arising from other grid cells.

The discretization chooses finite dimensional subspaces $\mathcal{M}_i \subset C^\infty(x_{i-1/2}, x_{i+1/2})$ and seeks numerical approximation $U(x,t) \in \mathcal{M}_i$ so that for all $t > 0$ and for all $x \in (x_{i-1/2}, x_{i+1/2})$ and for all $W \in \mathcal{M}_i$,

$$\begin{aligned} 0 = {} & \frac{d}{dt} \int_{x_{i-1/2}}^{x_{i+1/2}} U(x,t) W(x) \, dx \\ & + f(\mathcal{R}(U(x_{i+1/2} - 0, t), U(x_{i+1/2} + 0, t); 0)) W(x_{i+1/2} - 0) \\ & - f(\mathcal{R}(U(x_{i-1/2} - 0, t), U(x_{i-1/2} + 0, t); 0)) W(x_{i-1/2} + 0) \\ & - \int_{x_{i-1/2}}^{x_{i+1/2}} f(U(x,t)) \frac{\partial W}{\partial x} \, dx. \end{aligned} \tag{5.35}$$

Here $\mathcal{R}(u_L, u_R; \xi)$ is the state moving at speed ξ in the solution of the Riemann problem with left state u_L and right state u_R.

In order to implement the discontinuous Galerkin method, we need to choose appropriate finite dimensional subspaces $\mathcal{M}_i \subset C^\infty(x_{i-1/2}, x_{i+1/2})$. We also need to determine appropriately accurate numerical quadrature rules for the integrals in (5.35), to determine initial values for the numerical solution, and to select appropriate temporal integration schemes. It will also be useful to apply a limiter to the states that are used in the Riemann problems.

5.14.2 Basis Functions

The subspaces $\mathcal{M}_i$ will consist of polynomials of degree at most k. In order to simplify the computations, we will choose orthonormal basis functions for these polynomials, and map these basis functions from $\xi \in [-1, 1]$ to $x \in [x_{i-1/2}, x_{i+1/2}]$ by $\xi_i(x) = 2\frac{x-x_i}{\Delta x_i}$. These basis functions are the well-known Legendre polynomials. If we define $p_{-1}(\xi) \equiv 0$, $p_0(\xi) \equiv 1$ and

$$p_j(\xi) = \frac{2j-1}{j}\xi p_{j-1}(\xi) - \frac{j-1}{j}p_{j-2}(\xi), \tag{5.36}$$

then it is well-known [42] that

$$\int_{-1}^{1} p_j(\xi)p_\ell(\xi)\,d\xi = 0 \text{ for all } j \neq \ell,$$

$$\int_{-1}^{1} p_j(\xi)^2\,d\xi = \frac{2}{2j+1}.$$

The polynomials $p_j(x)$ are easily generated by the three-term recurrence (5.36). So that our basis functions for the discontinuous Galerkin method are orthonormal, we take

$$b_j(\xi) = \frac{p_j(\xi)}{\sqrt{\int_{-1}^{1} p_j(\xi)^2\,d\xi}} = p_j(\xi)\sqrt{j+\frac{1}{2}}. \tag{5.37}$$

Note that for all j, $b_{2j}(\xi)$ is an even function of ξ and $b_{2j+1}(\xi)$ is an odd function of ξ. Also note that the three-term recurrence (5.36) implies that $p_j(1) = 1$ for $j \geq 0$.

We will represent the numerical solution in the form

$$U(x,t) = \sum_{j=0}^{k} u_{i,j}(t)b_j(\xi_i(x)) \text{ for all } x \in (x_{i-1/2}, x_{i+1/2}), \text{ where } \xi_i(x) = 2\frac{x-x_i}{\Delta x_i}.$$

It is sufficient that the Galerkin equations (5.35) be satisfied with W chosen to be one of the basis functions. In order to simplify the resulting Galerkin equations, let us define

$$\mathbf{u}_i(t) = \begin{bmatrix} u_{i,0}(t) \\ \vdots \\ u_{i,k}(t) \end{bmatrix} \text{ and } \mathbf{b}(\xi) = \begin{bmatrix} b_0(\xi) \\ \vdots \\ b_k(\xi) \end{bmatrix}.$$

Then the Galerkin equations can be written in the form of a system of ordinary differential equations in each mesh interval:

$$0 = \frac{d}{dt} \int_{x_{i-1/2}}^{x_{i+1/2}} \mathbf{b}(\xi_i(x))\, \mathbf{b}(\xi_i(x))^\top \mathbf{u}_i(t)\, dx + \mathbf{b}(1) f(\mathcal{R}(\mathbf{b}(1)^\top \mathbf{u}_i(t), \mathbf{b}(-1)^\top \mathbf{u}_{i+1}(t); 0))$$

$$- \mathbf{b}(-1) f(\mathcal{R}(\mathbf{b}(1)^\top \mathbf{u}_{i-1}(t), \mathbf{b}(-1)^\top \mathbf{u}_i(t); 0)) - \int_{x_{i-1/2}}^{x_{i+1/2}} \frac{d\mathbf{b}(\xi_i(x))}{dx} f(\mathbf{b}(\xi_i(x))^\top \mathbf{u}_i(t))\, dx.$$

Note that the orthonormality of the basis functions greatly simplifies the term involving the time derivatives:

$$\frac{d}{dt} \int_{x_{i-1/2}}^{x_{i+1/2}} \mathbf{b}(\xi_i(x)) \mathbf{b}(\xi_i(x))^\top \mathbf{u}(t)\, dx = \int_{-1}^{1} \mathbf{b}(\xi) \mathbf{b}(\xi)^\top\, d\xi\, \frac{d\mathbf{u}_i}{dt} \frac{\triangle x_i}{2} = \frac{d\mathbf{u}_i}{dt} \frac{\triangle x_i}{2}.$$

Eventually, a limiter will be used to replace the states in the Riemann problems. Let us define

$$\begin{aligned} \mathbf{f}_i(t) \frac{\triangle x_i}{2} &= -\mathbf{b}(1) f(\mathcal{R}(\mathbf{b}(1)^\top \mathbf{u}_i(t), \mathbf{b}(-1)^\top \mathbf{u}_{i+1}(t); 0)) \\ &\quad + \mathbf{b}(-1) f(\mathcal{R}(\mathbf{b}(1)^\top \mathbf{u}_{i-1}(t), \mathbf{b}(-1)^\top \mathbf{u}_i(t); 0)) \\ &\quad + \int_{-1}^{1} \mathbf{b}'(\xi) f(\mathbf{b}(\xi)^\top \mathbf{u}_i(t))\, d\xi. \end{aligned} \tag{5.38}$$

Then the system of ordinary differential equations is $d\mathbf{u}_i/dt = \mathbf{f}_i(t)$.

5.14.3 Numerical Quadrature

Note that we will need to evaluate the flux at the two cell sides, where we solve Riemann problems, possibly approximately. For the spatial integral of the flux, it will be advantageous to use numerical quadrature rules that involve function values at the ends of the integration interval. These are called **Lobatto quadrature** rules [82]. In general, if $\xi_0 = -1$, $\xi_m = 1$ and $\xi_1, \ldots, \xi_{m-1}$ are the zeros of $b'_m(\xi)$, then the approximation

$$\int_{-1}^{1} \phi(\xi)\, d\xi \approx \sum_{j=0}^{m} \phi(\xi_j) \alpha_j$$

is exact for all $\phi \in \mathcal{P}^{2m-1}$, provided that the weights α_j are chosen appropriately. The first four rules are

$$\int_{-1}^{1} \phi(x)\,dx \approx \phi(-1) + \phi(1) \text{ (trapezoidal rule; exact for } \phi \in \mathcal{P}^1)$$

$$\approx \frac{1}{3}\{\phi(-1) + 4\phi(0) + \phi(1)\} \text{ (simpson's rule; exact for } \phi \in \mathcal{P}^3)$$

$$\approx \frac{1}{6}\left\{\phi(-1) + 5\phi\left(-\frac{1}{\sqrt{5}}\right) + 5\phi\left(\frac{1}{\sqrt{5}}\right) + \phi(1)\right\} \text{ (exact for } \phi \in \mathcal{P}^5)$$

$$\approx \frac{1}{90}\left\{9\phi(-1) + 49\phi\left(-\sqrt{\frac{3}{7}}\right) + 64\phi(0) + 49\phi\left(\sqrt{\frac{3}{7}}\right) + 9\phi(1)\right\} \text{ (exact for } \phi \in \mathcal{P}^7).$$

The quadrature rules are chosen so that the coefficients of the $\mathbf{u}_i'$ in the Galerkin equations are integrated exactly; in other words, we use $k+1$ Lobatto quadrature points for discontinuous Galerkin methods involving polynomials of degree at most k. These rules suggest that we pre-compute the values of $\mathbf{b}(\xi)$ and $\mathbf{b}'(\xi)$ at the quadrature points, since these will be the same for all grid cells.

For time integration of $\frac{d\mathbf{u}}{dt} = \mathbf{f}(u)$, we will use the same Runge–Kutta methods as in the ENO scheme (Section 5.13).

5.14.4 Initial Data

We will choose our initial function $U(x,t) = \sum_{j=0}^{k} u_{i,j}(t) b_j(\xi(x))$ at $t=0$ so that it has minimal $\mathcal{L}_2$ error in approximating the true initial data. The solution of this minimization problem has error orthogonal to the space of piecewise polynomials. In other words,

$$0 = \frac{2}{\triangle x_i}\int_{x_{i-1/2}}^{x_{i+1/2}} \mathbf{b}(\xi_i(x))\{u(x,0) - \mathbf{b}(\xi_i(x))^\top \mathbf{u}_i(0)\}dx$$

$$= \int_{-1}^{1} \mathbf{b}(\xi) u\left(x_i + \frac{1}{2}\xi \triangle x_i, 0\right) d\xi - \mathbf{u}_i(0).$$

This gives us a formula for $\mathbf{u}_i(0)$. If the initial data is sufficiently smooth, we can approximate $\mathbf{u}_i(0)$ by using numerical quadrature. Using the orthonormality of the basis functions and the fact that the first basis function is constant, it is easy to see that this value for $\mathbf{u}_i(0)$ conserves the initial data:

$$\begin{aligned}
\int_{x_{i-1/2}}^{x_{i+1/2}} \mathbf{b}(\xi_i(x))^\top \mathbf{u}_i(0)\,dx &= \int_{-1}^{1} \mathbf{b}(\xi')^\top \int_{-1}^{1} \mathbf{b}(\xi) u\left(x_i + \xi\frac{\triangle x_i}{2}, 0\right) d\xi\, d\xi' \frac{\triangle x_i}{2} \\
&= \int_{-1}^{1}\int_{-1}^{1} \mathbf{b}(\xi')^\top\, d\xi' \mathbf{b}(\xi) u\left(x_i + \xi\frac{\triangle x_i}{2}, 0\right) d\xi \frac{\triangle x_i}{2} \\
&= \int_{-1}^{1} \sqrt{2}\mathbf{e}_0^\top \mathbf{b}(\xi) u\left(x_i + \xi\frac{\triangle x_i}{2}, 0\right) d\xi \frac{\triangle x_i}{2} \\
&= \int_{-1}^{1} u\left(x_i + \xi\frac{\triangle x_i}{2}, 0\right) d\xi \frac{\triangle x_i}{2} \\
&= \int_{x_{i-1/2}}^{x_{i+1/2}} u(x,0)\,dx.
\end{aligned}$$

5.14.5 Limiters

Shu and Cockburn [29] suggest that a modification of the minmod limiter be used with the discontinuous Galerkin scheme in order to avoid degeneration of accuracy at local extrema. They suggest that we compute the cell averages

$$\overline{u}_i(t) \equiv \frac{1}{\triangle x_i}\int_{x_{i-1/2}}^{x_{i+1/2}} U(x,t)\,dx = u_{i,0}(t) b_0$$

and the values of the solution at the cell boundaries

$$\begin{aligned}
u_{i+1/2,L}(t) &\equiv U(x_{i+1/2} - 0, t) = \mathbf{b}(1)^\top \mathbf{u}_i(t), \\
u_{i-1/2,R}(t) &\equiv U(x_{i-1/2} + 0, t) = \mathbf{b}(-1)^\top \mathbf{u}_i(t).
\end{aligned}$$

Let $\triangle x = \max_i \triangle x_i$. If $u(x,0) \in \mathcal{C}^2$, let M be a bound on the second spatial derivative of the initial data near critical points. The idea is to pick some $\delta > 0$ and choose $M > 0$ so that for any critical point x_* of $u(x,0)$ we have

$$\text{for all } |x - x_*| < \delta, \left|\frac{d^2 u(x,0)}{dx^2}\right| \leq M.$$

In practice, people often experiment with M until they achieve acceptable results.

The values at the cell sides are then modified as follows. If the polynomial order k is greater than 1 and $|u_{i+1/2,L}(t) - \overline{u}_i(t)| \geq M\triangle x^2$, then $u_{i+1/2,L}(t)$ is replaced by $\sum_{j=0}^{1} u_{i,j}(t) b_j(1)$. If we still have $|u_{i+1/2,L}(t) - \overline{u}_i(t)| \geq M\triangle x^2$, then we use the minmod limiter, which involves the following calculations. Let

$\triangle u_{i,+} = u_{i+1/2,L}(t) - \overline{u}_i(t)$ and $s = \text{sign}(\triangle u_{i,+})$. Then

$$u_{i+1/2,L}(t) = \overline{u}_i(t) + \begin{cases} s \min\{\triangle\overline{u}_{i-1/2}(t), \triangle\overline{u}_{i+1/2}(t)\}, & s = \text{sign}(\triangle\overline{u}_{i-1/2}(t)) = \text{sign}(\triangle\overline{u}_{i+1/2}(t)) \\ 0, & \text{otherwise.} \end{cases}$$

Similarly, if the polynomial order k is greater than 1 and $\overline{u}_i(t) - |u_{i-1/2,R}(t)| \geq M\triangle x^2$, then $u_{i-1/2,R}(t)$ is replaced by $\sum_{j=0}^{1} u_{i,j}(t)b_j(-1)$. If we still have $\overline{u}_i(t) - |u_{i-1/2,R}(t)| \geq M\triangle x^2$, then we use the minmod limiter. Let $\triangle u_{i,-} = \overline{u}_i(t) - u_{i+1/2,R}(t)$ and $s = \text{sign}(\triangle u_{i,-})$. Then

$$u_{i-1/2,R}(t) = \overline{u}_i(t) - \begin{cases} s \min\{\triangle\overline{u}_{i-1/2}(t), \triangle\overline{u}_{i+1/2}(t)\}, & s = \text{sign}(\triangle\overline{u}_{i-1/2}(t)) = \text{sign}(\triangle\overline{u}_{i+1/2}(t)) \\ 0, & \text{otherwise.} \end{cases}$$

The resulting states at the cell sides are used to compute the fluxes at the solution of Riemann problems:

$$f_{i+1/2}(t) = f(\mathcal{R}(u_{i+1/2,L}(t), u_{i+1/2,R}(t), 0)).$$

Here the flux at the solution of the Riemann problem could be given either by the flux at the exact solution of the Riemann problem, the Engquist-Osher flux (see Example 5.7.4), the Rusanov flux (see Section 3.3.4), or the Harten–Hyman modification of the Roe solver (see Section 4.13.9). These fluxes are used in (5.38) to compute the right-hand side $\mathbf{f}_i(t)$ in the ordinary differential equations $d\mathbf{u}i/dt = \mathbf{f}_i(t)$.

5.14.6 Timestep Selection

Cockburn and Shu [29] prove that for $0 \leq k \leq 2$, the discontinuous Galerkin scheme described above is stable for linear advection with CFL $= 1/(2k+1)$.

A C^{++} program to implement the discontinuous Galerkin schemes for a variety of nonlinear scalar conservation laws can be found in **Program 5.14-80: Schemes2.C,** which calls several Fortran subroutines in **Program 5.14-81: dgm.f.** The same version of these schemes for linear advection on uniform grids appears in **Program 5.14-82: LinearAdvectionSchemes.C.** Students can exercise the former program by clicking on **Executable 5.14-39: guiriemann2** Students can also exercise the discontinuous Galerkin scheme for linear advection by clicking on **Executable 5.14-40: guilinearad2** By setting the number of cells to 0, the user will cause the program to perform a mesh refinement study. In this case, the errors are computed by summing the absolute values of the pointwise errors at the Lobatto quadrature points, times the appropriate mesh factor. Also, the numerical solution and analytical solution are plotted throughout each mesh interval, so that the students can see the behavior of the piecewise polynomials as time evolves, or as the mesh is refined.

Exercises for 5.14

5.14.1 Show that the first-order discontinuous Galerkin scheme, which uses piecewise constant polynomials, is the same as Godunov's method.

5.14.2 Write down the steps in the second-order version of the discontinuous Galerkin scheme, which uses piecewise linear polynomials.

5.15 Case Studies

In deciding which scheme to use in a particular application, it is reasonable to examine the accuracy and efficiency of the schemes, as well as ease of programming. For nonlinear problems, we might also want to know which of the numerical fluxes are best. We might also ask if higher-order schemes are more efficient than lower order schemes. The answers, of course, depend on the intended application.

For problems involving shocks, none of the schemes is better than first-order accurate. For Riemann problems, none of the schemes is better than second-order accurate. This is important, because Liu [108] has shown that in general the solution of hyperbolic conservation laws for initial data with compact support evolve to the solution of Riemann problems at large time. Thus, it might appear that the nominal high order of some schemes will be apparent in the numerical results only at small time for nonlinear conservation laws, or for problems also involving some reasonable amount of physical diffusion.

In the sections below, we will describe some results with the numerical methods in this chapter, for a variety of test problems. These results do not by any means determine the suitability of a scheme in general. The mesh refinement studies were performed with 100, 200, 400, 800, 1600, 3200 and 6400 grid cells. The timestep was chosen to be 90% of the stability limit for each scheme. For the discontinuous Galerkin scheme, the limiter factor M was chosen sufficiently large so that no limiting was performed for any of the problems; this was because smaller values of this factor lead to nonconvergence for Riemann problems. Also note that the errors are measured differently for the discontinuous Galerkin method. All other methods measure the $\mathcal{L}_1$ error in the cell averages, while the discontinuous Galerkin method measures the $\mathcal{L}_1$ norm of the error.

5.15.1 Case Study: Linear Advection

Consider the linear advection problem with periodic boundary conditions

$$\begin{aligned}
&\frac{\partial u}{\partial t} + \frac{\partial u}{\partial x} = 0 \text{ for all } x \in (0, 1) \text{ for all } t > 0 \\
&u(1, t) = u(0, t) \text{ for all } t > 0 \\
&u(x, 0) \text{ given, for all } x \in (0, 1).
\end{aligned}$$

Zalesak [180] proposed several initial values for testing schemes applied to this problem:

$$\textbf{square pulse}: u(x,0) = \begin{cases} 2, & 0.1 \le x \le 0.2 \\ 1, & \text{otherwise} \end{cases}$$

$$\textbf{triangular pulse}: u(x,0) = \begin{cases} 2 - 20|x - 0.15|, & 0.1 \le x \le 0.2 \\ 1, & \text{otherwise} \end{cases}$$

$$\textbf{smooth Gaussian pulse}: u(x,0) = \begin{cases} 1 + \exp(-10^4(x - 0.15)^2) - \exp(-25), & 0.1 \le x \le 0.2 \\ 1, & \text{otherwise.} \end{cases}$$

Zalesak specified that each problem should be solved with 100 cells on a uniform grid, so that the initial disturbance is described in a fixed number of grid cells. We have performed several numerical experiments with the schemes described in this chapter for these test problems.

For the square pulse, the upwind scheme was $O(\triangle x^{0.5})$, while the higher-order schemes were all roughly $O(\triangle x^{0.7})$. For a given (fine) mesh, the most accurate second-order schemes were the side-centered wave propagation, MUSCL, TVD and the Nessyahu–Tadmor scheme. However, the schemes that obtained a given accuracy for the least computer time were MUSCL, TVD and wave propagation. Second-order ENO was the least accurate and least efficient, and discontinuous Galerkin was not much better. We also compared second-order wave propagation with the third-order schemes in this chapter. The most accurate schemes were PPM, discontinuous Galerkin, wave propagation and the Liu–Tadmor scheme, while the most efficient schemes were PPM, wave propagation, Liu–Tadmor and then discontinuous Galerkin. The discontinuous Galerkin scheme took about 50 times more computer time than PPM for a given accuracy, while ENO was about 350 times more expensive.

For the triangular pulse, the upwind scheme was $O(\triangle x)$, while most of the higher-order schemes were $O(\triangle x^{1.4})$. All of the second-order schemes obtained roughly the same accuracy for a given mesh size, except the discontinuous Galerkin and ENO schemes, which were less accurate. The second-order discontinuous Galerkin and ENO schemes were significantly less efficient, as well. Of wave propagation and the third-order schemes, the most accurate were discontinuous Galerkin, followed by Liu-Tadmor, PPM, second-order wave propagation and third-order ENO. The most accurate third-order scheme was discontinuous Galerkin, and the least accurate was ENO. The most efficient of wave propagation and the third-order schemes were wave propagation, PPM, Liu–Tadmor and discontinuous Galerkin; ENO was the least efficient. ENO took roughly 1000 times as long as PPM for a given accuracy, and discontinuous Galerkin took about 16 times the computer work.

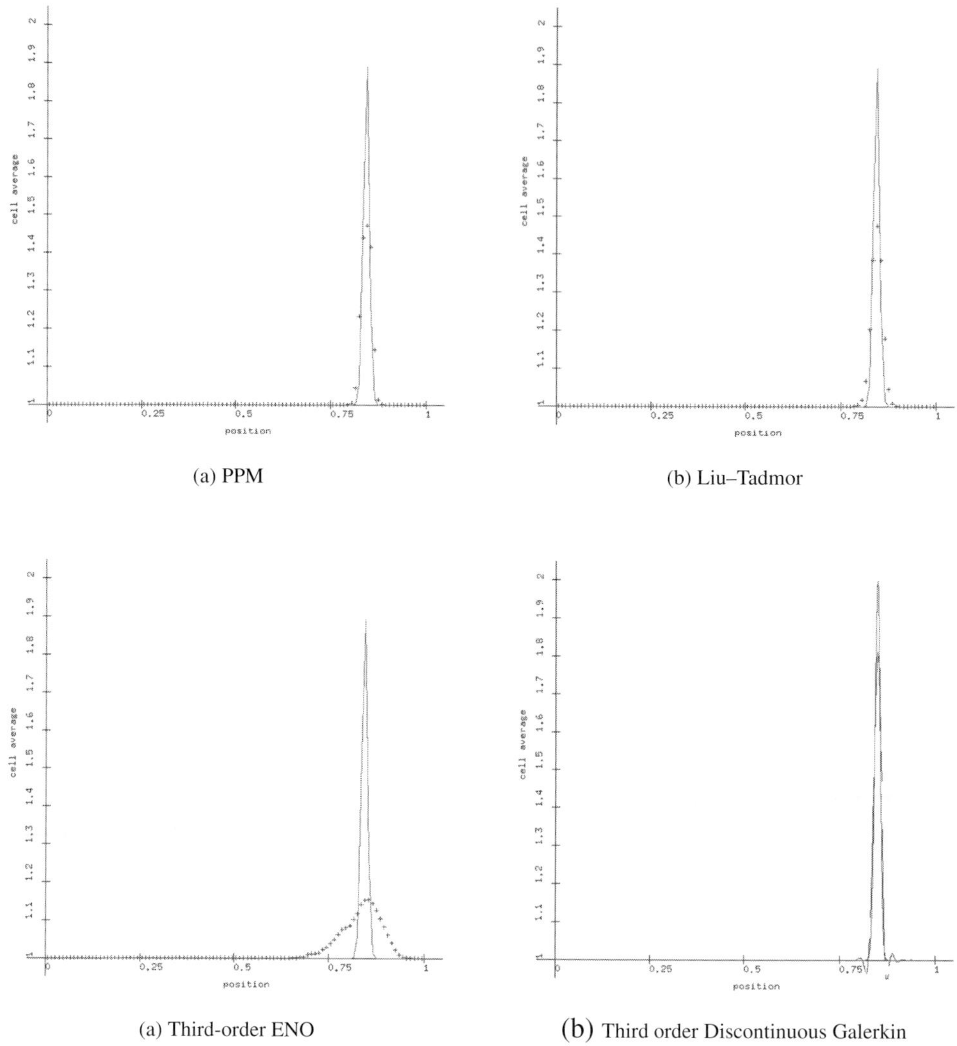

(a) PPM (b) Liu–Tadmor

(a) Third-order ENO (b) Third order Discontinuous Galerkin

Fig. 5.5 Comparison of schemes for linear advection Gaussian pulse, 100 grid cells

For the smooth Gaussian pulse, upwind was first-order, the second-order schemes were all second-order, and the third-order schemes were mostly somewhat over second order, typically around $O(\triangle x^{2.3})$. In this case, we could determine that the second derivative of the initial data at the critical point $x = 0.15$ is -2×10^4, so we took $M = 2 \times 10^4$ in the discontinuous Galerkin limiter. Figure 5.5 shows the results with several third-order schemes with 100 grid cells. Note that the peak values of the solution are reduced substantially with PPM and the Liu–Tadmor schemes; third-order ENO is far worse. The third-order discontinuous Galerkin maintains the peak value of the solution pretty well, but has some undershoots before

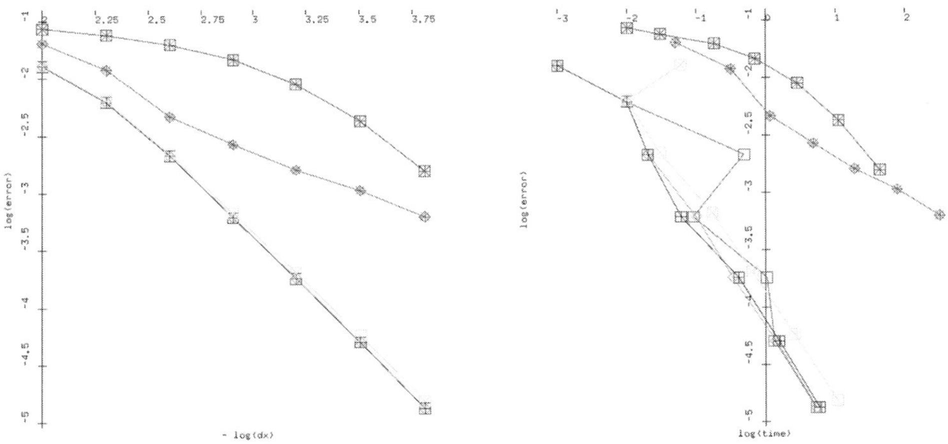

(a) Second order accuracy

(b) Second order efficiency

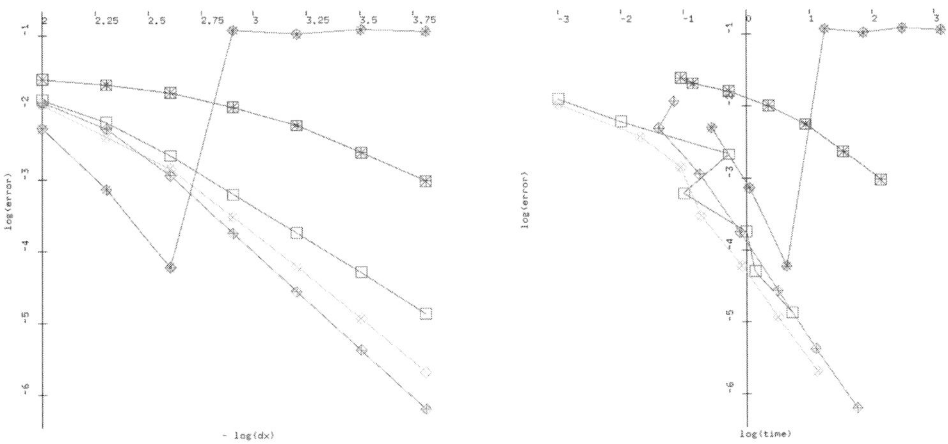

(a) Third-order accuracy

(b) Third-order efficiency

Fig. 5.6 Comparison of schemes for linear advection Gaussian Pulse (box = MUSCL, diamond = TVD, box plus = wave propagation, box cross = Nessyahu-Tadmor, diamond plus = Liu–Tadmor, diamond cross = PPM, box plus cross = ENO, diamond plus cross = discontinuous Galerkin)

and after the pulse. Here the discontinuous Galerkin method has a competitive advantage, in that it samples the initial data at Lobatto quadrature points in each grid cell; the third-order scheme samples the initial data at an average of three points per cell, versus one point per cell for the other schemes.

The least accurate and least efficient second-order schemes were ENO and discontinuous Galerkin, with the other schemes all roughly the same. The most accurate third-order schemes were Liu–Tadmor, PPM and MUSCL. Discontinuous Galerkin

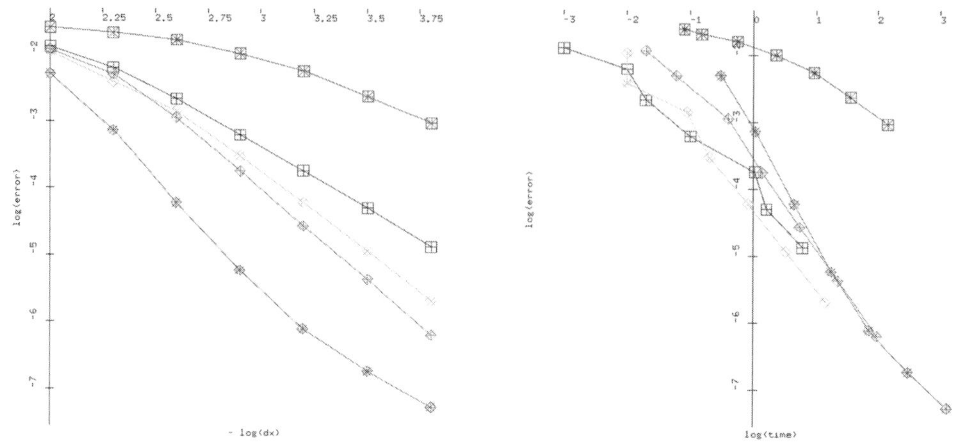

(a) Accuracy

(b) Efficiency

Fig. 5.7 Comparison of wave propagation and third-order schemes for linear advection Gaussian pulse, unlimited discontinuous Galerkin (box = MUSCL, box cross = Nessyahu–Tadmor, box plus cross = ENO, diamond plus cross = discontinuous Galerkin)

was most accurate at coarse meshes but then the limiter caused a convergence failure; ENO was generally the least accurate third-order scheme. PPM and MUSCL were the most efficient of these schemes. These results are displayed in Figure 5.6. Since it is possible that our implementation of the discontinuous Galerkin limiter has an error, to produce Figure 5.7 we took $M = 2 \times 10^9$ so that the limiter never activated, and achieved convergence for the discontinuous Galerkin method. For this problem, wave propagation had order 1.8, PPM had order 2.3, Liu–Tadmor had order 2.7, third-order ENO had order 1.3 and third-order discontinuous Galerkin had order 3 for coarse mesh and 1.8 on the finest mesh tested. In this case, the discontinuous Galerkin method was nearly as efficient as the best third-order schemes.

Students can reproduce these results or create their own by clicking on the following link, **Executable 5.15-41: guilinearad2** By setting the number of cells to 0, the user will cause the program to perform a mesh refinement study. Note that the discontinuous Galerkin errors are computed by summing the absolute values of the pointwise errors at the Lobatto quadrature points, times the quadrature weight and the appropriate mesh factor. The errors for the other schemes are the sums of the absolute values of the errors in the cell averages, times the mesh width.

5.15.2 Case Study: Burgers' Equation

We compared the schemes in this chapter for Burgers' equation,

$$\frac{\partial u}{\partial t} + \frac{\partial u^2/2}{\partial x} = 0 \text{ for all } x \in (0, 1) \text{ for all } t > 0.$$

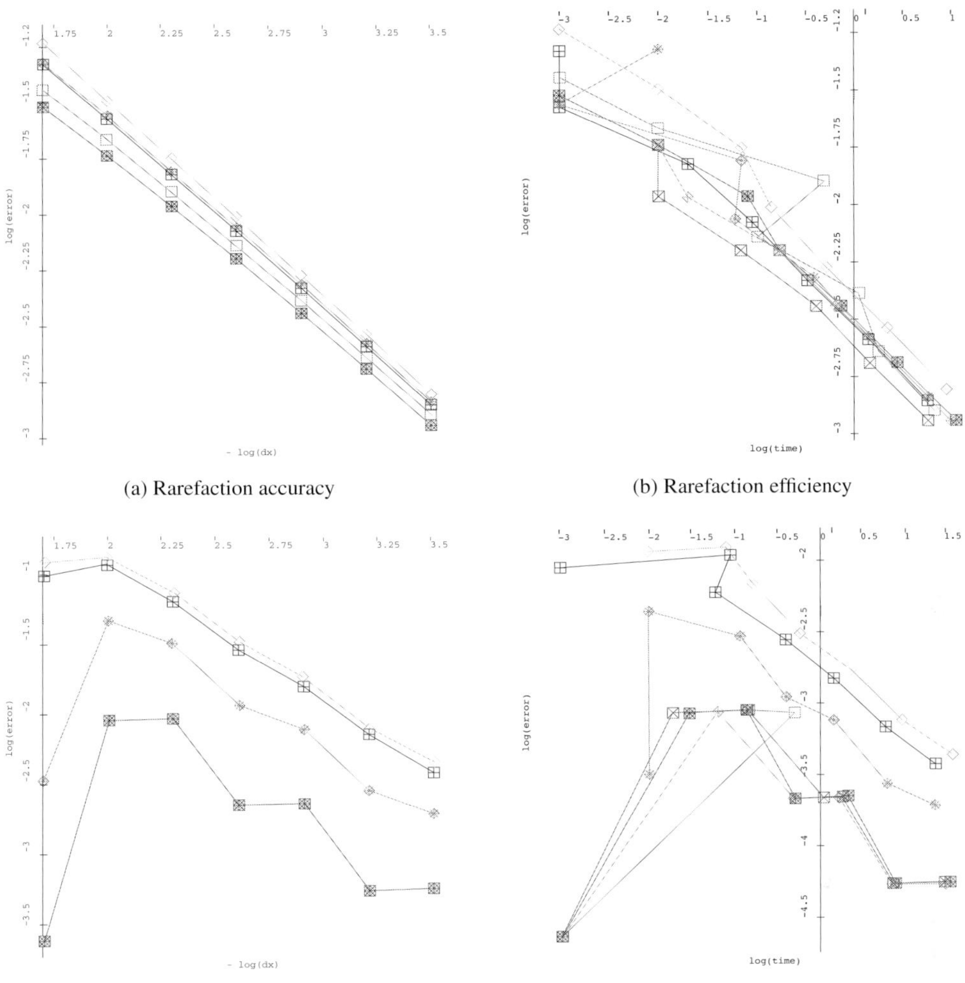

(a) Rarefaction accuracy

(b) Rarefaction efficiency

(c) Shock accuracy

(d) Shock efficiency

Fig. 5.8 Comparison of approximate Riemann solvers for Burgers' transonic rarefaction ($u_L = -1$, $u_R = 1$) and Shock ($u_L = 2$, $u_R = -1$) (box = exact, diamond = Rusanov, box plus = Marquina, box cross = Harten–Hyman, diamond cross = Harten–Lax-vanLeer, box plus cross = Linde, diamond plus cross = Engquist–Osher)

First, we considered a transonic rarefaction ($u_L = -1$, $u_R = 1$). Of the various choices for numerical fluxes in Godunov's method, the most efficient were Marquina's flux, followed by the Harten–Hyman flux and the Engquist–Osher flux (see the top of Figure 5.8). The most accurate numerical fluxes were Linde's flux, followed by the Engquist-Osher flux. These all performed better than the flux at the true solution of the Riemann problem. For a strong shock ($u_L = 2$, $u_R = -1$), the most efficient numerical fluxes were Harten-Hyman, Harten-Lax–vanLeer, Linde and the flux at the the solution of the Riemann problem (in that order). The most

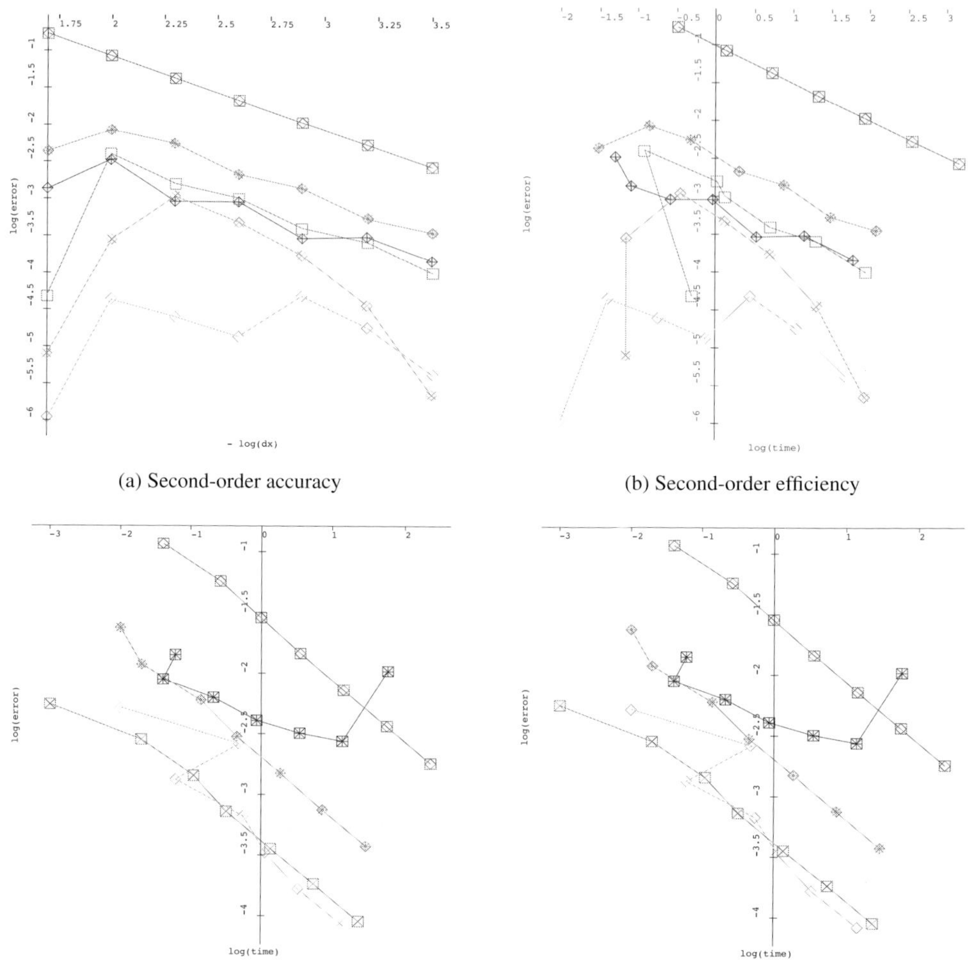

(a) Second-order accuracy (b) Second-order efficiency

(c) Third-order accuracy (d) Third-order efficiency

Fig. 5.9 Comparison of Schemes for Burgers' Transonic Rarefaction ($u_L = -1$, $u_R = 1$) (box = TVD, diamond = MUSCL, diamond cross = side-centered wave propagation, box cross = PPM, diamond plus = Nessyahu–Tadmor, box plus cross = Liu–Tadmor, diamond plus cross = ENO, box diamond = discontinuous Galerkin)

accurate fluxes were Linde, Engquist–Osher, Marquina and Rusanov. These results appear at the bottom of Figure 5.8.

Among the second-order schemes applied to the transonic rarefaction, the MUSCL scheme was most accurate and most efficient, followed closely by the Nessyahu–Tadmor scheme (see Figure 5.9). The ENO and discontinuous Galerkin schemes were the least accurate and least efficient. The discontinuous Galerkin scheme took about 3000 times longer to reach the same accuracy as the MUSCL scheme (remember that the errors are measured differently in these two cases,

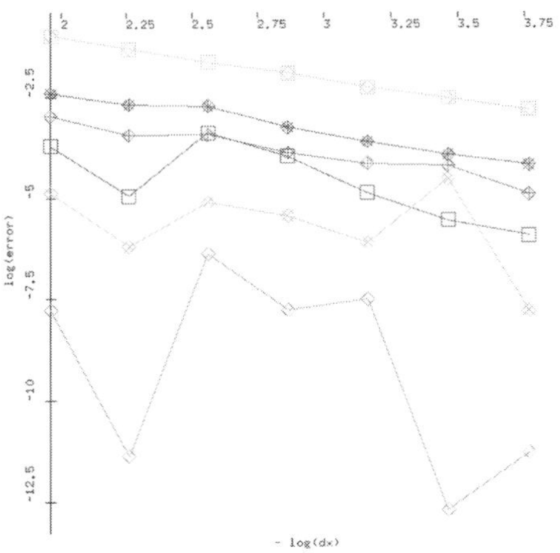

(a) Second-order accuracy

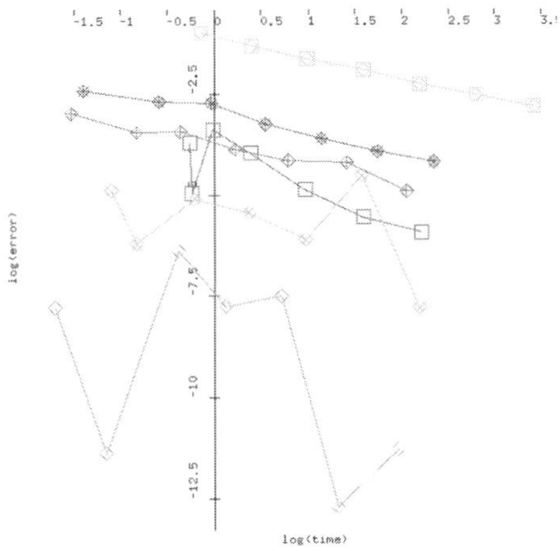

(b) Second-order efficiency

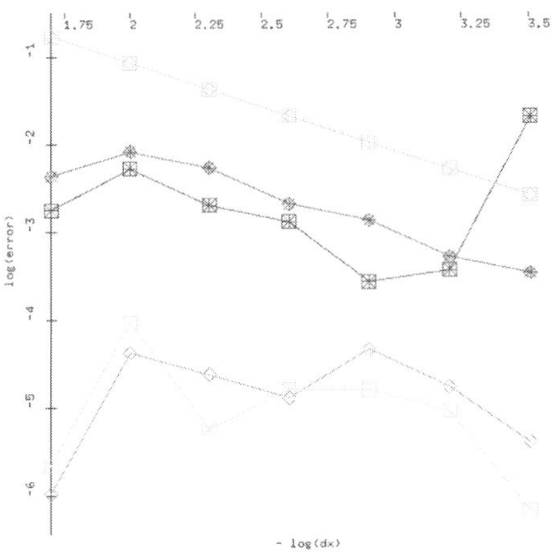

(a) Third-order accuracy

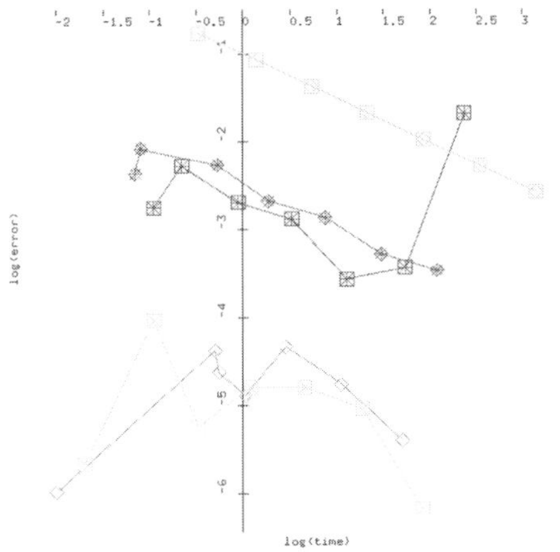

(b) Third-order efficiency

Fig. 5.10 Comparison of Schemes for Burgers' Strong Shock ($u_L = 2, u_R = -1$) (box = TVD, diamond = MUSCL, diamond cross = side-centered wave propagation, box cross = PPM, diamond plus = Nessyahu–Tadmor, box plus cross = Liu–Tadmor, diamond plus cross = ENO, box diamond = discontinuous Galerkin)

because of the differences in the schemes). ENO took about 35 times as long as MUSCL. Among MUSCL and the third-order schemes, MUSCL and PPM were most accurate and efficient, while discontinuous Galerkin and Liu–Tadmor were least accurate and efficient. In this case, the limiter destroys the accuracy of the Liu–Tadmor scheme; third-order convergence was verified with the same algorithm on other test problems.

Of the second-order schemes, the most accurate and efficient for a strong shock ($u_L = 2$, $u_R = -1$) were MUSCL and wave propagation, while the least accurate and efficient were discontinuous Galerkin and ENO (see Figure 5.10). In a

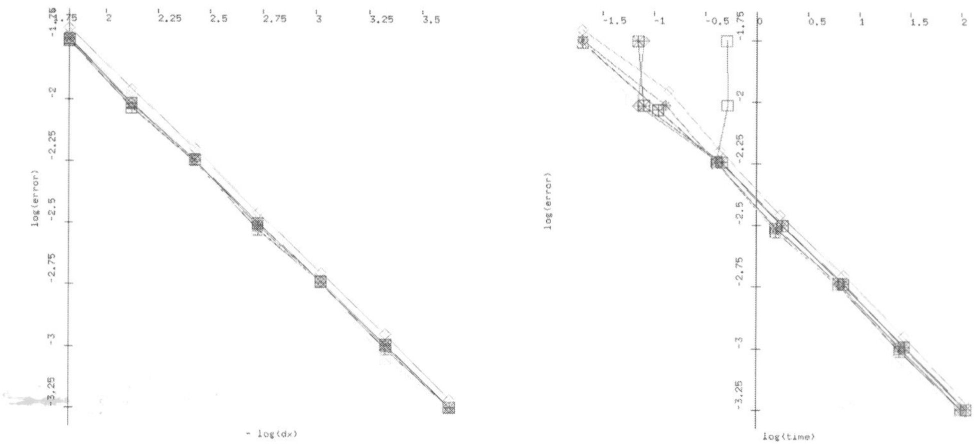

(a) Shock

(b) Rarefaction

Fig. 5.11 Comparison of approximate Riemann solver for Buckley–Leverett shock–rarefaction–shock ($u_L = 0.001$, $u_R = 0.999$) (box = exact, diamond = Rusanov, box plus = Marquina, box cross = Harten–Hyman, diamond cross = Harten–Lax–vanLeer, box plus cross = Linde, diamond plus cross = Engquist–Osher)

comparison of MUSCL with the third-order schemes, the most accurate and efficient were MUSCL and PPM. MUSCL and PPM both capture the shock in at most one grid cell, with essentially perfect resolution of the solution otherwise.

We also compared the schemes for a weak shock ($u_L = 1$, $u_R = 0.9$). All of the approximate Riemann solvers achieved essentially the same accuracy, and nearly the same efficiency.

The ENO and discontinuous Galerkin schemes are at a competitive disadvantage because of their use of Runge–Kutta time integration, involving several substeps. Runge–Kutta schemes have also been known to develop spurious oscillations and convergence to unphysical solutions for nonlinear problems [179]. The discontinuous Galerkin schemes are at a further disadvantage because of their more restrictive stability restriction on the timesteps. These schemes are accurate for problems with smooth initial data, but their performance on the Riemann problems we have tested is not competitive.

Students can reproduce these results by clicking on **Executable 5.15-42: guiriemann2** By setting the number of cells to 0, the user will cause the program to perform a mesh refinement study.

5.15.3 Case Study: Traffic Flow

The traffic flow problem in Section 3.2.1 can be interesting to students because it has more readily identifiable physical significance. In our experiments, the

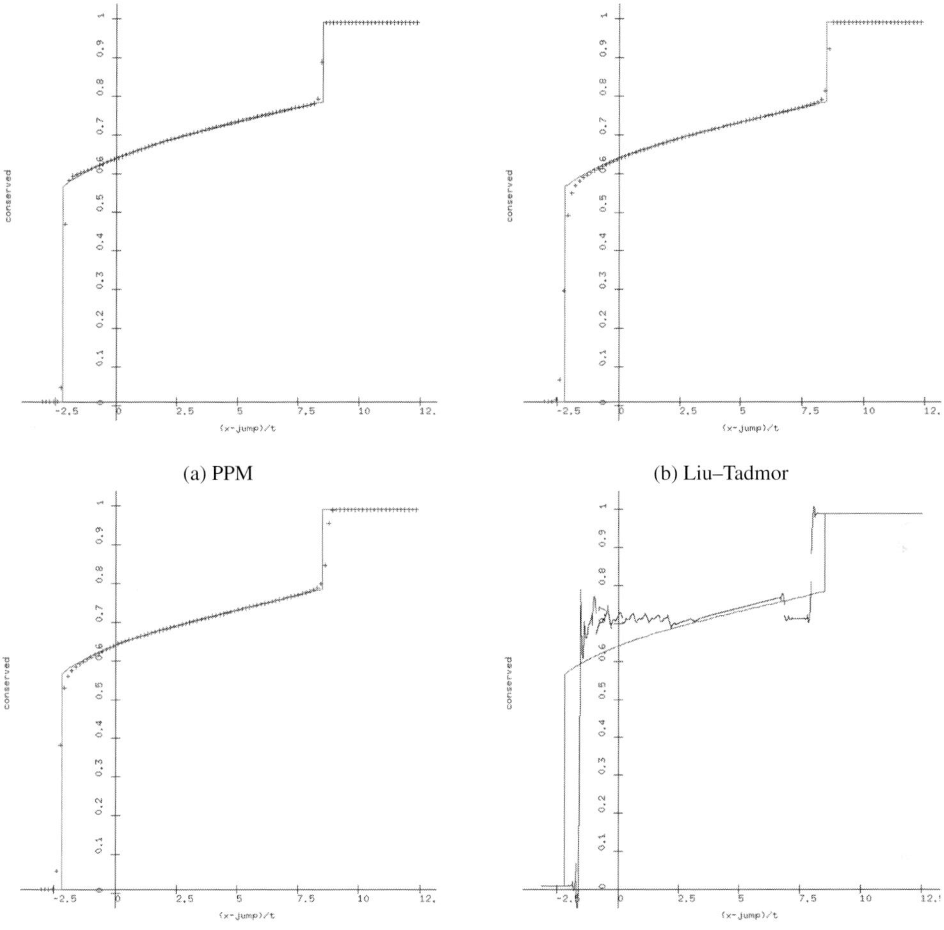

(a) PPM (b) Liu–Tadmor

(c) ENO (d) Discontinuous Galerkin

Fig. 5.12 Comparison of third-order schemes for Buckley–Leverett shock–rarefaction–shock ($u_L = 0.001$, $u_R = 0.999$)

numerical methods performed much the same for traffic flow problems as for Burgers' equation. Students should experiment on their own to verify this claim.

5.15.4 Case Study: Buckley–Leverett Model

We also made comparisons of the various numerical fluxes for a Buckley-Leverett shock-rarefaction-shock. In this comparison, all of the numerical fluxes performed similarly, with the Harten-Hyman flux slightly more accurate and efficient, and the Rusanov flux slightly less. These results appear in Figure 5.11 Comparisons of schemes appear in Figure 5.12. In this case, the accuracy of the discontinuous Galerkin scheme is poor because it appears to be converging to a solution involving two discontinuities and an intermediate constant state (see Figure 5.13.)

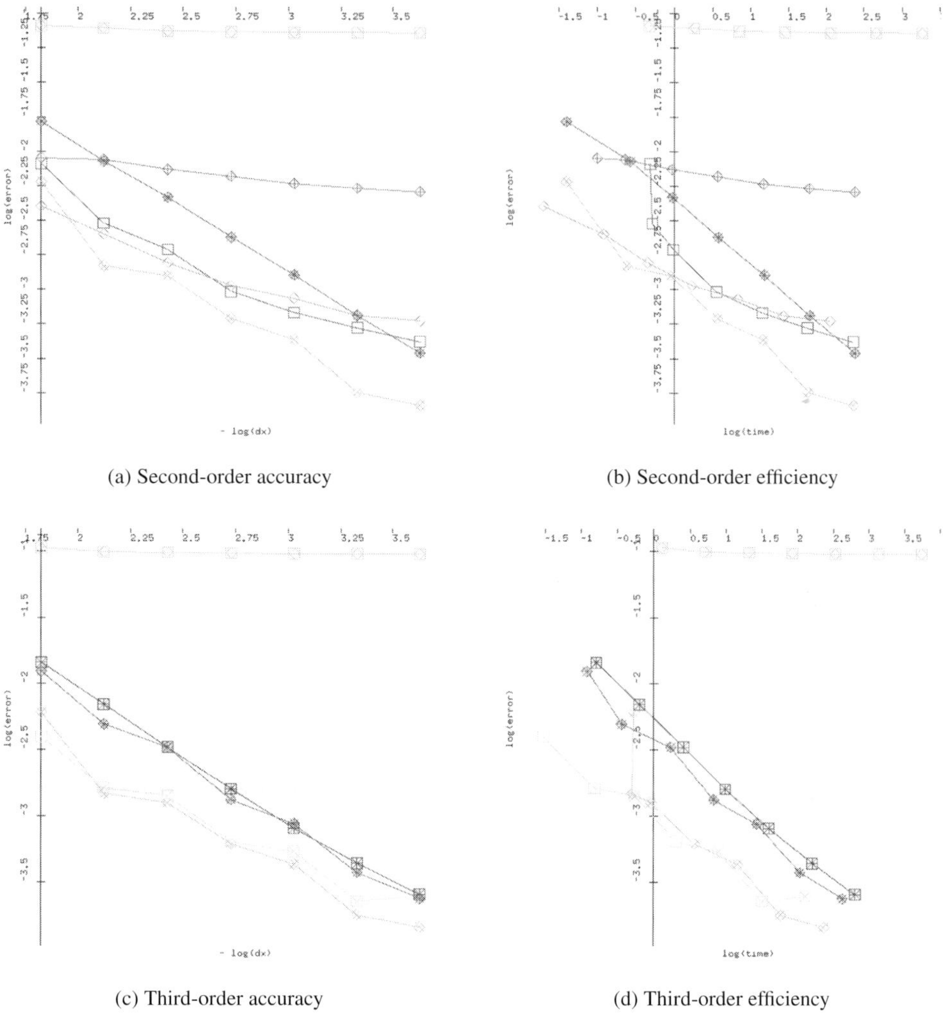

(a) Second-order accuracy

(b) Second-order efficiency

(c) Third-order accuracy

(d) Third-order efficiency

Fig. 5.13 Comparison of Schemes for Buckley–Leverett shock–rarefaction–Shock ($u_L = 0.001$, $u_R = 0.999$) (box = TVD, diamond = MUSCL, diamond cross = side-centered wave propagation, box cross = PPM, diamond plus = Nessyahu–Tadmor, box plus cross = Liu–Tadmor, diamond plus cross = ENO, box diamond = discontinuous Galerkin)

Exercises for 5.15

5.15.1 In order to verify that the schemes are operating properly, it is useful to have a smooth test problem. Zalesak's smooth Gaussian requires a large number of grid cells to look smooth to the numerical computation. Choose one of the numerical methods in this chapter, and verify that it achieves the correct order for initial data $u_0(x) = \sin(\pi x)$. Remember to use period boundary conditions.

5.15.2 Choose one of the schemes in this chapter and compare its performance on linear advection with Zalesak's smooth Gaussian, and with initial data

$$u_0(x) = 1 + \exp(-625(x/2 - 1)^2).$$

5.15.3 Goodman and LeVeque [59] suggested the Burgers' equation initial data

$$u_0(x) = \begin{cases} -0.5, & x < 0.5 \\ 0.2 + 0.7\cos(2\pi x), & x > 0.5 \end{cases}$$

for a problem with periodic boundary conditions.

(a) Use Equation (3.5) to determine the analytical solution for this problem. Your solution should be in the form of an algorithm.

(b) Program the MUSCL scheme for this problem, and run the method until the solution develops a shock.

(c) Plot the $\mathcal{L}_1$ error in the cell averages for your solution as a function of time. How does the development of the shock degrade the accuracy of the scheme?

5.15.4 Cockburn and Shu [29] suggested the initial data

$$u_0(x) = \frac{1}{4} + \frac{1}{2}\sin \pi x$$

for Burgers' equation. Repeat the exercises of the previous problem for the second-order discontinuous Galerkin method.

5.15.5 LeVeque [97, page 205] shows results for traffic flow with initial data apparently given by a smooth Gaussian

$$u_0(x) = 0.25 + 0.75\exp(-0.25x^2), \ |x| \leq 30.$$

(a) Use equation (3.5) to determine the analytical solution for this problem. Your solution should be in the form of an algorithm.

(b) Program the wave propagation scheme for this problem, and run the method until the solution develops a shock.

(c) Plot the $\mathcal{L}_1$ error in the cell averages for your solution as a function of time. How does the development of the shock degrade the accuracy of the scheme?

5.15.6 Repeat the previous exercise for a red light (LeVeque [97, page 206]) $\rho_L = 0.25$, $\rho_R = 1.$, using the Nessyahu–Tadmor scheme.

5.15.7 Repeat the previous exercise for a green light (LeVeque [97, page 207]) $\rho_L = 1.$, $\rho_R = 0.$ using the second-order ENO scheme.

5.15.8 Suppose that the traffic flow has a spatially varying speed limit (LeVeque [97, page 369]) with flux $f(u, x) = u_{\max}(x)u(1 - u)$

(a) Examine the ideas behind the analytical solution (3.5) of the conservation law for $f(u)$ to find an analytical solution for $f(u, x)$ in smooth flow. What are the stationary states in this conservation law?

(b) Develop a modification of Godunov's method to solve this problem on a periodic domain. Verify that your scheme is first-order accurate for $u_{\max}(x) = 2 + \sin(\pi x)$ and $u_0(x) = \sin^2(\pi x)$ where $x \in (-1, 1)$.

(c) Develop a modification of the MUSCL to solve this problem. Verify that your scheme is second-order accurate.

5.15.9 Suppose that the traffic flow has a source term due to an entrance ramp (LeVeque [97, page 396]):

$$\frac{\partial u}{\partial t} + \frac{\partial f(u)}{\partial x} = D\delta(x),$$

where $\delta(x)$ is the Dirac delta-function.

(a) Formulate the integral form of this conservation law.

(b) Examine the ideas behind the analytical solution (3.5) of the conservation law with no source terms to find an analytical solution for traffic flow with a source term.

(c) Develop a modification of the wave propagation scheme to solve this problem. Verify that your scheme is first-order accurate.

(d) Is it possible to design a second-order accurate scheme for this problem with a source term?

5.15.10 Suppose that a enclosed one-dimensional oil reservoir ($\mathrm{v}_T = 0$, $\mathbf{K}g(\rho_w - \rho_o) = 1$) is initialized with water on the top half and oil on the bottom half ($s_o = 1$ on the bottom or left, $s_o = 1$ on the top or right). Program Godunov's method for this problem, including reflecting boundary conditions. Use your numerical solution to examine how the fluid flow will evolve until waves bounce off a boundary and return to the middle of the reservoir. The steady-state solution will have oil on the top and water on the bottom. Use the model parameters described in Section 3.2.3, and ignore capillary pressure.

5.15.11 Oil is produced from reservoirs by wells. If water and oil are incompressible, then oil production requires at least two wells, a water injector and an oil producer. In the plane between the injector and producer, flow can look roughly one-dimensional, with the injector on the left and the producer on the right. Suppose that we produce fluid at a specified rate v_T (see Section 3.2.3) and inject water at the same rate. If the oil reservoir currently has oil saturation s_o and gravity number $\mathbf{K}g(\rho_w - \rho_o)/\mathrm{v}_T$ (with $\rho_w > \rho_o$), determine conditions on s_o so that fluid flows into the reservoir from the injector, and out of the reservoir into the producer. In other words, under what conditions is the characteristic speed positive at the injector ($s_o = 0$) and positive at the producer? Study the Oleinik chord condition (Lemma 3.1.19) to answer this question.

5.15.12 In order to produce oil at a specified rate, the pressure at the injector has to be adjusted so that the desired rate is maintained. Show that the incompressibility condition $\partial \mathrm{v}_T/\partial x = 0$ and the formula for v_T in section 3.2.3 gives us an ordinary differential equation for the pressure in water, with coefficients that depend on the oil saturation. Find the analytical solution of this pressure equation.

5.15.13 In the oil industry, the most common numerical method for solving the Buckley–Leverett equations is the **upstream weighting method**. This method depends on the special circumstances in the formulation of the Buckley-Leverett flux, described in section 3.2.3. For both of the oil and water phases, the potential gradient at cell side $i + \frac{1}{2}$ is given by

$$\psi_{j,i+1/2}(s) = -\frac{p_{j,i+1} - p_{j,i}}{x_{i+1} - x_i} + \left(\frac{\partial d}{\partial x}\right)_{i+1/2} g\frac{\rho_{j,i+1} + \rho j, i}{2}, \quad j = o, w$$

where $p_{j,i}$ is the phase pressure in grid cell i, d is the depth, g is the normal component of gravity and $\rho_{j,i}$ is the density of phase j in cell i. Phase pressures differ by the capillary pressure, which is ignored in this form of the Buckley–Leverett model. The upstream weighting method chooses the phase mobilities $\lambda_j = \kappa_{rj}(s_j)/\mu_j$, which are phase relative permeability divided by phase viscosity, to be given by

$$\lambda_{j,i+1/2} = \begin{cases} \lambda_j(s_{j,i}), & \psi_{j,i+1/2} \geq 0 \\ \lambda_j(s_{j,i+1}), & \psi_{j,i+1/2} < 0. \end{cases}$$

Although it is unimportant to this exercise, the permeability is harmonically averaged

$$\kappa_{i+1/2} = \frac{2\kappa_i \kappa_{i+1}}{\kappa_i + \kappa_{i+1}}.$$

The flux of oil is chosen to be

$$f_{o,i+1/2} = \kappa_{i+1/2} \sum_{j=o.w} \psi_{j,i+1/2} \lambda_{j,i+1/2}.$$

The oil saturation is updated by the conservative difference

$$s_i^{n+1} \phi_i = s_i^n \phi_i - \frac{\Delta t^n}{\Delta x_i} [f_{i+1/2} - f_{i-1/2}] \equiv H(s_{i-1}^n, s_i^n, s_{i+1}^n).$$

Show that upstream weighting is monotone (Section 5.2.2) if the timestep is chosen to be sufficiently small [141]. Determine how the timestep should be chosen to guarantee that upstream weighting is monotone.

6

Methods for Hyperbolic Systems

The most interesting conservation laws involve nonlinear systems. However, generalizing the numerical methods of Chapter 5 to hyperbolic systems poses a challenge, because there are multiple conserved quantities and multiple characteristic speeds. Another challenge is that the theory for hyperbolic systems is not as advanced as it is for scalar laws; for a survey of the recent state of this theory, see [41]. As a result, the theory of numerical methods for nonlinear systems of hyperbolic conservation laws is even more primitive. In this chapter we will concentrate on the description of numerical methods for hyperbolic systems, and ignore issues of convergence theory.

6.1 First-Order Schemes for Nonlinear Systems

We would like to develop methods to solve nonlinear systems of the form

$$\frac{\partial \mathbf{u}}{\partial t} + \frac{\partial \mathbf{f}(\mathbf{u})}{\partial x} = 0.$$

We will present three first-order schemes for this problem: the Lax–Friedrichs scheme, the random choice scheme and the Godunov (or upwind) scheme. Another possibility is front tracking [26, 56, 98, 135, 136]. The front tracking scheme typically requires substantially different data structures from the other schemes in this text, and its correct implementation for general multi-dimensional problems is very difficult.

6.1.1 Lax–Friedrichs Method

The general Lax–Friedrichs scheme was described previously in Section 4.2.1. This scheme involves two half-steps:

$$\mathbf{u}_{j+1/2}^{n+1/2} = \left\{\mathbf{u}_j^n \triangle x_j + \mathbf{u}_{j+1}^n \triangle x_{j+1} - \triangle t^{n+1/2}[\mathbf{f}(\mathbf{u}_{j+1}^n) - \mathbf{f}(\mathbf{u}_j^n)]\right\} \frac{1}{\triangle x_j + \triangle x_{j+1}}$$

and

$$\mathbf{u}_j^{n+1} = \left\{ \mathbf{u}_{j-1/2}^{n+1/2} + \mathbf{u}_{j+1/2}^{n+1/2} - \left[\mathbf{f}(\mathbf{u}_{j+1/2}^{n+1/2}) - \mathbf{f}(\mathbf{u}_{j-1/2}^{n+1/2}) \right] \frac{\triangle t^{n+1/2}}{\triangle x_j} \right\} \frac{1}{2}.$$

Both of these steps are conservative differences. The timestep should be chosen so that $\frac{\lambda \Delta t}{\Delta x} \leq 1$ for all characteristic speeds λ associated with problem. In practice, it is common to look only at the characteristic speeds associated with the discrete states, and reduce the timestep by a safety factor (say 0.9).

Note that the only characteristic information needed for this method is a bound on the maximum absolute value of the characteristic speeds, in order to determine a stable timestep. In particular, no Riemann problems are solved and no characteristic directions are used. However, the scheme is very diffusive and requires two half-steps on a staggered grid. A Fortran implementation of the Lax-Friedrichs scheme can be found in subroutine `laxfriedrichs` in **Program 6.1-83: schemes.f.**

6.1.2 Random Choice Method

The random choice method can be implemented in the following form. At each cell side $j + \frac{1}{2}$, compute

$$\mathbf{w}_{j+1/2}^{n+1/2} = \mathcal{R}(\mathbf{w}_j^n, \mathbf{w}_{j+1}^n, \xi) \text{ where } \xi \in \left(-\triangle x_j / \triangle t^{n+1/2}, \ \triangle x_{j+1} / \triangle t^{n+1/2}\right).$$

Here ξ is a uniformly distributed random number. Then in each cell j compute

$$\mathbf{w}_j^{n+1} = \mathcal{R}(\mathbf{w}_{j-1/2}^{n+1/2}, \mathbf{w}_{j+1/2}^{n+1/2}, \xi) \text{ where } \xi \in \left(-\triangle x_j / \triangle t^{n+1/2}, \ \triangle x_j / \triangle t^{n+1/2}\right)$$

where ξ is a uniformly distributed random number. The timestep is chosen as in the Lax–Friedrichs scheme. This scheme has been implemented as subroutine `random_choice` in **Program 6.1-84: schemes.f.**

One advantage of the random choice method is that it has no numerical diffusion. Further, it is the only scheme for which convergence to the solution of nonlinear systems of conservation laws has been proved [55]. On the other hand, it tends to approximate rarefactions in a staircase fashion. The performance of the algorithm can be improved with better sampling techniques. See [31, 151] for discussions of these better approaches.

6.1.3 Godunov's Method

Formally, Godunov's method is very simple. At each cell side, we find the flux at the solution of a Riemann problem:

$$\mathbf{f}_{j+1/2}^{n+1/2} = \mathbf{f}(\mathcal{R}(\mathbf{u}_j^n, \mathbf{u}_{j+1}^n; 0)).$$

The scheme is completed by a conservative difference

$$\mathbf{u}_j^{n+1} = \mathbf{u}_j^n - \left[\mathbf{f}_{j+1/2}^{n+1/2} - \mathbf{f}_{j-1/2}^{n+1/2}\right] \frac{\triangle t^{n+1/2}}{\triangle x_j}. \tag{6.1}$$

The timestep can be chosen as in the Lax–Friedrichs scheme. We discussed methods for solving the Riemann problem in the case studies of Chapter 4, and methods for approximating the solution of the Riemann problem in Section 4.13. We will review the approximate Riemann solvers below.

6.1.3.1 Godunov's Method with the Rusanov Flux

Rusanov's flux was described previously in Section 4.2.2. Let $\lambda_{j+1/2}$ be an upper bound on the maximum absolute value of the characteristic speeds for all states involved in the Riemann problem with left state $\mathbf{u}_j^n$ and right state $\mathbf{u}_{j+1}^n$. First, we compute the fluxes at the cell centers by

$$\mathbf{f_j^n} = \mathbf{f(u_j^n)}.$$

Next, we compute the speeds and fluxes at the cell sides by

$$\lambda_{j+1/2} = \max_{u \text{ between } u_j^n, u_{j+1}^n} \rho\left(\frac{\partial \mathbf{f}(\mathbf{u})}{\partial \mathbf{u}}\right)$$

$$\mathbf{f}_{j+1/2}^{n+1/2} = \left\{\mathbf{f}(\mathbf{u}_{j+1}^n) + \mathbf{f}(\mathbf{u}_j^n) - [\mathbf{u}_{j+1}^n - \mathbf{u}_j^n]\lambda_{j+1/2}\right\} \frac{1}{2}.$$

Here ρ refers to the spectral radius, which is the maximum absolute value of an eigenvalue of the matrix argument. Finally, we perform conservative difference in equation (6.1).

A Fortran version of the Rusanov flux computation is available as subroutine `rusanov_flux` in **Program 6.1-85: schemes.f.** This flux computation is combined with a conservative difference in procedure `runScheme` in **Program 6.1-86: GUIRiemannProblem.C.**

Alternatively, we could compute the fluxes at the cell centers as before, then the speeds and flux differences at the cell sides

$$\lambda_{j+1/2} = \max_{u \text{ between } u_j^n, u_{j+1}^n} \rho\left(\frac{\partial \mathbf{f}(\mathbf{u})}{\partial \mathbf{u}}\right)$$

$$\triangle\mathbf{f}_{j+1/2}^{-} = \left\{\mathbf{f}(\mathbf{u}_{j+1}^n) - \mathbf{f}(\mathbf{u}_j^n) - [\mathbf{u}_{j+1}^n - \mathbf{u}_j^n]\lambda_{j+1/2}\right\}/2,$$

$$\triangle\mathbf{f}_{j+1/2}^{+} = \left\{\mathbf{f}(\mathbf{u}_{j+1}^n) - \mathbf{f}(\mathbf{u}_j^n) + [\mathbf{u}_{j+1}^n - \mathbf{u}_j^n]\lambda_{j+1/2}\right\}/2.$$

Then we could perform the conservative difference in flux increment form

$$\mathbf{u}_j^{n+1} = \mathbf{u}_j^n - [\triangle\mathbf{f}_{j+1/2}^{-} + \triangle\mathbf{f}_{j-1/2}^{+}]\frac{\triangle t^{n+1/2}}{\triangle x_j}. \tag{6.2}$$

Note that a very modest amount of characteristic information is used (just to compute $\lambda_{j+1/2}$ for each cell side), no Riemann problems are solved, and no staggered mesh is used. However, the scheme is first order and diffusive (although usually not so diffusive as the Lax–Friedrichs scheme).

6.1.3.2 Godunov's Method with the Harten–Lax–vanLeer (HLL) Solver

In Section 4.13.10 we discussed a technique for approximating the solution of Riemann problems by using a single intermediate state. We presented this approximate Riemann solver in the form

$$\tilde{\mathcal{R}}(\mathbf{u}_L, \mathbf{u}_R, \xi) = \begin{cases} \mathbf{u}_L, & \xi < \underline{\lambda} \\ \mathbf{u}_{LR}, & \underline{\lambda} < \xi < \overline{\lambda} \\ \mathbf{u}_R, & \overline{\lambda} < \xi \end{cases}$$

where

$$\mathbf{u}_{LR} = \frac{\overline{\lambda}\mathbf{u}_R - \underline{\lambda}\mathbf{u}_L}{\overline{\lambda} - \underline{\lambda}} - \frac{\mathbf{f}(\mathbf{u}_R) - \mathbf{f}(\mathbf{u}_L)}{\overline{\lambda} - \underline{\lambda}}.$$

and $\underline{\lambda}$ and $\overline{\lambda}$ are lower and upper bounds on the characteristic speeds in the solution of the Riemann problem involving states $\mathbf{u}_L$ and $\mathbf{u}_R$. We can implement this scheme as follows. For each cell j, compute the flux

$$\mathbf{f}_j^n = \mathbf{f}(\mathbf{u}_j^n),$$

and compute the minimum characteristic speed $\lambda_{1,j}^n$ and the maximum characteristic speed $\lambda_{m,j}$. For each cell side $j + \frac{1}{2}$, compute minimum and maximum average characteristic speeds, $\lambda_{1,j+1/2}^n$ and $\lambda_{m,j+1/2}^n$ (for example, from a Roe matrix, or from a characteristic analysis at $\frac{1}{2}(\mathbf{u}_j^n + \mathbf{u}_{j+1}^n)$). Also compute the lower and upper bounds, $\underline{\lambda}_{j+1/2}^n$ and $\overline{\lambda}_{j+1/2}^n$; for gas dynamics, a good choice is

$$\begin{aligned} \underline{\lambda}_{j+1/2}^n &= \min\{\lambda_{1,j}^n, \lambda_{1,j+1/2}^n\} \\ \overline{\lambda}_{j+1/2}^n &= \max\{\lambda_{m,j+1}^n, \lambda_{m,j+1/2}^n\} \end{aligned}$$

Also at the cell side, compute the solution increment and flux

$$\begin{aligned} \triangle\mathbf{u}_{j+1/2}^n &= \mathbf{u}_{j+1}^n - \mathbf{u}_j^n \\ \mathbf{f}_{j+1/2}^n &= \begin{cases} \mathbf{f}_j^n, & \underline{\lambda}_{j+1/2}^n > 0 \\ \mathbf{f}_{j+1}^n, & \overline{\lambda}_{j+1/2}^n < 0 \\ \left[\mathbf{f}_j^n \overline{\lambda}_{j+1/2}^n - \mathbf{f}_{j+1}^n \underline{\lambda}_{j+1/2}^n + \triangle\mathbf{u}_{j+1/2}^n \underline{\lambda}_{j+1/2}^n \overline{\lambda}_{j+1/2}^n\right] \frac{1}{\overline{\lambda}_{j+1/2}^n - \underline{\lambda}_{j+1/2}^n}, & \text{otherwise.} \end{cases} \end{aligned}$$

Finally, for each cell j perform the conservative difference (6.1).

A Fortran version of the Harten–Lax–vanLeer flux computation for the shallow water equations is available as subroutine `harten_lax_vanleer_sw` in **Program 6.1-87: shallow_water.f**; the corresponding subroutine for gas dynamics is subroutine `harten_lax_vanleer_gd` in **Program 6.1-88: gas_dynamics.f**. The pointer to this function is passed to procedure `runScheme` in **Program 6.1-89: GUIRiemannProblem.C**, and combined with a conservative difference.

Alternatively, we could implement this algorithm in flux increment form. For each cell j, compute the minimum and maximum characteristic speeds $\lambda^n_{1,j}$ and $\lambda^n_{m,j}$, and flux $\mathbf{f}^n_j$ as before. For each cell side $j+\frac{1}{2}$, compute $\underline{\lambda}^n_{j+1/2}$ and $\overline{\lambda}^n_{j+1/2}$ as before. At the cell sides, also compute the increments

$$
\begin{aligned}
\triangle u^n_{j+1/2} &= \mathbf{u}^n_{j+1} - \mathbf{u}^n_j \\
\triangle \mathbf{f}^n_{j+1/2} &= \mathbf{f}^n_{j+1} - \mathbf{f}^n_j . \\
\triangle \mathbf{f}^-_{j+1/2} &= \begin{cases} 0, & \underline{\lambda}^n_{j+1/2} > 0 \\ \triangle \mathbf{f}^n_{j+1/2}, & \overline{\lambda}^n_{j+1/2} < 0 \\ -\left[\triangle \mathbf{f}^n_{j+1/2} - \triangle \mathbf{u}^n_{j+1/2}\overline{\lambda}^n_{j+1/2}\right] \frac{\underline{\lambda}^n_{j+1/2}}{\overline{\lambda}^n_{j+1/2} - \underline{\lambda}^n_{j+1/2}}, & \text{otherwise} \end{cases} \\
\triangle \mathbf{f}^+_{j+1/2} &= \begin{cases} \triangle \mathbf{f}^n_{j+1/2}, & \underline{\lambda}^n_{j+1/2} > 0 \\ 0, & \overline{\lambda}^n_{j+1/2} < 0 \\ \left[\triangle \mathbf{f}^n_{j+1/2} - \triangle \mathbf{u}^n_{j+1/2}\underline{\lambda}^n_{j+1/2}\right] \frac{\overline{\lambda}^n_{j+1/2}}{\underline{\lambda}^n_{j+1/2} - \underline{\lambda}^n_{j+1/2}}, & \text{otherwise.} \end{cases}
\end{aligned}
$$

Finally in each cell j perform the conservative difference (6.2).

6.1.3.3 Godunov's Method with the Harten–Hyman Fix for Roe's Solver

In the special cases where a Roe solver (see Section 4.13.8) is available, such as for shallow water or gas dynamics, we described a modification to prevent entropy violations in transonic rarefactions. The algorithm proceeds as follows. For each cell j, compute the fluxes $\mathbf{f}^n_j = \mathbf{f}(\mathbf{u}^n_j)$ and the characteristic speeds Λ^n_j. For each cell side $j+\frac{1}{2}$, find the Roe matrix $\mathbf{A}^n_{j+1/2}$ so that

$$\mathbf{f}(\mathbf{u}^n_{j+1}) - \mathbf{f}(\mathbf{u}^n_j) = \mathbf{A}^n_{j+1/2}(\mathbf{u}^n_{j+1} - \mathbf{u}^n_j),$$

and find the eigenvectors $\mathbf{X}^n_{j+1/2}$ and eigenvalues $\Lambda^n_{j+1/2}$ so that

$$\mathbf{A}^n_{j+1/2}\mathbf{X}^n_{j+1/2} = \mathbf{X}^n_{j+1/2}\Lambda^n_{j+1/2}.$$

Still at this cell side $j+\frac{1}{2}$, solve

$$\mathbf{X}^n_{j+1/2}\mathbf{y}^n_{j+1/2} = \mathbf{u}^n_{j+1} - \mathbf{u}^n_j$$

for the characteristic expansion coefficients $\mathbf{y}^n_{j+1/2}$. For each cell side $j+1/2$ and each transonic wave family i, meaning that wave i is not linearly degenerate and

$\mathbf{e}_i^\top \Lambda_j^n \mathbf{e}_i < 0 < \mathbf{e}_i^\top \Lambda_{j+1}^n \mathbf{e}_i$, compute the interpolation factor $\beta_{i,j+1/2}^n$ and component of the wave-field decomposition coefficients $\mathbf{e}_i^\top \mathbf{a}_{j+1/2}^n$ by

$$\beta_{i,j+1/2}^n = \max\left\{0, \min\left\{1, \frac{\mathbf{e}_i^\top \Lambda_{j+1}^n \mathbf{e}_i - \mathbf{e}_i^\top \Lambda_{j+1/2}^n \mathbf{e}_i}{\mathbf{e}_i^\top \Lambda_{j+1}^n \mathbf{e}_i - \mathbf{e}_i^\top \Lambda_j^n \mathbf{e}_i}\right\}\right\}$$

$$\mathbf{e}_i^\top \mathbf{a}_{j+1/2}^n = \left\{|\mathbf{e}_i^\top \Lambda_j^n \mathbf{e}_i|\beta_{i,j+1/2}^n + |\mathbf{e}_i^\top \Lambda_{j+1}^n \mathbf{e}_i|(1-\beta_{i,j+1/2}^n)\right\} \mathbf{e}_i^\top \mathbf{y}_{j+1/2}^n.$$

For all other waves, the component of the vector of wave-field decomposition coefficients $\mathbf{a}_{j+1/2}^n$ is given by

$$\mathbf{e}_i^\top \mathbf{a}_{j+1/2}^n = |\mathbf{e}_i^\top \Lambda_{j+1/2}^n \mathbf{e}_i| \mathbf{e}_i^\top \mathbf{y}_{j+1/2}^n.$$

Finally, at the cell side assemble the flux vector

$$\mathbf{f}_{j+1/2}^{n+1/2} = \left[\mathbf{f}_j^n + \mathbf{f}_{j+1}^n - \mathbf{X}_{j+1/2}^n \mathbf{a}_{j+1/2}^n\right] \frac{1}{2}.$$

Next, perform the conservative difference (6.1) in each cell j.

A Fortran version of the Harten–Hyman flux computation for the shallow water equations is available as subroutine `harten_hyman_sw` in **Program 6.1-90: shallow_water.f**; the corresponding subroutine for gas dynamics is subroutine `harten_hyman_gd` in **Program 6.1-91: gas_dynamics.f**. The pointer to this function is passed to procedure `runScheme` in **Program 6.1-92: GUIRiemannProblem.C**, and combined with a conservative difference.

Alternatively, we can implement the Harten–Hyman scheme in terms of flux increments. Let us add the subscript i for the wave-field component. For genuinely nonlinear transonic rarefaction waves, meaning that $\Lambda_{i,j}^n < 0 < \Lambda_{i,j+1}^n$, we have the following prescription for entries of the vector of wave-field decomposition coefficients:

$$\mathbf{a}_{i,j+1/2}^+ = \frac{1}{2}\left[\Lambda_{i,j+1/2}^n - \Lambda_{i,j}^n \beta_{i,j+1/2}^n + \Lambda_{i,j+1}^n(1-\beta_{i,j+1/2}^n)\right] \mathbf{y}_{i,j+1/2}^n \tag{6.3a}$$

$$\mathbf{a}_{i,j+1/2}^- = \frac{1}{2}\left[\Lambda_{i,j+1/2}^n + \Lambda_{i,j}^n \beta_{i,j+1/2}^n - \Lambda_{i,j+1}^n(1-\beta_{i,j+1/2}^n)\right] \mathbf{y}_{i,j+1/2}^n \tag{6.3b}$$

For all other entries, we have

$$\mathbf{a}_{i,j+1/2}^+ = \begin{cases} \Lambda_{i,j+1/2}^n \mathbf{y}_{i,j+1/2}^n, & \Lambda_{i,j+1/2}^n > 0 \\ 0, & \Lambda_{i,j+1/2}^n \le 0 \end{cases} \tag{6.4a}$$

$$\mathbf{a}_{i,j+1/2}^- = \begin{cases} \Lambda_{i,j+1/2}^n \mathbf{y}_{i,j+1/2}^n, & \Lambda_{i,j+1/2}^n < 0 \\ 0, & \Lambda_{i,j+1/2}^n \ge 0 \end{cases} \tag{6.4b}$$

These allow us to compute the positive and negative flux increments

$$\triangle\mathbf{f}_{j+1/2}^- = \mathbf{X}_{j+1/2}^n \mathbf{a}_{j+1/2}^- \quad \text{and} \quad \triangle\mathbf{f}_{j+1/2}^+ = \mathbf{X}_{j-1/2}^n \mathbf{a}_{j+1/2}^+ \tag{6.5}$$

and perform the conservative difference (6.2).

Exercises for 6.1

6.1.1 Describe the use of Linde's approximate Riemann solver (see Section 4.13.11) for nonlinear hyperbolic systems.

6.1.2 The generalization of the Marquina flux described in exercise 3.4.3 to nonlinear hyperbolic systems is described in [45]. The method begins by solving eigenvalue problems at the cell centers, and finding the wave-field decompositions of the conserved variables and the flux:

$$\begin{aligned}\frac{\partial \mathbf{f}}{\partial \mathbf{u}}(\mathbf{u}_j^n)\mathbf{X}_j^n &= \mathbf{X}_j^n \Lambda_j^n \\ \mathbf{X}_j^n \mathbf{y}_j^n &= \mathbf{u}_j^n \\ \mathbf{X}_j^n \mathbf{z}_j^n &= \mathbf{f}(u_j^n).\end{aligned}$$

Then at each cell side $j + \frac{1}{2}$ and for each wave family i we compute the maximum wave speed by

$$\alpha_i = \max\{|\lambda_{i,j}^n|, |\lambda_{i,j+1}^n|\}$$

and the wave-field components of the numerical flux by

$$\mathbf{e}_i \cdot \mathbf{z}_{j+1/2}^+ = \begin{cases} \mathbf{e}_i \cdot \mathbf{z}_j^n, & \lambda_{i,j}^n > 0 \text{ and } \lambda_{i,j+1}^n > 0 \\ 0, & \lambda_{i,j}^n < 0 \text{ and } \lambda_{i,j+1}^n < 0 \\ \frac{1}{2}(\mathbf{e}_i \cdot \mathbf{z}_j^n + \alpha_{i,j+1/2}\mathbf{e}_i \cdot \mathbf{y}_j^n), & \text{otherwise} \end{cases}$$

$$\mathbf{e}_i \cdot \mathbf{z}_{j+1/2}^- = \begin{cases} 0, & \lambda_{i,j}^n > 0 \text{ and } \lambda_{i,j+1}^n > 0 \\ \mathbf{e}_i \cdot \mathbf{z}_{j+1}^n, & \lambda_{i,j}^n < 0 \text{ and } \lambda_{i,j+1}^n < 0 \\ \frac{1}{2}(\mathbf{e}_i \cdot \mathbf{z}_{j+1}^n - \alpha_{i,j+1/2}\mathbf{e}_i \cdot \mathbf{y}_j^n), & \text{otherwise.} \end{cases}$$

The flux at the cell side is then given by

$$\mathbf{f}_{j+1/2}^{n+1/2} = \sum_i [\mathbf{X}_j^n \mathbf{e}_i \mathbf{e}_i \cdot \mathbf{z}_{j+1/2}^+ + \mathbf{X}_{j+1}^n \mathbf{e}_i \mathbf{e}_i \cdot \mathbf{z}_{j+1/2}^-].$$

Show that this flux is consistent. Test this scheme for various problems involving shallow water.

6.2 Second-Order Schemes for Nonlinear Systems

6.2.1 Lax–Wendroff Method

For nonlinear systems, it is useful to view the Lax–Wendroff scheme as a **predictor–corrector scheme**. In this case, the scheme corresponds to using Lax–Friedrichs as a predictor, and the Godunov scheme as a corrector. This gives us the following algorithm. For each cell j compute the flux $\mathbf{f}_j^n = \mathbf{f}(\mathbf{u}_j^n)$. For each cell side $j + 1/2$ compute the conserved quantities

$$\mathbf{u}_{j+1/2}^{n+1/2} = \left\{\mathbf{u}_{j+1}^n \triangle x_j + \mathbf{u}_j^n \triangle x_{j+1} - [\mathbf{f}_{j+1}^n - \mathbf{f}_j^n]\triangle t^{n+1/2}\right\} \frac{1}{\triangle x_j + \triangle x_{j+1}}$$

and the flux $\mathbf{f}^{n+1/2}_{j+1/2} = \mathbf{f}(\mathbf{u}^{n+1/2}_{j+1/2})$. Finally, for each cell j compute

$$\mathbf{u}^{n+1}_j = \mathbf{u}^n_j - \left[\mathbf{f}^{n+1/2}_{j+1/2} - \mathbf{f}^{n+1/2}_{j-1/2}\right]\frac{\triangle t}{\triangle x_j}.$$

The advantage of this scheme is its simplicity; the disadvantage is its lack of monotonicity.

This scheme has been implemented as subroutine `lax_wendroff` in **Program 6.2-93: schemes.f.**

6.2.2 MacCormack's Method

On even numbered steps, for each cell j compute the flux $\mathbf{f}^n_j = \mathbf{f}(\mathbf{u}^n_j)$, then for each cell j compute the provisional conserved quantities

$$\tilde{\mathbf{u}}^{n+1}_j = \mathbf{u}^n_j - \frac{\triangle t}{\triangle x_j}[\mathbf{f}^n_{j+1} - \mathbf{f}^n_j]$$

and the provisional flux $\tilde{\mathbf{f}}^{n+1}_j = \mathbf{f}(\tilde{\mathbf{u}}^{n+1}_j)$, and finally for each cell j compute

$$\tilde{\mathbf{u}}^{n+2}_j = \tilde{\mathbf{u}}^{n+1}_j - \frac{\triangle t}{\triangle x_j}[\tilde{\mathbf{f}}^{n+1}_j - \tilde{\mathbf{f}}^{n+1}_{j-1}]$$

$$\mathbf{u}^{n+1}_j = \frac{1}{2}[\mathbf{u}^n_j + \tilde{\mathbf{u}}^{n+2}_j].$$

On odd numbered steps, the differencing to the left is performed first. Like the Lax-Wendroff scheme, this scheme is simple to program, but numerical oscillations can be destructive. In practice, this method would be combined with a numerical diffusion (see Section 4.13.2).

This scheme has been implemented as subroutine `maccormack` in **Program 6.2-94: schemes.f.**

6.2.3 Higher-Order Lax–Friedrichs Schemes

The second-order version of the Lax–Friedrichs scheme for nonlinear systems of conservation laws is similar to the scheme for scalar laws [121]. The scheme begins by computing slopes $\triangle\mathbf{u}^n_{j+1/2} = \mathbf{u}^n_{j+1} - \mathbf{u}^n_j$ in the flux variables at the cell sides. Next, we compute the component-wise limited slopes in the conserved quantities in each cell j:

$$\mathbf{e}_i^\top \triangle\mathbf{u}^n_j = \text{limiter}(\mathbf{e}_i^\top \triangle\mathbf{u}^n_{j-1/2}, \mathbf{e}_i^\top \triangle\mathbf{u}^n_{j+1/2}).$$

We also compute the matrices of derivatives $\partial\mathbf{f}/\partial\mathbf{w}(\mathbf{w}_j^n)$ and $\partial\mathbf{u}/\partial\mathbf{w}(\mathbf{w}_j^n)$, and use these to compute the state

$$\mathbf{u}_j^{n+1/4} = \mathbf{u}_j^n - \frac{\partial\mathbf{f}}{\partial\mathbf{w}}\left(\frac{\partial\mathbf{u}}{\partial\mathbf{w}}\right)^{-1} \triangle\mathbf{u}_j^n \frac{\triangle t^{n+1/2}}{4\triangle x_j}.$$

Then we compute the flux variables $\mathbf{w}_j^{n+1/4}$ from $\mathbf{u}_j^{n+1/4}$, and the flux $\mathbf{f}_j^{n+1/4} = \mathbf{f}(\mathbf{w}_j^{n+1/4})$. For all cell sides $j+\frac{1}{2}$ we compute the solution at the half-time by

$$\mathbf{u}_{j+1/2}^{n+1/2} = \left\{\left[\mathbf{u}_j^n + \triangle\mathbf{u}_j^n\frac{1}{4}\right]\triangle x_j + \left[\mathbf{u}_{j+1}^n - \triangle\mathbf{u}_{j+1}^n\frac{1}{4}\right]\triangle x_{j+1} - \left[\mathbf{f}_{j+1}^{n+1/4} - \mathbf{f}_j^{n+1/4}\right]\triangle t^{n+1/2}\right\}\frac{1}{\triangle x_j + \triangle x_{j+1}}.$$

The second half-step is similar. We compute the slope $\triangle\mathbf{u}_j^{n+1/2} = \mathbf{u}_{j+1/2}^{n+1/2} - \mathbf{u}_{j-1/2}^{n+1/2}$ in each cell. Then at each cell side we limit the slopes component-wise by

$$\mathbf{e}_i^\top\triangle\mathbf{u}_{j+1/2}^{n+1/2} = \text{limiter}(\mathbf{e}_i^\top\triangle\mathbf{u}_j^{n+1/2}, \mathbf{e}_i^\top\triangle\mathbf{u}_{j+1}^{n+1/2}).$$

We compute the flux variables $\mathbf{w}_{j+1/2}^{n+1/2}$ from $\mathbf{u}_{j+1/2}^{n+1/2}$, use these to compute the matrices of derivatives $\partial\mathbf{f}/\partial\mathbf{w}(\mathbf{w}_{j+1/2}^{n+1/2})$ and $\partial\mathbf{u}/\partial\mathbf{w}(\mathbf{w}_{j+1/2}^{n+1/2})$, and use these to compute the advanced time state

$$\mathbf{u}_{j+1/2}^{n+3/4} = \mathbf{u}_{j+1/2}^{n+1/2} - \frac{\partial\mathbf{f}}{\partial\mathbf{w}}\left(\frac{\partial\mathbf{u}}{\partial\mathbf{w}}\right)^{-1} \triangle\mathbf{u}_{j+1/2}^{n+1/2}\frac{\triangle t^{n+1/2}}{2(\triangle x_j + \triangle x_{j+1})}.$$

Then we compute the flux variables $\mathbf{w}_{j+1/2}^{n+3/4}$ from $\mathbf{u}_{j+1/2}^{n+3/4}$, and the flux $\mathbf{f}_{j+1/2}^{n+3/4} = \mathbf{f}(\mathbf{w}_{j+1/2}^{n+3/4})$. Finally, at each cell center we compute the new solution by

$$\mathbf{u}_j^{n+1} = \frac{1}{2}\left\{\left[\mathbf{u}_{j-1/2}^{n+1/2} + \triangle\mathbf{u}_{j-1/2}^{n+1/2}\frac{1}{4}\right] + \left[\mathbf{u}_{j+1/2}^{n+1/2} - \triangle\mathbf{u}_{j+1/2}^{n+1/2}\frac{1}{4}\right] - \left[\mathbf{f}_{j+1/2}^{n+3/4} - \mathbf{f}_{j-1/2}^{n+3/4}\right]\frac{\triangle t^{n+1/2}}{\triangle x_j}\right\}.$$

This scheme has been implemented as subroutine `nessyahu_tadmor` in **Program 6.2-95: schemes.f.**

The extension by Liu and Tadmor [110] of the Lax–Friedrichs scheme to third-order is a bit more complicated. The piecewise quadratic reconstruction of the solution is determined for each component of the solution vector $\mathbf{u}$ as in Section 5.11, producing vectors of quadratic functions $\overline{\mathbf{q}}_j(x, t^n)$ in each grid cell $(x_{j-1/2}, x_{j+1/2})$.

Conservation requires that

$$\int_{x_j}^{x_{j+1}} \mathbf{u}\left(x, t^n + \frac{\triangle t^{n+1/2}}{2}\right) dx = \int_{x_j}^{x_{j+1/2}} \overline{\mathbf{q}}_j(x, t^n)\, dx + \int_{x_{j+1/2}}^{x_{j+1}} \overline{\mathbf{q}}_{j+1}(x, t^n)\, dx$$
$$- \int_{t^n}^{t^n+\triangle t^{n+1/2}/2} \mathbf{f}(\mathbf{u}(x_{j+1}, t))\, dt$$
$$+ \int_{t^n}^{t^n+\triangle t^{n+1/2}/2} \mathbf{f}(\mathbf{u}(x_j, t))\, dt.$$

The integrals of the reconstructions are computed component-wise as in Equations (5.33). The flux integrals are approximated by Simpson's rule, which requires values of the flux at $\mathbf{u}(x_j, t^n + \tau)$ for $\tau = \triangle t^{n+1/2}/4$ and $\tau = \triangle t^{n+1/2}/2$. These states are approximated by a Taylor expansion of the form

$$\mathbf{u}(x, t + \tau) \approx \mathbf{u} + \frac{\partial \mathbf{u}}{\partial t}\tau + \frac{\partial^2 \mathbf{u}}{\partial t^2}\frac{\tau^2}{2}. \tag{6.6}$$

The principal difficulty in applying the Liu–Tadmor scheme to hyperbolic systems has to do with evaluation of the time derivatives of $\mathbf{u}$ in this Taylor expansion.

Since $\mathbf{f}$ is a function of $\mathbf{u}$, and $\partial\mathbf{u}/\partial t + \partial\mathbf{f}/\partial x = 0$, the chain rule implies that

$$\frac{\partial \mathbf{u}}{\partial t} = -\frac{\partial \mathbf{f}}{\partial x} = -\frac{\partial \mathbf{f}}{\partial \mathbf{u}}\frac{\partial \mathbf{u}}{\partial x}.$$

Similarly, if $\mathbf{u}_i$ is the ith component of $\mathbf{u}$, then

$$\begin{aligned}
\frac{\partial^2 \mathbf{u}_i}{\partial t^2} &= -\frac{\partial}{\partial t}\left[\frac{\partial \mathbf{f}_i}{\partial x}\right] = -\sum_j \frac{\partial}{\partial t}\left[\frac{\partial \mathbf{f}_i}{\partial \mathbf{u}_j}\frac{\partial \mathbf{u}_j}{\partial x}\right] \\
&= -\sum_j\sum_k \frac{\partial^2 \mathbf{f}_i}{\partial \mathbf{u}_j \partial \mathbf{u}_k}\frac{\partial \mathbf{u}_k}{\partial t}\frac{\partial \mathbf{u}_j}{\partial x} - \sum_j \frac{\partial \mathbf{f}_i}{\partial \mathbf{u}_j}\frac{\partial}{\partial x}\left[\frac{\partial \mathbf{u}_j}{\partial t}\right] \\
&= \sum_j\sum_k \frac{\partial^2 \mathbf{f}_i}{\partial \mathbf{u}_j \partial \mathbf{u}_k}\frac{\partial \mathbf{f}_k}{\partial x}\frac{\partial \mathbf{u}_j}{\partial x} + \sum_j \frac{\partial \mathbf{f}_i}{\partial \mathbf{u}_j}\frac{\partial}{\partial x}\left[\frac{\partial \mathbf{f}_j}{\partial x}\right] \\
&= \sum_j\sum_k\sum_\ell \frac{\partial^2 \mathbf{f}_i}{\partial \mathbf{u}_j \partial \mathbf{u}_k}\frac{\partial \mathbf{f}_k}{\partial \mathbf{u}_\ell}\frac{\partial \mathbf{u}_\ell}{\partial x}\frac{\partial \mathbf{u}_j}{\partial x} + \sum_j\sum_k \frac{\partial \mathbf{f}_i}{\partial \mathbf{u}_j}\frac{\partial}{\partial x}\left[\frac{\partial \mathbf{f}_j}{\partial \mathbf{u}_k}\frac{\partial \mathbf{u}_k}{\partial x}\right] \\
&= \sum_j\sum_k\sum_\ell \frac{\partial^2 \mathbf{f}_i}{\partial \mathbf{u}_j \partial \mathbf{u}_k}\frac{\partial \mathbf{f}_k}{\partial \mathbf{u}_\ell}\frac{\partial \mathbf{u}_\ell}{\partial x}\frac{\partial \mathbf{u}_j}{\partial x} + \sum_j\sum_k\sum_\ell \frac{\partial \mathbf{f}_i}{\partial \mathbf{u}_j}\frac{\partial^2 \mathbf{f}_j}{\partial \mathbf{u}_k \partial \mathbf{u}_\ell}\frac{\partial \mathbf{u}_\ell}{\partial x}\frac{\partial \mathbf{u}_k}{\partial x} \\
&\quad + \sum_j\sum_k \frac{\partial \mathbf{f}_i}{\partial \mathbf{u}_j}\frac{\partial \mathbf{f}_j}{\partial \mathbf{u}_k}\frac{\partial^2 \mathbf{u}_k}{\partial x^2}.
\end{aligned}$$

Let us define the matrices

$$\mathbf{F} = \frac{\partial \mathbf{f}}{\partial \mathbf{u}} \quad \text{and} \quad \mathbf{G}_{(k)} = \frac{\partial^2 \mathbf{f}_k}{\partial \mathbf{u} \partial \mathbf{u}}.$$

Note that $\mathbf{G}_{(k)}$ is symmetric for each index k. Then

$$\frac{\partial \mathbf{u}}{\partial t} = -\mathbf{F}\frac{\partial \mathbf{u}}{\partial x} \tag{6.7a}$$

$$\frac{\partial^2 \mathbf{u}}{\partial t^2} = \sum_i \mathbf{e}_i \left(\frac{\partial \mathbf{u}}{\partial x}\right)^\top \mathbf{G}_{(i)} \mathbf{F} \frac{\partial \mathbf{u}}{\partial x} + \sum_j \mathbf{F}\mathbf{e}_j \left(\frac{\partial \mathbf{u}}{\partial x}\right)^\top \mathbf{G}_{(j)} \frac{\partial \mathbf{u}}{\partial x} + \mathbf{F}\mathbf{F}\frac{\partial^2 \mathbf{u}}{\partial x^2}. \tag{6.7b}$$

These expressions can be used to form Taylor expansions for $\mathbf{u}(x, t+\tau)$.

Of course, the conserved quantity vector $\mathbf{u}$ and the flux vector $\mathbf{f}$ are actually functions of the vector of flux variables $\mathbf{w}$. Let us define the matrices

$$\mathcal{A} = \frac{\partial \mathbf{u}}{\partial \mathbf{w}},\ \mathcal{F} = \frac{\partial \mathbf{f}}{\partial \mathbf{w}},\ \mathcal{G}_{(k)} = \frac{\partial \mathbf{f}_k}{\partial \mathbf{w} \partial \mathbf{w}},\ \mathcal{B}_{(k)} = \frac{\partial \mathbf{u}_k}{\partial \mathbf{w} \partial \mathbf{w}}.$$

Then the matrix of flux derivatives with respect to the conserved quantities is evaluated by

$$\mathbf{F} = \frac{\partial \mathbf{f}}{\partial \mathbf{u}} = \frac{\partial \mathbf{f}}{\partial \mathbf{w}} \left(\frac{\partial \mathbf{u}}{\partial \mathbf{w}}\right)^{-1} = \mathcal{F}\mathcal{A}^{-1}. \tag{6.8}$$

Since $\mathbf{I} = \mathcal{A}\mathcal{A}^{-1}$, we have

$$0 = \frac{\partial}{\partial \mathbf{w}_k} \left(\mathcal{A}\mathcal{A}^{-1}\right) = \frac{\partial \mathcal{A}}{\partial \mathbf{w}_k} \mathcal{A}^{-1} + \mathcal{A}\frac{\partial \mathcal{A}^{-1}}{\partial \mathbf{w}_k},$$

so

$$\frac{\partial \mathcal{A}^{-1}}{\partial \mathbf{w}_k} = -\mathcal{A}^{-1}\frac{\mathcal{A}}{\partial \mathbf{w}_k}\mathcal{A}^{-1}.$$

We can compute the components of the matrix $\mathbf{G}_{(k)}$ by

$$\begin{aligned}
\mathbf{e}_i^\top \mathbf{G}_{(k)} \mathbf{e}_j &= \frac{\partial \mathbf{f}_k}{\partial \mathbf{u}_i \partial \mathbf{u}_j} = \frac{\partial}{\partial \mathbf{u}_j}\left(\frac{\partial \mathbf{f}_k}{\partial \mathbf{u}_i}\right) \\
&= \sum_\ell \sum_m \frac{\partial}{\partial \mathbf{w}_\ell}\left[\frac{\partial \mathbf{f}_k}{\partial \mathbf{w}_m}\frac{\partial \mathbf{w}_m}{\partial \mathbf{u}_i}\right]\frac{\partial \mathbf{w}_\ell}{\partial \mathbf{u}_j} \\
&= \sum_\ell \sum_m \frac{\partial^2 \mathbf{f}_k}{\partial \mathbf{w}_m \partial \mathbf{w}_\ell}\frac{\partial \mathbf{w}_m}{\partial \mathbf{u}_i}\frac{\partial \mathbf{w}_\ell}{\partial \mathbf{u}_j} + \sum_\ell \sum_m \frac{\partial \mathbf{f}_k}{\partial \mathbf{w}_m}\mathbf{e}_m^\top \frac{\partial \mathcal{A}^{-1}}{\partial \mathbf{w}_\ell}\mathbf{e}_i \frac{\partial \mathbf{w}_\ell}{\partial \mathbf{u}_j} \\
&= \sum_\ell \sum_m \mathbf{e}_m^\top \mathcal{G}_{(k)} \mathbf{e}_\ell \mathbf{e}_m^\top \mathcal{A}^{-1} \mathbf{e}_i \mathbf{e}_\ell^\top \mathcal{A}^{-1} \mathbf{e}_j \\
&\quad - \sum_\ell \sum_m \sum_p \sum_q \mathbf{e}_k^\top \mathcal{F} \mathbf{e}_m \mathbf{e}_m^\top \mathcal{A}^{-1} \mathbf{e}_p \mathbf{e}_p^\top \frac{\partial \mathcal{A}}{\partial \mathbf{w}_\ell} \mathbf{e}_q \mathbf{e}_q^\top \mathcal{A}^{-1} \mathbf{e}_i \mathbf{e}_\ell^\top \mathcal{A}^{-1} \mathbf{e}_j.
\end{aligned}$$

In other words,

$$\mathbf{G}_{(k)} = \mathcal{A}^{-\top}\mathcal{G}_{(k)}\mathcal{A}^{-1} - \sum_p \mathcal{A}^{-\top}\mathcal{B}_{(p)}\mathcal{A}^{-1}\mathbf{e}_k^\top \mathcal{F}\mathcal{A}^{-1}\mathbf{e}_p. \tag{6.9}$$

We can now evaluate the Taylor series (6.6) by evaluating each of the time derivatives of $\mathbf{u}$ in (6.7) through the use of the formulas for the flux derivatives in Equations (6.8) and (6.9).

This scheme has been implemented as subroutine `liu_tadmor` in **Program 6.2-96: schemes.f.**

6.2.4 TVD Methods

Although the analytical solutions of scalar conservation laws are total variation diminishing, this is not the case for systems. Following LeVeque [97, pp. 341ff], we will demonstrate this fact in the next lemma.

Lemma 6.2.1 *For a constant coefficient hyperbolic system*

$$\frac{\partial \mathbf{u}}{\partial t} + \mathbf{A}\frac{\partial \mathbf{u}}{\partial x} = 0$$

with $\mathbf{A} = \mathbf{X}\Lambda\mathbf{X}^{-1}$, *the total variation of the characteristic expansion coefficients* $\mathbf{y} = \mathbf{X}^{-1}\mathbf{u}$ *has constant total variation, but the total variation of the solution* $\mathbf{u}$ *can grow by a factor of the condition number of* $\mathbf{X}$ *in any given time interval.*

Proof The individual characteristic expansion coefficients satisfy

$$\frac{\partial \mathbf{y}_j}{\partial t} + \lambda_j \frac{\partial \mathbf{y}_j}{\partial x} = 0.$$

We can easily solve these scalar equations to get $\mathbf{y}_j(x,t) = \mathbf{y}_j(x - \lambda_j t, 0)$. Then the total variation in the characteristic expansion coefficients satisfies

$$\begin{aligned}
TV(\mathbf{y}_j(x,t)) &= \sum_j \limsup_{\epsilon\to 0} \frac{1}{\epsilon}\int_{-\infty}^{\infty} \|\mathbf{y}_j(x+\epsilon,t) - \mathbf{y}_j(x,t)\|_1 \, dx \\
&= \limsup_{\epsilon\to 0} \frac{1}{\epsilon}\int_{-\infty}^{\infty} \|\mathbf{y}_j(x-\lambda_j t+\epsilon,0) - \mathbf{y}_j(x-\lambda_j t,0)\|_1 \, dx \\
&= \limsup_{\epsilon\to 0} \frac{1}{\epsilon}\int_{-\infty}^{\infty} \|\mathbf{y}_j(x+\epsilon,0) - \mathbf{y}_j(x,0)\|_1 \, dx \\
&= \sum_j TV(\mathbf{y}_j(x,0)).
\end{aligned}$$

However, the total variation in the solution $\mathbf{u}$ can increase:

$$\begin{aligned}
\sum_i TV(\mathbf{u}_i(x,t)) &= \limsup_{\epsilon\to 0} \frac{1}{\epsilon}\int_{-\infty}^{\infty} \|\mathbf{u}_i(x+\epsilon,t)-\mathbf{u}_i(x,t)\|_1 \, dx \\
&= \limsup_{\epsilon\to 0} \frac{1}{\epsilon}\int_{-\infty}^{\infty} \|\mathbf{X}[\mathbf{y}(x+\epsilon,t)-\mathbf{y}_j(x,t)]\|_1 \, dx \\
&\le \|\mathbf{X}\|_1 \limsup_{\epsilon\to 0} \frac{1}{\epsilon}\int_{-\infty}^{\infty} \|\mathbf{y}(x+\epsilon,t)-\mathbf{y}_j(x,t)\|_1 \, dx \\
&= \|\mathbf{X}\|_1 \sum_j \limsup_{\epsilon\to 0} \frac{1}{\epsilon}\int_{-\infty}^{\infty} |\mathbf{y}_j(x-\lambda_j t+\epsilon,t)-\mathbf{y}_j(x-\lambda_j t,t)| \, dx \\
&= \|\mathbf{X}\|_1 \sum_j \limsup_{\epsilon\to 0} \frac{1}{\epsilon}\int_{-\infty}^{\infty} |\mathbf{y}_j(x+\epsilon,0)-\mathbf{y}_j(x,0)| \, dx \\
&= \|\mathbf{X}\|_1 \limsup_{\epsilon\to 0} \frac{1}{\epsilon}\int_{-\infty}^{\infty} \|\mathbf{y}(x+\epsilon,0)-\mathbf{y}(x,0)\|_1 \, dx \\
&= \|\mathbf{X}\|_1 \limsup_{\epsilon\to 0} \frac{1}{\epsilon}\int_{-\infty}^{\infty} \|\mathbf{X}^{-1}\,[\mathbf{u}(x+\epsilon,0)-\mathbf{u}(x,0)]\,\|_1 \, dx \\
&\le \|\mathbf{X}\|_1 \|\mathbf{X}^{-1}\|_1 \limsup_{\epsilon\to 0} \frac{1}{\epsilon}\int_{-\infty}^{\infty} \|\mathbf{u}(x+\epsilon,0)-\mathbf{u}(x,0)\|_1 \, dx \\
&= \|\mathbf{X}\|_1 \|\mathbf{X}^{-1}\|_1 TV(\mathbf{u}(x,0)).
\end{aligned}$$

The growth in the total variation is bounded above by the condition number of the matrix of eigenvectors of A independent of time, but the total variation can still grow in time. □

In designing numerical schemes, it is common to limit variations in the characteristic expansion coefficients of nonlinear systems of conservation laws. This approach is the basis of the scheme due to Sweby [155]. At each cell side $j+\frac{1}{2}$, we assume that we have a Roe decomposition

$$\mathbf{f}_{j+1}^n - \mathbf{f}_j^n = \mathbf{A}_{j+1/2}^n[\mathbf{u}_{j+1}^n - \mathbf{u}_j^n],$$

where the Roe matrix is diagonalizable:

$$\mathbf{A}_{j+1/2}^n = \mathbf{X}_{j+1/2}^n \Lambda_{j+1/2}^n (\mathbf{X}_{j+1/2}^n)^{-1}.$$

We compute the characteristic expansion coefficients $\mathbf{y}_{j+1/2}^n$ of the jump by solving

$$\mathbf{X}_{j+1/2}^n \mathbf{y}_{j+1/2}^n = \mathbf{u}_{j+1}^n - \mathbf{u}_j^n.$$

Since the Roe flux can produce entropy-violating discontinuities for transonic rarefactions, we will use the Harten-Hyman flux (see Section 4.13.9). We compute

$$\lambda^-_{i,j+1/2} = \begin{cases} \lambda^n_{i,j} \max\left\{0, \min\left\{1, \frac{\lambda^n_{i,j+1}-\lambda^n_{i,j+1/2}}{\lambda^n_{i,j+1}-\lambda^n_{i,j}}\right\}\right\}, & \lambda^n_{i,j} < 0 < \lambda^n_{i,j+1} \\ \lambda^n_{i,j+1/2}, & (\lambda^n_{i,j} \geq 0 \text{ or } \lambda^n_{i,j+1} \leq 0) \text{ and } \lambda^n_{i,j+1/2} < 0 \\ 0, & (\lambda^n_{i,j} \geq 0 \text{ or } \lambda^n_{i,j+1} \leq 0) \text{ and } \lambda^n_{i,j+1/2} \geq 0 \end{cases}$$

$$\lambda^+_{i,j+1/2} = \begin{cases} \lambda^n_{i,j} \max\left\{0, \min\left\{1, \frac{\lambda^n_{i,j+1/2}-\lambda^n_{i,j}}{\lambda^n_{i,j+1}-\lambda^n_{i,j}}\right\}\right\}, & \lambda^n_{i,j} < 0 < \lambda^n_{i,j+1} \\ 0, & (\lambda^n_{i,j} \geq 0 \text{ or } \lambda^n_{i,j+1} \leq 0) \text{ and } \lambda^n_{i,j+1/2} < 0 \\ \lambda^n_{i,j+1/2}, & (\lambda^n_{i,j} \geq 0 \text{ or } \lambda^n_{i,j+1} \leq 0) \text{ and } \lambda^n_{i,j+1/2} \geq 0 \end{cases}$$

Then the Harten–Hyman flux is given by

$$\mathbf{f}^L_{j+1/2} = \mathbf{f}^n_j + \sum_i \mathbf{X}^n_{j+1/2}\mathbf{e}_i\lambda^-_{i,j+1/2}\mathbf{e}_i^\top \mathbf{y}^n_{j+1/2} = \mathbf{f}^n_{j+1} - \sum_i \mathbf{X}^n_{j+1/2}\mathbf{e}_i\lambda^+_{i,j+1/2}\mathbf{e}_i^\top \mathbf{y}^n_{j+1/2}.$$

Sweby prefers to work with the flux differences

$$\begin{aligned} \triangle\mathbf{f}^+_{j+1/2} &= \mathbf{f}^n_{j+1} - \mathbf{f}^L_{j+1/2} = \sum_i \mathbf{X}^n_{j+1/2}\mathbf{e}_i\lambda^+_{i,j+1/2}\mathbf{e}_i^\top \mathbf{y}^n_{j+1/2} \\ \triangle\mathbf{f}^-_{j+1/2} &= \mathbf{f}^L_{j+1/2} - \mathbf{f}^n_j = \sum_i \mathbf{X}^n_{j+1/2}\mathbf{e}_i\lambda^-_{i,j+1/2}\mathbf{e}_i^\top \mathbf{y}^n_{j+1/2}, \end{aligned}$$

and compute the first-order solution with the Harten–Hyman flux by computing the wave-field components by either (6.3) or (6.4), the flux increments by (6.5), and the new low-order solution by

$$\begin{aligned} \mathbf{u}^L_j &= \mathbf{u}^n_j - \left[(\mathbf{f}^L_{j+1/2} - \mathbf{f}^n_j) - (\mathbf{f}^L_{j-1/2} - \mathbf{f}^n_j)\right] \frac{\triangle t^{n+1/2}}{\triangle x_j} \\ &= \mathbf{u}^n_j - \left[\triangle\mathbf{f}^-_{j+1/2} + \triangle\mathbf{f}^+_{j-1/2}\right] \frac{\triangle t^{n+1/2}}{\triangle x_j}. \end{aligned}$$

The Lax–Wendroff process (see Section 3.5) applied to the Harten–Hyman flux would give us

$$\mathbf{f}_{j+1/2} = \left\{\mathbf{f}^n_j \triangle x_{j+1} + \mathbf{f}^n_{j+1}\triangle x_j - \frac{\partial \mathbf{f}^n}{\partial \mathbf{u}}_{j+1/2}\left[\mathbf{f}^n_{j+1} - \mathbf{f}^n_j\right]\triangle t^{n+1/2}\right\} \frac{1}{\triangle x_j + \triangle x_{j+1}}.$$

In order to use the intermediate state of the Harten–Hyman approximate Riemann solver, we will modify the Lax–Wendroff flux slightly:

$$\begin{aligned}
\mathbf{f}^H_{j+1/2} &= \Big\{\mathbf{f}^n_j \triangle x_{j+1} + \mathbf{f}^n_{j+1} \triangle x_j \\
&\quad - \mathbf{X}^n_{j+1/2}\Big[\lambda^+_{j+1/2}(\mathbf{X}^n_{j+1/2})^{-1}\triangle\mathbf{f}^+_{j+1/2} + \lambda^-_{j+1/2}(\mathbf{X}^n_{j+1/2})^{-1}\triangle\mathbf{f}^-_{j+1/2}\Big]\triangle t^{n+1/2}\Big\}\frac{1}{\triangle x_j + \triangle x_{j+1}} \\
&= \Big\{\Big[\mathbf{f}^L_{j+1/2} - \triangle\mathbf{f}^-_{j+1/2}\Big]\triangle x_{j+1} + \Big[\mathbf{f}^L_{j+1/2} + \triangle\mathbf{f}^+_{j+1/2}\Big]\triangle x_j \\
&\quad - \mathbf{X}^n_{j+1/2}\Big[\lambda^+_{j+1/2}(\mathbf{X}^n_{j+1/2})^{-1}\triangle\mathbf{f}^+_{j+1/2} + \lambda^-_{j+1/2}(\mathbf{X}^n_{j+1/2})^{-1}\triangle\mathbf{f}^-_{j+1/2}\Big]\triangle t_{n+1/2}\Big\}\frac{1}{\triangle x_j + \triangle x_{j+1}} \\
&= \mathbf{f}^L_{j+1/2} - \mathbf{X}^n_{j+1/2}\Big[\mathbf{I}\triangle x_{j+1} + \lambda^-_{j+1/2}\triangle t^{n+1/2}\Big](\mathbf{X}^n_{j+1/2})^{-1}\triangle\mathbf{f}^-_{j+1/2}\frac{1}{\triangle x_j + \triangle x_{j+1}} \\
&\quad + \mathbf{X}^n_{j+1/2}\Big[\mathbf{I}\triangle x_j - \lambda^+_{j+1/2}\triangle t^{n+1/2}\Big](\mathbf{X}^n_{j+1/2})^{-1}\triangle\mathbf{f}^+_{j+1/2}\frac{1}{\triangle x_j + \triangle x_{j+1}} \\
&= \mathbf{f}^L_{j+1/2} - \sum_i \mathbf{X}^n_{j+1/2}\mathbf{e}_i\Big[\triangle x_{j+1} + \lambda^-_{i,j+1/2}\triangle t^{n+1/2}\Big]\lambda^-_{i,j+1/2}\mathbf{e}_i^\top\mathbf{y}^n_{j+1/2}\frac{1}{\triangle x_j + \triangle x_{j+1}} \\
&\quad + \sum_i \mathbf{X}^n_{j+1/2}\mathbf{e}_i\Big[\triangle x_j - \lambda^+_{i,j+1/2}\triangle t^{n+1/2}\Big]\lambda^+_{i,j+1/2}\mathbf{e}_i^\top\mathbf{y}^n_{j+1/2}\frac{1}{\triangle x_j + \triangle x_{j+1}}.
\end{aligned}$$

Following the ideas for scalar laws in Section 5.8, we will form

$$\mathbf{g}^+_{j+1/2} = \Big[\mathbf{I}\triangle x_j - \Lambda^+_{j+1/2}\triangle t^{n+1/2}\Big]\Lambda^+_{j+1/2}\mathbf{y}^n_{j+1/2}\frac{1}{\triangle x_j + \triangle x_{j+1}} \text{ and}$$

$$\mathbf{g}^-_{j+1/2} = \Big[\mathbf{I}\triangle x_{j+1} - \Lambda^-_{j+1/2}\triangle t^{n+1/2}\Big]\Lambda^-_{j+1/2}\mathbf{y}^n_{j+1/2}\frac{1}{\triangle x_j + \triangle x_{j+1}}.$$

We will limit $\mathbf{g}^+_{j-1/2}$ and $\mathbf{g}^+_{j+1/2}$ component-wise to get $\mathbf{g}^+_j$; similar limiting will produce $\mathbf{g}^-_j$. Then the flux used by the TVD scheme is

$$\mathbf{f}^{n+1/2}_{j+1/2} = \mathbf{f}^L_{j+1/2} + \mathbf{X}^n_{j+1/2}\Big[\mathbf{g}^+_j - \mathbf{g}^-_{j+1}\Big].$$

Sweby prefers to use this flux in increment form:

$$\mathbf{u}^{n+1}_j = \mathbf{u}^n_j - \Big\{\mathbf{X}^n_{j+1/2}\Big[\mathbf{g}^+_j - \mathbf{g}^-_{j+1}\Big] - \mathbf{X}^n_{j-1/2}\Big[\mathbf{g}^+_{j-1} - \mathbf{g}^-_j\Big]\Big\}\frac{\triangle t^{n+1/2}}{\triangle x_j}.$$

When this scheme was first developed, it represented a major advance over the popular flux-corrected transport (FCT) scheme [22]. Flux-corrected transport used a diffusive step followed by an anti-diffusive step designed to steepen fronts. The TVD scheme performed much better than FCT, causing the latter to fall out of favor. Lately, the MUSCL and wave propagation schemes have been more popular than the TVD schemes.

The TVD scheme has been implemented as subroutine `tvd` in **Program 6.2-97: schemes.f.**

6.2.5 MUSCL

Next, let us generalize the MUSCL scheme to nonlinear hyperbolic systems of conservation laws. We will work with the flux variables $\mathbf{w}$, and compute the characteristic information at each cell center as follows:

$$\left(\frac{\partial \mathbf{f}}{\partial \mathbf{w}}\right)_j^n \mathbf{Y}_j^n = \left(\frac{\partial \mathbf{u}}{\partial \mathbf{w}}\right)_j^n \mathbf{Y}_j^n \Lambda_j^n.$$

Some authors suggest applying the slope limiting to the characteristic expansion coefficients $\mathbf{Y}^{-1}\triangle w$. The justification is that the characteristic expansion approximately decouples the system of conservation laws into separate scalar laws, for which it is known that the analytical solution does not produce new extrema. In this approach, we solve

$$\mathbf{Y}_j^n \mathbf{z}_{j+1/2}^n = \mathbf{w}_{j+1/2}^n \text{ and } \mathbf{Y}_j^n \mathbf{z}_{j-1/2}^n = \mathbf{w}_{j-1/2}^n,$$

and compute the cell-centered average characteristic expansion slopes

$$\tilde{\mathbf{z}}_j^n = \frac{\frac{\triangle x_j + 2\triangle x_{j-1}}{\triangle x_j + \triangle x_{j+1}} \mathbf{z}_{j+1/2}^n + \frac{\triangle x_j + 2\triangle x_{j+1}}{\triangle x_j + \triangle x_{j-1}} \mathbf{z}_{j-1/2}^n}{\triangle x_{j-1} + \triangle x_j + \triangle x_{j+1}} \triangle x_j.$$

Then we compute the limited characteristic expansion slopes component-wise

$$\mathbf{z}_{i,j}^n = \begin{cases} \operatorname{sign}(\tilde{\mathbf{z}}_{i,j}^n) \min\{2|\mathbf{z}_{i,j+1/2}^n|, 2|\mathbf{z}_{i,j-1/2}^n|, |\tilde{\mathbf{z}}_{i,j}^n|\}, & (\mathbf{z}_{i,j+1/2}^n)(\mathbf{z}_{i,j-1/2}^n) \geq 0 \\ 0, & \text{otherwise.} \end{cases}$$

These slopes are used to compute left and right states for Riemann problems:

$$\mathbf{w}_{j+1/2}^{n+1/2,L} = \mathbf{w}_j^n + \sum_i \mathbf{Y}_j^n \mathbf{e}_i \left[1 - \frac{(\lambda_i)_j^n \triangle t^{n+1/2}}{\triangle x_j}\right] \frac{1}{2} \mathbf{e}_i^\top \mathbf{z}_j^n,$$

$$\mathbf{w}_{j+1/2}^{n+1/2,R} = \mathbf{w}_{j+1}^n - \sum_i \mathbf{Y}_{j+1}^n \mathbf{e}_i \left[1 + \frac{(\lambda_i)_{j+1}^n \triangle t^{n+1/2}}{\triangle x_j}\right] \frac{1}{2} \mathbf{e}_i^\top \mathbf{z}_{j+1}^n.$$

This scheme has been implemented as subroutine `musclwave` in **Program 6.2-98: schemes.f.**

Alternatively, we can compute increments in the flux variables at the cell sides:

$$\mathbf{w}_{j+1/2}^n = \mathbf{w}_{j+1}^n - \mathbf{w}_j^n,$$

then we can compute a cell-centered average slope at each cell center

$$\tilde{\mathbf{s}}_j^n \triangle x_j = \left[\mathbf{w}_{j+1/2}^n \frac{\triangle x_j + 2\triangle x_{j-1}}{\triangle x_j + \triangle x_{j+1}} + \mathbf{w}_{j-1/2}^n \frac{\triangle x_j + 2\triangle x_{j+1}}{\triangle x_j + \triangle x_{j-1}}\right] \frac{\triangle x_j}{\triangle x_{j-1} + \triangle x_j + \triangle x_{j+1}}.$$

Slope limiting in the flux variables is performed component-wise. Afterward, we compute the characteristic expansion coefficients $\mathbf{z}_j^n$ of the slopes:

$$\mathbf{Y}_j^n \mathbf{z}_j^n = \mathbf{s}_j^n \triangle x_j .$$

Our next step is to compute flux variables at the cell sides:

$$\mathbf{w}_{j+1/2}^{n+1/2,L} = \mathbf{w}_j^n + \sum_i \mathbf{Y}_j^n \mathbf{e}_i \left[1 - \frac{(\lambda_i)_j^n \triangle t^{n+1/2}}{\triangle x_j} \right] \frac{1}{2} \mathbf{e}_i^\top \mathbf{z}_j^n ,$$

$$\mathbf{w}_{j-1/2}^{n+1/2,R} = \mathbf{w}_j^n - \sum_i \mathbf{Y}_j^n \mathbf{e}_i \left[1 + \frac{(\lambda_i)_j^n \triangle t^{n+1/2}}{\triangle x_j} \right] \frac{1}{2} \mathbf{e}_i^\top \mathbf{z}_j^n .$$

This scheme has been implemented as subroutine `musclvars` in **Program 6.2-99: schemes.f.**

Limiting slopes in characteristic expansion coefficients makes sense for problems with smooth flux functions and distinct characteristic speeds, such as gas dynamics. However, for other problems with discontinuities in the characteristic directions or speeds, or with nearly equal characteristic speeds and nearly singular $\mathbf{Y}$, the use of the characteristic expansion coefficients in the slope limiting can introduce other difficulties. For example, in polymer flooding the characteristic speeds can coalesce with a single characteristic direction, and so the characteristic expansion coefficients can become very large, corresponding to unphysical intermediate states in the Riemann problem. Generally, it is less diffusive but more expensive to limit the characteristic expansion coefficients.

No matter which way the limiting is performed, we use the left and right states to solve a Riemann problem and compute a flux:

$$\mathbf{f}_{j+1/2}^{n+1/2} = \mathbf{f}\left(\mathcal{R}(\mathbf{w}_{j+1/2}^{n+1/2,L}, \mathbf{w}_{j+1/2}^{n+1/2,R} \; ; \; 0) \right).$$

Afterward, we apply a conservative difference scheme:

$$\mathbf{u}_j^{n+1} = \mathbf{u}_j^n - \left\{ \mathbf{f}_{j+1/2}^{n+1/2} - \mathbf{f}_{j-1/2}^{n+1/2} \right\} \frac{\triangle t^{n+1/2}}{\triangle x_j} .$$

6.2.6 Wave Propagation Methods

LeVeque [97] suggests an approach that combines aspects of both slope limiter and TVD schemes. Suppose that at each cell side, we have a decomposition

$$\mathbf{f}(\mathbf{u}_{j+1}^n) - \mathbf{f}(\mathbf{u}_j^n) = \mathbf{X}_{j+1/2}^n \Lambda_{j+1/2}^n \mathbf{a}_{j+1/2}^n .$$

This means that the approximate Riemann solver has effectively provided a flux

$$\begin{aligned}\mathbf{f}^n_{j+1/2} &= \mathbf{f}(\mathbf{u}^n_j) + \mathbf{X}^n_{j+1/2}\left(\Lambda^n_{j+1/2}\right)^- \mathbf{a}^n_{j+1/2}\\ &= \mathbf{f}(\mathbf{u}^n_{j+1}) - \mathbf{X}^n_{j+1/2}\left(\Lambda^n_{j+1/2}\right)^+ \mathbf{a}^n_{j+1/2}.\end{aligned}$$

The conservative difference could be written in the form

$$\begin{aligned}\mathbf{u}^{n+1}_j &= \mathbf{u}^n_j - \left[(\mathbf{f}^n_{j+1/2} - \mathbf{f}(\mathbf{u}^n_j)) - (\mathbf{f}^n_{j-1/2} - \mathbf{f}(\mathbf{u}^n_j))\right]\frac{\triangle t^{n+1/2}}{\triangle x_j}\\ &= \mathbf{u}^n_j - \left[\mathbf{X}^n_{j+1/2}\left(\Lambda^n_{j+1/2}\right)^- \mathbf{a}^n_{j+1/2} + \mathbf{X}^n_{j-1/2}\left(\Lambda^n_{j-1/2}\right)^+ \mathbf{a}^n_{j-1/2}\right]\frac{\triangle t^{n+1/2}}{\triangle x_j}.\end{aligned}$$

In practice, we compute the first-order flux increment

$$\triangle\mathbf{f}^n_j = \mathbf{X}^n_{j-1/2}(\Lambda^n_{j-1/2})^+\mathbf{a}^n_{j-1/2} + \mathbf{X}^n_{j+1/2}(\Lambda^n_{j+1/2})^-\mathbf{a}^n_{j+1/2}$$

and perform the conservative difference as follows:

$$\mathbf{u}^{n+1}_j = \mathbf{u}^n_j - \triangle\mathbf{f}^n_j\frac{\triangle t^{n+1/2}}{\triangle x_j}.$$

The decomposition of the flux differences into waves varies with the method used to approximate the flux in the solution of the Riemann problem. The ideas behind the decomposition were discussed in Section 4.13. For Rusanov's method, we would use Equation (4.90), for Roe's flux we would use (4.13.8), for the Harten–Hyman modification of Roe's flux we would use (6.3) or (6.4) and (6.5), for the Harten–Lax–vanLeer flux we would use (4.97), and for Linde's flux we would use (4.102). These computations depend on the physical model.

Next, let us describe how to form a second-order correction to this algorithm. At each cell side $j+\frac{1}{2}$ and for each wave family i we compute a limited wave-field decomposition coefficient

$$\mathbf{e}_i^\top\tilde{\mathbf{a}}^n_{j+1/2} = \begin{cases}\phi\left(\frac{(\mathbf{X}^n_{j-1/2}\mathbf{e}_i)^\top(\mathbf{X}^n_{j+1/2}\mathbf{e}_i)}{\|\mathbf{X}^n_{j-1/2}\mathbf{e}_i\|^2}\mathbf{e}_i^\top\mathbf{a}^n_{j-1/2},\ \mathbf{e}_i^\top\mathbf{a}^n_{j+1/2}\right), & \mathbf{e}_i^\top\Lambda^n_{j+1/2}\mathbf{e}_i \ge 0\\ \phi\left(\frac{(\mathbf{X}^n_{j+3/2}\mathbf{e}_i)^\top(\mathbf{X}^n_{j+1/2}\mathbf{e}_i)}{\|\mathbf{X}^n_{j+3/2}\mathbf{e}_i\|^2}\mathbf{e}_i^\top\mathbf{a}^n_{j+3/2},\ \mathbf{e}_i^\top\mathbf{a}^n_{j+1/2}\right), & \mathbf{e}_i^\top\Lambda^n_{j+1/2}\mathbf{e}_i < 0\end{cases} \tag{6.10}$$

where $\phi(a, b)$ is some limiter. Then the second-order flux increments are computed at each cell side by

$$\triangle\mathbf{f}^{n+1/2}_{j+1/2} = \mathbf{X}^n_{j+1/2}|\Lambda^n_{j+1/2}|\left(I - |\Lambda^n_{j+1/2}|\frac{2\triangle t^{n+1/2}}{\triangle x_j + \triangle x_{j+1}}\right)\tilde{\mathbf{a}}^n_{j+1/2}.$$

The scheme is completed by performing a conservative difference at cell centers:

$$\mathbf{u}^{n+1}_j = \mathbf{u}^n_j - \frac{\triangle t^{n+1/2}}{\triangle x}\left[\triangle\mathbf{f}^n_j + \left(\triangle\mathbf{f}^{n+1/2}_{j+1/2} - \triangle\mathbf{f}^{n+1/2}_{j-1/2}\right)\frac{1}{2}\right].$$

We have implemented this scheme as subroutine `wave_propagation` in **Program 6.2-100: schemes.f.**

If a Roe matrix is not available, LeVeque suggests that we could use $\mathbf{A}^n_{j+1/2} = \frac{\partial \mathbf{f}}{\partial \mathbf{u}}(\frac{\mathbf{u}^n_j+\mathbf{u}^n_{j+1}}{2})$ together with entropy fixes. This is equivalent to using a form of the weak wave Riemann solver described in Section 4.13.4. Assuming that

$$\mathbf{A}^n_{j+1/2}\mathbf{X}^n_{j+1/2} = \mathbf{X}^n_{j+1/2}\Lambda^n_{j+1/2}.$$

we could solve

$$\mathbf{X}^n_{j+1/2}\mathbf{a}^n_{j+1/2} = \mathbf{u}^n_{j+1} - \mathbf{u}^n_j$$

for $\mathbf{a}^n_{j+1/2}$, and compute the first-order flux increment

$$\triangle \mathbf{f}^n_j = \sum \mathbf{X}^n_{j-1/2}(\Lambda^n_{j-1/2})^+\mathbf{a}^n_{j-1/2} + \sum \mathbf{X}^n_{j+1/2}(\Lambda^n_{j+1/2})^-\mathbf{a}^n_{j+1/2}.$$

At each cell side $j + 1/2$ and for each wave family i we compute a limited wave-field decomposition coefficient

$$\mathbf{e}_i^\top \tilde{\mathbf{a}}^n_{j+1/2} = \begin{cases} \phi\left(\frac{(\mathbf{X}^n_{j-1/2}\mathbf{e}_i)^\top(\mathbf{X}^n_{j+1/2}\mathbf{e}_i)}{\|\mathbf{X}^n_{j-1/2}\mathbf{e}_i\|^2}\mathbf{e}_i^\top\mathbf{a}^n_{j-1/2}, \mathbf{e}_i^\top\mathbf{a}^n_{j+1/2}\right), & \mathbf{e}_i^\top\Lambda^n_{j+1/2}\mathbf{e}_i \geq 0 \\ \phi\left(\frac{(\mathbf{X}^n_{j+3/2}\mathbf{e}_i)^\top(\mathbf{X}^n_{j+1/2}\mathbf{e}_i)}{\|\mathbf{X}^n_{j+3/2}\mathbf{e}_i\|^2}\mathbf{e}_i^\top\mathbf{a}^n_{j+3/2}, \mathbf{e}_i^\top\mathbf{a}^n_{j+1/2}\right), & \mathbf{e}_i^\top\Lambda^n_{j+1/2}\mathbf{e}_i < 0 \end{cases}$$

where $\phi(a, b)$ is some limiter. Then the second-order flux increments are computed in each cell by

$$\triangle \mathbf{f}^{n+1/2}_{j,R} = \mathbf{X}^n_{j+1/2}|\Lambda^n_{j+1/2}|\left(I - |\Lambda^n_{j+1/2}|\frac{\triangle t^{\,n+1/2}}{\triangle x_j}\right)\tilde{\mathbf{a}}^n_{j+1/2}$$

$$\triangle \mathbf{f}^{n+1/2}_{j,L} = \mathbf{X}^n_{j-1/2}|\Lambda^n_{j-1/2}|\left(I - |\Lambda^n_{j-1/2}|\frac{\triangle t^{n+1/2}}{\triangle x_j}\right)\tilde{\mathbf{a}}^n_{j-1/2}.$$

The scheme is completed by performing a conservative difference at cell centers:

$$\mathbf{u}^{n+1}_j = \mathbf{u}^n_j - \frac{\triangle t^{n+1/2}}{\triangle x}\left[\triangle \mathbf{f}^n_j + \left(\triangle \mathbf{f}^{n+1/2}_{j,R} + \triangle \mathbf{f}^{n+1/2}_{j,L}\right)\frac{1}{2}\right].$$

Randy LeVeque has developed a library of routines for solving hyperbolic conservation laws, called CLAWPACK. This library is available online from netlib at **Program 6.2-101: CLAWPACK.**

6.2.7 PPM

Although Colella and Woodward [35] describe the piecewise parabolic method only for scalar laws and gas dynamics, the ideas are easy to generalize to other hyperbolic systems. They compute the divided differences

$$\mathbf{w}^n[x_k, x_{k-1}] = \frac{\mathbf{w}^n_k - \mathbf{w}^n_{k-1}}{\triangle x_k + \triangle x_{k-1}}, \quad k = j, j+1$$

and the slopes

$$\tilde{\mathbf{s}}^n_j = \mathbf{w}^n[x_j, x_{j-1}] \frac{2\triangle x_{j+1} + \triangle x_j}{\triangle x_{j+1} + \triangle x_j + \triangle x_{j-1}} + \mathbf{w}^n[x_{j+1} - x_j] \frac{\triangle x_j + 2\triangle x_{j-1}}{\triangle x_{j+1} + \triangle x_j + \triangle x_{j-1}}$$

in the flux variables, then apply the MUSCL limiter component-wise as in Equation (5.32) to get the limited slopes $\mathbf{s}^n_j$. Initial values for the flux variables at the cell sides are given by

$$\begin{aligned}\mathbf{w}^n_{j+1/2} =& \mathbf{w}^n_j + \mathbf{w}^n[x_{j+1}, x_j]\triangle x_j \\ & - \mathbf{w}^n[x_{j+1}, x_j] \frac{2\triangle x_{j+1}\triangle x_j}{\triangle x_{j+2} + \triangle x_{j+1} + \triangle x_j + \triangle x_{j-1}} \left\{ \frac{\triangle x_{j+2} + \triangle x_{j+1}}{2\triangle x_{j+1} + \triangle x_j} - \frac{\triangle x_j + \triangle x_{j-1}}{\triangle x_{j+1} + 2\triangle x_j} \right\} \\ & + \left\{ \tilde{\mathbf{s}}^n_j \frac{\triangle x_{j+2} + \triangle x_{j+1}}{2\triangle x_{j+1} + \triangle x_j} - \tilde{\mathbf{s}}^n_{j+1} \frac{\triangle x_j + \triangle x_{j-1}}{\triangle x_{j+1} + 2\triangle x_j} \right\} \frac{\triangle x_{j+1}\triangle x_j}{\triangle x_{j+2} + \triangle x_{j+1} + \triangle x_j + \triangle x_{j-1}}.\end{aligned}$$

The values $\mathbf{w}^n_{j+1/2}$ of the flux variables are then limited component-wise as in the scalar case to produce values $\mathbf{w}^R_{j-1/2}$ and $\mathbf{w}^L_{j+1/2}$; these give the quadratic reconstruction

$$\mathbf{w}^n(x) = \mathbf{a}_j + \{\mathbf{b}_j + \mathbf{c}_j(1 - \xi(x))\}\xi(x) \quad \text{where} \quad \xi(x) = \frac{x - x_{j-1/2}}{\triangle x_j},$$

in which $\mathbf{a}_j = \mathbf{w}^R_{j-1/2}$, $\mathbf{b}_j = \mathbf{w}^L_{h+1/2} - \mathbf{w}^R_{j-1/2}$ and $\mathbf{c}_j = 6\mathbf{w}^n_j - 3(\mathbf{w}^L_{j+1/2} + \mathbf{w}^R_{j-1/2})$.

Next, we require values for the temporal averages of the flux variables at the cell sides for states in Riemann problems. Here the computations proceed as for a hyperbolic system with constant coefficients $\mathbf{A} = \mathbf{X}\Lambda\mathbf{X}^{-1}$. Lemma 4.1.7 showed that the analytical solution of $\partial\mathbf{u}/\partial t + \mathbf{X}\Lambda\mathbf{X}^{-1}\partial\mathbf{u}/\partial x = 0$ is

$$\mathbf{u}(x, t) = \sum_i \mathbf{X}\mathbf{e}_i\mathbf{e}_i^\top\mathbf{X}^{-1}\mathbf{u}(x - \lambda_i t, 0).$$

Our flux variables satisfy

$$\frac{\partial\mathbf{w}}{\partial t} + \mathbf{Y}\Lambda\mathbf{Y}^{-1}\frac{\partial\mathbf{w}}{\partial x} = 0$$

where

$$\frac{\partial \mathbf{f}}{\partial \mathbf{w}}\mathbf{Y} = \frac{\partial \mathbf{u}}{\partial \mathbf{w}}\mathbf{Y}\Lambda .$$

Colella and Woodward would take

$$\begin{aligned}
&\frac{1}{\triangle t^{n+1/2}}\int_0^{\triangle t^{n+1/2}} \mathbf{w}^n(x_{j+1/2}-0, t^n+\tau)\,d\tau \\
&\approx \overline{\mathbf{w}}^L_{j+1/2} \\
&= \mathbf{w}^L_{j+1/2} + \sum_{\mathbf{e}_i^\top \Lambda_j^n \mathbf{e}_i \ge 0} \mathbf{Y}_j^n \mathbf{e}_i \mathbf{e}_i^\top (\mathbf{Y}_j^n)^{-1}\left[\frac{1}{\triangle t^{n+1/2}}\int_0^{\triangle t^{n+1/2}} \mathbf{w}_j^n(x_{j+1/2} - \mathbf{e}_i^\top \Lambda_j^n \mathbf{e}_i \tau, 0)\,d\tau - \mathbf{w}^L_{j+1/2}\right] \\
&= \mathbf{w}^L_{j+1/2} + \sum_{\mathbf{e}_i^\top \Lambda_j^n \mathbf{e}_i \ge 0} \mathbf{Y}_j^n \mathbf{e}_i \mathbf{e}_i^\top (\mathbf{Y}_j^n)^{-1}\left\{-\frac{\mathbf{e}_i^\top \Lambda_j^n \mathbf{e}_i \triangle t^{n+1/2}}{2\triangle x_j}\left[\mathbf{b}_j - \mathbf{c}_j\left(1 - \frac{2\mathbf{e}_i^\top \Lambda_j^n \mathbf{e}_i \triangle t^{n+1/2}}{3\triangle x_j}\right)\right]\right\}
\end{aligned}$$

and

$$\begin{aligned}
&\frac{1}{\triangle t^{n+1/2}}\int_0^{\triangle t^{n+1/2}} \mathbf{w}^n(x_{j-1/2}+0, t^n+\tau)\,d\tau \\
&\approx \overline{\mathbf{w}}^R_{j-1/2} \\
&= \mathbf{w}^R_{j-1/2} + \sum_{\mathbf{e}_i^\top \Lambda_j^n \mathbf{e}_i \le 0} \mathbf{Y}_j^n \mathbf{e}_i \mathbf{e}_i^\top (\mathbf{Y}_j^n)^{-1}\left[\frac{1}{\triangle t^{n+1/2}}\int_0^{\triangle t^{n+1/2}} \mathbf{w}_j^n(x_{j-1/2} - \mathbf{e}_i^\top \Lambda_j^n \mathbf{e}_i \tau, 0)\,d\tau - \mathbf{w}^R_{j-1/2}\right] \\
&= \mathbf{w}^R_{j-1/2} + \sum_{\mathbf{e}_i^\top \Lambda_j^n \mathbf{e}_i \le 0} \mathbf{Y}_j^n \mathbf{e}_i \mathbf{e}_i^\top (\mathbf{Y}_j^n)^{-1}\left\{-\frac{\mathbf{e}_i^\top \Lambda_j^n \mathbf{e}_i \triangle t^{n+1/2}}{2\triangle x_j}\left[\mathbf{b}_j - \mathbf{c}_j\left(1 - \frac{2\mathbf{e}_i^\top \Lambda_j^n \mathbf{e}_i \triangle t^{n+1/2}}{3\triangle x_j}\right)\right]\right\}.
\end{aligned}$$

The flux at the solution to the Riemann problem with these two states is used in a conservative difference to complete the scheme. We have implemented this scheme as subroutine ppm in **Program 6.2-102: schemes.f.**

6.2.8 ENO

Throughout the **ENO scheme**, Osher and Shu compute eigenvectors and eigenvalues of $\mathbf{A}^n_{j+1/2}$, which is an average value for $\partial\mathbf{f}/\partial\mathbf{u}$. This could be done by using Roe's approximate Riemann solver (see Section 4.13.8), or by computing the flux derivatives at the average of the the states in cell j and $j+1$. They use $\mathbf{A}^n_{j+1/2}$ to find the characteristic directions and speeds so that

$$\mathbf{A}^n_{j+1/2}\mathbf{X}^n_{j+1/2} = \mathbf{X}^n_{j+1/2}\Lambda^n_{j+1/2}.$$

Throughout the algorithm, they compute the divided difference table for each component of $(\mathbf{X}^n_{j+1/2})^{-1}\mathbf{f}^n_{j+k}$, with the divided difference table being appropriately upwinded based on the sign of the corresponding eigenvalue in $\Lambda^n_{j+1/2}$. The processing of the divided difference table for each wave-field component is the same as in the scalar algorithm, discussed in Section 5.13 above. This divided difference table produces values for $(\mathbf{X}^n_{j+1/2})^{-1}\mathbf{f}^n_{j+1/2}$, and the flux is computed by

$$\mathbf{f}^n_{j+1/2} = \mathbf{X}^n_j \left\{(\mathbf{X}^n_{j+1/2})^{-1}\mathbf{f}^n_{j+1/2}\right\}.$$

The Runge–Kutta steps for time integration are the same as in the scalar algorithm. We have implemented this scheme inside procedure `runScheme` in **Program 6.2-103: GUIRiemannProblem.C.** The ENO divided difference table is computed in subroutine `eno_rf` in **Program 6.2-104: schemes.f.**

6.2.9 Discontinuous Galerkin Method

Application of the discontinuous Galerkin method to a system of hyperbolic conservation laws is similar to the scalar case described in Section 5.14. For all $b(x) \in C^\infty(x_L, x_R)$, the weak formulation of the conservation law is

$$\begin{aligned} 0 &= \int_{x_L}^{x_R} \left[\frac{\partial \mathbf{u}}{\partial t} + \frac{\partial \mathbf{f}(\mathbf{u})}{\partial x}\right] b\, dx = \sum_k \int_{x_{k-1/2}}^{x_{k+1/2}} \frac{\partial \mathbf{u}}{\partial t} b + \frac{\partial \mathbf{f}(\mathbf{u}) b}{\partial x} - \mathbf{f}(\mathbf{u}) \frac{db}{dx}\, dx \\ &= \sum_k \left[\frac{d}{dt}\int_{x_{k-1/2}}^{x_{k+1/2}} \mathbf{u}\, b\, dx + \mathbf{f}(\mathbf{u}) b|_{x_{k-1/2}}^{x_{k+1/2}} - \int_{x_{k-1/2}}^{x_{k+1/2}} \mathbf{f}(\mathbf{u}) \frac{db}{dx}\, dx\right]. \end{aligned}$$

Let $\mathbf{b}(\xi)$ be the vector of orthonormal Legendre polynomials on $\xi \in (-1, 1)$, defined in equation (5.37). The Galerkin approximation will approximate the solution in the form

$$\text{for all } x \in (x_{k-1/2}, x_{k+1/2}),\ \mathbf{u}(x,t) = \mathbf{U}_k(t)\mathbf{b}(\xi_k(x)) \quad \text{where} \quad \xi_k(x) = 2\frac{x - x_k}{\triangle x_k}.$$

Here $\mathbf{U}_k(t)$ is an array of coefficients for each conserved quantity with respect to each basis function. The weak form of the conservation law is replaced by a weak form with $b(x)$ replaced by an arbitrary polynomial of degree at most k. If we let the $\mathbf{b}(\xi)$ be the vector of orthonormal Legrendre polynomials, then we obtain the Galerkin equations

$$\begin{aligned} 0 &= \frac{d}{dt}\int_{x_{k-1/2}}^{x_{k+1/2}} \mathbf{U}_k(t)\mathbf{b}\left(\xi_k(x)\right)\mathbf{b}\left(\xi_k(x)\right)^\top dx \\ &+ \mathbf{f}\left(\mathcal{R}\left(\mathbf{U}_k(t)\mathbf{b}(1), \mathbf{U}_{k+1}(t)\mathbf{b}(-1); 0\right)\right)\mathbf{b}(1)^\top - \mathbf{f}\left(\mathcal{R}\left(\mathbf{U}_{k-1}(t)\mathbf{b}(1), \mathbf{U}_k(t)\mathbf{b}(-1); 0\right)\right)\mathbf{b}(-1)^\top \\ &- \int_{x_{k-1/2}}^{x_{k+1/2}} \mathbf{f}\left(\mathbf{U}_k(t)\mathbf{b}(\xi_k(x))\right) \frac{d\mathbf{b}\left(\xi_k(x)\right)^\top}{dx}\, dx. \end{aligned}$$

The orthonormality of the entries of $\mathbf{b}(\xi)$ then gives us the system of ordinary differential equations

$$\frac{d\mathbf{U}_k}{dt} = \frac{2}{\triangle x_k}\left\{-\mathbf{f}(\mathcal{R}(\mathbf{U}_k(t)\mathbf{b}(1), \mathbf{U}_{k+1}(t)\mathbf{b}(-1);0)) + \mathbf{f}(\mathcal{R}(\mathbf{U}_{k-1}(t)\mathbf{b}(1), \mathbf{U}_k(t)\mathbf{b}(-1);0)) + \int_{-1}^{1} \mathbf{f}(\mathbf{U}_k(t)\mathbf{b}(\xi))\mathbf{b}'(\xi)^\top \, d\xi\right\}.$$

The initial value for $\mathbf{U}_k$ is determined by the orthogonality condition

$$0 = \int_{x_{k-1/2}}^{x_{k+1/2}} [\mathbf{u}(x,0) - \mathbf{U}_k(0)\mathbf{b}(\xi_k(x))]\,\mathbf{b}(\xi_k(x))^\top \, dx$$
$$= \int_{-1}^{1} \mathbf{u}\left(x_k + \frac{\xi \triangle x_k}{2}, 0\right) \mathbf{b}(\xi)^\top \, d\xi \, \frac{\triangle x_k}{2} - \mathbf{U}_k(0)\int_{-1}^{1} \mathbf{b}(\xi)\mathbf{b}(\xi)^\top \, d\xi \, \frac{\triangle x_k}{2},$$

which implies that

$$\mathbf{U}_k(0) = \int_{-1}^{1} \mathbf{u}\left(x_k + \frac{\xi \triangle x_k}{2}, 0\right)\mathbf{b}(\xi)^\top \, d\xi.$$

As in the scalar case, the integrals are replaced by Lobatto quadrature rules:

$$\int_{-1}^{1} \phi(\xi)\, d\xi \approx \sum_{q=0}^{Q} \phi(\xi_q)\alpha_q, \quad -1 = \xi_0 < \xi_1 < \ldots < \xi_Q = 1.$$

Thus our quadrature rule approximation gives us initial values

$$\mathbf{U}_k(0) = \sum_{q=0}^{Q} \mathbf{u}\left(x_k + \frac{\xi_q \triangle x_k}{2}, 0\right)\alpha_q \mathbf{b}(\xi_q)^\top,$$

and ordinary differential equations

$$\frac{d\mathbf{U}_k}{dt} = \frac{2}{\triangle x_k}\left\{ -\mathbf{f}(\mathcal{R}(\mathbf{U}_k(t)\mathbf{b}(1), \mathbf{U}_{k+1}(t)\mathbf{b}(-1);0)) + \mathbf{f}(\mathcal{R}(\mathbf{U}_{k-1}(t)\mathbf{b}(1), \mathbf{U}_k(t)\mathbf{b}(-1);0)) + \sum_{q=0}^{Q} \mathbf{f}\left(\mathbf{U}_k(t)\mathbf{b}(\xi_q)\right)\alpha_q \mathbf{b}'(\xi_q)^\top \right\}.$$

Of course, most nonlinear hyperbolic systems are formulated in terms of flux variables $\mathbf{w}(x,t)$, with the conserved quantities given by $\mathbf{u}(\mathbf{w}(x,t))$ and the fluxes given by $\mathbf{f}(\mathbf{w}(x,t))$. This requires several modifications of the discontinuous Galerkin formulation. Initial conditions commonly provide values for the flux variables. Let

$$\tilde{\mathbf{W}}_k(t) = \left[\mathbf{w}\left(x_k + \frac{\xi_0 \triangle x_k}{2}, t\right), \ldots, \mathbf{w}\left(x_k + \frac{\xi_Q \triangle x_k}{2}, t\right)\right]$$

be the array of flux variables at the Lobatto quadrature points for grid cell k. Then the initial values for the discontinuous Galerkin method are

$$\mathbf{U}_k(0) = \sum_{q=0}^{Q} \mathbf{u}(\tilde{\mathbf{W}}_k(0)\mathbf{e}_q)\alpha_q \mathbf{b}(\xi_q)^\top .$$

The values of the initial solution at the Lobatto quadrature points are $\mathbf{U}_k(0)\mathbf{b}(\xi_q)$; these are possibly different from the initial values of the conserved quantities evaluated as functions of the flux variables at the quadrature points. Thus we must solve

$$\mathbf{u}(\mathbf{W}_k(0)\mathbf{e}_q) = \mathbf{U}_k(0)\mathbf{b}(\xi_q)$$

to get values of the flux variables that are consistent with the point values of the initial solution. Also, the ordinary differential equations must be modified to provide flux variables for arguments to the flux:

$$\begin{aligned}\frac{d\mathbf{U}_k}{dt} = \frac{2}{\triangle x_k}\Bigg\{ & -\mathbf{f}(\mathcal{R}(\mathbf{W}_k(t)\mathbf{e}_Q, \mathbf{W}_{k+1}(t)\mathbf{e}_0; 0)) \\ & + \mathbf{f}(\mathcal{R}(\mathbf{W}_{k-1}(t)\mathbf{e}_Q, \mathbf{W}_k(t)\mathbf{e}_0; 0)) \\ & + \sum_{q=0}^{Q} \mathbf{f}(\mathbf{W}_k(t)\mathbf{e}_q)\alpha_q \mathbf{b}'(\xi_q)^\top \Bigg\} .\end{aligned}$$

After advancing the discontinuous Galerkin method in time, the flux variables can be determined by solving

$$\mathbf{u}(\mathbf{W}_k(t)\mathbf{e}_q) = \mathbf{U}_k(t)\mathbf{b}(\xi_q)$$

for the vector of flux variables at each quadrature point.

Limiting is performed on the conserved quantities component-wise, as in the scalar case. These produce vectors $\mathbf{u}_{k+1/2,L}(t)$ and $\mathbf{u}_{k-1/2,R}(t)$ in each cell. From these limited conserved quantities we can determine vectors of flux variables $\mathbf{w}_{k+1/2,L}(t)$ and $\mathbf{w}_{k-1/2,R}(t)$ for use in evaluating the fluxes.

We have implemented the discontinuous Galerkin scheme for nonlinear hyperbolic systems inside `discontinuous_galerkin` in **Program 6.2-105: dgm.f.**

Exercise for 6.2

6.2.1 Another version of the weak wave solver involves decomposing the flux difference: $\mathbf{X}^n_{j+1/2}\mathbf{z}^n_{j+1/2} = \mathbf{f}(\mathbf{u}^n_{j+1}) - \mathbf{f}(\mathbf{u}^n_j)$. Describe how to modify the weak wave form of wave propagation for this wave-field decomposition. Compare its performance to the weak wave solver described above.

6.3 Case Studies

6.3.1 Wave Equation

The wave equation

$$\frac{\partial^2 \mathbf{u}}{\partial t^2} - c^2 \frac{\partial^2 \mathbf{u}}{\partial x^2} = 0 \quad \text{for all } x \in \mathbf{R} \quad \text{for all } t > 0$$
$$\mathbf{u}(x,0) = \mathbf{u}_0(x), \frac{\partial \mathbf{u}}{\partial t}(x,0) = v_0(x) \quad \text{for all } x \in \mathbf{R}$$

is a simple system of linear hyperbolic conservation laws, for which the analytical solution is well known (see exercise 4.1.1). The numerical methods in this chapter will perform well for this problem, but will not be especially efficient. Because this problem has no shocks, it is advantageous to use high-order numerical methods for its solution. Some useful numerical methods include spectral methods [61] and multi-pole expansions [62]. Similar comments apply to linear elasticity (discussed in section 4.7) and Maxwell's equations (discussed in section 4.3).

6.3.2 Shallow Water

The shallow water equations were presented in Example 4.1.1 and analyzed in Section 4.1. These equations are a very simple nonlinear system with practical implications. For example, the dam break problem is a Riemann problem with initial conditions

$$h(x,0) = \begin{cases} h_L, & x < 0 \\ h_R, & x < 0 \end{cases} \tag{6.11a}$$

$$v(x,0) = 0. \tag{6.11b}$$

If gravity has the value $g = 1$, then the "parting of the sea" problem has initial data

$$v(x,0) = \begin{cases} -2, & x < 0 \\ 2, & x < 0 \end{cases} \tag{6.12a}$$

$$h(x,0) = 1. \tag{6.12b}$$

This Riemann problem leads to two rarefactions with an intermediate state with zero height. This problem requires some care in programming the numerical methods, because the characteristic speed computations involve taking the square root of h; oscillations in the numerical method produce unphysical values of h. Reducing the initial velocities to $\pm\frac{1}{2}$ produces two rarefactions that are much easier for numerical methods to resolve.

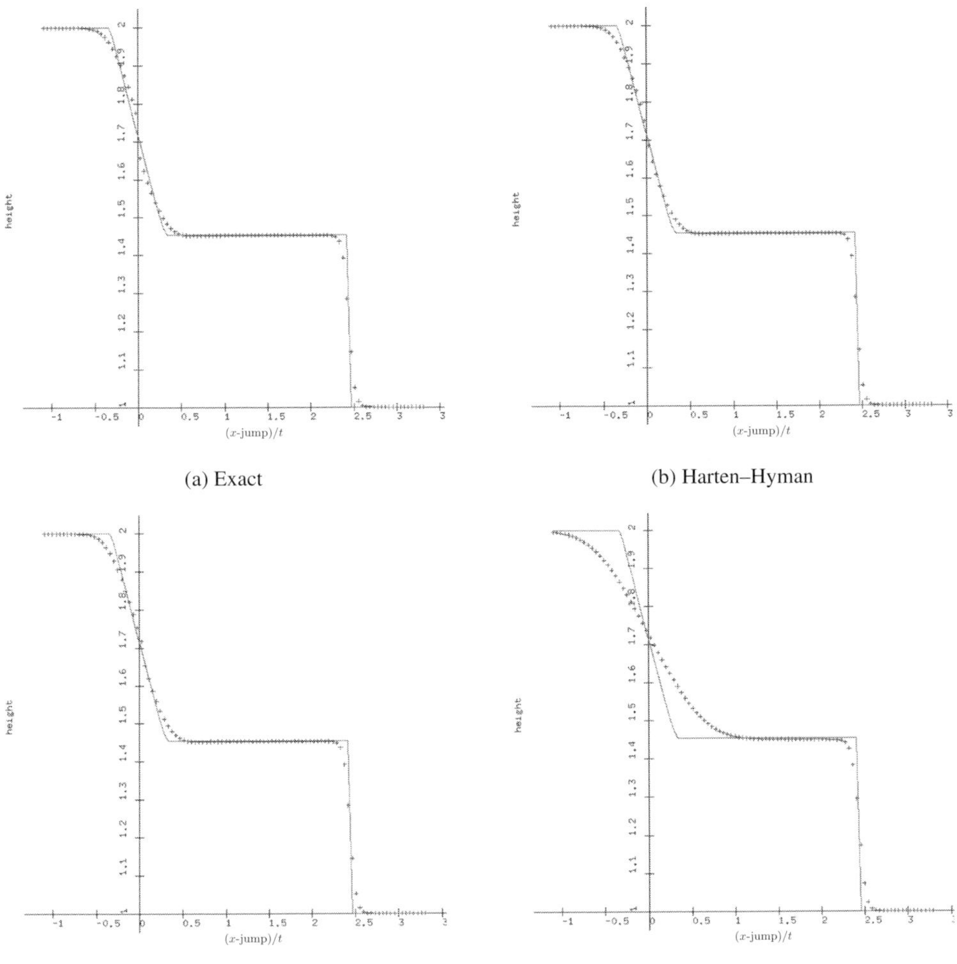

(a) Exact

(b) Harten–Hyman

(c) Harten–Lax–van Leer

(d) Rusanov

Fig. 6.1 Various Riemann Solvers for Transonic Dam Break: Height vs. x/t

For a problem with two shocks, we can solve the Riemann problem

$$v(x,0) = \begin{cases} 1, & x < 0 \\ -1, & x < 0 \end{cases} \tag{6.13a}$$

$$h(x,0) = 1. \tag{6.13b}$$

Finally, to see a shock moving left and a rarefaction moving right, solve

$$h(x,0) = \begin{cases} 1, & x < 0 \\ 2, & x < 0 \end{cases} \tag{6.14a}$$

$$v(x,0) = 1. \tag{6.14b}$$

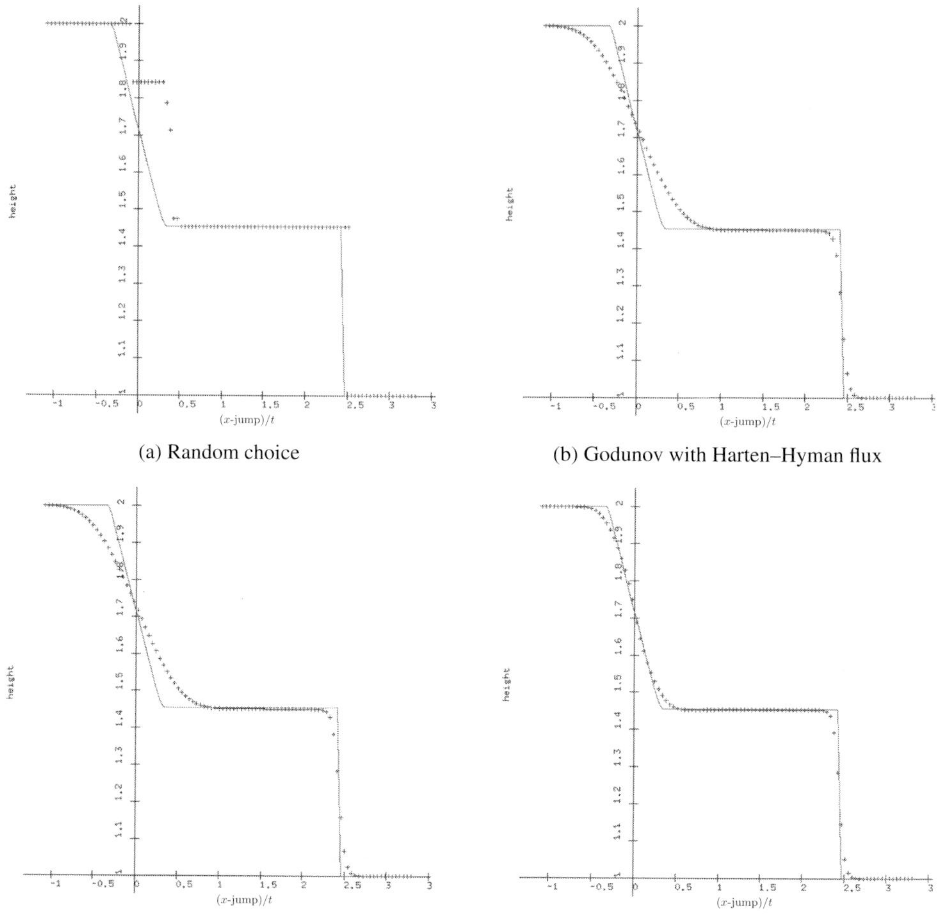

(a) Random choice (b) Godunov with Harten–Hyman flux

(c) Lax–Friedrichs (d) ENO 1

Fig. 6.2 Various first-order schemes for transonic dam break: Height vs. x/t

Students can create their own interesting Riemann problems by clicking on **Executable 6.3-43: guiShallowWater.** Afterward, students can experiment with a variety of numerical methods by clicking on **Executable 6.3-44: guiRiemannProblem.**

All of the approximate Riemann problem solvers (except the unmodified Roe solver) performed well with Godunov's method. Figure 6.1 shows the water height for the dam break problem in a moving frame of reference so that the rarefaction is transonic. The exact solution of the Riemann problem produces a small jump at the sonic point, and the Rusanov solver smears the waves quite a bit more than the other Riemann solvers. The Linde solver and the Harten-Lax-van Leer solver produce essentially the same results. The first-order schemes performed reasonably well for this problem, as shown in Figure 6.2. Various second-order results are shown in Figure 6.3.

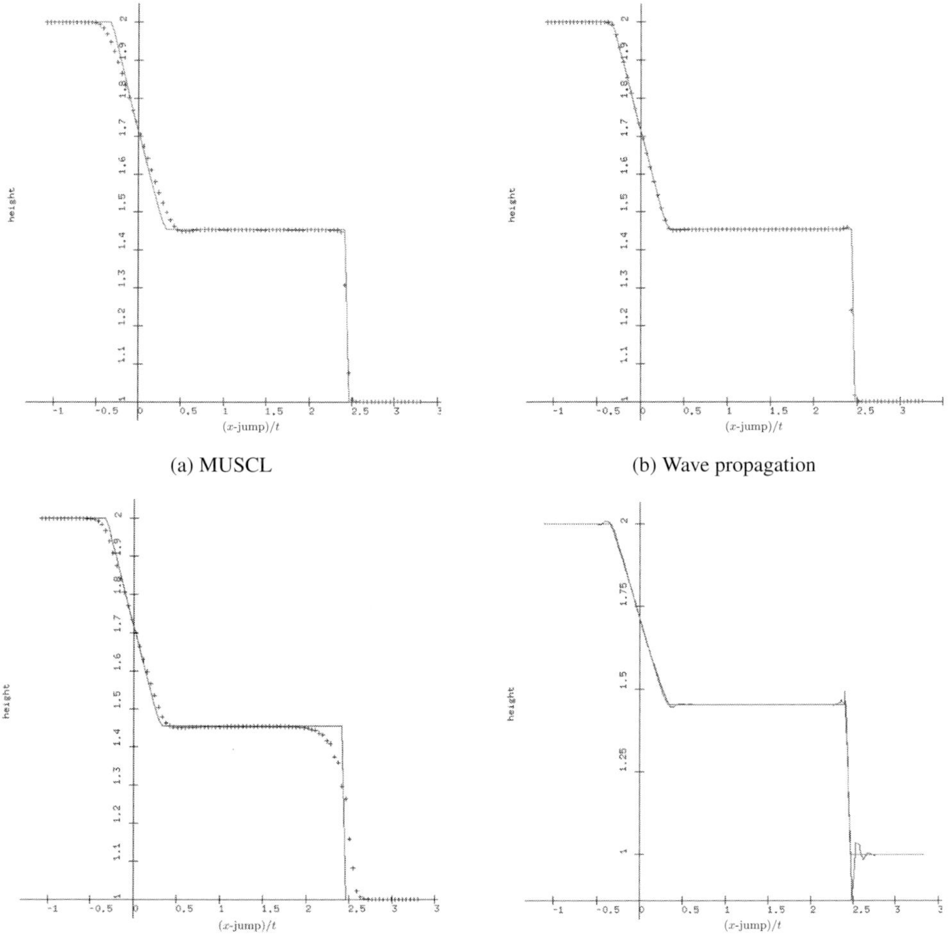

(a) MUSCL

(b) Wave propagation

(c) ENO 2

(d) DGM 2

Fig. 6.3 Various second-order schemes for transonic dam break: height vs. x/t

6.3.3 Gas Dynamics

The gas dynamics equations were studied in Section 4.4. Some interesting gas dynamics test problems be found in [106], as well as comparisons between many of the schemes described in this chapter, and other interesting schemes. This paper also describes how the test codes were obtained from available sources.

Almost all of the gas dynamics test problems are for a polytropic gas with $\gamma = 1.4$, representing air. One of the famous Riemann problems for this system is the Sod shock tube, which involves a rarefaction and a shock:

$$\rho(x,0) = \begin{cases} 1, & x < 0 \\ 1/8, & x > 0, \end{cases} \quad v(x,0) = 0, \quad p(x,0) = \begin{cases} 1, & x < 0 \\ 1/10, & x > 0. \end{cases} \tag{6.15}$$

Here are some other test problems to consider when debugging code. For two rarefactions, try the Riemann problem

$$\rho(x,0) = 1, \quad v(x,0) = \begin{cases} -1, & x < 0 \\ 1, & x > 0, \end{cases} \quad p(x,0) = 1; \tag{6.16}$$

for two shocks, try

$$\rho(x,0) = 1, \quad v(x,0) = \begin{cases} 1, & x < 0 \\ -1, & x > 0, \end{cases} \quad p(x,0) = 1; \tag{6.17}$$

and for a shock and a rarefaction, try

$$\rho(x,0) = 1, \; v(x,0) = 1, \; p(x,0) = \begin{cases} 1, & x < 0 \\ 2, & x > 0. \end{cases} \tag{6.18}$$

Students can shift the velocities in some of these problems in order to produce transonic rarefactions that cause trouble for techniques such as the Roe approximate Riemann solver. Finally, we mention the Colella-Woodward interacting blast wave problem [177]

$$\rho(x,0) = 1, \quad v(x,0) = 0, \quad p(x,0) = \begin{cases} 1000, & 0 < x < 0.1 \\ 0.01, & 0.1 < x < 0.9 \\ 100, & 0.9 < x < 1.0, \end{cases} \tag{6.19}$$

in which the gas is confined between two reflecting walls at $x = 0$ and $x = 1$. (As we discussed in section 4.4.8, at a reflecting wall the normal component of velocity is an odd function of position, while density, pressure and tangential components of velocity are even functions.) This problem is particularly difficult to solve accurately with first-order (and even second-order) methods.

In numerical experiments with the Sod shock tube problem, we found that Godunov's method combined with the exact Riemann solver, Roe, Harten–Hyman, Harten–Lax van Leer and Linde solvers all produced acceptable results for timesteps around 0.9 times the CFL timestep. Among first-order schemes using the Harten-Hyman modification of the Roe approximate Riemann solver, Godunov was most efficient, followed by first-order ENO, Lax–Friedrichs and discontinuous Galerkin. Among second-order schemes, the most efficient was MUSCL, followed by wave propagation, Lax–Friedrichs, TVD and ENO. Second-order discontinuous Galerkin (with $M = \infty$ for a Riemann problem) required timesteps around 0.2 times the stable timestep in order to avoid negative density and pressure. The most efficient third-order scheme was PPM, followed by ENO and Lax–Friedrichs (which were roughly equivalent).

Students can create their own test problems for gas dynamics with Riemann problem initial data by clicking on the following link: **Executable 6.3-45: guiGas Dynamics.**

6.3.4 MHD

The equations of magnetohydrodynamics were studied previously in section 4.5. The most famous Riemann problem for these equations is the Brio and Wu shock tube in air [23]:

$$\rho(x,0) = \begin{cases} 1, & x < 0 \\ 1/8, & x > 0 \end{cases} \tag{6.20a}$$

$$\mathbf{v}(x,0) = 0 \tag{6.20b}$$

$$\mathbf{B}(x,0) = \begin{cases} \sqrt{4\pi}, & x < 0 \\ -\sqrt{4\pi}, & x > 0 \end{cases} \tag{6.20c}$$

$$p(x,0) = \begin{cases} 1, & x < 0 \\ 1/10, & x > 0. \end{cases} \tag{6.20d}$$

This problem generalizes the Sod shock tub problem (6.15). The solution involves a rarefaction, a compound wave consisting of an over-compressive shock and a rarefaction, a contact discontinuity, a shock and a rarefaction.

The Brio and Wu shock tube is the default test problem for MHD in **Executable 6.3-46: guiRiemannProblem.** Students can also create their own test problems for MHD.

6.3.5 Nonlinear Elasticity

We discussed nonlinear elasticity previously in Section 4.6. The application of shock-capturing schemes to nonlinear elasticity has been developed less than other applications. To some extent, this is because there are so many different constitutive models that no one of them has captured the full audience. There are some analytical solutions for Goursat problems described in [157]. Wilkins [176] described a problem involving the impact of an elastic-plastic aluminum plate described by a nonlinear elastic response and a von Mises yield surface. Most of the interesting problems in elasticity are multi-dimensional.

6.3.6 Cristescu's Vibrating String

The equations for a vibrating string were studied in section 4.8. The model in that section was modified from the one in Keyfitz and Kranzer [81] in order to provide

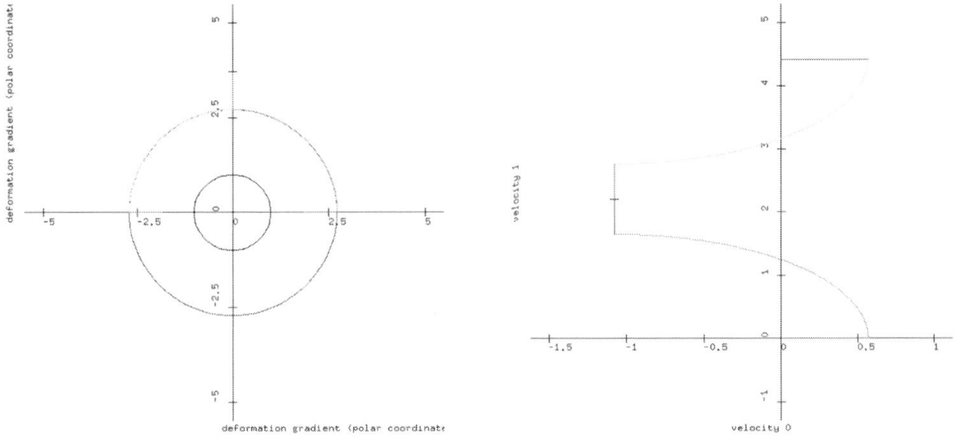

(a) Deformation gradient

(b) Velocity

Fig. 6.4 Riemann problem solution for vibrating string: Rarefaction–contact–rarefaction and rarefaction–contact–rarefaction ($\mathbf{v}_L = [0, 0]$, $\phi_L = 1.85914$, $\theta_L = 0$; $\mathbf{v}_R = [0, 4.4157]$, $\phi_R = 1.85914$, $\theta_R = 3.14159$)

a more physically realistic tension. This problem is not a common test problem for numerical methods. However, it is possible to generate rarefactions with contact discontinuities in their middle for this problem:

$$\mathbf{v}(x,0) = \begin{cases} \begin{bmatrix} 0 \\ 0 \end{bmatrix}, & x < 0 \\ \begin{bmatrix} 0 \\ 4.4157 \end{bmatrix}, & x > 0 \end{cases} \tag{6.21a}$$

$$\phi(x,0) = 1.85914 \tag{6.21b}$$

$$\theta(x,0) = \begin{cases} 0, & x < 0 \\ 3.14159, & x > 0. \end{cases} \tag{6.21c}$$

Figure 6.4 shows the solution of this Riemann problem in state space.

The analytical solution of the Riemann problem involves two contact discontinuities moving at speeds in the middle of rarefactions. As a result, this Riemann problem is not resolved well by the methods that use characteristic directions. Figure 6.5 shows that the Godunov scheme with either the exact Riemann solver or the Harten–Lax–van Leer approximate Riemann solver fails to capture the velocity peaks near the left and right states with 100 grid cells. The Lax–Friedrichs scheme has a similar difficulty. However, Figure 6.6 shows that the second-order Lax–Friedrichs scheme of Nessyahu and Tadmor and the second-order discontinuous Galerkin method do a better job of resolving this peak. Second-order Godunov (the

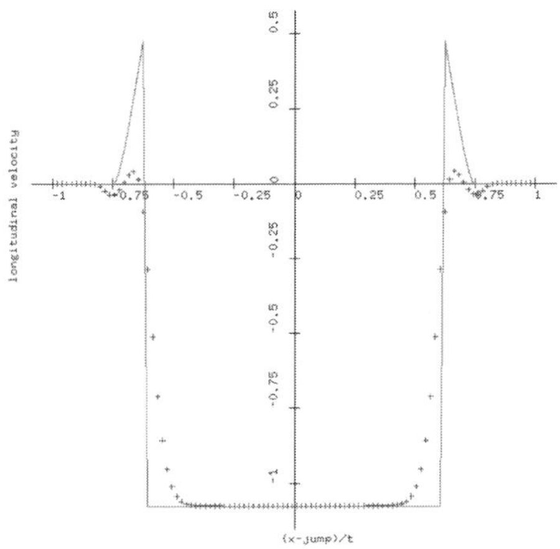

(a) Godunov with exact Riemann solve

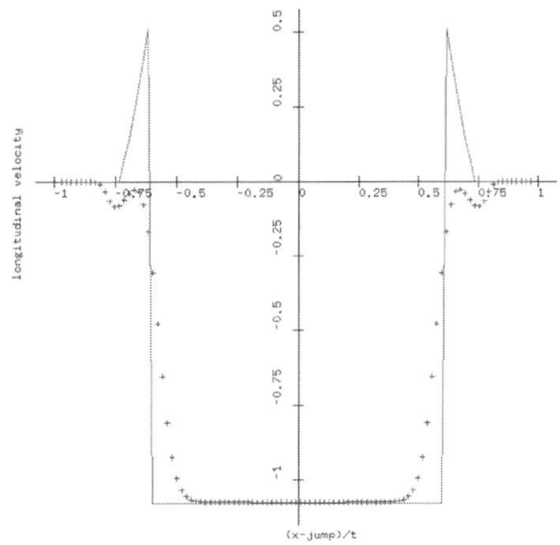

(b) Godunov with Harten–Lax–van Leer

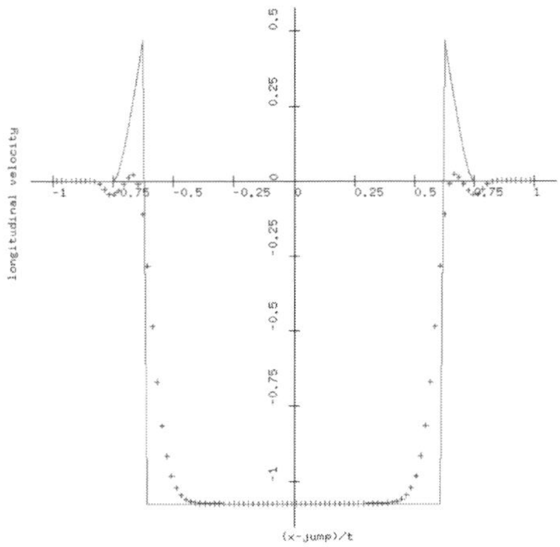

(c) Lax–Friedrichs

Fig. 6.5 First-order schemes for vibrating string: longitudinal velocity vs. x/t

MUSCL scheme) and wave propagation do a poor job in resolving this contact discontinuity.

Since the vibrating string problem does not have a Roe solver, the ENO schemes use the average of the left and right states to compute characteristic speeds and directions. In this case, the average of the left and right deformation gradients is zero, which represents a string compressed to zero volume. Such a string is not under tension, and characteristic speeds for the vibrating string are meaningless. As a result, the ENO schemes abort.

Several additional test problems can be found in the comments in **Program 6.3-106: string.f.** Students can develop their own test problems with Riemann initial data by clicking on the link **Executable 6.3-47: guiString.**

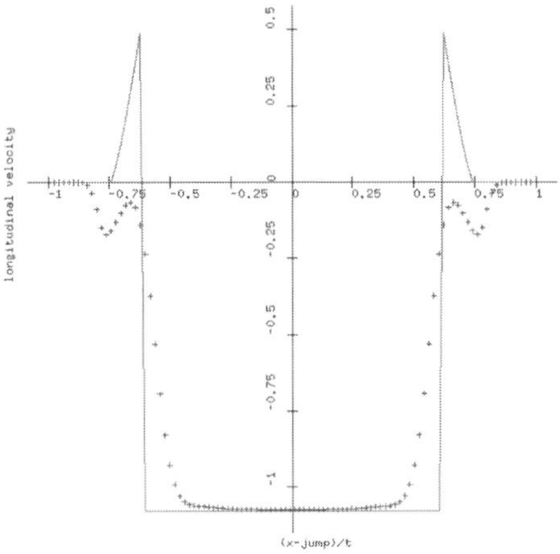

(a) MUSCL with Harten–Lax–van Leer

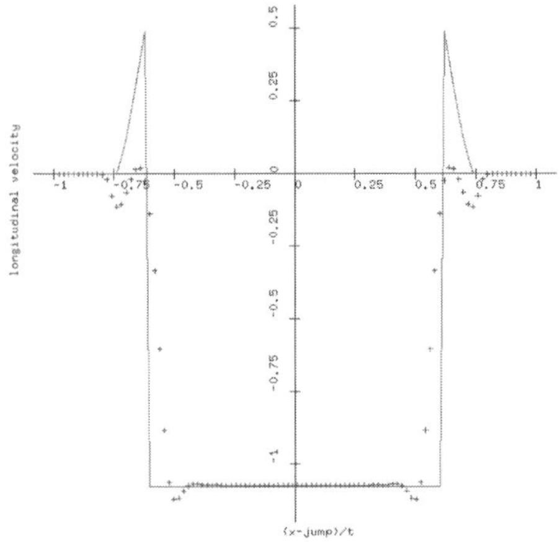

(b) Wave propagation

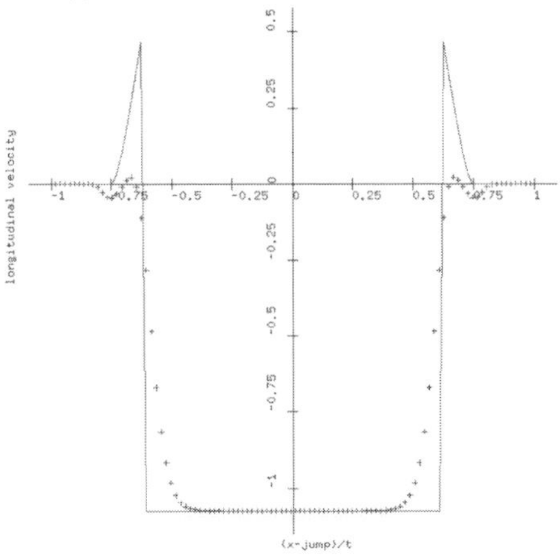

(c) Nessyahu–Tadmor

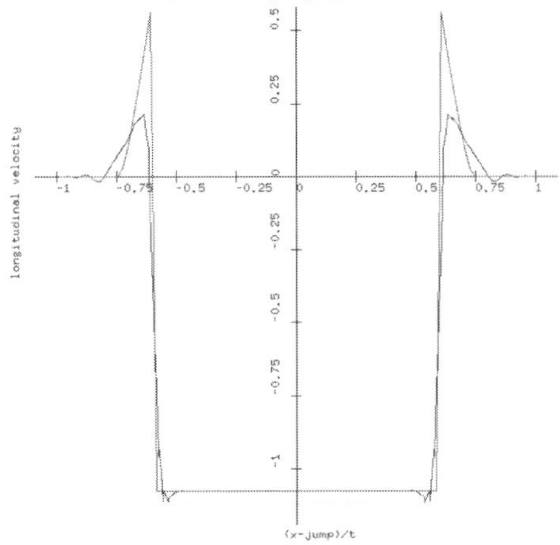

(d) Discontinuous Galerkin 2

Fig. 6.6 Second-order schemes for vibrating string: longitudinal velocity vs. x/t

6.3.7 Plasticity

The Antman–Szymczak model for plasticity was analyzed in Section 4.9. The solution of the Riemann problem was described by Trangenstein and Pember in [163]. This paper contains a table of 21 Riemann problems with structurally different solutions. Here is one of the more challenging of those problems:

$$v(x,0 = \begin{cases} -0.1, & x < 0 \\ -5.081, & x > 0, \end{cases} \quad \epsilon(x,0) = \begin{cases} 3.5, & x < 0 \\ 0.12111, & x > 0, \end{cases}$$
$$\pi(x,0) = \begin{cases} 3.25, & x < 0 \\ 0.217, & x > 0. \end{cases} \tag{6.22}$$

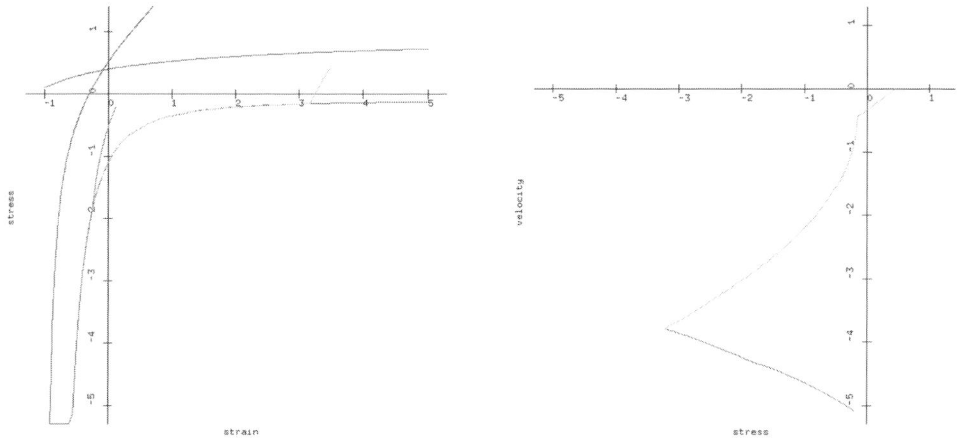

(a) Stress vs strain (b) Velocity vs stress

Fig. 6.7 Riemann problem solution for plasticity ($v_L = -0.1, \epsilon_L = 3.5, \pi_i = 3.25$; $v_R = -5.081, \epsilon_R = 0.12111, \pi_R = 0.217$)

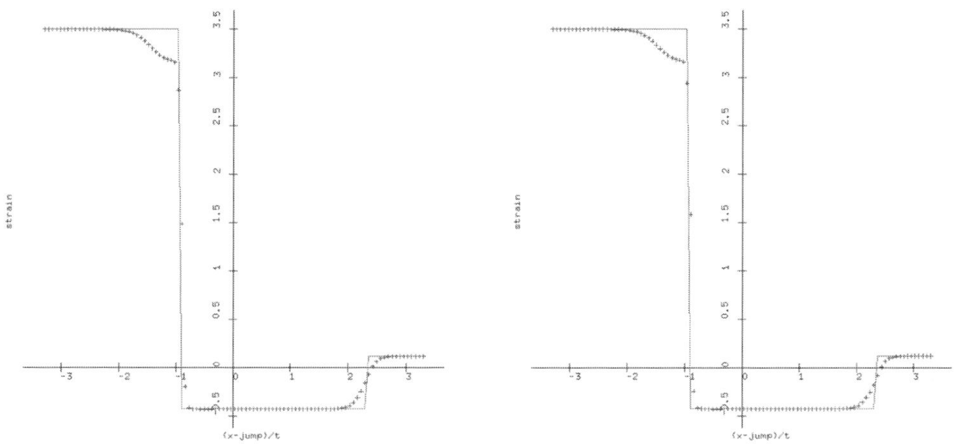

(a) Godunov with exact Riemann solve (b) Godunov with Harten–Lax–van Leer

Fig. 6.8 First-order schemes for plasticity: strain vs. x/t

This problem involves a shock from an elastic state to a well-compressed plastic state, as shown in Figure 6.7.

Note that we cannot easily apply the Lax–Friedrichs scheme to the plasticity problem. The difficulty is that the plastic strain, which acts as a hysteresis parameter, has no natural definition during the mesh staggering. The plastic strain is integrated along particle paths, and the Lax–Friedrichs scheme tangles the particle paths each

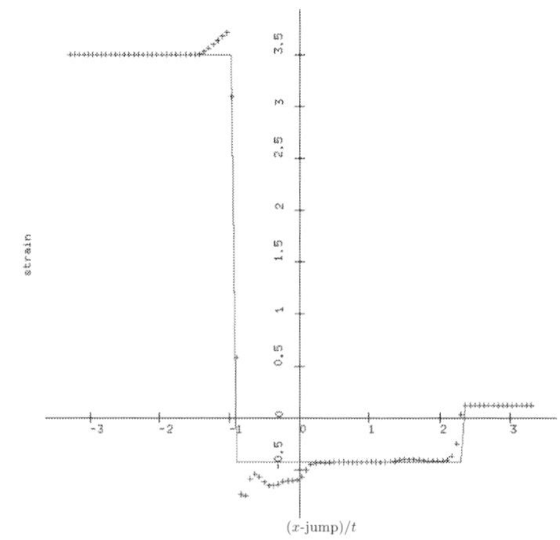

(a) MUSCL, exact Riemann solve, CFL = 0.5

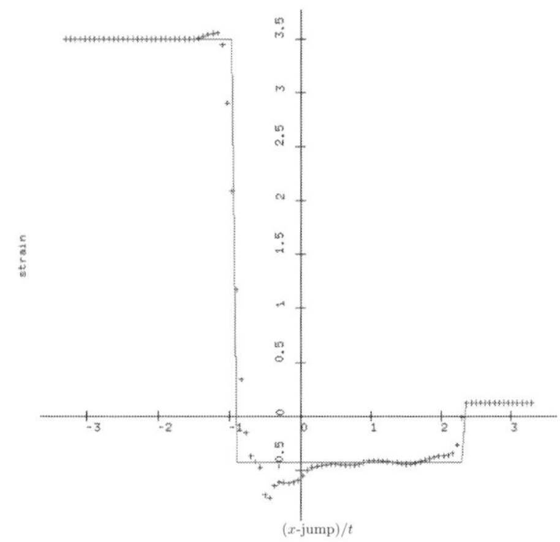

(b) MUSCL, Harten–Lax–van Leer, CFL = 0.5

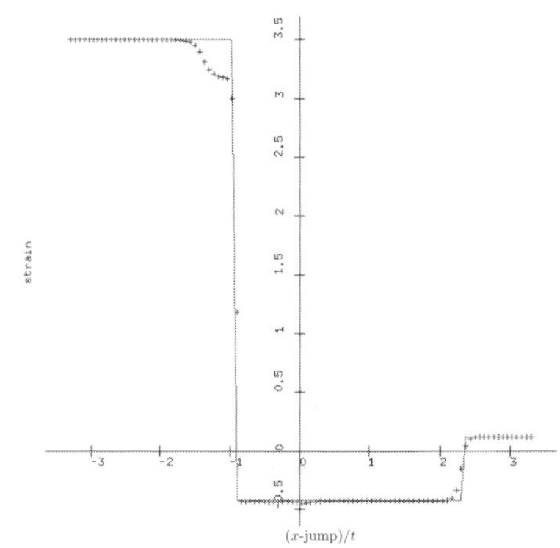

(c) ENO 2

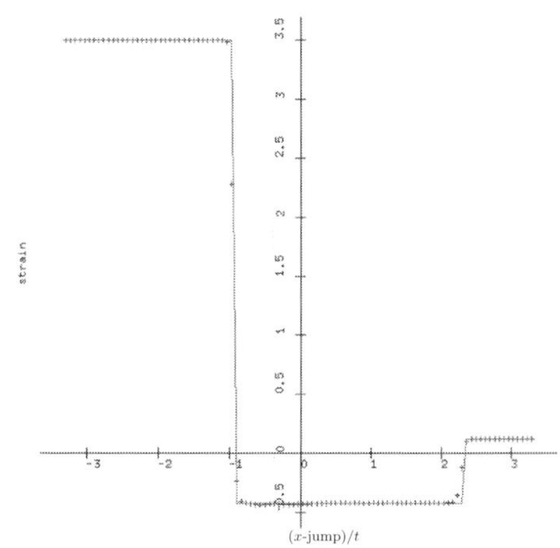

(d) Wave propagation

Fig. 6.9 Second-order schemes for plasticity: strain vs. x/t

half timestep. This scheme might take advantage of ideas for schemes for plasticity in the Eulerian frame of reference [118, 164].

Some of the schemes perform well for this problem, and some do not. The MUSCL schemes and discontinuous Galerkin tend to develop numerical oscillations, while wave propagation and ENO do not. It is interesting to note that restoring the characteristic projection step in the MUSCL scheme (see Section 5.9.4) does not remedy this problem. Results with first-order schemes are shown in Figure 6.8 and second-order schemes are shown in Figure 6.9.

All 21 of the Trangenstein-Pember Riemann problems can be found in the comments of subroutine `initplasticity` in **Program 6.3-107: plasticity.f.** Students

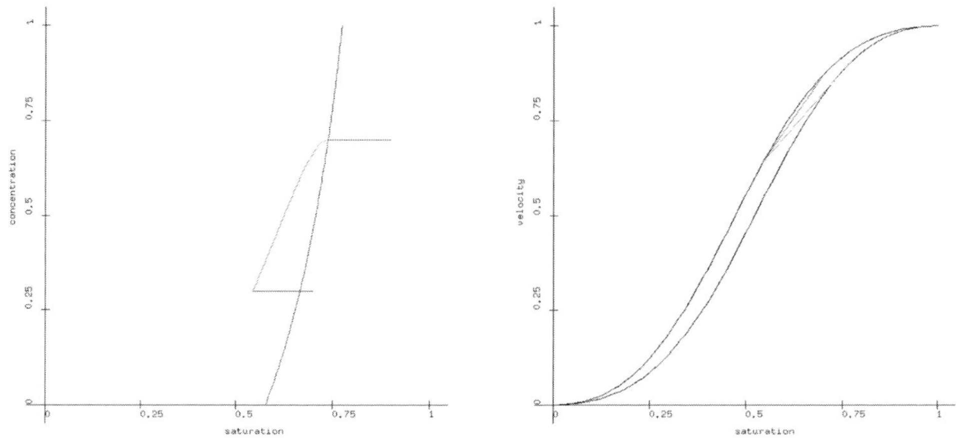

(a) Concentration vs. saturation (b) Velocity vs. saturation

Fig. 6.10 Riemann problem solution for polymer oil bank ($s_L = 0.9$, $c_L = 0.7$; $s_R = 0.7$, $c_R = 0$)

can develop their own test problems with Riemann initial data by clicking on the link **Executable 6.3-48: guiPlasticity.**

6.3.8 Polymer Model

The polymer model was studied in Section 4.10. The analytical solution of the Riemann problem for this model was described in general terms by Keyfitz and Kranzer in [81] and Pope in [132]. Allen *et al.* [3] present pictures of the structurally different solutions of the Riemann problem with and without gravity. Of the different Riemann problems, the following water flooding problem [75] is interesting because it develops an "oil bank" just ahead of a contact discontinuity moving at the same speed as the leading edge of a rarefaction:

$$s(x,0) = \begin{cases} 0.9, & x < 0 \\ 0.7, & x > 0, \end{cases} \quad c(x,0) = \begin{cases} 0.7, & x < 0 \\ 0.3, & x > 0. \end{cases} \tag{6.23}$$

In this model, the mobilities are $\lambda_w(s_w, c) = s_w^2/(\mu_0[1/2 + c])$ and $\lambda_o(s_o) = s_o^2/\mu_o$ where the viscosity of oil is $\mu_o = 0.35$. A second interesting test case [75] which roughly corresponds to water-flooding after polymer injection is

$$s(x,0) = \begin{cases} 1.0, & x < 0 \\ 0.3, & x > 0, \end{cases} \quad c(x,0) = \begin{cases} 0.1, & x < 0 \\ 0.9, & x > 0. \end{cases} \tag{6.24}$$

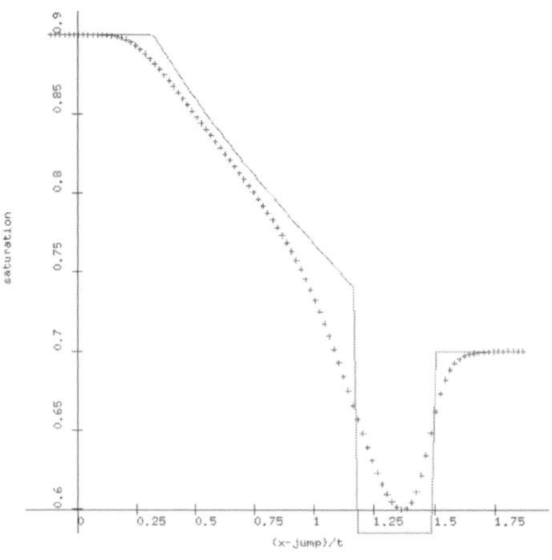

(a) Godunov with exact Riemann solve

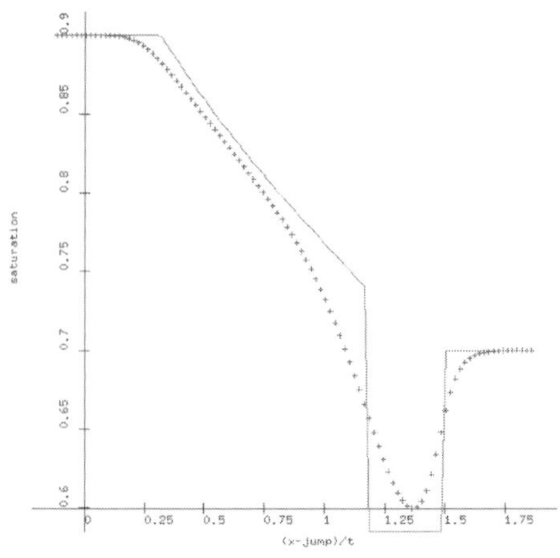

(b) Godunov with Harten–Lax–van Leer

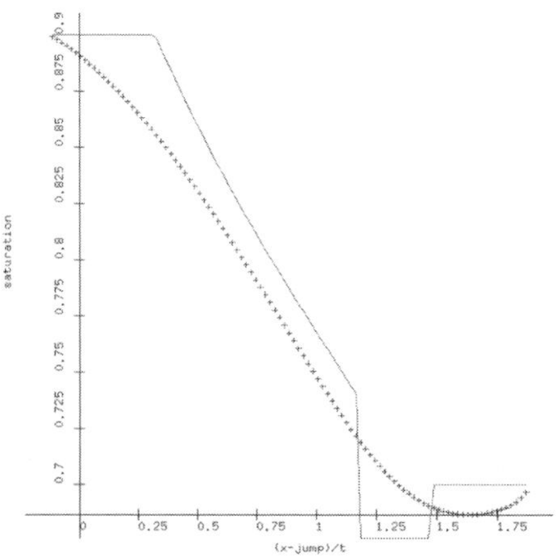

(c) Lax–Friedrichs

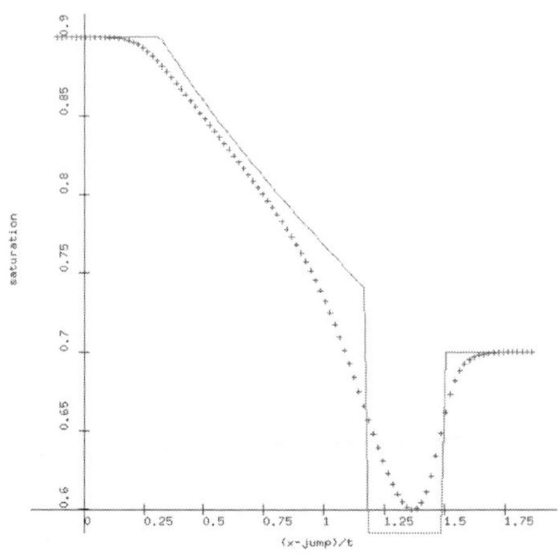

(d) ENO 1

Fig. 6.11 First-order schemes for polymer oil bank: saturation vs. x/t

Most schemes develop an overshoot at the leading edge of the slow rarefaction, and the size of the overshoot is not reduced during mesh refinement.

The solution to the polymer oil bank Riemann problem is shown in Figure 6.10. This problem has a rarefaction with leading edge traveling at the same speed as a contact discontinuity, with a shock moving ahead of both. The water saturation dips between the shock and contact discontinuity, corresponding to an oil bank that is being pushed toward a production well.

Figure 6.11 shows numerical results with several first-order schemes for the polymer oil bank problem. The Lax–Friedrichs scheme performs the worst, but

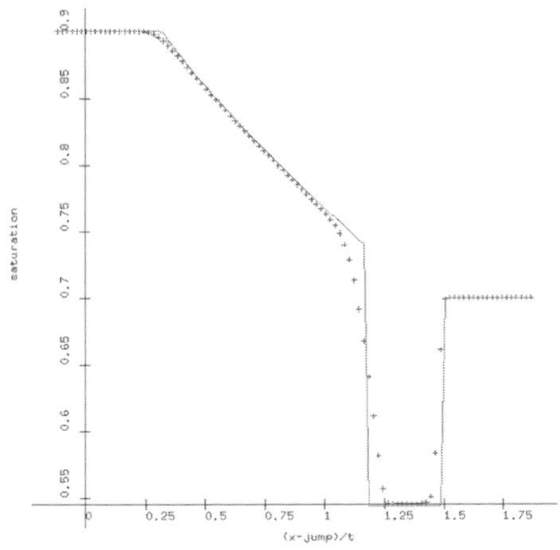

(a) MUSCL

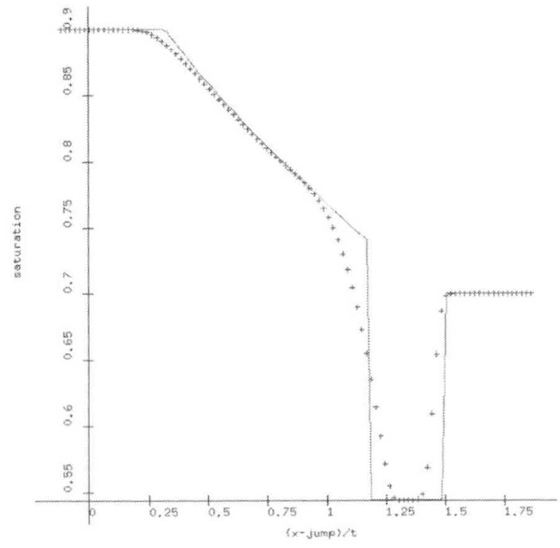

(b) Nessyahu–Tadmor

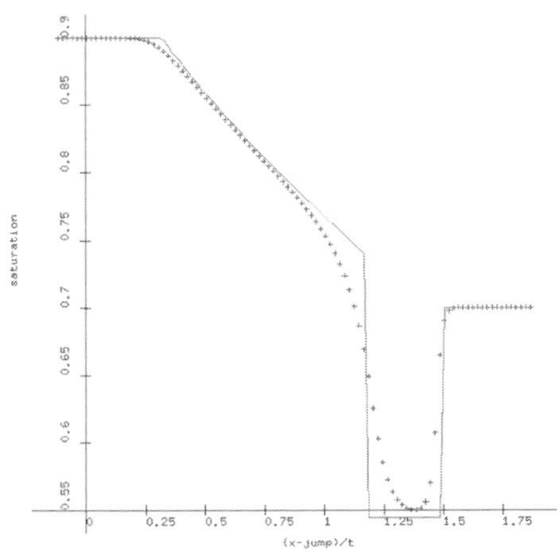

(c) ENO 2

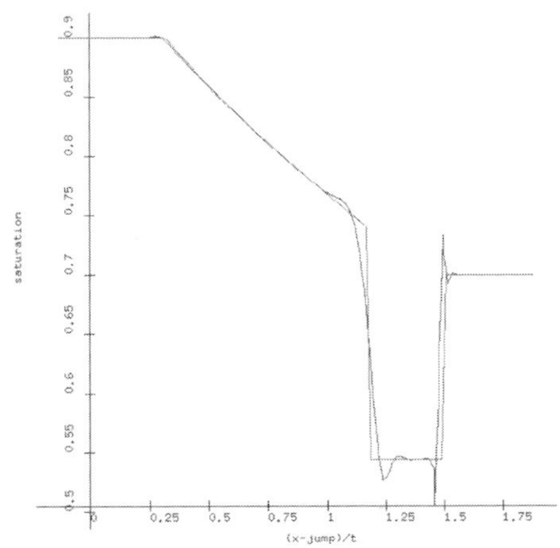

(d) Discontinuous Galerkin 2

Fig. 6.12 Second-order schemes for polymer oil bank: saturation vs. x/t

none of the schemes do a good job on resolving the contact discontinuity. Results with second-order schemes are shown in Figure 6.12. The MUSCL scheme does the best job of resolving the oil bank, but all of the second-order schemes show significant improvement over the first-order schemes. MUSCL is less expensive per step than the other schemes, which involve substeps (and reduced stability restrictions on the timestep for the discontinuous Galerkin method).

Details of the implementation of the polymer model can be found in **Program 6.3-108: polymer.f.** Students can develop their own test problems with Riemann initial data by clicking on the link **Executable 6.3-49: guiPolymer**.

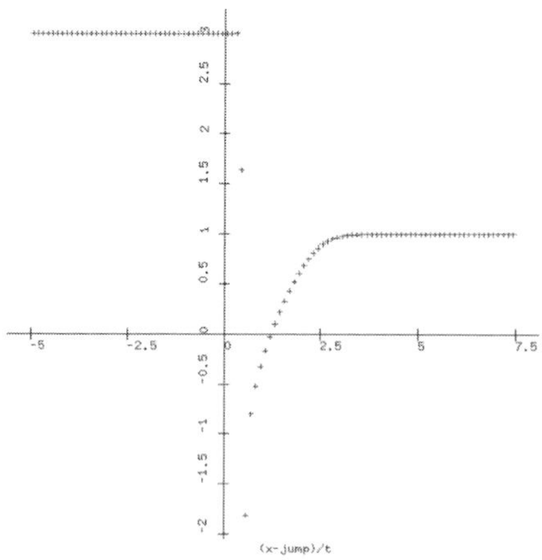

(a) Godunov with Harten–Lax–van Leer, $t = 1$

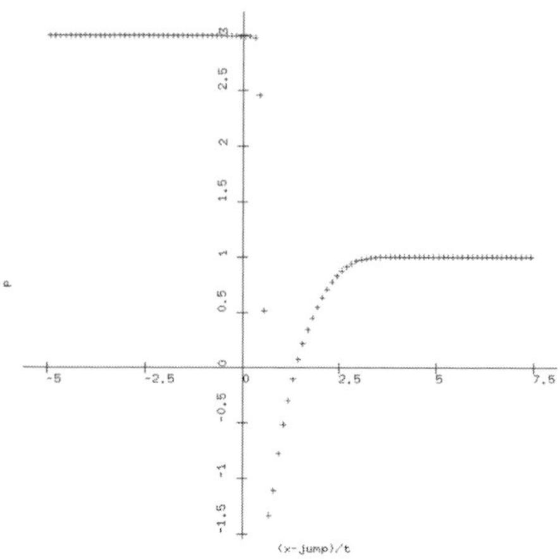

(b) Godunov with Rusanov, $t = 1$

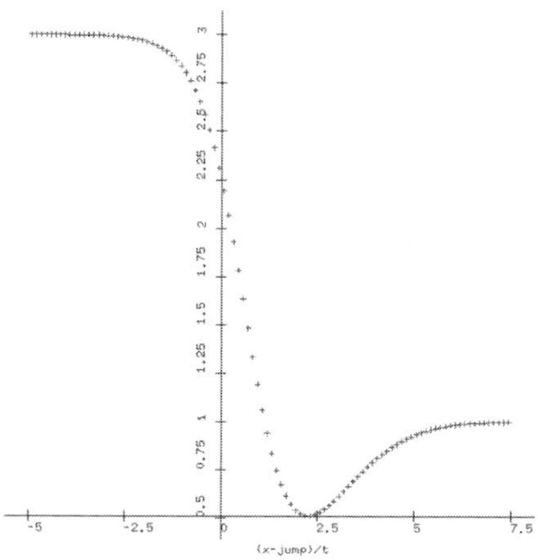

(c) Lax-Friedrichs, $t = 1$

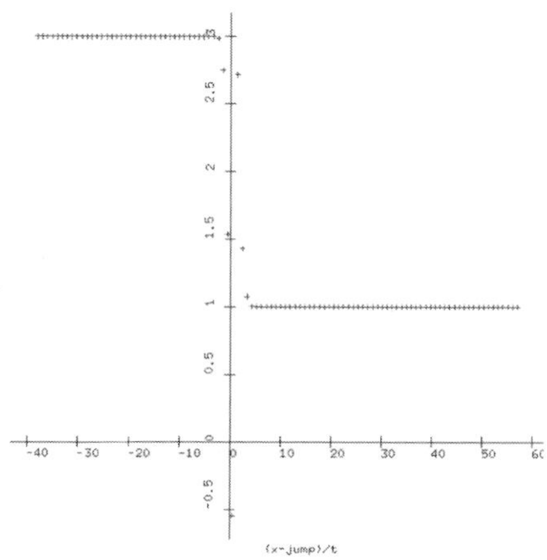

(d) ENO 1, $t = 0.13$

Fig. 6.13 First-order schemes for Schaeffer–schechter–shearer model: P vs. x/t

6.3.9 Schaeffer–Schechter–Shearer Model

Finally, let us discuss a computation involving the Schaeffer-Schechter-Shearer model of Section 4.12. The Riemann problem

$$p(x,0) = \begin{cases} 3, & x < 0 \\ 1, & x > 0, \end{cases} \qquad q(x,0) = \begin{cases} 10, & x < 0 \\ 1, & x > 0 \end{cases} \tag{6.25}$$

has no solution involving waves studied in Chapter 4. Numerical methods applied to such a problem will show a variety of results, none of which appear to represent known behavior. The computational results could be interpreted as failures of the

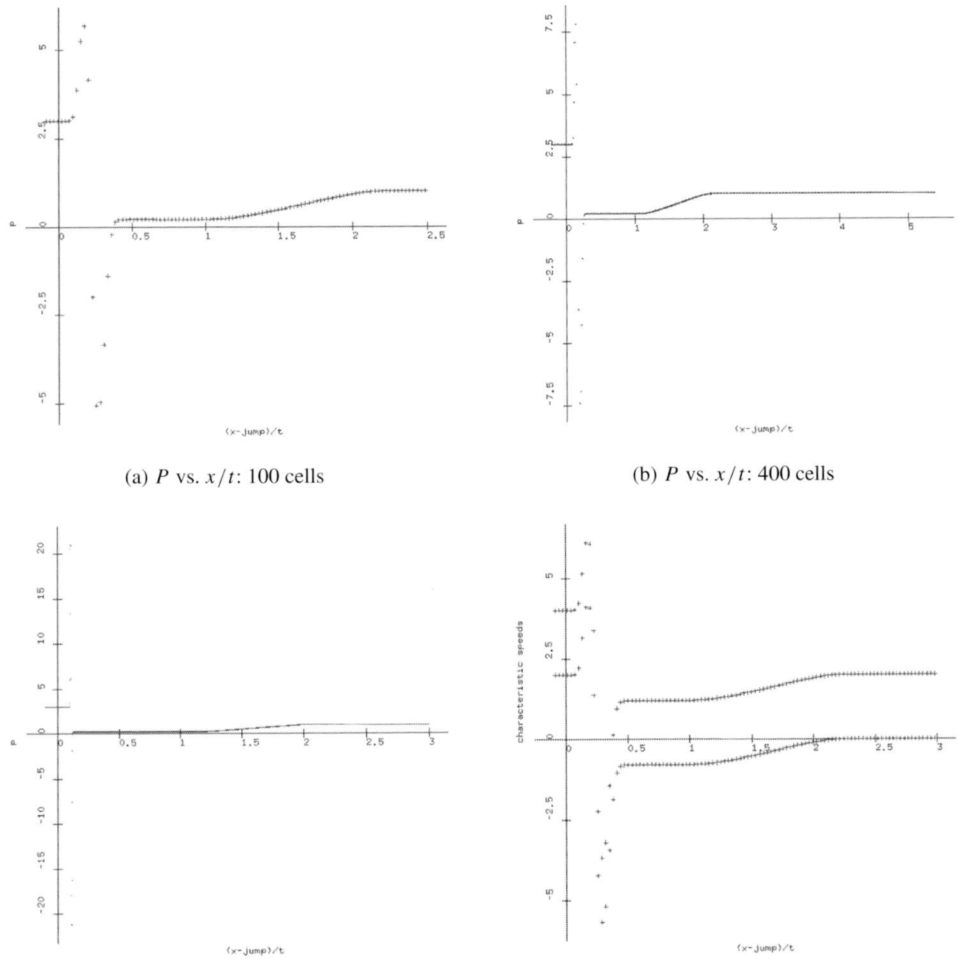

(a) P vs. x/t: 100 cells

(b) P vs. x/t: 400 cells

(c) P vs. x/t: 1600 cells

(d) Characteristic speeds vs. x/t: 100 cells

Fig. 6.14 Convergence study using the Nessyahu–Tadmor scheme for the Schaeffer–Schechter–Shearer model

numerical method, rather than problems with the model. Indeed, the initial reaction of some people to the elliptic regions in three-phase flow in porous media, reported in [15], was to assume that the problem was in the numerical method, not the model.

Figure 6.13 shows some numerical results with various first-order methods for this problem. The Godunov scheme with either the Harten–Lax–van Leer or the Rusanov flux shows some numerical oscillation in the discontinuity, while the Lax–Friedrichs scheme smears the wave so severely (even at CFL $= 0.9$) that no problem is readily apparent. The first-order ENO scheme, however, effectively blows up just after time 0.13; this result is perhaps more honest than the previous three results.

In Figure 6.14 we show results with the second-order Nessyahu–Tadmor scheme for this problem. Note that the characteristic speeds in part (d) of this figure indicate a discontinuity of some sort in the slow wave family and a rarefaction in the fast wave family. However, the characteristic speeds in the viscous profile of the discontinuity go higher than the value at the left state and lower than the value at the intermediate state. In addition, mesh refinement leads to ever-larger values in the viscous profile of the discontinuity. This could appear to be a numerical instability, but the real problem is that the slow wave family through the left state does not intersect the fast wave family through the right state at any finite state.

Exercises for 6.3

6.3.1 Use the results in Example 4.1.12 to determine the left state for a shallow water Riemann problem involving a shock moving right with speed 1 into water with height 1 and velocity 0. Assume that $g = 1$. Construct a numerical method to solve this Riemann problem. Does your numerical solution show evidence of a wave in the other wave family? How do you explain this?

6.3.2 Suppose that you have a tank of water of some fixed height and zero velocity that is given an external velocity ("shaking the pan"). Equivalently, consider a shallow water problem in which the water has a fixed height and nonzero initial velocity, but is is confined between two reflecting walls. Construct a numerical scheme to solve this problem. What kind of waves occur as a result of the first reflection from the walls? How many reflections can you model before the results are no longer trustworthy?

6.3.3 Nonlinear resonance problems in the wave propagation scheme can occur when waves in the wrong wave family are excited due to the use of inner products between the characteristic directions in the limiting. LeVeque suggested a modification of the wave propagation scheme to overcome the nonlinear resonance problem. For the Roe solver, the limited wave-field decomposition would take the revised form

$$\mathbf{e}_i \cdot \tilde{\mathbf{a}}^n_{j+1/2} = \begin{cases} \phi(\mathbf{e}_i \cdot (\mathbf{X}^n_{j+1/2})^{-1}\mathbf{a}^n_{j-1/2}, \mathbf{e}_i \cdot (\mathbf{X}^n_{j+1/2})^{-1}\mathbf{a}^n_{j+1/2}), & \mathbf{e}_i \cdot \Lambda^n_{j+1/2}\mathbf{e}_i \geq 0 \\ \phi(\mathbf{e}_i \cdot (\mathbf{X}^n_{j+1/2})^{-1}\mathbf{a}^n_{j+3/2}, \mathbf{e}_i \cdot (\mathbf{X}^n_{j+1/2})^{-1}\mathbf{a}^n_{j+1/2}), & \mathbf{e}_i \cdot \Lambda^n_{j+1/2}\mathbf{e}_i < 0. \end{cases}$$

Program this scheme and test it on the "parting sea" Riemann problem with $h(x, 0) = 1$ and $v(x, 0) = \pm 0.5$.

6.3.4 How would you change the Harten–Hyman modification of the Roe solver to provide the information needed for the modified wave-field decomposition in the previous exercise?

6.3.5 Perform the previous exercise for the Harten–Lax-van Leer approximate Riemann solver. This scheme would apply to more general nonlinear systems of hyperbolic conservation laws, for which Roe solvers are not available.

6.3.6 Solve the Colella–Woodward interacting blast wave problem using one of the schemes discussed above. Plot the numerical results (ρ, $\mathbf{v}$, p versus x) for times 0.01, 0.016, 0.026, 0.028, 0.030,

0.032, 0.034 and 0.038. Perform a mesh refinement study to determine whether your solution is well resolved.

6.3.7 Colella and Woodward [35] suggested several techniques specific to gas dynamics for improving the resolution of shocks and contact discontinuities. Read about "discontinuity detection," modify the PPM scheme to incorporate this approach, and test its performance on the Sod shock tube problem.

6.3.8 Colella [32] has suggested using fourth-order slopes in the MUSCL reconstruction step.

(a) For a scalar law, determine the average of the second derivative of the *quintic* interpolation to the cell averages, instead of the cubic interpolation discussed in section 5.9.2. Test these fourth-order slopes on the Zalesak test problems for linear advection in exercise 2.2.5.

(b) Examine the effect of fourth-order slopes on the resolution of the contact discontinuity in the Sod shock tube problem.

6.3.9 Harten [65] has suggested a modification of the ENO scheme to improve the resolution of contact discontinuities. Read about "sub-cell resolution," modify the ENO scheme to incorporate this approach, and test its performance on the Sod shock tube problem.

7

Methods in Multiple Dimensions

Most interesting physical problems involve multiple spatial dimensions. Unfortunately, the theory of hyperbolic conservation laws in multiple dimensions is not very well developed. Very little is known about the appropriate norms to use in discussing the stability and uniqueness of conservation laws in multiple dimensions. As a result, the theory for numerical methods is also very primitive.

Nevertheless, people need to solve multidimensional problems. For complicated nonlinear problems, the common approach is to compute numerical solutions. In this chapter, we will discuss several numerical approaches.

7.1 Numerical Methods in Two Dimensions

7.1.1 Operator Splitting

Suppose that we want to solve the two-dimensional system of partial differential equations

$$\frac{\partial \mathbf{u}}{\partial t} + \frac{\partial \mathbf{f}_1}{\partial x_1} + \frac{\partial \mathbf{f}_2}{\partial x_2} = 0.$$

Also suppose that for the one-dimensional problem

$$\frac{\partial \mathbf{u}}{\partial t} + \frac{\partial \mathbf{f}}{\partial x} = 0$$

we would use the numerical method

$$\mathbf{u}_j^{n+1} = \mathbf{u}_j^n - \left[\mathbf{f}_{j+\frac{1}{2}}(\mathbf{u}^n) - \mathbf{f}_{j-\frac{1}{2}}(\mathbf{u}^n)\right] \frac{\triangle t}{\triangle x_j}.$$

Then in two dimensions we can apply the following computational strategy, called first-order **operator splitting**:

$$\mathbf{u}_{ij}^{n+1,n} = \mathbf{u}_{ij}^{n} - \left[(\mathbf{f}_1)_{i+1/2,j}(\mathbf{u}^n) - (\mathbf{f}_1)_{i-1/2,j}(\mathbf{u}^n)\right] \frac{\triangle t^{n+1/2}}{\triangle x_{1,i}} \tag{7.1a}$$

$$\mathbf{u}_{ij}^{n+1} = \mathbf{u}_{ij}^{n+1,n} - \left[(\mathbf{f}_2)_{i,j+1/2}(\mathbf{u}^{n+1,n}) - (\mathbf{f}_2)_{i,j-1/2}(\mathbf{u}^{n+1,n})\right] \frac{\triangle t^{n+1/2}}{\triangle x_{2,j}}. \tag{7.1b}$$

This approach is based on approximating the exact evolution operator for the differential equation by a product of one-dimensional evolution operators, and then approximating the evolution operators by numerical methods. Operator splitting of the evolution operators is first-order accurate in time [38, 158]. Spatial errors (and additional temporal errors) are due to the approximation of the one-dimensional evolution operators.

Second-order operator splitting is often called **Strang splitting** [152]. This method takes the form

$$\mathbf{u}_{ij}^{n,n+1/2} = \mathbf{u}_{ij}^{n} - \left[(\mathbf{f}_2)_{i,j+1/2}(\mathbf{u}^n) - (\mathbf{f}_2)_{i,j-1/2}(\mathbf{u}^n)\right] \frac{\triangle t^{n+1/2}}{2\triangle x_{2,j}} \tag{7.2a}$$

$$\mathbf{u}_{ij}^{n+1,n+1/2} = \mathbf{u}_{ij}^{n,n+1/2} - \left[(\mathbf{f}_1)_{i+1/2,j}(\mathbf{u}^{n,n+1/2}) - (\mathbf{f}_1)_{i-1/2,j}(\mathbf{u}^{n,n+1/2})\right] \frac{\triangle t^{n+1/2}}{\triangle x_{1,i}} \tag{7.2b}$$

$$\mathbf{u}_{ij}^{n+1} = \mathbf{u}_{ij}^{n+1,n+1/2} - \left[(\mathbf{f}_2)_{i,j+1/2}(\mathbf{u}^{n+1,n+1/2}) - (\mathbf{f}_2)_{i,j-\frac{1}{2}}(\mathbf{u}^{n+1,n+1/2})\right] \frac{\triangle t^{n+1/2}}{2\triangle x_{2,j}}. \tag{7.2c}$$

Any appropriate numerical flux from Chapter 6 on methods in one dimension can be used to evaluate the numerical fluxes in these steps. It is also common to combine the last half step of one timestep with the first half-step of the next timestep in order to save work.

For problems on rectangular domains with simple boundary conditions, either form of operator splitting can be very easy to implement, if a trustworthy one-dimensional method is already available. Peyret and Taylor [131, page 73] point out that operator splitting can take advantage of different stability restrictions in the coordinate directions by taking multiple split steps in the direction with the more restrictive stability condition. They also point out that the treatment of boundary conditions for the intermediate steps of the splitting are tricky, so it is common to use the analytical values for the boundary conditions when available. Additional information about operator splitting techniques in general can be found in [117] and [178].

We have implemented operator splitting in **Program 7.1-109: GUIRectangle.C** Procedures `runOnce` and `runSimulation` can perform either first-order or second-order splitting by calling procedure `runSchemeSplit` in the appropriate ways. This procedure is designed to loop over grid lines and call appropriate modifications of the one-dimensional schemes for operator splitting. Of course, some modification of the functions for the various models are necessary in multiple dimensions. For example, the shallow water equations in program **Program 7.1-110: shallow_water.m4** were modified to deal with a vector fluid velocity and appropriate normal directions. Similar modifications were performed to **Program 7.1-111: burgers.m4** and **Program 7.1-112: gas_dynamics.m4.** In fact, these files were designed to use macro processing so that the fundamental code could be written once and implemented in different coordinate dimensions and directions as needed. Essentially all of the computational routines were written in Fortran for easier array addressing.

Students can execute operator splitting programs in two dimensions by clicking on **Executable 7.1-50: guiRectangle.** The student can select either of Burgers' equations, shallow water or gas dynamics, first- or second-order operator splitting, and either of a variety of numerical schemes. In two dimensions, the computational results for scalar fields (such as water height in shallow water, or pressure in gas dynamics) can be displayed either as 2D contours, 2D color fills or 3D surface plots. The 3D graphics in the surface plot uses a trackball for rotation with the left mouse button. The figure can be sliced along the ends of any coordinate axis using the middle mouse button. Values at points in the figure can be determined using the right mouse button.

By setting the number of grid cells in one of the coordinate directions to zero, the student can perform a convergence analysis. In this case, the program performs a fine grid computation with the given method, and measures the errors in coarser grid results compared to the fine grid result.

Some computational results for the 2D Riemann problem with Burgers' equation are shown in Figure 7.1.

7.1.2 Donor Cell Methods

The conservation law

$$\frac{\partial \mathbf{u}}{\partial t} + \sum_{k=1}^{2} \frac{\partial \mathbf{F}(\mathbf{u})\mathbf{e}_k}{\partial \mathbf{x}_k} = 0$$

can be written in integral form as

$$\int_{\Omega_{ij}} \mathbf{u}(\mathbf{x}, t^{n+1})\, d\mathbf{x} = \int_{\Omega_{ij}} \mathbf{u}(\mathbf{x}, t^n)\, d\mathbf{x} - \int_{t^n}^{t^{n+1}} \int_{\partial\Omega_{ij}} \mathbf{F}(\mathbf{u})\mathbf{n}\, ds\, dt.$$

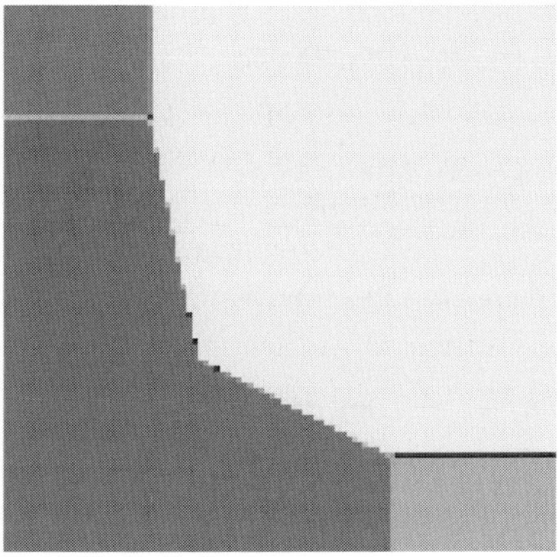

(a) Color fill plot

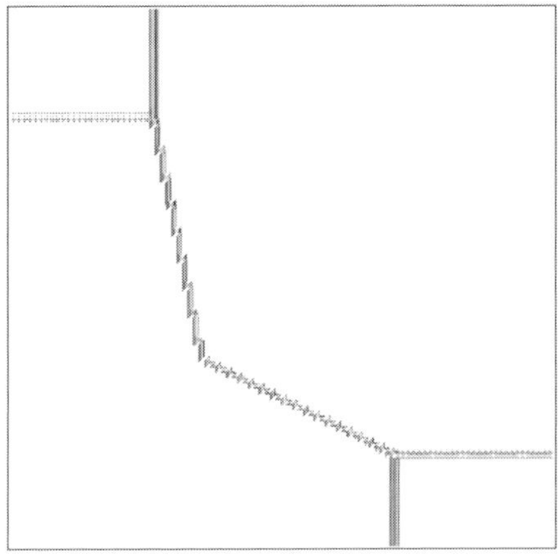

(b) Contour plot

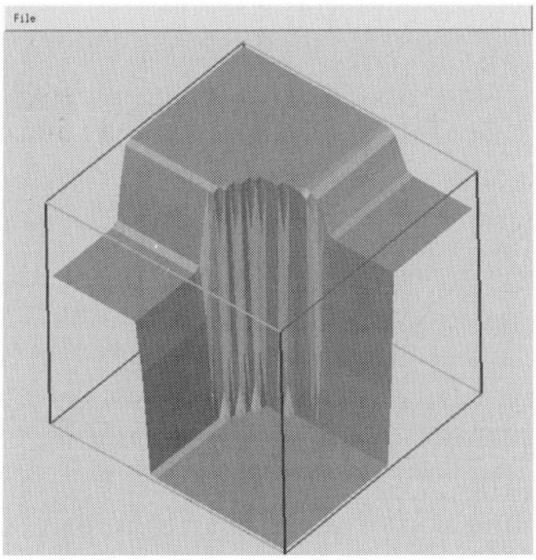

(c) Surface plot

Fig. 7.1 2D Riemann problem for Burgers' equation: Second-order operator splitting with MUSCL on 100×100 grid (40×40 for surface plot); initial condition is $u = -1$ for $x_0 > 0, x_1 > 0$, $u = .67$ for $x_0 < 0, x_1 > 0$, $u = .33$ for $x_0 > 0, x_1 < 0$, $u = 1$ for $x_0 < 0, x_1 < 0$.

On a rectangular grid cell $\Omega_{ij} = (\mathbf{x}_{1,i-1/2}, \mathbf{x}_{1,i+1/2}) \times (\mathbf{x}_{2,j-1/2}, \mathbf{x}_{2,j+1/2})$ this integral form can be written

$$\int_{\mathbf{x}_{2,j-1/2}}^{\mathbf{x}_{2,j+1/2}} \int_{\mathbf{x}_{1,i-1/2}}^{\mathbf{x}_{1,i+1/2}} \mathbf{u}(\mathbf{x}_1, \mathbf{x}_2, t^{n+1})\, d\mathbf{x}_1\, d\mathbf{x}_2$$
$$= \int_{\mathbf{x}_{2,j-1/2}}^{\mathbf{x}_{2,j+1/2}} \int_{\mathbf{x}_{1,i-1/2}}^{\mathbf{x}_{1,i+1/2}} \mathbf{u}(\mathbf{x}_1, \mathbf{x}_2, t^{n})\, d\mathbf{x}_1\, d\mathbf{x}_2$$

$$
\begin{aligned}
&- \int_{t^n}^{t^{n+1}} \int_{\mathbf{x}_{2,j-1/2}}^{\mathbf{x}_{2,j+1/2}} \mathbf{F}(\mathbf{u}(\mathbf{x}_{1,i+1/2}, \mathbf{x}_2, t))\mathbf{e}_1 \, d\mathbf{x}_2 \, dt \\
&+ \int_{t^n}^{t^{n+1}} \int_{\mathbf{x}_{2,j-1/2}}^{\mathbf{x}_{2,j+1/2}} \mathbf{F}(\mathbf{u}(\mathbf{x}_{1,i-1/2}, \mathbf{x}_2, t))\mathbf{e}_1 \, d\mathbf{x}_2 \, dt \\
&- \int_{t^n}^{t^{n+1}} \int_{\mathbf{x}_{1,i-1/2}}^{\mathbf{x}_{1,i+1/2}} \mathbf{F}(\mathbf{u}(\mathbf{x}_1, \mathbf{x}_{2,j+1/2}, t))\mathbf{e}_2 \, d\mathbf{x}_1 \, dt \\
&+ \int_{t^n}^{t^{n+1}} \int_{\mathbf{x}_{1,i-1/2}}^{\mathbf{x}_{1,i+1/2}} \mathbf{F}(\mathbf{u}(\mathbf{x}_1, \mathbf{x}_{2,j-1/2}, t))\mathbf{e}_2 \, d\mathbf{x}_1 \, dt.
\end{aligned} \tag{7.3}
$$

We will typically compute cell averages

$$\mathbf{u}_{ij}^n \approx \frac{1}{\triangle x_{1,i} \triangle x_{2,j}} \int_{\mathbf{x}_{2,j-1/2}}^{\mathbf{x}_{2,j+1/2}} \int_{\mathbf{x}_{1,i-1/2}}^{\mathbf{x}_{1,i+1/2}} \mathbf{u}(\mathbf{x}_1, \mathbf{x}_2, t^{(n)}) \, d\mathbf{x}_1 \, d\mathbf{x}_2$$

and flux side and time integrals

$$
\begin{aligned}
\mathbf{f}_{i+1/2,j}^{n+1/2} &\approx \int_{t^n}^{t^{n+1}} \int_{\mathbf{x}_{2,j-1/2}}^{\mathbf{x}_{2,j+1/2}} \mathbf{F}(\mathbf{u}(\mathbf{x}_{1,i+1/2}, \mathbf{x}_2, t))\mathbf{e}_1 \, d\mathbf{x}_2 \, dt \\
\mathbf{f}_{i,j+1/2}^{n+1/2} &\approx \int_{t^n}^{t^{n+1}} \int_{\mathbf{x}_{1,i-1/2}}^{\mathbf{x}_{1,i+1/2}} \mathbf{F}(\mathbf{u}(\mathbf{x}_1, \mathbf{x}_{2,j+1/2}, t))\mathbf{e}_2 \, d\mathbf{x}_1 \, dt
\end{aligned}
$$

to perform a conservative difference

$$\mathbf{u}_{ij}^{n+1} \triangle x_{1,i} \triangle x_{2,j} = \mathbf{u}_{ij}^n \triangle x_{1,i} \triangle x_{2,j} - \mathbf{f}_{i+1/2,j}^{n+1/2} + \mathbf{f}_{i-1/2,j}^{n+1/2} - \mathbf{f}_{i,j+1/2}^{n+1/2} + \mathbf{f}_{i,j-1/2}^{n+1/2}. \tag{7.4}$$

7.1.2.1 Traditional Donor Cell Upwind Method

Early work on unsplit methods for hyperbolic conservation laws took particularly simple forms, as in [131, 137]. For example, the **donor cell scheme** form of Godunov's method would compute flux side and time integrals

$$\mathbf{f}_{i+1/2,j}^{n+1/2} = \mathbf{F}(\mathcal{R}(\mathbf{u}(\mathbf{x}_{1,i}, \mathbf{x}_{2,j}, t^n), \mathbf{u}(\mathbf{x}_{1,i+1}, \mathbf{x}_{2,j}, t^n); 0))\mathbf{e}_1 \, \triangle x_{2,j} \triangle t^{n+1/2} \tag{7.5a}$$

$$\mathbf{f}_{i,j+1/2}^{n+1/2} = \mathbf{F}(\mathcal{R}(\mathbf{u}(\mathbf{x}_{1,i}, \mathbf{x}_{2,j}, t^n), \mathbf{u}(\mathbf{x}_{1,i}, \mathbf{x}_{2,j+1}, t^n); 0))\mathbf{e}_2 \, \triangle x_{1,i} \triangle t^{n+1/2} \tag{7.5b}$$

and then perform a conservative difference. Unfortunately, this very natural scheme has a more restrictive stability restriction on its timestep than an operator split scheme.

To understand the reason for the stability restriction, we will consider linear advection with a constant velocity $\mathbf{v}$ in two dimensions:

$$\frac{\partial \mathbf{u}}{\partial t} + \mathbf{v}_1 \frac{\partial \mathbf{u}}{\partial \mathbf{x}_1} + \mathbf{v}_2 \frac{\partial \mathbf{u}}{\partial \mathbf{x}_2} = 0.$$

The donor cell scheme for this problem operates as follows. For each cell side

$i + \frac{1}{2}, j$ we compute a flux integral over the cell side and timestep by

$$\begin{aligned}\mathbf{f}^n_{i+1/2,j} &= [\mathbf{u}^n_{ij}\mathbf{v}_1^+ + \mathbf{u}^n_{i+1,j}\mathbf{v}_1^-]\triangle x_{2,j}\triangle t^{n+1/2} \\ &= \mathbf{F}(\mathcal{R}(\mathbf{u}^n_{ij}, \mathbf{u}^n_{i+1,j}; 0))\mathbf{e}_1\triangle x_{2,j}\triangle t^{n+1/2}\end{aligned}$$

and for each cell side $i, j + \frac{1}{2}$ we compute

$$\begin{aligned}\mathbf{f}^n_{i,j+1/2} &= [\mathbf{u}^n_{ij}\mathbf{v}_2^+ + \mathbf{u}^n_{i,j+1}\mathbf{v}_2^-]\triangle x_{1,i}\triangle t^{n+1/2} \\ &= \mathbf{F}(\mathcal{R}(\mathbf{u}^n_{ij}, \mathbf{u}^n_{i,j+1}; 0))\mathbf{e}_2\triangle x_{1,i}\triangle t^{n+1/2}.\end{aligned}$$

(As usual, the plus/minus superscripts on the velocity components denote the positive/negative parts of their values, and $\mathcal{R}$ denotes the solution of a Riemann problem.) Then we perform the conservative difference (7.4). Note that the donor cell scheme for linear advection can be rewritten

$$\begin{aligned}\mathbf{u}^{n+1}_{ij}\triangle x_{1,i}\triangle x_{2,j} &= \mathbf{u}^n_{ij}\left(\triangle x_{1,i}\triangle x_{2,j} - |\mathbf{v}_1|\triangle x_{2,j}\triangle t^{n+1/2} - |\mathbf{v}_2|\triangle x_{1,i}\triangle t^{n+1/2}\right) \\ &+ \mathbf{u}^n_{i-1,j}\mathbf{v}_1^+\triangle x_{2,j}\triangle t^{n+1/2} - \mathbf{u}^n_{i+1,j}\mathbf{v}_1^-\triangle x_{2,j}\triangle t^{n+1/2} \\ &+ \mathbf{u}^n_{i,j-1}\mathbf{v}_2^+\triangle x_{1,i}\triangle t^{n+1/2} - \mathbf{u}^n_{i,j+1}\mathbf{v}_2^-\triangle x_{1,i}\triangle t^{n+1/2}.\end{aligned}$$

By examining the coefficients in this equation, we see that the new solution is a weighted average of values of the previous solution provided that for all values of the cell indices i, j the timestep satisfies

$$\left[\frac{|\mathbf{v}_1|}{\triangle x_{1,i}} + \frac{|\mathbf{v}_2|}{\triangle x_{2,j}}\right]\triangle t^{n+1/2} \leq 1. \tag{7.6}$$

If this inequality is violated, then the scheme is unstable. This stability restriction caused early researchers to favor operator-split methods over unsplit methods.

7.1.2.2 First-Order Corner Transport Upwind Method

The donor cell timestep restriction can be improved by modifying the scheme to include the effects of diagonal flow. This approach, called **corner transport upwind**, is due to Colella [33]. First, we will develop this scheme for linear advection. Recall that the solution of the linear advection problem is $\mathbf{u}(\mathbf{x}, t) = \mathbf{u}_0(\mathbf{x} - \mathbf{v}t)$. If we integrate the conservation laws over the grid cell $\Omega_{ij} = (\mathbf{x}_{1,i-1/2}, \mathbf{x}_{1,i+1/2}) \times (\mathbf{x}_{2,j-1/2}, \mathbf{x}_{2,j+1/2})$, then the corner transport upwind scheme computes

$$\int_{\Omega_{ij}} \mathbf{u}^{n+1}(\mathbf{x})\, d\mathbf{x} = \int_{\Omega_{ij}} \mathbf{u}^n(\mathbf{x} - \mathbf{v}\triangle t^{n+1/2})\, d\mathbf{x} = \int_{R_{ij}} \mathbf{u}^n(\mathbf{x})\, d\mathbf{x}$$

where

$$\begin{aligned}R_{ij} &= (\mathbf{x}_{1,i-1/2} - \mathbf{v}_1\triangle t^{n+1/2}, \mathbf{x}_{1,i+1/2} - \mathbf{v}_1\triangle t^{n+1/2}) \\ &\times (\mathbf{x}_{2,j-1/2} - \mathbf{v}_2\triangle t^{n+1/2}, \mathbf{x}_{2,j+1/2} - \mathbf{v}_2\triangle t^{n+1/2})\end{aligned}$$

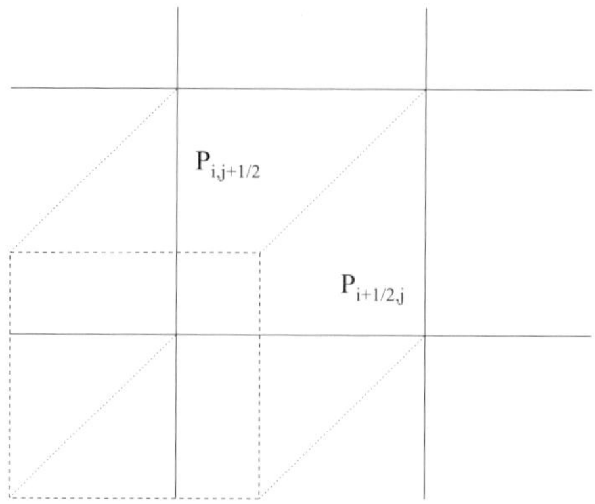

Fig. 7.2 Corner transport upwind: solid lines enclose grid cell Ω_{ij}, dashed lines enclose transported grid cell R_{ij}, dotted lines mark parallelograms $P_{i+1/2,j}$ and $P_{i,j+1/2}$

represents Ω_{ij} translated back in time along the velocity field. If $P_{i\pm1/2,j}$ and $P_{i,j\pm1/2}$ are signed parallelograms associated with the velocity field at the cell sides, then

$$R_{ij} = \Omega_{ij} - P_{i+1/2,j} + P_{i-1/2,j} - P_{i,j+1/2} + P_{i,j-1/2}.$$

(See Figure 7.2.) As a result,

$$\begin{aligned}\int_{\Omega_{ij}} \mathbf{u}^{n+1}(\mathbf{x})\, d\mathbf{x} = \int_{\Omega_{ij}} \mathbf{u}^{n}(\mathbf{x})\, d\mathbf{x} - \int_{P_{i+1/2,j}} \mathbf{u}^{n}(\mathbf{x})\, d\mathbf{x} + \int_{P_{i-1/2,j}} \mathbf{u}^{n}(\mathbf{x})\, d\mathbf{x} \\ - \int_{P_{i,j+1/2}} \mathbf{u}^{n}(\mathbf{x})\, d\mathbf{x} + \int_{P_{i,j-1/2}} \mathbf{u}^{n}(\mathbf{x})\, d\mathbf{x}.\end{aligned} \tag{7.7}$$

In order to compute these integrals, we will use the fact that $\mathbf{u}^n$ is piecewise constant on the individual grid cells. The equation $\int_{\Omega_{ij}} \mathbf{u}^{n+1}(\mathbf{x})\, d\mathbf{x} = \int_{R_{ij}} \mathbf{u}^{n}(\mathbf{x})\, d\mathbf{x}$ can be written as

$$\begin{aligned}\mathbf{u}_{ij}^{n+1} \triangle x_{1,i} \triangle x_{2,j} = {} & \mathbf{u}_{ij}^{n} \left(\triangle x_{1,i} - |\mathbf{v}_1| \triangle t^{n+1/2}\right) \left(\triangle x_{2,j} - |\mathbf{v}_2| \triangle t^{n+1/2}\right) \\ & + \mathbf{u}_{i-1,j}^{n} \mathbf{v}_1^{+} \triangle t^{n+1/2} \left(\triangle x_{2,j} - |\mathbf{v}_2| \triangle t^{n+1/2}\right) \\ & + \mathbf{u}_{i+1,j}^{n} \left(-\mathbf{v}_1^{-} \triangle t^{n+1/2}\right) \left(\triangle x_{2,j} - |\mathbf{v}_2| \triangle t^{n+1/2}\right) \\ & + \mathbf{u}_{i,j-1}^{n} \left(\triangle x_{1,i} - |\mathbf{v}_1| \triangle t^{n+1/2}\right) \mathbf{v}_2^{+} \triangle t^{n+1/2} \\ & + \mathbf{u}_{i,j+1}^{n} \left(\triangle x_{1,i} - |\mathbf{v}_1| \triangle t^{n+1/2}\right) \left(-\mathbf{v}_2^{-} \triangle t^{n+1/2}\right) \\ & + \mathbf{u}_{i-1,j-1}^{n} \mathbf{v}_1^{+} \triangle t^{n+1/2} \mathbf{v}_2^{+} \triangle t^{n+1/2} \\ & + \mathbf{u}_{i+1,j-1}^{n} \left(-\mathbf{v}_1^{-} \triangle t^{n+1/2}\right) \mathbf{v}_2^{+} \triangle t^{n+1/2} \\ & + \mathbf{u}_{i-1,j+1}^{n} \mathbf{v}_1^{+} \triangle t^{n+1/2} \left(-\mathbf{v}_2^{-} \triangle t^{n+1/2}\right) \\ & + \mathbf{u}_{i+1,j+1}^{n} \left(-\mathbf{v}_1^{-} \triangle t^{n+1/2}\right) \left(-\mathbf{v}_2^{-} \triangle t^{n+1/2}\right).\end{aligned} \tag{7.8}$$

It is easy to see that the new solution is a weighted average of old solution values if and only if the timestep is chosen so that in every grid cell Ω_{ij}

$$\max\left\{\frac{|\mathbf{v}_1|}{\triangle x_{1,i}}, \frac{|\mathbf{v}_2|}{\triangle x_{2,j}}\right\} \triangle t^{n+1/2} \leq 1.$$

This stability condition for corner transport upwind is less restrictive than the donor cell condition (7.6), and similar to the stability restriction for operator splitting.

In order to write corner transport upwind in the form (7.7) we can expand and collect terms in (7.8) to get

$$\begin{aligned}
\mathbf{u}_{ij}^{n+1} \triangle x_{1,i} \triangle x_{2,j} &= \mathbf{u}_{ij}^{n} \triangle x_{1,i} \triangle x_{2,j} \\
&- \left\{\left[\mathbf{u}_{ij}^{n}\left(\triangle x_{2,j} - |\mathbf{v}_2|\frac{\triangle t^{n+1/2}}{2}\right) + \mathbf{u}_{i,j-1}^{n}\mathbf{v}_2^{+}\frac{\triangle t^{n+1/2}}{2} - \mathbf{u}_{i,j+1}^{n}\mathbf{v}_2^{-}\frac{\triangle t^{n+1/2}}{2}\right]\mathbf{v}_1^{+}\triangle t^{n+1/2}\right. \\
&+ \left.\left[\mathbf{u}_{i+1j}^{n}\left(\triangle x_{2,j} - |\mathbf{v}_2|\frac{\triangle t^{n+1/2}}{2}\right) + \mathbf{u}_{i+1,j-1}^{n}\mathbf{v}_2^{+}\frac{\triangle t^{n+1/2}}{2} - \mathbf{u}_{i+1,j+1}^{n}\mathbf{v}_2^{-}\frac{\triangle t^{n+1/2}}{2}\right]\mathbf{v}_1^{-}\triangle t^{n+1/2}\right\} \\
&+ \left\{\left[\mathbf{u}_{i-1,j}^{n}\left(\triangle x_{2,j} - |\mathbf{v}_2|\frac{\triangle t^{n+1/2}}{2}\right) + \mathbf{u}_{i-1,j-1}^{n}\mathbf{v}_2^{+}\frac{\triangle t^{n+1/2}}{2} - \mathbf{u}_{i-1,j+1}^{n}\mathbf{v}_2^{-}\frac{\triangle t^{n+1/2}}{2}\right]\mathbf{v}_1^{+}\triangle t^{n+1/2}\right. \\
&+ \left.\left[\mathbf{u}_{ij}^{n}\left(\triangle x_{2,j} - |\mathbf{v}_2|\frac{\triangle t^{n+1/2}}{2}\right) + \mathbf{u}_{i,j-1}^{n}\mathbf{v}_2^{+}\frac{\triangle t^{n+1/2}}{2} - \mathbf{u}_{i,j+1}^{n}\mathbf{v}_2^{-}\frac{\triangle t^{n+1/2}}{2}\right]\mathbf{v}_1^{-}\triangle t^{n+1/2}\right\} \\
&- \left\{\left[\mathbf{u}_{ij}^{n}\left(\triangle x_{1,i} - |\mathbf{v}_1|\frac{\triangle t^{n+1/2}}{2}\right) + \mathbf{u}_{i-1,j}^{n}\mathbf{v}_1^{+}\frac{\triangle t^{n+1/2}}{2} - \mathbf{u}_{i+1,j}^{n}\mathbf{v}_1^{-}\frac{\triangle t^{n+1/2}}{2}\right]\mathbf{v}_2^{+}\triangle t^{n+1/2}\right. \\
&+ \left.\left[\mathbf{u}_{i,j+1}^{n}\left(\triangle x_{1,i} - |\mathbf{v}_1|\frac{\triangle t^{n+1/2}}{2}\right) + \mathbf{u}_{i-1,j+1}^{n}\mathbf{v}_1^{+}\frac{\triangle t^{n+1/2}}{2} - \mathbf{u}_{i+1,j+1}^{n}\mathbf{v}_1^{-}\frac{\triangle t^{n+1/2}}{2}\right]\mathbf{v}_2^{-}\triangle t^{n+1/2}\right\} \\
&+ \left\{\left[\mathbf{u}_{i,j-1}^{n}\left(\triangle x_{1,i} - |\mathbf{v}_1|\frac{\triangle t^{n+1/2}}{2}\right) + \mathbf{u}_{i-1,j-1}^{n}\mathbf{v}_1^{+}\frac{\triangle t^{n+1/2}}{2} - \mathbf{u}_{i+1,j-1}^{n}\mathbf{v}_1^{-}\frac{\triangle t^{n+1/2}}{2}\right]\mathbf{v}_2^{+}\triangle t^{n+1/2}\right. \\
&+ \left.\left[\mathbf{u}_{ij}^{n}\left(\triangle x_{1,i} - |\mathbf{v}_1|\frac{\triangle t^{n+1/2}}{2}\right) + \mathbf{u}_{i-1,j}^{n}\mathbf{v}_1^{+}\frac{\triangle t^{n+1/2}}{2} - \mathbf{u}_{i+1,j}^{n}\mathbf{v}_1^{-}\frac{\triangle t^{n+1/2}}{2}\right]\mathbf{v}_2^{-}\triangle t^{n+1/2}\right\}.
\end{aligned} \tag{7.9}$$

This equation allows us to interpret the corner transport upwind scheme as a conservative difference. We define the flux side and time integrals in the first coordinate direction by

$$\begin{aligned}
&\mathbf{f}_{i+1/2,j}^{n+1/2} \Delta X_{z,j} \Delta t^{n+1/2} \\
&= \left[\mathbf{u}_{ij}^{n}\left(\triangle x_{2,j} - |\mathbf{v}_2|\frac{\triangle t^{n+1/2}}{2}\right) + \mathbf{u}_{i,j-1}^{n}\mathbf{v}_2^{+}\frac{\triangle t^{n+1/2}}{2} - \mathbf{u}_{i,j+1}^{n}\mathbf{v}_2^{-}\frac{\triangle t^{n+1/2}}{2}\right]\mathbf{v}_1^{+}\triangle t^{n+1/2} \\
&+ \left[\mathbf{u}_{i+1,j}^{n}\left(\triangle x_{2,j} - |\mathbf{v}_2|\frac{\triangle t^{n+1/2}}{2}\right) + \mathbf{u}_{i+1,j-1}^{n}\mathbf{v}_2^{+}\frac{\triangle t^{n+1/2}}{2} - \mathbf{u}_{i+1,j+1}^{n}\mathbf{v}_2^{-}\frac{\triangle t^{n+1/2}}{2}\right]\mathbf{v}_1^{-}\triangle t^{n+1/2}
\end{aligned}$$

$$
\begin{aligned}
&= \left(\mathbf{u}^n_{ij} - \left[\left\{ \mathbf{u}^n_{ij} \mathbf{v}_2^+ + \mathbf{u}^n_{i,j+1} \mathbf{v}_2^- \right\} \triangle x_{1,i} \triangle t^{n+1/2} \right.\right. \\
&\qquad \left.\left. - \left\{ \mathbf{u}^n_{i,j-1} \mathbf{v}_2^+ + \mathbf{u}^n_{ij} \mathbf{v}_2^- \right\} \triangle x_{1,i} \triangle t^{n+1/2} \right] \frac{1}{2 \triangle x_{1,i} \triangle x_{2,j}} \right) \mathbf{v}_1^+ \triangle t^{n+1/2} \triangle x_{2,j} \\
&\quad + \left(\mathbf{u}^n_{i+1,j} - \left[\left\{ \mathbf{u}^n_{i+1,j} \mathbf{v}_2^+ + \mathbf{u}^n_{i+1,j+1} \mathbf{v}_2^- \right\} \triangle x_{1,i+1} \triangle t^{n+1/2} \right.\right. \\
&\qquad \left.\left. - \left\{ \mathbf{u}^n_{i+1,j-1} \mathbf{v}_2^+ + \mathbf{u}^n_{i+1,j} \mathbf{v}_2^- \right\} \triangle x_{1,i+1} \triangle t^{n+1/2} \right] \frac{1}{2 \triangle x_{1,i+1} \triangle x_{2,j}} \right) \mathbf{v}_1^- \triangle t^{n+1/2} \triangle x_{2,j} .
\end{aligned}
$$

Note that the final expression shows that $\mathbf{f}^{n+1/2}_{i+1/2,j}$ is the flux side and time integral with flux evaluated at the solution of a Riemann problem. Similar expressions hold for the fluxes in the second coordinate direction.

In summary, the corner transport upwind scheme for linear advection involves the following steps. First, we compute transverse flux side and time integrals by solving Riemann problems at the cell sides, using the cell averages:

$$
\begin{aligned}
\mathbf{f}^n_{i+1/2,j} &= \mathbf{F}\left(\mathcal{R}\left(\mathbf{u}^n_{ij}, \mathbf{u}^n_{i+1,j}; 0 \right) \right) \mathbf{e}_1 \triangle x_{2,j} \triangle t^{n+1/2} \\
\mathbf{f}^n_{i,j+1/2} &= \mathbf{F}\left(\mathcal{R}\left(\mathbf{u}^n_{ij}, \mathbf{u}^n_{i,j+1}; 0 \right) \right) \mathbf{e}_2 \triangle x_{1,i} \triangle t^{n+1/2} .
\end{aligned}
$$

Then we compute left and right states at the cell sides by

$$
\mathbf{u}^{n+1/2,L}_{i+1/2,j} = \mathbf{u}^n_{ij} - \left[\mathbf{f}^n_{i,j+1/2} - \mathbf{f}^n_{i,j-1/2} \right] \frac{1}{2 \triangle x_{1,i} \triangle x_{2,j}} \tag{7.10a}
$$

$$
\mathbf{u}^{n+1/2,R}_{i+1/2,j} = \mathbf{u}^n_{i+1,j} - \left[\mathbf{f}^n_{i+1,j+1/2} - \mathbf{f}^n_{i+1,j-1/2} \right] \frac{1}{2 \triangle x_{i+1} \triangle x_{2,j}} \tag{7.10b}
$$

$$
\mathbf{u}^{n+1/2,L}_{i,j+1/2} = \mathbf{u}^n_{ij} - \left[\mathbf{f}^n_{i+1/2,j} - \mathbf{f}^n_{i-1/2,j} \right] \frac{1}{2 \triangle x_{1,i} \triangle x_{2,j}} \tag{7.10c}
$$

$$
\mathbf{u}^{n+1/2,R}_{i,j+1/2} = \mathbf{u}^n_{i,j+1} - \left[\mathbf{f}^n_{i+1/2,j+1} - \mathbf{f}^n_{i-1/2,j+1} \right] \frac{1}{2 \triangle x_{1,i} \triangle x_{2,j}} . \tag{7.10d}
$$

These allow us to compute flux side and time integrals associated with Riemann problems at the cell sides:

$$
\begin{aligned}
\mathbf{f}^{n+1/2}_{i+1/2,j} &= \mathbf{F}\left(\mathcal{R}\left(\mathbf{u}^{n+1/2,L}_{i+1/2,j}, \mathbf{u}^{n+1/2,R}_{i+1/2,j}; 0 \right) \right) \mathbf{e}_1 \triangle x_{2,j} \triangle t^{n+1/2} \\
\mathbf{f}^{n+1/2}_{i,j+1/2} &= \mathbf{F}\left(\mathcal{R}\left(\mathbf{u}^{n+1/2,L}_{i,j+1/2}, \mathbf{u}^{n+1/2,R}_{i,j+1/2}; 0 \right) \right) \mathbf{e}_2 \triangle x_{1,i} \triangle t^{n+1/2} .
\end{aligned}
$$

Finally, we can perform the conservative difference (7.4).

7.1.2.3 Wave Propagation Form of First-Order Corner Transport Upwind

LeVeque [94] has described an alternative form of the corner transport upwind scheme. He would rewrite (7.8) in the form

$$
\begin{aligned}
\mathbf{u}_{ij}^{n+1} \triangle x_{1,i} \triangle x_{2,j} = \mathbf{u}_{ij}^{n} \triangle x_{1,i} \triangle x_{2,j} \\
& - \Big[\left(\mathbf{u}_{ij}^{n} - \mathbf{u}_{i-1,j}^{n}\right) \mathbf{v}_1^{+} \left(\triangle x_{2,j} - |\mathbf{v}_2| \frac{\triangle t^{n+1/2}}{2} \right) \triangle t^{n+1/2} \\
& + \left(\mathbf{u}_{i-1,j}^{n} - \mathbf{u}_{i-1,j-1}^{n}\right) \mathbf{v}_2^{+} \left(\mathbf{v}_1^{+} \frac{\triangle t^{n+1/2}}{2} \right) \triangle t^{n+1/2} \\
& + \left(\mathbf{u}_{i-1,j+1}^{n} - \mathbf{u}_{i-1,j}^{n}\right) \mathbf{v}_2^{+} \left(\mathbf{v}_1^{+} \frac{\triangle t^{n+1/2}}{2} \right) \triangle t^{n+1/2} \Big] \\
& - \Big[\left(\mathbf{u}_{i+1,j}^{n} - \mathbf{u}_{ij}^{n}\right) \mathbf{v}_1^{-} \left(\triangle x_{2,j} - |\mathbf{v}_2| \frac{\triangle t^{n+1/2}}{2} \right) \triangle t^{n+1/2} \\
& - \left(\mathbf{u}_{i+1,j}^{n} - \mathbf{u}_{i+1,j-1}^{n}\right) \mathbf{v}_2^{+} \left(\mathbf{v}_1^{-} \frac{\triangle t^{n+1/2}}{2} \right) \triangle t^{n+1/2} \\
& - \left(\mathbf{u}_{i+1,j+1}^{n} - \mathbf{u}_{i+1,j}^{n}\right) \mathbf{v}_2^{-} \left(\mathbf{v}_1^{-} \frac{\triangle t^{n+1/2}}{2} \right) \triangle t^{n+1/2} \Big] \\
& - \Big[\left(\mathbf{u}_{ij}^{n} - \mathbf{u}_{i,j-1}^{n}\right) \mathbf{v}_2^{+} \left(\triangle x_{1,i} - |\mathbf{v}_1| \frac{\triangle t^{n+1/2}}{2} \right) \triangle t^{n+1/2} \\
& + \left(\mathbf{u}_{i,j-1}^{n} - \mathbf{u}_{i-1,j-1}^{n}\right) \mathbf{v}_1^{+} \left(\mathbf{v}_2^{+} \frac{\triangle t^{n+1/2}}{2} \right) \triangle t^{n+1/2} \\
& + \left(\mathbf{u}_{i+1,j-1}^{n} - \mathbf{u}_{i,j-1}^{n}\right) \mathbf{v}_1^{-} \left(\mathbf{v}_2^{+} \frac{\triangle t^{n+1/2}}{2} \right) \triangle t^{n+1/2} \Big] \\
& - \Big[\left(\mathbf{u}_{i,j+1}^{n} - \mathbf{u}_{ij}^{n}\right) \mathbf{v}_2^{-} \left(\triangle x_{1,i} - |\mathbf{v}_1| \frac{\triangle t^{n+1/2}}{2} \right) \triangle t^{n+1/2} \\
& + \left(\mathbf{u}_{i,j+1}^{n} - \mathbf{u}_{i-1,j+1}^{n}\right) \mathbf{v}_1^{+} \left(\mathbf{v}_2^{-} \frac{\triangle t^{n+1/2}}{2} \right) \triangle t^{n+1/2} \\
& + \left(\mathbf{u}_{i+1,j+1}^{n} - \mathbf{u}_{i,j+1}^{n}\right) \mathbf{v}_1^{-} \left(\mathbf{v}_2^{-} \frac{\triangle t^{n+1/2}}{2} \right) \triangle t^{n+1/2} \Big]. \qquad (7.11)
\end{aligned}
$$

The terms in the first set of square brackets have the interpretation as the wave through side $i - \frac{1}{2}$, j times the area in cell ij swept by this wave, plus corrections for waves from sides $i-1$, $j - \frac{1}{2}$ and $i-1$, $j + \frac{1}{2}$. The other terms in the square brackets have similar interpretations for the other sides of cell i, j. The computation is actually performed as a modification of the donor cell scheme:

$$\begin{aligned}\mathbf{u}_{ij}^{n+1} &= \mathbf{u}_{ij}^{n} - \triangle(\mathbf{Fe}_1)_{ij}^{n}\frac{\triangle t^{n+1/2}}{\triangle x_{1,i}} - \triangle(\mathbf{Fe}_2)_{ij}^{n}\frac{\triangle t^{n+1/2}}{\triangle x_{2,j}}\\ &\quad + \triangle^2(\mathbf{Fe}_1)_{i+1/2,j}\frac{\triangle t^{n+1/2}}{\triangle x_{1,i}} - \triangle^2(\mathbf{Fe}_1)_{i-1/2,j}\frac{\triangle t^{n+1/2}}{\triangle x_{1,i}}\\ &\quad + \triangle^2(\mathbf{Fe}_2)_{i,j+1/2}\frac{\triangle t^{n+1/2}}{\triangle x_{2,j}} - \triangle^2(\mathbf{Fe}_2)_{i,j-1/2}\frac{\triangle t^{n+1/2}}{\triangle x_{2,j}}.\end{aligned}$$

Here the donor cell flux differences are

$$\begin{aligned}\triangle(\mathbf{Fe}_1)_{ij}^{n} &= \left(\mathbf{u}_{i+1,j}^{n} - \mathbf{u}_{ij}^{n}\right)\mathbf{v}_1^- + \left(\mathbf{u}_{ij}^{n} - \mathbf{u}_{i-1,j}^{n}\right)\mathbf{v}_1^+\\ &= \left[\mathbf{F}\left(\mathcal{R}\left(\mathbf{u}_{i+1,j}^{n}, \mathbf{u}_{ij}^{n}; 0\right)\right)\mathbf{e}_1 - \mathbf{F}\left(\mathbf{u}_{ij}^{n}\right)\mathbf{e}_1\right] + \left[\mathbf{F}\left(\mathbf{u}_{ij}^{n}\right)\mathbf{e}_1 - \mathbf{F}\left(\mathcal{R}\left(\mathbf{u}_{ij}^{n}, \mathbf{u}_{i-1,j}^{n}; 0\right)\right)\mathbf{e}_1\right]\\ \triangle(\mathbf{Fe}_2)_{ij}^{n} &= \left(\mathbf{u}_{i,j+1}^{n} - \mathbf{u}_{ij}^{n}\right)\mathbf{v}_2^- + \left(\mathbf{u}_{ij}^{n} - \mathbf{u}_{i,j-1}^{n}\right)\mathbf{v}_2^+\\ &= \left[\mathbf{F}\left(\mathcal{R}\left(\mathbf{u}_{i,j+1}^{n}, \mathbf{u}_{ij}^{n}; 0\right)\right)\mathbf{e}_2 - \mathbf{F}\left(\mathbf{u}_{ij}^{n}\right)\mathbf{e}_2\right] + \left[\mathbf{F}\left(\mathbf{u}_{ij}^{n}\right)\mathbf{e}_2 - \mathbf{F}\left(\mathcal{R}\left(\mathbf{u}_{ij}^{n}, \mathbf{u}_{i,j-1}^{n}; 0\right)\right)\mathbf{e}_2\right].\end{aligned}$$

These are interpreted as waves (in this case differences in $\mathbf{u}$) times speeds (in this case plus/minus parts of the velocity components). The waves are computed at the cell sides, and stored as flux differences in the grid cells. The flux corrections are

$$\begin{aligned}\triangle^2(\mathbf{Fe}_1)_{i+1/2,j} &= \left(\mathbf{u}_{ij}^{n} - \mathbf{u}_{i,j-1}^{n}\right)\mathbf{v}_1^+\mathbf{v}_2^+\frac{\triangle t^{n+1/2}}{2\triangle x_{2,j}} + \left(\mathbf{u}_{i+1,j}^{n} - \mathbf{u}_{i+1,j-1}^{n}\right)\mathbf{v}_1^-\mathbf{v}_2^+\frac{\triangle t^{n+1/2}}{2\triangle x_{2,j}}\\ &\quad + \left(\mathbf{u}_{i,j+1}^{n} - \mathbf{u}_{ij}^{n}\right)\mathbf{v}_1^+\mathbf{v}_2^-\frac{\triangle t^{n+1/2}}{2\triangle x_{2,j}} + \left(\mathbf{u}_{i+1,j+1}^{n} - \mathbf{u}_{i+1,j}^{n}\right)\mathbf{v}_1^-\mathbf{v}_2^-\frac{\triangle t^{n+1/2}}{2\triangle x_{2,j}}\\ &= \left[\left\{\mathbf{F}\left(\mathbf{u}_{ij}^{n}\right)\mathbf{e}_2 - \mathbf{F}\left(\mathcal{R}\left(\mathbf{u}_{ij}^{n}, \mathbf{u}_{i,j-1}^{n}; 0\right)\right)\mathbf{e}_2\right\}\right.\\ &\quad \left. + \left\{\mathbf{F}\left(\mathcal{R}\left(\mathbf{u}_{i,j+1}^{n}, \mathbf{u}_{ij}^{n}; 0\right)\right)\mathbf{e}_2 - \mathbf{F}\left(\mathbf{u}_{ij}^{n}\right)\mathbf{e}_2\right\}\right]\frac{\triangle t^{n+1/2}}{2\triangle x_{2,j}}\mathbf{v}_1^+\\ &\quad + \left[\left\{\mathbf{F}\left(\mathbf{u}_{i+1,j}^{n}\right)\mathbf{e}_2 - \mathbf{F}\left(\mathcal{R}\left(\mathbf{u}_{i+1,j}^{n}, \mathbf{u}_{i+1,j-1}^{n}; 0\right)\right)\mathbf{e}_2\right\}\right.\\ &\quad \left. + \left\{\mathbf{F}\left(\mathcal{R}\left(\mathbf{u}_{i+1,j+1}^{n}, \mathbf{u}_{i+1,j}^{n}; 0\right)\right)\mathbf{e}_2 - \mathbf{F}\left(\mathbf{u}_{i+1,j}^{n}\right)\mathbf{e}_2\right\}\right]\frac{\triangle t^{n+1/2}}{2\triangle x_{2,j}}\mathbf{v}_1^-\\ \triangle^2(\mathbf{Fe}_2)_{i,j+1/2} &= \left(\mathbf{u}_{ij}^{n} - \mathbf{u}_{i-1,j}^{n}\right)\mathbf{v}_1^+\mathbf{v}_2^+\frac{\triangle t^{n+1/2}}{2\triangle x_{1,i}} + \left(\mathbf{u}_{i,j+1}^{n} - \mathbf{u}_{i-1,j+1}^{n}\right)\mathbf{v}_1^+\mathbf{v}_2^-\frac{\triangle t^{n+1/2}}{2\triangle x_{1,i}}\\ &\quad + \left(\mathbf{u}_{i+1,j}^{n} - \mathbf{u}_{ij}^{n}\right)\mathbf{v}_1^-\mathbf{v}_2^+\frac{\triangle t^{n+1/2}}{2\triangle x_{1,i}} + \left(\mathbf{u}_{i+1,j+1}^{n} - \mathbf{u}_{i,j+1}^{n}\right)\mathbf{v}_1^-\mathbf{v}_2^-\frac{\triangle t^{n+1/2}}{2\triangle x_{1,i}}\\ &= \left[\left\{\mathbf{F}\left(\mathbf{u}_{ij}^{n}\right)\mathbf{e}_1 - \mathbf{F}\left(\mathcal{R}\left(\mathbf{u}_{ij}^{n}, \mathbf{u}_{i-1,j}^{n}; 0\right)\right)\mathbf{e}_1\right\}\right.\\ &\quad \left. + \left\{\mathbf{F}\left(\mathcal{R}\left(\mathbf{u}_{i+1,j}^{n}, \mathbf{u}_{ij}^{n}; 0\right)\right)\mathbf{e}_1 - \mathbf{F}\left(\mathbf{u}_{ij}^{n}\right)\mathbf{e}_1\right\}\right]\frac{\triangle t^{n+1/2}}{2\triangle x_{1,i}}\mathbf{v}_2^+\\ &\quad + \left[\left\{\mathbf{F}\left(\mathbf{u}_{i,j+1}^{n}\right)\mathbf{e}_1 - \mathbf{F}\left(\mathcal{R}\left(\mathbf{u}_{i,j+1}^{n}, \mathbf{u}_{i-1,j+1}^{n}; 0\right)\right)\mathbf{e}_1\right\}\right.\\ &\quad \left. + \left\{\mathbf{F}\left(\mathcal{R}\left(\mathbf{u}_{i+1,j+1}^{n}, \mathbf{u}_{i,j+1}^{n}; 0\right)\right)\mathbf{e}_1 - \mathbf{F}\left(\mathbf{u}_{i,j+1}^{n}\right)\mathbf{e}_1\right\}\right]\frac{\triangle t^{n+1/2}}{2\triangle x_{1,i}}\mathbf{v}_2^-.\end{aligned}$$

These flux corrections are assembled by looping over the cell sides, computing the waves, and storing the effects in the appropriate side locations. A Fortran implementation of this method appears in subroutine `step2` of CLAWPACK.

7.1.2.4 Second-Order Corner Transport Upwind Method

In order to form a second-order corner transport upwind method, it suffices to compute second-order accurate approximations to the flux side and time integrals in the integral form of the conservation law (7.3). Consider, for example, the integral

$$\mathbf{f}^{n+1/2}_{i+1/2,j} = \int_{t^n}^{t^{n+1}} \int_{\mathbf{x}_{2,j-1/2}}^{\mathbf{x}_{2,j+1/2}} \mathbf{F}\left(\mathbf{u}\left(\mathbf{x}_{1,i+1/2}, \mathbf{x}_2, t\right)\right) \mathbf{e}_1 \, d\mathbf{x}_2 \, dt.$$

We can use midpoint rule quadrature for both integrals, provided that we can determine a second-order approximation to $\mathbf{F}(\mathbf{u}(\mathbf{x}_{1,i+1/2}, \mathbf{x}_{2,j}, t^n + \triangle t^{n+1/2}/2))\mathbf{e}_1$. To this end, we consider the Taylor expansion

$$\begin{aligned}
\mathbf{u}\left(\mathbf{x}_1 + \frac{\triangle x_1}{2}, \mathbf{x}_2, t + \frac{\triangle t}{2}\right) &\approx \mathbf{u} + \frac{\partial \mathbf{u}}{\partial \mathbf{x}_1}\frac{\triangle x_1}{2} + \frac{\partial \mathbf{u}}{\partial t}\frac{\triangle t}{2} \\
&= \mathbf{u} + \frac{\partial \mathbf{u}}{\partial \mathbf{x}_1}\frac{\triangle x_1}{2} - \left(\frac{\partial \mathbf{u}}{\partial \mathbf{x}_1}\mathbf{v}_1 + \frac{\partial \mathbf{u}}{\partial \mathbf{x}_2}\mathbf{v}_2\right)\frac{\triangle t}{2} \\
&= \mathbf{u} - \left(\frac{\mathbf{v}_2 \triangle t}{\triangle x_2}\right)\frac{\partial \mathbf{u}}{\partial \mathbf{x}_2}\frac{\triangle x_2}{2} + \left(1 - \frac{\mathbf{v}_1 \triangle t}{\triangle x_1}\right)\frac{\partial \mathbf{u}}{\partial \mathbf{x}_1}\frac{\triangle x_1}{2}.
\end{aligned}$$

The first term on the right is the value used by the donor cell scheme in Equation (7.5). The first two terms provide the values used for the first-order corner transport upwind states in Equation (7.10). The third term must be added to achieve second-order accuracy in the corner transport upwind scheme.

Suppose that we want to solve the system of two-dimensional nonlinear conservation laws

$$\frac{\partial \mathbf{u}}{\partial t} + \frac{\partial \mathbf{F}\mathbf{e}_1}{\partial \mathbf{x}_1} + \frac{\partial \mathbf{F}\mathbf{e}_2}{\partial \mathbf{x}_2} = 0,$$

where $\mathbf{u}(\mathbf{w})$ and $\mathbf{F}(\mathbf{w})$ are functions of the flux variables $\mathbf{w}$. The MUSCL implementation of the second-order corner transport upwind scheme is the following. At the cell sides $i + \frac{1}{2}, j$ and $i, j + \frac{1}{2}$ compute the flux variable increments

$$\triangle \mathbf{w}^n_{i+1/2,j} = \mathbf{w}^n_{i+1,j} - \mathbf{w}^n_{ij} \quad \text{and} \quad \triangle \mathbf{w}^n_{i,j+1/2} = \mathbf{w}^n_{i,j+1} - \mathbf{w}^n_{ij}.$$

In each grid cell we compute the eigenvectors Y and eigenvalues Λ of the flux derivatives:

$$\frac{\partial \mathbf{Fe}_1}{\partial \mathbf{w}}\mathbf{Y}_1 = \frac{\partial \mathbf{u}}{\partial \mathbf{w}}\mathbf{Y}_1\Lambda_1 \quad \text{and} \quad \frac{\partial \mathbf{Fe}_2}{\partial \mathbf{w}}\mathbf{Y}_2 = \frac{\partial \mathbf{u}}{\partial \mathbf{w}}\mathbf{Y}_2\Lambda_2.$$

Of course, the matrices $\partial\mathbf{Fe}_1/\partial\mathbf{w}$, $\partial\mathbf{Fe}_2/\partial\mathbf{w}$, $\partial\mathbf{u}/\partial\mathbf{w}$, $\mathbf{Y}_1$, $\mathbf{Y}_2$ and Λ_1, Λ_2 vary with time and grid cell; these dependencies have been suppressed in the notation. Also in this grid cell, solve

$$\begin{aligned}
\mathbf{Y}_1\mathbf{z}_{i+1/2,j} &= \triangle\mathbf{w}^n_{i+1/2,j} \quad \text{and} \quad \mathbf{Y}_1\mathbf{z}_{i-1/2,j} = \triangle\mathbf{w}^n_{i-1/2,j}\\
\mathbf{Y}_2\mathbf{z}_{i,j+1/2} &= \triangle\mathbf{w}^n_{i,j+1/2} \quad \text{and} \quad \mathbf{Y}_2\mathbf{z}_{i,j-1/2} = \triangle\mathbf{w}^n_{i,j-1/2}.
\end{aligned}$$

Compute cell-centered slopes

$$\begin{aligned}
\tilde{\mathbf{z}}_{1,ij} &= \left\{\mathbf{z}_{i+1/2,j}\frac{\triangle x_{1,i}+2\triangle x_{1,i-1}}{\triangle x_{1,i}+\triangle x_{1,i+1}} + \mathbf{z}_{i-1/2,j}\frac{\triangle x_{1,i}+2\triangle x_{1,i+1}}{\triangle x_{1,i}+\triangle x_{1,i-1}}\right\}\frac{\triangle x_{1,i}}{\triangle x_{1,i-1}+\triangle x_{1,i}+\triangle x_{1,i+1}}\\
\tilde{\mathbf{z}}_{2,ij} &= \left\{\mathbf{z}_{i,j+1/2}\frac{\triangle x_{2,j}+2\triangle x_{2,j-1}}{\triangle x_{2,j}+\triangle x_{2,j+1}} + \mathbf{z}_{i,j-1/2}\frac{\triangle x_{2,j}+2\triangle x_{2,j+1}}{\triangle x_{2,j}+\triangle x_{2,j-1}}\right\}\frac{\triangle x_{2,j}}{\triangle x_{2,j-1}+\triangle x_{2,j}+\triangle x_{2,j+1}}
\end{aligned}$$

and componentwise the limited slopes

$$\mathbf{z}^n_{1,ij} = \mathrm{muscl}(\tilde{\mathbf{z}}_{1,ij}, \mathbf{z}_{i-1/2,j}, \mathbf{z}_{i+1/2,j}) \quad \text{and} \quad \mathbf{z}^n_{2,ij} = \mathrm{muscl}(\tilde{\mathbf{z}}_{2,ij}, \mathbf{z}_{i,j-1/2}, \mathbf{z}_{i,j+1/2}).$$

Still in cell i, j, use characteristic information to compute the transverse flux states at the cell sides:

$$\begin{aligned}
\mathbf{w}^{n,L}_{i+\frac{1}{2},j} &= \mathbf{w}^n_{ij} + \frac{1}{2}(\mathbf{Y}_1)_{ij}\left[\mathbf{I} - (\Lambda_1)_{ij}\frac{\triangle t^{n+1/2}}{\triangle x_{1,i}}\right]\mathbf{z}^n_{1,ij}\\
\mathbf{w}^{n,R}_{i-\frac{1}{2},j} &= \mathbf{w}^n_{ij} - \frac{1}{2}(\mathbf{Y}_1)_{ij}\left[\mathbf{I} + (\Lambda_1)_{ij}\frac{\triangle t^{n+1/2}}{\triangle x_{1,i}}\right]\mathbf{z}^n_{1,ij}\\
\mathbf{w}^{n,L}_{i,j+\frac{1}{2}} &= \mathbf{w}^n_{ij} + \frac{1}{2}(\mathbf{Y}_2)_{ij}\left[\mathbf{I} - (\Lambda_2)_{ij}\frac{\triangle t^{n+1/2}}{\triangle x_{2,j}}\right]\mathbf{z}^n_{2,ij}\\
\mathbf{w}^{n,R}_{i,j-\frac{1}{2}} &= \mathbf{w}^n_{ij} - \frac{1}{2}(\mathbf{Y}_2)_{ij}\left[\mathbf{I} + (\Lambda_2)_{ij}\frac{\triangle t^{n+1/2}}{\triangle x_{2,j}}\right]\mathbf{z}^n_{2,ij}.
\end{aligned}$$

Next, we use these corner transport upwind states to compute transverse flux integrals by solving Riemann problems at cell sides $i+\frac{1}{2}$, j and i, $j+\frac{1}{2}$:

$$\begin{aligned}
\mathbf{f}^n_{i+\frac{1}{2},j} &= \mathbf{F}\left(\mathcal{R}\left(\mathbf{w}^{n,L}_{i+\frac{1}{2},j}, \mathbf{w}^{n,R}_{i+\frac{1}{2},j}; 0\right)\right)\mathbf{e}_1\triangle x_{2,j}\triangle t^{n+1/2},\\
\mathbf{f}^n_{i,j+\frac{1}{2}} &= \mathbf{F}\left(\mathcal{R}\left(\mathbf{w}^{n,L}_{i,j+\frac{1}{2}}, \mathbf{w}^{n,R}_{i,j+\frac{1}{2}}; 0\right)\right)\mathbf{e}_2\triangle x_{1,i}\triangle t^{n+1/2}.
\end{aligned}$$

In each grid cell i, j, we correct the previous left and right states by incorporating the transverse fluxes:

$$\mathbf{w}_{i+\frac{1}{2},j}^{n+\frac{1}{2},L} = \mathbf{w}_{i+\frac{1}{2},j}^{n,L} - \left(\frac{\partial \mathbf{u}}{\partial \mathbf{w}}\right)_{ij}^{-1} \left[\mathbf{f}_{i,j+\frac{1}{2}}^{n} - \mathbf{f}_{i,j-\frac{1}{2}}^{n}\right] \frac{1}{2\triangle x_{1,i}\triangle x_{2,j}} \tag{7.12a}$$

$$\mathbf{w}_{i-\frac{1}{2},j}^{n+\frac{1}{2},R} = \mathbf{w}_{i-\frac{1}{2},j}^{n,R} - \left(\frac{\partial \mathbf{u}}{\partial \mathbf{w}}\right)_{ij}^{-1} \left[\mathbf{f}_{i,j+\frac{1}{2}}^{n} - \mathbf{f}_{i,j-\frac{1}{2}}^{n}\right] \frac{1}{2\triangle x_{1,i}\triangle x_{2,j}} \tag{7.12b}$$

$$\mathbf{w}_{i,j+\frac{1}{2}}^{n+\frac{1}{2},L} = \mathbf{w}_{i,j+\frac{1}{2}}^{n,L} - \left(\frac{\partial \mathbf{u}}{\partial \mathbf{w}}\right)_{ij}^{-1} \left[\mathbf{f}_{i+\frac{1}{2},j}^{n} - \mathbf{f}_{i-\frac{1}{2},j}^{n}\right] \frac{1}{2\triangle x_{1,i}\triangle x_{2,j}} \tag{7.12c}$$

$$\mathbf{w}_{i,j-\frac{1}{2}}^{n+\frac{1}{2},R} = \mathbf{w}_{i,j-\frac{1}{2}}^{n,R} - \left(\frac{\partial \mathbf{u}}{\partial \mathbf{w}}\right)_{ij}^{-1} \left[\mathbf{f}_{i+\frac{1}{2},j}^{n} - \mathbf{f}_{i-\frac{1}{2},j}^{n}\right] \frac{1}{2\triangle x_{1,i}\triangle x_{2,j}}. \tag{7.12d}$$

The conservative fluxes are computed by solving another set of Riemann problems at the sides $i+\frac{1}{2}, j$ and $i, j+\frac{1}{2}$:

$$\mathbf{f}_{i+\frac{1}{2},j}^{n+\frac{1}{2}} = \mathbf{F}\left(\mathcal{R}\left(\mathbf{w}_{i+\frac{1}{2},j}^{n+\frac{1}{2},L}, \mathbf{w}_{i+\frac{1}{2},j}^{n+\frac{1}{2},R}; 0\right)\right) \mathbf{e}_1 \triangle x_{2,j} \triangle t^{n+1/2},$$
$$\mathbf{f}_{i,j+\frac{1}{2}}^{n+\frac{1}{2}} = \mathbf{F}\left(\mathcal{R}\left(\mathbf{w}_{i,j+\frac{1}{2}}^{n+\frac{1}{2},L}, \mathbf{w}_{i,j+\frac{1}{2}}^{n+\frac{1}{2},R}; 0\right)\right) \mathbf{e}_2 \triangle x_{1,i} \triangle t^{n+1/2}.$$

Finally, we perform the conservative difference (7.4).

This second-order corner transport upwind scheme is subject to the stability restriction

$$\triangle t \leq \min_{ij}\{\triangle x_{1,i}/\|\Lambda_1\|_\infty, \ \triangle x_{2,j}/\|\Lambda_2\|_\infty\}.$$

The difficulty is that the second-order corner transport upwind scheme requires an average of four Riemann problems solutions per cell. In contrast, second-order operator splitting has the same order and stability restriction, but requires only an average of two Riemann problem solutions per cell, provided that the last half-step of the prior operator split step is combined with the first half-step of the subsequent split step.

We have implemented the second-order corner transport upwind scheme in **Program 7.1-113: ctu2d.m4.** Students can execute this scheme in two dimensions by clicking on **Executable 7.1-51: guiRectangle.** The student can select either of Burgers' equation, shallow water or gas dynamics under '`Riemann Problem Parameters`. Students should select `unsplit` for the `splitting` under `Numerical Method parameters`, and `Godunov` should have the value `True`. In two dimensions, the computational results for scalar fields (such as water height in shallow water, or pressure in gas dynamics) can be displayed either as 2D contours, 2D color fills or 3D surface plots. The 3D graphics in the surface plot uses a trackball for rotation with the left mouse button. The figure can be sliced along the ends of any coordinate axis using the middle mouse button. Values at points in the figure can be determined using the right mouse button.

7.1.3 Wave Propagation

It would be straightforward to implement the MUSCL scheme above as a wave propagation scheme, by replacing the flux differences in (7.12) and (7.4) with the wave-field decompositions from a Riemann problem solver, as in Section 6.2.6 above. LeVeque [97] has adopted a different approach for the two-dimensional conservation law. For each cell side he decomposes the flux differences into waves, typically using a Roe decomposition:

$$\begin{aligned}(\mathbf{Fe}_1)^n_{i+1,j} - (\mathbf{Fe}_1)^n_{ij} &= \mathbf{X}_{i+1/2,j}\Lambda_{i+1/2,j}\mathbf{X}^{-1}_{i+1/2,j}\left(\mathbf{u}^n_{i+1,j} - \mathbf{u}^n_{ij}\right)\\(\mathbf{Fe}_2)^n_{i,j+1} - (\mathbf{Fe}_2)^n_{ij} &= \mathbf{X}_{i,j+1/2}\Lambda_{i,j+1/2}\mathbf{X}^{-1}_{i,j+1/2}\left(\mathbf{u}^n_{i,j+1} - \mathbf{u}^n_{ij}\right)\end{aligned}$$

These equations imply that Riemann problem solutions would provide flux differences

$$\begin{aligned}\mathbf{F}\left(\mathcal{R}\left(\mathbf{u}^n_{ij},\mathbf{u}^n_{i+1,j};0\right)\right)\mathbf{e}_1 - \mathbf{F}\left(\mathbf{u}^n_{ij}\right)\mathbf{e}_1 &= \mathbf{X}_{i+1/2,j}\left(\Lambda_{i+1/2,j}\right)^-\left(\mathbf{X}_{i+1/2,j}\right)^{-1}\left(\mathbf{u}^n_{i+1,j} - \mathbf{u}^n_{ij}\right)\\&\equiv \left(\mathbf{A}^n_{i+1/2,j}\right)^-\left(\mathbf{u}^n_{i+1,j} - \mathbf{u}^n_{ij}\right)\\\mathbf{F}\left(\mathbf{u}^n_{ij}\right)\mathbf{e}_1 - \mathbf{F}\left(\mathcal{R}\left(\mathbf{u}^n_{i-1,j},\mathbf{u}^n_{ij};0\right)\right)\mathbf{e}_1 &= \mathbf{X}_{i-1/2,j}\left(\Lambda_{i-1/2,j}\right)^+\left(\mathbf{X}_{i-1/2,j}\right)^{-1}\left(\mathbf{u}^n_{ij} - \mathbf{u}^n_{i-1,j}\right)\\&\equiv \left(\mathbf{A}^n_{i-1/2,j}\right)^+\left(\mathbf{u}^n_{ij} - \mathbf{u}^n_{i-1,j}\right)\end{aligned}$$

with similar results in the second coordinate direction. In this form of the wave propagation scheme, the matrices $(\mathbf{A}^n_{i+1/2,j})^\pm$ are most easily understood for the Roe approximate Riemann solver. We have provided information about how to use interpret the fluctuation matrices $\mathbf{A}$ for other approximate Riemann solvers, such as the Rusanov flux (see equation (4.91)), or the Harten–Lax–van Leer flux (see Equation (4.98)). Note that the Harten–Hyman modification of Roe's flux requires special interpretation of the plus/minus parts of the eigenvalue matrix Λ; see Section 4.13.9 for more details.

The wave decompositions at the cell sides are used to compute the donor cell flux differences

$$\begin{aligned}\triangle(\mathbf{Fe}_1)^n_{ij} &= \mathbf{A}^-_{i+1/2,j}\left(\mathbf{u}^n_{i+1,j} - \mathbf{u}^n_{ij}\right) + \mathbf{A}^+_{i-1/2,j}\left(\mathbf{u}^n_{ij} - \mathbf{u}^n_{i-1,j}\right)\\\triangle(\mathbf{Fe}_2)^n_{ij} &= \mathbf{A}^-_{i,j+1/2}\left(\mathbf{u}^n_{i,j+1} - \mathbf{u}^n_{ij}\right) + \mathbf{A}^+_{i,j-1/2}\left(\mathbf{u}^n_{ij} - \mathbf{u}^n_{i,j-1}\right).\end{aligned}$$

Slope limiting is similar to the 1D wave propagation scheme in section 6.2.6. In the first coordinate direction, the wave-field decomposition coefficients

$$\mathbf{a}^n_{i+1/2,j} = \left(\mathbf{X}_{i+1/2,j}\right)^{-1}\left(\mathbf{u}^n_{i+1,j} - \mathbf{u}^n_{ij}\right)$$

are limited as in equation (6.10) to produce $\tilde{\mathbf{a}}^n_{i+1/2,j}$. The flux corrections are computed as follows:

$$\begin{aligned}
\triangle^2(\mathbf{Fe}_1)_{i+1/2,j} &= \mathbf{X}_{i+1/2,j}|\Lambda_{i+1/2,j}|\left(I - |\Lambda_{i+1/2,j}|\frac{2\triangle t^{n+1/2}}{\triangle x_{1,i} + \triangle x_{1,i+1}}\right)\tilde{\mathbf{a}}^n_{i+1/2,j} \\
&+ \Big[\mathbf{A}^+_{i+1/2,j}\mathbf{A}^+_{i,j-1/2}(\mathbf{u}^n_{ij} - \mathbf{u}^n_{i,j-1}) + \mathbf{A}^+_{i+1/2,j}\mathbf{A}^-_{i,j+1/2}(\mathbf{u}^n_{i,j+1} - \mathbf{u}^n_{ij}) \\
&+ \mathbf{A}^-_{i+1/2,j}\mathbf{A}^+_{i+1,j-1/2}(\mathbf{u}^n_{i+1,j} - \mathbf{u}^n_{i+1,j-1}) \\
&+ \mathbf{A}^-_{i+1/2,j}\mathbf{A}^-_{i+1,j+1/2}(\mathbf{u}^n_{i+1,j+1} - \mathbf{u}^n_{i+1,j})\Big]\frac{\triangle t^{n+1/2}}{2\triangle x_{2,j}} \\
\triangle^2(\mathbf{Fe}_2)_{i,j+1/2} &= \mathbf{X}_{i,j+1/2}|\Lambda_{i,j+1/2}|\left(I - |\Lambda_{i,j+1/2}|\frac{2\triangle t^{n+1/2}}{\triangle x_{2,j} + \triangle x_{2,j+1}}\right)\tilde{\mathbf{a}}^n_{i,j+1/2} \\
&+ \Big[\mathbf{A}^+_{i,j+1/2}\mathbf{A}^+_{i-1/2,j}(\mathbf{u}^n_{ij} - \mathbf{u}^n_{i-1,j}) + \mathbf{A}^+_{i,j+1/2}\mathbf{A}^-_{i+1/2,j}(\mathbf{u}^n_{i+1,j} - \mathbf{u}^n_{ij}) \\
&+ \mathbf{A}^-_{i,j+1/2}\mathbf{A}^+_{i-1/2,j+1}(\mathbf{u}^n_{i,j+1} - \mathbf{u}^n_{i-1,j+1}) \\
&+ \mathbf{A}^-_{i,j+1/2}\mathbf{A}^-_{i+1/2,j+1}(\mathbf{u}^n_{i+1,j+1} - \mathbf{u}^n_{i,j+1})\Big]\frac{\triangle t^{n+1/2}}{2\triangle x_{2,j}}.
\end{aligned}$$

The scheme is completed with the conservative difference

$$\begin{aligned}
\mathbf{u}^{n+1}_{ij} &= \mathbf{u}^n_{ij} - \triangle(\mathbf{Fe}_1)^n_{ij}\frac{\triangle t^{n+1/2}}{\triangle x_{1,i}} - \triangle(\mathbf{Fe}_2)^n_{ij}\frac{\triangle t^{n+1/2}}{\triangle x_{2,j}} \\
&+ \triangle^2(\mathbf{Fe}_1)_{i+1/2,j}\frac{\triangle t^{n+1/2}}{\triangle x_{1,i}} - \triangle^2(\mathbf{Fe}_1)_{i-1/2,j}\frac{\triangle t^{n+1/2}}{\triangle x_{1,i}} \\
&+ \triangle^2(\mathbf{Fe}_2)_{i,j+1/2}\frac{\triangle t^{n+1/2}}{\triangle x_{2,j}} - \triangle^2(\mathbf{Fe}_2)_{i,j-1/2}\frac{\triangle t^{n+1/2}}{\triangle x_{2,j}}.
\end{aligned}$$

The stability condition for this method is the same as for the corner transport upwind scheme.

We have not implemented the wave propagation scheme in two dimensions. Interested students can obtain Randy LeVeque's code for wave propagation in CLAWPACK, available from netlib.

7.1.4 2D Lax–Friedrichs

The Lax–Friedrichs scheme in two dimensions uses a staggered grid in two half-steps, similar to the one-dimensional scheme in Section 6.1.1. The two half-steps are based on the following integral forms of the conservation law:

$$\begin{aligned}
&\int_{\mathbf{x}_{2,j}}^{\mathbf{x}_{2,j+1}}\int_{\mathbf{x}_{1,i}}^{\mathbf{x}_{1,i+1}} \mathbf{u}\left(\mathbf{x}_1,\mathbf{x}_2,t^{n+1/2}\right) d\mathbf{x}_1\, d\mathbf{x}_2 \\
&= \int_{\mathbf{x}_{2,j}}^{\mathbf{x}_{2,j+1}}\int_{\mathbf{x}_{1,i}}^{\mathbf{x}_{1,i+1}} \mathbf{u}\left(\mathbf{x}_1,\mathbf{x}_2,t^{n}\right) d\mathbf{x}_1\, d\mathbf{x}_2 \\
&\quad - \int_{t^n}^{t^{n+1/2}}\int_{\mathbf{x}_{2,j}}^{\mathbf{x}_{2,j+1}} \mathbf{F}\left(\mathbf{w}\left(\mathbf{x}_{1,i+1},\mathbf{x}_2,t^{n}\right)\right)\mathbf{e}_1\, d\mathbf{x}_2\, dt \\
&\quad + \int_{t^n}^{t^{n+1/2}}\int_{\mathbf{x}_{2,j}}^{\mathbf{x}_{2,j+1}} \mathbf{F}\left(\mathbf{w}\left(\mathbf{x}_{1,i},\mathbf{x}_2,t^{n}\right)\right)\mathbf{e}_1\, d\mathbf{x}_2\, dt \\
&\quad - \int_{t^n}^{t^{n+1/2}}\int_{\mathbf{x}_{1,i}}^{\mathbf{x}_{1,i+1}} \mathbf{F}\left(\mathbf{w}\left(\mathbf{x}_1,\mathbf{x}_{2,j+1},t^{n}\right)\right)\mathbf{e}_2\, d\mathbf{x}_1\, dt \\
&\quad + \int_{t^n}^{t^{n+1/2}}\int_{\mathbf{x}_{1,i}}^{\mathbf{x}_{1,i+1}} \mathbf{F}\left(\mathbf{w}\left(\mathbf{x}_1,\mathbf{x}_{2,j},t^{n}\right)\right)\mathbf{e}_2\, d\mathbf{x}_1\, dt,
\end{aligned} \tag{7.13a}$$

$$\begin{aligned}
&\int_{\mathbf{x}_{2,j-1/2}}^{\mathbf{x}_{2,j+1/2}}\int_{\mathbf{x}_{1,i-1/2}}^{\mathbf{x}_{1,i+1/2}} \mathbf{u}\left(\mathbf{x}_1,\mathbf{x}_2,t^{n+1}\right) d\mathbf{x}_1\, d\mathbf{x}_2 \\
&= \int_{\mathbf{x}_{2,j-1/2}}^{\mathbf{x}_{2,j+1/2}}\int_{\mathbf{x}_{1,i-1/2}}^{\mathbf{x}_{1,i+1/2}} \mathbf{u}\left(\mathbf{x}_1,\mathbf{x}_2,t^{n+1/2}\right) d\mathbf{x}_1\, d\mathbf{x}_2 \\
&\quad - \int_{t^{n+1/2}}^{t^{n+1}}\int_{\mathbf{x}_{2,j-1/2}}^{\mathbf{x}_{2,j+1/2}} \mathbf{F}\left(\mathbf{w}\left(\mathbf{x}_{1,i+1/2},\mathbf{x}_2,t^{n}\right)\right)\mathbf{e}_1\, d\mathbf{x}_2\, dt \\
&\quad + \int_{t^{n+1/2}}^{t^{n+1}}\int_{\mathbf{x}_{2,j-1/2}}^{\mathbf{x}_{2,j+1/2}} \mathbf{F}\left(\mathbf{w}\left(\mathbf{x}_{1,i-1/2},\mathbf{x}_2,t^{n}\right)\right)\mathbf{e}_1\, d\mathbf{x}_2\, dt \\
&\quad - \int_{t^{n+1/2}}^{t^{n+1}}\int_{\mathbf{x}_{1,i-1/2}}^{\mathbf{x}_{1,i+1/2}} \mathbf{F}\left(\mathbf{w}\left(\mathbf{x}_1,\mathbf{x}_{2,j+1/2},t^{n}\right)\right)\mathbf{e}_2\, d\mathbf{x}_1\, dt \\
&\quad + \int_{t^{n+1/2}}^{t^{n+1}}\int_{\mathbf{x}_{1,i-1/2}}^{\mathbf{x}_{1,i+1/2}} \mathbf{F}\left(\mathbf{w}\left(\mathbf{x}_1,\mathbf{x}_{2,j-1/2},t^{n}\right)\right)\mathbf{e}_2\, d\mathbf{x}_1\, dt.
\end{aligned} \tag{7.13b}$$

Unlike the one-dimensional case, the fluxes in the integrals are not necessarily constant in time. Instead, these integrals will be approximated by appropriate quadratures.

7.1.4.1 First-Order Lax–Friedrichs

In the first-order Lax–Friedrichs scheme in 2D, we assume that the flux variables $\mathbf{w}$ are piecewise constant on the grid cells. Flux integrals in (7.13) are approximated by forward Euler in time. This leads to the following scheme for the conserved quantities $\mathbf{u}$:

$$
\begin{aligned}
&\mathbf{u}^{n+1}_{i+1/2,j+1/2}\frac{\triangle x_{1,i}+\triangle x_{1,i+1}}{2}\frac{\triangle x_{2,j}+\triangle x_{2,j+1}}{2}\\
&= \mathbf{u}^n_{i,j}\frac{\triangle x_{1,i}}{2}\frac{\triangle x_{2,j}}{2} + \mathbf{u}^n_{i+1,j}\frac{\triangle x_{1,i+1}}{2}\frac{\triangle x_{2,j}}{2} + \mathbf{u}^n_{i,j+1}\frac{\triangle x_{1,i}}{2}\frac{\triangle x_{2,j+1}}{2} + \mathbf{u}^n_{i+1,j+1}\frac{\triangle x_{1,i+1}}{2}\frac{\triangle x_{2,j+1}}{2}\\
&\quad - \left[\mathbf{F}(\mathbf{w}^n_{i+1,j+1})\mathbf{e}_1\frac{\triangle x_{2,j+1}}{2} + \mathbf{F}(\mathbf{w}^n_{i+1,j})\mathbf{e}_1\frac{\triangle x_{2,j}}{2}\right]\frac{\triangle t^{n+1/2}}{2}\\
&\quad + \left[\mathbf{F}(\mathbf{w}^n_{i,j+1})\mathbf{e}_1\frac{\triangle x_{2,j+1}}{2} + \mathbf{F}(\mathbf{w}^n_{i,j})\mathbf{e}_1\frac{\triangle x_{2,j}}{2}\right]\frac{\triangle t^{n+1/2}}{2}\\
&\quad - \left[\mathbf{F}(\mathbf{w}^n_{i+1,j+1})\mathbf{e}_2\frac{\triangle x_{1,i+1}}{2} + \mathbf{F}(\mathbf{w}^n_{i,j+1})\mathbf{e}_2\frac{\triangle x_{1,i}}{2}\right]\frac{\triangle t^{n+1/2}}{2}\\
&\quad + \left[\mathbf{F}(\mathbf{w}^n_{i+1,j})\mathbf{e}_1\frac{\triangle x_{1,i+1}}{2} + \mathbf{F}(\mathbf{w}^n_{i,j})\mathbf{e}_1\frac{\triangle x_{1,i}}{2}\right]\frac{\triangle t^{n+1/2}}{2}\\
&\mathbf{u}^{n+1}_{i,j}\triangle x_{1,i}\triangle x_{2,j}\\
&= \left[\mathbf{u}^{n+1/2}_{i-1/2,j-1/2} + \mathbf{u}^{n+1/2}_{i+1/2,j-1/2} + \mathbf{u}^{n+1/2}_{i-1/2,j+1/2} + \mathbf{u}^{n+1/2}_{i+1/2,j+1/2}\right]\frac{\triangle x_{1,i}}{2}\frac{\triangle x_{2,j}}{2}\\
&\quad - \left[\mathbf{F}(\mathbf{w}^n_{i+1/2,j+1/2})\mathbf{e}_1 + \mathbf{F}(\mathbf{w}^n_{i+1/2,j-1/2})\mathbf{e}_1\right]\frac{\triangle x_{2,j}}{2}\frac{\triangle t^{n+1/2}}{2}\\
&\quad + \left[\mathbf{F}(\mathbf{w}^n_{i-1/2,j+1/2})\mathbf{e}_1 + \mathbf{F}(\mathbf{w}^n_{i-1/2,j-1/2})\mathbf{e}_1\right]\frac{\triangle x_{2,j}}{2}\frac{\triangle t^{n+1/2}}{2}\\
&\quad - \left[\mathbf{F}(\mathbf{w}^n_{i+1/2,j+1/2})\mathbf{e}_2 + \mathbf{F}(\mathbf{w}^n_{i-1/2,j+1/2})\mathbf{e}_2\right]\frac{\triangle x_{1,i}}{2}\frac{\triangle t^{n+1/2}}{2}\\
&\quad + \left[\mathbf{F}(\mathbf{w}^n_{i+1/2,j-1/2})\mathbf{e}_2 + \mathbf{F}(\mathbf{w}^n_{i-1/2,j-1/2})\mathbf{e}_2\right]\frac{\triangle x_{1,i}}{2}\frac{\triangle t^{n+1/2}}{2}.
\end{aligned}
$$

Note that it is necessary to determine the flux variables $\mathbf{w}$ from the conserved quantities $\mathbf{u}$ after each half-step. The stability restriction for this scheme is the same as for the corner transport upwind scheme.

7.1.4.2 Second-Order Lax–Friedrichs

The second-order version of the Lax–Friedrichs scheme due to Nessyahu and Tadmor [121] was extended to 2D by Jiang and Tadmor in [78]. In this scheme, we assume that the current solution is piecewise linear

$$\mathbf{u}^n(\mathbf{x}) = \mathbf{u}^n_{ij} + \mathbf{S}^n_{ij}\begin{bmatrix}\frac{\mathbf{x}_1-\mathbf{x}_{1,i}}{\triangle x_{1,i}}\\ \frac{\mathbf{x}_2-\mathbf{x}_{2,j}}{\triangle x_{2,j}}\end{bmatrix}.$$

The columns of the slope matrix $\mathbf{S}^n_{ij}$ are computed in the same way as the limited slopes $\triangle\mathbf{u}$ in the 1D scheme of Section 6.2.3:

$$
\begin{aligned}
\mathbf{S}^n_{ij}\mathbf{e}_1 &= \text{limiter}\left\{\left(\mathbf{u}^n_{ij} - \mathbf{u}^n_{i-1,j}\right), \left(\mathbf{u}^n_{i+1,j} - \mathbf{u}^n_{ij}\right)\right\}\\
\mathbf{S}^n_{ij}\mathbf{e}_2 &= \text{limiter}\left\{\left(\mathbf{u}^n_{ij} - \mathbf{u}^n_{i,j-1}\right), \left(\mathbf{u}^n_{i,j+1} - \mathbf{u}^n_{ij}\right)\right\}.
\end{aligned}
$$

Flux integrals are approximated by the midpoint rule in time and the trapezoidal rule in space. This implies that the first half-step takes the form

$$\begin{aligned}
&\mathbf{u}_{i+1/2,j+1/2}^{n+1/2}\frac{\triangle x_{1,i}+\triangle x_{1,i+1}}{2}\frac{\triangle x_{2,j}+\triangle x_{2,j+1}}{2}\\
&=\left\{u_{ij}^{n}+\mathbf{S}_{ij}^{n}\begin{bmatrix}\triangle x_{1,i}\\ \triangle x_{2,j}\end{bmatrix}\frac{1}{4}\right\}\frac{\triangle x_{1,i}}{2}\frac{\triangle x_{2,j}}{2}+\left\{u_{i+1,j}^{n}+\mathbf{S}_{i+1,j}^{n}\begin{bmatrix}-\triangle x_{1,i+1}\\ \triangle x_{2,j}\end{bmatrix}\frac{1}{4}\right\}\frac{\triangle x_{1,i+1}}{2}\frac{\triangle x_{2,j}}{2}\\
&+\left\{u_{ij+1}^{n}+\mathbf{S}_{i,j+1}^{n}\begin{bmatrix}\triangle x_{1,i}\\ -\triangle x_{2,j+1}\end{bmatrix}\frac{1}{4}\right\}\frac{\triangle x_{1,i}}{2}\frac{\triangle x_{2,j+1}}{2}\\
&+\left\{u_{i+1,j+1}^{n}+\mathbf{S}_{i+1,j+1}^{n}\begin{bmatrix}-\triangle x_{1,i+1}\\ -\triangle x_{2,j+1}\end{bmatrix}\frac{1}{4}\right\}\frac{\triangle x_{1,i+1}}{2}\frac{\triangle x_{2,j+1}}{2}\\
&-\left\{(\mathbf{Fe}_1)(\mathbf{w}_{i+1,j}^{n+1/4})+(\mathbf{Fe}_1)(\mathbf{w}_{i+1,j+1}^{n+1/4})\right\}\frac{\triangle x_{2,j}+\triangle x_{2,j+1}}{4}\frac{\triangle t^{n+1/2}}{2}\\
&+\left\{(\mathbf{Fe}_1)(\mathbf{w}_{ij}^{n+1/4})+(\mathbf{Fe}_1)(\mathbf{w}_{i,j+1}^{n+1/4})\right\}\frac{\triangle x_{2,j}+\triangle x_{2,j+1}}{4}\frac{\triangle t^{n+1/2}}{2}\\
&-\left\{(\mathbf{Fe}_2)(\mathbf{w}_{i,j+1}^{n+1/4})+(\mathbf{Fe}_2)(\mathbf{w}_{i+1,j+1}^{n+1/4})\right\}\frac{\triangle x_{1,i}+\triangle x_{1,i+1}}{4}\frac{\triangle t^{n+1/2}}{2}\\
&+\left\{(\mathbf{Fe}_2)(\mathbf{w}_{ij}^{n+1/4})+(\mathbf{Fe}_2)(\mathbf{w}_{i+1,j}^{n+1/4})\right\}\frac{\triangle x_{1,i}+\triangle x_{1,i+1}}{4}\frac{\triangle t^{n+1/2}}{2},
\end{aligned}$$

where the states for the midpoint rule temporal integrals of the fluxes are given by the following Taylor's rule approximation:

$$\begin{aligned}
\mathbf{u}\left(\mathbf{x}_1,\mathbf{x}_2,t+\frac{\triangle t}{4}\right)&\approx\mathbf{u}+\frac{\partial\mathbf{u}}{\partial t}\frac{\triangle t}{4}=\mathbf{u}-\left(\frac{\partial\mathbf{Fe}_1}{\partial\mathbf{x}_1}+\frac{\partial\mathbf{Fe}_2}{\partial\mathbf{x}_2}\right)\frac{\triangle t}{4}\\
&=\mathbf{u}-\frac{\partial\mathbf{Fe}_1}{\partial\mathbf{w}}\left(\frac{\partial\mathbf{u}}{\partial\mathbf{w}}\right)^{-1}\frac{\partial\mathbf{u}}{\partial\mathbf{x}_1}\frac{\triangle t}{4}-\frac{\partial\mathbf{Fe}_2}{\partial\mathbf{w}}\left(\frac{\partial\mathbf{u}}{\partial\mathbf{w}}\right)^{-1}\frac{\partial\mathbf{u}}{\partial\mathbf{x}_2}\frac{\triangle t}{4}\\
&=\mathbf{u}-\frac{\partial\mathbf{Fe}_1}{\partial\mathbf{w}}\left(\frac{\partial\mathbf{u}}{\partial\mathbf{w}}\right)^{-1}\mathbf{Se}_1\frac{\triangle t}{4}-\frac{\partial\mathbf{Fe}_2}{\partial\mathbf{w}}\left(\frac{\partial\mathbf{u}}{\partial\mathbf{w}}\right)^{-1}\mathbf{Se}_2\frac{\triangle t}{4}.
\end{aligned}$$

For the second half-step, we compute slopes

$$\begin{aligned}
\mathbf{S}_{i+1/2,j+1/2}^{n+1/2}\mathbf{e}_1&=\text{limiter}\left\{\left(\mathbf{u}_{i+1/2,j+1/2}^{n+1/2}-\mathbf{u}_{i-1/2,j+1/2}^{n+1/2}\right),\left(\mathbf{u}_{i+3/2,j+1/2}^{n+1/2}-\mathbf{u}_{i+1/2,j+1/2}^{n+1/2}\right)\right\}\\
\mathbf{S}_{i+1/2,j+1/2}^{n+1/2}\mathbf{e}_2&=\text{limiter}\left\{\left(\mathbf{u}_{i+1/2,j+1/2}^{n+1/2}-\mathbf{u}_{i+1/2,j-1/2}^{n+1/2}\right),\left(\mathbf{u}_{i+1/2,j+3/2}^{n+1/2}-\mathbf{u}_{i+1/2,j+1/2}^{n+1/2}\right)\right\}
\end{aligned}$$

and states for temporal integrals

$$\begin{aligned}
\mathbf{u}_{i+1/2,j+1/2}^{n+3/4}=\mathbf{u}_{i+1/2,j+1/2}^{n+1/2}&-\frac{\partial\mathbf{Fe}_1}{\partial\mathbf{w}}\left(\frac{\partial\mathbf{u}}{\partial\mathbf{w}}\right)^{-1}\mathbf{S}_{i+1/2,j+1/2}^{n+1/2}\mathbf{e}_1\frac{\triangle t}{4}\\
&-\frac{\partial\mathbf{Fe}_2}{\partial\mathbf{w}}\left(\frac{\partial\mathbf{u}}{\partial\mathbf{w}}\right)^{-1}\mathbf{S}_{i+1/2,j+1/2}^{n+1/2}\mathbf{e}_2\frac{\triangle t}{4}.
\end{aligned}$$

Then the solution at the new time is given by

$$
\begin{aligned}
&\mathbf{u}_{i,j}^{n+1} \triangle x_{1,i} \triangle x_{2,j} \\
&= \left[\left\{ u_{i-1/2,j-1/2}^{n+1/2} + \mathbf{S}_{i-1/2,j-1/2}^{n+1/2} \begin{bmatrix} \triangle x_{1,i} \\ \triangle x_{2,j} \end{bmatrix} \frac{1}{4} \right\} \right. \\
&+ \left\{ u_{i+1/2,j-1/2}^{n+1/2} + \mathbf{S}_{i+1/2,j-1/2}^{n+1/2} \begin{bmatrix} -\triangle x_{1,i} \\ \triangle x_{2,j} \end{bmatrix} \frac{1}{4} \right\} \\
&+ \left\{ u_{i-1/2,j+1/2}^{n+1/2} + \mathbf{S}_{i-1/2,j+1/2}^{n+1/2} \begin{bmatrix} \triangle x_{1,i} \\ -\triangle x_{2,j} \end{bmatrix} \frac{1}{4} \right\} \\
&+ \left. \left\{ u_{i+1/2,j+1/2}^{n+1/2} + \mathbf{S}_{i+1/2,j+1/2}^{n+1/2} \begin{bmatrix} -\triangle x_{1,i} \\ -\triangle x_{2,j} \end{bmatrix} \frac{1}{4} \right\} \right] \frac{\triangle x_{1,i}}{2} \frac{\triangle x_{2,j}}{2} \\
&- \left\{ (\mathbf{Fe}_1)(\mathbf{w}_{i+1/2,j-1/2}^{n+3/4} + (\mathbf{Fe}_1)(\mathbf{w}_{i+1/2,j+1/2}^{n+3/4} \right\} \frac{\triangle x_{2,j}}{2} \frac{\triangle t^{n+1/2}}{2} \\
&+ \left\{ (\mathbf{Fe}_1)(\mathbf{w}_{i-1/2,j-1/2}^{n+3/4} + (\mathbf{Fe}_1)(\mathbf{w}_{i-1/2,j+1/2}^{n+3/4} \right\} \frac{\triangle x_{2,j}}{2} \frac{\triangle t^{n+1/2}}{2} \\
&- \left\{ (\mathbf{Fe}_2)(\mathbf{w}_{i-1/2,j+1/2}^{n+3/4} + (\mathbf{Fe}_2)(\mathbf{w}_{i+1/2,j+1/2}^{n+3/4} \right\} \frac{\triangle x_{1,i}}{2} \frac{\triangle t^{n+1/2}}{2} \\
&+ \left\{ (\mathbf{Fe}_2)(\mathbf{w}_{i-1/2,j-1/2}^{n+3/4} + (\mathbf{Fe}_2)(\mathbf{w}_{i+1/2,j-1/2}^{n+3/4} \right\} \frac{\triangle x_{1,i}}{2} \frac{\triangle t^{n+1/2}}{2} .
\end{aligned}
$$

It is helpful to note that Jiang and Tadmor include a copy of their scheme in their paper. Also, note that the development of a third-order extension of the Lax–Friedrichs scheme suffers from the same difficulty facing the generalizations of the MUSCL and wave propagation schemes: the development of appropriate piecewise quadratic reconstructions in multiple dimensions. Some work in this direction appears in [100, 101].

We have implemented the first- and second-order corner Lax–Friedrichs scheme in **Program 7.1-114: lf2d.m4.** Students can execute this scheme in two dimensions by clicking on **Executable 7.1-52: guiRectangle.** The student can select either of Burgers' equations, shallow water or gas dynamics under '`Riemann Problem Parameters`. Students should select `unsplit` for the `splitting` under `Numerical Method parameters`, and `Lax-Friedrichs` should be the first scheme with value `True`. In two dimensions, the computational results for scalar fields (such as water height in shallow water, or pressure in gas dynamics) can be displayed either as 2D contours, 2D color fills or 3D surface plots. The 3D graphics in the surface plot uses a trackball for rotation with the left mouse button. The figure can be sliced along the ends of any coordinate axis using the middle mouse button. Values at points in the figure can be determined using the right mouse button.

7.1.5 Multidimensional ENO

The multidimensional form of the **ENO scheme** is straightforward. Recall that ENO views the conservation law in terms of the method of lines:

$$\frac{\partial \mathbf{u}}{\partial t} = -\frac{\partial \mathbf{f}_1}{\partial x_1} - \frac{\partial \mathbf{f}_2}{\partial x_2}.$$

Whenever the ordinary differential equation integrator requires the value of $\partial \mathbf{f}_1/\partial x_1$ and $\partial \mathbf{f}_2/\partial x_2$, we would apply the standard one-dimensional ENO scheme to each of $\mathbf{f}_1$ and $\mathbf{f}_2$ in the separate coordinate directions. Since the ordinary differential equation solvers all involve multiple sub-steps, the overall scheme involves significant coupling between grid cells, even diagonal neighbors.

We have implemented the ENO scheme in **Program 7.1-115: eno2d.m4.** Students can execute this scheme in two dimensions by clicking on **Executable 7.1-53: guiRectangle.** The student can select either of Burgers' equations, shallow water or gas dynamics under `Riemann Problem Parameters`. Students should select `unsplit` for the `splitting` under `Numerical Method parameters`, and `ENO` should be the first scheme with value `True`. In two dimensions, the computational results for scalar fields (such as water height in shallow water, or pressure in gas dynamics) can be displayed either as 2D contours, 2D color fills or 3D surface plots. The 3D graphics in the surface plot uses a trackball for rotation with the left mouse button. The figure can be sliced along the ends of any coordinate axis using the middle mouse button. Values at points in the figure can be determined using the right mouse button.

7.1.6 Discontinuous Galerkin Method on Rectangles

The application of the discontinuous Galerkin method to a 2D rectangular grid is a natural extension of the 1D ideas in Section 6.2.9. If $\beta(\mathbf{x})$ is an arbitrary smooth function, we note that the weak form of a scalar conservation law on a grid cell is

$$\begin{aligned}
0 &= \int_{\mathbf{x}_{2,j-1/2}}^{\mathbf{x}_{2,j+1/2}} \int_{\mathbf{x}_{1,i-1/2}}^{\mathbf{x}_{1,i+1/2}} \beta \left[\frac{\partial u}{\partial t} + \frac{\partial f_1}{\partial \mathbf{x}_1} + \frac{\partial f_2}{\partial \mathbf{x}_2} \right] d\mathbf{x}_1 \, d\mathbf{x}_2 \\
&= \int_{\mathbf{x}_{2,j-1/2}}^{\mathbf{x}_{2,j+1/2}} \int_{\mathbf{x}_{1,i-1/2}}^{\mathbf{x}_{1,i+1/2}} \frac{\partial \beta u}{\partial t} - \frac{\partial \beta}{\partial \mathbf{x}_1} f_1 - \frac{\partial \beta}{\partial \mathbf{x}_2} f_2 \, d\mathbf{x}_1 \, d\mathbf{x}_2 \\
&+ \int_{x_{2,j-1/2}}^{x_{2,j+1/2}} \beta(\mathbf{x}_{1,i+1/2}, \mathbf{x}_2) f_1(\mathbf{x}_{1,i+1/2}, \mathbf{x}_2, t) \, d\mathbf{x}_2 \\
&- \int_{x_{2,j-1/2}}^{x_{2,j+1/2}} \beta(\mathbf{x}_{1,i-1/2}, \mathbf{x}_2) f_1(\mathbf{x}_{1,i-1/2}, \mathbf{x}_2, t) \, d\mathbf{x}_2 \\
&+ \int_{x_{1,i-1/2}}^{x_{1,i+1/2}} \beta(\mathbf{x}_1, \mathbf{x}_{2,j+1/2}) f_2(\mathbf{x}_1, \mathbf{x}_{2,j+1/2}, t) \, d\mathbf{x}_1 \\
&- \int_{x_{1,i-1/2}}^{x_{1,i+1/2}} \beta(\mathbf{x}_1, \mathbf{x}_{2,j-1/2}) f_2(\mathbf{x}_1, \mathbf{x}_{2,j-1/2}, t) \, d\mathbf{x}_1 .
\end{aligned}$$

Since these equations hold componentwise in the conserved quantities, it suffices to consider a scalar law u.

We approximate our scalar conserved quantity by a linear combination of orthonormal basis functions,

$$u(\mathbf{x}_1, \mathbf{x}_2, t) = \mathbf{b}\left(\xi_{1,i}(\mathbf{x}_1)\right)^\top \mathbf{U}_{ij}(t)\mathbf{b}\left(\xi_{2,j}(\mathbf{x}_2)\right),$$

where the coordinate transformations are

$$\xi_{1,i}(\mathbf{x}_1) = 2\frac{\mathbf{x}_1 - \mathbf{x}_{1,i}}{\triangle x_{1,i}}, \quad \mathbf{x}_1 \in \left(\mathbf{x}_{1,i-1/2}, \mathbf{x}_{1,i+1/2}\right)$$

$$\xi_{2,j}(\mathbf{x}_1) = 2\frac{\mathbf{x}_2 - \mathbf{x}_{2,j}}{\triangle x_{2,j}}, \quad \mathbf{x}_2 \in \left(\mathbf{x}_{2,j-1/2}, \mathbf{x}_{2,j+1/2}\right),$$

and $\mathbf{U}_{ij}(t)$ is the array of unknown coefficients of the basis functions for each conserved quantity. Taking $\beta(\mathbf{x})$ to be any component of the array $\mathbf{b}(\xi_{1,i}(\mathbf{x}_1))\mathbf{b}(\xi_{2,j}(\mathbf{x}_2))^\top$ allows us to write the weak form of the discontinuous Galerkin method as follows:

$$\begin{aligned}
0 = &\int_{x_{2,j-1/2}}^{x_{2,j+1/2}} \int_{x_{1,i-1/2}}^{x_{1,i+1/2}} \mathbf{b}\big(\xi_{1,i}(\mathbf{x}_1)\big)\mathbf{b}\big(\xi_{1,i}(\mathbf{x}_1)\big)^\top \frac{d\mathbf{U}_{ij}}{dt}\mathbf{b}\big(\xi_{2,j}(\mathbf{x}_2)\big)\mathbf{b}\big(\xi_{2,j}(\mathbf{x}_2)\big)^\top \, d\mathbf{x}_1 \, d\mathbf{x}_2 \\
&- \int_{x_{2,j-1/2}}^{x_{2,j+1/2}} \int_{x_{1,i-1/2}}^{x_{1,i+1/2}} \frac{\partial \mathbf{b}\big(\xi_{1,i}(\mathbf{x}_1)\big)}{\partial \mathbf{x}_1} f_1\Big(\mathbf{b}\big(\xi_{1,i}(\mathbf{x}_1)\big)^\top \mathbf{U}_{ij}(t)\mathbf{b}\big(\xi_{2,j}(\mathbf{x}_2)\big)\Big)\mathbf{b}\big(\xi_{2,j}(\mathbf{x}_2)\big)^\top \, d\mathbf{x}_1 \, d\mathbf{x}_2 \\
&- \int_{x_{2,j-1/2}}^{x_{2,j+1/2}} \int_{x_{1,i-1/2}}^{x_{1,i+1/2}} \mathbf{b}\big(\xi_{1,i}(\mathbf{x}_1)\big) f_2\Big(\mathbf{b}\big(\xi_{1,i}(\mathbf{x}_1)\big)^\top \mathbf{U}_{ij}(t)\mathbf{b}\big(\xi_{2,j}(\mathbf{x}_2)\big)\Big)\frac{\partial \mathbf{b}\big(\xi_{2,j}(\mathbf{x}_2)\big)}{\partial \mathbf{x}_2}^\top \, d\mathbf{x}_1 \, d\mathbf{x}_2 \\
&+ \int_{x_{2,j-1/2}}^{x_{2,j+1/2}} \mathbf{b}(1) f_1\Big(\mathcal{R}\big(\mathbf{b}(1)^\top \mathbf{U}_{i,j}(t)\mathbf{b}\big(\xi_{2,j}(\mathbf{x}_2)\big), \mathbf{b}(-1)^\top \mathbf{U}_{i+1,j}(t)\mathbf{b}\big(\xi_{2,j}(\mathbf{x}_2)\big); 0\big)\Big)\mathbf{b}\big(\xi_{2,j}(\mathbf{x}_2)\big)^\top \, d\mathbf{x}_2 \\
&- \int_{x_{2,j-1/2}}^{x_{2,j+1/2}} \mathbf{b}(-1) f_1\Big(\mathcal{R}\big(\mathbf{b}(1)^\top \mathbf{U}_{i-1,j}(t)\mathbf{b}\big(\xi_{2,j}(\mathbf{x}_2)\big), \mathbf{b}(-1)^\top \mathbf{U}_{i,j}(t)\mathbf{b}\big(\xi_{2,j}(\mathbf{x}_2)\big); 0\big)\Big)\mathbf{b}\big(\xi_{2,j}(\mathbf{x}_2)\big)^\top \, d\mathbf{x}_2 \\
&+ \int_{x_{1,i-1/2}}^{x_{1,i+1/2}} \mathbf{b}\big(\xi_{1,i}(\mathbf{x}_1)\big) f_2\Big(\mathcal{R}\big(\mathbf{b}\big(\xi_{1,i}(\mathbf{x}_1)\big)^\top \mathbf{U}_{i,j}(t)\mathbf{b}(-1), \mathbf{b}\big(\xi_{1,i}(\mathbf{x}_1)\big)^\top \mathbf{U}_{i,j+1}(t)\mathbf{b}(1); 0\big)\Big)\mathbf{b}(1)^\top \, d\mathbf{x}_1 \\
&- \int_{x_{1,i-1/2}}^{x_{1,i+1/2}} \mathbf{b}\big(\xi_{1,i}(\mathbf{x}_1)\big) f_2\Big(\mathcal{R}\big(\mathbf{b}\big(\xi_{1,i}(\mathbf{x}_1)\big)^\top \mathbf{U}_{i,j-1}(t)\mathbf{b}(1), \mathbf{b}\big(\xi_{1,i}(\mathbf{x}_1)\big)^\top \mathbf{U}_{i,j}(t)\mathbf{b}(1); 0\big)\Big)\mathbf{b}(-1)^\top \, d\mathbf{x}_1.
\end{aligned}$$

Changing integration variables and using the orthonormality of the basis functions produces

$$\begin{aligned}
0 = &\frac{d\mathbf{U}_{ij}}{dt}\frac{\triangle x_{1,i}}{2}\frac{\triangle x_{2,j}}{2} \\
&- \int_{-1}^{1}\int_{-1}^{1} \mathbf{b}'(\xi_1) f_1\left(\mathbf{b}(\xi_1)^\top \mathbf{U}_{ij}(t)\mathbf{b}(\xi_2)\right)\mathbf{b}(\xi_2)^\top \, d\xi_1 \, \frac{\triangle x_{2,j}}{2} \, d\xi_2 \\
&- \int_{-1}^{1}\int_{-1}^{1} \mathbf{b}(\xi_1) f_2\left(\mathbf{b}(\xi_1)^\top \mathbf{U}_{ij}(t)\mathbf{b}(\xi_2)\right)\mathbf{b}'(\xi_2)^\top \, \frac{\triangle x_{1,i}}{2} \, d\xi_1 \, d\xi_2
\end{aligned}$$

$$+\int_{-1}^{1} \mathbf{b}(1) f_1\left(\mathcal{R}(\mathbf{b}(1)^\top \mathbf{U}_{i,j}(t)\mathbf{b}(\xi_2), \mathbf{b}(-1)^\top \mathbf{U}_{i+1,j}(t)\mathbf{b}(\xi_2); 0)\right) \mathbf{b}(\xi_2)^\top \frac{\triangle x_{2,j}}{2}\, d\xi_2$$
$$-\int_{-1}^{1} \mathbf{b}(-1) f_1\left(\mathcal{R}(\mathbf{b}(1)^\top \mathbf{U}_{i-1,j}(t)\mathbf{b}(\xi_2), \mathbf{b}(-1)^\top \mathbf{U}_{i,j}(t)\mathbf{b}(\xi_2); 0)\right) \mathbf{b}(\xi_2)^\top \frac{\triangle x_{2,j}}{2}\, d\xi_2$$
$$+\int_{-1}^{1} \mathbf{b}(\xi_1) f_2\left(\mathcal{R}(\mathbf{b}(\xi_1)^\top \mathbf{U}_{i,j}(t)\mathbf{b}(1), \mathbf{b}(\xi_1)^\top \mathbf{U}_{i,j+1}(t)\mathbf{b}(-1); 0)\right) \mathbf{b}(1)^\top \frac{\triangle x_{1,i}}{2}\, d\mathbf{x}_1$$
$$-\int_{-1}^{1} \mathbf{b}(\xi_1) f_2\left(\mathcal{R}(\mathbf{b}(\xi_1)^\top \mathbf{U}_{i,j-1}(t)\mathbf{b}(1), \mathbf{b}(\xi_1)^\top \mathbf{U}_{i,j}(t)\mathbf{b}(-1); 0)\right) \mathbf{b}(-1)^\top \frac{\triangle x_{1,i}}{2}\, d\mathbf{x}_1.$$

The remaining integrals are approximated by applying product Lobatto quadrature rules arising from the 1D quadrature rule $\int_{-1}^{1} \phi(\xi)\, d\xi \approx \sum_{q=0}^{Q} \phi(\xi_q)\alpha_q$. This gives us the continuous-in-time form of the discontinuous Galerkin scheme for scalar laws in 2D:

$$\begin{aligned}
0 =\; & \frac{d\mathbf{U}_{ij}}{dt}\frac{\triangle x_{1,i}\triangle x_{2,j}}{4} \\
& - \sum_{q_2=0}^{Q}\sum_{q_1=0}^{Q} \mathbf{b}'(\xi_{q_1}) f_1\left(\mathbf{b}(\xi_{q_1})^\top \mathbf{U}_{ij}(t)\mathbf{b}(\xi_{q_2})\right) \mathbf{b}(\xi_{q_2})^\top \alpha_{q_1}\alpha_{q_2}\frac{\triangle x_{2,j}}{2} \\
& - \sum_{q_2=0}^{Q}\sum_{q_1=0}^{Q} \mathbf{b}(\xi_{q_1}) f_2\left(\mathbf{b}(\xi_{q_1})^\top \mathbf{U}_{ij}(t)\mathbf{b}(\xi_{q_2})\right) \mathbf{b}'(\xi_{q_2})^\top \alpha_{q_1}\alpha_{q_2}\frac{\triangle x_{1,i}}{2} \\
& - \sum_{q_2=0}^{Q} \mathbf{b}(1) f_1\left(\mathcal{R}\left(\mathbf{b}(1)^\top \mathbf{U}_{i,j}(t)\mathbf{b}(\xi_{q_2}), \mathbf{b}(-1)^\top \mathbf{U}_{i+1,j}(t)\mathbf{b}(\xi_{q_2}); 0\right)\right) \mathbf{b}(\xi_{q_2})^\top \alpha_{q_2}\frac{\triangle x_{2,j}}{2} \\
& + \sum_{q_2=0}^{Q} \mathbf{b}(-1) f_1\left(\mathcal{R}\left(\mathbf{b}(1)^\top \mathbf{U}_{i-1j}(t)\mathbf{b}(\xi_{q_2}), \mathbf{b}(-1)^\top \mathbf{U}_{i,j}(t)\mathbf{b}(\xi_{q_2}); 0\right)\right) \mathbf{b}(\xi_{q_2})^\top \alpha_{q_2}\frac{\triangle x_{2,j}}{2} \\
& - \sum_{q_1=0}^{Q} \mathbf{b}(\xi_{q_1}) f_2\left(\mathcal{R}\left(\mathbf{b}(\xi_{q_1})^\top \mathbf{U}_{i,j}(t)\mathbf{b}(1), \mathbf{b}(\xi_{q_1})^\top \mathbf{U}_{i,j+1}(t)\mathbf{b}(-1); 0\right)\right) \mathbf{b}(1)^\top \alpha_{q_1}\frac{\triangle x_{1,i}}{2} \\
& + \sum_{q_1=0}^{Q} \mathbf{b}(\xi_{q_1}) f_2\left(\mathcal{R}\left(\mathbf{b}(\xi_{q_1})^\top \mathbf{U}_{i,j-1}(t)\mathbf{b}(1), \mathbf{b}(\xi_{q_1})^\top \mathbf{U}_{i,j}(t)\mathbf{b}(-1); 0\right)\right) \mathbf{b}(-1)^\top \alpha_{q_1}\frac{\triangle x_{1,i}}{2}.
\end{aligned}$$

Time integration is performed by the same Runge–Kutta schemes as in the ENO scheme of Section 5.13.

Limiting in two dimensions is more complicated than in one dimension. It is easy to see that the cell average is $\overline{u}_{ij}(t) = \mathbf{e}_0^\top \mathbf{U}_{ij}(t)\mathbf{e}_0\mathbf{b}_0^2$. In the first coordinate direction, we compute

$$u(\mathbf{x}_{1,i-1/2}, \mathbf{x}_{2,j}, t) - \overline{u}_{ij}(t) = \mathbf{b}(-1)^\top \mathbf{U}_{ij}(t)\mathbf{b}(0) - \mathbf{b}_0\mathbf{e}_0^\top \mathbf{U}_{ij}(t)\mathbf{e}_0\mathbf{b}_0,$$
$$u(\mathbf{x}_{1,i+1/2}, \mathbf{x}_{2,j}, t) - \overline{u}_{ij}(t) = \mathbf{b}(1)^\top \mathbf{U}_{ij}(t)\mathbf{b}(0) - \mathbf{b}_0\mathbf{e}_0^\top \mathbf{U}_{ij}(t)\mathbf{e}_0\mathbf{b}_0.$$

If

$$\max\left\{|u(\mathbf{x}_{1,i-1/2}, \mathbf{x}_{2,j}, t) - \overline{u}_{ij}(t)|,\ |u(\mathbf{x}_{1,i+1/2}, \mathbf{x}_{2,j}, t) - \overline{u}_{ij}(t)|\right\} \leq M \triangle x_{1,i}^2$$

then no limiting is done in the first coordinate direction. Otherwise, we compute

$$\triangle u_{1,ij} = \operatorname{minmod}(\overline{u}_{i+1,j}(t) - \overline{u}_{ij}(t),\ \overline{u}_{i,j}(t) - \overline{u}_{i-1,j}(t))$$

where

$$\operatorname{minmod}(a, b) = \begin{cases} \min\{|a|, |b|\}\operatorname{sign}(a), & ab > 0 \\ 0, & ab \leq 0. \end{cases}$$

Then for all polynomial orders $0 < p_0 < Q$ and $0 \leq p_1 < Q$ we set

$$\mathbf{e}_{p_0}^\top \mathbf{u}_{ij}(t)\mathbf{e}_{p_1} = \begin{cases} \triangle u_{1,ij}, & p_0 = 1 \text{ and } p_1 = 0 \\ 0, & \text{otherwise.} \end{cases}$$

In other words, the first row of $\mathbf{u}_{ij}(t)$ is unchanged, and the first entry of the second row is the only nonzero below the first row. In the second coordinate direction, we compute $u(\mathbf{x}_{1,i}, \mathbf{x}_{2,j+1/2}, t) - \overline{u}_{ij}(t)$ and $u(\mathbf{x}_{1,i}, \mathbf{x}_{2,j-1/2}, t) - \overline{u}_{ij}(t)$ using similar expressions, to see if limiting should be done. If so, we compute $\triangle u_{2,ij}$ using similar expressions and zero all but the first column of $\mathbf{U}_{ij}(t)$ and the first entry of its second column. For a system of conservation laws in multiple dimensions, the limiting is done in characteristic expansion coefficients in each coordinate direction. This limiting is performed just after the initial data is determined, and just after each Runge–Kutta step for time integration.

Note that for hyperbolic systems, the discontinuous Galerkin equations are coupled through the flux function evaluations. Each cell side requires the solution of $Q + 1$ Riemann problems, where $Q + 1$ is the number of quadrature points. As we saw in Section 5.14.3, the number of Lobatto quadrature points should be at least one plus the highest degree of the basis polynomials. The number of Riemann problems is roughly two times the order of the method times the number of grid cells, and the number of flux function evaluations in the interior of the grid cells is four times the square of the order times the number of grid cells. In addition, the number of Runge–Kutta steps increases with the order of the scheme, and the size of the stable timestep decreases with the order (see Section 5.14.6).

We have implemented the discontinuous Galerkin scheme in **Program 7.1-116: dgm2d.m4.** Students can execute this scheme in two dimensions by clicking on **Exe-**

cutable 7.1-54: guiRectangle. The student can select either Burgers' equation, shallow water or gas dynamics under 'Riemann Problem Parameters. Students should select `unsplit` for the `splitting` under `Numerical Method parameters`, and `Discontinuous Galerkin` should be the first scheme with value `True`. In two dimensions, the computational results for scalar fields (such as water height in shallow water, or pressure in gas dynamics) can be displayed either as 2D contours, 2D color fills or 3D surface plots. The 3D graphics in the surface plot uses a trackball for rotation with the left mouse button. The figure can be sliced along the ends of any coordinate axis using the middle mouse button. Values at points in the figure can be determined using the right mouse button.

Exercises for 7.1

7.1.1 Verify that the donor cell scheme is **free-stream-preserving**: if $\mathbf{u}^n_{ijk} = \mathbf{u}$ for all grid cells Ω_{ijk}, then $\mathbf{u}^{n+1}_{ijk} = \mathbf{u}$ for all Ω_{ijk}.

7.1.2 Verify that the first-order corner transport upwind scheme is free-stream-preserving.

7.2 Riemann Problems in Two Dimensions

In order to generate interesting test problems for two-dimensional calculations, we will consider a generalization of the one-dimensional Riemann problem, consisting of four constant states in the quadrants of the plane. In order to reduce the number of possible cases, we will assume that neighboring constant states are associated with one-dimensional Riemann problems in which a single wave is involved in the evolution.

7.2.1 Burgers' Equation

The solution of two-dimensional scalar Riemann problems was examined by Lindquist [104, 105]. For Burgers' equation, the flux is given by $\mathbf{F}(u) = \frac{1}{2}u^2\mathbf{n}^\top$. In order to simplify the discussion, we will assume that $\mathbf{n} > 0$. We will order the states counterclockwise in the plane, beginning with the upper right-hand quadrant. There are 24 possible orderings of these states. Along each positive or negative axis it is possible to have either a rarefaction or a shock, giving a total of 16 combinations. Two of these combinations ($\mathbf{R}_{21}\mathbf{R}_{32}\mathbf{S}_{3}4\mathbf{S}_{41}$ and $\mathbf{S}_{21}\mathbf{S}_{32}\mathbf{R}_{3}4\mathbf{R}_{41}$) are impossible due to conflicts among the Lax admissibility conditions. Because both components of $\mathbf{n}$ are positive, five of these combinations are related by interchanging the axes ($\mathbf{R}_{21}\mathbf{S}_{32}\mathbf{R}_{34}\mathbf{R}_{41} \Longrightarrow \mathbf{R}_{21}\mathbf{R}_{32}\mathbf{S}_{34}\mathbf{R}_{41}$, $\mathbf{S}_{21}\mathbf{R}_{32}\mathbf{R}_{34}\mathbf{R}_{41} \Longrightarrow \mathbf{R}_{21}\mathbf{R}_{32}\mathbf{R}_{34}\mathbf{S}_{41}$, $\mathbf{S}_{21}\mathbf{R}_{32}\mathbf{S}_{34}\mathbf{R}_{41} \Longrightarrow \mathbf{R}_{21}\mathbf{S}_{32}\mathbf{R}_{34}\mathbf{S}_{41}$, $\mathbf{S}_{21}\mathbf{S}_{32}\mathbf{R}_{34}\mathbf{S}_{41} \Longrightarrow \mathbf{S}_{21}\mathbf{R}_{32}\mathbf{S}_{34}\mathbf{S}_{41}$

and $\mathbf{S}_{21}\mathbf{S}_{32}\mathbf{S}_{34}\mathbf{R}_{41} \Longrightarrow \mathbf{R}_{21}\mathbf{S}_{32}\mathbf{S}_{34}\mathbf{S}_{41}$). This leaves nine distinct cases:

$$
\begin{aligned}
\mathbf{R}_{21}\mathbf{R}_{32}\mathbf{R}_{34}\mathbf{R}_{41} &: u_1 > u_2 > u_4 > u_3 \text{ or } u_1 > u_4 > u_2 > u_3\\
\mathbf{R}_{21}\mathbf{R}_{32}\mathbf{R}_{34}\mathbf{S}_{41} &: u_4 > u_1 > u_2 > u_3\\
\mathbf{R}_{21}\mathbf{R}_{32}\mathbf{S}_{34}\mathbf{R}_{41} &: u_1 > u_2 > u_3 > u_4\\
\mathbf{R}_{21}\mathbf{S}_{32}\mathbf{R}_{34}\mathbf{R}_{41} &: u_4 > u_1 > u_3 > u_2 \text{ or } u_4 > u_3 > u_1 > u_2\\
\mathbf{R}_{21}\mathbf{S}_{32}\mathbf{S}_{34}\mathbf{R}_{41} &: u_1 > u_3 > u_2 > u_4 \text{ or } u_1 > u_3 > u_4 > u_2 \text{ or}\\
&\quad u_3 > u_1 > u_2 > u_4 \text{ or } u_3 > u_1 > u_4 > u_2\\
\mathbf{R}_{21}\mathbf{S}_{32}\mathbf{S}_{34}\mathbf{S}_{41} &: u_3 > u_4 > u_1 > u_2\\
\mathbf{S}_{21}\mathbf{R}_{32}\mathbf{R}_{34}\mathbf{S}_{41} &: u_2 > u_4 > u_1 > u_3 \text{ or } u_2 > u_4 > u_3 > u_1 \text{ or}\\
&\quad u_4 > u_2 > u_1 > u_3 \text{ or } u_4 > u_2 > u_3 > u_1\\
\mathbf{S}_{21}\mathbf{R}_{32}\mathbf{S}_{34}\mathbf{S}_{41} &: u_2 > u_3 > u_4 > u_1\\
\mathbf{S}_{21}\mathbf{S}_{32}\mathbf{S}_{34}\mathbf{S}_{41} &: u_3 > u_2 > u_4 > u_1 \text{ or } u_3 > u_4 > u_2 > u_1.
\end{aligned}
$$

Within each of these cases, it is possible to generate transonic rarefactions, or rarefactions involving all positive speeds or all negative speeds. Similarly, shocks could have negative or positive speeds. The actual possibilities vary with the cases.

We have already displayed some numerical results for this Riemann problem in Figure 7.1. This Riemann problem used the initial data

$u_2 = 0.67$	$u_1 = -1$
$u_3 = 1.0$	$u_4 = 0.33$

and $\mathbf{n} = [1, 1]$. The solution involves only shocks.

We have implemented the 2D Riemann problem for Burgers' equation in **Program 7.2-117: GUIRectangle.C.** Students can execute a variety of schemes for this problem by clicking on **Executable 7.2-55: guiRectangle.** The student can select Burgers' equation under `Riemann Problem Parameters`. Students should select the desired `splitting` under `Numerical Method parameters`, and set the boolean flag for their desired method to `True`. In two dimensions, the computational results for the solution can be displayed either as 2D contours, 2D color fills or 3D surface plots. The 3D graphics in the surface plot uses a trackball for rotation with the left mouse button. The figure can be sliced along the ends of any coordinate axis using the middle mouse button. Values at points in the figure can be determined using the right mouse button. Students can perform an error analysis, including a comparison of schemes, by setting the number of cells in one of the coordinate directions to 0 under `Numerical Method parameters`. These comparisons can be very time-consuming.

7.2.2 Shallow Water

As we showed in section 4.1.10, the shallow water model in example 4.1.1 has four possible elementary waves with associated admissibility conditions:

$$\begin{aligned}
\mathbf{R}^- &: \mathbf{v}_R - \mathbf{v}_L = \mathbf{n}2(\sqrt{gh_L} - \sqrt{gh_R}), && h_R \le h_L \\
\mathbf{R}^+ &: \mathbf{v}_R - \mathbf{v}_L = \mathbf{n}2(\sqrt{gh_R} - \sqrt{gh_L}), && h_L \le h_R \\
\mathbf{S}^- &: \mathbf{v}_R - \mathbf{v}_L = -\mathbf{n}(h_R - h_L)\sqrt{\frac{g}{2}\left(\frac{1}{h_L} + \frac{1}{h_R}\right)}, && h_R > h_L \\
\mathbf{S}^+ &: \mathbf{v}_R - \mathbf{v}_L = -\mathbf{n}(h_L - h_R)\sqrt{\frac{g}{2}\left(\frac{1}{h_R} + \frac{1}{h_L}\right)}, && h_L > h_R.
\end{aligned}$$

In the shallow water model, there is no inherent direction associated with the flux function, so without loss of generality we may rotate the 2D Riemann problem for shallow water so that the largest value of h occurs in the upper right-hand quadrant. Again, we will order the states counterclockwise in the plane, beginning with the upper right-hand quadrant. This leaves six possible orderings of the states:

$$h_1 > h_2 > h_3 > h_4 \quad \text{and} \quad h_1 > h_4 > h_3 > h_2 \tag{7.14a}$$
$$h_1 > h_2 > h_4 > h_3 \quad \text{and} \quad h_1 > h_4 > h_2 > h_3 \tag{7.14b}$$
$$h_1 > h_3 > h_2 > h_4 \quad \text{and} \quad h_1 > h_3 > h_4 > h_2. \tag{7.14c}$$

By switching the coordinate axes, we may obtain the second set of inequalities on each of these lines from the first set. This leaves us with three distinct orderings of the states.

The 2D Riemann problem for shallow water may have either a rarefaction or a shock on either the positive or negative branches of each coordinate axis. The wave curves and admissibility conditions imply that

$$\begin{aligned}
\mathbf{R}_{21} &\Longrightarrow \mathbf{e}_1^\top(\mathbf{v}_1 - \mathbf{v}_2) > 0 \\
\mathbf{S}_{21} &\Longrightarrow \mathbf{e}_1^\top(\mathbf{v}_1 - \mathbf{v}_2) < 0 \\
\mathbf{R}_{32} &\Longrightarrow \mathbf{e}_2^\top(\mathbf{v}_2 - \mathbf{v}_3) > 0 \\
\mathbf{S}_{32} &\Longrightarrow \mathbf{e}_2^\top(\mathbf{v}_2 - \mathbf{v}_3) < 0 \\
\mathbf{R}_{34} &\Longrightarrow \mathbf{e}_1^\top(\mathbf{v}_4 - \mathbf{v}_3) > 0 \\
\mathbf{S}_{34} &\Longrightarrow \mathbf{e}_1^\top(\mathbf{v}_4 - \mathbf{v}_3) < 0 \\
\mathbf{R}_{41} &\Longrightarrow \mathbf{e}_2^\top(\mathbf{v}_1 - \mathbf{v}_4) > 0 \\
\mathbf{S}_{41} &\Longrightarrow \mathbf{e}_2^\top(\mathbf{v}_1 - \mathbf{v}_4) < 0.
\end{aligned}$$

For each of the three distinct orderings of the states in (7.14), there are two admissible waves on each positive or negative axis, giving a total of 16 cases for each ordering. By comparing the signs of the components of $(\mathbf{v}_1 - \mathbf{v}_2) + (\mathbf{v}_2 - \mathbf{v}_3)$ with $(\mathbf{v}_1 - \mathbf{v}_4) + (\mathbf{v}_4 - \mathbf{v}_3)$, we see that only four of the 16 cases are possible for each ordering. This gives us a total of 12 situations to consider more carefully:

$$\mathbf{R}_{21}^{+}\mathbf{R}_{32}^{+}\mathbf{R}_{34}^{-}\mathbf{R}_{41}^{+},\ h_1 > h_2 > h_3 > h_4 \tag{7.15a}$$
$$\mathbf{R}_{21}^{+}\mathbf{S}_{32}^{-}\mathbf{R}_{34}^{-}\mathbf{S}_{41}^{-},\ h_1 > h_2 > h_3 > h_4 \tag{7.15b}$$
$$\mathbf{S}_{21}^{-}\mathbf{R}_{32}^{+}\mathbf{S}_{34}^{+}\mathbf{R}_{41}^{+},\ h_1 > h_2 > h_3 > h_4 \tag{7.15c}$$
$$\mathbf{S}_{21}^{-}\mathbf{S}_{32}^{-}\mathbf{S}_{34}^{+}\mathbf{S}_{41}^{-},\ h_1 > h_2 > h_3 > h_4 \tag{7.15d}$$
$$\mathbf{R}_{21}^{+}\mathbf{R}_{32}^{+}\mathbf{R}_{34}^{+}\mathbf{R}_{41}^{+},\ h_1 > h_2 > h_4 > h_3 \tag{7.15e}$$
$$\mathbf{R}_{21}^{+}\mathbf{S}_{32}^{-}\mathbf{R}_{34}^{+}\mathbf{S}_{41}^{-},\ h_1 > h_2 > h_4 > h_3 \tag{7.15f}$$
$$\mathbf{S}_{21}^{-}\mathbf{R}_{32}^{+}\mathbf{S}_{34}^{-}\mathbf{R}_{41}^{+},\ h_1 > h_2 > h_4 > h_3 \tag{7.15g}$$
$$\mathbf{S}_{21}^{-}\mathbf{S}_{32}^{-}\mathbf{S}_{34}^{-}\mathbf{S}_{41}^{-},\ h_1 > h_2 > h_4 > h_3 \tag{7.15h}$$
$$\mathbf{R}_{21}^{+}\mathbf{R}_{32}^{-}\mathbf{R}_{34}^{-}\mathbf{R}_{41}^{+},\ h_1 > h_3 > h_4 > h_2 \tag{7.15i}$$
$$\mathbf{R}_{21}^{+}\mathbf{S}_{32}^{+}\mathbf{R}_{34}^{-}\mathbf{S}_{41}^{-},\ h_1 > h_3 > h_4 > h_2 \tag{7.15j}$$
$$\mathbf{S}_{21}^{-}\mathbf{R}_{32}^{-}\mathbf{S}_{34}^{+}\mathbf{R}_{41}^{+},\ h_1 > h_3 > h_4 > h_2 \tag{7.15k}$$
$$\mathbf{S}_{21}^{-}\mathbf{S}_{32}^{+}\mathbf{S}_{34}^{+}\mathbf{S}_{41}^{-},\ h_1 > h_3 > h_4 > h_2. \tag{7.15l}$$

Before examining these cases in detail, we note that $\sqrt{h_R} - \sqrt{h_L}$ is a strictly increasing function of h_R and a strictly decreasing function of h_L. It is also easy to show that on a shock locus $\mathbf{S}^+$

$$(\mathbf{v}_L - \mathbf{v}_R) \cdot \mathbf{n}\sqrt{\frac{2}{g}} = (h_L - h_R)\sqrt{\frac{1}{h_R} + \frac{1}{h_L}}$$

is a strictly increasing function of h_L and a strictly decreasing function of h_R.

All but four of the cases in (7.15) are impossible. For example, the first case (7.15a), namely $\mathbf{R}_{21}^{+}\mathbf{R}_{32}^{+}\mathbf{R}_{34}^{-}\mathbf{R}_{41}^{+}$ and $h_1 > h_2 > h_3 > h_4$ implies that

$$\begin{aligned}\begin{bmatrix} 2(\sqrt{gh_1} - \sqrt{gh_2}) \\ 2(\sqrt{gh_2} - \sqrt{gh_3}) \end{bmatrix} &= (\mathbf{v}_1 - \mathbf{v}_2) + (\mathbf{v}_2 - \mathbf{v}_3) \\ &= (\mathbf{v}_1 - \mathbf{v}_4) + (\mathbf{v}_4 - \mathbf{v}_3) \\ &= \begin{bmatrix} 2(\sqrt{gh_1} - \sqrt{gh_4}) \\ 2(\sqrt{gh_3} - \sqrt{gh_4}) \end{bmatrix}.\end{aligned}$$

The first component of this equation implies that $\sqrt{h_2} = \sqrt{h_4}$, which violates the assumed ordering of the states. In the case (7.15b), the second component of the same vector sum implies that

$$(h_2 - h_3)\sqrt{\frac{1}{h_3} + \frac{1}{h_2}} = (h_1 - h_4)\sqrt{\frac{1}{h_4} + \frac{1}{h_1}}\ ; \tag{7.16}$$

since $h_1 > h_2 > h_3 > h_4$ this equation is impossible. Case (7.15c) is impossible because the first vector component implies that

$$\sqrt{h_1} - \sqrt{h_4} = \sqrt{h_2} - \sqrt{h_3}$$

under the assumption $h_1 > h_2 > h_3 > h_4$. Case (7.15d) is impossible because the second vector component is the impossible equation (7.16). Case (7.15i) is impossible because the first vector component implies that

$$\sqrt{h_1} - \sqrt{h_2} = \sqrt{h_3} - \sqrt{h_4}$$

under the assumption $h_1 > h_3 > h_4 > h_2$. Case (7.15j) is impossible because the first vector component implies the same impossible equation. Case (7.15k) is impossible because the first vector component implies

$$(h_3 - h_4)\sqrt{\frac{1}{h_3} + \frac{1}{h_4}} = (h_1 - h_2)\sqrt{\frac{1}{h_1} + \frac{1}{h_2}}$$

under the assumption $h_1 > h_3 > h_4 > h_2$. Case (7.15l) is impossible for the same reason that the previous case was impossible.

The four possible cases are the following. In case (7.15e) we could have four rarefactions, $\mathbf{R}_{21}^{+}\mathbf{R}_{32}^{+}\mathbf{R}_{34}^{+}\mathbf{R}_{41}^{+}$, under the conditions that $h_1 > h_2 > h_4 > h_3$ and

$$\sqrt{h_1} - \sqrt{h_2} = \sqrt{h_4} - \sqrt{h_3}.$$

Given shallow water heights satisfying this equation and the velocity at one of the states, the velocities at all of the states are determined. Secondly, in case (7.15f) we could have two rarefactions and two shocks, $\mathbf{R}_{21}^{+}\mathbf{S}_{32}^{-}\mathbf{R}_{34}^{+}\mathbf{S}_{41}^{-}$, under the conditions that $h_1 > h_2 > h_4 > h_3$ and

$$\sqrt{h_1} - \sqrt{h_2} = \sqrt{h_4} - \sqrt{h_3}$$

$$(h_2 - h_3)\sqrt{\frac{1}{h_3} + \frac{1}{h_2}} = (h_1 - h_4)\sqrt{\frac{1}{h_4} + \frac{1}{h_1}}.$$

Thirdly, in case (7.15g) we could have two shocks and two rarefactions, $\mathbf{S}_{21}^{-}\mathbf{R}_{32}^{+}\mathbf{S}_{34}^{-}\mathbf{R}_{41}^{+}$, under the conditions that $h_1 > h_2 > h_4 > h_3$ and

$$(h_1 - h_2)\sqrt{\frac{1}{h_2} + \frac{1}{h_1}} = (h_4 - h_3)\sqrt{\frac{1}{h_3} + \frac{1}{h_4}}$$

$$\sqrt{h_2} - \sqrt{h_3} = \sqrt{h_1} - \sqrt{h_4}.$$

Finally, in case (7.15h) we could have four shocks, $\mathbf{S}_{21}^{-}\mathbf{S}_{32}^{-}\mathbf{S}_{34}^{-}\mathbf{S}_{41}^{-}$, under the conditions that $h_1 > h_2 > h_4 > h_3$ and

$$(h_1 - h_2)\sqrt{\frac{1}{h_2} + \frac{1}{h_1}} = (h_4 - h_3)\sqrt{\frac{1}{h_3} + \frac{1}{h_4}}$$

$$(h_2 - h_3)\sqrt{\frac{1}{h_3} + \frac{1}{h_2}} = (h_1 - h_4)\sqrt{\frac{1}{h_4} + \frac{1}{h_1}}.$$

For example, the initial conditions

$\mathbf{w}_2 = \begin{bmatrix} 0.5625 \\ -0.5 \\ 0 \end{bmatrix}$	$\mathbf{w}_1 = \begin{bmatrix} 1 \\ 0 \\ 0 \end{bmatrix}$
$\mathbf{w}_3 = \begin{bmatrix} .0625 \\ -0.5 \\ -1 \end{bmatrix}$	$\mathbf{w}_4 = \begin{bmatrix} 0.25 \\ 0.0 \\ -1 \end{bmatrix}$

will produce four rarefactions.

We have implemented the 2D Riemann problem for shallow water in **Program 7.2-118: GUIRectangle.C.** Students can execute a variety of schemes for this problem by clicking on **Executable 7.2-56: guiRectangle.** The student can select shallow water under `Riemann Problem Parameters`. Students should select the desired `splitting` under `Numerical Method parameters`, and set the boolean flag for their desired method to `True`. In two dimensions, the computational results for water height can be displayed either as 2D contours, 2D color fills or 3D surface plots. The 3D graphics in the surface plot uses a trackball for rotation with the left mouse button. The figure can be sliced along the ends of any coordinate axis using the middle mouse button. Values at points in the figure can be determined using the right mouse button. Students can perform an error analysis, including a comparison of schemes, by setting the number of cells in one of the coordinate directions to 0 under `Numerical Method parameters`. These comparisons can be very time-consuming.

7.2.3 Gas Dynamics

The 2D Riemann problem for gas dynamics was described by Schulz-Rinne [144]. Following [106], we will recommend six of these Riemann problems for student

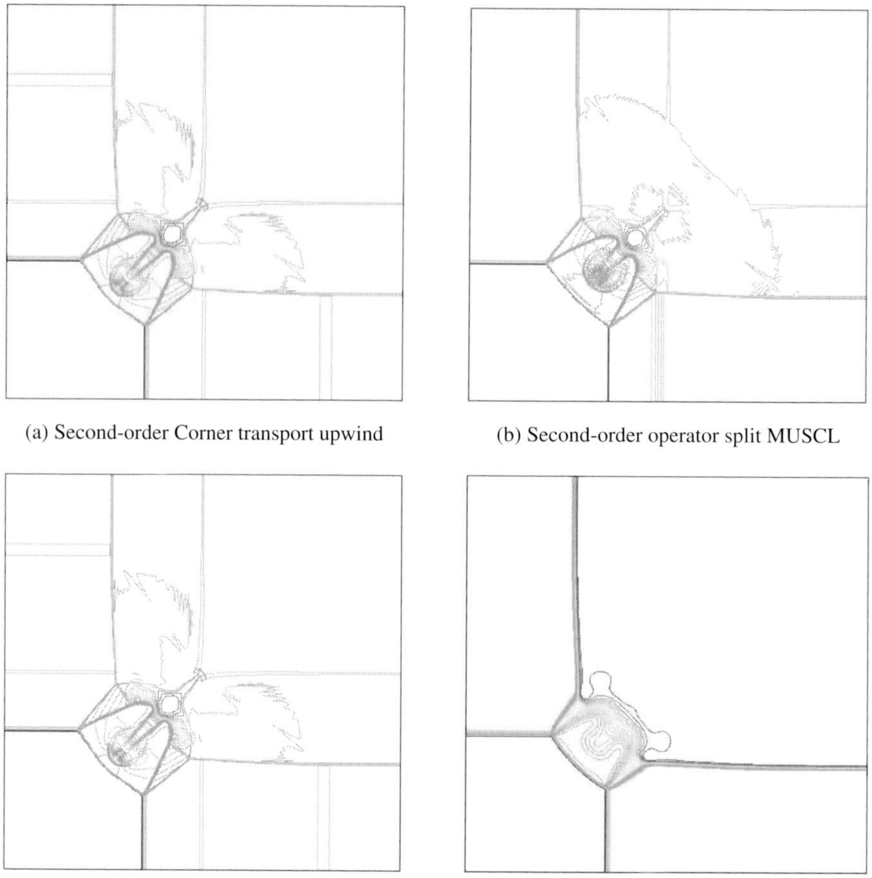

(a) Second-order Corner transport upwind

(b) Second-order operator split MUSCL

(c) Second-order ENO

(d) Second-order Lax-Friedrichs

Fig. 7.3 2D Riemann problem for gas dynamics: 400×400 grid, initial data in Equation (7.17)

consideration. For four shocks producing a narrow jet, try

$$\begin{array}{c|c} \mathbf{w}_2 = \begin{bmatrix} 0.5323 \\ 1.206 \\ 0 \\ 0.3 \end{bmatrix} & \mathbf{w}_1 = \begin{bmatrix} 1.5 \\ 0 \\ 0 \\ 1.5 \end{bmatrix} \\ \hline \mathbf{w}_3 = \begin{bmatrix} 0.138 \\ 1.206 \\ 1.206 \\ 0.029 \end{bmatrix} & \mathbf{w}_4 = \begin{bmatrix} 0.5323 \\ 0.0 \\ 1.206 \\ 0.3 \end{bmatrix} \end{array}. \tag{7.17}$$

(Recall that the flux variable vector $\mathbf{w}$ is ordered ρ, $\mathbf{v}$ then p.) Some numerical results for this problem are shown in Figure 7.3.

Another problem involving four shocks is the following

$$\begin{array}{c|c} \mathbf{w}_2 = \begin{bmatrix} 0.5065 \\ 0.8939 \\ 0 \\ 0.35 \end{bmatrix} & \mathbf{w}_1 = \begin{bmatrix} 1.1 \\ 0 \\ 0 \\ 1.1 \end{bmatrix} \\ \hline \mathbf{w}_3 = \begin{bmatrix} 1.1 \\ 0.8939 \\ 0.8939 \\ 1.1 \end{bmatrix} & \mathbf{w}_4 = \begin{bmatrix} 0.5065 \\ 0.0 \\ 0.8939 \\ 0.35 \end{bmatrix} \end{array}. \tag{7.18}$$

The following problem involves four contact discontinuities:

$$\begin{array}{c|c} \mathbf{w}_2 = \begin{bmatrix} 2 \\ 0.75 \\ 0.5 \\ 1 \end{bmatrix} & \mathbf{w}_1 = \begin{bmatrix} 1 \\ 0.75 \\ -0.5 \\ 1 \end{bmatrix} \\ \hline \mathbf{w}_3 = \begin{bmatrix} 1 \\ -0.75 \\ 0.5 \\ 1 \end{bmatrix} & \mathbf{w}_4 = \begin{bmatrix} 3 \\ -0.75 \\ -0.5 \\ 1 \end{bmatrix} \end{array}. \tag{7.19}$$

The next problem involves two shocks and two contact discontinuities:

$$\begin{array}{c|c} \mathbf{w}_2 = \begin{bmatrix} 1 \\ 0.7276 \\ 0 \\ 1 \end{bmatrix} & \mathbf{w}_1 = \begin{bmatrix} 0.5313 \\ 0 \\ 0 \\ 0.4 \end{bmatrix} \\ \hline \mathbf{w}_3 = \begin{bmatrix} 0.8 \\ 0 \\ 0 \\ 1 \end{bmatrix} & \mathbf{w}_4 = \begin{bmatrix} 1 \\ 0 \\ 0.7276 \\ 1 \end{bmatrix} \end{array}. \tag{7.20}$$

The final two problems each involve two contact discontinuities, a rarefaction and a shock; both produce interesting vortices:

$$\begin{array}{c|c} \mathbf{w}_2 = \begin{bmatrix} 0.5197 \\ -0.6259 \\ -0.3 \\ 0.4 \end{bmatrix} & \mathbf{w}_1 = \begin{bmatrix} 1 \\ 0.1 \\ -0.3 \\ 1 \end{bmatrix} \\ \hline \mathbf{w}_3 = \begin{bmatrix} 0.8 \\ 0.1 \\ -0.3 \\ 0.4 \end{bmatrix} & \mathbf{w}_4 = \begin{bmatrix} 0.4 \\ 0.5313 \\ 0.1 \\ 0.4276 \end{bmatrix} \end{array} \tag{7.21}$$

$$
\begin{array}{c|c}
\mathbf{w}_2 = \begin{bmatrix} 2 \\ 0 \\ -0.3 \\ 1 \end{bmatrix} & \mathbf{w}_1 = \begin{bmatrix} 1 \\ 0 \\ -0.4 \\ 1 \end{bmatrix} \\
\hline
\mathbf{w}_3 = \begin{bmatrix} 1.0625 \\ 0 \\ 0.2145 \\ 0.4 \end{bmatrix} & \mathbf{w}_4 = \begin{bmatrix} 0.4 \\ 0.5197 \\ 0 \\ -1.1259 \end{bmatrix}
\end{array}. \tag{7.22}
$$

Exercises for 7.2

7.2.1 Determine initial conditions for the Burgers' equation 2D Riemann problem so that all waves are rarefactions, and the rarefactions between u_3 and its neighbors are all transonic. Test your favorite numerical scheme in 2D on your Riemann problem.

7.2.2 Determine initial conditions for the Burgers' equation 2D Riemann problem so that the wave pattern is $\mathbf{S}_{21}\mathbf{R}_{32}\mathbf{R}_{34}\mathbf{S}_{41}$, with both rarefactions transonic. Verify your results numerically.

7.2.3 Determine initial conditions for the 2D Riemann problem for shallow water so that all waves are shocks. Verify your solution numerically.

7.2.4 Test your favorite numerical method on problem (7.19). Perform a mesh refinement study. What order of accuracy is your numerical method actually achieving?

7.2.5 We did not discuss MHD in multiple dimensions because there are special issues regarding the treatment of the $\nabla \cdot \mathbf{B} = 0$ condition. Read Tóth [160] and report to the class on the suggested numerical approaches to this problem.

7.3 Numerical Methods in Three Dimensions

7.3.1 Operator Splitting

Suppose that we want to solve the three-dimensional system of partial differential equations

$$\frac{\partial \mathbf{u}}{\partial t} + \frac{\partial \mathbf{f}_1}{\partial \mathbf{x}_1} + \frac{\partial \mathbf{f}_2}{\partial \mathbf{x}_2} + \frac{\partial \mathbf{f}_3}{\partial \mathbf{x}_3} = 0.$$

With spatial operator splitting, we would select a method for the one-dimensional problem

$$\frac{\partial \mathbf{u}}{\partial t} + \frac{\partial \mathbf{f}}{\partial x} = 0,$$

which we will write as

$$\mathbf{u}_i^{n+} = \mathbf{u}_i^n - [\mathbf{f}_{i+1/2}(\mathbf{u}^n) - \mathbf{f}_{i-1/2}(\mathbf{u}^n)]\frac{\triangle t}{\triangle x_i}.$$

A first-order operator splitting scheme for the partial differential equation using the one-dimensional method would take the form

$$\mathbf{u}_{ijk}^{n+1,n,n} = \mathbf{u}_{ijk}^{n} - \left[(\mathbf{f}_1)_{i+1/2,jk}(\mathbf{u}^n) - (\mathbf{f}_1)_{i-1/2,jk}(\mathbf{u}^n)\right] \tag{7.23a}$$

$$\mathbf{u}_{ijk}^{n+1,n+1,n} = \mathbf{u}_{ijk}^{n+1,n,n} - \left[(\mathbf{f}_2)_{i,j+1/2,k}(\mathbf{u}^{n+1,n,n}) - (\mathbf{f}_2)_{i,j-1/2,k}(\mathbf{u}^{n+1,n,n})\right] \tag{7.23b}$$

$$\mathbf{u}_{ijk}^{n+1} = \mathbf{u}_{ijk}^{n+1,n+1,n} - \left[(\mathbf{f}_3)_{i,j,k+1/2}(\mathbf{u}^{n+1,n+1,n}) - (\mathbf{f}_3)_{i,j,k-1/2}(\mathbf{u}^{n+1,n+1,n})\right]. \tag{7.23c}$$

No matter what the order of the one-dimensional scheme might be, the resulting operator splitting scheme would be at most first-order accurate in time. Its spatial order would be determined by the spatial order of the one-dimensional scheme. The operator splitting scheme is typically stable if each of its individual steps is stable.

A second-order operator splitting of a three-dimensional conservation law might consider the differential equation in the form

$$\frac{\partial u}{\partial t} + \left(\frac{\partial \mathbf{f}_1}{\partial \mathbf{x}_1} + \frac{\partial \mathbf{f}_2}{\partial \mathbf{x}_2}\right) + \frac{\partial \mathbf{f}_3}{\partial \mathbf{x}_3} = 0.$$

Other pairings of the partial derivatives would also work. Thus one timestep with second-order operator splitting might look like

$$\begin{aligned}
\mathbf{u}_{ijk}^{n,n,n+1/2} &= \mathbf{u}_{ijk}^{n} - \left[(\mathbf{f}_3)_{i,j,k+1/2}(\mathbf{u}^n) - (\mathbf{f}_3)_{i,j,k-1/2}(\mathbf{u}^n)\right] \frac{\triangle t^{n+1/2}}{2\triangle x_{3,k}}, \\
\mathbf{u}_{ijk}^{n,n+1/2,n+1/2} &= \mathbf{u}_{ijk}^{n,n,n+1/2} - \left[(\mathbf{f}_2)_{i,j+1/2,k}(\mathbf{u}^{n,n,n+1/2}) \right. \\
&\qquad \left. - (\mathbf{f}_2)_{i,j-1/2,k}(\mathbf{u}^{n,n,n+1/2})\right] \frac{\triangle t^{n+1/2}}{2\triangle x_{2,j}}, \\
\mathbf{u}_{ijk}^{n+1,n+1/2,n+1/2} &= \mathbf{u}_{ijk}^{n,n+1/2,n+1/2} - \left[(\mathbf{f}_1)_{i+1/2,j,k}(\mathbf{u}^{n,n+1/2,n+1/2}) \right. \\
&\qquad \left. - (\mathbf{f}_1)_{i-1/2,j,k}(\mathbf{u}^{n,n+1/2,n+1/2})\right] \frac{\triangle t^{n+1/2}}{\triangle x_{1,i}}, \\
\mathbf{u}_{ijk}^{n+1,n+1,n+1/2} &= \mathbf{u}_{ijk}^{n+1,n+1/2,n+1/2} - \left[(\mathbf{f}_2)_{i,j+1/2,k}(\mathbf{u}^{n+1,n+1/2,n+1/2}) \right. \\
&\qquad \left. - (\mathbf{f}_2)_{i,j-1/2,k}(\mathbf{u}^{n+1,n+1/2,n+1/2})\right] \frac{\triangle t^{n+1/2}}{2\triangle x_{2,j}}, \\
\mathbf{u}_{ijk}^{n+1} &= \mathbf{u}_{ijk}^{n+1,n+1,n+1/2} - \left[(\mathbf{f}_3)_{i,j,k+1/2}(\mathbf{u}^{n+1,n+1,n+1/2}) \right. \\
&\qquad \left. - (\mathbf{f}_3)_{i,j,k-1/2}(\mathbf{u}^{n+1,n+1,n+1/2})\right] \frac{\triangle t^{n+1/2}}{2\triangle x_{3,k}}.
\end{aligned}$$

As with the two-dimensional scheme, it is possible to combine the last update in the third coordinate direction at the end of one timestep with the first update at the beginning of the next timestep.

Operator splittings of even higher order are possible [58]. Such splittings necessarily involve negative splitting coefficients, which complicate upwinding issues and stability considerations. A different approach might be to use deferred correction [46]. This approach can reach a pre-determined temporal order using lower-order schemes, such as first- or second-order operator splitting. However, this approach requires more data storage than higher-order operator splitting schemes. Neither of these two approaches are useful for increasing spatial order.

7.3.2 Donor Cell Methods

The integral form of the three-dimensional conservation law

$$\frac{\partial \mathbf{u}}{\partial t} + \sum_{\ell=1}^{3} \frac{\partial \mathbf{F}(\mathbf{u})\mathbf{e}_\ell}{\partial \mathbf{x}_\ell} = 0$$

is

$$\int_{\Omega_{ijk}} \mathbf{u}(\mathbf{x}, t^{n+1})\, d\mathbf{x} = \int_{\Omega_{ijk}} \mathbf{u}(\mathbf{x}, t^{n})\, d\mathbf{x} - \int_{t^n}^{t^{n+1}} \int_{\partial\Omega_{ijk}} \mathbf{F}(\mathbf{u})\mathbf{n}\, ds\, dt,$$

where $\mathbf{n}$ is the outer normal and s is surface area. Let us denote the intervals

$$\begin{aligned} I_{1,i} &= (\mathbf{x}_{1,i-1/2}, \mathbf{x}_{1,i+1/2}) \\ I_{2,j} &= (\mathbf{x}_{2,j-1/2}, \mathbf{x}_{2,j+1/2}) \\ I_{3,k} &= (\mathbf{x}_{3,k-1/2}, \mathbf{x}_{3,k+1/2}). \end{aligned}$$

On the rectangular grid cell $\Omega_{ijk} = I_{1,i} \times I_{2,j} \times I_{3,k}$ the integral form can be written

$$\begin{aligned} &\int_{I_{3,k}} \int_{I_{2,j}} \int_{I_{1,i}} \mathbf{u}(\mathbf{x}_1, \mathbf{x}_2, \mathbf{x}_3, t^{n+1})\, d\mathbf{x}_1\, d\mathbf{x}_2\, d\mathbf{x}_3 \\ &\quad = \int_{I_{3,k}} \int_{I_{2,j}} \int_{I_{1,i}} \mathbf{u}(\mathbf{x}_1, \mathbf{x}_2, \mathbf{x}_3, t^{n})\, d\mathbf{x}_1\, d\mathbf{x}_2\, d\mathbf{x}_3 \\ &\qquad - \int_{t^n}^{t^{n+1}} \int_{I_{3,k}} \int_{I_{2,j}} \mathbf{F}(\mathbf{u}(\mathbf{x}_{1,i+1/2}, \mathbf{x}_2, \mathbf{x}_3, t))\mathbf{e}_1\, d\mathbf{x}_2\, d\mathbf{x}_3\, dt \\ &\qquad + \int_{t^n}^{t^{n+1}} \int_{I_{3,k}} \int_{I_{2,j}} \mathbf{F}(\mathbf{u}(\mathbf{x}_{1,i-1/2}, \mathbf{x}_2, \mathbf{x}_3, t))\mathbf{e}_2\, d\mathbf{x}_2\, d\mathbf{x}_3\, dt \\ &\qquad - \int_{t^n}^{t^{n+1}} \int_{I_{1,i}} \int_{I_{3,k}} \mathbf{F}(\mathbf{u}(\mathbf{x}_1, \mathbf{x}_{2,j+1/2}, \mathbf{x}_3, t))\mathbf{e}_2\, d\mathbf{x}_3\, d\mathbf{x}_1\, dt \\ &\qquad + \int_{t^n}^{t^{n+1}} \int_{I_{1,i}} \int_{I_{3,k}} \mathbf{F}(\mathbf{u}(\mathbf{x}_1, \mathbf{x}_{2,j-1/2}, \mathbf{x}_3, t))\mathbf{e}_2\, d\mathbf{x}_3\, d\mathbf{x}_1\, dt \\ &\qquad - \int_{t^n}^{t^{n+1}} \int_{I_{2,j}} \int_{I_{1,i}} \mathbf{F}(\mathbf{u}(\mathbf{x}_1, \mathbf{x}_2, \mathbf{x}_{3,k+1/2}, t))\mathbf{e}_3\, d\mathbf{x}_1\, d\mathbf{x}_2\, dt \\ &\qquad + \int_{t^n}^{t^{n+1}} \int_{I_{2,j}} \int_{I_{1,i}} \mathbf{F}(\mathbf{u}(\mathbf{x}_1, \mathbf{x}_2, \mathbf{x}_{3,k-1/2}, t))\mathbf{e}_3\, d\mathbf{x}_1\, d\mathbf{x}_2\, dt. \end{aligned} \tag{7.24}$$

We will define cell averages

$$\mathbf{u}^n_{ijk} \triangle x_{1,i} \triangle x_{2,j} \triangle x_{3,k} \approx \int_{I_{3,k}} \int_{I_{2,j}} \int_{I_{1,i}} \mathbf{u}(\mathbf{x}_1, \mathbf{x}_2, \mathbf{x}_3, t^n) \, d\mathbf{x}_1 \, d\mathbf{x}_2 \, d\mathbf{x}_3$$

and flux side and time integrals

$$\mathbf{f}^{n+1/2}_{i+1/2,jk} \approx \int_{t^n}^{t^{n+1}} \int_{I_{3,k}} \int_{I_{2,j}} \mathbf{F}(\mathbf{u}(\mathbf{x}_{1,i+1/2}, \mathbf{x}_2, \mathbf{x}_3, t))\mathbf{e}_1 \, d\mathbf{x}_2 \, d\mathbf{x}_3 \, dt$$

$$\mathbf{f}^{n+1/2}_{i,j+1/2,k} \approx \int_{t^n}^{t^{n+1}} \int_{I_{1,i}} \int_{I_{3,k}} \mathbf{F}(\mathbf{u}(\mathbf{x}_1, \mathbf{x}_{2,j+1/2}, \mathbf{x}_3, t))\mathbf{e}_2 \, d\mathbf{x}_3 \, d\mathbf{x}_1 \, dt$$

$$\mathbf{f}^{n+1/2}_{ij,k+1/2} \approx \int_{t^n}^{t^{n+1}} \int_{I_{2,j}} \int_{I_{1,i}} \mathbf{F}(\mathbf{u}(\mathbf{x}_1, \mathbf{x}_2, \mathbf{x}_{3,k+1/2}, t))\mathbf{e}_3 \, d\mathbf{x}_1 \, d\mathbf{x}_2 \, dt$$

in order to perform a conservative difference

$$\begin{aligned}\mathbf{u}^{n+1}_{ijk} \triangle x_{1,i} \triangle x_{2,j} \triangle x_{3,k} = \mathbf{u}^n_{ijk} \triangle x_{1,i} \triangle x_{2,j} \triangle x_{3,k} + \mathbf{f}^{n+1/2}_{i-1/2,jk} - \mathbf{f}^{n+1/2}_{i+1/2,jk} + \mathbf{f}^{n+1/2}_{i,j-1/2,k} \\ - \mathbf{f}^{n+1/2}_{i,j+1/2,k} + \mathbf{f}^{n+1/2}_{ij,k-1/2} - \mathbf{f}^{n+1/2}_{ij,k+1/2}.\end{aligned} \tag{7.25}$$

Individual schemes will differ in how the flux integrals are computed.

In the traditional donor cell upwind scheme, the flux integrals are given by

$$\mathbf{f}^{n+1/2}_{i+1/2,jk} = \mathbf{F}\left(\mathcal{R}\left(\mathbf{u}^n_{ijk}, \mathbf{u}^n_{i+1,jk}\right)\right) \mathbf{e}_1 \triangle x_{2,j} \triangle x_{3,k} \triangle t^{n+1/2} \tag{7.26a}$$

$$\mathbf{f}^{n+1/2}_{i,j+1/2,k} = \mathbf{F}\left(\mathcal{R}\left(\mathbf{u}^n_{ijk}, \mathbf{u}^n_{i,j+1,k}\right)\right) \mathbf{e}_2 \triangle x_{3,k} \triangle x_{1,i} \triangle t^{n+1/2} \tag{7.26b}$$

$$\mathbf{f}^{n+1/2}_{ij,k+1/2} = \mathbf{F}\left(\mathcal{R}\left(\mathbf{u}^n_{ijk}, \mathbf{u}^n_{ij,k+1}\right)\right) \mathbf{e}_3 \triangle x_{1,i} \triangle x_{2,j} \triangle t^{n+1/2}. \tag{7.26c}$$

As in the two-dimensional case, this choice leads to restricted stability, as can easily be seen by considering the linear advection equation. If the velocity vector $\mathbf{v}$ is constant and $\mathbf{F}(\mathbf{u}) = \mathbf{u}\mathbf{v}^\top$, then the donor cell upwind flux integrals are

$$\mathbf{f}^{n+1/2}_{i+1/2,jk} = \left[\mathbf{u}^n_{ijk}\mathbf{v}_1^+ + \mathbf{u}^n_{i+1,jk}\mathbf{v}_1^-\right] \triangle x_{2,j} \triangle x_{3,k} \triangle t^{n+1/2}$$

$$\mathbf{f}^{n+1/2}_{i,j+1/2,k} = \left[\mathbf{u}^n_{ijk}\mathbf{v}_2^+ + \mathbf{u}^n_{i,j+1,k}\mathbf{v}_2^-\right] \triangle x_{3,k} \triangle x_{1,i} \triangle t^{n+1/2}$$

$$\mathbf{f}^{n+1/2}_{ij,k+1/2} = \left[\mathbf{u}^n_{ijk}\mathbf{v}_3^+ + \mathbf{u}^n_{ij,k+1}\mathbf{v}_3^-\right] \triangle x_{1,i} \triangle x_{2,j} \triangle t^{n+1/2}.$$

Then the cell averages are given by

$$\begin{aligned}&\mathbf{u}^{n+1}_{ijk} \triangle x_{1,i} \triangle x_{2,j} \triangle x_{3,k} \\ &\quad = \mathbf{u}^n_{ijk} \triangle x_{1,i} \triangle x_{2,j} \triangle x_{3,k} \\ &\qquad + \left\{\mathbf{u}^n_{i-1,jk}\mathbf{v}_1^+ + \mathbf{u}^n_{ijk}\mathbf{v}_1^- - \mathbf{u}^n_{ijk}\mathbf{v}_1^+ + \mathbf{u}_{i+1,jk}\mathbf{v}_1^-\right\} \triangle x_2 \triangle x_3 \triangle t^{n+1/2} \\ &\qquad + \left\{\mathbf{u}^n_{i,j-1,k}\mathbf{v}_2^+ + \mathbf{u}^n_{ijk}\mathbf{v}_2^- - \mathbf{u}^n_{ijk}\mathbf{v}_2^+ + \mathbf{u}_{i,j+1,k}\mathbf{v}_2^-\right\} \triangle x_3 \triangle x_1 \triangle t^{n+1/2} \\ &\qquad + \left\{\mathbf{u}^n_{ij,k-1}\mathbf{v}_3^+ + \mathbf{u}^n_{ijk}\mathbf{v}_3^- - \mathbf{u}^n_{ijk}\mathbf{v}_3^+ + \mathbf{u}_{ij,k+1}\mathbf{v}_3^-\right\} \triangle x_1 \triangle x_2 \triangle t^{n+1/2}\end{aligned}$$

$$
\begin{aligned}
&= \mathbf{u}^n_{ijk}\left(\triangle x_{1,i}\triangle x_{2,j}\triangle x_{3,k} - |\mathbf{v}_1|\triangle x_{2,j}\triangle x_{3,k}\triangle t^{n+1/2}\right.\\
&\quad \left. - |\mathbf{v}_2|\triangle x_{3,k}\triangle x_{1,i}\triangle t^{n+1/2} - |\mathbf{v}_3|\triangle x_{1,i}\triangle x_{2,j}\triangle t^{n+1/2}\right)\\
&\quad + \left[\mathbf{u}^n_{i-1,jk}\mathbf{v}_1^+ - \mathbf{u}^n_{i+1,jk}\mathbf{v}_1^-\right]\triangle x_{2,j}\triangle x_{3,k}\triangle t^{n+1/2}\\
&\quad + \left[\mathbf{u}^n_{i,j-1,k}\mathbf{v}_2^+ - \mathbf{u}^n_{i,j+1,k}\mathbf{v}_2^-\right]\triangle x_{3,k}\triangle x_{1,i}\triangle t^{n+1/2}\\
&\quad + \left[\mathbf{u}^n_{ij,k-1}\mathbf{v}_3^+ - \mathbf{u}^n_{ij,k+1}\mathbf{v}_3^-\right]\triangle x_{1,i}\triangle x_{2,j}\triangle t^{n+1/2}.
\end{aligned}
$$

The coefficients in this equation are nonnegative if and only if the timestep satisfies

$$
\left[\frac{|\mathbf{v}_1|}{\triangle x_{1,i}} + \frac{|\mathbf{v}_2|}{\triangle x_{2,j}} + \frac{|\mathbf{v}_3|}{\triangle x_{3,k}}\right]\triangle t^{n+1/2} \le 1. \tag{7.27}
$$

If this inequality is violated, then the donor cell upwind scheme is unstable.

7.3.3 Corner Transport Upwind Scheme

The 3D corner transport upwind scheme is described in [140]. This scheme is most easily developed for linear advection, for which the analytical solution is $\mathbf{u}(\mathbf{x}, t) = \tilde{\mathbf{u}}_0(\mathbf{x} - \mathbf{v}t)$. If we integrate the conservation law over the grid cell Ω_{ijk}, then the corner transport upwind scheme computes

$$
\int_{\Omega_{ijk}} \mathbf{u}^{n+1}(\mathbf{x})\, d\mathbf{x} = \int_{\Omega_{ijk}} \mathbf{u}^n(\mathbf{x} - \mathbf{v}\triangle t^{n+1/2})\, d\mathbf{x} = \int_{R_{ijk}} \mathbf{u}^n(\mathbf{x})\, d\mathbf{x}
$$

where

$$
\begin{aligned}
R_{ijk} &= (\mathbf{x}_{1,i-1/2} - \mathbf{v}_1\triangle t^{n+1/2}, \mathbf{x}_{1,i+1/2} - \mathbf{v}_1\triangle t^{n+1/2})\\
&\quad \times (\mathbf{x}_{2,j-1/2} - \mathbf{v}_2\triangle t^{n+1/2}, \mathbf{x}_{2,j+1/2} - \mathbf{v}_2\triangle t^{n+1/2})\\
&\quad \times (\mathbf{x}_{3,k-1/2} - \mathbf{v}_3\triangle t^{n+1/2}, \mathbf{x}_{3,k+1/2} - \mathbf{v}_3\triangle t^{n+1/2})
\end{aligned}
$$

is the rectangle formed by tracing Ω_{ijk} back in time along the velocity field. The equation $\int_{\Omega_{ijk}} \mathbf{u}^{n+1}(\mathbf{x})\, d\mathbf{x} = \int_{R_{ijk}} \mathbf{u}^n(\mathbf{x})\, d\mathbf{x}$ can be written

$$
\begin{aligned}
&\mathbf{u}^{n+1}_{ijk}\triangle x_{1,i}\triangle x_{2,j}\triangle x_{3,k}\\
&= \mathbf{u}^n_{ijk}\left(\triangle x_{1,i} - |\mathbf{v}_1|\triangle t^{n+1/2}\right)\left(\triangle x_{2,j} - |\mathbf{v}_2|\triangle t^{n+1/2}\right)\left(\triangle x_{3,k} - |\mathbf{v}_3|\triangle t^{n+1/2}\right)\\
&\quad + \left[\mathbf{u}^n_{i-1,jk}\mathbf{v}_1^+ + \mathbf{u}^n_{i+1,jk}(-\mathbf{v}_1^-)\right]\triangle t^{n+1/2}\left(\triangle x_{2,j} - |\mathbf{v}_2|\triangle t^{n+1/2}\right)\left(\triangle x_{3,k} - |\mathbf{v}_3|\triangle t^{n+1/2}\right)\\
&\quad + \left[\mathbf{u}^n_{i,j-1,k}\mathbf{v}_2^+ + \mathbf{u}^n_{i,j+1,k}(-\mathbf{v}_2^-)\right]\triangle t^{n+1/2}\left(\triangle x_{3,k} - |\mathbf{v}_3|\triangle t^{n+1/2}\right)\left(\triangle x_{1,i} - |\mathbf{v}_1|\triangle t^{n+1/2}\right)\\
&\quad + \left[\mathbf{u}^n_{ij,k-1}\mathbf{v}_3^+ + \mathbf{u}^n_{ij,k+1}(-\mathbf{v}_3^-)\right]\triangle t^{n+1/2}\left(\triangle x_{1,i} - |\mathbf{v}_1|\triangle t^{n+1/2}\right)\left(\triangle x_{2,j} - |\mathbf{v}_2|\triangle t^{n+1/2}\right)\\
&\quad + \left[\mathbf{u}_{i-1,j-1,k}\mathbf{v}_1^+\mathbf{v}_2^+ + \mathbf{u}_{i+1,j-1,k}(-\mathbf{v}_1^-)\mathbf{v}_2^+ + \mathbf{u}_{i-1,j+1,k}\mathbf{v}_1^+(-\mathbf{v}_2^-) + \mathbf{u}_{i+1,j+1,k}(-\mathbf{v}_1^-)(-\mathbf{v}_2^-)\right]\\
&\qquad \left(\triangle t^{n+1/2}\right)^2\left(\triangle x_{3,k} - |\mathbf{v}_3|\triangle t^{n+1/2}\right)
\end{aligned}
$$

$$
\begin{aligned}
&+ \left[\mathbf{u}_{i-1,j,k-1}\mathbf{v}_3^+\mathbf{v}_1^+ + \mathbf{u}_{i-1,j,k+1}(-\mathbf{v}_3^-)\mathbf{v}_1^+ + \mathbf{u}_{i+1,j,k-1}\mathbf{v}_3^+(-\mathbf{v}_1^-) + \mathbf{u}_{i+1,j,k+1}(-\mathbf{v}_3^-)(-\mathbf{v}_1^-)\right] \\
&\quad \left(\triangle t^{n+1/2}\right)^2 \left(\triangle x_{2,j} - |\mathbf{v}_2|\triangle t^{n+1/2}\right) \\
&+ \left[\mathbf{u}_{i,j-1,k-1}\mathbf{v}_2^+\mathbf{v}_3^+ + \mathbf{u}_{i,j+1,k-1}(-\mathbf{v}_2^-)\mathbf{v}_3^+ + \mathbf{u}_{i,j-1,k+1}\mathbf{v}_2^+(-\mathbf{v}_3^-) + \mathbf{u}_{i,j+1,k+1}(-\mathbf{v}_2^-)(-\mathbf{v}_3^-)\right] \\
&\quad \left(\triangle t^{n+1/2}\right)^2 \left(\triangle x_{1,i} - |\mathbf{v}_1|\triangle t^{n+1/2}\right) \\
&+ \mathbf{u}_{i-1,j-1,k-1}\mathbf{v}_1^+\mathbf{v}_2^+\mathbf{v}_3^+ \left(\triangle t^{n+1/2}\right)^3 + \mathbf{u}_{i+1,j-1,k-1}(-\mathbf{v}_1^-)\mathbf{v}_2^+\mathbf{v}_3^+ \left(\triangle t^{n+1/2}\right)^3 \\
&+ \mathbf{u}_{i-1,j+1,k-1}\mathbf{v}_1^+(-\mathbf{v}_2^-)\mathbf{v}_3^+ \left(\triangle t^{n+1/2}\right)^3 + \mathbf{u}_{i+1,j+1,k-1}(-\mathbf{v}_1^-)(-\mathbf{v}_2^-)\mathbf{v}_3^+ \left(\triangle t^{n+1/2}\right)^3 \\
&+ \mathbf{u}_{i-1,j-1,k+1}\mathbf{v}_1^+\mathbf{v}_2^+(-\mathbf{v}_3^-) \left(\triangle t^{n+1/2}\right)^3 + \mathbf{u}_{i+1,j-1,k+1}(-\mathbf{v}_1^-)\mathbf{v}_2^+(-\mathbf{v}_3^-) \left(\triangle t^{n+1/2}\right)^3 \\
&+ \mathbf{u}_{i-1,j+1,k+1}\mathbf{v}_1^+(-\mathbf{v}_2^-)(-\mathbf{v}_3^-) \left(\triangle t^{n+1/2}\right)^3 + \mathbf{u}_{i+1,j+1,k+1}(-\mathbf{v}_1^-)(-\mathbf{v}_2^-)(-\mathbf{v}_3^-) \left(\triangle t^{n+1/2}\right)^3 .
\end{aligned}
$$

From this equation, it is easy to see that the new solution is a weighted average of old solution values if and only if the timestep is chosen so that in every grid cell

$$
\max\left\{\frac{|\mathbf{v}_1|}{\triangle x_{1,i}}, \frac{|\mathbf{v}_2|}{\triangle x_{2,j}}, \frac{|\mathbf{v}_3|}{\triangle x_{3,k}}\right\} \triangle t^{n+1/2} \le 1.
$$

This stability restriction is considerably better than the donor cell stability restriction, and identical to the first- and second-order operator splitting stability restrictions.

If $P_{i\pm1/2,jk}$, $P_{i,j\pm1/2,k}$ and $P_{ij,k\pm1/2}$ are signed parallelepipeds associated with the velocity fields at the cell sides, then

$$
R_{ijk} = \Omega_{ijk} - P_{i+1/2,jk} + P_{i-1/2,jk} - P_{i,j+1/2,k} + P_{i,j-1/2,k} - P_{ij,k+1/2} + P_{ij,k-1/2}.
$$

(See Figure 7.4.) The donor cell fluxes are associated with integrating the initial data $\mathbf{u}^n$ over the signed regions

$$
\begin{aligned}
S_{i+1/2,jk} &= (\mathbf{x}_{1,i+1/2} - \mathbf{v}_1\triangle t^{n+1/2}, \mathbf{x}_{1,i+1/2}) \times I_{2,j} \times I_{3,k} \\
S_{i,j+1/2,k} &= I_{1,i} \times (\mathbf{x}_{2,j+1/2} - \mathbf{v}_2\triangle t^{n+1/2}, \mathbf{x}_{2,j+1/2}) \times I_{3,k} \\
S_{ij,k+1/2} &= I_{1,i} \times I_{2,j} \times (\mathbf{x}_{3,k+1/2} - \mathbf{v}_3\triangle t^{n+1/2}, \mathbf{x}_{3,k+1/2}).
\end{aligned}
$$

These are each completely contained within a single grid cell. Correspondingly, each parallelepiped can in turn be decomposed as a donor cell region plus or minus prisms associated with the cell edges (see Figure 7.4):

$$
\begin{aligned}
P_{i+1/2,jk} &= S_{i+1/2,jk} - Q_{i+1/2,j+1/2,k} + Q_{i+1/2,j-1/2,k} \\
&\quad - Q_{i+1/2,j,k+1/2} + Q_{i+1/2,j,k-1/2} \\
P_{i,j+1/2,k} &= S_{i,j+1/2,k} - Q_{i,j+1/2,k+1/2} + Q_{i,j+1/2,k-1/2} \\
&\quad - Q_{i+1/2,j+1/2,k} + Q_{i-1/2,j+1/2,k} \\
P_{ij,k+1/2} &= S_{ij,k+1/2} - Q_{i+1/2,j,k+1/2} + Q_{i-1/2,j,k+1/2} \\
&\quad - Q_{i,j+1/2,k+1/2} + Q_{i,j-1/2,k+1/2}.
\end{aligned}
$$

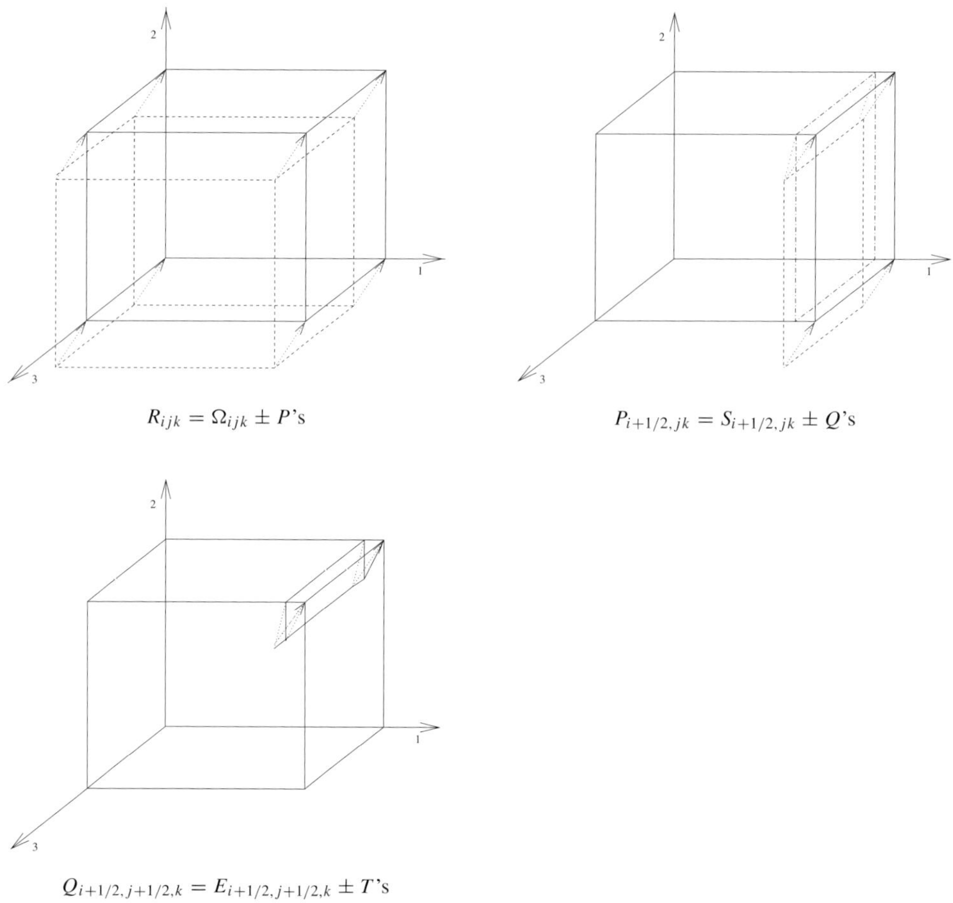

Fig. 7.4 Conservative difference involves fluxes at cell sides

The prisms lie primarily in either of four grid cells around an edge, depending on the sign of the velocity components associated with coordinate directions other than the cell index for the edge. Further, there are two prisms at each cell edge, each determined by having one side perpendicular to one of the two coordinate axes that are perpendicular to the edge. The average of $\mathbf{u}^n$ is the same for both of these prisms at a given cell edge within a given cell, but differs for all four edges within a given cell due to the velocity field. The edge prisms can be further decomposed in terms of prisms contained within the grid cells, and tetrahedrons associated with the cell corners (see Figure 7.4):

$$
\begin{aligned}
Q_{i,j+1/2,k+1/2} &= E_{i,j+1/2,k+1/2} - T_{i+1/2,j+1/2,k+1/2} + T_{i-1/2,j+1/2,k+1/2} \\
Q_{i+1/2,j,k+1/2} &= E_{i+1/2,j,k+1/2} - T_{i+1/2,j+1/2,k+1/2} + T_{i+1/2,j-1/2,k+1/2} \\
Q_{i+1/2,j+1/2,k} &= E_{i+1/2,j+1/2,k} - T_{i+1/2,j+1/2,k+1/2} + T_{i+1/2,j+1/2,k-1/2}.
\end{aligned}
$$

Of course, there are three tetrahedrons at each cell corner, determined by extending a coordinate axis from a corner of Ω_{ijk} to a side of R_{ijk}. The average of $\mathbf{u}^n$ is the same for all three of these, so we do not distinguish them in our notation.

7.3.3.1 Linear Advection with Positive Velocity

The expressions for the corner transport upwind scheme with general velocity field are too long to present here. In order to develop the scheme in finite difference form, we will assume that the velocity components are all positive and use upwinding considerations to lead us to the general expressions. The averages over the prisms $Q_{i,j\pm1/2,k\pm1/2}$, $Q_{i\pm1/2,j,k\pm1/2}$ and $Q_{i\pm1/2,j\pm1/2,k}$ primarily within cell Ω_{ijk} are (respectively)

$$\mathbf{u}_{ijk}^{n+1/3,1} = \mathbf{u}_{ijk}^n \left(1 - \frac{\mathbf{v}_1 \triangle t^{n+1/2}}{3\triangle x_{1,i}}\right) + \mathbf{u}_{i-1,jk} \frac{\mathbf{v}_1 \triangle t^{n+1/2}}{3\triangle x_{1,i}} \tag{7.28a}$$

$$\mathbf{u}_{ijk}^{n+1/3,2} = \mathbf{u}_{ijk}^n \left(1 - \frac{\mathbf{v}_2 \triangle t^{n+1/2}}{3\triangle x_{2,j}}\right) + \mathbf{u}_{i,j-1,k} \frac{\mathbf{v}_2 \triangle t^{n+1/2}}{3\triangle x_{2,j}} \tag{7.28b}$$

$$\mathbf{u}_{ijk}^{n+1/3,3} = \mathbf{u}_{ijk}^n \left(1 - \frac{\mathbf{v}_3 \triangle t^{n+1/2}}{3\triangle x_{3,k}}\right) + \mathbf{u}_{ij,k-1} \frac{\mathbf{v}_3 \triangle t^{n+1/2}}{3\triangle x_{3,k}}. \tag{7.28c}$$

Using these definitions, we see that

$$\begin{aligned}
&\mathbf{u}_{i-1,j-1,k-1}^n \mathbf{v}_1 \triangle t^{n+1/2} \mathbf{v}_2 \triangle t^{n+1/2} \mathbf{v}_3 \triangle t^{n+1/2} \\
&= \left[\mathbf{u}_{i,j-1,k-1}^{n+1/3,1} - \mathbf{u}_{i,j-1,k-1}^n \left(1 - \frac{\mathbf{v}_1 \triangle t^{n+1/2}}{3\triangle x_{1,i}}\right)\right] \triangle x_{1,i} \mathbf{v}_2 \triangle t^{n+1/2} \mathbf{v}_3 \triangle t^{n+1/2} \\
&\quad + \left[\mathbf{u}_{i-1,j,k-1}^{n+1/3,2} - \mathbf{u}_{i-1,j,k-1}^n \left(1 - \frac{\mathbf{v}_2 \triangle t^{n+1/2}}{3\triangle x_{2,j}}\right)\right] \triangle x_{2,j} \mathbf{v}_3 \triangle t^{n+1/2} \mathbf{v}_1 \triangle t^{n+1/2} \\
&\quad + \left[\mathbf{u}_{i-1,j-1,k}^{n+1/3,3} - \mathbf{u}_{i-1,j-1,k}^n \left(1 - \frac{\mathbf{v}_3 \triangle t^{n+1/2}}{3\triangle x_{3,k}}\right)\right] \triangle x_{3,k} \mathbf{v}_1 \triangle t^{n+1/2} \mathbf{v}_2 \triangle t^{n+1/2}.
\end{aligned}$$

Geometrically, this equation says that the sum of the integrals over the corner regions in $R_{ijk} - \Omega_{ijk}$ is the sum of the integrals over the prisms minus the sum of the integrals over the edge regions. Algebraically, we are using equations (7.28) to replace $\mathbf{u}_{i-1,j-1,k-1}^n$ at the corner of the stencil with states associated with the edges. Using the definitions of the prism averages again, we obtain

$$\begin{aligned}
&\mathbf{u}_{i-1,j-1,k}^n \mathbf{v}_1 \triangle t^{n+1/2} \mathbf{v}_2 \triangle t^{n+1/2} (\triangle x_{3,k} - \mathbf{v}_3 \triangle t^{n+1/2}) \\
&\quad + \mathbf{u}_{i-1,j,k-1}^n \mathbf{v}_3 \triangle t^{n+1/2} \mathbf{v}_1 \triangle t^{n+1/2} (\triangle x_{2,j} - \mathbf{v}_2 \triangle t^{n+1/2}) \\
&\quad + \mathbf{u}_{i,j-1,k-1}^n \mathbf{v}_2 \triangle t^{n+1/2} \mathbf{v}_3 \triangle t^{n+1/2} (\triangle x_{1,i} - \mathbf{v}_1 \triangle t^{n+1/2}) \\
&\quad + \mathbf{u}_{i-1,j-1,k-1}^n \mathbf{v}_1 \triangle t^{n+1/2} \mathbf{v}_2 \triangle t^{n+1/2} \mathbf{v}_3 \triangle t^{n+1/2}
\end{aligned}$$

$$
\begin{aligned}
&= \mathbf{u}^{n+1/3,1}_{i,j-1,k-1} \triangle x_{1,i} \mathbf{v}_2 \triangle t^{n+1/2} \mathbf{v}_3 \triangle t^{n+1/2} \\
&\quad + \mathbf{u}^{n+1/3,2}_{i-1,j,k-1} \triangle x_{2,j} \mathbf{v}_3 \triangle t^{n+1/2} \mathbf{v}_1 \triangle t^{n+1/2} \\
&\quad + \mathbf{u}^{n+1/3,3}_{i-1,j-1,k} \triangle x_{3,k} \mathbf{v}_1 \triangle t^{n+1/2} \mathbf{v}_2 \triangle t^{n+1/2} \\
&\quad - \left[\mathbf{u}^{n}_{i-1,j-1,k} + \mathbf{u}^{n}_{i-1,j,k-1} + \mathbf{u}^{n}_{i,j-1,k-1}\right] \mathbf{v}_1 \triangle t^{n+1/2} \mathbf{v}_2 \triangle t^{n+1/2} \mathbf{v}_3 \triangle t^{n+1/2} \frac{2}{3} \\
&= \left[\mathbf{u}^{n+1/3,1}_{i,j-1,k-1} - \mathbf{u}^{n+1/3,1}_{i,j-1,k} - \mathbf{u}^{n+1/3,1}_{ij,k-1}\right] \triangle x_{1,i} \mathbf{v}_2 \triangle t^{n+1/2} \mathbf{v}_3 \triangle t^{n+1/2} \\
&\quad + \left[\mathbf{u}^{n+1/3,2}_{i-1,j,k-1} - \mathbf{u}^{n+1/3,2}_{i-1,jk} - \mathbf{u}^{n+1/3,2}_{ij,k-1}\right] \triangle x_{2,j} \mathbf{v}_3 \triangle t^{n+1/2} \mathbf{v}_1 \triangle t^{n+1/2} \\
&\quad + \left[\mathbf{u}^{n+1/3,3}_{i-1,j-1,k} - \mathbf{u}^{n+1/3,3}_{i-1,jk} - \mathbf{u}^{n+1/3,3}_{i,j-1,k}\right] \triangle x_{3,k} \mathbf{v}_1 \triangle t^{n+1/2} \mathbf{v}_2 \triangle t^{n+1/2} \\
&\quad + \mathbf{u}^{n}_{i-1,jk} \mathbf{v}_1 \triangle t^{n+1/2} \left[\left(1 - \frac{\mathbf{v}_2 \triangle t^{n+1/2}}{3 \triangle x_{2,j}}\right) \triangle x_{2,j} \mathbf{v}_3 \triangle t^{n+1/2} + \left(1 - \frac{\mathbf{v}_3 \triangle t^{n+1/2}}{3 \triangle x_{3,k}}\right) \triangle x_{3,k} \mathbf{v}_2 \triangle t^{n+1/2}\right] \\
&\quad + \mathbf{u}^{n}_{i,j-1,k} \mathbf{v}_2 \triangle t^{n+1/2} \left[\left(1 - \frac{\mathbf{v}_1 \triangle t^{n+1/2}}{3 \triangle x_{1,i}}\right) \triangle x_{1,i} \mathbf{v}_3 \triangle t^{n+1/2} + \left(1 - \frac{\mathbf{v}_3 \triangle t^{n+1/2}}{3 \triangle x_{3,k}}\right) \triangle x_{3,k} \mathbf{v}_1 \triangle t^{n+1/2}\right] \\
&\quad + \mathbf{u}^{n}_{ij,k-1} \mathbf{v}_3 \triangle t^{n+1/2} \left[\left(1 - \frac{\mathbf{v}_1 \triangle t^{n+1/2}}{3 \triangle x_{1,i}}\right) \triangle x_{1,i} \mathbf{v}_2 \triangle t^{n+1/2} + \left(1 - \frac{\mathbf{v}_2 \triangle t^{n+1/2}}{3 \triangle x_{2,j}}\right) \triangle x_{2,j} \mathbf{v}_1 \triangle t^{n+1/2}\right].
\end{aligned}
$$

Geometrically, we are relating integrals over corner and edge regions in R_{ijk} to edge prisms E and Q. Algebraically, we are using (7.28) to replace values of $\mathbf{u}$ associated with edges with values associated with the sides. Next, the sum of the integrals in $R_{ijk} - \Omega_{ijk}$ is

$$
\begin{aligned}
&\mathbf{u}^{n}_{i-1,jk} \mathbf{v}_1 \triangle t^{n+1/2} (\triangle x_{2,j} - \mathbf{v}_2 \triangle t^{n+1/2})(\triangle x_{3,k} - \mathbf{v}_3 \triangle t^{n+1/2}) \\
&\quad + \mathbf{u}^{n}_{i,j-1,k} \mathbf{v}_2 \triangle t^{n+1/2} (\triangle x_{3,k} - \mathbf{v}_3 \triangle t^{n+1/2})(\triangle x_{1,i} - \mathbf{v}_1 \triangle t^{n+1/2}) \\
&\quad + \mathbf{u}^{n}_{ij,k-1} \mathbf{v}_3 \triangle t^{n+1/2} (\triangle x_{1,i} - \mathbf{v}_1 \triangle t^{n+1/2})(\triangle x_{2,j} - \mathbf{v}_2 \triangle t^{n+1/2}) \\
&\quad + \mathbf{u}^{n}_{i-1,j-1,k} \mathbf{v}_1 \triangle t^{n+1/2} \mathbf{v}_2 \triangle t^{n+1/2} (\triangle x_{3,k} - \mathbf{v}_3 \triangle t^{n+1/2}) \\
&\quad + \mathbf{u}^{n}_{i-1,j,k-1} \mathbf{v}_3 \triangle t^{n+1/2} \mathbf{v}_1 \triangle t^{n+1/2} (\triangle x_{2,j} - \mathbf{v}_2 \triangle t^{n+1/2}) \\
&\quad + \mathbf{u}^{n}_{i,j-1,k-1} \mathbf{v}_2 \triangle t^{n+1/2} \mathbf{v}_3 \triangle t^{n+1/2} (\triangle x_{1,i} - \mathbf{v}_1 \triangle t^{n+1/2}) \\
&\quad + \mathbf{u}^{n}_{i-1,j-1,k-1} \mathbf{v}_1 \triangle t^{n+1/2} \mathbf{v}_2 \triangle t^{n+1/2} \mathbf{v}_3 \triangle t^{n+1/2} \\
&= \left[\mathbf{u}^{n+1/3,1}_{i,j-1,k-1} - \mathbf{u}^{n+1/3,1}_{i,j-1,k} - \mathbf{u}^{n+1/3,1}_{ij,k-1}\right] \triangle x_{1,i} \mathbf{v}_2 \triangle t^{n+1/2} \mathbf{v}_3 \triangle t^{n+1/2} \\
&\quad + \left[\mathbf{u}^{n+1/3,2}_{i-1,j,k-1} - \mathbf{u}^{n+1/3,2}_{i-1,jk} - \mathbf{u}^{n+1/3,2}_{ij,k-1}\right] \triangle x_{2,j} \mathbf{v}_3 \triangle t^{n+1/2} \mathbf{v}_1 \triangle t^{n+1/2} \\
&\quad + \left[\mathbf{u}^{n+1/3,3}_{i-1,j-1,k} - \mathbf{u}^{n+1/3,3}_{i-1,jk} - \mathbf{u}^{n+1/3,3}_{i,j-1,k}\right] \triangle x_{3,k} \mathbf{v}_1 \triangle t^{n+1/2} \mathbf{v}_2 \triangle t^{n+1/2}
\end{aligned}
$$

$$+ \mathbf{u}^n_{i-1,jk} \mathbf{v}_1 \triangle t^{n+1/2} \left[\triangle x_{2,j} \triangle x_{3,k} + \mathbf{v}_2 \triangle t^{n+1/2} \mathbf{v}_3 \triangle t^{n+1/2} \frac{1}{3} \right]$$

$$+ \mathbf{u}^n_{i,j-1,k} \mathbf{v}_2 \triangle t^{n+1/2} \left[\triangle x_{1,i} \triangle x_{3,k} + \mathbf{v}_1 \triangle t^{n+1/2} \mathbf{v}_3 \triangle t^{n+1/2} \frac{1}{3} \right]$$

$$+ \mathbf{u}^n_{ij,k-1} \mathbf{v}_3 \triangle t^{n+1/2} \left[\triangle x_{1,i} \triangle x_{2,j} + \mathbf{v}_1 \triangle t^{n+1/2} \mathbf{v}_2 \triangle t^{n+1/2} \frac{1}{3} \right].$$

The integral over $R_{ijk} \cap \Omega_{ijk}$ is

$$\begin{aligned}
&\mathbf{u}^n_{ijk} \left(\triangle x_{1,i} - \mathbf{v}_1 \triangle t^{n+1/2} \right) \left(\triangle x_{2,j} - \mathbf{v}_2 \triangle t^{n+1/2} \right) \left(\triangle x_{3,k} - \mathbf{v}_3 \triangle t^{n+1/2} \right) \\
&\quad = \mathbf{u}^n_{ijk} \triangle x_{1,i} \triangle x_{2,j} \triangle x_{3,k} \\
&\qquad - \mathbf{u}^n_{ijk} \left[\mathbf{v}_1 \triangle t^{n+1/2} \triangle x_{2,j} \triangle x_{3,k} + \mathbf{v}_2 \triangle t^{n+1/2} \triangle x_{3,k} \triangle x_{1,i} + \mathbf{v}_3 \triangle t^{n+1/2} \triangle x_{1,i} \triangle x_{2,j} \right] \\
&\qquad + \mathbf{u}^n_{ijk} \left(1 - \frac{\mathbf{v}_1 \triangle t^{n+1/2}}{3 \triangle x_{1,i}} \right) \triangle x_{1,i} \mathbf{v}_2 \triangle t^{n+1/2} \mathbf{v}_3 \triangle t^{n+1/2} \\
&\qquad + \mathbf{u}^n_{ijk} \left(1 - \frac{\mathbf{v}_2 \triangle t^{n+1/2}}{3 \triangle x_{2,j}} \right) \mathbf{v}_1 \triangle t^{n+1/2} \triangle x_{2,j} \mathbf{v}_3 \triangle t^{n+1/2} \\
&\qquad + \mathbf{u}^n_{ijk} \left(1 - \frac{\mathbf{v}_3 \triangle t^{n+1/2}}{3 \triangle x_{3,k}} \right) \mathbf{v}_1 \triangle t^{n+1/2} \mathbf{v}_2 \triangle t^{n+1/2} \triangle x_{3,k} \\
&\quad = \mathbf{u}^n_{ijk} \triangle x_{1,i} \triangle x_{2,j} \triangle x_{3,k} \\
&\qquad - \mathbf{u}^n_{ijk} \left[\mathbf{v}_1 \triangle t^{n+1/2} \triangle x_{2,j} \triangle x_{3,k} + \mathbf{v}_2 \triangle t^{n+1/2} \triangle x_{3,k} \triangle x_{1,i} + \mathbf{v}_3 \triangle t^{n+1/2} \triangle x_{1,i} \triangle x_{2,j} \right] \\
&\qquad + \mathbf{v}_1 \triangle t^{n+1/2} \triangle x_{2,j} \triangle x_{3,k} \left\{ \left[\mathbf{u}^{n+1/3,2}_{ijk} - \mathbf{u}^n_{i,j-1,k} \frac{\mathbf{v}_2 \triangle t^{n+1/2}}{3 \triangle x_{2,j}} \right] \frac{\mathbf{v}_3 \triangle t^{n+1/2}}{2 \triangle x_{3,k}} \right. \\
&\qquad\qquad \left. + \left[\mathbf{u}^{n+1/3,3}_{ijk} - \mathbf{u}^n_{ij,k-1} \frac{\mathbf{v}_3 \triangle t^{n+1/2}}{3 \triangle x_{3,k}} \right] \frac{\mathbf{v}_2 \triangle t^{n+1/2}}{2 \triangle x_{2,j}} \right\} \\
&\qquad + \mathbf{v}_2 \triangle t^{n+1/2} \triangle x_{3,k} \triangle x_{1,i} \left\{ \left[\mathbf{u}^{n+1/3,1}_{ijk} - \mathbf{u}^n_{i-1,jk} \frac{\mathbf{v}_1 \triangle t^{n+1/2}}{3 \triangle x_{1,i}} \right] \frac{\mathbf{v}_3 \triangle t^{n+1/2}}{2 \triangle x_{3,k}} \right. \\
&\qquad\qquad \left. + \left[\mathbf{u}^{n+1/3,3}_{ijk} - \mathbf{u}^n_{ij,k-1} \frac{\mathbf{v}_3 \triangle t^{n+1/2}}{3 \triangle x_{3,k}} \right] \frac{\mathbf{v}_1 \triangle t^{n+1/2}}{2 \triangle x_{1,i}} \right\} \\
&\qquad + \mathbf{v}_3 \triangle t^{n+1/2} \triangle x_{1,i} \triangle x_{2,j} \left\{ \left[\mathbf{u}^{n+1/3,1}_{ijk} - \mathbf{u}^n_{i-1,jk} \frac{\mathbf{v}_1 \triangle t^{n+1/2}}{3 \triangle x_{1,i}} \right] \frac{\mathbf{v}_2 \triangle t^{n+1/2}}{2 \triangle x_{2,j}} \right. \\
&\qquad\qquad \left. + \left[\mathbf{u}^{n+1/3,2}_{ijk} - \mathbf{u}^n_{i,j-1,k} \frac{\mathbf{v}_2 \triangle t^{n+1/2}}{3 \triangle x_{2,j}} \right] \frac{\mathbf{v}_1 \triangle t^{n+1/2}}{2 \triangle x_{1,i}} \right\}
\end{aligned}$$

Putting all of the integrals over the pieces of R_{ijk} together, we obtain

$$
\begin{aligned}
&\mathbf{u}_{ijk}^{n+1} \triangle x_{1,i} \triangle x_{2,j} \triangle x_{3,k} \\
&\quad = \mathbf{u}_{ijk}^{n} \triangle x_{1,i} \triangle x_{2,j} \triangle x_{3,k} \\
&\qquad + \mathbf{v}_1 \triangle t^{n+1/2} \triangle x_{2,j} \triangle x_{3,k} \left\{ -\mathbf{u}_{ijk}^{n} + \left[\mathbf{u}_{ijk}^{n+1/3,2} - \mathbf{u}_{ij,k-1}^{n+1/3,2} \right] \frac{\mathbf{v}_3 \triangle t^{n+1/2}}{2 \triangle x_{3,k}} \right. \\
&\qquad\qquad \left. + \left[\mathbf{u}_{ijk}^{n+1/3,3} - \mathbf{u}_{i,j-1,k}^{n+1/3,3} \right] \frac{\mathbf{v}_2 \triangle t^{n+1/2}}{2 \triangle x_{2,j}} \right\} \\
&\qquad + \mathbf{v}_1 \triangle t^{n+1/2} \triangle x_{2,j} \triangle x_{3,k} \left\{ \mathbf{u}_{i-1,jk}^{n} + \left[\mathbf{u}_{i-1,j,k-1}^{n+1/3,2} - \mathbf{u}_{i-1,jk}^{n+1/3,2} \right] \frac{\mathbf{v}_3 \triangle t^{n+1/2}}{2 \triangle x_{3,k}} \right. \\
&\qquad\qquad \left. + \left[\mathbf{u}_{i-1,j-1,k}^{n+1/3,3} - \mathbf{u}_{i-1,jk}^{n+1/3,3} \right] \frac{\mathbf{v}_2 \triangle t^{n+1/2}}{2 \triangle x_{2,j}} \right\} \\
&\qquad + \mathbf{v}_2 \triangle t^{n+1/2} \triangle x_{3,k} \triangle x_{1,i} \left\{ -\mathbf{u}_{ijk}^{n} + \left[\mathbf{u}_{ijk}^{n+1/3,1} - \mathbf{u}_{i-1,jk}^{n+1/3,1} \right] \frac{\mathbf{v}_3 \triangle t^{n+1/2}}{2 \triangle x_{3,k}} \right. \\
&\qquad\qquad \left. + \left[\mathbf{u}_{ijk}^{n+1/3,3} - \mathbf{u}_{i,j-1,k}^{n+1/3,3} \right] \frac{\mathbf{v}_1 \triangle t^{n+1/2}}{2 \triangle x_{1,i}} \right\} \\
&\qquad + \mathbf{v}_2 \triangle t^{n+1/2} \triangle x_{3,k} \triangle x_{1,i} \left\{ \mathbf{u}_{i,j-1,k}^{n} + \left[\mathbf{u}_{i,j-1,k-1}^{n+1/3,1} - \mathbf{u}_{i,j-1,k}^{n+1/3,1} \right] \frac{\mathbf{v}_3 \triangle t^{n+1/2}}{2 \triangle x_{3,k}} \right. \\
&\qquad\qquad \left. + \left[\mathbf{u}_{i-1,j-1,k}^{n+1/3,3} - \mathbf{u}_{i,j-1,k}^{n+1/3,3} \right] \frac{\mathbf{v}_1 \triangle t^{n+1/2}}{2 \triangle x_{1,i}} \right\} \\
&\qquad + \mathbf{v}_3 \triangle t^{n+1/2} \triangle x_{1,i} \triangle x_{2,j} \left\{ -\mathbf{u}_{ijk}^{n} + \left[\mathbf{u}_{ijk}^{n+1/3,1} - \mathbf{u}_{i,j-1,k}^{n+1/3,1} \right] \frac{\mathbf{v}_2 \triangle t^{n+1/2}}{2 \triangle x_{2,j}} \right. \\
&\qquad\qquad \left. + \left[\mathbf{u}_{ijk}^{n+1/3,2} - \mathbf{u}_{i-1,jk}^{n+1/3,2} \right] \frac{\mathbf{v}_1 \triangle t^{n+1/2}}{2 \triangle x_{1,i}} \right\} \\
&\qquad + \mathbf{v}_3 \triangle t^{n+1/2} \triangle x_{1,i} \triangle x_{2,j} \left\{ \mathbf{u}_{ij,k-1}^{n} + \left[\mathbf{u}_{i,j-1,k-1}^{n+1/3,1} - \mathbf{u}_{ij,k-1}^{n+1/3,1} \right] \frac{\mathbf{v}_2 \triangle t^{n+1/2}}{2 \triangle x_{2,j}} \right. \\
&\qquad\qquad \left. + \left[\mathbf{u}_{i-1,j-1,k}^{n+1/3,2} - \mathbf{u}_{ij,k-1}^{n+1/3,2} \right] \frac{\mathbf{v}_1 \triangle t^{n+1/2}}{2 \triangle x_{1,i}} \right\} \\
&\quad \equiv \mathbf{u}_{ijk}^{n} \triangle x_{1,i} \triangle x_{2,j} \triangle x_{3,k} + \left[-\mathbf{u}_{i+1/2,jk}^{n+1/2,L} + \mathbf{u}_{i-1/2,jk}^{n+1/2,L} \right] \mathbf{v}_1 \triangle t^{n+1/2} \triangle x_{2,j} \triangle x_{3,k} \\
&\qquad + \left[-\mathbf{u}_{i,j+1/2,j}^{n+1/2,L} + \mathbf{u}_{i,j-1/2,k}^{n+1/2,L} \right] \mathbf{v}_2 \triangle t^{n+1/2} \triangle x_{3,k} \triangle x_{1,i} \\
&\qquad + \left[-\mathbf{u}_{ij,k+1/2}^{n+1/2,L} + \mathbf{u}_{ij,k-1/2}^{n+1/2,L} \right] \mathbf{v}_3 \triangle t^{n+1/2} \triangle x_{1,i} \triangle x_{2,j} .
\end{aligned}
$$

The terms in the braces in this equation are the averages over the side parallelepipeds $P_{i\pm1/2,jk}$, $P_{i,j\pm1/2,k}$ and $P_{ij,k\pm1/2}$.

Let us rewrite the averages over the side parallelepipeds:

$$
\begin{aligned}
\mathbf{u}_{i+1/2,jk}^{n+1/2,L} &= \mathbf{u}_{ijk}^{n} - \left[\mathbf{u}_{ijk}^{n+1/3,2} - \mathbf{u}_{ij,k-1}^{n+1/3,2}\right] \frac{\mathbf{v}_3 \triangle t^{n+1/2}}{2\triangle x_{3,k}} - \left[\mathbf{u}_{ijk}^{n+1/3,3} - \mathbf{u}_{i,j-1,k}^{n+1/3,3}\right] \frac{\mathbf{v}_2 \triangle t^{n+1/2}}{2\triangle x_{2,j}} \\
&= \mathbf{u}_{ijk}^{n} - \left[\mathbf{F}(\mathbf{u}_{ijk}^{n+1/3,2}) - \mathbf{F}(\mathbf{u}_{ij,k-1}^{n+1/3,2})\right] \mathbf{e}_3 \frac{\triangle t^{n+1/2}}{2\triangle x_{3,k}} - \left[\mathbf{F}(\mathbf{u}_{ijk}^{n+1/3,3}) - \mathbf{F}(\mathbf{u}_{i,j-1,k}^{n+1/3,3})\right] \mathbf{e}_2 \frac{\triangle t^{n+1/2}}{2\triangle x_{2,j}} \\
\mathbf{u}_{i,j+1/2,k}^{n+1/2,L} &= \mathbf{u}_{ijk}^{n} - \left[\mathbf{u}_{ijk}^{n+1/3,1} - \mathbf{u}_{ij,k-1}^{n+1/3,1}\right] \frac{\mathbf{v}_3 \triangle t^{n+1/2}}{2\triangle x_{3,k}} - \left[\mathbf{u}_{ijk}^{n+1/3,3} - \mathbf{u}_{i-1,jk}^{n+1/3,3}\right] \frac{\mathbf{v}_1 \triangle t^{n+1/2}}{2\triangle x_{1,i}} \\
&= \mathbf{u}_{ijk}^{n} - \left[\mathbf{F}(\mathbf{u}_{ijk}^{n+1/3,1} - \mathbf{F}(\mathbf{u}_{ij,k-1}^{n+1/3,1})\right] \mathbf{e}_3 \frac{\triangle t^{n+1/2}}{2\triangle x_{3,k}} - \left[\mathbf{F}(\mathbf{u}_{ijk}^{n+1/3,3}) - \mathbf{F}(\mathbf{u}_{i-1,jk}^{n+1/3,3})\right] \mathbf{e}_1 \frac{\triangle t^{n+1/2}}{2\triangle x_{1,i}} \\
\mathbf{u}_{ij,k+1/2}^{n+1/2,L} &= \mathbf{u}_{ijk}^{n} - \left[\mathbf{u}_{ijk}^{n+1/3,1} - \mathbf{u}_{i,j-1,k}^{n+1/3,1}\right] \frac{\mathbf{v}_2 \triangle t^{n+1/2}}{2\triangle x_{2,j}} - \left[\mathbf{u}_{ijk}^{n+1/3,2} - \mathbf{u}_{i-1,jk}^{n+1/3,2}\right] \frac{\mathbf{v}_1 \triangle t^{n+1/2}}{2\triangle x_{1,i}} \\
&= \mathbf{u}_{ijk}^{n} - \left[\mathbf{F}(\mathbf{u}_{ijk}^{n+1/3,1}) - \mathbf{F}(\mathbf{u}_{i,j-1,k}^{n+1/3,1})\right] \mathbf{e}_2 \frac{\triangle t^{n+1/2}}{2\triangle x_{2,j}} - \left[\mathbf{F}(\mathbf{u}_{ijk}^{n+1/3,2}) - \mathbf{F}(\mathbf{u}_{i-1,jk}^{n+1/3,2})\right] \mathbf{e}_1 \frac{\triangle t^{n+1/2}}{2\triangle x_{1,i}}.
\end{aligned}
$$

Similarly, the averages over the edge prisms can be written

$$
\begin{aligned}
\mathbf{u}_{ijk}^{n+1/3,1} &= \mathbf{u}_{ijk}^{n} - \left[\mathbf{u}_{ijk}^{n} - \mathbf{u}_{i-1,jk}^{n}\right] \frac{\mathbf{v}_1 \triangle t^{n+1/2}}{3\triangle x_{1,i}} = \mathbf{u}_{ijk}^{n} - \left[\mathbf{F}(\mathbf{u}_{ijk}^{n}) - \mathbf{F}(\mathbf{u}_{i-1,jk}^{n})\right] \mathbf{e}_1 \frac{\triangle t^{n+1/2}}{3\triangle x_{1,i}} \\
\mathbf{u}_{ijk}^{n+1/3,2} &= \mathbf{u}_{ijk}^{n} - \left[\mathbf{u}_{ijk}^{n} - \mathbf{u}_{i,j-1,k}^{n}\right] \frac{\mathbf{v}_2 \triangle t^{n+1/2}}{3\triangle x_{2,j}} = \mathbf{u}_{ijk}^{n} - \left[\mathbf{F}(\mathbf{u}_{ijk}^{n}) - \mathbf{F}(\mathbf{u}_{i,j-1,k}^{n})\right] \mathbf{e}_2 \frac{\triangle t^{n+1/2}}{3\triangle x_{2,j}} \\
\mathbf{u}_{ijk}^{n+1/3,3} &= \mathbf{u}_{ijk}^{n} - \left[\mathbf{u}_{ijk}^{n} - \mathbf{u}_{ij,k-1}^{n}\right] \frac{\mathbf{v}_3 \triangle t^{n+1/2}}{3\triangle x_{3,k}} = \mathbf{u}_{ijk}^{n} - \left[\mathbf{F}(\mathbf{u}_{ijk}^{n}) - \mathbf{F}(\mathbf{u}_{ij,k-1}^{n})\right] \mathbf{e}_3 \frac{\triangle t^{n+1/2}}{3\triangle x_{3,k}}.
\end{aligned}
$$

7.3.3.2 Linear Advection with Arbitrary Velocity

Next, let us extend this scheme to linear advection problems with arbitrary velocity fields. First we compute

$$
\begin{aligned}
\mathbf{u}_{ijk}^{n+1/3,1} &= \mathbf{u}_{ijk}^{n} - \left[\left(\mathbf{u}_{ijk}^{n}\mathbf{v}_1^{+} + \mathbf{u}_{i+1,jk}^{n}\mathbf{v}_1^{-}\right) - \left(\mathbf{u}_{i-1,jk}^{n}\mathbf{v}_1^{+} + \mathbf{u}_{ijk}^{n}\mathbf{v}_1^{-}\right)\right] \frac{\triangle t^{n+1/2}}{3\triangle x_{1,i}} \\
&= \mathbf{u}_{ijk}^{n} - \left[\mathbf{F}\left(\mathcal{R}\left(\mathbf{u}_{ijk}^{n}, \mathbf{u}_{i+1,jk}^{n}; 0\right)\right) - \mathbf{F}\left(\mathcal{R}\left(\mathbf{u}_{i-1,jk}^{n}, \mathbf{u}_{ijk}^{n}; 0\right)\right)\right] \mathbf{e}_1 \frac{\triangle t^{n+1/2}}{3\triangle x_{1,i}}.
\end{aligned}
$$

The expressions for $\mathbf{u}_{ijk}^{n+1/3,2}$ and $\mathbf{u}_{ijk}^{n+1/3,3}$ are similar. Then we compute

$$\begin{aligned}
\mathbf{u}_{i+1/2,jk}^{n+1/2,L} &= \mathbf{u}_{ijk}^{n} - \left[\left(\mathbf{u}_{ijk}^{n+1/3,2}\mathbf{v}_3^{+} + \mathbf{u}_{ij,k+1}^{n+1/3,2}\mathbf{v}_3^{-}\right) - \left(\mathbf{u}_{ij,k-1}^{n+1/3,2}\mathbf{v}_3^{+} + \mathbf{u}_{ijk}^{n+1/3,2}\mathbf{v}_3^{-}\right)\right]\frac{\triangle t^{n+1/2}}{2\triangle x_{3,k}}\\
&\quad - \left[\left(\mathbf{u}_{ijk}^{n+1/3,3}\mathbf{v}_2^{+} + \mathbf{u}_{i,j+1,k}^{n+1/3,3}\mathbf{v}_2^{-}\right) - \left(\mathbf{u}_{i,j-1,k}^{n+1/3,3}\mathbf{v}_2^{+} + \mathbf{u}_{ijk}^{n+1/3,3}\mathbf{v}_2^{-}\right)\right]\frac{\triangle t^{n+1/2}}{2\triangle x_{2,j}}\\
&= \mathbf{u}_{ijk}^{n} - \left[\mathbf{F}\left(\mathcal{R}\left(\mathbf{u}_{ijk}^{n+1/3,2}, \mathbf{u}_{ij,k+1}^{n+1/3,2}; 0\right)\right) - \mathbf{F}\left(\mathcal{R}\left(\mathbf{u}_{ij,k-1}^{n+1/3,2}, \mathbf{u}_{ijk}^{n+1/3,2}; 0\right)\right)\right]\mathbf{e}_3\frac{\triangle t^{n+1/2}}{2\triangle x_{3,k}}\\
&\quad - \left[\mathbf{F}\left(\mathcal{R}\left(\mathbf{u}_{ijk}^{n+1/3,3}, \mathbf{u}_{i,j+1,k}^{n+1/3,3}; 0\right)\right) - \mathbf{F}\left(\mathcal{R}\left(\mathbf{u}_{i,j-1,k}^{n+1/3,3}, \mathbf{u}_{ijk}^{n+1/3,3}; 0\right)\right)\right]\mathbf{e}_2\frac{\triangle t^{n+1/2}}{2\triangle x_{2,j}}\\
\mathbf{u}_{i+1/2,jk}^{n+1/2,R} &= \mathbf{u}_{i+1,jk}^{n} - \left[\left(\mathbf{u}_{i+1,jk}^{n+1/3,2}\mathbf{v}_3^{+} + \mathbf{u}_{i+1,j,k+1}^{n+1/3,2}\mathbf{v}_3^{-}\right) - \left(\mathbf{u}_{i+1,j,k-1}^{n+1/3,2}\mathbf{v}_3^{+} + \mathbf{u}_{i+1,jk}^{n+1/3,2}\right)\right]\frac{\triangle t^{n+1/2}}{2\triangle x_{3,k}}\\
&\quad - \left[\left(\mathbf{u}_{i+1,jk}^{n+1/3,3}\mathbf{v}_2^{+} + \mathbf{u}_{i+1,j+1,k}^{n+1/3,3}\mathbf{v}_2\right) - \left(\mathbf{u}_{i+1,j-1,k}^{n+1/3,3}\mathbf{v}_2^{+} + \mathbf{u}_{i+1,jk}^{n+1/3,3}\mathbf{v}_2^{-}\right)\right]\frac{\triangle t^{n+1/2}}{2\triangle x_{2,j}}\\
&= \mathbf{u}_{ijk}^{n} - \left[\mathbf{F}\left(\mathcal{R}\left(\mathbf{u}_{i+1,jk}^{n+1/3,2}, \mathbf{u}_{i+1,j,k+1}^{n+1/3,2}; 0\right)\right) - \mathbf{F}\left(\mathcal{R}\left(\mathbf{u}_{i+1,j,k-1}^{n+1/3,2}, \mathbf{u}_{i+1,jk}^{n+1/3,2}; 0\right)\right)\right]\mathbf{e}_3\frac{\triangle t^{n+1/2}}{2\triangle x_{3,k}}\\
&\quad - \left[\mathbf{F}\left(\mathcal{R}\left(\mathbf{u}_{i+1,jk}^{n+1/3,3}, \mathbf{u}_{i+1,j+1,k}^{n+1/3,3}; 0\right)\right) - \mathbf{F}\left(\mathcal{R}\left(\mathbf{u}_{i+1,j-1,k}^{n+1/3,3}, \mathbf{u}_{i+1,jk}^{n+1/3,3}; 0\right)\right)\right]\mathbf{e}_2\frac{\triangle t^{n+1/2}}{2\triangle x_{2,j}}.
\end{aligned}$$

The expressions for $\mathbf{u}_{i,j+1/2,k}^{n+1/2,L}$, $\mathbf{u}_{i,j+1/2,k}^{n+1/2,R}$, $\mathbf{u}_{i,j,k+1/2}^{n+1/2,L}$ and $\mathbf{u}_{i,j,k+1/2}^{n+1/2,R}$ are similar. Afterward, we compute the flux integrals

$$\begin{aligned}
\mathbf{f}_{i+1/2,jk}^{n+1/2} &= \left[\mathbf{u}_{i+1/2,jk}^{n+1/2,L}\mathbf{v}_1^{+} + \mathbf{u}_{i+1/2,jk}^{n+1/2,R}\mathbf{v}_1^{-}\right]\triangle x_{2,j}\triangle x_{3,k}\triangle t^{n+1/2}\\
&= \mathbf{F}\left(\mathcal{R}\left(\mathbf{u}_{i+1/2,jk}^{n+1/2,L}, \mathbf{u}_{i+1/2,jk}^{n+1/2,R}; 0\right)\right)\mathbf{e}_1\triangle x_{2,j}\triangle x_{3,k}\triangle t^{n+1/2}
\end{aligned}$$

and the flux integrals $\mathbf{f}_{i,j+1/2,k}^{n+1/2}$ and $\mathbf{f}_{ij,k+1/2}^{n+1/2}$ using similar expressions. The corner transport upwind scheme for general linear advection is completed with the conservative difference (7.25).

7.3.3.3 General Nonlinear Problems

Now it is easy to describe the first-order corner transport upwind scheme for general nonlinear systems in three dimensions. First, we compute the flux integrals

$$(\mathbf{f}_1)_{i+\frac{1}{2},jk}^{n} = \mathbf{F}\left(\mathcal{R}\left(\mathbf{w}_{ijk}^{n}, \mathbf{w}_{i+1,jk}^{n}; 0\right)\right)\mathbf{e}_1\triangle x_{2,j}\triangle x_{3,k}\triangle t^{n+1/2},$$

and the flux integrals $(\mathbf{f}_2)_{i,j+\frac{1}{2},k}^{n}$ and $(\mathbf{f}_3)_{ij,k+\frac{1}{2}}^{n}$ using similar expressions. This step costs a total of three Riemann problems per cell. Then we compute the states

$$\mathbf{w}_{ijk}^{n+\frac{1}{3},1} = \mathbf{w}_{ijk}^{n} - \left(\frac{\partial \mathbf{u}}{\partial \mathbf{w}}\right)^{-1} \left[(\mathbf{f}_1)_{i+\frac{1}{2},jk} - (\mathbf{f}_1)_{i-\frac{1}{2},jk}\right] \frac{1}{3\triangle x_{1,i}\triangle x_{2,j}\triangle x_{3,k}}$$

and the states $\mathbf{w}_{ijk}^{n+\frac{1}{3},2}$ and $\mathbf{w}_{ijk}^{n+\frac{1}{3},3}$ using similar expressions. An alternative approach would be to compute $\mathbf{u}_{ijk}^{n+1/3,1}$ in an obvious fashion, then decode $\mathbf{w}_{ijk}^{n+\frac{1}{3},1}$ from it. Next, we compute the flux integrals

$$(\mathbf{f}_1)_{i+\frac{1}{2},jk}^{n+\frac{1}{3},2} = \mathbf{F}\left(\mathcal{R}\left(\mathbf{w}_{ijk}^{n+\frac{1}{3},2}, \mathbf{w}_{i+1,jk}^{n+\frac{1}{3},2}; 0\right)\right) \mathbf{e}_1 \triangle x_{2,j}\triangle x_{3,k}\triangle t^{n+1/2}$$
$$(\mathbf{f}_1)_{i+\frac{1}{2},jk}^{n+\frac{1}{3},3} = \mathbf{F}\left(\mathcal{R}\left(\mathbf{w}_{ijk}^{n+\frac{1}{3},3}, \mathbf{w}_{i+1,jk}^{n+\frac{1}{3},3}; 0\right)\right) \mathbf{e}_1 \triangle x_{2,j}\triangle x_{3,k}\triangle t^{n+1/2}$$

with similar expressions for the flux integrals $(\mathbf{f}_2)_{i,j+\frac{1}{2},k}^{n+\frac{1}{3},3}$, $(\mathbf{f}_2)_{i,j+\frac{1}{2},k}^{n+\frac{1}{3},1}$, $(\mathbf{f}_3)_{ij,k+\frac{1}{2}}^{n+\frac{1}{3},1}$ and $(\mathbf{f}_3)_{ij,k+\frac{1}{2}}^{n+\frac{1}{3},2}$. This costs a total of another six Riemann problems per cell. Afterward, we compute the transverse correction

$$\mathbf{h}_{ijk}^{n+\frac{1}{2},1} = \left(\frac{\partial \mathbf{u}}{\partial \mathbf{w}}\right)^{-1} \left\{\left[(\mathbf{f}_2)_{i,j+\frac{1}{2},k}^{n+\frac{1}{3},3} - (\mathbf{f}_2)_{i,j-\frac{1}{2},k}^{n+\frac{1}{3},3}\right] + \left[(\mathbf{f}_3)_{ij,k+\frac{1}{2}}^{n+\frac{1}{3},2} - (\mathbf{f}_3)_{ij,k-\frac{1}{2}}^{n+\frac{1}{3},2}\right]\right\} \frac{1}{2\triangle x_{1,i}\triangle x_{2,j}\triangle x_{3,k}}$$

and the states

$$\mathbf{w}_{i+\frac{1}{2},jk}^{n+\frac{1}{2},L} = \mathbf{w}_{i-\frac{1}{2},jk}^{n+\frac{1}{2},R} = \mathbf{w}_{ijk}^{n} - \mathbf{h}_{ijk}^{n+\frac{1}{2},1}. \tag{7.29}$$

Similar expressions hold in the other coordinate directions. Then we compute the flux integrals

$$(\mathbf{f}_1)_{i+\frac{1}{2},jk}^{n+\frac{1}{2}} = \mathbf{F}\left(\mathcal{R}\left(\mathbf{w}_{i+\frac{1}{2},jk}^{n+\frac{1}{2},L}, \mathbf{w}_{i+\frac{1}{2},jk}^{n+\frac{1}{2},R}; 0\right)\right) \mathbf{e}_1 \triangle x_{2,j}\triangle x_{3,k}\triangle t^{n+1/2}$$

and the flux integrals $(\mathbf{f}_2)_{i,j+\frac{1}{2},k}^{n+\frac{1}{2}}$ and $(\mathbf{f}_3)_{ij,k+\frac{1}{2}}^{n+\frac{1}{2}}$ using similar expressions. This costs another three Riemann problems per cell. Finally we perform the conservative difference (7.25). The total algorithm costs 12 Riemann problems per cell, while second-order operator splitting costs about four Riemann problems per step.

7.3.3.4 Second-Order Corner Transport Upwind

In order to improve the corner-transport upwind scheme to second-order accuracy, we can perform a modified equation analysis to determine the leading terms in the error, and then approximate those terms by finite difference approximations. If we use midpoint-rule quadratures for the flux integrals, then we evaluate the flux in the first coordinate direction at

$$\begin{aligned}
&\mathbf{u}\left(\mathbf{x}_1 + \frac{\triangle x_1}{2}, \mathbf{x}_2, \mathbf{x}_3, t + \frac{\triangle t}{2}\right) \\
&= \mathbf{u} + \frac{\partial \mathbf{u}}{\partial \mathbf{x}_1}\frac{\triangle x_1}{2} + \frac{\partial \mathbf{u}}{\partial t}\frac{\triangle t}{2} + O(\triangle x_1^2) + O(\triangle t^2) \\
&= \mathbf{u} + \frac{\partial \mathbf{u}}{\partial \mathbf{x}_1}\frac{\triangle x_1}{2} - \left[\frac{\partial \mathbf{Fe}_1}{\partial \mathbf{x}_1} + \frac{\partial \mathbf{Fe}_2}{\partial \mathbf{x}_2} + \frac{\partial \mathbf{Fe}_3}{\partial \mathbf{x}_3}\right]\frac{\triangle t}{2} + O(\triangle x_1^2) + O(\triangle t^2).
\end{aligned}$$

The first-order corner transport upwind scheme essentially computes

$$\begin{aligned}
\mathbf{u}_{i+1/2,jk}^{n+1/2,L} = \mathbf{u}_{ijk}^{n} - \frac{\partial \mathbf{Fe}_2}{\partial \mathbf{x}_2}\left(\mathbf{u}_{ijk}^{n+1/3,3}\right)\frac{\triangle t^{n+1/2}}{2} - \frac{\partial \mathbf{Fe}_3}{\partial \mathbf{x}_3}\left(\mathbf{u}_{ijk}^{n+1/3,2}\right)\frac{\triangle t^{n+1/2}}{2} \\
+ O(\triangle x_{2,j}\triangle t^{n+1/2}) + O(\triangle x_{3,k}\triangle t^{n+1/2}).
\end{aligned}$$

This expression needs to add an approximation for

$$\frac{\partial \mathbf{u}}{\partial \mathbf{x}_1}\frac{\triangle x_1}{2} - \frac{\partial \mathbf{Fe}_1}{\partial \mathbf{x}_1}\frac{\triangle t}{2}$$

to reach second-order accuracy. This suggests that a second-order corner transport upwind scheme can be obtained by a simple modification of the first-order scheme. If a characteristic analysis provides us with

$$\frac{\partial \mathbf{Fe}_1}{\partial \mathbf{w}}\mathbf{Y}_1 = \frac{\partial \mathbf{u}}{\partial \mathbf{w}}\mathbf{Y}_1\Lambda_1$$

then we replace the computation of $\mathbf{u}_{i+1/2,jk}^{n+1/2,L}$ and $\mathbf{u}_{i+1/2,jk}^{n+1/2,R}$ in (7.29) with

$$\begin{aligned}
\mathbf{w}_{i+\frac{1}{2},jk}^{n+\frac{1}{2},L} &= \mathbf{w}_{ijk}^{n} + \mathbf{Y}_1\left(\mathbf{I} - \Lambda_1\frac{\triangle t^{n+1/2}}{\triangle x_{1,i}}\right)\mathbf{Y}_1^{-1}\frac{\partial \mathbf{w}}{\partial \mathbf{x}_1}\frac{\triangle x_{1,i}}{2} - \mathbf{h}_{ijk}^{n+\frac{1}{2},1} \\
\mathbf{w}_{i-\frac{1}{2},jk}^{n+\frac{1}{2},R} &= \mathbf{w}_{ijk}^{n} - \mathbf{Y}_1\left(\mathbf{I} - \Lambda_1\frac{\triangle t^{n+1/2}}{\triangle x_{1,i}}\right)\mathbf{Y}_1^{-1}\frac{\partial \mathbf{w}}{\partial \mathbf{x}_1}\frac{\triangle x_{1,i}}{2} - \mathbf{h}_{ijk}^{n+\frac{1}{2},1}.
\end{aligned}$$

The slope $\frac{\partial \mathbf{w}}{\partial \mathbf{x}_1}\triangle x_{1,i}$ can be provided by the standard MUSCL slope limiting. Similar equations hold in the other coordinate directions.

The finite difference scheme presented above does not reduce to the form used by Colella [33] in two dimensions, because the characteristic tracing is used only in the determination of the flux variables $\mathbf{w}_{i+\frac{1}{2},jk}^{n+\frac{1}{2},L,R}$. Numerical experiments in [14] indicate that the Colella form is preferable for miscible displacement computations in two dimensions. Thus, it might be desirable to incorporate this extra tracing in three dimensions. The only modification would be to the states at which the fluxes $(\mathbf{f}_1)_{i+\frac{1}{2},jk}^{n+\frac{1}{3},2}$ and $(\mathbf{f}_1)_{i+\frac{1}{2},jk}^{n+\frac{1}{3},3}$ are evaluated. If characteristic tracing is added to these states, then we would obtain four distinct values, associated with the four cell edges in the first coordinate direction, replacing each of the former states $\mathbf{w}_{ijk}^{n+1/3,2}$ and $\mathbf{w}_{ijk}^{n+1/3,3}$. This would lead to a significant increase in the number of Riemann problems being solved. Saltzman [140] also described some alternative forms of his 3D corner transport upwind scheme.

7.3.4 Wave Propagation

The three-dimensional version of the wave propagation algorithm is described in [88]. We will attempt to describe it succinctly here. The flux increments in the first coordinate direction are

$$
\begin{aligned}
(\mathbf{F}\mathbf{e}_1)_{i+1/2,jk}^{n+1/2} - (\mathbf{F}\mathbf{e}_1)_{ijk}^{n} &= \mathbf{A}_{i+1/2,jk}^{-} \triangle \mathbf{u}_{i+1/2,jk}^{n} - \mathbf{h}_{i+1/2,jk} \\
(\mathbf{F}\mathbf{e}_1)_{i+1/2,jk}^{n+1/2} - (\mathbf{F}\mathbf{e}_1)_{i+1,jk}^{n} &= -\mathbf{A}_{i+1/2,jk}^{+} \triangle \mathbf{u}_{i+1/2,jk}^{n} - \mathbf{h}_{i+1/2,jk}
\end{aligned}
$$

where the transverse flux corrections are

$$
\begin{aligned}
\mathbf{h}_{i+1/2,jk} = &\Big[\mathbf{A}_{i+1/2,jk}^{-}\left(\mathbf{A}_{i+1,j,k+1/2}^{-} \triangle \mathbf{u}_{i+1,j,k+1/2}^{n+1/3,2} + \mathbf{A}_{i+1,j,k-1/2}^{+} \triangle \mathbf{u}_{i+1,j,k-1/2}^{n+1/3,2}\right) \\
&+ \mathbf{A}_{i+1/2,jk}^{+}\left(\mathbf{A}_{ij,k+1/2}^{-} \triangle \mathbf{u}_{ij,k+1/2}^{n+1/3,2} + \mathbf{A}_{ij,k-1/2}^{+} \triangle \mathbf{u}_{ij,k-1/2}^{n+1/3,2}\right)\Big] \frac{\triangle t^{n+1/2}}{2 \triangle x_{3,k}} \\
&+ \Big[\mathbf{A}_{i+1/2,jk}^{-}\left(\mathbf{A}_{i+1,j+1/2,k}^{-} \triangle \mathbf{u}_{i+1,j+1/2,k}^{n+1/3,3} + \mathbf{A}_{i+1,j-1/2,k}^{+} \triangle \mathbf{u}_{i+1,j-1/2,k}^{n+1/3,3}\right) \\
&+ \mathbf{A}_{i+1/2,jk}^{+}\left(\mathbf{A}_{i,j+1/2,k}^{-} \triangle \mathbf{u}_{i,j+1/2,k}^{n+1/3,3} + \mathbf{A}_{i,j+1/2,k}^{+} \triangle \mathbf{u}_{i,j-1/2,k}^{n+1/3,3}\right)\Big] \frac{\triangle t^{n+1/2}}{2 \triangle x_{2,j}} .
\end{aligned}
$$

These expressions require the evaluation of

$$
\begin{aligned}
\triangle \mathbf{u}_{ij,k+1/2}^{n+1/3,2} = \triangle \mathbf{u}_{ij,k+1/2}^{n} - &\Big[\mathbf{A}_{i,j+1/2,k+1}^{-} \triangle \mathbf{u}_{i,j+1/2,k+1}^{n} + \mathbf{A}_{i,j-1/2,k+1}^{+} \triangle \mathbf{u}_{i,j-1/2,k+1}^{n} \\
&- \mathbf{A}_{i,j+1/2,k}^{-} \triangle \mathbf{u}_{i,j+1/2,k}^{n} - \mathbf{A}_{i,j-1/2,k}^{+} \triangle \mathbf{u}_{i,j-1/2,k}^{n}\Big] \frac{\triangle t^{n+1/2}}{3 \triangle x_{2,j}}
\end{aligned}
$$

and a similar expression for $\triangle \mathbf{u}_{i,j+1/2,k}^{n+1/3,3}$. Slope information is incorporated in much the same way as in 2D. Similar expressions are used for the flux increments in the other coordinate directions. An implementation can be found in file `step3.f` of CLAWPACK.

7.4 Curvilinear Coordinates

Sometimes it is useful to employ curvilinear coordinates in numerical computation. Two common curvilinear coordinate systems are spherical and cylindrical coordinates. Other cases arise in specific problems, such as stream functions in incompressible flow [11, 127]. For problems with appropriate symmetry, we can use low-dimensional methods to compute solutions to higher-dimensional problems. The discussion that follows is adopted from [116].

7.4.1 Coordinate Transformations

Suppose that $\mathbf{a}$ denotes the vector of Cartesian (rectangular) coordinates, and $\mathbf{y}$ denotes the vector of curvilinear coordinates. Let $\mathbf{H}$ be the matrix of **scale factors**

$$\mathbf{H}^2 \equiv \left(\frac{\partial \mathbf{a}}{\partial \mathbf{y}}\right)^\top \left(\frac{\partial \mathbf{a}}{\partial \mathbf{y}}\right).$$

Since $\mathbf{H}^2$ is symmetric and nonnegative, it has a square root $\mathbf{H}$. As a result, the matrix

$$\mathbf{Q} \equiv \frac{\partial \mathbf{a}}{\partial \mathbf{y}} \mathbf{H}^{-1}$$

is orthogonal. The columns of $\mathbf{Q}$ are called the **unit base vectors**.

Suppose that we are given a vector $\mathbf{x}$ in the Cartesian coordinate system. In order to write this vector in terms of the unit base vectors, we must solve $\mathbf{Q}\mathbf{w} = \mathbf{x}$ for the **physical components** $\mathbf{w}$ of the vector $\mathbf{x}$. Since $\mathbf{Q}$ is orthogonal, this is easy:

$$\mathbf{w} = \mathbf{Q}^\top \mathbf{x}.$$

Similarly, if $\mathbf{M}$ is a matrix that operates on Cartesian vectors $\mathbf{x} = \mathbf{Q}\mathbf{w}$, then the physical components of $\mathbf{M}\mathbf{x}$ are $\mathbf{Q}^\top\mathbf{M}\mathbf{x}$. Thus the matrix representation of the linear transformation $\mathbf{x} \to \mathbf{M}\mathbf{x}$ in the curvilinear coordinate system is $\mathbf{w} \to (\mathbf{Q}^\top\mathbf{M}\mathbf{Q})\mathbf{w}$.

Suppose that we are given a scalar $\omega(\mathbf{y})$, and we want to compute its gradient $\nabla_{\mathbf{a}}\omega$ in the Cartesian coordinate system. Since

$$\frac{\partial \omega}{\partial \mathbf{a}} = \frac{\partial \omega}{\partial \mathbf{y}} \frac{\partial \mathbf{y}}{\partial \mathbf{a}} = \frac{\partial \omega}{\partial \mathbf{y}} \left(\frac{\partial \mathbf{a}}{\partial \mathbf{y}}\right)^{-1} = \frac{\partial \omega}{\partial \mathbf{y}} (\mathbf{Q}\mathbf{H})^{-1} = \frac{\partial \omega}{\partial \mathbf{y}} \mathbf{H}^{-1}\mathbf{Q}^\top,$$

we have

$$\nabla_{\mathbf{a}}\omega = \mathbf{Q}\mathbf{H}^{-1}\nabla_{\mathbf{y}}\omega. \tag{7.30}$$

Similarly, suppose that we want to compute the matrix of Cartesian spatial derivatives of a vector $\mathbf{x} = \mathbf{Q}\mathbf{w}$, where the physical components $\mathbf{w}$ are known functions of the curvilinear coordinates $\mathbf{y}$. We compute

$$\frac{\partial \mathbf{x}}{\partial \mathbf{a}} = \frac{\partial \mathbf{x}}{\partial \mathbf{y}} \frac{\partial \mathbf{y}}{\partial \mathbf{a}} = \frac{\partial \mathbf{Q}\mathbf{w}}{\partial \mathbf{y}} \mathbf{H}^{-1}\mathbf{Q}^\top.$$

The physical components of this matrix are

$$\mathbf{Q}^\top \frac{\partial \mathbf{x}}{\partial \mathbf{a}} \mathbf{Q} = \mathbf{Q}^\top \frac{\partial \mathbf{Q}\mathbf{w}}{\partial \mathbf{y}} \mathbf{H}^{-1}.$$

In order to compute the divergence of $\mathbf{x}$, we take the trace of $\frac{\partial \mathbf{x}}{\partial \mathbf{a}}$:

$$\nabla_{\mathbf{a}} \cdot \mathbf{x} = \operatorname{tr}\left(\frac{\partial \mathbf{x}}{\partial \mathbf{a}}\right) = \operatorname{tr}\left(\frac{\partial \mathbf{x}}{\partial \mathbf{y}}\frac{\partial \mathbf{y}}{\partial \mathbf{a}}\right) = \operatorname{tr}\left(\frac{\partial \mathbf{y}}{\partial \mathbf{a}}\frac{\partial \mathbf{Q}\mathbf{w}}{\partial \mathbf{y}}\right).$$

Now, recall from Equation (4.18) that the matrix of minors is divergence-free. In other words,

$$\nabla_{\mathbf{y}} \cdot \left(\frac{\partial \mathbf{y}}{\partial \mathbf{a}}\left|\frac{\partial \mathbf{a}}{\partial \mathbf{y}}\right|\right) = 0.$$

We can add this zero to the previous expression to obtain

$$\nabla_{\mathbf{a}} \cdot \mathbf{x} = \operatorname{tr}\left(\frac{\partial \mathbf{y}}{\partial \mathbf{a}}\frac{\partial \mathbf{Q}\mathbf{w}}{\partial \mathbf{y}}\right) + \nabla_{\mathbf{y}} \cdot \left(\frac{\partial \mathbf{y}}{\partial \mathbf{a}}\left|\frac{\partial \mathbf{a}}{\partial \mathbf{y}}\right|\right)\mathbf{Q}\mathbf{w}.$$

Now, the product rule for differentiation shows that

$$\nabla_{\mathbf{a}} \cdot \mathbf{x} = \left|\frac{\partial \mathbf{y}}{\partial \mathbf{a}}\right| \nabla_{\mathbf{y}} \cdot \left(\frac{\partial \mathbf{y}}{\partial \mathbf{a}}\mathbf{Q}\mathbf{w}\left|\frac{\partial \mathbf{a}}{\partial \mathbf{y}}\right|\right) = \left|\frac{\partial \mathbf{y}}{\partial \mathbf{a}}\right| \nabla_{\mathbf{y}} \cdot \left(\mathbf{H}^{-1}\mathbf{w}\left|\frac{\partial \mathbf{a}}{\partial \mathbf{y}}\right|\right). \tag{7.31}$$

Finally, we will occasionally need to compute the physical components of the divergence of an array. If $\tilde{\mathbf{S}} \equiv \mathbf{Q}^\top \mathbf{S}\mathbf{Q}$, then

$$\begin{aligned}
\mathbf{Q}^\top(\nabla_{\mathbf{a}} \cdot \mathbf{S})^\top &= \mathbf{Q}^\top\left\{\nabla_{\mathbf{y}} \cdot \left(\mathbf{H}^{-1}\tilde{\mathbf{S}}\mathbf{Q}^\top\left|\frac{\partial \mathbf{a}}{\partial \mathbf{y}}\right|\right)\right\}^\top \left|\frac{\partial \mathbf{y}}{\partial \mathbf{a}}\right| \\
&= \sum_i \mathbf{e}_i\mathbf{e}_i^\top \mathbf{Q}^\top\left\{\nabla_{\mathbf{y}} \cdot \left(\mathbf{H}^{-1}\tilde{\mathbf{S}}\mathbf{Q}^\top\left|\frac{\partial \mathbf{a}}{\partial \mathbf{y}}\right|\right)\right\}^\top \left|\frac{\partial \mathbf{y}}{\partial \mathbf{a}}\right| \\
&= \sum_i \mathbf{e}_i\left\{\nabla_{\mathbf{y}} \cdot \left(\mathbf{H}^{-1}\tilde{\mathbf{S}}\mathbf{Q}^\top\left|\frac{\partial \mathbf{a}}{\partial \mathbf{y}}\right|\right)\mathbf{Q}\mathbf{e}_i\right\} \left|\frac{\partial \mathbf{y}}{\partial \mathbf{a}}\right| \\
&= \sum_i \mathbf{e}_i\left\{\nabla_{\mathbf{y}} \cdot \left(\mathbf{H}^{-1}\tilde{\mathbf{S}}\mathbf{Q}^\top\mathbf{Q}\mathbf{e}_i\left|\frac{\partial \mathbf{a}}{\partial \mathbf{y}}\right|\right) - \operatorname{tr}\left(\mathbf{H}^{-1}\tilde{\mathbf{S}}\mathbf{Q}^\top\frac{\partial \mathbf{Q}\mathbf{e}_i}{\partial \mathbf{y}}\left|\frac{\partial \mathbf{a}}{\partial \mathbf{y}}\right|\right)\right\} \left|\frac{\partial \mathbf{y}}{\partial \mathbf{a}}\right| \\
&= \sum_i \mathbf{e}_i\left\{\nabla_{\mathbf{y}} \cdot \left(\mathbf{H}^{-1}\tilde{\mathbf{S}}\mathbf{e}_i\left|\frac{\partial \mathbf{a}}{\partial \mathbf{y}}\right|\right) - \operatorname{tr}\left(\mathbf{H}^{-1}\tilde{\mathbf{S}}\mathbf{Q}^\top\frac{\partial \mathbf{Q}\mathbf{e}_i}{\partial \mathbf{y}}\left|\frac{\partial \mathbf{a}}{\partial \mathbf{y}}\right|\right)\right\} \left|\frac{\partial \mathbf{y}}{\partial \mathbf{a}}\right|.
\end{aligned} \tag{7.32}$$

7.4.2 Spherical Coordinates

In spherical coordinates, the vector of curvilinear coordinates is

$$\mathbf{y}^\top = \begin{bmatrix} r, & \theta, & \phi \end{bmatrix},$$

where r is the distance from the center of the sphere, ϕ is a latitudinal angle, and θ is a longitudinal angle. It follows that the Cartesian coordinate vector is

$$\mathbf{a} = \begin{bmatrix} r\sin\theta\cos\phi, & r\sin\theta\sin\phi, & r\cos\theta \end{bmatrix}.$$

From this, it is easy to compute

$$\frac{\partial \mathbf{a}}{\partial \mathbf{y}} = \begin{bmatrix} \sin\theta\cos\phi & r\cos\theta\cos\phi & -r\sin\theta\sin\phi \\ \sin\theta\sin\phi & r\cos\theta\sin\phi & r\sin\theta\cos\phi \\ \cos\theta & -r\sin\theta & 0 \end{bmatrix}$$

and

$$\mathbf{H}^2 = \left(\frac{\partial \mathbf{a}}{\partial \mathbf{y}}\right)^\top \left(\frac{\partial \mathbf{a}}{\partial \mathbf{y}}\right) = \begin{bmatrix} 1 & & \\ & r^2 & \\ & & r^2\sin^2\theta \end{bmatrix}.$$

Similarly, the orthogonal matrix of unit base vectors is

$$\mathbf{Q} = \frac{\partial \mathbf{a}}{\partial \mathbf{y}}\mathbf{H}^{-1} = \begin{bmatrix} \sin\theta\cos\phi & \cos\theta\cos\phi & -\sin\phi \\ \sin\theta\sin\phi & \cos\theta\sin\phi & \cos\phi \\ \cos\theta & -\sin\theta & 0 \end{bmatrix}.$$

Note that the curvilinear derivatives of the columns of $\mathbf{Q}$ are

$$\begin{aligned} \frac{\partial \mathbf{Q}\mathbf{e}_r}{\partial \mathbf{y}} &= \mathbf{Q}\mathbf{e}_\theta\mathbf{e}_\theta^\top + \mathbf{Q}\mathbf{e}_\phi \sin\theta \mathbf{e}_\phi^\top, \\ \frac{\partial \mathbf{Q}\mathbf{e}_\theta}{\partial \mathbf{y}} &= -\mathbf{Q}\mathbf{e}_r\mathbf{e}_\theta^\top + \mathbf{Q}\mathbf{e}_\phi \cos\theta \mathbf{e}_\phi^\top, \\ \frac{\partial \mathbf{Q}\mathbf{e}_\phi}{\partial \mathbf{y}} &= -\begin{bmatrix} \cos\phi \\ \sin\phi \\ 0 \end{bmatrix} \mathbf{e}_\phi^\top. \end{aligned}$$

Given a vector $\mathbf{x}$ in Cartesian coordinates, we can compute the physical components of $\mathbf{x}$ to be

$$\begin{bmatrix} \mathbf{w}_r \\ \mathbf{w}_\theta \\ \mathbf{w}_\phi \end{bmatrix} = \mathbf{w} = \mathbf{Q}^\top\mathbf{x} = \begin{bmatrix} \mathbf{x}_1\sin\theta\cos\phi + \mathbf{x}_2\sin\theta\sin\phi + \mathbf{x}_3\cos\theta \\ \mathbf{x}_1\cos\theta\cos\phi + \mathbf{x}_2\cos\theta\sin\phi - \mathbf{x}_3\sin\theta \\ -\mathbf{x}_1\sin\phi + \mathbf{x}_2\cos\phi \end{bmatrix}.$$

Next, we compute the gradient of a scalar using Equation (7.30)

$$\begin{aligned} \nabla_{\mathbf{a}}\omega &= \mathbf{Q}\mathbf{H}^{-1}\nabla_{\mathbf{y}}\omega \\ &= \begin{bmatrix} \sin\theta\cos\phi & \cos\theta\cos\phi & -\sin\phi \\ \sin\theta\sin\phi & \cos\theta\sin\phi & \cos\phi \\ \cos\theta & -\sin\theta & 0 \end{bmatrix} \begin{bmatrix} 1 & & \\ & 1/r & \\ & & 1/(r\sin\theta) \end{bmatrix} \begin{bmatrix} \frac{\partial\omega}{\partial r} \\ \frac{\partial\omega}{\partial\theta} \\ \frac{\partial\omega}{\partial\phi} \end{bmatrix} \\ &= \begin{bmatrix} \frac{\partial\omega}{\partial r}\sin\theta\cos\phi + \frac{\partial\omega}{\partial\theta}\frac{\cos\theta\cos\phi}{r} - \frac{\partial\omega}{\partial\phi}\frac{\sin\phi}{r\sin\theta} \\ \frac{\partial\omega}{\partial r}\sin\theta\sin\phi + \frac{\partial\omega}{\partial\theta}\frac{\cos\theta\sin\phi}{r} + \frac{\partial\omega}{\partial\phi}\frac{\cos\phi}{r\sin\theta} \\ \frac{\partial\omega}{\partial r}\cos\theta - \frac{\partial\omega}{\partial\theta}\frac{\sin\theta}{r} \end{bmatrix}. \end{aligned} \quad (7.33)$$

The deformation gradient $\mathbf{J}$ has physical components

$$\mathbf{Q}^\top \mathbf{J} \mathbf{Q} \equiv \mathbf{Q}^\top \frac{\partial \mathbf{x}}{\partial \mathbf{a}} \mathbf{Q} = \mathbf{Q}^\top \frac{\partial \mathbf{Q} \mathbf{w}}{\partial \mathbf{y}} \mathbf{H}^{-1}$$
$$= \begin{bmatrix} \frac{\partial \mathbf{w}_r}{\partial r} & \frac{1}{r}(\frac{\partial \mathbf{w}_r}{\partial \theta} - \mathbf{w}_\theta) & \frac{1}{r \sin\theta}\frac{\partial \mathbf{w}_r}{\partial \phi} - \frac{\mathbf{w}_\phi \cos\phi}{r \sin\theta} \\ \frac{\partial \mathbf{w}_\theta}{\partial r} & \frac{1}{r}(\frac{\partial \mathbf{w}_\theta}{\partial \theta} + \mathbf{w}_r) & \frac{1}{r \sin\theta}\frac{\partial \mathbf{w}_\theta}{\partial \phi} - \frac{\mathbf{w}_\phi \sin\phi}{r \sin\theta} \\ \frac{\partial \mathbf{w}_\phi}{\partial r} & \frac{1}{r}\frac{\partial \mathbf{w}_\phi}{\partial \theta} & \frac{1}{r \sin\theta}\frac{\partial \mathbf{w}_\phi}{\partial \phi} + \frac{\mathbf{w}_r}{r} + \frac{\mathbf{w}_\theta \cos\theta}{r \sin\theta} \end{bmatrix} = \begin{bmatrix} \mathbf{J}_{rr} & \mathbf{J}_{r\theta} & \mathbf{J}_{rz} \\ \mathbf{J}_{\theta r} & \mathbf{J}_{\theta\theta} & \mathbf{J}_{\theta z} \\ \mathbf{J}_{zr} & \mathbf{J}_{z\theta} & \mathbf{J}_{zz} \end{bmatrix}. \tag{7.34}$$

Similarly, the divergence of a vector $\mathbf{x}$ can be computed from Equation (7.31):

$$\nabla_{\mathbf{a}} \cdot \mathbf{x} = \left|\frac{\partial \mathbf{y}}{\partial \mathbf{a}}\right| \nabla_{\mathbf{y}} \cdot \left(\mathbf{H}^{-1}\mathbf{w}\left|\frac{\partial \mathbf{a}}{\partial \mathbf{y}}\right|\right)$$
$$= \frac{1}{r^2}\frac{\partial r^2 \mathbf{w}_r}{\partial r} + \frac{1}{r \sin\theta}\frac{\partial \mathbf{w}_\theta \sin\theta}{\partial \theta} + \frac{1}{r \sin\theta}\frac{\partial \mathbf{w}_\phi}{\partial \phi}. \tag{7.35}$$

Finally, the physical components of the divergence of an array can be computed from Equation (7.32):

$$\mathbf{Q}^\top(\nabla_{\mathbf{a}} \cdot S)^\top = \mathbf{Q}^\top \left[\nabla_{\mathbf{y}} \cdot \left(\mathbf{H}^{-1} T \mathbf{Q}^\top \left|\frac{\partial \mathbf{a}}{\partial \mathbf{y}}\right|\right)\right]^\top \left|\frac{\partial \mathbf{y}}{\partial \mathbf{a}}\right|$$
$$= \begin{bmatrix} \frac{\partial r^2 \tilde{S}_{rr}}{\partial r}\frac{1}{r^2} + \frac{\partial \tilde{S}_{\theta r}\sin\theta}{\partial \theta}\frac{1}{r\sin\theta} + \frac{\partial \tilde{S}_{\phi r}}{\partial \phi}\frac{1}{r\sin\theta} - \frac{\tilde{S}_{\theta\theta}+\tilde{S}_{\phi\phi}}{r} \\ \frac{\partial r^2 \tilde{S}_{r\theta}}{\partial r}\frac{1}{r^2} + \frac{\partial \tilde{S}_{\theta\theta}\sin\theta}{\partial \theta}\frac{1}{r\sin\theta} + \frac{\partial \tilde{S}_{\phi\theta}}{\partial \phi}\frac{1}{r\sin\theta} + \frac{\tilde{S}_{\theta r}}{r} - \frac{\tilde{S}_{\phi\phi}\cos\theta}{r\sin\theta} \\ \frac{\partial r^2 \tilde{S}_{r\phi}}{\partial r}\frac{1}{r^2} + \frac{\partial \tilde{S}_{\theta\phi}\sin\theta}{\partial \theta}\frac{1}{r} + \frac{\partial \tilde{S}_{\phi\phi}}{\partial \phi}\frac{1}{r\sin\theta} + \frac{\tilde{S}_{\phi r}+\tilde{S}_{\phi\theta}}{r} \end{bmatrix}. \tag{7.36}$$

Suppose that we are given a scalar conservation law

$$\frac{\partial \mathbf{u}}{\partial t} + \nabla_{\mathbf{x}} \cdot \mathbf{f}(\mathbf{u}) = 0.$$

The physical components of the flux are

$$\begin{bmatrix} \mathbf{f}_r \\ \mathbf{f}_\theta \\ \mathbf{f}_\phi \end{bmatrix} = \begin{bmatrix} \mathbf{f}_1 \sin\theta\cos\phi + \mathbf{f}_2 \sin\theta\sin\phi + \mathbf{f}_3\cos\theta \\ \mathbf{f}_1 \cos\theta\cos\phi + \mathbf{f}_2\cos\theta\sin\phi - \mathbf{f}_3\sin\theta \\ -\mathbf{f}_1\sin\phi + \mathbf{f}_2\cos\phi \end{bmatrix}.$$

Thus a scalar conservation law in spherical coordinates can be written in the conservation form

$$\frac{\partial \mathbf{u} r}{\partial t} + \frac{\partial \mathbf{f}_r r^2}{\partial r}\frac{1}{r^2} + \frac{\partial \mathbf{f}_\theta \sin\theta}{\partial \theta}\frac{1}{r\sin\theta} + \frac{\partial \mathbf{f}_\phi}{\partial \phi}\frac{1}{r\sin\theta} = 0.$$

Of course, partial differential equations for conservation laws are not as complete a description as the integral form of the law. If we integrate the spherical coordinate form of the conservation law over a spherical element, we obtain

$$
\begin{aligned}
0 &= \int_{t_1}^{t_2} \int_{r_1}^{r_2} \int_{\theta_1}^{\theta_2} \int_{\phi_1}^{\phi_2} \left\{ \frac{\partial \mathbf{u}}{\partial t} + \frac{1}{r^2 \sin\theta} \left[\frac{\partial \mathbf{f}_r r^2 \sin\theta}{\partial r} + \frac{\partial \mathbf{f}_\theta r \sin\theta}{\partial \theta} + \frac{\partial \mathbf{f}_\phi r}{\partial \phi} \right] \right\} r^2 \sin\theta \, d\phi \, d\theta dr dt \\
&= \int_{r_1}^{r_2} \int_{\theta_1}^{\theta_2} \int_{\phi_1}^{\phi_2} [\mathbf{u}(r,\theta,\phi,t_2) - \mathbf{u}(r,\theta,\phi,t_1)] \, d\phi \, \sin\theta \, d\theta r^2 \, dr \\
&+ \int_{t_1}^{t_2} \int_{\theta_1}^{\theta_2} \int_{\phi_1}^{\phi_2} \left[\mathbf{f}_r(r_2,\theta,\phi,t) r_2^2 - \mathbf{f}_r(r_1,\theta,\phi,t) r_1^2 \right] d\phi \, \sin\theta \, d\theta \, dt \\
&+ \int_{t_1}^{t_2} \int_{r_1}^{r_2} \int_{\phi_1}^{\phi_2} [\mathbf{f}_\theta(r,\theta_2,\phi,t) \sin\theta_2 - \mathbf{f}_\theta(r,\theta_1,\phi,t) \sin\theta_1] \, d\phi \, r dr \, dt \\
&+ \int_{t_1}^{t_2} \int_{r_1}^{r_2} \int_{\theta_1}^{\theta_2} \left[\mathbf{f}_\phi(r,\theta,\phi_2,t) - \mathbf{f}_\phi(r,\theta,\phi_1,t) \right] d\theta \, r dr \, dt.
\end{aligned}
$$

Note that the volume measure cancels the denominators in the partial differential equations.

It is common to assume that the motion of a material is such that the physical components are functions of r only, and there are no θ or ϕ component of arrays. Thus in spherical symmetry, $\mathbf{w}_\theta = 0$ and $\mathbf{w}_\phi = 0$, so the Cartesian coordinates are related to the physical components by

$$
\mathbf{x} = \begin{bmatrix} \mathbf{w}_r \sin\theta \cos\phi \\ \mathbf{w}_r \sin\theta \sin\phi \\ \mathbf{w}_r \cos\theta \end{bmatrix}.
$$

The gradient of a scalar simplifies to

$$
\nabla_{\mathbf{a}} \omega = \mathbf{Q}\mathbf{H}^{-1} \nabla_{\mathbf{y}} \omega = \begin{bmatrix} \frac{\partial \omega}{\partial r} \sin\theta \cos\phi \\ \frac{\partial \omega}{\partial r} \sin\theta \sin\phi \\ \frac{\partial \omega}{\partial r} \cos\theta \end{bmatrix}.
$$

Similarly, the divergence of a vector $\mathbf{x}$ in spherical symmetry is

$$
\nabla_a \cdot \mathbf{x} = \left| \frac{\partial \mathbf{y}}{\partial \mathbf{a}} \right| \nabla_{\mathbf{y}} \cdot \left(\mathbf{H}^{-1} \mathbf{w} \left| \frac{\partial \mathbf{a}}{\partial \mathbf{y}} \right| \right) = \frac{1}{r^2} \frac{\partial r^2 \mathbf{w}_r}{\partial r}.
$$

Consider a scalar conservation law in Cartesian coordinates:

$$
\frac{\partial \mathbf{u}}{\partial t} + \nabla_{\mathbf{x}} \cdot \mathbf{f}(\mathbf{u}) = 0.
$$

If the motion is spherically symmetric, then $\mathbf{f}_\theta = 0$ and $\mathbf{f}_\phi = 0$. In this case, the conservation law simplifies to

$$
\frac{\partial \mathbf{u} r^2}{\partial t} + \frac{\partial \mathbf{f}_r r^2}{\partial r} = 0.
$$

If we integrate the spherically symmetric form of the conservation law over a spherical shell, we obtain

$$\begin{aligned} 0 &= \frac{1}{4\pi}\int_{t_1}^{t_2}\int_{r_1}^{r_2}\int_0^{\pi}\int_0^{2\pi}\left[\frac{\partial \mathbf{u}}{\partial t} + \frac{1}{r^2\sin\theta}\frac{\partial \mathbf{f}_r r^2 \sin\theta}{\partial r}\right] r^2 \sin\theta \, d\phi \, d\theta \, dr \, dt \\ &= \int_{r_1}^{r_2} [\mathbf{u}(r,t_2) - \mathbf{u}(r,t_1)] \, r^2 dr + \int_{t_1}^{t_2} \mathbf{f}_r(r_2,t) r_2^2 - \mathbf{f}_r(r_1,t) r_1{}^2 \, dt \end{aligned}$$

7.4.2.1 Case Study: Eulerian Gas Dynamics in Spherical Coordinates

Recall that Eulerian conservation of mass in gas dynamics takes the Cartesian form

$$\frac{\partial \rho}{\partial t} + \nabla_{\mathbf{x}} \cdot (\mathbf{v}\rho) = 0.$$

We can use equation (7.35) to obtain the conservation of mass in spherical coordinates:

$$\frac{\partial \rho}{\partial t} + \frac{\partial \mathbf{v}_r \rho r^2}{\partial r}\frac{1}{r^2} + \frac{\partial \mathbf{v}_\theta \rho \sin\theta}{\partial \theta}\frac{1}{r\sin\theta} + \frac{\partial \mathbf{v}_z \rho}{\partial z}\frac{1}{r\sin\theta} = 0.$$

Eulerian conservation of momentum in Cartesian coordinates is

$$\frac{\partial \mathbf{v}\rho}{\partial t} + \nabla_{\mathbf{x}} \cdot (\mathbf{v}\rho\mathbf{v}^\top + Ip) = g\rho.$$

We can use equation (7.36) to obtain conservation of momentum in spherical coordinates:

$$\begin{aligned} \begin{bmatrix} g_r\rho \\ g_\theta\rho \\ g_\phi\rho \end{bmatrix} &\equiv \mathbf{Q}^\top g\rho = \frac{\partial \mathbf{Q}^\top \mathbf{v}\rho}{\partial t} + \mathbf{Q}^\top \left\{\nabla_{\mathbf{a}} \cdot (\mathbf{v}\rho\mathbf{v}^\top + Ip)\right\}^\top \\ &= \frac{\partial}{\partial t}\begin{bmatrix} \mathbf{v}_r\rho \\ \mathbf{v}_\theta\rho \\ \mathbf{v}_\phi\rho \end{bmatrix} + \begin{bmatrix} \frac{\partial(\mathbf{v}_r^2\rho+p)r^2}{\partial r}\frac{1}{r^2} + \frac{\partial \mathbf{v}_\theta\mathbf{v}_r\rho\sin\theta}{\partial\theta}\frac{1}{r\sin\theta} + \frac{\partial \mathbf{v}_\phi\mathbf{v}_r\rho}{\partial\phi}\frac{1}{r\sin\theta} - \frac{(\mathbf{v}_\theta^2+\mathbf{v}_\phi^2)\rho+2p}{r} \\ \frac{\partial \mathbf{v}_r\mathbf{v}_\theta\rho r^2}{\partial r}\frac{1}{r^2} + \frac{\partial(\mathbf{v}_\theta^2\rho+p)\sin\theta}{\partial\theta}\frac{1}{r\sin\theta} + \frac{\partial \mathbf{v}_\phi\mathbf{v}_\theta\rho}{\partial\phi}\frac{1}{r\sin\theta} + \frac{\mathbf{v}_\phi\mathbf{v}_r\rho}{r} - \frac{(\mathbf{v}_\phi^2\rho+p)\cos\theta}{\sin\theta} \\ \frac{\partial \mathbf{v}_r\mathbf{v}_\phi\rho r^2}{\partial r}\frac{1}{r^2} + \frac{\partial \mathbf{v}_\theta\mathbf{v}_\phi\rho\sin\theta}{\partial\theta}\frac{1}{r\sin\theta} + \frac{\partial(\mathbf{v}_\phi^2\rho+p)r}{\partial\phi}\frac{1}{r\sin\theta} + \frac{\mathbf{v}_\phi\mathbf{v}_r\rho}{r} + \frac{\mathbf{v}_\phi\mathbf{v}_\theta\rho\cos\theta}{r\sin\theta} \end{bmatrix}. \end{aligned}$$

Finally, conservation of energy in spherical coordinates becomes

$$\begin{aligned} &\frac{\partial\rho(e+\frac{1}{2}[\mathbf{v}_r^2+\mathbf{v}_\theta^2+\mathbf{v}_\phi^2])}{\partial t} + \frac{\partial r^2\mathbf{v}_r[p+\rho(e+\frac{1}{2}\mathbf{v}_r^2+\mathbf{v}_\theta^2+\mathbf{v}_\phi^2)]}{\partial r}\frac{1}{r^2} \\ &\quad + \frac{\partial \mathbf{v}_\theta[p+\rho(e+\frac{1}{2}\mathbf{v}_r^2+\mathbf{v}_\theta^2+\mathbf{v}_\phi^2)]\sin\theta}{\partial\theta}\frac{1}{r\sin\theta} + \frac{\partial \mathbf{v}_\phi[p+\rho(e+\frac{1}{2}\mathbf{v}_r^2+\mathbf{v}_\theta^2+\mathbf{v}_\phi^2)]}{\partial\phi}\frac{1}{r\sin\theta} \\ &= \rho(g_r\mathbf{v}_r + g_\theta\mathbf{v}_\theta + g_\phi\mathbf{v}_\phi). \end{aligned}$$

For spherically symmetric gas dynamics, source terms due to gravity are zero. Eulerian conservation of mass in spherical symmetry is

$$\frac{\partial \rho}{\partial t} + \frac{\partial \mathbf{v}_r \rho r^2}{\partial r}\frac{1}{r^2} = 0.$$

Eulerian conservation of momentum in spherical symmetry is

$$0 = \frac{\partial \mathbf{v}_r \rho}{\partial t} + \frac{\partial (\mathbf{v}_r^2 \rho + p) r^2}{\partial r}\frac{1}{r^2} - \frac{2p}{r} = \frac{\partial \mathbf{v}_r \rho}{\partial t} + \frac{\partial \mathbf{v}_r^2 \rho r^2}{\partial r}\frac{1}{r^2} + \frac{\partial p}{\partial r}.$$

Finally, conservation of energy in spherical coordinates becomes

$$\frac{\partial \rho (e + \frac{1}{2}\mathbf{v}_r^2)}{\partial t} + \frac{\partial r^2 \mathbf{v}_r [p + \rho(e + \frac{1}{2}\mathbf{v}_r^2)]}{\partial r}\frac{1}{r^2} = 0.$$

We can write this system in the form of a system of partial differential equations

$$\frac{\partial}{\partial t}\begin{bmatrix} \rho \\ \mathbf{v}_r \rho \\ (e + \frac{1}{2}\mathbf{v}_r^2)\rho \end{bmatrix} + \frac{1}{r^2}\frac{\partial}{\partial r}\begin{bmatrix} \mathbf{v}_r \rho r^2 \\ \mathbf{v}_r^2 \rho r^2 \\ \mathbf{v}_r [(e + \frac{1}{2}\mathbf{v}_r^2)\rho + p] r^2 \end{bmatrix} + \begin{bmatrix} 0 \\ \frac{\partial p}{\partial r} \\ 0 \end{bmatrix} = 0,$$

or in integral form

$$\begin{aligned}
\int_{r_{j-1/2}}^{r_{j+1/2}} r^2 \begin{bmatrix} \rho \\ \mathbf{v}_r \rho \\ (e + \frac{1}{2}\mathbf{v}_r^2)\rho \end{bmatrix}^{n+1} dr &= \int_{r_{j-1/2}}^{r_{j+1/2}} r^2 \begin{bmatrix} \rho \\ \mathbf{v}_r \rho \\ (e + \frac{1}{2}\mathbf{v}_r^2)\rho \end{bmatrix}^{n} dr \\
&\quad - \int_{t^n}^{t^{n+1}} \begin{bmatrix} \mathbf{v}_r \rho \\ \mathbf{v}_r^2 \rho \\ \mathbf{v}_r [(e + \frac{1}{2}\mathbf{v}_r^2)\rho + p] \end{bmatrix}_{j+1/2} dt \; r_{j+1/2}^2 \\
&\quad + \int_{t^n}^{t^{n+1}} \begin{bmatrix} \mathbf{v}_r \rho \\ \mathbf{v}_r^2 \rho \\ \mathbf{v}_r [(e + \frac{1}{2}\mathbf{v}_r^2)\rho + p] \end{bmatrix}_{j-1/2} dt \; r_{j-1/2}^2 \\
&\quad + \int_{t^n}^{t^{n+1}} \int_{r_{j-1/2}}^{r_{j+1/2}} \begin{bmatrix} 0 \\ \frac{\partial p}{\partial r} \\ 0 \end{bmatrix} r^2 \, dr \, dt.
\end{aligned}$$

In order to develop a discretization of these equations, let us assume that we are given cell averages

$$\mathbf{u}_j^n \equiv \begin{bmatrix} \rho \\ \mathbf{v}_r \rho \\ (e + \frac{1}{2}\mathbf{v}_r^2)\rho \end{bmatrix}_j^n \approx \frac{3}{r_{j+1/2}^3 - r_{j-1/2}^3} \int_{r_{j-1/2}}^{r_{j+1/2}} \begin{bmatrix} \rho \\ \mathbf{v}_r \rho \\ (e + \frac{1}{2}\mathbf{v}_r^2)\rho \end{bmatrix} r^2 \, dr.$$

We could compute flux integrals

$$\mathbf{f}^{n+1/2}_{j+1/2} \approx \int_{t^n}^{t^{n+1}} \begin{bmatrix} \rho \mathbf{v}_r \\ \mathbf{v}_r^2 \rho \\ \mathbf{v}_r[(e + \frac{1}{2}\mathbf{v}_r^2)\rho + p] \end{bmatrix} dt$$

by a slope limiter scheme, as in section 6.2.5. The characteristic tracing and Riemann problem solution would also provide a values for the pressures $p^{n+1/2}_{j\pm\frac{1}{2}}$ in the pressure derivative for the momentum equation. Even the approximate Riemann solvers in Section 4.13 can be designed so that they provide values for the conserved quantities at the solution of the Riemann problem, from which the flux variables can be decoded. Then our finite difference takes the form

$$\mathbf{u}_j^{n+1} = \mathbf{u}_j^n - \left[\mathbf{f}^{n+1/2}_{j+1/2} r^2_{j+1/2} - \mathbf{f}^{n+1/2}_{j-1/2} r^2_{j-1/2}\right] \frac{3}{r^3_{j+1/2} - r^3_{j-1/2}}$$
$$+ \begin{bmatrix} 0 \\ p^{n+1/2}_{j+\frac{1}{2}} - p^{n+1/2}_{j+1/2} \\ 0 \end{bmatrix} \frac{\triangle t^{n+1/2}}{r_{j+1/2} - r_{j-1/2}}.$$

This scheme is similar to the approach in [35]. It is designed to be **free-stream-preserving**, meaning that if $\mathbf{v}_r = 0$ and ρ and p are constant at time t^n, then the scheme produces the same values at t^{n+1}. LeVeque [97, page 376] suggests that in wave propagation schemes the source terms due to curvilinear coordinates can be treated via operator splitting. This can also be free-stream-preserving if done carefully.

7.4.2.2 Case Study: Lagrangian Solid Mechanics in Spherical Coordinates

Equality of mixed partial derivatives in spherical coordinates gives us

$$\frac{\partial}{\partial t}\begin{bmatrix} \mathbf{J}_{rr} & \mathbf{J}_{r\theta} & \mathbf{J}_{rz} \\ \mathbf{J}_{\theta r} & \mathbf{J}_{\theta\theta} & \mathbf{J}_{\theta z} \\ \mathbf{J}_{zr} & \mathbf{J}_{z\theta} & \mathbf{J}_{zz} \end{bmatrix} = \mathbf{Q}^\top \frac{\partial \mathbf{v}}{\partial \mathbf{a}} \mathbf{Q} = \begin{bmatrix} \frac{\partial \mathbf{v}_r}{\partial r} & \frac{1}{r}(\frac{\partial \mathbf{v}_r}{\partial \theta} - \mathbf{v}_\theta) & \frac{1}{r\sin\theta}\frac{\partial \mathbf{v}_r}{\partial \phi} - \frac{\mathbf{v}_\phi}{r} \\ \frac{\partial \mathbf{v}_\theta}{\partial r} & \frac{1}{r}(\frac{\partial \mathbf{v}_\theta}{\partial \theta} + \mathbf{v}_r) & \frac{1}{r\sin\theta}\frac{\partial \mathbf{v}_\theta}{\partial \phi} - \frac{\mathbf{v}_\phi \cos\phi}{r\sin\phi} \\ \frac{\partial \mathbf{v}_\phi}{\partial r} & \frac{1}{r}\frac{\partial \mathbf{v}_\phi}{\partial \theta} & \frac{1}{r\sin\theta}\frac{\partial \mathbf{v}_\phi}{\partial \phi} + \frac{\mathbf{v}_r}{r} + \frac{\mathbf{v}_\theta \cos\theta}{r\sin\theta} \end{bmatrix}.$$

The physical components of the Cauchy stress are

$$\begin{bmatrix} \mathbf{S}_{rr} & \mathbf{S}_{r\theta} & \mathbf{S}_{r\phi} \\ \mathbf{S}_{\theta r} & \mathbf{S}_{\theta\theta} & \mathbf{S}_{\theta\phi} \\ \mathbf{S}_{\phi r} & \mathbf{S}_{\phi\theta} & \mathbf{S}_{\phi\phi} \end{bmatrix} \equiv \tilde{\mathbf{S}} = \mathbf{Q}^\top \mathbf{S}\mathbf{Q}.$$

Since $\mathbf{S}$ is symmetric, so is $\tilde{\mathbf{S}}$. The physical components of the first Piola–Kirchhoff stress are

$$\begin{bmatrix} \mathbf{T}_{rr} & \mathbf{T}_{r\theta} & \mathbf{T}_{r\phi} \\ \mathbf{T}_{\theta r} & \mathbf{T}_{\theta\theta} & \mathbf{T}_{\theta\phi} \\ \mathbf{T}_{\phi r} & \mathbf{T}_{\phi\theta} & \mathbf{T}_{\phi\phi} \end{bmatrix}$$
$$\equiv \mathbf{T} = \mathbf{Q}^\top \mathbf{S}\,\mathbf{J}^{-\top}\mathbf{Q}|\mathbf{J}|$$
$$= \begin{bmatrix} \mathbf{S}_{rr} & \mathbf{S}_{r\theta} & \mathbf{S}_{r\phi} \\ \mathbf{S}_{\theta r} & \mathbf{S}_{\theta\theta} & \mathbf{S}_{\theta\phi} \\ \mathbf{S}_{\phi r} & \mathbf{S}_{\phi\theta} & \mathbf{S}_{\phi\phi} \end{bmatrix} \begin{bmatrix} \mathbf{J}_{\theta\theta}\mathbf{J}_{\phi\phi} - \mathbf{J}_{\phi\theta}\mathbf{J}_{\theta\phi} & -\mathbf{J}_{\theta r}\mathbf{J}_{\phi\phi} + \mathbf{J}_{\phi r}\mathbf{J}_{\theta\phi} & \mathbf{J}_{\theta r}\mathbf{J}_{\phi\theta} - \mathbf{J}_{\phi r}\mathbf{J}_{\theta\theta} \\ -\mathbf{J}_{r\theta}\mathbf{J}_{\phi\phi} + \mathbf{J}_{\phi\theta}\mathbf{J}_{r\phi} & \mathbf{J}_{rr}\mathbf{J}_{\phi\phi} - \mathbf{J}_{\phi r}\mathbf{J}_{\theta\phi} & -\mathbf{J}_{rr}\mathbf{J}_{\phi\theta} + \mathbf{J}_{\phi r}\mathbf{J}_{r\theta} \\ \mathbf{J}_{r\theta}\mathbf{J}_{\phi\theta} - \mathbf{J}_{\theta\theta}\mathbf{J}_{r\phi} & -\mathbf{J}_{rr}\mathbf{J}_{\theta\phi} + \mathbf{J}_{\theta r}\mathbf{J}_{r\phi} & \mathbf{J}_{rr}\mathbf{J}_{\theta\theta} - \mathbf{J}_{\theta r}\mathbf{J}_{r\theta} \end{bmatrix}.$$

Note that $\mathbf{T}$ is not necessarily symmetric. Thus conservation of momentum in spherical coordinates is

$$\begin{bmatrix} \mathbf{g}_r\rho \\ \mathbf{g}_\theta\rho \\ \mathbf{g}_\phi\rho \end{bmatrix} = \mathbf{Q}^\top\mathbf{g}\rho = \frac{\partial \mathbf{Q}^\top\mathbf{v}\rho}{\partial t} - \mathbf{Q}^\top\left[\nabla_{\mathbf{a}} \cdot (\mathbf{J}^{-1}|\mathbf{J}|\mathbf{S})\right]^\top$$
$$= \frac{\partial}{\partial t}\begin{bmatrix} \mathbf{v}_r\rho \\ \mathbf{v}_\theta\rho \\ \mathbf{v}_z\rho \end{bmatrix} - \begin{bmatrix} \frac{\partial r^2\mathbf{T}_{rr}}{\partial r}\frac{1}{r^2} + \frac{\partial \mathbf{T}_{\theta r}\sin\theta}{\partial\theta}\frac{1}{r\sin\theta} + \frac{\partial\mathbf{T}_{\phi r}}{\partial\phi}\frac{1}{r\sin\theta} - \frac{\mathbf{T}_{\theta\theta}+\mathbf{T}_{\phi\phi}}{r} \\ \frac{\partial r^2\mathbf{T}_{r\theta}}{\partial r}\frac{1}{r^2} + \frac{\partial\mathbf{T}_{\theta\theta}\sin\theta}{\partial\theta}\frac{1}{r\sin\theta} + \frac{\partial\mathbf{T}_{\phi\theta}}{\partial\phi}\frac{1}{r\sin\theta} + \frac{\mathbf{T}_{\theta r}}{r} - \frac{\mathbf{T}_{\phi\phi}\cos\theta}{r\sin\theta} \\ \frac{\partial r^2\mathbf{T}_{r\phi}}{\partial r}\frac{1}{r^2} + \frac{\partial\mathbf{T}_{\theta\phi}\sin\theta}{\partial\theta}\frac{1}{r} + \frac{\partial\mathbf{T}_{\phi\phi}}{\partial\phi}\frac{1}{r\sin\theta} + \frac{\mathbf{T}_{\phi r}+\mathbf{T}_{\phi\theta}}{r} \end{bmatrix}.$$

Finally, Cartesian conservation of energy

$$\frac{\partial(\epsilon + \frac{1}{2}\mathbf{v}\cdot\mathbf{v})\rho}{\partial t} - \nabla_{\mathbf{a}} \cdot (\mathbf{J}^{-1}\mathbf{S}|\mathbf{J}|\mathbf{v}) = (\omega + \mathbf{g}\cdot\mathbf{v})\rho$$

becomes

$$\begin{aligned} &\frac{\partial(\epsilon + \frac{1}{2}\mathbf{v}\cdot\mathbf{v})\rho}{\partial t} - \frac{1}{r^2}\frac{\partial(\mathbf{T}_{rr}\mathbf{v}_r + \mathbf{T}_{\theta r}\mathbf{v}_\theta + \mathbf{T}_{zr}\mathbf{v}_z)r^2}{\partial r} \\ &\quad - \frac{1}{r\sin\theta}\frac{\partial(\mathbf{T}_{r\theta}\mathbf{v}_r + \mathbf{T}_{\theta\theta}\mathbf{v}_\theta + \mathbf{T}_{z\theta}\mathbf{v}_z)\sin\theta}{\partial\theta} \\ &\quad + \frac{1}{r\sin\theta}\frac{\partial\mathbf{T}_{rz}\mathbf{v}_r + \mathbf{T}_{\theta z}\mathbf{v}_\theta + \mathbf{T}_{zz}\mathbf{v}_z}{\partial z} \\ &= (\omega + \mathbf{g}_r\mathbf{v}_r + \mathbf{g}_\theta\mathbf{v}_\theta + \mathbf{g}_z\mathbf{v}_z)\rho. \end{aligned}$$

In spherical symmetry, the deformation gradient $\mathbf{J} = \frac{\partial\mathbf{x}}{\partial\mathbf{a}}$ has physical components

$$\mathbf{Q}^\top\frac{\partial\mathbf{x}}{\partial\mathbf{a}}\mathbf{Q} = \begin{bmatrix} \frac{\partial\mathbf{w}_r}{\partial r} & 0 & 0 \\ 0 & \frac{\mathbf{w}_r}{r} & 0 \\ 0 & 0 & \frac{\mathbf{w}_r}{r} \end{bmatrix}.$$

The physical components of a spherically symmetric Cauchy stress tensor are

$$\mathbf{Q}^\top \mathbf{S}\mathbf{Q} = \begin{bmatrix} \tilde{\mathbf{S}}_{rr} & 0 & 0 \\ 0 & \tilde{\mathbf{S}}_{\theta\theta} & 0 \\ 0 & 0 & \tilde{\mathbf{S}}_{\phi\phi} \end{bmatrix}.$$

Because of the form of the deformation gradient, we typically have $\tilde{\mathbf{S}}_{\theta\theta} = \tilde{\mathbf{S}}_{\phi\phi}$; this is the case, for example, in linear elasticity. Thus the physical components of the stress divergence are

$$\mathbf{Q}^\top (\nabla_{\mathbf{a}} \cdot \mathbf{S})^\top = \begin{bmatrix} \frac{1}{r^2}\frac{\partial r^2 \tilde{\mathbf{S}}_{rr}}{\partial r} - \frac{\tilde{\mathbf{S}}_{\theta\theta}+\tilde{\mathbf{S}}_{\phi\phi}}{r} \\ 0 \\ 0 \end{bmatrix}.$$

Equality of mixed partial derivatives in spherical symmetry gives us

$$\frac{\partial}{\partial t}\begin{bmatrix} \mathbf{J}_{rr} & 0 & 0 \\ 0 & \mathbf{J}_{\theta\theta} & 0 \\ 0 & 0 & \mathbf{J}_{\phi\phi} \end{bmatrix} = \begin{bmatrix} \frac{\partial \mathrm{v}_r}{\partial r} & 0 & 0 \\ 0 & \frac{\mathrm{v}_r}{r} & 0 \\ 0 & 0 & \frac{\mathrm{v}_r}{r} \end{bmatrix}.$$

We can also write

$$\frac{\partial |\mathbf{J}|}{\partial t} = \frac{\partial}{\partial t}\left(\frac{\mathbf{w}_r^2}{r^2}\frac{\partial \mathbf{w}_r}{\partial r}\right) = \frac{2\mathrm{v}_r \mathbf{w}_r}{r^2}\frac{\partial \mathbf{w}_r}{\partial r} + \frac{\mathbf{w}_r^2}{r^2}\frac{\partial \mathrm{v}_r}{\partial r} = \frac{1}{r^2}\frac{\partial \mathrm{v}_r \mathbf{w}_r^2}{\partial r}.$$

The physical components of the first Piola–Kirchhoff stress are

$$\begin{aligned} \begin{bmatrix} \mathbf{T}_{rr} & 0 & 0 \\ 0 & \mathbf{T}_{\theta\theta} & 0 \\ 0 & 0 & \mathbf{T}_{\phi\phi} \end{bmatrix} &\equiv \mathbf{T} = \mathbf{Q}^\top \tilde{\mathbf{S}} \mathbf{J}^{-\top} \mathbf{Q} |\mathbf{J}| \\ &= \begin{bmatrix} \tilde{\mathbf{S}}_{rr}\left(\frac{\mathrm{v}_r}{r}\right)^2 & 0 & 0 \\ 0 & \tilde{\mathbf{S}}_{\theta\theta}\frac{\mathbf{v}_r}{r}\frac{\partial \mathbf{v}_r}{\partial r} & 0 \\ 0 & 0 & \tilde{\mathbf{S}}_{\phi\phi}\frac{\mathbf{v}_r}{r}\frac{\partial \mathbf{v}_r}{\partial r} \end{bmatrix}. \end{aligned}$$

Thus conservation of momentum in spherical symmetry is

$$\mathbf{g}_r \rho = \frac{\partial \mathrm{v}_r \rho}{\partial t} - \frac{\partial \mathbf{T}_{rr} r^2}{\partial r}\frac{1}{r^2} - \frac{\mathbf{T}_{\theta\theta} + \mathbf{T}_{\phi\phi}}{r}.$$

Finally, cylindrically symmetric conservation of energy is

$$\frac{\partial(\epsilon + \frac{1}{2}\mathrm{v}_r^2)\rho}{\partial t} - \frac{1}{r^2}\frac{\partial \mathbf{T}_{rr}\mathrm{v}_r r^2}{\partial r} = (\omega + \mathbf{g}_r \mathrm{v}_r)\rho.$$

Let us describe the computations involved in updating the equations of motion. Assuming that $\mathbf{T}_{\phi\phi} = \mathbf{T}_{\theta\theta}$, we have

$$
\begin{aligned}
0 &= \int_{t^n}^{t^{n+1}} \int_{r_{j-1/2}}^{r_{j+1/2}} \left[\frac{\partial \mathrm{v}_r \rho}{\partial t} - \frac{\partial \mathbf{T}_{rr}}{\partial r} - \frac{2\mathbf{T}_{rr} - \mathbf{T}_{\theta\theta} - \mathbf{T}_{\phi\phi}}{r} - \mathbf{f}_r \rho \right] r^2 dr\, dt \\
&= \int_{r_{j-1/2}}^{r_{j+1/2}} \left[\rho \mathrm{v}_r(r, t^{n+1}) - \rho \mathrm{v}_r(r, t^n) \right] r^2 dr - \int_{t^n}^{t^{n+1}} \int_{r_{j-1/2}}^{r_{j+1/2}} \frac{\partial \mathbf{T}_{rr}}{\partial r} r^2 dr\, dt \\
&\quad - \int_{t^n}^{t^{n+1}} \int_{r_{j-1/2}}^{r_{j+1/2}} \left[(2\mathbf{T}_{rr} - \mathbf{T}_{\theta\theta} - \mathbf{T}_{\phi\phi}) r + \mathbf{f}_r \rho r^2 \right] dr\, dt \\
&\approx \left[(\mathrm{v}_r)_j^{n+1} - (\mathrm{v}_r)_j^n \right] \rho_j \frac{r_{j+1/2}^3 - r_{j-1/2}^3}{3} \\
&\quad - \frac{(\mathbf{T}_{rr})_{j+1/2}^{n+1/2} - (\mathbf{T}_{rr})_{j-1/2}^{n+1/2}}{\triangle r_j} \frac{r_{j+1/2}^3 - r_{j-1/2}^3}{3} \triangle t^{n+1/2} \\
&\quad - \left[(2\mathbf{T}_{rr} - \mathbf{T}_{\theta\theta} - \mathbf{T}_{\phi\phi})_j^{n+1} + (2\mathbf{T}_{rr} - \mathbf{T}_{\theta\theta} - \mathbf{T}_{\phi\phi})_j^n \right] \frac{r_{j+1/2}^2 - r_{j-1/2}^2}{4} \triangle t^{n+1/2} \\
&\quad - \mathbf{f}_j^{n+1/2} \rho_j \frac{r_{j+1/2}^3 - r_{j-1/2}^3}{3} \triangle t^{n+1/2}.
\end{aligned}
$$

The radial stresses $(\mathbf{T}_{rr})_{j+1/2}^{n+1/2}$ can be computed by a slope limiter algorithm, as in Section 6.2.5. If the constitutive law can be written in the form

$$
\frac{d\mathbf{T}_{rr}}{dt} = h_r \frac{\partial \mathrm{v}_r}{\partial r} + h_\theta \frac{\mathrm{v}_r}{r}
$$

then the quasilinear form of isothermal Lagrangian solid mechanics is

$$
\frac{\partial}{\partial t} \begin{bmatrix} \mathrm{v}_r \\ \mathbf{T}_{rr} \end{bmatrix} + \begin{bmatrix} 0 & -\frac{1}{\rho} \\ -h_r & 0 \end{bmatrix} \frac{\partial}{\partial r} \begin{bmatrix} \mathrm{v}_r \\ \mathbf{T}_{rr} \end{bmatrix} = \begin{bmatrix} \mathbf{f}_r + \frac{2(\mathbf{T}_{rr} - \mathbf{T}_{\theta\theta})}{r\rho} \\ h_\theta \frac{\mathrm{v}_r}{r} \end{bmatrix}.
$$

Note that

$$
\begin{bmatrix} 0 & -\frac{1}{\rho} \\ -h_r & 0 \end{bmatrix} \begin{bmatrix} 1 & 1 \\ \lambda\rho & -\lambda\rho \end{bmatrix} = \begin{bmatrix} 1 & 1 \\ \lambda\rho & -\lambda\rho \end{bmatrix} \begin{bmatrix} -\lambda & 0 \\ 0 & \lambda \end{bmatrix}
$$

where

$$
\lambda = \sqrt{h_r / \rho}.
$$

We solve

$$
Y c_{j+1/2}^n = \begin{bmatrix} \mathrm{v}_{j+1}^n - \mathrm{v}_j^n \\ (\mathbf{T}_{rr})_{j+1}^n - (\mathbf{T}_{rr})_j^n \end{bmatrix}
$$

for the characteristic expansion coefficients, apply a limiter to determine c_j^n, and compute

$$\begin{bmatrix} \mathbf{v}_r \\ \mathbf{T}_{rr} \end{bmatrix}_{j+1/2}^{n+1/2,L} = \begin{bmatrix} \mathbf{v}_r \\ \mathbf{T}_{rr} \end{bmatrix}_j^n + \begin{bmatrix} 1 \\ -\lambda\rho \end{bmatrix}_j^n \left(1 - \frac{\lambda_j^n \Delta t^{n+1/2}}{\Delta r_j}\right) \frac{\mathbf{e}_\theta^\top c_j^n}{2} + \begin{bmatrix} \mathbf{f}_r + \frac{2(\mathbf{T}_{rr} - \mathbf{T}_{\theta\theta})}{r\rho} \\ h_\theta \frac{\mathbf{v}_r}{r} \end{bmatrix}_j^n \frac{\Delta t^{n+1/2}}{2}$$

$$\begin{bmatrix} \mathbf{v}_r \\ \mathbf{T}_{rr} \end{bmatrix}_{j-1/2}^{n+1/2,R} = \begin{bmatrix} \mathbf{v}_r \\ \mathbf{T}_{rr} \end{bmatrix}_j^n - \begin{bmatrix} 1 \\ \lambda\rho \end{bmatrix}_j^n \left(1 - \frac{\lambda_j^n \Delta t^{n+1/2}}{\Delta r_j}\right) \frac{\mathbf{e}_r^\top c_j^n}{2} + \begin{bmatrix} \mathbf{f}_r + \frac{2(\mathbf{T}_{rr} - \mathbf{T}_{\theta\theta})}{r\rho} \\ h_\theta \frac{\mathbf{v}_r}{r} \end{bmatrix}_j^n \frac{\Delta t^{n+1/2}}{2}$$

In summary, we can perform limiting and characteristic tracing to determine velocity and radial stress at the cell sides and half-time. The particle positions can be updated by

$$(\mathbf{w}_r)_{j+1/2}^{n+1} = (\mathbf{w}_r)_{j+1/2}^n + (\mathbf{v}_r)_{j+1/2}^{n+1/2} \Delta t^{n+1/2}.$$

The determinant of the deformation gradient is given by

$$|\mathbf{Q}^\top \mathbf{J} \mathbf{Q}| = \frac{(\mathbf{w}_{j+1/2}^n)_r^3 - (\mathbf{w}_{j-1/2}^n)_r^3}{r_{j+1/2}^3 - r_{j-1/2}^3}$$

and its r, r component is given by

$$\left(\frac{\partial \mathbf{w}_r}{\partial r}\right)_j^n = \frac{(\mathbf{w}_{j+1/2}^n)_r - (\mathbf{w}_{j-1/2}^n)_r}{r_{j+1/2} - r_{j-1/2}}.$$

This gives us the information about the deformation gradient needed to determine the full stress tensor at the new time. Then we can use the momentum equation to update the velocity.

7.4.3 Cylindrical Coordinates

In cylindrical coordinates, the vector of curvilinear coordinates is

$$\mathbf{y}^\top = \begin{bmatrix} r, & \theta, & z \end{bmatrix},$$

where r is the distance from the center of the cylinder, z is the distance along the axis of the cylinder, and θ is the angle around the axis of the cylinder. It follows that the Cartesian coordinate vector is

$$\mathbf{a}^\top = \begin{bmatrix} r\cos\theta, & r\sin\theta, & z \end{bmatrix}.$$

From this, it is easy to compute

$$\frac{\partial \mathbf{a}}{\partial \mathbf{y}} = \begin{bmatrix} \cos\theta & -r\sin\theta & 0 \\ \sin\theta & r\cos\theta & 0 \\ 0 & 0 & 1 \end{bmatrix}$$

and

$$\mathbf{H}^2 = \left(\frac{\partial \mathbf{a}}{\partial \mathbf{y}}\right)^\top \left(\frac{\partial \mathbf{a}}{\partial \mathbf{y}}\right) = \begin{bmatrix} 1 & & \\ & r^2 & \\ & & 1 \end{bmatrix}.$$

Thus the orthogonal matrix of unit base vectors is

$$\mathbf{Q} = \frac{\partial \mathbf{a}}{\partial \mathbf{y}} \mathbf{H}^{-1} = \begin{bmatrix} \cos\theta & -\sin\theta & 0 \\ \sin\theta & \cos\theta & 0 \\ 0 & 0 & 1 \end{bmatrix}.$$

Note that

$$\frac{\partial \mathbf{Q}\mathbf{e}_r}{\partial \mathbf{y}} = \mathbf{Q}\mathbf{e}_\theta \mathbf{e}_\theta^\top, \quad \frac{\partial \mathbf{Q}\mathbf{e}_\theta}{\partial \mathbf{y}} = -\mathbf{Q}\mathbf{e}_r \mathbf{e}_\theta^\top, \quad \frac{\partial \mathbf{Q}\mathbf{e}_z}{\partial \mathbf{y}} = 0.$$

Given a vector $\mathbf{x}$ in Cartesian coordinates, we can compute the physical components of $\mathbf{x}$ to be

$$\begin{bmatrix} \mathbf{w}_r \\ \mathbf{w}_\theta \\ \mathbf{w}_z \end{bmatrix} = \mathbf{w} = \mathbf{Q}^\top \mathbf{x} = \begin{bmatrix} \cos\theta & \sin\theta & 0 \\ -\sin\theta & \cos\theta & 0 \\ 0 & 0 & 1 \end{bmatrix} \begin{bmatrix} \mathbf{x}_1 \\ \mathbf{x}_2 \\ \mathbf{x}_3 \end{bmatrix} = \begin{bmatrix} \mathbf{x}_1 \cos\theta + \mathbf{x}_2 \sin\theta \\ -\mathbf{x}_1 \sin\theta + \mathbf{x}_2 \cos\theta \\ \mathbf{x}_3 \end{bmatrix}.$$

Next, we compute the gradient of a scalar ω using Equation (7.30)

$$\nabla_{\mathbf{a}} \omega = \mathbf{Q}\mathbf{H}^{-1} \nabla_{\mathbf{y}} \omega = \begin{bmatrix} \cos\theta & -\frac{\sin\theta}{r} & 0 \\ \sin\theta & \frac{\cos\theta}{r} & 0 \\ 0 & 0 & 1 \end{bmatrix} \begin{bmatrix} \frac{\partial\omega}{\partial r} \\ \frac{\partial\omega}{\partial\theta} \\ \frac{\partial\omega}{\partial z} \end{bmatrix} = \begin{bmatrix} \frac{\partial\omega}{\partial r}\cos\theta - \frac{\partial\omega}{\partial\theta}\frac{\sin\theta}{r} \\ \frac{\partial\omega}{\partial r}\sin\theta + \frac{\partial\omega}{\partial\theta}\frac{\cos\theta}{r} \\ \frac{\partial\omega}{\partial z} \end{bmatrix}.$$

The deformation gradient $\mathbf{J}$ has physical components

$$\mathbf{Q}^\top \mathbf{J} \mathbf{Q} \equiv \mathbf{Q}^\top \frac{\partial \mathbf{x}}{\partial \mathbf{a}} \mathbf{Q} = \mathbf{Q}^\top \frac{\partial \mathbf{Q}\mathbf{w}}{\partial \mathbf{y}} \mathbf{H}^{-1} = \begin{bmatrix} \frac{\partial \mathbf{w}_r}{\partial r} & \frac{1}{r}\left(\frac{\partial \mathbf{w}_r}{\partial\theta} - \mathbf{w}_\theta\right) & \frac{\partial \mathbf{w}_r}{\partial z} \\ \frac{\partial \mathbf{w}_\theta}{\partial r} & \frac{1}{r}\left(\frac{\partial \mathbf{w}_\theta}{\partial\theta} + \mathbf{w}_r\right) & \frac{\partial \mathbf{w}_\theta}{\partial z} \\ \frac{\partial \mathbf{w}_z}{\partial r} & \frac{1}{r}\frac{\partial \mathbf{w}_z}{\partial\theta} & \frac{\partial \mathbf{w}_z}{\partial z} \end{bmatrix} = \begin{bmatrix} \mathbf{J}_{rr} & \mathbf{J}_{r\theta} & \mathbf{J}_{rz} \\ \mathbf{J}_{\theta r} & \mathbf{J}_{\theta\theta} & \mathbf{J}_{\theta z} \\ \mathbf{J}_{zr} & \mathbf{J}_{z\theta} & \mathbf{J}_{zz} \end{bmatrix}. \tag{7.37}$$

Similarly, the divergence of a vector $\mathbf{x}$ can be computed from equation (7.31):

$$\nabla_{\mathbf{a}} \cdot \mathbf{x} = \left|\frac{\partial \mathbf{y}}{\partial \mathbf{a}}\right| \nabla_{\mathbf{y}} \cdot \left(\mathbf{H}^{-1} \mathbf{w} \left|\frac{\partial \mathbf{a}}{\partial \mathbf{y}}\right|\right) = \frac{1}{r}\left[\frac{\partial r \mathbf{w}_r}{\partial r} + \frac{\partial \mathbf{w}_\theta}{\partial\theta} + \frac{\partial r \mathbf{w}_z}{\partial z}\right]. \tag{7.38}$$

We will also need to compute the physical components of the divergence of an array $\mathbf{S}$. Let $\tilde{\mathbf{S}} = \mathbf{Q}^\top\mathbf{S}\mathbf{Q}$. Then equation (7.32) shows us that

$$\mathbf{Q}^\top(\nabla_{\mathbf{a}}\cdot\mathbf{S})^\top = \left|\frac{\partial\mathbf{y}}{\partial\mathbf{a}}\right|\sum_i e_i\left\{\nabla_{\mathbf{y}}\cdot\left(\mathbf{H}^{-1}\tilde{\mathbf{S}}e_i\left|\frac{\partial\mathbf{a}}{\partial\mathbf{y}}\right|\right)-\mathrm{tr}\left(\mathbf{H}^{-1}\tilde{\mathbf{S}}\mathbf{Q}^\top\frac{\partial\mathbf{Q}e_i}{\partial\mathbf{y}}\left|\frac{\partial\mathbf{a}}{\partial\mathbf{y}}\right|\right)\right\}$$

$$= \begin{bmatrix} \frac{\partial\tilde{\mathbf{S}}_{rr}r}{\partial r}+\frac{\partial\tilde{\mathbf{S}}_{\theta r}}{\partial\theta}+\frac{\partial\tilde{\mathbf{S}}_{zr}r}{\partial z}-\tilde{\mathbf{S}}_{\theta\theta} \\ \frac{\partial\tilde{\mathbf{S}}_{r\theta}r}{\partial r}+\frac{\partial\tilde{\mathbf{S}}_{\theta\theta}}{\partial\theta}+\frac{\partial\tilde{\mathbf{S}}_{z\theta}r}{\partial z}+\tilde{\mathbf{S}}_{\theta r} \\ \frac{\partial\tilde{\mathbf{S}}_{rz}r}{\partial r}+\frac{\partial\tilde{\mathbf{S}}_{\theta z}}{\partial\theta}+\frac{\partial\tilde{\mathbf{S}}_{zz}r}{\partial z}\end{bmatrix}\frac{1}{r}. \tag{7.39}$$

Suppose that we are given a scalar conservation law

$$\frac{\partial\mathbf{u}}{\partial t}+\nabla_{\mathbf{x}}\cdot\mathbf{f}(\mathbf{u})=0.$$

The physical components of the flux are

$$\begin{bmatrix}\mathbf{f}_r\\ \mathbf{f}_\theta\\ \mathbf{f}_z\end{bmatrix}=\begin{bmatrix}\mathbf{f}_1\cos\theta+\mathbf{f}_2\sin\theta\\ -\mathbf{f}_1\sin\theta+\mathbf{f}_2\cos\theta\\ \mathbf{f}_3\end{bmatrix}.$$

Thus a scalar conservation law in cylindrical coordinates can be written in the conservation form

$$\frac{\partial\mathbf{u}}{\partial t}+\frac{1}{r}\left[\frac{\partial\mathbf{f}_r r}{\partial r}+\frac{\partial\mathbf{f}_\theta}{\partial\theta}+\frac{\partial\mathbf{f}_z r}{\partial z}\right]=0.$$

However, the more useful expression is the integral form of the law. If we integrate the cylindrical coordinate form of the conservation law over a cylindrical element, we obtain

$$\begin{aligned}0 &= \int_{t_1}^{t_2}\int_{r_1}^{r_2}\int_{\theta_1}^{\theta_2}\int_{z_1}^{z_2}\left\{\frac{\partial\mathbf{u}}{\partial t}+\frac{1}{r}\left[\frac{\partial\mathbf{f}_r r}{\partial r}+\frac{\partial\mathbf{f}_\theta}{\partial\theta}+\frac{\partial\mathbf{f}_z r}{\partial z}\right]\right\}r\,dz\,d\theta\,dr\,dt\\
&= \int_{r_1}^{r_2}\int_{\theta_1}^{\theta_2}\int_{z_1}^{z_2}[\mathbf{u}(r,\theta,z,t_2)-\mathbf{u}(r,\theta,z,t_1)]\,dz\,d\theta\,r dr\\
&\quad+\int_{t_1}^{t_2}\int_{\theta_1}^{\theta_2}\int_{z_1}^{z_2}[\mathbf{f}_r(r_2,\theta,z,t)r_2-\mathbf{f}_r(r_1,\theta,z,t)r_1]\,dz\,d\theta\,dt\\
&\quad+\int_{t_1}^{t_2}\int_{r_1}^{r_2}\int_{z_1}^{z_2}[\mathbf{f}_\theta(r,\theta_2,z,t)-\mathbf{f}_\theta(r,\theta_1,z,t)]\,dz\,dr\,dt\\
&\quad+\int_{t_1}^{t_2}\int_{r_1}^{r_2}\int_{\theta_1}^{\theta_2}[\mathbf{f}_z(r,\theta,z_2,t)-\mathbf{f}_z(r,\theta,z_1,t)]\,d\theta\,r dr\,dt.\end{aligned}$$

It is common to assume that the motion of a material is such that the physical components are functions of r and z only, and there is no θ component of arrays.

Thus in cylindrical symmetry $\mathbf{w}_\theta = 0$ and Cartesian coordinates are related to physical components by

$$\mathbf{x}^\top = \left[\mathbf{w}_r \cos\theta, \quad \mathbf{w}_r \sin\theta, \quad \mathbf{w}_z\right].$$

The physical components of a deformation gradient in cylindrical symmetry simplify to

$$\mathbf{Q}^\top \frac{\partial \mathbf{x}}{\partial \mathbf{a}} \mathbf{Q} = \begin{bmatrix} \frac{\partial \mathbf{w}_r}{\partial r} & 0 & \frac{\partial \mathbf{w}_r}{\partial z} \\ 0 & \frac{\mathbf{w}_r}{r} & 0 \\ \frac{\partial \mathbf{w}_z}{\partial r} & 0 & \frac{\partial \mathbf{w}_z}{\partial z} \end{bmatrix}.$$

The physical components of the Cauchy stress are

$$\mathbf{Q}^\top \mathbf{S} \mathbf{Q} = \begin{bmatrix} \mathbf{S}_{rr} & 0 & \mathbf{S}_{rz} \\ 0 & \mathbf{S}_{\theta\theta} & 0 \\ \mathbf{S}_{zr} & 0 & \mathbf{S}_{zz} \end{bmatrix}.$$

Similarly, the stress divergence simplifies to

$$\mathbf{Q}^\top (\nabla_{\mathbf{a}} \cdot \mathbf{S})^\top = \begin{bmatrix} \frac{1}{r}\frac{\partial r \mathbf{T}_{rr}}{\partial r} + \frac{\partial \mathbf{T}_{zr}}{\partial z} - \frac{\mathbf{T}_{\theta\theta}}{r} \\ 0 \\ \frac{1}{r}\frac{\partial r \mathbf{T}_{rr}}{\partial r} + \frac{\partial \mathbf{T}_{zz}}{\partial z} \end{bmatrix}$$

If the scalar conservation law is cylindrically symmetric, then $\mathbf{f}_\theta = 0$ and the remaining quantities are assumed to be independent of θ. The conservation law simplifies to

$$\frac{\partial \mathbf{u}}{\partial t} + \frac{1}{r}\left[\frac{\partial \mathbf{f}_r r}{\partial r} + \frac{\partial \mathbf{f}_z r}{\partial z}\right] = 0.$$

If we integrate the cylindrical coordinate form of the conservation law over a cylindrical shell, we obtain

$$\begin{aligned} 0 &= \frac{1}{2\pi}\int_{t_1}^{t_2}\int_{r_1}^{r_2}\int_0^{2\pi}\int_{z_1}^{z_2}\left\{\frac{\partial \mathbf{u}}{\partial t} + \frac{1}{r}\left[\frac{\partial \mathbf{f}_r r}{\partial r} + \frac{\partial \mathbf{f}_z r}{\partial z}\right]\right\} r\,dz\,d\theta\,dr\,dt \\ &= \int_{r_1}^{r_2}\int_{z_1}^{z_2} [\mathbf{u}(r,z,t_2) - \mathbf{u}(r,z,t_1)]\,dz\,rdr \\ &\quad + \int_{t_1}^{t_2}\int_{z_1}^{z_2} [\mathbf{f}_r(r_2,z,t)r_2 - \mathbf{f}_r(r_1,z,t)r_1]\,dz\,dt \\ &\quad + \int_{t_1}^{t_2}\int_{r_1}^{r_2} [\mathbf{f}_z(r,z_2,t) - \mathbf{f}_z(r,z_1,t)]\,rdr\,dt. \end{aligned}$$

7.4.3.1 Case Study: Eulerian Gas Dynamics in Cylindrical Coordinates

Recall that Eulerian conservation of mass in gas dynamics takes the Cartesian form

$$\frac{\partial \rho}{\partial t} + \nabla_{\mathbf{x}} \cdot (\mathbf{v}\rho) = 0.$$

We can use Equation (7.38) to obtain the conservation of mass in cylindrical coordinates:

$$\frac{\partial \rho}{\partial t} + \frac{1}{r}\frac{\partial \mathbf{v}_r \rho r}{\partial r} + \frac{1}{r}\frac{\partial \mathbf{v}_\theta \rho}{\partial \theta} + \frac{\partial \mathbf{v}_z \rho}{\partial z} = 0.$$

Eulerian conservation of momentum in Cartesian coordinates is

$$\frac{\partial \mathbf{v}\rho}{\partial t} + \nabla_{\mathbf{x}} \cdot (\mathbf{v}\rho\mathbf{v}^\top + Ip) = \mathbf{g}\rho.$$

We can use equation (7.39) to obtain conservation of momentum in cylindrical coordinates:

$$\begin{bmatrix} \mathbf{g}_r \rho \\ \mathbf{g}_\theta \rho \\ \mathbf{g}_z \rho \end{bmatrix} \equiv \mathbf{Q}^\top \mathbf{g}\rho = \frac{\partial \mathbf{Q}^\top \mathbf{v}\rho}{\partial t} + \mathbf{Q}^\top \left\{ \nabla_{\mathbf{a}} \cdot \left(\mathbf{v}\rho\mathbf{v}^\top + Ip \right) \right\}^\top$$

$$= \frac{\partial}{\partial t}\begin{bmatrix} \mathbf{v}_r \rho \\ \mathbf{v}_\theta \rho \\ \mathbf{v}_z \rho \end{bmatrix} + \begin{bmatrix} \frac{\partial \mathbf{v}_r^2 \rho r + p}{\partial r}\frac{1}{r} + \frac{\partial \mathbf{v}_\theta \mathbf{v}_r \rho}{\partial \theta}\frac{1}{r} + \frac{\partial \mathbf{v}_z \mathbf{v}_r \rho}{\partial z} - \frac{\mathbf{v}_\theta^2 \rho}{r} \\ \frac{\partial \mathbf{v}_r \mathbf{v}_\theta \rho r}{\partial r}\frac{1}{r} + \frac{\partial (\mathbf{v}_\theta^2 \rho + p)}{\partial \theta}\frac{1}{r} + \frac{\partial \mathbf{v}_z \mathbf{v}_\theta \rho}{\partial z} + \frac{\mathbf{v}_\theta \rho \mathbf{v}_r}{r} \\ \frac{\partial \mathbf{v}_r \mathbf{v}_z \rho r}{\partial r}\frac{1}{r} + \frac{\partial \mathbf{v}_\theta \mathbf{v}_z \rho}{\partial \theta}\frac{1}{r} + \frac{\partial \mathbf{v}_z^2 \rho + p}{\partial z} \end{bmatrix}.$$

Finally, conservation of energy in Cartesian coordinates takes the form

$$\frac{\partial \rho(e + \frac{1}{2}\mathbf{v}\cdot\mathbf{v})}{\partial t} + \nabla_{\mathbf{x}} \cdot \left[\mathbf{v}\rho \left(e + \frac{1}{2}\mathbf{v}\cdot\mathbf{v} \right) + \mathbf{v}p \right] = \rho\mathbf{g}\cdot\mathbf{v}.$$

In cylindrical coordinates, this becomes

$$\frac{\partial \rho(e + \frac{1}{2}[\mathbf{v}_r^2 + \mathbf{v}_\theta^2 + \mathbf{v}_z^2])}{\partial t} + \frac{1}{r}\frac{\partial r\mathbf{v}_r[p + \rho(e + \frac{1}{2}\mathbf{v}_r^2 + \mathbf{v}_\theta^2 + \mathbf{v}_z^2)]}{\partial r} + \frac{1}{r}\frac{\partial \mathbf{v}_\theta[p + \rho(e + \frac{1}{2}\mathbf{v}_r^2 + \mathbf{v}_\theta^2 + \mathbf{v}_z^2)]}{\partial \theta}$$
$$+ \frac{\partial \mathbf{v}_z[p + \rho(e + \frac{1}{2}\mathbf{v}_r^2 + \mathbf{v}_\theta^2 + \mathbf{v}_z^2)]}{\partial z} = \rho(\mathbf{g}_r\mathbf{v}_r + \mathbf{g}_\theta\mathbf{v}_\theta + \mathbf{g}_z\mathbf{v}_z).$$

In cylindrical symmetry, mass conservation is

$$\frac{\partial \rho}{\partial t} + \frac{\partial \mathbf{v}_r \rho r}{\partial r}\frac{1}{r} + \frac{\partial \mathbf{v}_z \rho}{\partial z} = 0.$$

Conservation of momentum is

$$\begin{bmatrix} \mathbf{g}_r \rho \\ \mathbf{g}_z \rho \end{bmatrix} = \begin{bmatrix} \frac{\partial \mathbf{v}_r \rho}{\partial t} \\ \frac{\partial \mathbf{v}_z \rho}{\partial t} \end{bmatrix} + \begin{bmatrix} \frac{1}{r}\frac{\partial r\rho\mathbf{v}_r^2 r}{\partial r} + \frac{\partial \rho\mathbf{v}_r\mathbf{v}_z}{\partial z} + \frac{\partial p}{\partial r} \\ \frac{1}{r}\frac{\partial \rho\mathbf{v}_z\mathbf{v}_r}{\partial r} + \frac{\partial \rho\mathbf{v}_z^2}{\partial z} + \frac{\partial p}{\partial z} \end{bmatrix}.$$

Energy conservation is

$$\frac{\partial \rho(e+\frac{1}{2}[\mathbf{v}_r^2+\mathbf{v}_z^2])}{\partial t}+\frac{1}{r}\frac{\partial r\mathbf{v}_r[p+\rho(e+\frac{1}{2}\mathbf{v}_r^2+\mathbf{v}_z^2)]}{\partial r}+\frac{\partial \mathbf{v}_z[p+\rho(e+\frac{1}{2}\mathbf{v}_r^2+\mathbf{v}_z^2)]}{\partial z}=\rho(\mathbf{g}_r\mathbf{v}_r+\mathbf{g}_z\mathbf{v}_z).$$

In order to develop a discretization of these equations, let us work with cell averages

$$\mathbf{u}_{ij}^n \approx \frac{2}{(r_{i+1/2}^2-r_{i-1/2}^2)\triangle z_j}\int_{z_{j-1/2}}^{z_{j+1/2}}\int_{r_{i-1/2}}^{r_{i+1/2}}\begin{bmatrix}\rho\\ \mathbf{v}_r\rho\\ \mathbf{v}_z\rho\\ (e+\frac{1}{2}[\mathbf{v}_r^2+\mathbf{v}_z^2])\rho\end{bmatrix} r dr\, dz.$$

We could compute flux integrals

$$\mathbf{f}_{i+1/2,j}^{n+1/2}\approx\int_{t^n}^{t^{n+1}}\int_{z_{j-1/2}}^{z_{j+1/2}}\begin{bmatrix}\rho\mathbf{v}_r\\ \rho\mathbf{v}_r^2\\ \rho\mathbf{v}_z\mathbf{v}_r\\ \mathbf{v}_r\left[p+\rho(e+\frac{1}{2}[\mathbf{v}_r^2+\mathbf{v}_z^2])\right]\end{bmatrix}(r_{i+1/2},z,t)\,dz\,dt\,r_{i+1/2}$$

$$\mathbf{f}_{i,j+1/2}^{n+1/2}\approx\int_{t^n}^{t^{n+1}}\int_{r_{i-1/2}}^{r_{i+1/2}}\begin{bmatrix}\rho\mathbf{v}_z\\ \rho\mathbf{v}_r\mathbf{v}_z\\ \rho\mathbf{v}_z^2\\ \mathbf{v}_z\left[p+\rho(e+\frac{1}{2}[\mathbf{v}_r^2+\mathbf{v}_z^2])\right]\end{bmatrix}(r,z_{j+1/2},t)\,rdr\,dt$$

using the corner transport upwind scheme with slope limiting. The Riemann problem solutions would also provide values for the pressures at the cell sides. The cell averages could be updated by

$$\begin{aligned}\mathbf{u}_{ij}^{n+1}&=\mathbf{u}_{ij}^n-\left[\mathbf{f}_{i+1/2,j}^{n+1/2}-\mathbf{f}_{i-1/2,j}^{n+1/2}+\mathbf{f}_{i,j+1/2}^{n+1/2}-\mathbf{f}_{i,j-1/2}^{n+1/2}\right]\frac{2}{(r_{i+1/2}^2-r_{i-1/2}^2)\triangle z_{j+1/2}}\\
&-\begin{bmatrix}0\\ p_{i+1/2,j}^{n+1/2}-p_{i-1/2,j}^{n+1/2}\\ 0\\ 0\end{bmatrix}\frac{\triangle t^{n+1/2}}{\triangle r_i}-\begin{bmatrix}0\\ 0\\ p_{i,j+1/2}^{n+1/2}-p_{i,j-1/2}^{n+1/2}\\ 0\end{bmatrix}\frac{\triangle t^{n+1/2}}{\triangle z_j}\\
&+\left\{\begin{bmatrix}0\\ \mathbf{g}_r\rho\\ \mathbf{g}_z\rho\\ (\mathbf{g}_r\mathbf{v}_r+\mathbf{g}_z\mathbf{v}_z)\rho\end{bmatrix}_j^n+\begin{bmatrix}0\\ \mathbf{g}_r\rho\\ \mathbf{g}_z\rho\\ (\mathbf{g}_r\mathbf{v}_r+\mathbf{g}_z\mathbf{v}_z)\rho\end{bmatrix}_j^{n+1}\right\}\frac{\triangle t^{n+1/2}}{2}.\end{aligned}$$

This equation is not implicit. The density can be updated at the new time, then combined with gravity to update the momentum. Afterward, the new density and velocity can be combined with gravity to update the energy. Of course, if $\mathbf{g}$ is due to the force of gravity, then cylindrical symmetry would require that $\mathbf{g}_r = 0$.

7.4.3.2 Case Study: Lagrangian Solid Mechanics in Cylindrical Coordinates

Lagrangian conservation of mass

$$\frac{\partial \rho}{\partial t} = 0$$

is unchanged in cylindrical coordinates. Equality of mixed partial derivatives in cylindrical coordinates gives us

$$\frac{\partial}{\partial t}\begin{bmatrix} \mathbf{J}_{rr} & \mathbf{J}_{r\theta} & \mathbf{J}_{rz} \\ \mathbf{J}_{\theta r} & \mathbf{J}_{\theta\theta} & \mathbf{J}_{\theta z} \\ \mathbf{J}_{zr} & \mathbf{J}_{z\theta} & \mathbf{J}_{zz} \end{bmatrix} = \mathbf{Q}^\top \frac{\partial \mathbf{J}}{\partial t}\mathbf{Q} = \mathbf{Q}^\top \frac{\partial \mathbf{v}}{\partial \mathbf{a}}\mathbf{Q} = \mathbf{Q}^\top \frac{\partial \mathbf{Q}\mathbf{Q}^\top \mathbf{v}}{\partial \mathbf{y}}\mathbf{H}^{-1}$$
$$= \begin{bmatrix} \frac{\partial \mathbf{v}_r}{\partial r} & \frac{1}{r}(\frac{\partial \mathbf{v}_r}{\partial \theta} - \mathbf{v}_\theta) & \frac{\partial \mathbf{v}_r}{\partial z} \\ \frac{\partial \mathbf{v}_\theta}{\partial r} & \frac{1}{r}(\frac{\partial \mathbf{v}_\theta}{\partial \theta} + \mathbf{v}_r) & \frac{\partial \mathbf{v}_\theta}{\partial z} \\ \frac{\partial \mathbf{v}_z}{\partial r} & \frac{1}{r}\frac{\partial \mathbf{v}_z}{\partial \theta} & \frac{\partial \mathbf{v}_z}{\partial z} \end{bmatrix}.$$

The physical components of the Cauchy stress are

$$\tilde{\mathbf{S}} = \mathbf{Q}^\top \mathbf{S}\mathbf{Q} \equiv \begin{bmatrix} \mathbf{S}_{rr} & \mathbf{S}_{r\theta} & \mathbf{S}_{rz} \\ \mathbf{S}_{\theta r} & \mathbf{S}_{\theta\theta} & \mathbf{S}_{\theta z} \\ \mathbf{S}_{zr} & \mathbf{S}_{z\theta} & \mathbf{S}_{zz} \end{bmatrix}.$$

Since $\mathbf{S}$ is symmetric, so is $\tilde{\mathbf{S}}$. The physical components of the first Piola–Kirchhoff stress are

$$\begin{bmatrix} \mathbf{T}_{rr} & \mathbf{T}_{r\theta} & \mathbf{T}_{rz} \\ \mathbf{T}_{\theta r} & \mathbf{T}_{\theta\theta} & \mathbf{T}_{\theta z} \\ \mathbf{T}_{zr} & \mathbf{T}_{z\theta} & \mathbf{T}_{zz} \end{bmatrix} \equiv \mathbf{T} = \mathbf{Q}^\top \mathbf{S}\mathbf{J}^{-\top}\mathbf{Q}|\mathbf{J}|$$
$$= \begin{bmatrix} \mathbf{S}_{rr} & \mathbf{S}_{r\theta} & \mathbf{S}_{rz} \\ \mathbf{S}_{\theta r} & \mathbf{S}_{\theta\theta} & \mathbf{S}_{\theta z} \\ \mathbf{S}_{zr} & \mathbf{S}_{z\theta} & \mathbf{S}_{zz} \end{bmatrix} \begin{bmatrix} \mathbf{J}_{\theta\theta}\mathbf{J}_{zz} - \mathbf{J}_{z\theta}\mathbf{J}_{\theta z} & -\mathbf{J}_{\theta r}\mathbf{J}_{zz} + \mathbf{J}_{zr}\mathbf{J}_{\theta z} & \mathbf{J}_{\theta r}\mathbf{J}_{z\theta} - \mathbf{J}_{zr}\mathbf{J}_{\theta\theta} \\ -\mathbf{J}_{r\theta}\mathbf{J}_{zz} + \mathbf{J}_{z\theta}\mathbf{J}_{rz} & \mathbf{J}_{rr}\mathbf{J}_{zz} - \mathbf{J}_{zr}\mathbf{J}_{\theta z} & -\mathbf{J}_{rr}\mathbf{J}_{z\theta} + \mathbf{J}_{zr}\mathbf{J}_{r\theta} \\ \mathbf{J}_{r\theta}\mathbf{J}_{z\theta} - \mathbf{J}_{\theta\theta}\mathbf{J}_{rz} & -\mathbf{J}_{rr}\mathbf{J}_{\theta z} + \mathbf{J}_{\theta r}\mathbf{J}_{rz} & \mathbf{J}_{rr}\mathbf{J}_{\theta\theta} - \mathbf{J}_{\theta r}\mathbf{J}_{r\theta} \end{bmatrix}.$$

Note that $\mathbf{T}$ is not necessarily symmetric. Thus conservation of momentum in cylindrical coordinates is

$$\begin{bmatrix} \mathbf{g}_r\rho \\ \mathbf{g}_\theta\rho \\ \mathbf{g}_z\rho \end{bmatrix} = \mathbf{Q}^\top \mathbf{g}\rho = \frac{\partial \mathbf{Q}^\top \mathbf{v}\rho}{\partial t} - \mathbf{Q}^\top \left[\nabla_{\mathbf{a}} \cdot (\mathbf{J}^{-1}|\mathbf{J}|\mathbf{S})\right]^\top$$
$$= \frac{\partial}{\partial t}\begin{bmatrix} \mathbf{v}_r\rho \\ \mathbf{v}_\theta\rho \\ \mathbf{v}_z\rho \end{bmatrix} - \begin{bmatrix} \frac{r\partial \mathbf{T}_{rr}}{\partial r} + \frac{\partial \mathbf{T}_{\theta r}}{\partial \theta} + \frac{\partial r\mathbf{T}_{zr}}{\partial z} - \mathbf{T}_{\theta\theta} \\ \frac{\partial r\mathbf{T}_{r\theta}}{\partial r} + \frac{\partial \mathbf{T}_{\theta\theta}}{\partial \theta} + \frac{\partial r\mathbf{T}_{z\theta}}{\partial z} + \mathbf{T}_{\theta r} \\ \frac{\partial r\mathbf{T}_{rz}}{\partial r} + \frac{\partial \mathbf{T}_{\theta z}}{\partial \theta} + \frac{\partial r\mathbf{T}_{zz}}{\partial z} \end{bmatrix}\frac{1}{r}.$$

Finally, Cartesian conservation of energy

$$\frac{\partial(\epsilon + \frac{1}{2}\mathbf{v}\cdot\mathbf{v})\rho}{\partial t} - \nabla_{\mathbf{a}} \cdot (\mathbf{J}^{-1}\mathbf{S}|\mathbf{J}|\mathbf{v}) = (\omega + \mathbf{g}\cdot\mathbf{v})\rho$$

becomes

$$\frac{\partial(\epsilon + \frac{1}{2}\mathbf{v}\cdot\mathbf{v})\rho}{\partial t} - \frac{1}{r}\left\{\frac{\partial(\mathbf{T}_{rr}\mathbf{v}_r + \mathbf{T}_{\theta r}\mathbf{v}_\theta + \mathbf{T}_{zr}\mathbf{v}_z)r}{\partial r} + \frac{\partial \mathbf{T}_{r\theta}\mathbf{v}_r + \mathbf{T}_{\theta\theta}\mathbf{v}_\theta + \mathbf{T}_{z\theta}\mathbf{v}_z}{\partial \theta}\right\}$$
$$+ \frac{\partial \mathbf{T}_{rz}\mathbf{v}_r + \mathbf{T}_{\theta z}\mathbf{v}_\theta + \mathbf{T}_{zz}\mathbf{v}_z}{\partial z} = (\omega + \mathbf{g}_r\mathrm{v}_r + \mathbf{g}_\theta\mathbf{v}_\theta + \mathbf{g}_z\mathbf{v}_z)\rho.$$

In cylindrical symmetry, equality of mixed partial derivatives gives us

$$\frac{\partial}{\partial t}\begin{bmatrix}\mathbf{J}_{rr} & 0 & \mathbf{J}_{rz}\\ 0 & \mathbf{J}_{\theta\theta} & 0\\ \mathbf{J}_{zr} & 0 & \mathbf{J}_{zz}\end{bmatrix} = \mathbf{Q}^\top\frac{\partial \mathbf{J}}{\partial t}\mathbf{Q} = \mathbf{Q}^\top\frac{\partial \mathbf{v}}{\partial \mathbf{a}}\mathbf{Q} = \begin{bmatrix}\frac{\partial \mathbf{v}_r}{\partial r} & 0 & \frac{\partial \mathbf{v}_r}{\partial z}\\ 0 & \frac{\mathbf{v}_r}{r} & 0\\ \frac{\partial \mathbf{v}_z}{\partial r} & 0 & \frac{\partial \mathbf{v}_z}{\partial z}\end{bmatrix}.$$

The physical components of the Cauchy stress are

$$\begin{bmatrix}\mathbf{S}_{rr} & 0 & \mathbf{S}_{rz}\\ 0 & \mathbf{S}_{\theta\theta} & 0\\ \mathbf{S}_{zr} & 0 & \mathbf{S}_{zz}\end{bmatrix} \equiv \tilde{\mathbf{S}} = \mathbf{Q}^\top\mathbf{S}\mathbf{Q}.$$

The physical components of the first Piola–Kirchhoff stress are

$$\mathbf{T} = \mathbf{Q}^\top\mathbf{S}\mathbf{J}^{-\top}\mathbf{Q}|\mathbf{J}| = \begin{bmatrix}\mathbf{S}_{rr} & 0 & \mathbf{S}_{rz}\\ 0 & \mathbf{S}_{\theta\theta} & 0\\ \mathbf{S}_{zr} & 0 & \mathbf{S}_{zz}\end{bmatrix}\begin{bmatrix}\mathbf{J}_{\theta\theta}\mathbf{J}_{zz} & 0 & -\mathbf{J}_{zr}\mathbf{J}_{\theta\theta}\\ 0 & \mathbf{J}_{rr}\mathbf{J}_{zz} & 0\\ -\mathbf{J}_{\theta\theta}\mathbf{J}_{rz} & 0 & \mathbf{J}_{rr}\mathbf{J}_{\theta\theta}\end{bmatrix}$$
$$\equiv \begin{bmatrix}\mathbf{T}_{rr} & 0 & \mathbf{T}_{rz}\\ 0 & \mathbf{T}_{\theta\theta} & 0\\ \mathbf{T}_{zr} & 0 & \mathbf{T}_{zz}\end{bmatrix}.$$

Conservation of momentum in cylindrical symmetry is

$$\begin{bmatrix}\mathbf{g}_r\rho\\ \mathbf{g}_z\rho\end{bmatrix} = \frac{\partial}{\partial t}\begin{bmatrix}\mathbf{v}_r\rho\\ \mathbf{v}_z\rho\end{bmatrix} - \begin{bmatrix}\frac{\partial r\mathbf{T}_{rr}}{\partial r} + \frac{\partial r\mathbf{T}_{zr}}{\partial z} - \mathbf{T}_{\theta\theta}\\ \frac{\partial r\mathbf{T}_{rz}}{\partial r} + \frac{\partial r\mathbf{T}_{zz}}{\partial z}\end{bmatrix}\frac{1}{r}.$$

Finally, conservation of energy is

$$\frac{\partial(\epsilon + \frac{1}{2}\mathbf{v}\cdot\mathbf{v})\rho}{\partial t} - \frac{1}{r}\frac{\partial(\mathbf{T}_{rr}\mathbf{v}_r + \mathbf{T}_{zr}\mathbf{v}_z)r}{\partial r} - \frac{\partial \mathbf{T}_{rz}\mathbf{v}_r + \mathbf{T}_{zz}\mathbf{v}_z}{\partial z} = (\omega + \mathbf{g}_r\mathrm{v}_r + \mathbf{g}_z\mathbf{v}_z)\rho.$$

A numerical scheme for cylindrically symmetric isothermal solid mechanics might proceed as follows. We could compute the fluxes at the cell sides using a slope limiter scheme. This would determine values for the velocity and the first

Piola–Kirchhoff stress. Then we could update the deformation gradient by

$$\mathbf{J}_{ij}^{n+1} = \mathbf{J}_{ij}^{n} + \left\{ \begin{bmatrix} \mathbf{v}_r \\ 0 \\ \mathbf{v}_z \end{bmatrix}_{i+1/2,j}^{n+1/2} r_{i+1/2} - \begin{bmatrix} \mathbf{v}_r \\ 0 \\ \mathbf{v}_z \end{bmatrix}_{i-1/2,j}^{n+1/2} r_{i-1/2} \right\} \mathbf{e}_r^\top \frac{2\triangle t^{n+1/2}}{r_{i+1/2}^2 - r_{i-1/2}^2}$$
$$+ \left\{ \begin{bmatrix} \mathbf{v}_r \\ 0 \\ \mathbf{v}_z \end{bmatrix}_{i+1/2,j}^{n+1/2} - \begin{bmatrix} \mathbf{v}_r \\ 0 \\ \mathbf{v}_z \end{bmatrix}_{i-1/2,j}^{n+1/2} \right\} \mathbf{e}_z^\top \frac{\triangle t^{n+1/2}}{dz_j}$$
$$+ \begin{bmatrix} 0 \\ (\mathbf{v}_r)_{i+1/2,j} + (\mathbf{v}_r)_{i-1/2,j} + (\mathbf{v}_r)_{i,j+1/2} + (\mathbf{v}_r)_{i,j-1/2} \\ 0 \end{bmatrix} \frac{\triangle t^{n+1/2}}{2(r_{i+1/2} + r_{i-1/2})}.$$

Afterward, we can update the stress. Then the momentum could be updated by

$$\begin{bmatrix} \rho \mathbf{v}_r \\ \rho \mathbf{v}_z \end{bmatrix}_{ij}^{n+1} = \begin{bmatrix} \rho \mathbf{v}_r \\ \rho \mathbf{v}_z \end{bmatrix}_{ij}^{n} + \left\{ \begin{bmatrix} \mathbf{T}_{rr} \\ \mathbf{T}_{rz} \end{bmatrix}_{i+1/2,j}^{n+1/2} r_{i+1/2} - \begin{bmatrix} \mathbf{T}_{rr} \\ \mathbf{T}_{rz} \end{bmatrix}_{i-1/2,j}^{n+1/2} r_{i-1/2} \right\} \frac{2\triangle t^{n+1/2}}{r_{i+1/2}^2 - r_{i-1/2}^2}$$
$$+ \left\{ \begin{bmatrix} \mathbf{T}_{zr} \\ \mathbf{T}_{zz} \end{bmatrix}_{i,j+1/2}^{n+1/2} - \begin{bmatrix} \mathbf{T}_{zr} \\ \mathbf{T}_{zz} \end{bmatrix}_{i,j-1/2}^{n+1/2} \right\} \frac{2\triangle t^{n+1/2}}{\triangle z_j}$$
$$- \left\{ \begin{bmatrix} \mathbf{T}_{\theta\theta} \\ 0 \end{bmatrix}_{ij}^{n} + \begin{bmatrix} \mathbf{T}_{\theta\theta} \\ 0 \end{bmatrix}_{ij}^{n+1} \right\} \frac{2\triangle t^{n+1/2}}{r_{i+1/2} + r_{i-1/2}} + \begin{bmatrix} \mathbf{g}_r \\ \mathbf{g}_z \end{bmatrix} \rho_{ij} \triangle t^{n+1/2}.$$

Exercises for 7.4

7.4.1 The spherically symmetric Burgers' equation has Cartesian flux $f(u, \mathbf{x}) = \frac{1}{2} u^2 \mathbf{x} / \|\mathbf{x}\|$.
(a) How would this problem be formulated in spherical coordinates?
(b) How would you formulate a free-stream-preserving numerical scheme for this problem?

7.4.2 Determine the equations describing spherically symmetric shallow water, and formulate a free-stream-preserving numerical scheme for their numerical solution.

7.4.3 The natural formulation of the gas dynamics momentum equation in spherical coordinates is

$$0 = \frac{\partial \mathbf{v}_r \rho}{\partial t} + \frac{1}{r^2} \frac{\partial r^2 (\mathbf{v}_r^2 \rho + p)}{\partial r} - \frac{2p}{r}.$$

Why did we rewrite this in the form

$$0 = \frac{\partial \mathbf{v}_r \rho}{\partial t} + \frac{1}{r^2} \frac{\partial r^2 \mathbf{v}_r^2 \rho}{\partial r} - \frac{\partial p}{\partial r}?$$

7.4.4 Consider linear elasticity in spherical symmetry.
(a) How would we compute the strain tensor in the numerical method described above?
(b) How would we compute the stress tensor?
(c) How would we update the velocity, using a slope limiter scheme?

7.4.5 Polar coordinates would represent a further simplification of the cylindrically symmetric problem, in which there would be no dependence on z. Formulate gas dynamics in polar coordinates, and describe a numerical method for its solution.

7.4.6 Determine the equations for cylindrically symmetric shallow water.

7.5 Source Terms

Because of the publication constraints, we did not have time to discuss several advanced topics in detail. Instead, we will provide some references to guide the reader.

In section 7.4 we saw that the use of various kinds of curvilinear coordinate systems introduces source terms into conservation laws. Source terms arise quite naturally in physical systems using Cartesian coordinates as well. For example, ramps are important to modeling traffic flow [97][page 167], wells are essential in oil recovery [6, 128, 129, 165], shallow water flow often occurs over variable topography [86], and combustion problems are important applications of gas dynamics [27, 79, 93, 130].

In many cases, these source terms can be treated accurately by operator splitting. For example, capillary pressure terms in Buckley–Leverett flow can often be treated in this way. With stiff source terms, however, such an approach can produce fictitious waves. Within the context of a MUSCL scheme, Pember [130] treated the stiff source terms implicitly in the determination of the states for the Riemann problem, and implicitly in the update of the new solution. Jin and Levermore [79] introduced the local equilibrium flux and an additional equation for relaxation to local equilibrium. The additional equation involves a user-defined relaxation parameter which must be chosen to be appropriately large. We invite interested readers to examine these approaches in more detail.

7.6 Geometric Flexibility

Many important physical problems involve flow around obstacles (such as gas flow around airplanes, water flow around ships, ocean currents around land masses). These problems require more sophisticated numerical schemes than we have had space to discuss in this text.

If an appropriate coordinate transformation is available, sometimes an irregular flow region can be transformed into a rectangular region, or a union of rectangles. Stream functions can often provide these transformations. If the transformations are sufficiently smooth, then high-order numerical methods (such as ENO and discontinuous Galerkin) can be used on the transformed system. For linear advection problems, the use of streamline methods [43] can convert systems of conservation laws into ordinary differential equations within stream-tubes, and self-similarity can be used to map the solution on a generic stream-tube onto an arbitrary stream-tube. Three-dimensional problems can be solved quite rapidly in this way, with no numerical diffusion across stream surfaces.

Alternatively, we can use quadrilateral grids in two dimensions [12, 47, 95, 166]. Limiting on such grids is easiest if the mesh is reasonably smooth and the quadrilateral indexing is "logically rectangular," in which case the limiting is performed between quadrilaterals differing in only one cell index much as limiting is performed on rectangular grids. Alternatively, we could use triangles or tetrahedrons [2, 9, 30, 111]. In these cases, limiting is necessary but much more difficult [10, 30].

An alternative to gridding an irregular flow domain is the use of Cartesian grids [16]. Here the idea is to make a first computation on a regular grid that ignores the obstacles, then correct those results to account for the flow around the obstacles. The corrections can usually be performed easily to achieve second-order accuracy. If only a small fraction of the grid cells need such a correction, then this approach can be must faster than using body-fitted grids. The new book by Li and Ito [102] provides significant detail regarding these methods.

8
Adaptive Mesh Refinement

8.1 Localized Phenomena

Many hyperbolic problems involve moving features of the solution that require additional resolution. For example, most schemes resolve shocks and contact discontinuities with lower accuracy than smooth features of the flow. In order to obtain good accuracy with such methods near the discontinuities, it is necessary to refine the mesh. However, if the mesh is refined everywhere, the increase in computational cost can be substantial.

Many of the methods we have examined are second-order accurate for smooth solutions, but only first-order accurate at discontinuities. This means that if we want to halve the error in the resolution of a shock by a formally second-order method, we must double the number of cells in each coordinate direction. The formally first-order methods are even worse: because of the large numerical diffusion, their order of accuracy at discontinuities is typically $O(\sqrt{\triangle x})$. If we want to halve the error in the resolution of a shock by a first-order method, we must quadruple the number of cells in each direction.

Suppose that we are computing the solution to a system of hyperbolic equations in d dimensions via an explicit method on a grid with n cells in each direction. Then the cost per timestep is proportional to n^d. Since the CFL condition requires that $\triangle t$ be proportional to $\triangle x$, which is in turn inversely proportional to n, the cost to compute the solution over the time required for a wave to cross the grid is proportional to n^{d+1}. In three dimensions, this means that if we double the number of cells in each direction, we have to do 16 times as much work. With a method that is formally second-order but actually first-order near discontinuities, we do 2^{d+1} times as much work to halve the error in the resolution of the discontinuity. On the other hand, if we use a first-order method, in order to reduce the error by a factor of 2, we must increase the work by a factor of $2^{2(d+1)}$; in three dimensions, this factor is 256.

One approach to this problem is to go to even higher order methods. Suppose that we could construct, say, an ENO scheme of order p. The work with such a scheme is still $O(n^d)$ per timestep, but the error is improved by a factor of 2^p by doubling the number of cells in each direction. This means that we need to increase the number of cells in each direction by a factor of $2^{1/p}$ to reduce the error by a factor of 2. In d dimensions, the work increases by a factor of $2^{(d+1)/p}$. This, of course, looks very attractive, when compared to the results in the previous paragraph. The difficulty is that the presence of a discontinuity often leads to a very large factor multiplying the power of 2 in the work estimate.

General experience in dealing with ordinary differential equations found that it is more efficient to use low-order methods and fine mesh near rough behavior, and higher-order methods on coarse meshes for smooth behavior. Since we are now dealing with partial differential equations, we can apply this experience only by varying the mesh resolution in space and time near regions of rough behavior.

Suppose that we use a coarse mesh with N cells in each direction to compute the solution, except near regions of rough behavior. In order to compute the solution on this coarse mesh, the work is proportional to N^{d+1}. Suppose that we have a propagating discontinuity surface, around which we place a mesh with a fixed number of cells normal to the surface, but refined by a factor of 2^r in the other coordinate directions. This means that there are an order of $(2^r N)^{d-1}$ cells in the locally refined mesh. The CFL condition requires that we take an order of $2^r N$ timesteps on the locally refined grid. The total work with the locally refined grid is proportional to $(2^r N)^d$. This means that we can increase the accuracy by a factor 2^r at a cost similar to solving a problem with one less coordinate dimension.

There are alternatives to adaptive mesh refinement. One alternative is to obtain maximal accuracy for a fixed cost [119]. Our approach is to seek a desired level of accuracy with nearly minimal cost.

Our approach will use nested arrays of (logically) rectangular grid patches, following the ideas developed by Marsha Berger [17, 18, 19, 20]. Other approaches use variable numbers of computational cells on unstructured meshes or meshes refined cell-by-cell; see, for example, [112].

The principal difficulty with all forms of adaptive mesh refinement is programming complexity. Our computations are carried out on arrays of logically rectangular patches defined in recursively finer index spaces. Each array of patches is designed so that its union is contained in the union of the next coarser array of patches. A fair amount of programming is required to handle the communication between the patches on the same and coarser or finer levels of refinement. In some problems, the behavior of this communication between scales is interesting in its own right [28].

The principal reason for working on logically rectangular grid patches is that we will generate a relatively small number of significant computational assignments, namely to integrate the solution of the differential equations on a grid patch. Typically, it is far easier to generate a robust and reliable numerical method on a regular grid. It is also easier to arrange the computations to take advantage of the machine memory hierarchy, particularly to make good use of the cache. Our strategy also makes it easier to make efficient use of pipelining and distributed memory machines.

8.2 Basic Assumptions

We assume that we are given some initial coarse mesh and an integer refinement ratio r. We will call this mesh the coarsest **level of refinement**, and denote it by $\mathcal{L}_0$. Imagine finer levels $\mathcal{L}_\ell$ defined recursively from the coarsest level $\mathcal{L}_0$ according to several rules.

Finite Termination In order to guarantee that the algorithm terminates, we require that there be a maximum number of levels, and a maximum number of timesteps on each level.

Logical Rectangularity We assume that each level $\mathcal{L}_\ell$ consists of some number of **patches**, each of which is a logically rectangular array of cells. A "logically rectangular" array of cells means that the array of cells can be mapped to a rectangular grid by a coordinate transformation. The grid itself can be non-rectangular in space; we only require that the grid be rectangular in the data arrays.

Grid Alignment We assume that if a coarse cell is refined in any part of its physical space, then it is refined everywhere. In other words, the boundary of a fine patch coincides with the boundary of a logically rectangular array of coarse grid cells.

Fixed Refinement Ratio We assume that we are given an integer **refinement ratio** r. Whenever a coarse cell on level $\mathcal{L}_\ell$ is refined, it is subdivided into r cells in each coordinate direction on level $\mathcal{L}_{\ell+1}$. Normally, the refinement ratio is a power of 2. Note that the assumptions of a fixed refinement ratio and of grid alignment imply that on any level $\mathcal{L}_\ell$ with $\ell > 0$, in any patch the number of cells in any coordinate direction is an integer multiple of the refinement ratio.

Proper Nesting We assume that the union of fine patches on level $\mathcal{L}_{\ell+1}$ is contained in the interior of the union of coarse patches on level $\mathcal{L}_\ell$. However, an individual fine patch is not required to lie inside any single coarse patch. Note that this assumption implies that the coarsest level $\mathcal{L}_0$ must completely cover the entire physical domain. This in turn implies that we must be able to provide a logically rectangular grid on the coarsest domain.

Synchronization After we advance the data on patches in level $\mathcal{L}_\ell$ to some time, we assume that the data on patches in level $\mathcal{L}_{\ell+1}$ are advanced by as many timesteps as required by stability and accuracy to reach exactly the same time as the coarser level $\mathcal{L}_\ell$. This assumption implies that coarse patches are integrated before fine patches; it also implies that the timestepping algorithm must be applied recursively within each timestep on all but the finest level.

Fine Preference We assume that the numerical scheme for our differential equation produces better results on the fine grid than on the coarse, in regions of the problem where both fine and coarse grid cells overlap.

Conservation We assume that the numerical scheme for our conservation law is conservative. Where fine and coarse grids overlap, we replace the coarse grid results with conservative coarsenings of the fine grid results; this process is called **upscaling**.

Regridding We assume that we are given a **regrid interval** which is a predetermined number of timesteps between regridding events on a coarse level. At the end of each regrid interval, the coarse level $\mathcal{L}_\ell$ selects coarse cells that need refinement at the new time, organizes these cells into some number of logically rectangular arrays of cells, and refines them to form the new cells on the finer level $\mathcal{L}_{\ell+1}$. On any level $\mathcal{L}_{\ell+1}$ that is not the finest level, the number of timesteps used for for synchronization with the coarser level $\mathcal{L}_\ell$ must be an integer multiple of the regrid interval. This is required so that regridding can be applied recursively.

8.3 Outline of the Algorithm

The adaptive mesh refinement process can be represented by the following "pseudo C^{++}" algorithm:

```
void Level::advance(double dt˙max) {
  bool time_to_sync_with_coarser_level=
    (coarserLevelExists() ? coarserLevel()->timeToRegrid() : false);
  double dt_sum=0.;
  int step=0;

  dt=findStableStepSize(dt_max,dt_sum);
  while (dt_sum<dt_max) {
    patch_arr->advance(dt);
    step++;
    dt_sum+=dt;
    if (finerLevelExists()) finerLevel()->advance(dt);
    if (canBeRefined() && step%regrid_interval==0) {
      if (dt_sum<dt_max || !time_to_sync_with_coarser_level) {
        regridFinerLevels();
      }
    }
```

```
      dt=findStableStepSize(dt_max,dt_sum);
    }
    if (coarserLevelExists()) {
      patch_arr->coarsenFluxSums(coarserLevel()->patch_arr);
      coarserLevel()->patch_arr->repeatConservativeDifference();
      patch_arr->coarsenConservedQuantities();
    }
  }
```

We will describe these parts of the timestepping algorithm in the sections below.

8.3.1 Timestep Selection

The first task in integrating the data on the patches belonging to an arbitrary level $\mathcal{L}_\ell$ is to select the current timestep $\triangle t_{c,\ell}$. If a coarser level exists, then the cell widths on the coarser level are a factor of the refinement ratio r times the cell widths on this level. Thus, if a coarser level exists, the maximum number of steps on this level should be roughly r, so we should have that $\triangle t_{c,\ell} \approx \triangle t_{c,\ell-1}/r$. Because of the extra resolution of the fine grid, calculations of the stable timestep may suggest that we take somewhat more than r timesteps. We assume that we can use appropriate stability conditions to determine the largest stable timestep $\triangle t_{s,\ell}$ over all the cells in all the patches on a given level $\mathcal{L}_\ell$.

It is generally a good idea to avoid rapid increases in the size of the timestep. Rapid decreases in $\triangle t$ may be required for stability, but rapid increases allow for the algorithm to jump too far past the time when discontinuities develop. This suggests that we require that the new timestep satisfy

$$\triangle t \leq \min\{\triangle t_{c,\ell}, \triangle t_{s,\ell}\} f$$

where f is some predetermined growth factor. Typically, $f = 1.1$ is a good choice.

If a coarser level does not exist, we have no synchronization to perform. Assuming that this coarsest level $\mathcal{L}_0$ can be refined, the number of timesteps we take to reach the desired time must be an integer multiple of the regrid interval. If necessary, we reduce $\triangle t_{c,0}$ to reach the desired time at a regridding interval.

If a coarser level exists and we have already advanced a total time increment of $\triangle t_{t,\ell}$ on this level, then the time remaining until synchronization is $\triangle t_{c,\ell-1} - \triangle t_{t,\ell}$. Thus the number of timesteps remaining until synchronization with the coarser level is at least $(\triangle t_{c,\ell-1} - \triangle t_{t,\ell})/\triangle t$. The maximum number of steps on this level is at least the number of steps already taken plus this number of steps remaining. If the current level can be refined, then the maximum number of steps must be an integer multiple of the regridding interval. After this adjustment, the updated value of the current timestep is the time remaining until synchronization divided by the number of steps remaining.

8.3.2 Advancing the Patches

The algorithm to advance the data on the patches takes the form of the following "pseudo C^{++}" code:

```
double PatchArr::advance(double dt) {
  for (int i=0;i<getNumber();i++) {
    Patch *p=(*this)[i];
    p->makeSpaceForData(time+dt);
    level->fillBoundaryData(p,time);
    p->advance(dt);
  }
  return getStableDt();
}
```

First, we make available sufficient computer memory for the computational results at the new time. This need not involve actual memory allocation at each timestep; rather existing memory associated with the patch is cleared and marked with the new time. The final step is to compute a timestep that is stable for the new data on all the cells of all the patches. This depends on the integration scheme, and is typically based on the CFL condition. The other intermediate tasks require more elaboration.

8.3.2.1 Boundary Data

Our adaptive mesh refinement algorithm is designed so that the data on an individual grid patch can be advanced in time without the user worrying about the current arrangement of the grid hierarchy. This allows the integration scheme to perform a very regular and efficient algorithm. In order to achieve this goal, we provide **ghost cells** for the data on each grid patch, and fill these extra cells with the best available information before advancing the data in time.

The number of ghost cells is determined by the stencil of the integration scheme. A simple scheme such as Rusanov's method would require a single ghost cell. A complicated scheme such as fourth-order ENO would require 16–20 ghost cells in each direction, since the flux stencil for each of the four Runge–Kutta steps requires 4–5 cells to either side of the intended location of the flux. Schemes with large stencils are not typically used with adaptive mesh refinement: the large number of ghost cells requires a large amount of communication between patches, relative to the cost of advancing the data on a patch. Note that discontinuous Galerkin methods do not use any ghost cells, because higher-order accuracy is achieved by carrying information about derivatives of the solution within each grid cell.

The synchronization assumption requires that coarse patches are integrated before fine patches, and the proper nesting assumption requires that the union of fine patches is contained in the interior of the union of the coarse patches. Thus fine patches cannot provide boundary data for coarser patches. Instead, boundary data are sought from the following sources in the following order of priority:

(i) physical boundary conditions at the outer sides of the computational domain;
(ii) data in cells on other patches in the same level of refinement;
(iii) space and time interpolation from coarser patches.

Note that the union of patches and ghost cells on level $\mathcal{L}_{\ell+1}$ may extend beyond the union of patches on level $\mathcal{L}_\ell$. Thus, it may be necessary to use an algorithm that recurses over coarser levels of refinement to find all the needed boundary data. The alternative is to expand coarse grid patches enough to contain the fine grid patches and their ghost cells.

Figure 8.1 illustrates the determination of boundary data for a patch. The patch of interest sits between the boundary of the physical domain (the heavy vertical line to the left) and another patch on the same level of refinement. Boundary data must be found for the ghost cells (in this case four) around the outside of the patch; the boundary of the ghost cells is drawn as a thin solid line around the patch. Part of these boundary data are determined by the boundary conditions, and part are provided by the data on the other patch. The remainder of the boundary data must be obtained by refining data from the overlying coarse patches, which are illustrated by dashed lines where no fine patches occur. Although Figure 8.1 indicates that the remaining boundary data can be provided by the coarser level of patches, in general it is possible that some of these refined boundary data could come from even coarser levels.

Note that when data are provided from coarser levels, then those data must be refined appropriately. For conserved quantities, this means that the space and time interpolation should preserve the overall accuracy of the integration scheme.

The strategies used to fill boundary data and to regrid are interrelated. If we required that fine patches and their ghost cells must be contained the union of the patches on the next coarser level, then we could avoid recursion in filling the ghost cells. However, we would require more cells on the coarser level, which would in turn require even coarser levels to be larger.

The safe strategy is to program the adaptive mesh refinement algorithm so that recursive filling of ghost cells is possible. By selection of a sufficiently large proper nesting buffer, it is then possible to prevent the recursion from occurring.

8.3.2.2 Flux Computation

Once the ghost cell data are available, the fluxes can be computed with the same subroutine used in a non-adaptive code. The choice of scheme used to integrate the conservation law dictates the form of this subroutine.

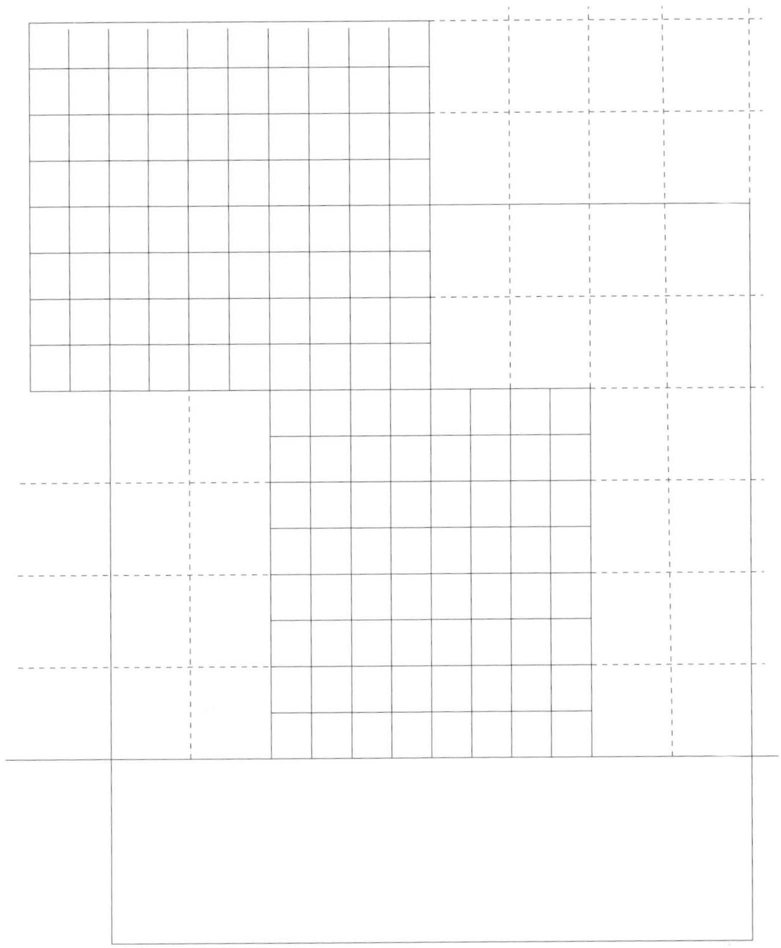

Fig. 8.1 Sources of boundary data during adaptive mesh refinement. The physical boundary is the very thick vertical line at bottom; the fine patch currently seeking boundary data is next to the physical boundary and has a boundary represented by lines of intermediate thickness. The fine patch requires data in "ghost cells" inside a larger rectangle (two coarse cells wider in both directions) surrounding the patch, represented by solid lines both inside and outside the physical domain.

We have a choice of providing each patch with sufficient storage for its own cells and the ghost cells, or of copying the data from the patch to some work space and copying the new results back from the work space. If we choose to avoid the copies to the work space, then the extra ghost cells on each patch represent a redundant use of computer memory, wherever ghost cells from one patch overlap cells on another patch in the same level. Another consequence of not having a work space is that all patches would have to store temporary variables (such as the sound speed in gas dynamics), leading to even greater memory requirements. An alternative is to do all work in the flux computation and time integration within the work space, at the extra cost of copying the necessary input and output. An intermediate strategy in which temporary variables are stored in a work space while others are stored on the patch with ghost cells is also possible. We have chosen not to store any ghost cells with the patches.

8.3.2.3 Time Integration

Essentially all schemes for hyperbolic conservation laws involve a conservative difference for time integration. In two dimensions, this takes the form

$$\begin{aligned}\mathbf{u}_{ij}^{n+1} = \mathbf{u}_{ij}^{n} - \frac{1}{(\triangle x_1)_i(\triangle x_2)_j} &\left\{\left[\mathbf{f}_{i+1/2,j}^{n+1/2} - \mathbf{f}_{i-1/2,j}^{n+1/2}\right](\triangle x_2)_j \triangle t^{n+1/2}\right.\\ &\left. - \left[\mathbf{f}_{i,j+1/2}^{n+1/2} - \mathbf{f}_{i,j-1/2}^{n+1/2}\right](\triangle x_2)_j \triangle t^{n+1/2}\right\}. \end{aligned} \tag{8.1}$$

This is a simple calculation, and should use the same subroutine that would be used in a non-adaptive algorithm.

However, if there are even finer patches, then the fine results should be preferable to these coarse results. There are two ways in which the fine results replace the coarse results. One is discussed in Section 8.3.5, in which the numerical solution in a coarse cell is replaced by an appropriate upscaling of the fine results. The other issue is discussed in Section 8.3.4, in which the coarse fluxes are replaced by an appropriate upscaling of fine fluxes, and the coarse conservative difference is repeated with the upscaled fluxes.

Suppose that on the next coarser level we are advancing by $\triangle T$ in time. In order to synchronize the current level with the coarser, we will take some sequence of timesteps $\triangle t$. Symbolically, we will write $\triangle t \in \triangle T$ to denote that the fine timesteps are contained within the coarse timestep. Similarly, suppose that some coarse cell side $I + 1/2, J$ coincides with the boundary of a fine patch. We will denote that a fine cell side $i + 1/2, j$ is contained within this coarse cell side by writing $i + 1/2, j \in I + 1/2, J$. Similar notation could be used in other coordinate directions, and in one or three dimensions.

Note that the fluxes $\mathbf{f}_{i+1/2,j}^{n+1/2}(\triangle x_2)_j \triangle t^{n+1/2}$ are numerical approximations to the integrals $\int_t^{t+\triangle t^{n+1/2}} \int_{s_{i+1/2,j}} \mathbf{fn}\, ds\, dt$ for a side $s_{i+1/2,j}$ of patch with logical indices ij. Then the total flux on the coarse cell side $S_{I+1/2,J}$ should be replaced with the sum of fine flux integrals $\sum_{\triangle t^{n+1/2} \in \triangle T} \sum_{i+1/2,j \in I+1/2,J} \mathbf{f}_{i+1/2,j}^{n+1/2}(\triangle x_2)_j \triangle t^{n+1/2}$.

As a result, we compute time integrals of the flux at outer sides of each patch. When we reflux, we will integrate these time integrals in space over all fine cell sides contained in a coarse cell side. The result will replace the original coarse flux integral in the refluxing step below.

8.3.3 Regridding

Because we are interested in solving time-dependent problems, we allow the mesh refinement to move in time. There are several design principles in this process:

Necessity: We want to locate the new fine mesh only where it is needed.
Proper Nesting: We want the union of patches on each level of refinement to be contained in the interior of the union of patches on the next coarser level.
Infrequency: we move the mesh after a fixed number of coarser timesteps, rather than after every timestep.

With respect to the Necessity principle, we shall use an error estimation procedure (described below) to determine where the unacceptably large errors occur on this level. By using an error estimator, rather than a gradient detector, we are able to place mesh refinement where discontinuities in the motion variables are about to form, or where the algorithm is not able to produce second-order accuracy for some other reason, such as a lack of smoothness in the equation of state.

With respect to the Infrequency principle, we obviously do not want to move the mesh every timestep on each level. Instead we shall move the mesh after a fixed number of timesteps, call the **regrid interval**, in order to keep the cost of error estimation within acceptable limits.

A "pseudo C^{++}" algorithm describing the regridding process consists of the following two procedures:

```
//called from Level::advance
  void Level::regridFinerLevels(double efficiency_tolerance) {
    findProperNestingList();
    if (finerLevelExists()) {
      finerLevel()->findProperNestingListFromCoarserLevel();
    }

    TagBoxArr dummy;
    updateFinePatchArr(lplot,efficiency_tolerance,dummy);
  }
```

In the remainder of this subsection, we will discuss these ideas in greater detail.

8.3.3.1 Proper Nesting

We want the union of the fine patches to be contained in the interior of the union of the coarser patches. This implies that the boundary of the union of the fine patches nowhere coincides with the boundary of the union of the coarser patches; adjacent cells can differ by at most one refinement level. As a result, all interfaces between coarse and fine grids are related by a fixed ratio. Furthermore, the flux integrals computed on the outer sides of the fine patches are used on the next coarser level only.

We require that the boundary of the union of the fine patches be a fixed number of coarse cells from the boundary of the union of the coarser patches. This number of cells is called the **proper nesting buffer**. If chosen properly, it can be reasonably small and still guarantee that the interpolation needed to provide initial data on new fine cells does not need to recurse over coarser levels.

Two "pseudo C^{++}" procedures to compute the proper nesting lists are as follows:

```
  void Level::findProperNestingList() {
    proper_nesting_list->clear();
    complement_list->clear();
    patch_arr->complement(physical_box,*complement_list);
    complement_list->
      bufferAndReplace(physical_box,proper_nesting_buffer);
    complement_list->complement(physical_box,*proper_nesting_list);
  }

//recursive:
  void Level::findProperNestingListFromCoarserLevel() {
    if (!canBeRefined()) return;
    proper_nesting_list->clear();
    complement_list->clear();
    const BoxList &coarse_complement_list=
      *(coarserLevel()->complement_list);
    int proper_nesting_buffer=getProperNestingBuffer();
    complement_list->bufferAndAppendFromCoarser(coarse_complement_list,
      physical_box,proper_nesting_buffer);
    complement_list->complement(physical_box,*proper_nesting_list);
    if (finerLevelExists()) {
      finerLevel()->findProperNestingListFromCoarserLevel();
    }
  }
```

In these procedures, think of the proper nesting list and the complement list as a list of boxes (*i.e.*, patches without data). The complement list is initialized by finding a list of boxes that exactly cover the complement of the union of the patches with respect to the physical domain. Each of the boxes in this list is buffered to make them larger by a fixed number of cells, namely the proper nesting buffer. The proper nesting list is the list of boxes that forms the complement of the union of theses buffered boxes.

Since we are moving the mesh after a fixed number of coarser timesteps, called the **regrid interval**, it is necessary to provide a buffer region around the cells currently requiring refinement. The purpose of the buffering is to prevent the waves of interest from moving off the refined mesh before the next regridding step. Here, we take advantage of the CFL condition: the explicit integration method we are using is designed so that a wave can travel across at most one cell in a timestep.

Another purpose of the proper nesting list is to guide the selection of the patches on finer levels. The patches on the current level have been selected so that the

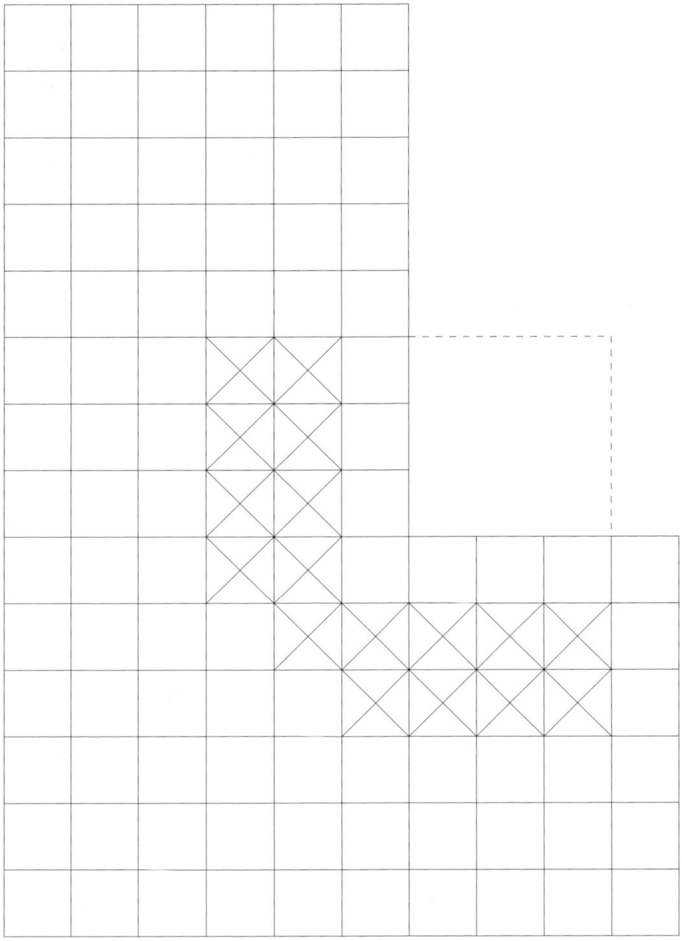

Fig. 8.2 The role of the proper nesting list during regridding: boxing of cells tagged for refinement must lie inside the union of current patches

interesting waves will lie inside the union of the patches until the patches have been advanced in time to synchronization with the next coarser level. We want to construct a list of patches on each of the finer levels, so that each fine patch lies inside the refinement of the proper nesting list belonging to its coarser level. When the union of patches on the current level is not convex, the proper nesting list prevents patches on the next finer level that straddle an interior corner of the union. As an example, Figure 8.2 shows a collection of tagged cells (which are each marked with an X) on two patches. The smallest box containing these tagged cells (which is illustrated with a heavy dotted line) lies partly outside the union of the patches. If this containing box were used to generate a refined patch, then some of the fine cells would not be properly nested.

8.3.3.2 Tagging Cells for Refinement

The next regridding step is to update the list of patches on the finer level. A "pseudo C^{++}" procedure to implement the ideas is the following:

```
//recursive:
  void Level::updateFinePatchArr(double efficiency_tolerance,
  TagBoxArr &coarse_tag_box_arr) {
    patch_arr->tagAllPoints(FALSE);
    if (finerLevelExists()) {
      finerLevel()->patch_arr->tagCoarserCells(*patch_arr,TRUE);
    }
    findErrorCells();
    TagBoxArr tag_box_arr;
    patch_arr->makeBufferedTagBoxArr(error_buffer,tag_box_arr);

    if (finerLevelExists()) {
      if (finerLevel()->canBeRefined()) {
        finerLevel()->
          updateFinePatchArr(efficiency_tolerance,tag_box_arr);
      }
    }
    if (tag_box_arr.getNumber()>0) {
      tag_box_arr.bufferTags(error_buffer);
      tag_box_arr.makeTagsUnique();
      BoxList box_list;
      tag_box_arr.findBoxesContainingTags(getShortestSide(),
        max_interior_cells,refinement_ratio,max_ghost_cells,
        efficiency_tolerance,box_list);
      makeIntegrable(tag_box_arr,box_list);
      if (finerLevelExists()) {
        PatchArr *new_patch_arr=newPatchArr(finerLevel());
        delete (finerLevel()->patch_arr);
        finerLevel()->patch_arr=new_patch_arr;
      } else {
        finer_level=newLevel(this);
      }
      finerLevel()->patch_arr->coarsenAndTagAll(tag_box_arr);
      if (getEosModel().usesRichardsonExtrapolation()) {
        tag_box_arr.bufferTags(1);
      }
      tag_box_arr.coarsenAndCopyTagsTo(coarse_tag_box_arr);
    } else if (finerLevelExists()) {
      delete finer_level; finer_level=0;
    }
```

First, cells in the patches belonging to the current level are tagged if their global integration error is too large. This procedure uses both Richardson extrapolation to estimate the local truncation error in the integration, and a simple device to estimate the number of timesteps to be performed on this level of refinement. This error estimation procedure is a standard procedure in the numerical integration of ordinary differential equations [42]. Suppose that at each timestep we commit an error of magnitude ϵ (principally the local truncation error); further, suppose that the computation permits a bound M on the growth of these errors. Note that the conservative difference (8.1) shows that the computational solution essentially amounts to applying a perturbation of the identity operator to the previous solution; it is therefore reasonable to expect M to be close to 1 for smooth flow and sufficiently fine mesh. Then the error e_n at step n satisfies

$$e_1 \leq \epsilon, \quad e_n \leq \epsilon + M e_{n-1} \quad \text{for } n > 1.$$

Then an argument by induction shows that

$$e_n \leq \epsilon \sum_{j=0}^{n-1} M^j = \epsilon \frac{M^n - 1}{M - 1}.$$

If M is close to one, then for small n the error bound will be approximately $n\epsilon$. Suppose that the local truncation error satisfies

$$\epsilon = C \Delta t^{k+1},$$

where k is the expected global order of the scheme. (We are assuming that spatial and temporal error orders are equal.) Then the error in taking one coarse step of size $r\Delta t$ is

$$e_n^c \approx C r^{k+1} \Delta t^{k+1}.$$

On the other hand, if we take r fine timesteps of size Δt, the error is

$$e_n^f \approx C r \Delta t^{k+1}.$$

This allows us to estimate the local error of a fine timestep by

$$\epsilon \approx \frac{e_n^c - e_n^f}{r^{k+1} - r} = \frac{w_n^c - w_n^f}{r^{k+1} - r},$$

where w is the quantity being monitored for errors. This gives us a computable estimate for the local truncation error. The global error can be estimated by multiplying the local error by the anticipated number of timesteps N. Here,

$$N \approx \frac{L}{s \Delta t},$$

where L is some length scale associated with the problem, s is some important wave speed, and Δt is the current timestep. Thus, cells are tagged if the relative error

$$\frac{|w_n^c - w_n^f|}{\max |w_n^f|} \frac{L}{(r^{k+1} - r) s \Delta t} > \text{tolerance}.$$

If the regrid interval has the value k, the steps in the error estimation procedure are the following:

(i) At the current level of refinement,
  (a) advance the data for one timestep, Δt, and
  (b) coarsen the results by a factor of k.
(ii) On a patch coarsened by a factor of k,
  (a) coarsen the data from the patch at the current time *minus* $(k - 1)$ timesteps, and
  (b) advance the data for one timestep, $k\Delta t$.
(iii) Compare the results of the two time integrations.

Note that the error estimation is performed on pseudo-patches that potentially lie in index spaces between the current level of refinement and the next coarser level (since the pseudo-patches are coarsened by a factor of the regrid interval, and mesh refinement uses an integer multiple of the regrid interval). This reduces the work in comparing the errors; in particular, we would not want to increase the work by integrating the fine patches and comparing to the values on a coarser level, since the results on the fine patches would have to be discarded when the patches are moved. The ordering of the coarsening and comparison operations also makes the algorithm simpler; if we were to compare errors on the current level of refinement by coarsening, integrating one coarse step and then refining, we would have to construct a high-order conservative interpolation. In particular, this interpolation would have to be of a higher order than that used to construct predictor values for the numerical flux computation.

It is interesting to consider the implementation of this error estimation strategy on a recursively refined mesh. Note that errors on coarse and fine meshes are estimated at different times. At first glance, this would appear to be undesirable. However, the alternative of comparing errors on all levels at the same time actually leads to much wasted work, and larger refined regions. This is because the error estimation on the coarse mesh places the refined cells where the disturbance will be moving, plus or minus buffer cells. Thus it is only necessary to buffer by regrid interval minus 1 cells, since that is the number of timesteps that will be taken between the times when the errors are estimated, and when the mesh will next be moved. If the errors had been computed at the same times, then it would be necessary to buffer by an additional cell on each level. Furthermore, the error estimation would have to proceed through more than one timestep, with recursive calls to integration on finer levels in order to provide data for finer grids. Since the mesh is going to be moved, this is extra work being performed for data that are mostly going to be discarded.

We want to prevent cells from being alternately coarsened and refined. If a cell is not currently tagged, it is tagged if its global error estimate exceeds some specified tolerance. On the other hand, if a cell is currently tagged, then it remains tagged unless its global error estimate falls below the specified tolerance divided by the regrid interval. This serves to prevent much of the chatter that occurs at the edges of refinements, where the error estimates hover around the error tolerance.

A cell is also tagged if it has an underlying fine cell. In order to determine these cells, we recursively update the patch lists on finer levels, and tag cells on the current level overlaying the tagged cells on the finer level.

8.3.3.3 Tag Buffering

Next, all cells sufficiently near the tagged cells are also tagged. The width of this buffer is called the "error buffer," and has been chosen to be equal to the regrid interval minus 1. This is the number of timesteps between the point where the errors were estimated and where the level will next be regridded.

8.3.3.4 Logically Rectangular Organization

The next step is to determine a list of boxes that contain the tagged cells. First, a large box containing all of the tags is found; then "cut points" are sought in order to split this big box into smaller boxes that cover the tagged cells more efficiently. A histogram of the cell tags in each coordinate direction helps us to determine how to cut the boxes. Cut points are selected according to goodness: a zero histogram entry near the center is best; an inflection point in the histogram near the center is next best; the mid-point is the choice of last resort. The best of these over all coordinate directions is chosen. This procedure is due to Berger, and differs from that in [18].

After this initial list of boxes is determined, it is further massaged. First, each box is further subdivided, if necessary, so that it lies inside the proper nesting list. Next, if the edge of a box falls too near a physical boundary, it is extended all the way to the boundary. Afterward, the list of boxes is searched to see if any two share a common side, so that they can be coalesced to make a larger box. Boxes that are too large (*i.e.*, they require more temporary space than we are willing to provide) are subdivided. Once these boxes are found, they are shrunk to the smallest size needed to contain the tagged cells, then expanded within their former boundaries to avoid physical boundaries among the ghost cells.

8.3.3.5 Initializing Data after Regridding

The final step is to make the new patches. If there are no tagged cells, then the current finer level is destroyed if it exists. If there are tagged cells and no current finer level, then a new finer level is created; the data on these new fine patches are obtained by refining the data on the overlying patches in the current level of refinement. If there are tagged cells and a finer level already exists, then the data on the new fine patches are determined by copying the data from the old fine patches where they are available, and refining them from the current level otherwise. In this case, a temporary memory bulge can occur while the finest level data are copied from old patches to new.

8.3.4 Refluxing

Numerical integration on an adaptively refined mesh involves communication between coarse and fine mesh patches. We have already seen that boundary data for fine patches may be obtained from refinement of data on coarser patches. Conversely, fine patches produce more accurate results that can be used to correct the data on coarser patches.

This coarsening of data takes two forms. First of all, when a coarse cell overlies finer cells, the data on the fine cells are coarsened and replace the coarse data. The computational form of this coarsening procedure varies from one flow variable to another, and will be discussed in section 8.3.5. However, we remark that the coarsening of conserved quantities needs to be conservative; that is, the volume-weighted average of the conserved variables on the fine grid replaces the conserved variables on the overlying coarse cells. This coarsening process, by itself, could not preserve conservation; it is also necessary to replace the coarse fluxes around the boundary of the fine patches with the boundary and time integral of the fluxes determined on the fine patches. In this way, the change in the values of conserved quantities in cells overlying finer patches is compensated by changes in conserved quantities on cells neighboring the refinement.

Because of the peculiarities of the hypoelastic equations of state for solids, we perform this refluxing process in a form different from that in [18]. We compute the time integrals of the fine fluxes, then coarsen them in space to provide improved values of the coarse fluxes; afterward, we repeat the conservative difference (8.1). This amounts to more work than the equivalent process in [18]. However, other solid mechanics variables (such as stress) may have a *nonlinear* dependence on the fluxes and need re-computation. This is discussed in more detail in section 8.3.5.

8.3.5 Upscaling

Suppose that we have a coarse cell $(A_J, A_J + \triangle x_J)$ on level $\mathcal{L}_k$, and we compute a numerical result

$$\mathbf{u}_J \approx \frac{1}{\triangle x_J} \int_{A_J}^{A_J+\triangle x_J} \mathbf{u}(x, t + \triangle t_K)\, dx.$$

If this cell is refined, then after R timesteps on the fine grid, we also have fine grid results

$$\mathbf{u}_j \approx \frac{1}{\triangle x_{k+1}} \int_{a_j}^{a_j+\triangle x_j} \mathbf{u}(x, t + \triangle t_K)\, dx, \quad (a_j, a_j + \triangle x_j)\, C\, (A_J, A_J + \triangle x_J)$$

on the fine cells contained in the coarse cell $(A_J, A_J + \triangle x_k)$. We assume that the fine grid results are more accurate than the coarse grid results, so we replace $\mathbf{u}_J$ by

$$\mathbf{u}_J \leftarrow \frac{1}{\triangle x_J} \sum_{(a_j, a_j+\triangle x_j)\, C\, (A_J, A_J+\triangle x_J)} [\mathbf{u}_j \triangle x_J].$$

8.3.6 Initialization

The beginning of the computation is somewhat different from the process presented above. On the coarsest level, the patches are determined by subdividing the physical domain into pieces that are not too big for the integrator; on finer levels, the patches are assumed to be determined by the initialization process on the next coarser level. Then, a user-supplied procedure is called to place the initial data on the patches; this procedure also determines a stable timestep.

If the current level can be refined, then a special error estimation process is conducted. Here, the initial data are advanced regrid interval timesteps and coarsened; they are also coarsened and advanced one timestep. The two results are compared in order to estimate the global truncation errors, and cells with unacceptably large errors are tagged. These cells are buffered by regrid interval cells in each coordinate direction, since that is the number of timesteps that will be taken before the first regridding is performed.

The tagged cells are organized into patches on a new fine level just as described above. Then the user-supplied procedure is called to determine the initial data on the new fine patches. If the new fine level can be refined, we advance it forward one timestep in order to provide boundary data for the recursive initialization process on finer levels. After the finer levels have been initialized, we return to the initial data, forgetting the results of the integrations on all of the levels, since the size of the first timestep on each level may be affected by the work on coarser levels during the normal integration process.

8.4 Object Oriented Programming

Adaptive mesh refinement involves much more complicated programming than straightforward integration on regular meshes. This program complexity is reflected not only in the data structures used to represent the patches on various levels of refinement, but also in the communication between the patches.

Other aspects of material models also add to the program complexity. Oftentimes the computations in the equations of state involve a large number of temporary variables, which could lead to large memory requirements if not handled carefully. Users also need to perform computations in one, two or three dimensions.

It is desirable to implement adaptive mesh refinement so that the mesh hierarchy and patch communication are independent of the equation of state and the number of physical dimensions. In this way, adaptive mesh refinement code could be debugged in one dimension and used in any number of dimensions.

It is also desirable that the treatment of various kinds of flow variables be driven by requests from the equation of state. The equation of state should decide how the variable is represented on the grid; for example, pressure in gas dynamics might be associated with cell centers, while momentum fluxes might be associated with cell sides. The equation of state should also decide how the variable should be coarsened and refined. If these goals are achieved, then no modification of the adaptive mesh refinement program is required to make changes in the equation of state.

8.4.1 Programming Languages

Languages such as Fortran 77, which are highly efficient for computations on rectangular arrays of data, do not provide the variety of data or programming structures that would make the implementation of adaptive mesh refinement easy, either for implementation or to maintenance. Fortran 90 still does not offer all the features we need, for reasons we will describe later. As a result, we decided to implement the adaptive mesh refinement program structure in C^{++}, with the numerically intensive routines written in Fortran 77.

Fortran 90 is attractive because it allows for dynamic memory allocation, and for the development of more complex data structures (called modules). It is possible to bundle array dimensions together with the array itself, thereby removing potential programming errors in passing arrays to subroutines.

However, Fortran 90 modules are fundamentally different from C^{++} classes. Fortran 90 does not associate subroutines with modules, in the way that C^{++} classes have member functions. Fortran 90 does not have access permissions for module data members. Further, although Fortran 90 does permit modules to be "inherited" from other modules, Fortran 90 does not allow for run-time binding of virtual functions. This point will be especially important in our development of the `EosModel` class below.

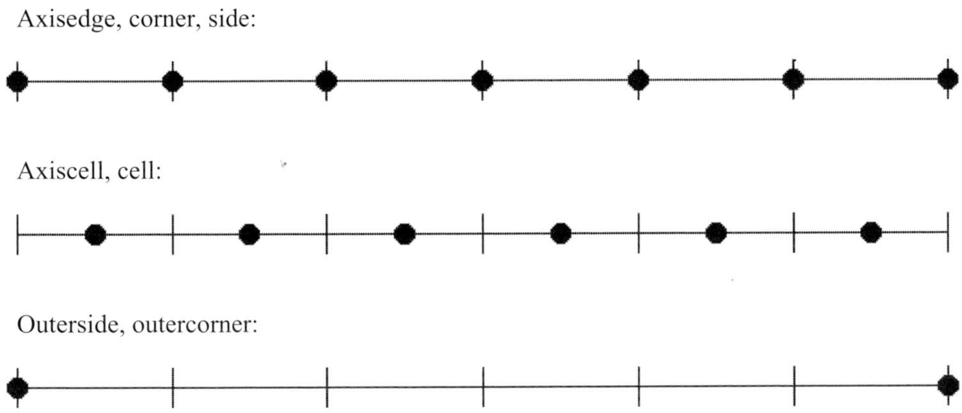

Fig. 8.3 One-dimensional geometries for variables

8.4.2 AMR Classes

There is no commonly accepted set of C^{++} classes for adaptive mesh refinement. Scott Baden [7] adopted notions of `floorplans` for his grid classes. Some of his notions and the ideas in this book were merged into the SAMRAI code written by Richard Hornung (PhD student of John Trangenstein) and Scott Kohn (PhD student of Scott Baden) at Lawrence Livermore National Laboratory. The SAMRAI code now exceeds by far the scope of the code we will describe below.

There are several basic classes we will use in our C^{++} implementation of adaptive mesh refinement. More complicated classes will build on these classes through class inheritance.

8.4.2.1 Geometric Indices

Variables used in the numerical solution of partial differential equations on logically rectangular grids can be associated with various spatial locations. For example, the conserved quantities in the conservation laws are commonly associated with the cell centers, since their numerical values are approximations to the average value of the exact conserved quantity in the cell. On the other hand, fluxes for the conservation laws are generally associated with the cell sides, since these are numerical approximations to the time integrals of the normal flux integrated over the cell sides.

Figure 8.3 shows the geometries we use in one-dimensional calculations. On a grid with cells indexed from `first(0)` to `last(0)`, cell-centered variables would be dimensioned

```
cell(first(0):last(0),nvar)
```

while corner-centered variables would be dimensioned

```
corner(first(0):last(0)+1,nvar)
```

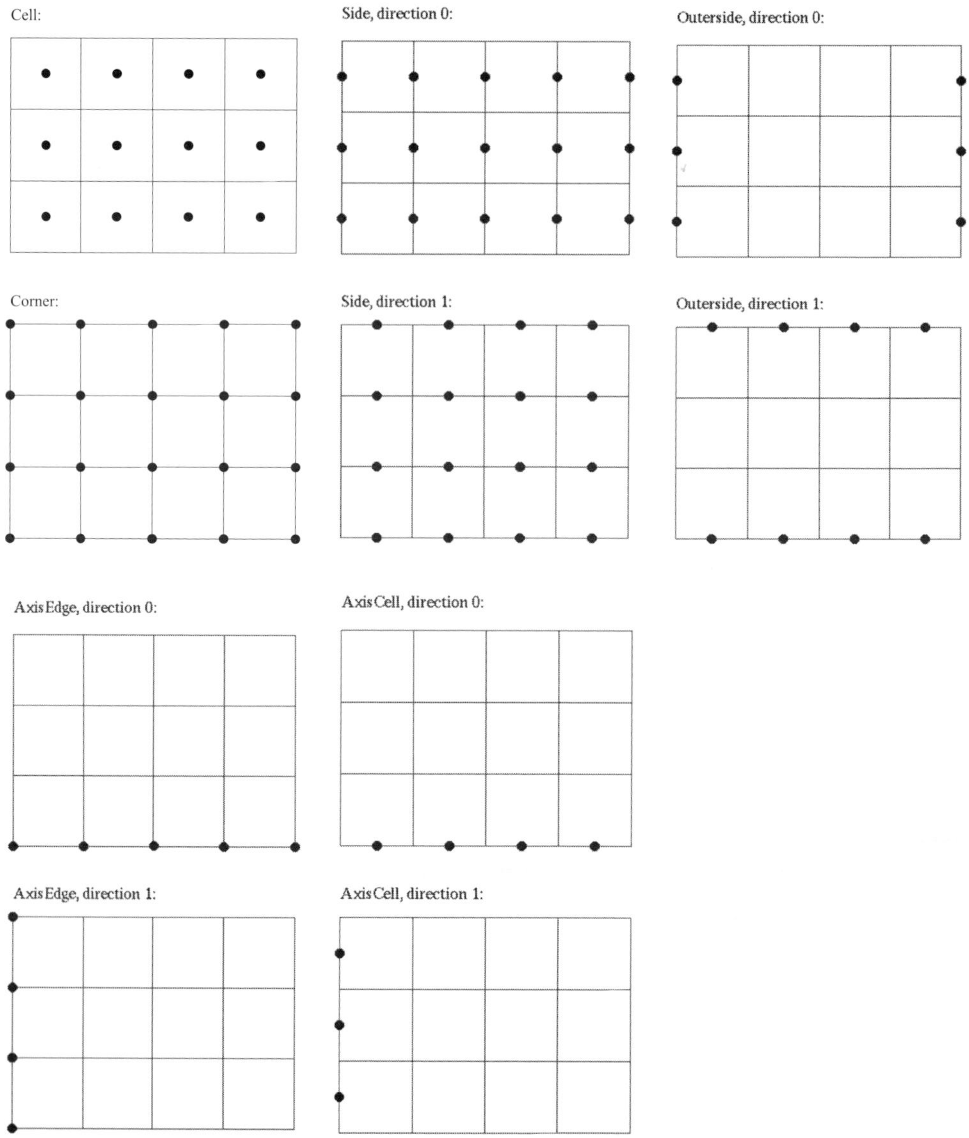

Fig. 8.4 Two-dimensional geometries

Variables with outerside geometry would be dimensioned

```
outerside(0:1,nvar)
```

Here, a first subscript of "0" refers to the left side and "1" refers to the right side.

The two-dimensional situation is more interesting. Cell-centered and corner-centered variables are easy to understand from Figure 8.4. Cell-centered variables (such as conserved quantities) in two dimensions would be dimensioned

```
cell(first(0):last(0),first(1):last(1),nvar)
```

while corner-centered variables (such as curvilinear grid coordinates) would be dimensioned

```
corner(first(0):last(0)+1,first(1):last(1)+1,nvar)
```

Side-centered variables (such as fluxes) are more complicated. Variables associated with sides in the first coordinate direction would be dimensioned

```
side0(first(0):last(0)+1,first(1):last(1),nvar)
```

while variables associated with sides in the second coordinate direction would be dimensioned

```
side1(first(1):last(1)+1,first(0):last(0),nvar)
```

In the latter case, we reverse the order of the first two subscripts in order to perform approximate Riemann problem solutions with unit stride. Outerside-centered variables (such as upscaled fluxes) are dimensioned

```
outerside0(first(1):last(1),0:1,nvar)
```

and

```
outerside1(first(0):last(0),0:1,nvar)
```

respectively. The axisedge geometry (useful for rectangular mesh data) is dimensioned

```
axisedge0(first(0):last(0)+1,nvar)
```

and

```
axisedge1(first(1):last(1)+1,nvar)
```

We can also identify three-dimensional geometries, illustrated in Figure 8.5. The cell-centered and corner-centered arrays are the easiest to understand. Cell-centered variables in three dimensions would be dimensioned

```
cell(first(0):last(0),first(1):last(1),first(2):last(2),nvar)
```

while corner-centered variables would be dimensioned

```
corner(first(0):last(0)+1,first(1):last(1)+1,first(2):last(2)+1,nvar)
```

Variables associated with sides in the first coordinate direction would be dimensioned

```
side0(first(0):last(0)+1,first(1):last(1),first(2):last(2),nvar)
```

variables associated with sides in the second coordinate direction would be dimensioned

```
side1(first(1):last(1)+1,first(2):last(2),first(0):last(0),nvar)
```

and variables associated with sides in the third coordinate direction would be dimensioned

```
side2(first(2):last(2)+1,first(0):last(0),first(1):last(1),nvar).
```

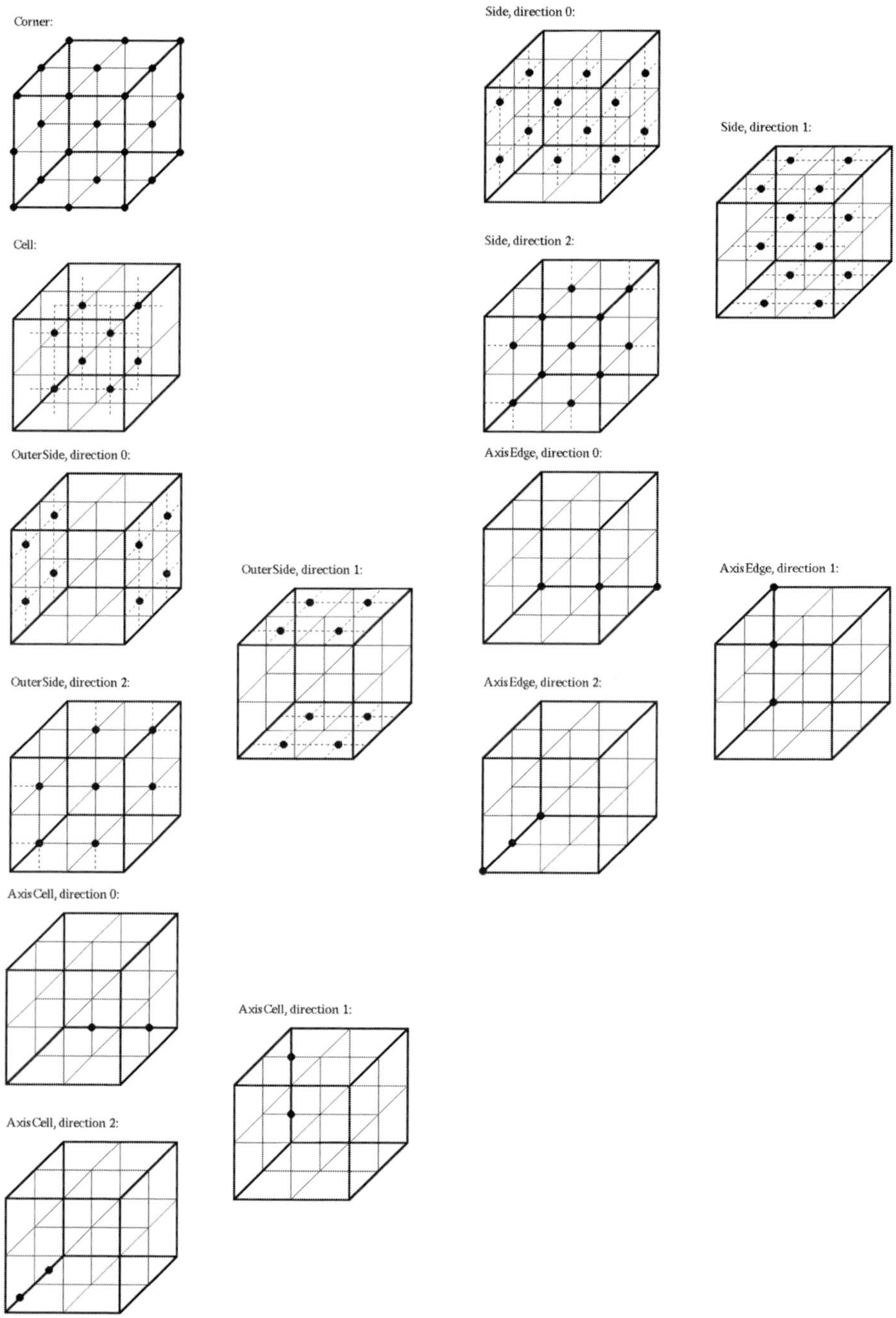

Fig. 8.5 Three-dimensional geometries

Outerside geometries are similar:

```
outerside0(first(1):last(1),first(2):last(2),0:1,nvar)
outerside1(first(2):last(2),first(0):last(0),0:1,nvar)
outerside2(first(0):last(0),first(1):last(1),0:1,nvar).
```

Axisedge geometries are also easy to understand:

```
axisedge0(first(0):last(0)+1,nvar)
axisedge1(first(1):last(1)+1,nvar)
axisedge2(first(2):last(2)+1,nvar).
```

There are several C^{++} classes designed to implement these geometries. In general, we want to avoid array addressing via C^{++}, because it is potentially much slower than Fortran array addressing. However, in some cases it is useful to form C^{++} arrays associated with an individual geometry, and access array entries via geometric indices. In this way, C^{++} can prevent the addressing of a cell-centered array at a corner index, for example.

In order to implement the various indices in C^{++}, we will use a generic dimensionally-dependent vector of integers in **Program 8.4-1: IntVect Class.** The data for this class consists of one, two or three integers, depending on the number of dimensions. Note that the implementation of `IntVect` is necessarily dimensionally-dependent.

The data for a `CellIndex`, `CornerIndex`, `SideIndex`, `OuterSideIndex` or `EdgeIndex` all involves an `IntVect`. For example, the definition of the `CornerIndex` class is described in **Program 8.4-120: CornerIndex Class** and the definition of the `CellIndex` class is described in **Program 8.4-121: CellIndex Class.** Although the data for both of these classes consist of an `IntVect`, we cannot perform arithmetic mixing these two classes because they are not derived from `IntVect`. However, nearly all of the dimensionally-dependent operations on these index classes are encapsulated in the `IntVect` class.

8.4.2.2 Boxes

A `Box` is a C^{++} class designed to represent a rectangular array of grid cells. As we can see from **Program 8.4-122: Box Class**, the data for a `Box` consists of two `CornerIndex` members, representing corners at the far ends of a diagonal of the `Box`. Because a `Box` uses the `CornerIndex` class to describe its data members, very few of the `Box` member functions need to be given dimensionally-dependent definitions. Boxes in various dimensions are shown in Figure 8.6.

A `LevelBox` is a `Box` that contains extra data to determine its index space; see **Program 8.4-123: LevelBox Class**. This extra information consists of two integers: a `level_number` and a `multilevel_number`. The `level_number` corresponds to the level of refinement in adaptive mesh refinement. The

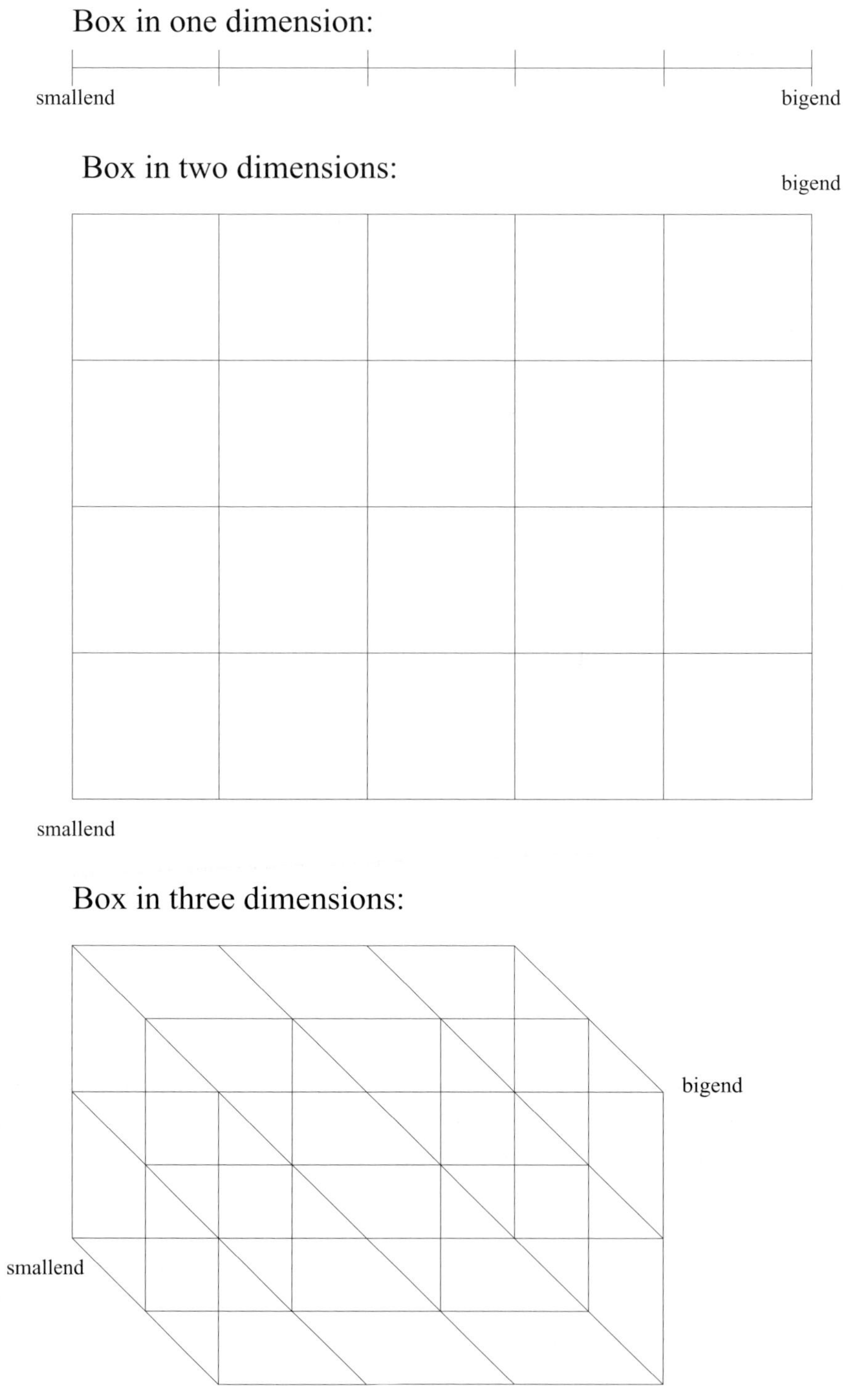
Box in one dimension:
smallend
bigend
Box in two dimensions:
bigend
smallend
Box in three dimensions:
bigend
smallend

Fig. 8.6 Boxes

`multilevel_number` corresponds to intermediate index spaces used, for example, for Richardson error estimation.

Essentially all arrays in the adaptive mesh refinement code are dimensioned by describing their geometry and a `Box`. The `Box` typically corresponds to the patch that is storing the data for the variable.

8.4.2.3 Data Pointers

Memory allocation for arrays in *C* or C^{++} returns a pointer, but leaves it to the programmer to remember the number of data entries involved in the memory allocation. In order to eliminate programming errors in passing these pointers around, we have developed a **Program 8.4-124: NumPtr**. class. This templated class consists of the data pointer plus an integer that contains the number of entries in the array.

If a `NumPtr` is a copy of an original memory allocation, it can be used just like a regular pointer, including operations such as `operator++`.

8.4.2.4 Lists

Lists can be **singly-linked** or **doubly-linked**, and **intrusive** or **non-intrusive**. We use combinations of these two choices for various purposes in the adaptive mesh refinement code. We do not use the standard template library lists in this code for two reasons. First, this code was developed before the standard template libraries were available. Second, our list functions operate faster, with controllable memory allocation.

A singly-linked list involves members that store pointers to the next item on the list. On the other hand, doubly-linked list members have pointers to both the previous and next items on the list. Singly-linked lists involve less storage, but they have difficulty with list traversal in reverse order. It is also difficult to remove an item from the middle of a singly-linked list, unless the previous item is known.

Intrusive lists assume that every item placed on the list already contains a data member to point to the next item on the list. Intrusive doubly-linked list members would have pointers to both the previous and the next items. On the other hand, non-intrusive lists form new list members by combining the data to be placed on the list with appropriate pointers to the next and previous list members. In other words, intrusive lists require that members already have pointers to next and previous, while non-intrusive lists add extra storage for this information.

In our adaptive mesh refinement library, all lists are templated. For the description of the intrusive singly-linked list, see **Program 8.4-125: Intrusive Singly-Linked List Template.** The intrusive doubly-linked list is implemented in **Program 8.4-126: Intrusive Doubly-Linked List Template.** The non-intrusive lists are implemented in **Program 8.4-127: Non-Intrusive Singly-Linked List Template** and **Program 8.4-128: Non-Intrusive Doubly-Linked List Template.**

The selection of intrusive or non-intrusive lists for specific purposes often depends on other issues. Classes derived from two base classes designed for intrusive lists can get confused about the pointers to next and previous members. On the other hand, in order to operate with items on a non-intrusive list, we need to get the item from the list member before operating on the item.

Currently, we use intrusive singly-linked lists for graphics (`GraphTool` lists and `PaletteList`), program timing (`TimedObjectList`), equation of states `EosModelList` and grids (`GridList`). We use non-intrusive singly-linked lists for graphics (`InputParameterList` and `XColormapList`). We use intrusive doubly-linked lists for `FlowVariableList` and keeping track of memory allocation in `MemoryDebugger`. We use non-intrusive doubly-linked lists for `BoxList`.

8.4.2.5 FlowVariables

Adaptive mesh refinement involves dynamically changing data structures. As the simulation time evolves, the computer memory requirements change, and the interaction among the data varies with the changes in the grid patches. As a result, it is useful for the user to describe in general ways how the data should be allocated and communicated, once an instance of the grid patches is determined.

A `FlowVariable` describes the information by adaptive mesh refinement in order to perform memory allocation and inter-patch communication. This information consists of the variable name, the number of variables (*e.g.*, one for pressure and the number of spatial dimensions for velocity), the geometry of the variable (*e.g.*, cell-centered or side-centered) the `IOSTATUS`, the refinement strategy, and the coarsening strategy. Here, **Program 8.4-129: IOSTATUS** is an enumeration. Some common (self-explanatory) values of `IOSTATUS` are `INPUT`, `INOUT`, `TEMP`, `FLUX`, `FLUXSUM`, `PLOT` and `MESH`. This basic functionality of a flow variable is implemented in the **Program 8.4-130: FlowVariableBase** class. Some of this information is further encapsulated in the **Program 8.4-131: FlowVariableDimensions** class.

A `FlowVariableBase` is derived from `IDLListNode` so that it can be placed on an intrusive doubly-linked list. Lists of `FlowVariableBases` can determine the total amount of data required to store everything with a specific `IOSTATUS` on a `Box`.

To associate a data type with a flow variable, we use a **Program 8.4-132: FlowVariable** template. This class is derived from `FlowVariableBase`, and parameterized by type. Data types of `double`, `int` or `bool` are common.

Recall that some geometries (such as `SIDE`), necessarily involve sub-arrays associated with the different coordinate directions in multiple dimensions. Thus the memory required to represent a `FlowVariable` on a `Box` involves an array of integers for the number of data values in each sub-array, and an array of data pointers for each of the sub-arrays. This extra information is contained in the **Program 8.4-133: FlowVariablePointer** class, derived from `FlowVariableBase`.

Occasionally, we would like to address the individual entries of the memory for a `FlowVariable` on a `Box`. For this purpose, we have constructed **Program 8.4-134: Array** template. Each `Array` owns a `FlowVariablePointer` to hold the data pointers for some `FlowVariable`. The `Array` class has several member functions that allow us to address individual numbers within the data storage for the `FlowVariable`.

8.4.2.6 Timesteps

We allocate the memory for all `FlowVariables` with a given IOSTATUS on a patch at the same time. The information required to and store the memory for all `FlowVariables` with a given `IOSTATUS` on a `Box` is contained in a **Program 8.4-135: TimestepBase**. This information consists of the `IOSTATUS`, the array of `FlowVariables` and the corresponding array of `FlowVariablePointers`.

A `Timestep` uses a specific `LevelBox` to allocate all the needed space for an array of `FlowVariables` with a given `IOSTATUS`. Thus **Program 8.4-136: Timestep** is derived from `TimestepBase`. It is also possible to allocate work space regions for flow variables without pre-specifying the `Box` used for addressing the data.

8.4.2.7 TagBoxes

In the regridding process within adaptive mesh refinement, it is necessary to flag those cells needing refinement, and to organize them into a list of `Boxes`. This process has already been described in section 8.3.3. At this point, we would like to focus on the design of the program to handle this process.

The **Program 8.4-137: TagBox** class consists of an `Array` of `bools` for each cell in the `Box`. In the regridding process, we might use a patch to make a `TagBox`, then set the boolean values to true for individual cells that need refinement. Afterward, we can call a number of `TagBox` functions to operate on the tags.

We also have a `TagBoxArr` class, to work with an array of `TagBoxes`. The `TagBoxArr` class has a member function, `TagBoxArr::findBoxesContainingTags`, that performs a kind of pattern recognition to determine a list of `Boxes` that contain the tagged cells efficiently.

8.4.2.8 DataBoxes

User problems can involve a variety of `FlowVariables` of different geometries and `IOSTATUSes`. However, the `Timestep` class is designed to hold the data for variables of the same `IOSTATUS`. Thus, in order to hold all the data for all the `FlowVariables` associated with some patch, we need a data structure that owns an array of `Timesteps`. This is implemented in **Program 8.4-138: DataBox**. A

DataBox owns an IOStatusArray, and a corresponding array of Timestep pointers. DataBoxes have useful member functions for copying and debugging.

Some DataBoxes are also used for regridding operations, and therefore need tags. For this reason, we have defined **Program 8.4-139: DataTagBox.** This class is derived from DataBox but not from TagBox, in order to avoid multiple inheritance of the base class LevelBox.

8.4.2.9 EOSModels

We have tried to isolate the problem-specific features of the code into a class designed by the user. All user models are derived from the library class **Program 8.4-140: EosModel.** An EosModel owns lists of FlowVariables defined by the user in the derived class. The EosModel also has a number of virtual functions. Many of these are **pure virtual functions**, meaning that the function pointer in the EosModel class is zero. As a result, the user is required to define specific instances of these functions in the derived class.

These virtual functions are very useful. When it is time to advance the data on a patch, we use a pointer to the EosModel to invoke **run-time binding** of the integration techniques appropriate to the material model. Because of this design principle, the adaptive mesh refinement code does not need to know anything about the integration techniques, the tests for error estimation, the determination of variables intended for graphical display, or even the individual FlowVariables themselves.

This means that the user can insert new equations of state into the adaptive mesh refinement library without modifying the library. In fact, there are only two places where it is necessary to identify the specific user equation of state to the adaptive mesh refinement library. The first place is where the name of the specific user equation of state is read by the input data and used to call the correct derived class constructor. The second place is where the problem model name is read from a restart file before constructing the derived equation of state. For convenience, the library provides a macro, DEFINE_MODEL_PROCEDURES to automatically construct the user's model; see the end of **Program 8.4-141: EosModel.H.**

8.4.2.10 Patch

We have already discussed the concept of the grid patch. Now it is time to make this concept concrete, by discussing the corresponding data structure. A Patch is derived from a DataTagBox; see **Program 8.4-142: Patch.H.** In addition, it owns a pointer to an EosModel. As a result, a Patch can invoke model-specific operations by calling the virtual functions in the EosModel class. Thus the Patch class performs such important operations as Patch::initialize and Patch::advance.

Within a given level of refinement, we can use several patches to cover the region of physical space requiring refinement. For this purpose, we have designed the `PatchArr` class; see **Program 8.4-143: PatchArr.H.** A `PatchArr` is derived from an array of `Patch` pointers. It has member functions to perform a variety of operations on `Patches`.

All loops over `Patches` are encapsulated in the `PatchArr` class. This is a natural point at which to distribute the adaptive mesh refinement computations over some collection of computer processors. The calculations on an individual `Patch` involve very similar operations on regular data arrays, with communication between `Patches` needed only for boundary values. It is possible to see the special code for distributed computing in **Program 8.4-144: PatchArr.C.** Almost all of the special programming for distributed computing is contained in `PatchArr.C` and `TagBoxArr.C`. There is no need for special distributed programming in individual user model classes, derived from `EosModel`.

8.4.2.11 Level

The final class of interest is `Level`; see **Program 8.4-145: Level.H.** This class encapsulates the data structures needed to perform the adaptive mesh refinement computations on an individual level of refinement. A `Level` is essentially a doubly-linked list member, because it contains pointers to the next coarser and finer `Levels`. More importantly, a `Level` contains a pointer to a `PatchArr`, which holds the array of `Patches` on that level of refinement. A `Level` also contains pointers to `BoxLists` for the proper nesting list and its complement.

`Level` class member functions contain all recursions over levels of refinement, while `PatchArr` member functions contain all loops over `Patches` on some level of refinement. This encapsulation is useful to collect all similar operations into a common location; however, it does mean that when we follow program execution, we jump between member functions in a number of C^{++} classes. This makes it difficult to follow the code in many editors.

8.5 ScalarLaw Example

It should not be necessary for a user of the adaptive mesh refinement code to be familiar with all of the various C^{++} classes described above. Instead, users typically want to know how to apply the adaptive mesh refinement code to their problem. In this subsection, we will describe how to apply adaptive mesh refinement to a scalar conservation, especially the Buckley–Leverett model.

If we were not writing code for adaptive mesh refinement, we might have a fairly simple main program that calls several Fortran subroutines. Without being specific about the arguments to these routines, the outline of the code might look like the following:

```
extern "C" {
  void initsl_(...);
  void stabledt_(...);
  void bccells_(...);
  void fluxderv_(...);
  void method_(...);
  void consdiff_(...);
}
void main(int argc, char *argv[]) {
  int ncells=...;
  int nghosts=...;
  double conserved[ncells+2*nghosts};
  double mesh[ncells+2*nghosts};

  initsl_(ncells,nghosts,  conserved,mesh); // initial values
  stabledt_(...);                           // CFL stability
  for (int step=0;step<max_steps;step++) {
    bccells_(...);                          // physical boundary values
    fluxderv_(...);                         // characteristic speeds
    method_(...);                           // scheme
    consdiff_(...);                         // conservative difference
    stabledt_(...);                         // CFL stability
  }
}
```

When we use the adaptive mesh refinement code, we will basically call these same routines from the the adaptive mesh refinement library. In this case, the structure of the code looks more like

```
void main(int argc, char *argv[]) {
  GlobalMain::run(...) {
    GridList:initialize(...) {
      Grid::initialize(...) {
        Level::initialize(...) {
          if (!coarserLevelExists()) {
            PatchArr::makePatchesFrom(...) {...}
          }
          PatchArr::initialize(...) {
            for (int i=0;i<getLength();i++) {
              operator[](i)->initialize() {
                EosModel::initialize() {
                  initsl_(...);
                }
              }
            }
          }
          if (Level::canBeRefined()) {
            Level::findInitialErrorCells() {...}
            if (cell are tagged) {
              EosModel::newLevel(...);
              Level::finerLevel()->initialize(...);
            }
```

```
          }
        }
      }
    }
    while ( sim_time < tmax && lmore_steps ) {
      GlobalMain::advance() {
        GridList::advance(...) {
          Grid::advance(...) {
            Level::advance(...) {
              while (!Level::isLastStep()) {
                PatchArr::advance(...) {
                 for (int i=0;i<getLength();i++) {
                  Level::fillModel(...);
                    Patch::advance(...) {
                      EosModel::stuffModelGhost(...) {
                        bccells_(...);
                      }
                      EosModel::computeFluxes(...) {
                        fluxderv_(...);
                        method_(...);
                      }
                      EosModel::conservativeDifference(...) {
                        consdiff_(...);
                        method_(...);
                      }
                    }
                  }
                }
                if (Level::finerLevelExists()) {
                  Level::finerLevel()->advance(...);
                }
                if (Level::timeToRegrid()) {
                  if (!isLastStep() || !time_to_sync_with_coarser_level)
                  {
                    Level::regridFinerLevels(...);
                  }
                }
                Level::findStableStepSize(...);
              } // end loop over steps
              if (Level::coarserLevelExists()) {
                PatchArr::coarsenFluxSums(...);
                coarserLevel()->patch_arr->conservativeDifference(...);
                PatchArr::coarsenOldToNew();
              }
            } // end Level::advance
          }
        }
      }
    }
}
```

Thus the same user-defined Fortran routines get called in adaptive mesh refinement. The difficulty for new users is that the call stack involves procedures from a number of different C^{++} classes.

We encapsulate the user-dependent code into a single `ScalarLaw` class, derived from `EosModel`; see **Program 8.5-146: ScalarLaw.H.** A `ScalarLaw` owns several `FlowVariable` pointers, in addition to the data needed to describe a computational problem. The purpose of each of these individual `FlowVariables` will become clearer as we describe the member functions in the `ScalarLaw` class.

8.5.1 ScalarLaw Constructor

When we construct a `ScalarLaw`, we declare each of the `FlowVariables` we will use in the computations; see **Program 8.5-147: ScalarLaw::ScalarLaw.** For example, the `conserved` quantity consists of one variable with `INOUT` status using `CELL`-centered data. Further, when we refine the `conserved` quantity, we use conservative linear interpolation, and we use conservative averaging to coarsen this variable. This `FlowVariable` constructor specifies all of the information needed to allocate the data on any patch at any point in the simulation, and to address that data.

The `ScalarLaw` constructor also specifies several `InputParameters` for a Riemann problem, and for controlling the numerical method.

8.5.2 initialize

At some point in the execution of the adaptive mesh refinement code, we call `Level::initialize` to initialize all of the data on a given level of refinement. This in turn calls `PatchArr::initialize`, which loop over its `Patches` and calls the `EosModel::initialize` procedure. Run-time binding of C^{++} virtual functions causes the `ScalarLaw::initialize` procedure to be called at that point.

The `ScalarLaw::initialize` procedure is basically a C^{++} wrapper around a Fortran subroutine that we have used previously, name `initsl`. In order to call the Fortran routine, we need to determine the dimensions of the arrays for `conserved` and the `mesh_var`. These array dimensions are provided by macros operating on the `Patch`.

In this particular implementation of `ScalarLaw::initialize`, we are using the `EosModel` workspace to hold the output from the Fortran initialization routine `initsl`. Thus, we use the `EosModel::getPtr` function to get the data pointer in the work space for both of the two `FlowVariables`. Afterward, we use `EosModel::postProcessInitialize` to copy the data from the model workspace to the `Patch`. For models in which ghost cell information is not needed for initialization, we could initialize the data directly on the `Patch`.

The maximum size of the `EosModel` work space determines the maximum size of the `Patches`. This is so that all of the data for an individual `Patch` and its extra boundary data can be contained in the work space. The number of cells in any coordinate direction in the workspace `Box` is determined by the macro `NSTRIP`. This must be set separately in Fortran, in the `mym4.i` file. `EosModel::setSizes` checks that the C++ and Fortran values for these parameters are the same.

8.5.3 *stableDt*

After initializing the data, and during the time stepping procedures of adaptive mesh refinement, we need to compute a stable time step using the CFL condition. This is performed in `ScalarLaw::stableDt`. This C++ procedure is a wrapper around two Fortran subroutines designed to compute the flux derivatives (`fluxderv`) and to apply the CFL condition (`stabledt`).

In this case, we work with data for `conserved` on the `Patch`; this data pointer is obtained by the `DataBoxFlowVariable::getPtrFrom` procedure. Since `dfdu` has `IOSTATUS TEMPORARY`, it is not stored on the `Patch`. As a result, the computer memory for `dfdu` is taken from the `EosModel` workspace.

8.5.4 *stuffModelGhost*

A timestep consists of using the conserved quantities to compute the numerical fluxes, and then applying a conservative difference. Since the computation of the fluxes depends on the numerical scheme, and high-order schemes typically have large computational stencils, we need information on cells beyond the interior of the `Patch` in order to compute all of the fluxes on the `Patch`. These extra cells are called **ghost cells**.

The number of ghost cells needed depends on the computational method, and on the choice of `FlowVariable`. For simplicity, we will give the same number of ghost cells to all `FlowVariables` with the same geometry. An even simpler strategy is to give all `FlowVariables` the same number of ghost cells. The number of ghost cells for each geometry are defined by macros that appear in ScalarLaw.C just before `ScalarLaw::ScalarLaw`. The Fortran values are set in `mym4.i`. In order to be sure that C++ and Fortran use the same values, we call `EosModel::setSizes` in the `ScalarLaw` constructor.

The numerical treatment of ghost cells is the most difficult aspect of developing the user model code. On non-adaptive meshes, the physical domain occurs in easily-identifiable locations. As a result, it is common for users to mix the code that assigns boundary values with the code that applies the numerical integrator. These operations need to be kept separate for adaptive mesh refinement.

Before calling `ScalarLaw::computeFluxes`, we need to provide values for the ghost cells used in the flux computation. Because the flux computation can involve temporary variables and almost surely involves ghost cells beyond the `Patch`, we perform the flux computations in the `EosModel` workspace.

For those `Patches` that have ghost cells interior to the physical domain, the ghost cell information is provided by `Level:fillModel`, called from `PatchArr::advance` just before we call `Patch::advance`. The adaptive mesh refinement library provided the best values for these ghost cells, in the manner described in Section 8.3.2. Since physical boundary conditions are problem-dependent, the adaptive mesh refinement library expects the user to provide these values.

In order to fill the ghost cells outside the physical domain for computations in the workspace, we use `ScalarLaw::stuffModelGhost`. There are two separate activities in this procedure. The first (subroutine `bcmesh`) provides values for the mesh outside the physical domain, and the second (`bccells`) provided values for `conserved` outside the physical domain.

8.5.5 stuffBoxGhost

In order to refine data, the adaptive mesh refinement library typically needs data on ghost cells in order to implement higher-order interpolation. Data might be refined to provide ghost cell information on a fine patch at the interface with the coarse grid, or to provide initial values for new fine patches after regridding. Since the refinement process may be recursive, we cannot use the `EosModel` workspace for the computations with the ghost cells.

Instead, the adaptive mesh refinement library will use `DataBoxes` constructed in the `Level::fillBox` procedure. Since some of the ghost cells for these `DataBoxes` may lie outside the physical domain, we have to provide `Scalar-Law::stuffBoxGhost` to set these values. If we are careful with array dimension, we can use the same Fortran routines (`bcmesh` and `bccells`) to determine these boundary values.

8.5.6 computeFluxes

After we define the data on the `Patch` and its ghost cells at the beginning of a time step, we are ready to compute the values of `conserved` at the end of the time step. We do this in `ScalarLaw::computeFluxes`, which is called from `Patch::advance`. This routine is just a C^{++} wrapper around two Fortran function, namely `fluxderv`, which computes the characteristic speeds, and `method`, which in this case computes the fluxes using a second-order MUSCL scheme. Users can change the integration technique merely by changing the code in subroutine `method`.

8.5.7 conservativeDifference

After we compute the fluxes, we compute the new `conserved` values via a conservative difference. This is done in `ScalarLaw::conservativeDifference`, which is merely a C++ wrapper around the Fortran routine `consdiff`. `ScalarLaw::conservativeDifference` is called from `Patch::advance` during time stepping on some level of refinement. It is also called from `Patch::conservativeDifference`, which is called by `PatchrArr::conservativeDifference` from `Level::advance` during refluxing; see Section 8.3.4.

8.5.8 findErrorCells

In order to decide where to place mesh refinement, the adaptive mesh refinement depends on the user to make the ultimate decision. The adaptive mesh refinement library will help the user by providing the results from coarse and fine integrations for user comparison. It is up to the user to examine these two integrations and tag cells for refinement in `ScalarLaw::findErrorCells`.

There are two versions of this procedure. One is a C++ wrapper around the Fortran subroutine `locshock`. This routine uses a gradient detector to tag cells for refinement. It is slightly less expensive to use this version of `ScalarLaw::findErrorCells`, but the results are not always acceptable.

If the `EosModel` parameter `use_richardson` is true, then `EosModel::usesRichardsonExtrapolation` will cause the adaptive mesh refinement library to call the second form of `ScalarLaw::findErrorCells`. This procedure is basically a C++ wrapper around the Fortran routine `errestsl`. This Fortran routine compares the coarse and fine integration results from the adaptive mesh refinement library and decides where to tag cells for refinement. The ideas in this routine are described in Section 8.3.3.

8.5.9 Numerical Example

In Figure 8.7 we show some numerical results for adaptive mesh refinement applied to the Buckley–Leverett model. The size of the data markers represents the size of the finest grid cell in that spatial region of the computation. Note that the mesh is adaptively refined around both the shock and the left-hand edge of the rarefaction. These are two locations where the MUSCL scheme fails to achieve second-order

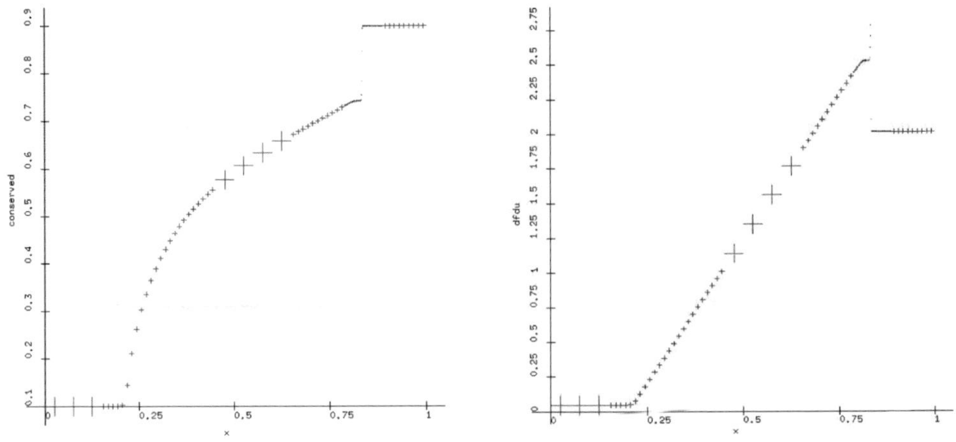

(a) Saturation (b) Characteristic speed

Fig. 8.7 Adaptive mesh refinement for Buckley–Leverett model

accuracy. Students can perform their own adaptive mesh refinement calculations by clicking on **Executable 8.5-57: cpphog.** The program will open up a graphical user interface in which the student can adjust a variety of parameters. For example, the student could vary `max_levels` to control how many levels of mesh refinement are used, `refinement_ratio` to control the refinement ration between the grid levels.

8.6 Linear Elasticity Example

Next, let us briefly discuss the application of adaptive mesh refinement to linear elasticity. We are particularly interested in solving Lamb's problem [1] for a transient line load. Because the model is linear, the problem does not involve shocks. However, it does involve interesting localized phenomena, including a p-wave, and s-wave and a head wave.

In order to implement the linear elasticity model in the adaptive mesh refinement code, we develop a C^{++} class called `LinearElasticity`. As expected, this class is derived from `EosModel`; see **Program 8.6-148: LE.H.** This model involves `FlowVariables` for displacement, deviatoric stress, pressure and velocity. The organization of this C^{++} class is much the same as the `ScalarLaw` class, but the Fortran routines involved in the integration are very different.

Figure 8.8 shows some numerical results for adaptive mesh refinement applied to Lamb's problem. We show 30 equally spaced contours of the second invariant of the deviatoric stress, and the corresponding adaptively refined grid. The contour plot shows the boundaries of the grid patches superimposed on the contour levels.

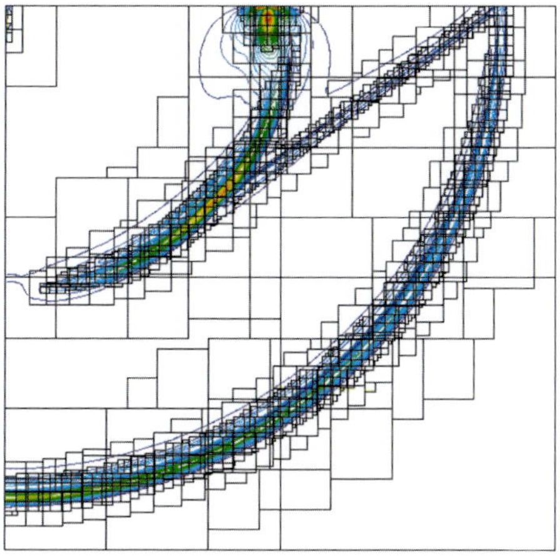

(a) J_2'

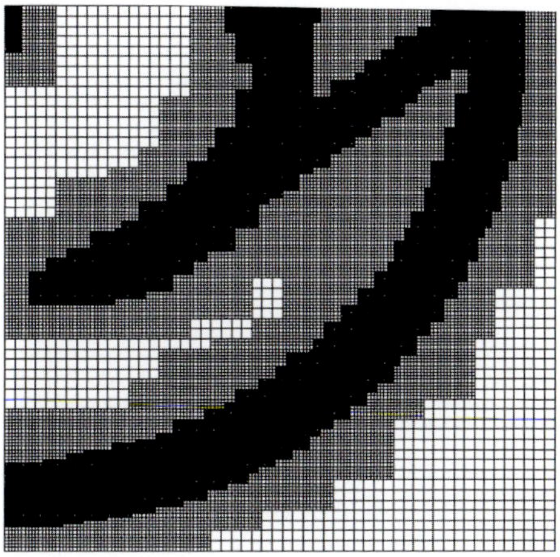

(b) Mesh

Fig. 8.8 Adaptive mesh refinement for Lamb's problem

Students can perform their own adaptive mesh refinement calculations by clicking on **Executable 8.6-58: cpphog.** The program will open up a graphical user interface in which the student can adjust a variety of parameters.

8.7 Gas Dynamics Examples

Finally, let us discuss the application of adaptive mesh refinement to gas dynamics. We will focus on the Colella–Woodward interacting blast wave problem in one dimension [177], and on the Schulz–Rinne two-dimensional Riemann problem case number 3 [144, 145]. Both problems are solved via the C^{++} class `GasDynamics`. Students can view the definition of this class by clicking on **Program 8.7-149: GD.H.** Of course, the program calls different Fortran routines to integrate the equations in one dimension and two dimensions. The one-dimensional code uses the MUSCL scheme, and the two-dimensional code uses the corner transport upwind scheme.

Figure 8.9 shows some numerical results for the blast wave problem using AMR. In this case, there is a variety of interesting behavior over a large portion of the domain at late time. However, a careful study of the results will see that the finest grid is concentrated at the discontinuities. Students can perform their own adaptive mesh refinement calculations by clicking on **Executable 8.7-59: cpphog.** The program will open up a graphical user interface in which the student can adjust a variety of parameters.

Figure 8.10 shows some numerical results for a 2D Riemann problem using AMR. At early time, the mesh is primarily along the coordinate axes, where

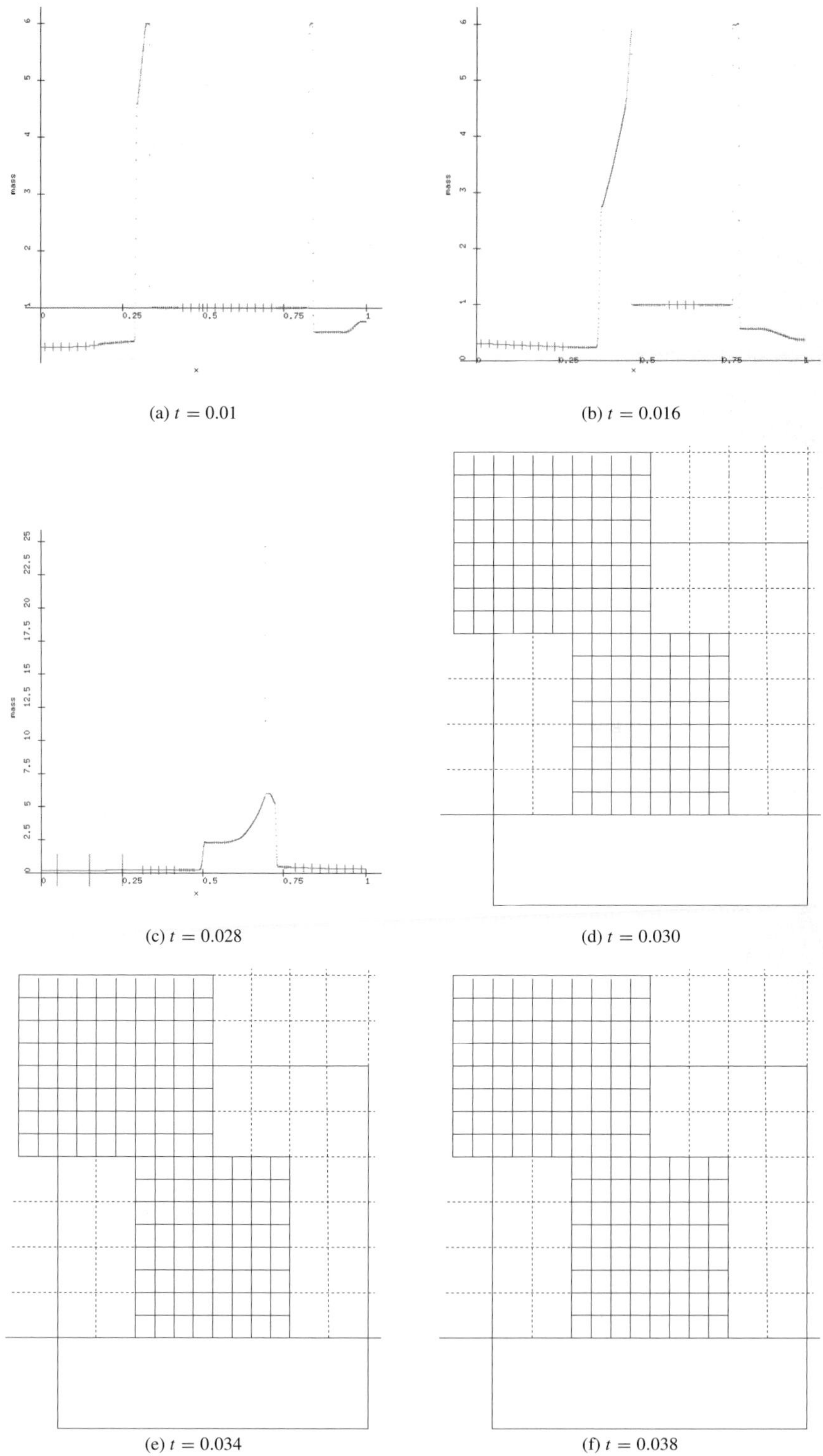

(a) $t = 0.01$ (b) $t = 0.016$

(c) $t = 0.028$ (d) $t = 0.030$

(e) $t = 0.034$ (f) $t = 0.038$

Fig. 8.9 Adaptive mesh refinement for blast wave problem: density vs. position. AMR uses 10 cells on coarse grid, a refinement ratio of 4, and a maximum of 5 levels of mesh

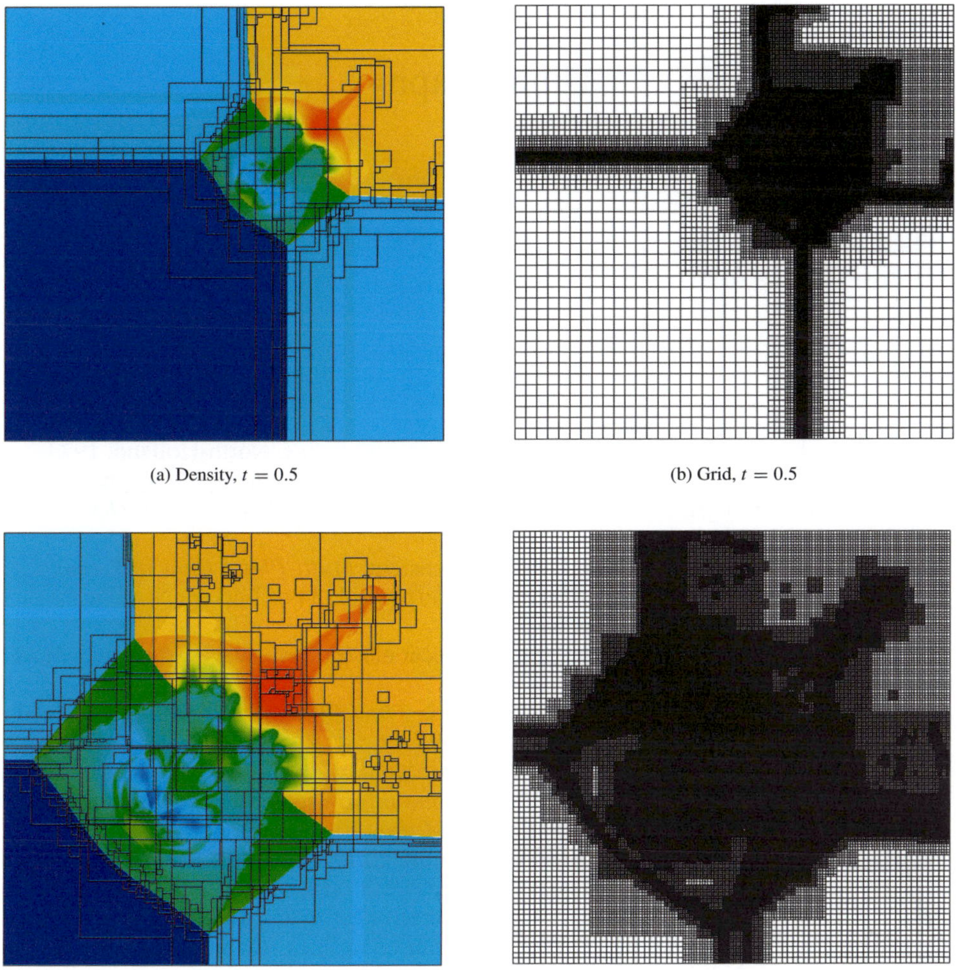

(a) Density, $t = 0.5$

(b) Grid, $t = 0.5$

(c) Density, $t = 1$

(d) Grid, $t = 1$

Fig. 8.10 Adaptive mesh refinement for 2D gas dynamics Riemann problem: density. AMR uses 40×40 grid, a refinement ratio of 2, and a maximum of five levels of mesh refinement.

the initial discontinuities are located. At later time, the solution develops some interesting shocks, and contact discontinuities that produce a variety of physical instabilities. The AMR algorithm refines over a substantial portion of the grid in order to capture all of this behavior. Students can perform their own adaptive mesh refinement calculations by clicking on **Executable 8.7-60: cpphog.** The program will open up a graphical user interface in which the student can adjust a variety of parameters.

Bibliography

[1] J. D. Achenbach, ed. *Wave Propagation in Elastic Solids*. North-Holland, 1973.

[2] N. Ahmad, B. Zafer, and R. Löhner. A godunov-type scheme for atmospheric flows on unstructured grids. scalar transport. *Pure and Applied Geosciences*, **163**: 1699–1735, 2005.

[3] M. B. Allen III, A. Behie, and J. A. Trangenstein. *Multi-Phase Flow in Porous Media: Mechanics, Mathematics and Numerics*, vol. 34 of Lecture Notes in Engineering. Springer-Verlag, 1988.

[4] S. S. Antman and W. G. Szymczak. Nonlinear elastoplastic waves. *Contemp. Math.*, **100**:27–54, 1989.

[5] R. E. G. Ayala. *Mathematics of Three-Phase Flow*. PhD thesis, Stanford University, Department of Petroleum Engineering, 1995.

[6] K. Aziz and A. Settari. *Petroleum Reservoir Simulation*. Applied Science, 1979.

[7] S. B. Baden. Software infrastructure for non-uniform scientific computations on parallel processors. *Appl. Comp. Rev., ACM*, **4**:7–10, 1996.

[8] L. E. Baker. Three-phase relative permeability correlations. In *SPE/DOE Enhanced Oil Recovery Symposium*, Tulsa, Oklahoma, 1988. SPE 17369.

[9] T. J. Barth and Mats G. Larson. A posterior error estimates for higher order Godunov finite volume methods on unstructured meshes. Technical Report NAS-02-001, NASA Ames, 2002.

[10] T. J. Barth and D. C. Jespersen. The design and application of upwind schemes on unstructured meshes. In *AIAA Paper 89-0366*, January, 1989.

[11] J. Bear, ed. *Dynamics of Fluids in Porous Media*. Dover, 1972.

[12] J. B. Bell, P. Colella, J. Trangenstein, and M. Welcome. Adaptive mesh refinement on moving quadrilateral grids. In *Proceedings of the AIAA 9th Computational Fluid Dynamics Conference*, Buffalo, June, 1988.

[13] J. B. Bell, P. Colella, and J. A. Trangenstein. Higher-order Godunov methods for general systems of hyperbolic conservation laws. *J. Comp. Phys.*, **82**:362–397, 1989.

[14] J. B. Bell, C. N. Dawson, and G. R. Shubin. An unsplit, higher order Godunov method for scalar conservation laws in multiple dimensions. *J. Comp. Phys.*, **74**:1–24, 1988.

[15] J. B. Bell, J. A. Trangenstein, and G. R. Shubin. Conservation laws of mixed type describing three-phase flow in a porous media. *SIAM J. Appl Math.*, **46**:1000–1017, 1986.

[16] M. Berger and J. Melton. An accuracy test of a cartesian grid method for steady flow in complex geometries. In *Proceedings of the 5th International Conference on Hyperbolic Problems*, Stonybook, NY, June, 1994.

[17] M. J. Berger and S. Bokhari. A partitioning strategy for non-uniform problems on multiprocessors. *IEEE Trans. Comp.*, C-36:570–580, 1987.

[18] M. J. Berger and P. Colella. Local adaptive mesh refinement for shock hydrodynamics. *J. Comp. Phys.*, **82**:64–84, 1989.

[19] M. J. Berger and J. Saltzman. AMR on the CM-2. *Appl. Numer. Math.*, **14**:239–253, 1994.

[20] M. J. Berger and J. Oliger. Adaptive mesh refinement for hyperbolic partial differential equations. *J. Comp. Phys.*, **53**:484–512, 1984.

[21] M. J. Blunt and P. R. King. Relative permeabilities from two and three dimensional pore scale network modeling. *Transport in Porous Media*, **4**:407–434, 1991.

[22] J. P. Boris and D. L. Book. Flux-corrected transport. III. minimal-error FCT algorithms. *J. Comp. Phys.*, **20**:397–431, 1976.

[23] M. Brio and C. C. Wu. An upwind differencing scheme for the equations of ideal magnetohydrodynamics. *J. Comp. Phys.*, **75**:400–422, 1988.

[24] J. M. Burgers. Application of a model system to illustrate some points of the statistical theory of free turbulence. *Nederl. Akad. Wetensch. Proc.*, **43**:2–12, 1940.

[25] M. A. Celia, T. F. Russell, I. Herrera, and R. E. Ewing. An Eulerian–Lagrangian localized adjoint method for the advection–diffussion equation. *Adv. Water Resources*, **13**:187–296, 1990.

[26] I-L. Chern and P. Colella. A conservative front tracking method for hyperbolic conservation laws. "Lawrence Livermore National Laboratory", LLNL report UCRL-97258, 1987.

[27] A. J. Chorin and J. E. Marsden. *A Mathematical Introduction to Fluid Mechanics*. Springer-Verlag, 1979.

[28] R. B. Clelland. *Simulation of Granular and Fluid Systems Using Combined Continuous and Discrete Methods*. PhD thesis, Duke University, Department of Mathematics, 1996.

[29] B. Cockburn and C. Shu. TVB Runge–Kutta local projection discontinuous galerkin finite element method for the consevation laws II: General framework. *Math. Comp.*, **52**:411–435, 1989.

[30] B. Cockburn, S. Hou, and C. Shu. The Runge–Kutta local projection discontinuous Galerkin finite element method for the conservation laws IV: The multidimensional case. *Math. Comp.*, **54**:545–581, 1990.

[31] P. Colella. Glimm's method for gas dynamics. *SIAM J. Sci. Stat. Comp.*, **3**:76–110, 1982.

[32] P. Colella. A direct Eulerian MUSCL scheme for gas dynamics. *SIAM J. Sci. Stat. Comput.*, **6**:104–117, 1985.

[33] P. Colella. Multidimensional unsplit methods for hyperbolic conservation laws. *J. Comp. Phys.*, **87**:171–200, 1990.

[34] P. Colella and H. M. Glaz. Efficient solution algorithms for the Riemann problem for real gases. *J. Comp. Phys.*, **59**:264–289, 1985.

[35] P. Colella and P. R. Woodward. The piecewise parabolic method (PPM) for gas dynamical simulations. *J. Comp. Phys.*, **54**:174–201, 1984.

[36] A. T. Corey, C. H. Rathjens, J. H. Henderson, and M. R. J. Wyllie. Three-phase relative permeability. *Pet. Trans., AIME*, **207**:349–351, 1956.

[37] R. Courant and K. O. Friedrichs. *Supersonic Flow and Shock Waves*. Springer, 1948.

[38] M. G. Crandall and A. Majda. Monotone difference approximations for scalar conservation laws. *Math. Comp.*, **34**:1–21, 1980.

[39] N. Cristescu. *Dynamic Plasticity*. North-Holland, 1967.

[40] C. M. Dafermos. The entropy rate admissibility criterion for solutions of hyperbolic conservation laws. *J. Differential Equations*, **14**:202–212, 1973.

[41] C. M. Dafermos. *Hyperbolic Conservation Laws in Continuum Physics*. Springer, 2000.

[42] G. Dahlquist and Å. Björck. *Numerical Methods*. Prentice-Hall, 1974. Translated by N. Anderson.

[43] Akhil Datta-Gupta and M. J. King. *Streamline Simulation: Theory and Practice*. Society of Petroleum Engineers, Textbook Series, 2006.

[44] J. E. Dennis and R. B. Schnabel. *Numerical Methods for Unconstrained Optimization and Nonlinear Equations*. Prentice-Hall, 1983.

[45] R. Donat and A. Marquina. Capturing shock reflections: An improved flux formula. *J. Comput. Phys*, **125**:42–58, 1996.

[46] A. Dutt, L. Greengard, and Vladimir Rokhlin. Spectral deferred correction methods for ordinary differential equations. *BIT*, **40**:241–266, 2000.

[47] M. G. Edwards and C. F. Rogers. Finite volume discretization with imposed flux continuity for the general tensor pressure equation. *Comp. Geosci.*, **2**:259–290, 1998.

[48] B. Einfeldt. On Godunov-type methods for gas dynamics. *SIAM J. Numer. Anal.*, **25**:294–318, 1988.

[49] B. Engquist and S. Osher. Stable and entropy satisfying approximations for transonic flow calculations. *Math. Comp.*, **34**:45–75, 1980.

[50] L. C. Evans and R. F. Gariepy. *Measure Theory and Fine Properties of Functions*. CRC Press, 1992.

[51] J. E. Fromm. A method for reducing dispersion in convective difference schemes. *J. Comp. Phys.*, **3**:176–189, 1968.

[52] Y. C. Fung. *Foundations of Solid Mechanics*. Prentice-Hall, 1965.

[53] K. Furati. The Riemann problem for polymer flooding with hysteresis. In *Proceedings of the Fifth International Conference on Hyperbolic Problems, Theory, Numerics, and Applications*, Stony Book, NY, 1994.

[54] P. R. Garabedian. *Partial Differential Equations*. John Wiley & Sons, 1964. S. Glimm *et al.* (eds) World Scientific, Singapore, 1996.

[55] J. Glimm. Solutions in the large for nonlinear hyperbolic systems of equations. *Comm. Pure Appl. Math.*, **18**:697–715, 1965.

[56] J. Glimm, E. Isaacson, D. Marchesin, and O. McBryan. Front tracking for hyperbolic systems. *Adv. Appl. Math.*, **2**:91–119, 1981.

[57] S. K. Godunov. Finite difference methods for numerical computation of discontinuous solutions of equations of fluid dynamics. *Mat. Sb.*, **47**:271–295, 1959. In Russian.

[58] D. Goldman and T. J. Kaper. *n*th-order operator splitting schemes and nonreversible systems. *SIAM J. Numer. Anal.*, **33**:349–367, 1996.

[59] J. Goodman and R. J. LeVeque. A geometric approach to high resolution TVD schemes. *SIAM J. Numer. Anal.*, **25**:268–284, 1988.

[60] J. Goodman and A. Majda. The validity of the modified equation for nonlinear shock waves. *J. Comp. Phys.*, **58**:336–348, 1985.

[61] D. Gottlieb and S. A. Orszag. *Numerical Analysis of Spectral Methods: Theory and Applications*, vol. 26 of *Regional Conference Series in Applied Mathematics*. SIAM, 1977.

[62] L. Greengard and J.-R. Li. High-order marching schemes for the wave equation in complex geometry. *J. Comput. Phys.*, **198**:295–309, 2004.

[63] A. Hanyga. *Mathematical Theory of Non-Linear Elasticity*. Halsted Press, 1985.

[64] A. Harten. On a class of high resolution total-variation-stable finite-difference schemes. *SIAM J. Numer. Anal.*, **21**:174–201, 1984.

[65] A. Harten. ENO schemes with subcell resolution. *J. Comp. Phys.*, **83**:148–184, 1989.

[66] A. Harten and J. M. Hyman. Self-adjusting grid methods for one-dimensional hyperbolic conservation laws. *J. Comput. Phys.*, **50**:235–269, 1983.

[67] A. Harten, J. M. Hyman, and P. D. Lax. On finite-difference approximations and entropy conditions for shocks. *Comm. Pure Appl Math.*, **29**:297–322, 1976. (with appendix by Barbara Keyfitz).

[68] A. Harten, P. D. Lax, and B. van Leer. On upstream differencing and Godunov-type schemes for hyperbolic conservation laws. *SIAM Review*, **25**:35–61, 1983.

[69] A. Harten and S. Osher. Uniformly high-order accurate nonoscillatory schemes I. *SIAM J. Numer. Anal.*, **24**:279–309, 1987.

[70] G. Hedstrom. Models of difference schemes for $u_t + u_x = 0$ by partial differential equations. *Math. Comp.*, **29**:969–977, 1975.

[71] F. G. Helfferich. Theory of multicomponent multiphase displacement in porous media. *Soc. Pet. Eng. J.*, **21**:51–62, 1981.

[72] C. Hirsch. *Numerical Computation of Internal and External Flows*, vol. **1**: *Fundamentals of Numerical Discretization*. John Wiley, 1988.

[73] C. Hirsch. *Numerical Computation of Internal and External Flows*, vol. **2**: *Computational Methods for Inviscid and Viscous Flows*. John Wiley, 1992.

[74] L. Holden. On the strict hyperbolicity of the Buckley-Leverett equations for three-phase flow in a porous medium. Norwegian Computing Center, P.O. Box 114 Blindern, 0314 Oslo 3, Norway, 1988.

[75] K. Holing. *A Conservative Front-Tracking Method for Two-Dimension Polymer Flooding*. PhD thesis, Norges Tekniske Høgskole Trondheim, 1990.

[76] G. Jennings. Discrete shocks. *Comm. Pure Appl. Math.*, XXVII:25–37, 1974.

[77] G.-S. Jiang and C. W. Shu. Efficient implementation of weighted ENO schemes. *J. Comput. Phys.*, **126**:202–228, 1996.

[78] G.-S. Jiang and E. Tadmor. Nonoscillatory central schemes for multidimensional hyperbolic conservation laws. *SIAM J. Sci. Comput.*, **19**:1892–1917, 1998.

[79] S. Jin and C. D. Levermore. Numerical schemes for hyperbolic conservation laws with stiff relaxation terms. *J. Comput. Phys.*, **126**:955–970, 1996.

[80] T. Johansen and R. Winther. The solution of the Riemann problem for a hyperbolic system of conservation laws modeling polymer flooding. *SIAM J. Math. Anal.*, **19**:541–566, 1988.

[81] B. L. Keyfitz and H. C. Kranzer. A system of non-strictly hyperbolic conservation laws arising in elasticity theory. *Arch. Rat. Mech. Anal.*, **72**:220–241, 1980.

[82] Z. Kopal. *Numerical Analysis, with Emphasis on the Application of Numerical Techniques to Problems of Infinitesimal Calculus in a Single Variable*. John Wiley, 1961.

[83] E. Kreyszig. *Introductory Functional Analysis with Applications*. John Wiley, 1978.

[84] S. N. Kruzkov. Results on the character of continuity of solutions of parabolic equations and some of their applications. *Math. Zametky*, **6**:97–108, 1969. In Russian.

[85] S. N. Kruzkov. First order quasi-linear equations in several independent variables. *Math. USSR Sb.*, **10**:217–243, 1970.

[86] A. Kurganov and D. Levy. Central-upwind schemes for the saint-venant system. *Math. Model. Numer. Anal.*, **3**:397–425, 2002.

[87] N. N. Kuznetsov. Accuracy of some approximate methods for computing the weak solutions of a first-order quasi-linear equation. *USSR Comp. Math. and Math. Phys.*, **16**:105–119, 1976.

[88] J. O. Langseth and R. J. LeVeque. A wave propagation method for three-dimensional hyperbolic conservation laws. *J. Comp. Phys.*, 2000.

[89] P. D. Lax. *Hyperbolic Systems of Conservation Laws and the Mathematical Theory of Shock Waves*, vol. 11 of *Regional Conference Series in Applied Mathematics*. SIAM, 1973.

[90] P. D. Lax and B. Wendroff. Systems of conservation laws. *Comm. Pure Appl. Math.*, **13**:217–237, 1960.

[91] P. D. Lax and B. Wendroff. On the stability of difference approximations to solutions of hyperbolic equations with variable coefficients. *Comm. Pure Appl. Math.*, **14**:497–520, 1961.

[92] B. P. Leonard. Simple high accuracy resolution program for convective modeling of discontinuities. *Int J. Numer. Methods Fluids*, **8**:1291–1318, 1979.

[93] J. LeVeque and H. C. Yee. A study of numerical methods for hyperbolic conservation laws with stiff source terms. *J. Comput. Phys.*, **86**:187–210, 1990.

[94] R. J. LeVeque. Wave propagation algorithms for multi-dimensional hyperbolic systems. *J. Comp. Phys.*, **131**:327–353, 1997.

[95] R. J. LeVeque. High resolution finite volume methods on arbitrary grids via wave propagation. *J. Comp. Phys.*, **78**:36–63, 1988.

[96] R. J. LeVeque. *Numerical Methods for Conservation Laws*. Birkhauser, 1990.

[97] R. J. LeVeque. *Finite Volume Methods for Hyperbolic Problems*. Cambridge Texts in Applied Mathematics. Cambridge University Press, 2002.

[98] R. J. LeVeque and K. M. Shyue. Two-dimensional front tracking based on high resolution wave propagation methods. *J. Comp. Phys.*, **123**:354–368, 1994.

[99] D. Levy, G. Puppo, and G. Russo. Central WENO schemes for hyperbolic systems of conservation laws. *Math. Model. Numer. Anal.*, **33**:547–571, 1999.

[100] D. Levy, G. Puppo, and G. Russo. Compact central WENO schemes for multidimensional conservation laws. *SIAM J. Sci. Comput.*, **22**:656–672, 2000.

[101] D. Levy, G. Puppo, and G. Russo. A third order central WENO scheme for 2D conservation laws. *Appl. Numer. Math.*, **33**:407–414, 2000.

[102] Z. Li and K. Ito. *The Immersed Interface Method – Numerical Solution of PDE's Involving Interfaces and Irregular Domains*, vol. 33 of *Frontiers in Applied Mathematics*. SIAM, 2006.

[103] Timur Linde. A practical, general-purpose, two-state HLL riemann solver for hyperbolic conservation laws. *Int. J. Numer. Math. Fluids*, **40**:391–402, 2002.

[104] W. B. Lindquist. Construction of solutions for two-dimensional Riemann problems. *Comput. Math. Appl.*, 12A:614–630, 1986.

[105] W. B. Lindquist. The scalar Riemann problem in two spatial dimensions: Piecewise smoothness of solutions and its breakdown. *SIAM J. Math. Anal.*, **17**:1178–1197, 1986.

[106] R. Liska and B. Wendroff. Comparison of several difference schemes on 1d and 2d test problems for the Euler equations. *SIAM J. Sci. Comput.*, **25**:995–1017, 2003.

[107] T. P. Liu. The Riemann problem for general systems of conservation laws. *J. Diff. Eqns.*, **18**:218–234, 1975.

[108] T. P. Liu. Linear and nonlinear large-time behavior of solutions of general systems of hyperbolic conservation laws. *Comm. Pure Appl. Math.*, **30**:767–796, 1977.

[109] X. D. Liu, S. Osher, and T. Chan. Weighted essentially nonoscillatory schemes. *J. Comput. Phys.*, **115**:200–212, 1994.

[110] X.-D. Liu and E. Tadmor. Third order nonoscillatory central scheme for hyperbolic conservation laws. *Numerische Mathematik*, **79**:397–425, 1998.

[111] R. Löhner. Extensions and improvements of the advancing front grid generation technique. *Comm. Numer. Meth. Engineering*, **12**:683–702, 1996.

[112] R. Löhner, K. Morgan, J. Peraire, and M. Vahdati. Finite element flux-correct transport (FEM-FCT) for the Euler and Navier–Stokes equations. *Int. J. Num. Meth. Fluids*, **7**:1093–1109, 1987.

[113] B. J. Lucier. Error bounds for the methods of Glimm, Godunov and LeVeque. *SIAM J. Numer. Anal.*, **22**:1074–1081, 1985.

[114] R. W. MacCormack. The effects of viscosity in hypervelocity impact cratering. Technical Report 69-354, AIAA, 1969.

[115] A. Majda and A. Ralston. Discrete shock profiles for system of conservation laws, 1979.

[116] L. E. Malvern. *Introduction to the Mechanics of a Continuous Medium.* Prentice-Hall, 1969.

[117] G. I. Marchuk. Splitting and alternating direction methods. In P. G. Ciarlet and J. L. Lions, eds., *Handbook of Numerical Analysis*, vol. 1. North-Holland, 1990.

[118] G. H. Miller and P. Colella. A high-order eulerian godunov method for elastic-plastic flow in solids. *J. Comp. Phys.*, **167**:131–176, 2001.

[119] R. N. Miller, K. and. Miller. Moving finite elements I. *SIAM J. Numer. Anal.*, **18**:1019–1032, 1981.

[120] E. M. Murman. Analysis of embedded shock waves calculated by relaxation methods. *AIAA J.*, **12**:626–633, 1974.

[121] H. Nessyahu and E. Tadmor. Non-oscillatory central differencing for hyperbolic conservation laws. *J. Comp. Phys.*, **87**:408–448, 1990.

[122] O. A. Oleinik. On the uniqueness of the generalized solution of Cauchy problem for nonlinear system of equations occurring in mechanics. *Uspekhi Mat. Nauk. (N.S.)*, 12(3(75)):3–73, 1957.

[123] S. Osher and Solomon. Upwind difference schemes for hyperbolic systems of conservation laws. *Math. Comp.*, **38**:339–374, 1982.

[124] S. Osher. Riemann solvers, the entropy condition, and difference approximations. *SIAM J. Numer. Anal.*, **21**:217–235, 1984.

[125] S. Osher. Convergence of generalized MUSCL schemes. *SIAM J. Numer. Anal.*, 22(5):947–961, 1985.

[126] S. Osher and S. Chakravarthy. High resolution schemes and the entropy condition. *SIAM J. Numer. Anal.*, **21**:955–984, 1984.

[127] Ronald L. Panton, ed. *Incompressible Flow*. John Wiley, 1984.

[128] D. W. Peaceman. *Fundamentals of Numerical Reservoir Simulation*. Elsevier Scientific Publishing, 1977.

[129] D.W. Peaceman. Interpretation of well-block pressure in numerical reservoir simulation with non-square grid blocks and anisotropic permeability. *Trans. AIME*, **275**:531–543, 1983. SPE 10528.

[130] R. B. Pember. Numerical methods for hyperbolic conservation laws with stiff relaxation II. Higher-order Godunov methods. *SIAM J. Sci. Comput.*, **14**:955–970, 1993.

[131] R. Peyret and T. D. Taylor. *Computational Methods for Fluid Flow*. Springer-Verlag, 1983.
[132] G. A. Pope. The application of fractional flow theory to enhanced oil recovery. *Society of Petroleum Engineers J.*, **20**:191–205, 1980.
[133] B. Keyfitz Quinn. Solutions with shocks: an example of an l_1-contractive semigroup. *Comm. Pure Appl. Math.*, XXIV:125–132, 1971.
[134] J.-F. Remacle, J. E. Flaherty, and M. S. Shephard. A adaptive discontinuous galerkin technique with an orthogonal basis applied to compressible flow problems. *SIAM Review*, **45**:53–72, 2003.
[135] N. H. Risebro. A front tracking alternative to the random choice method. *Proc. Amer. Math. Soc.*, 117 (No. 4):1125–1139, 1993.
[136] N. H. Risebro and A. Tveito. Front tracking applied to a non-strictly hyperbolic system of conservation laws. *SIAM J. Sci. Stat. Comput.*, **12**:1401–1419, 1991.
[137] P. J. Roache. *Computational Fluid Dynamics*. Hermosa, 1972.
[138] P. L. Roe. Approximate Riemann solvers, parameter vectors, and difference schemes. *J. Comput. Phys.*, **43**:357–372, 1981.
[139] W. Rudin. *Real and Complex Analysis*. McGraw-Hill, 1987.
[140] J. Saltzman. An unsplit 3-D upwind method for hyperbolic conservation laws. *J. Comput. Phys.*, **115**:153–168, 1994.
[141] P. H. Sammon. An analysis of upstream differencing. *Soc. Pet. Eng. J. Res. Engrg.*, **3**:1053–1056, 1988.
[142] R. Sanders. On convergence of monotone finite difference schemes with variable spatial differencing. *Math. Comp.*, **40**:91–106, 1983.
[143] D. G. Schaeffer, M. Shearer, and S. Schecter. Non-strictly hyperbolic conservation laws with a parabolic line. *J. Diff. Eqns.*, **103**:94–126, 1993.
[144] C. W. Schulz-Rinne. Classication of the Riemann problem for two-dimensional gas dynamics. *SIAM J. Math. Anal.*, **24**:76–88, 1993.
[145] C. W. Schulz-Rinne, J. P. Collins, and H. M. Glaz. Numerical solution of the Riemann problem for two-dimensional gas dynamics. *SIAM J. Sci. Comput.*, **14**:1394–1414, 1993.
[146] C. Werner Schulz-Rinne. *The Riemann Problem for Two-Dimensional Gas Dynamics and New Limters for High-Order Schemes*. PhD thesis, Swiss Federal Institute of Technology, Zurich, Switzerland, 1993.
[147] M. Shearer. Loss of strict hyperbolicity of the Buckley–Leverett equations for three-phase flow in a porous medium. In *Proc. I.M.A. Workshop on Oil Reservoir Simulation*, 1986.
[148] M. Shearer and J. A. Trangenstein. Loss of real characteristics for models of three-phase flow in a porous medium. *Transport in Porous Media*, **4**:499–525, 1989.
[149] C.-W. Shu and S. Osher. Efficient implementation of essentially non-oscillatory shock capturing schemes. II. *J. Comp. Phys.*, **83**:32–78, 1989.
[150] Joel Smoller. *Shock Waves and Reaction–Diffusion Equations*. Springer-Verlag, 1982.
[151] G. A. Sod. *Numerical Methods in Fluid Dynamics*. Cambridge University Press, 1985.
[152] G. Strang. On the construction and comparison of difference schemes. *SIAM J. Numer. Anal.*, **5**:506–517, 1968.
[153] W. Strauss. ed. *Partial Differential Equations: An Introduction*. John Wiley, 1992.
[154] J. C. Strikwerda. *Finite Difference Schemes and Partial Differential Equations*. Wadsworth & Borrks/Cole, 1989.

[155] P. K. Sweby. High resolution schemes using flux limiters for hyperbolic conservation laws. *SIAM J. Numer. Anal.*, **21**:995–1011, 1984.

[156] E. Tadmor. Numerical viscosity and the entropy condition for conservative difference schemes. *Math. Comp.*, **43**:369–381, 1984.

[157] A. Tang and T. C. T. Ting. Wave curves for the Riemann problem of plane waves in isotropic elastic solids. *Internat. J. Engng. Sci.*, **25**:1343–1381, 1987.

[158] M. Taylor. *Partial Differential Equations III*. Applied Mathematical Sciences No. 117. Springer-Verlag, 1996.

[159] E. F. Toro. *Riemann Solvers and Numerical Methods for Fluid Dynamics*. Springer, 1997.

[160] Gábor Tóth. The $\nabla \cdot b = 0$ constraint in shock-capturing magnetohydrodynamics codes. *J. Comp. Phys.*, **161**:605–652, 2000.

[161] J. A. Trangenstein. Three-phase flow with gravity. *Contemp. Math.*, **100**:147–160, 1988.

[162] J. A. Trangenstein and P. Colella. A higher-order Godunov method for modeling finite deformation in elastic-plastic solids. *Comm. Pure Appl. Math.*, **XLIV**:41–100, 1991.

[163] J. A. Trangenstein and R. B. Pember. The Riemann problem for longitudinal motion in an elastic-plastic bar. *SIAM J. Sci. Stat. Comput.*, **12**:180–207, 1991.

[164] J. A. Trangenstein and R. B. Pember. Numerical algorithms for strong discontinuities in elastic-plastic solids. *J. Comp. Phys.*, **103**:63–89, 1992.

[165] J. A. Trangenstein and Z. Bi. Multi-scale iterative techniques and adaptive mesh refinement for flow in porous media. *Adv. Water Resources*, **25**:1175–1213, 2002.

[166] D. P. Trebotich and P. Colella. A projection method for incompressible viscous flow on moving quadrilateral grids. *J. Comput. Phys.*, **166**:191–217, 2001.

[167] C. Truesdell. *The Elements of Continuum Mechanics*. Springer-Verlag, 1966.

[168] B. van Leer. *Springer Lecture Notes in Physics*, vol. 18 Towards the Ultimate Conservative Difference Scheme. I. The Quest of Monotonicity. Springer-Verlag, 1973.

[169] B. van Leer. Towards the ultimate conservative difference scheme. II. Monotonicity and conservation combined in a second-order scheme. *J. Comp. Phys.*, **14**:361–370, 1974.

[170] B. van Leer. Towards the ultimate conservative difference scheme. III. Upstream-centered finite-difference schemes for ideal compressible flow. *J. Comp. Phys.*, **23**:263–275, 1977.

[171] B. van Leer. Towards the ultimate conservative difference scheme. IV. A new approach to numerical convection. *J. Comp. Phys.*, **23**:276–299, 1977.

[172] H. Wang, D. Liang, R. E. Ewing, S. E. Lyons, and G. Qin. An ELLAM-MFEM solution technique for compressible fluid flows in porous media with point sources and sinks. *J. Comput. Phys.*, **159**:344–376, 2000.

[173] R. F. Warming and R. W. Beam. Upwind second order difference schemes and applications in aerodynamic flows. *AIAA Journal*, **24**:1241–1249, 1976.

[174] B. Wendroff. The Riemann problem for materials with nonconvex equations of state. I. Isentropic flow. *J. Math. Anal. Appl.*, **38**:454–466, 1972.

[175] G. Whitham. *Linear and Nonlinear Waves*. Wiley-Interscience, 1974.

[176] M. L. Wilkins. Calculation of elastic-plastic flow. *Meth. Comp. Phys.*, **3**:211–263, 1964.

[177] P. Woodward and P. Colella. The numerical simulation of two-dimensional fluid flow with strong shocks. *J. Comp. Phys.*, **54**:115–173, 1984.

[178] N. N. Yanenko. *The Method of Fractional Steps*. Springer-Verlag, 1971.

[179] H. C. Yee, K. Sweby, P, and D. F. Griffiths. Dynamical approach study of spurious steady-state numerical solutions of nonlinear differential equations. 1. The dynamics of time discretization and its implications for algorithm development in computational fluid dynamics. *J. Comput. Phys.*, **97**:249–310, 1991.

[180] S. T. Zalesak. Fully multidimensional flux corrected transport algorithms for fluids. *J. Comput. Phys.*, **31**:335–362, 1979.

Index

$\mathcal{L}_1$-contracting, 357

accuracy, 71, 73–75, 78, 80, 341, 360, 409, 418, 544
acoustic tensor, 189, 229, 235
adaptive mesh refinement, 17, 545, 547, 549, 550, 560, 562, 563, 568, 570–574, 576–580
admissibility, 90–92, 98, 150–153, 155, 157, 159, 498, 500
advection
  –diffusion, 1, 9–11, 31, 32, 129, 359
  linear, 6, 14, 55, 81, 326, 330, 331, 362, 418, 478, 510
Alfvén speed, 214
amplitude, 40, 43, 44
anomalous, 238–240, 254
aquifer, 103

Beam, 133, 363, 364, 366, 367, 376, 377
Berger, 545
bounded, 353
Buckley, 107, 427, 542
buffering, 554
bulk modulus, 224, 234
Burgers, 82–84, 87, 88, 92, 97, 291, 319, 382, 389, 390, 395, 401, 407, 422, 498

capillary pressure, 106, 108, 268, 430, 542
Cartesian
  coordinates, 522
  grid, 543
Cauchy
  stress, 222–224, 529, 536
cell, 12, 477, 508
centered
  difference, 26, 32, 119, 312
  rarefaction, 97, 98, 156–159, 192, 193, 201, 207, 218–220, 258, 259, 266, 267, 272, 280
CFL
  condition, 15, 16, 395, 410, 417, 544, 545, 549, 554, 577
  factor, 389
  number, 15, 24, 25, 115, 118, 119, 128, 134, 341, 371, 460
characteristic
  cone, 140
  coordinate, 8, 81, 140
  direction, 137, 151, 154, 155, 157, 159, 293, 298, 452
  expansion coefficient, 139, 161, 323, 447, 497
  line, 7, 84, 121
  projection, 388, 389, 466
  speed, 82, 98, 101, 120, 136–138, 153–155, 157, 158, 161, 176, 432, 433, 579
  tracing, 388, 520, 529, 533
CLAWPACK, 450
closed, 353
Colella, 208, 295, 298, 322, 402, 450, 479
Colella–Woodward interacting blast wave problem, 208, 323, 324
compact, 353
  set, 333, 337, 353, 356
  support, 172, 328, 354–356, 418
concentration, 103, 104, 111, 129, 268–271, 273
conservation law, 8, 10, 81, 99, 135, 136, 326, 474
conservative
  approximate Riemann solver, 285–287, 289, 292, 294, 295, 304, 305, 313, 314, 316, 317, 322, 323
  flux, 377, 487
  scheme, 14, 286, 289, 332, 356, 364, 367, 368, 370, 477, 547
consistent
  approximate Riemann solver, 285–289, 304, 305, 313
  numerical flux, 285, 287, 288, 331, 332, 343, 347, 355, 438
  scheme, 43, 57, 58, 62, 327, 330, 353, 368
constitutive law, 222–225, 227, 229, 259
contact discontinuity, 138, 159, 297, 314, 473, 544
continuity equation, 222
contractive, 351
convection–diffusion, 40, 80
convergence, 25, 37, 44, 63, 66, 69, 90, 170, 180, 286, 289, 329, 331, 332, 343, 353, 355, 359, 368, 432, 433
coordinate
  Cartesian, 522
  curvilinear, 521–523, 529
  cylindrical, 521, 533, 535–537, 539
  Eulerian, 145
  Lagrangian, 140, 145, 235
  polar, 542
  self-similar, 98, 202
  spherical, 521, 523, 525, 527–529, 541
corner transport upwind, 479
Courant, 187, 188
  CFL condition, 15, 395, 410, 417, 544, 545, 549, 554, 577
  CFL number, 15, 115, 118, 119, 128, 134, 341, 460

d'Alembert, 174
Darcy
  law, 103, 106, 107, 268, 273, 274
  velocity, 108, 268, 270
decomposition, 437, 488
deferred correction, 508
deformation
  finite, 221, 233, 255
  gradient, 141, 142, 145, 222–225, 228, 236, 253, 255, 324, 463, 525, 530, 533, 534, 536, 541
  Green tensor, 223, 236

deformation (*cont.*)
infinitesimal, 233
plastic, 255, 258
rate, 233
diagonalizable, 139, 182, 304, 444
diffusion
anti-, 32, 33
artificial, 119, 133
numerical, 26, 31, 34, 39, 117, 129, 290–292, 295, 312–314, 316, 319, 359, 360, 369, 373, 375, 387, 433, 439, 543, 544
diffusive
anti-, 32, 33, 382, 446
flux, 10, 299–301, 378
scheme, 33, 119, 124, 125, 129, 369, 433, 435, 446
diffusivity, 103, 129
discontinuity
admissible, 240
contact, 138, 159, 208, 297, 314, 473, 544
development, 84, 85, 87
entropy-violating, 312
normal, 146, 227
propagation, 85, 86, 89–92, 96, 145, 146, 169, 175, 197, 207, 239, 359, 545
speed, 87, 89, 92, 146–150, 194–197, 207, 220, 221, 227, 239, 242, 259, 271, 279, 280
stationary, 101, 227
surface, 145, 146, 173
discretization, 12
dispersion, 40, 41, 43, 44
dissipation, 43, 44, 51

E-scheme, 378
efficiency, 69, 80, 341
elasticity
hyper-, 223
hypo-, 224
linear, 233, 234
elliptic
PDE, 1, 2, 104, 269, 274
regions, 276, 471
Engquist, 33, 297–301, 369, 370, 372, 373, 381, 417, 423
Engquist–Osher flux, 301
ENO scheme, 452, 494
enthalpy
specific, 188
entropy, 191, 207, 218, 291
flux, 93, 95, 165, 170–172, 218
function, 93, 95, 165, 170–172
inequality, 94, 173, 287
specific, 187, 188, 190, 192, 193, 218, 219
violation, 312
equivelocity, 241, 253, 270
Eulerian
coordinate, 140, 145
Eulerian–Lagrangian localized adjoint, 52, 53
frame of reference, 140, 142, 143, 145, 175, 187, 188, 190–192, 194, 206, 207, 221, 223, 227, 229, 232, 233, 235, 254, 466, 537
normal, 146

Fick's law, 103
finite
difference, 12, 14, 15, 18, 21, 23, 24, 26, 28, 30, 38, 39, 42, 59, 70, 127, 130, 131, 286, 289, 327, 331, 370, 372, 519, 520, 529
element, 221
flux
consistent, 285, 287, 331, 332, 343, 355, 438
corrected transport (FCT), 374, 446
diffusive, 10, 299–301, 378
Engquist–Osher, 297–301, 369, 370, 372, 373, 381, 417, 423
Godunov, 122, 346, 370
Rusanov, 119, 291, 293, 301, 316, 369, 417, 424, 427, 434, 471, 488
splitting, 371
variable, 136, 156, 159, 176
Fourier
analysis, 32, 38, 39, 41, 47–49, 51, 53–55, 114, 118, 119, 129, 133, 134
finite transform, 42
interpolation, 62
inversion formula, 39, 42, 54, 61
series, 32
transform, 6, 39, 41, 42, 44, 53, 55, 56, 60, 68
frame of reference
Eulerian, 140, 142, 143, 145, 175, 187, 188, 190–192, 194, 206, 207, 221, 223, 227, 229, 232, 233, 235, 254, 466, 537
Lagrangian, 140, 143, 145, 146, 175, 206, 207, 222, 223, 225–227, 232, 233, 236, 255, 319, 529, 532, 539
free-stream-preserving, 498, 529
frequency, 40
Friedrichs, 187, 188
CFL condition, 15, 395, 410, 417, 544, 545, 549, 554, 577
CFL number, 15, 115, 118, 119, 128, 134, 341, 460
Lax–Friedrichs diffusion, 292
Lax–Friedrichs scheme, 112, 114, 115, 128, 132, 176, 331, 353, 360, 361, 395, 397, 438, 440, 489–491
Fromm
monotonization, 364
scheme, 133, 364–367

Galerkin
discontinuous, 53, 412, 453, 494, 549
Galilean transformation, 99, 101, 112
gas constant, 188
genuinely nonlinear, 138, 150, 154, 157, 159, 190, 238, 257, 272, 277, 302
ghost cell, 127, 549, 577
Godunov
flux, 122, 346, 370
scheme, 120, 121, 180, 285, 319, 402, 433, 438, 477
theorem, 359, 362
Godunov scheme, 208, 255
gradient detector, 553, 579
Green's function, 9, 10

Harten, 283, 312, 314, 370, 408
Harten–Hyman–Lax theorem, 342, 353, 359
Harten–Lax–van Leer scheme, 314, 316, 319, 320, 322, 340, 423, 435, 436, 449, 457, 460, 462, 472, 488
heat capacity, 187, 188
heat equation, 9
Helmholtz free energy, 223
Hooke's law, 233
hull
concave, 98
convex, 98
Hyman, 312
Harten–Hyman–Lax theorem, 342, 353, 359
hyperbolic, 1, 2, 136–138, 186, 228, 230, 258, 267, 270, 275, 276
strictly, 136
hyperelastic, 224
hysteresis, 255, 257, 258, 272, 465

ideal gas, 187
immiscible, 105, 273
implicit, 24, 26, 83, 538, 542
incompressible, 103, 105, 273, 430, 521
interfacial tension, 106
interpolation operator, 60
inviscid, 91

Jensen's inequality, 182

Kirchhoff, 174, 222–224, 236, 255, 530, 531, 539–541

Lagrangian
coordinate, 140, 145, 235
frame of reference, 140, 143, 145, 146, 175, 206, 207, 222, 223, 225–227, 232, 233, 236, 255, 319, 529, 532, 539
normal, 145, 146
Lamé constant, 224
Laplace equation, 2

Lax
  admissibility, 90–92, 98, 150–153, 155, 157, 159, 498, 500
  convergence theorem, 37
  equivalence theorem, 62, 65, 69, 331
  Harten–Hyman–Lax theorem, 342, 353, 359
  Harten–Lax–van Leer scheme, 314, 316, 319, 320, 322, 340, 423, 435, 436, 449, 457, 460, 462, 472, 488
  Lax equivalence theorem, 353
  Lax–Friedrichs diffusion, 292, 360
  Lax–Friedrichs scheme, 112, 114, 115, 128, 132, 176, 331, 353, 360, 361, 395, 397, 438, 440, 489–491
  Lax–Richtmyer stable, 329
  Lax–Wendroff process, 129, 445
  Lax–Wendroff scheme, 49, 53, 74, 78, 80, 131, 133, 134, 291, 331, 360–364, 366, 367, 374, 376, 377, 438, 439
  Lax–Wendroff theorem, 332, 339, 345, 353, 356, 367
  monograph, 81, 152, 169, 172
Lax–Friedrichs scheme, 128
Lax–Wendroff scheme, 131
leap-frog scheme, 50, 51, 53
level of refinement, 546
Leverett, 107, 427, 542
Levy
  CFL condition, 15, 395, 410, 417, 544, 545, 549, 554, 577
  CFL number, 15, 115, 118, 119, 128, 134, 341, 460
limiter, 375, 376
  flux-, 377
  minmod, 377, 416
  MUSCL, 377
  slope-, 383, 386, 391
  superbee, 377
  van Leer, 377, 382
linear method, 328
linearly degenerate, 138, 150, 154, 157, 159, 190, 212–214, 217, 237, 238, 269, 272, 301, 314, 316, 437
Lobatto, 414, 415, 417, 421, 422, 454, 455, 496, 497
local truncation error, 37, 57, 327, 329

MacCormack, 132–134, 439
Mach, 197–199, 207, 208, 323, 324
magnetohydrodynamics, 208
makefile, 17
Marquina, 125, 128, 370, 411, 423, 438
material derivative, 142
Maxwell, 139, 183, 184, 186, 209, 456
method of lines, 408
MHD, 208
minmod, 387
miscible displacement, 103, 104, 110, 111, 129
model
  Buckley–Leverett, 107, 427, 542
  gas, 186
  magnetohydrodynamics (MHD), 208
  miscible displacement, 103, 520
  Mooney–Rivlin, 224
  plasticity, 255, 464, 465
  polymer, 267
  shallow water, 136
  solid mechanics, 221
  traffic, 102
  vibrating string, 235
modified
  equation, 30, 35, 115, 119, 129, 347
  Roe, 312, 316
monotone, 341–343, 345, 347, 351, 359
monotonic, 361–365
monotonicity-preserving, 357–359, 361, 402
monotonization, 364
Mooney, 224, 232
MUSCL scheme, 385, 447

Newton
  interpolation, 385, 403
  iteration, 295, 323
  second law, 222, 223, 229
nonreflecting boundary, 125
normal
  Eulerian, 146
  Lagrangian, 145, 146
numerical diffusion, 292, 360

Oleinik, 95, 96, 98, 357, 368, 430
order
  of PDE, 1
  of scheme, 23, 26, 29, 37, 39, 43, 129, 328, 347, 353, 359–361
Osher, 33, 292, 297–301, 322, 368–370, 372, 373, 381, 408, 409, 417, 423, 452
over-compressive, 461

parabolic PDE, 1, 2
Parseval, 53, 59–63
particle velocity, 270
patch, 546
Peclet, 41, 110, 129
permeability, 103
phase error, 43, 45, 46, 51, 52, 118
physical components, 522
piecewise parabolic method (PPM), 402
Piola, 222–224, 236, 255, 530, 531, 539–541
plasticity, 255, 267, 464, 465
pointwise error, 326
polytropic gas, 188
predictor–corrector scheme, 438
proper nesting, 546, 549, 550
proper nesting buffer, 554
pure virtual function, 572

quadrature, 414
quasilinear form, 82, 83, 136, 227, 230, 237, 270

Rankine–Hugoniot
  jump condition, 86, 90, 147, 151, 152
  locus, 150, 155, 159, 163, 199, 201, 272, 277, 278, 283, 296
rarefaction, 159, 204, 205, 208, 254, 266, 295–297, 312, 323, 324
  centered, 91, 97–99, 115, 123, 156–159, 192, 193, 201, 207, 218–220, 258, 259, 266, 267, 272, 280
  curve, 159, 163, 201, 219, 283, 295, 297
  elastic, 266, 267
  transonic, 98, 122, 128, 129, 131, 290, 295, 297, 340
refinement
  adaptive, 17, 545
  ratio, 546
  study, 72, 341, 472
reflecting, 114, 125, 205, 390, 459
refluxing, 560
regrid interval, 547, 553
regridding, 547
Richardson, 556, 568
Richtmyer
  Lax–Richtmyer stable, 329
  two-step Lax–Wendroff scheme, 131
Riemann
  invariant, 158, 163, 193, 220, 254, 273, 280, 297, 298
  problem, 33, 97, 98, 101, 116, 121, 153, 159, 180, 182, 283, 286, 290, 297, 299, 303, 340, 359, 433, 476, 482, 498, 503, 519, 542
  solver, 174, 180, 284–286, 292–295, 297–300, 304, 306, 312–314, 316, 317, 319, 320, 322–324, 340, 353, 368, 390, 395, 401, 407, 426, 434, 435, 438, 446, 449, 450, 452, 457, 458, 460, 462, 472, 488
Rivlin, 224, 232
Roe
  flux, 305, 308, 311, 340, 445, 449, 488
  matrix, 298, 303–308, 310–312, 314, 324, 435, 436, 444, 450
  solver, 306, 308, 311–314, 319, 322–324, 340, 346, 417, 436, 452, 457, 460, 463, 472, 488

Runge–Kutta, 402, 411, 415, 426, 453, 496, 497, 549
Rusanov
flux, 119, 291, 293, 301, 316, 369, 417, 424, 427, 434, 471, 488
scheme, 118–120, 178, 292, 549
solver, 292, 299, 300, 314, 320, 457

saturation, 105, 268, 273
scale factors, 522
Schaeffer, 277, 278, 280, 281, 311, 319, 321, 322, 470
Schechter, 277, 278, 280, 281, 311, 319, 321, 322, 470
scheme
anti-dissipative, 47, 48
Beam–Warming, 133, 363, 364, 366, 367, 376, 377
conservative, 14, 15, 17, 18, 23, 24, 26, 28, 34, 70, 286, 289, 353, 355, 356, 364, 367, 368, 370, 477, 547
consistent, 43, 57, 58, 62, 327, 330, 353, 368
diffusive, 33, 119, 124, 125, 129, 369, 433, 435, 446
discontinuous Galerkin, 53, 412, 453, 494, 549
dispersive, 43
dissipative, 43, 44, 48–50, 114, 120, 122
donor cell upwind, 477, 509
E-, 368, 371
essentially non-oscillatory (ENO), 408, 409, 452, 494
Eulerian–Lagrangian localized adjoint, 52, 53
explicit centered, 52
explicit centered difference, 26, 32, 119
explicit downwind, 23, 24, 46
explicit upwind, 14–16, 23, 28, 31, 34, 44, 52, 111, 112, 129
finite difference, 12, 14, 15, 18, 21, 23, 24, 26, 28, 30, 38, 39, 42, 59, 70, 127, 130, 131, 286, 289, 327, 331, 370, 372, 519, 520, 529
flux corrected transport (FCT), 374, 446
Fromm, 133, 364–367
front tracking, 432
Godunov, 120, 121, 180, 208, 255, 285, 319, 402, 433, 438, 477
Harten–Lax–van Leer, 314, 316, 319, 320, 322, 340, 423, 435, 436, 449, 457, 460, 462, 472, 488
implicit downwind, 24, 25, 34, 49
implicit upwind, 26, 29, 34, 48
Lax–Friedrichs, 112, 114, 115, 128, 132, 176, 331, 353, 360, 361, 395, 397, 438, 440, 489–491
Lax–Wendroff, 49, 53, 74, 78, 80, 131, 133, 134, 291, 331, 360–364, 366, 367, 374, 376, 377, 438, 439
leap-frog, 50, 51, 53, 133, 134
Leonard, 35
linear explicit two-step, 35, 55
MacCormack, 132–134, 439
Marquina, 125, 128, 370, 411, 423, 438
monotone, 341–343, 345, 347, 351, 359
monotonic, 361–365
monotonicity-preserving, 357–359, 361, 402
MUSCL, 364, 447, 485, 542
operator splitting, 475, 481, 487, 506, 507, 511, 519, 529, 542
order, 23, 26, 29, 37, 39, 43, 47, 129, 328, 347, 353, 359–361
pseudo-spectral, 81
random choice, 433
Richtmyer two-step Lax–Wendroff, 131, 132
Rusanov, 118–120, 178, 292, 549
stream-tube, 543
total variation diminishing (TVD), 357–359, 370
unsplit, 477
upstream, 289
upstream weighting, 430
upwind, 176
wave propagation, 391, 448, 488, 521
WENO, 401
self-similar, 97, 98, 101, 156, 159
shear modulus, 224, 233, 234
Shearer, 277, 278, 280, 281, 311, 319, 321, 322, 470
shock, 33, 91, 92, 98, 115, 123, 128, 134, 150, 151, 153, 156, 159, 204, 205, 207, 208, 254, 267, 291, 296, 323, 324
capturing, 183, 221, 461
curve, 155, 159
elastic, 267
reflection, 153
speed, 151, 155, 156
shock strength, 196
similarity, 157
slope
harmonically averaged, 367, 377
minmod, 387
monotonized, 377
MUSCL, 366, 386
side-centered, 366
superbee, 387
smoothness monitor, 361, 375
Sod, 208, 324, 457
Solomon, 292, 297, 298, 322
solution ratio, 56
sound speed, 207
spin tensor, 232
stability, 91, 117, 152, 188, 197, 328, 329, 353, 354, 357, 359
Lax–Richtmyer, 329
stable, 37
finite difference approximation, 59
staggered, 116, 395, 401, 433, 435, 489
stationary, 90, 91, 101, 121, 122, 152, 227, 290, 340
stencil, 15, 127, 376, 410, 411, 513, 549, 577
strain, 236, 256–258, 267
energy, 240
plastic, 256–258, 267
Strang splitting, 475
streamline, 542
stress, 223, 225, 234, 256, 561
Cauchy, 222–224, 529, 536
Jaumann rate, 232
Piola–Kirchhoff, 222, 223, 224, 236, 255, 530, 531, 539–541
strictly hyperbolic, 136
substantial derivative, 142
superbee, 387
support, 353
symbol
of finite difference approximation, 56
of partial differential operator, 55
of scheme, 57, 58
symmetry
cylindrical, 536, 540
spherical, 526, 528

tension, 236
thermodynamics, 187, 222
timestep, 12, 15, 176, 179, 433, 434
total variation, 354
total variation diminishing, 357
transformation
coordinate, 1, 2, 99, 101, 112, 495, 542
transonic, 98, 122, 131, 290, 295, 340
traveling wave, 89, 151
truncation error, 37, 57
truncation operator, 61
TV stable, 354

umbilic point, 276
unit base vectors, 522

upscaling, 547, 561
upstream
  scheme, 289
  weighting, 430

van Leer, 291, 361, 387, 390
  Harten–Lax–van Leer scheme, 314, 316, 319, 320, 322, 340, 423, 435, 436, 449, 457, 460, 462, 472, 488
  smoothness monitor, 361, 375
velocity, 142
viscous, 89, 92–94, 96, 151, 170, 172, 472

Warming, 133, 363, 364, 366, 367, 376, 377
wave
  equation, 174
  length, 44
  number, 39, 40, 42, 44, 45, 52
weak
  form, 412
  solution, 332
  wave, 293
  wave approximation, 254, 255, 322, 323
Wendroff
  Lax–Wendroff process, 129, 445
  Lax–Wendroff scheme, 49, 53, 74, 78, 80, 131, 133, 134, 291, 331, 360–364, 366, 367, 374, 376, 377, 438, 439
  Lax–Wendroff theorem, 332, 339, 345, 353, 356, 367
Woodward, 208, 402, 450

Zalesak test problem, 29, 366, 367, 377, 382, 395, 402, 407, 419–422, 428, 429, 473